549/ENV

AF443259
Libraries

European Mineralogical Union
Notes in Mineralogy

Commissioning Editor: R. Oberti

Volume 13

ENVIRONMENTAL MINERALOGY II

UNIVERSITY TEXTBOOK

Edited by

D.J. Vaughan and R.A. Wogelius

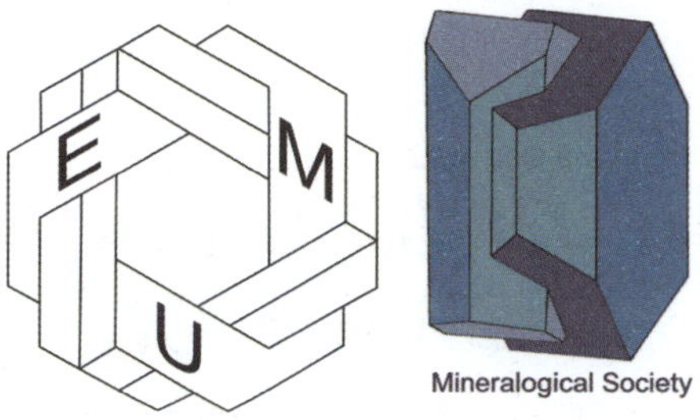

Published by the European Mineralogical Union and the
Mineralogical Society of Great Britain & Ireland,
London, 2013

The publication of this textbook is supported by the European Mineralogical Union

EMU Notes in Mineralogy

A series published under the auspices of the European Mineralogical Union (EMU).

Initiator of the EMU Schools and the EMU Notes in Mineralogy
Giovanni Ferraris, Torino, President of the EMU 1992–1996

Commissioning Editor: R. Oberti, Pavia (previous editors: Giovanni Ferraris, Torino,
Tamás G. Weiszburg and Gábor Papp, Budapest)

Editors of this Volume
D.J. Vaughan and R.A. Wogelius, University of Manchester, UK

Managing Editor and Indexer: Kevin Murphy, London

Front cover design: Michel H. Guay

Executive Committee of the EMU (2012–2016)
Wilhelm Heinrich, Germany (President), Roberta Oberti, Italy (Past-President), Corina
Ionescu, Romania (Vice President), F. Gervilla, Spain (Vice President), Juraj Majzlan,
Germany (Secretary) and Wim van Westrenen, The Netherlands (Treasurer)

On the front cover: Minewastes at the abandoned Berkeley Pit, Butte, Montana (USA); the open
pit is now a deep artificial lake of very acid water contaminated with toxic metals (photo courtesy
of D.J. Vaughan).

ISSN: 1417 2917
ISBN: 978-0-903056-32-8

© 2013 EMU and the Mineralogical Society of Great Britain & Ireland
Published by the European Mineralogical Union and the Mineralogical Society of
Great Britain & Ireland (12, Baylis Mews, Amyand Park Road,
Twickenham TW1 3HQ, UK)
Printed by Cambrian Press, Aberystwyth, UK

08377839

UNIVERSITY OF
BRIGHTON

2 1 OCT 2013

INFORMATION
SERVICES

The EMU Notes in Mineralogy Series

Published volumes

Volume	Year	Editors	Title
1	1997	S. Merlino	*Modular aspects of minerals*
2	2000	D.J. Vaughan R. Wogelius	*Environmental mineralogy*
3	2001	C.A. Geiger	*Solid solutions in silicate and oxide systems*
4	2002	C. Gramaccioli	*Energy modelling in minerals*
5	2003	D.A. Carswell R. Compagnoni	*Ultrahigh pressure metamorphism*
6	2004	A. Beran E. Libowitzky	*Spectroscopic methods in mineralogy*
7	2005	R. Miletich	*Mineral behaviour at extreme conditions*
8	2010	F.E. Brenker G. Jordan	*Nanoscopic approaches in Earth and Planetary Sciences*
9	2010	G. Christidis	*Advances in the characterization of Industrial minerals*
10	2010	M. Prieto	*Ion-partitioning in ambient-temperature aqueous systems*
11	2011	M.F. Brigatti A. Mottana	*Layered mineral structures and their application in advanced technologies*
12	2012	J. Dubessy M.-C. Caumon F. Rull	*Applications of Raman Spectroscopy to Earth Sciences and Cultural Heritage*
13	2013	D.J. Vaughan R.A. Wogelius	*Environmental Mineralogy II*

Copies of the EMU Notes (volumes 1–7) are distributed in Europe by the larger member societies of the European Mineralogical Union:

Società Italiana di Mineralogia e Petrologia:
www.socminpet.it

Mineralogical Society of Great Britain & Ireland:
www.minersoc.org

Société Française de Minéralogie et de Cristallographie:
www.sfmc-fr.org

and by the Secretary of the EMU, Prof. Juraj Majzlan:
juraj.majzlan@uni-jena.de

in America by the Mineralogical Society of America:
www.minsocam.org

Institutional orders as well as individual requests from outside Europe and America should be sent to the Secretary of EMU, Prof. Juraj Majzlan:
juraj.majzlan@uni-jena.de

For volumes 8 onwards, the Mineralogical Society of Great Britain acts as co-publisher and copies may be ordered from www.minersoc.org

or from:
Mineralogical Society
12 Baylis Mews, Amyand Park Road
Twickenham TW1 3HQ
UK
E-mail: admin@minersoc.org
Tel. + 44 (0)20 8891 6600
Fax: + 44 (0)20 8891 6599

Contents

Part II. Mineralogy of key environmental systems

Chapter 3. Minerals and soil development
by DAVID A.C. MANNING

**Chapter 4. Mineralogy of marine sediment systems: a geochemical
framework**
by ANDREW C. APLIN and KEVIN G. TAYLOR

X

Part III. Mineralogy and specific environmental problems

Chapter 9. Mineralogy in long-term nuclear waste management 383
by CHARLES CURTIS and KATHERINE MORRIS

0863778839

Foreword

In 1997, the European Mineralogical Union (EMU) began organizing a series of Short Courses ('Schools') with the associated publication of a series of review volumes (the 'EMU Notes in Mineralogy') on topics of interest to mineral scientists. The second School was held in Budapest in May 2000, along with the publication of Volume 2 of the Notes, on the then emerging subject of *Environmental Mineralogy*. This volume (edited by D.J. Vaughan and R.A. Wogelius) was well received and has sold well in the 12 years since it appeared (such that very few copies remain available for purchase). Given the continuing demand for books in this field, the President and Council of EMU approached the editors and asked them to consider several options. These were: (1) simply to reprint the original volume; (2) to produce a new volume using the original chapter authors as far as possible; or (3) to organize a new School and accompanying volume. It was decided to take the second of these options and to publish what might be thought of as a 'second edition', although the extensive revisions undertaken in what we have entitled *Environmental Mineralogy II* justify regarding it as a new book, and hence in making it 'Volume 13' of the Notes.

The layout and organization of this new book follow closely that used in the old volume. I was delighted to find that the great majority of contributors to the earlier volume were keen to take on the job of producing a new, up-to-date chapter on their chosen subject. Again, therefore, the book consists of 11 chapters, eight of which retain the same authors, and two of which have added one author (Kevin Taylor joins Andy Aplin for Chapter 4, and Kath Morris joins Charles Curtis for Chapter 9). In the case of Chapter 3, the previous authors are no longer actively involved in the area of soil science, and the challenge of that topic area has been ably taken up by David Manning. All of the chapters were subject to external peer review and I wish to thank Nick Bryan, Linda Campbell, Hugh Coe, Ian Freestone, Barry Johnson, Francis Livens, Jon Lloyd, Richard Pattrick, Claire Robinson, Eva Valsami-Jones and Dave Wray for their help with this important process. At the same time, it should be emphasized that any errors and imperfections that remain in this volume are solely the responsibility of the authors and editors.

The publication of these volumes was formerly undertaken by Eötvös University Press but now that role is being undertaken by The Mineralogical Society of Great Britain and Ireland. In that context, I would like to thank Kevin Murphy, the Executive Director of the Society, who has taken on the formidable task of being the Managing Editor for the series. Without his hard work, patience and attention to detail, this volume would never have been completed. I would also like to thank the EMU and its Officers and Council, particularly Giovanni Ferraris and Roberta Oberti, for their encouragement and support in this venture.

Finally, as our overall aims in producing this book remain the same. I think it appropriate to quote from the Foreword to this volume's predecessor by saying "Environmental mineralogy is a new field, and one of great international importance, not least to the countries of Europe. It is my hope that this volume will contribute to the development of this field, and lead to collaborative ventures involving mineral scientists from European countries and beyond who are concerned with scientific solutions to our many environmental problems."

David J. Vaughan
(former President of the European Mineralogical Union)

Introduction: The nature and scope of environmental mineralogy

DAVID J. VAUGHAN

*Williamson Research Centre for Molecular Environmental Science, and School
of Earth, Atmospheric and Environmental Sciences, University of Manchester
Oxford Road, Manchester, M13 9PL, England, UK
E-mail: David.Vaughan@manchester.ac.uk*

Minerals, as the inorganic solids that comprise the rocks, sediments and soils of the Earth, are an essential part of our environment. So, in a sense, all mineralogy is 'environmental mineralogy'. However, the term 'environmental' has come to be employed (particularly in combination with terms such as 'science', 'issue' or 'problem') to refer to those systems at or near the surface of the Earth where the geosphere comes into contact with the hydrosphere, atmosphere and biosphere. This is, of course, the 'environment' upon which the human race depends for survival and, hence, is now sometimes referred to as the 'critical zone'. It can be subject to disruptions due to human activity, particularly activity associated with the exploitation and utilization of Earth's resources. This is the sense in which we use the term 'environmental' in this book. Thus, we consider here those systems containing minerals that constitute the most important or key environments: soils, modern sediments, atmospheric aerosols, and the interior or exterior parts of certain micro- and macro-organisms. Particularly important are the roles that minerals play in processes that act over time to control or influence the environment at various scales of observation. Both pure systems and those contaminated as a result of human activity are considered. We also focus on certain specific problems that arise from resource exploitation or utilization and that involve minerals in some way; either, or both, in creating the problem or ameliorating it. These include problems associated with the waste generated by mining, particularly mining of metals, industrial and domestic wastes, and those wastes produced by the nuclear industry. Particular problems can arise from use of minerals and rocks in buildings and monuments and other cultural artefacts. The relationship between minerals and human health constitutes a special case where the environment includes the human body itself.

Environmental mineralogy is still a relatively new field. At the time of writing of the first edition of this book, it was described as having no definite boundaries nor any established and accepted body of literature. That volume has helped to define the nature and scope of the new field, alongside books, monographs and journal articles produced in the decade that followed. In the Appendix to this first chapter are listed some of the key texts of relevance to environmental mineralogy, and some of the most relevant journals that constitute the emergent literature of environmental mineralogy.

© Copyright 2013 the European Mineralogical Union and the Mineralogical Society of Great Britain & Ireland
DOI: 10.1180/EMU-notes.13.1

The objectives of the first edition, which remain essentially the same for this new volume, were to help to define the subject of environmental mineralogy, and to provide an initial source of information both for mineralogists and other scientists who wish to understand or work in this field. It was hoped that it might also provide a text for use by those teaching courses in the subject at advanced undergraduate or graduate student level. In no sense was that first edition, nor is the present volume, intended to be comprehensive; that would be impossible, given the constraints involved. However, it is intended to point the way for those who wish to pursue scholarship and research in the field.

The layout of this new volume follows closely that of the first edition but, of course, the content has been fully revised and updated to take into account the advances of the past 12 years. Hence, following this brief introductory chapter, we begin with a review of the analytical, experimental and computational methods that are of importance in environmental mineralogy (R.A. Wogelius and D.J. Vaughan). These include long-established techniques used to characterize minerals and mineral associations, and the newer methods that can be used to study very fine-particle solids, mineral surfaces, and interactions between mineral surfaces and fluids. Examples are given of analytical, experimental and computational approaches applied to important mineral systems.

There follows a major section addressing what we term "key environmental systems". Firstly, soils are reviewed by D.A.C. Manning in regard to mineralogical aspects of soil formation and anthropogenic impacts on soils. Then the mineralogy of modern sediments is discussed by A. Aplin and K. Taylor, including the processes of sediment formation and modification both through natural mechanisms and those associated with pollutants. Microbial interactions with minerals are discussed by S. Welch and J. Banfield, and the mineralogy of atmospheric aerosol particles considered by M. Pósfai and Á. Molnár.

The five other major chapters are each devoted to discussing a specific environmental problem in the context of the relevant mineralogy. These are: metalliferous and related minewastes (D. Blowes, C.J. Ptacek and J. Jambor); industrial and domestic waste disposal and containment (R. Hermanns Stengele and M. Plötze); nuclear waste (C.D. Curtis and K. Morris); the buildings and monuments that constitute our collective cultural heritage (G. Chiari), and the particular relationship between minerals and human health (H.C.W. Skinner). As noted above, this volume cannot possibly be comprehensive. Rather, we aim to help in defining and advancing the field by using carefully selected examples.

As for the first edition, three important points needed to be made in closing this introductory chapter:

(1) Mineralogy (and therefore mineralogists) are important to environmental science – the techniques of mineralogy, which are aimed at characterizing key natural solid materials, their relationships and interactions both with each other and with fluids and gases, offer unique insights into very important parts of our environment. The corollary is also true. Mineralogists need environmental science as some of their more traditional areas of study and research decline in importance.

(2) The molecular-scale understanding that is at the core of modern mineralogy is vital to the advancement of environmental mineralogy and hence environmental

science. In this sense, environmental mineralogy is part of the larger field which has become known as 'molecular environmental science'. It will only be through such fundamental understanding that we will be able to model and predict the behaviour of environmental systems containing minerals.

(3) The interlinking study of examples of natural systems with carefully chosen model laboratory systems, and with computer modelling from molecular to macroscopic scales, will be the key to advances in the field.

Environmental mineralogy is one of the most exciting, rapidly developing, socially and economically relevant areas of study and research in modern science. It is hoped that this extensively revised and updated new edition of what has now become an established volume will further encourage developments in this field.

Appendix

A. Some key texts of relevance to environmental mineralogy

Alpers, C.N., Jambor, J.L. & Nordstrom, D.K. (editors) (2000) *Sulfate Minerals – Crystallography, Geochemistry and Environmental Significance.* Reviews in Mineralogy and Geochemistry, **40**. Mineralogical Society of America, Washington, D.C. 608 pp.

Banfield, J.F. & Navrotsky, A. (eds.) (2001) *Nanoparticles and the Environment.* Reviews in Mineralogy and Geochemistry, **44**. Mineralogical Society of America, Washington, D.C. 349 pp.

Banfield, J.F. & Nealson, K.H. (editors) (1998) *Geomicrobiology: Interactions between Microbes and Minerals.* Reviews in Mineralogy and Geochemistry, **35**. Mineralogical Society of America, Washington, D.C., 448 pp.

Banfield, J.F., Cervina-Silva, J. & Nealson, K.M. (editors) (2005) *Molecular Geomicrobiology.* Reviews in Mineralogy and Geochemistry, **59**. Mineralogical Society of America, Washington, D.C. 294 pp.

Brady, P.V. (editor) (1996) *Physics and Chemistry of Mineral Surfaces.* CRC Series on the Chemistry and Physics of Surfaces and. Interfaces. CRC Press, Boca Raton, Florida, USA, 268 p.

Cabri, L.J. & Vaughan, D.J. (editors) (1998) *Modern Approaches to Ore and Environmental Mineralogy.* MAC Short Course Series, **27**. Mineralogical Association of Canada, Ottawa, Canada, 421 pp.

Calas, G. & Brown, G.E. Jr. (editors) (2011) Environmental Mineralogy. *Comptes Rendus Geoscience,* Thematic Issue, **343**, 83–259.

Cornell, R.M. & Schwertmann, U. (2003) *The Iron Oxides: Structure, Properties, Reactions, Occurrences and Uses,* 2nd edition. Wiley-VCH.

Cotter-Howells, J., Campbell, L., Batchelder, M. & Valsami-Jones, E. (editors) (2000) *Environmental Mineralogy: Microbial Interactions, Anthropogenic Influences, Contaminated Land and Waste Disposal.* Mineralogical Society Series, **9**. Mineralogical Society, London, 414 pp.

Dove, P.M., De Yoreo, J.J. & Weiner, S. (editors) (2003) *Biomineralization.* Reviews in Mineralogy and Geochemistry, **54**. Mineralogical Society of America, Chantilly, Virginia, USA, 381 pp.

Drever, J.I. (1997) *The Geochemistry of Natural Waters,* 3rd edition. Prentice Hall, New Jersey, USA, 436 pp.

Fleet, M.E. & Fyfe, W.S. (editors) (1984) *Environmental Geochemistry.* MAC Short Course Handbook, **10**. Mineralogical Association of Canada, Ottawa, Canada, 306 pp.

Guthrie, G.D. & Mossman, B.T. (editor) (1993) *Health Effects of Mineral Dusts.* Reviews in Mineralogy and Geochemistry, **28**. Mineralogical Society of America, Washington, D.C. 584 pp.

Henderson, P. (1982) *Inorganic Geochemistry.* Pergamon Press, Oxford, 353 pp.

Hochella, M.F., Jr. & White, A.F. (editors) (1990) *Mineral–Water Interface Geochemistry.* Reviews in Mineralogy and Geochemistry, **23**. Mineralogical Society of America, Washington, D.C. 603 pp.

Hudson-Edwards, K. & Vaughan, D.J. (2012) *Environmental Minerals.* Wiley-Blackwell, London, 384 pp.

Jambor, J.L. & Blowes, D.W. (editors) (1994) *The Environmental Geochemistry of Sulfide Minewastes*. MAC Short Course Handbook, **22**. Mineralogical Association of Canada, Ottawa, Canada, 438 pp.

Jambor, J.L., Blowes, D.W. & Ritchie, A.I.M. (2003) *Environmental Aspects of Mine Wastes*. MAC Short Course Handbook, **31**. Mineralogical Association of Canada, Ottawa, Canada, 400 pp.

Keller, E.A. (1992) *Environmental Geology*, 6th edition. MacMillian, New York.

Knoll, A., Canfield, D. & Konhauser, K. (2012) *Fundamentals of Geobiology*. Wiley-Blackwell, UK, 443 pp.

Langmuir, D. (1997) *Aqueous Environmental Geochemistry*. Prentice Hall, New Jersey, USA 600 pp.

Marfunin, A.S. (editor) (1998) *Advanced Mineralogy, Vol. 3. Mineral Matter in Space, Mantle, Ocean Floor, Biosphere, Environmental Management, and Jewelry*. Springer-Verlag, Berlin, 444 pp.

McIntosh, J.M. & Groat, L.A. (editors) (1997) *Biological–Mineralogical Interactions* MAC Short Course Handbook, **25**. Mineralogical Association of Canada, Ottawa, Canada, 239 pp.

Sahai, N. & Schoonen, M.A.A. (2006) *Medical Mineralogy and Geochemistry*. Reviews in Mineralogy and Geochemistry, **64**. Mineralogical Society of America, Chantilly, Virginia, USA, 332 p.

Sposito, G. (1989) *The Chemistry of Soils*. Oxford University Press, New York, 277 pp.

Stumm, W. (1992) *Chemistry of the Solid–Water Interface*. Wiley-Interscience, New York 448 pp.

Vaughan, D.J. (editor) (2006) *Sulfide Mineralogy and Geochemistry*. Reviews in Mineralogy and Geochemistry, **61**. Mineralogical Society of America, Chantilly, Virginia, USA, 714 pp.

Vaughan, D.J. & Pattrick, R.A.D. (editors) (1995) *Mineral Surfaces*. Mineralogical Society Series, **5**. Chapman & Hall, London, 384 pp.

White, A.F. & Brantley, S.L. (editors) (1995) *Chemical Weathering Rates of Silicate Minerals*. Reviews in Mineralogy and Geochemistry, **31**. Mineralogical Society of America, Washington, D.C. 583 pp.

B. Some key journals publishing articles relevant to environmental mineralogy

American Mineralogist
Applied Geochemistry
Applied and Environmental Microbiology
Chemical Geology
Clays and Clay Minerals
Clay Minerals
Elements
Environmental Pollution
Environmental Science & Technology
European Journal of Mineralogy
Geobiology
Geoderma
Geochimica et Cosmochimica Acta
Geomicrobiological Journal
Journal of Colloid and Interface Science
Journal of Nuclear Materials
Langmuir
Mineralogical Magazine
Science of the Total Environment
Soil Science Society of America Journal
The Canadian Mineralogist

Analytical, experimental and computational methods in environmental mineralogy

Roy A. WOGELIUS[1,*] and David J. VAUGHAN[2]

*Williamson Research Centre for Molecular Environmental Science,
and School of Earth, Atmospheric and Environmental Science, University of
Manchester, Oxford Road, Manchester M13 9PL, U.K.*
e-mail: [1]*Roy.Wogelius@manchester.ac.uk,* [2]*David.Vaughan@manchester.ac.uk*
**Corresponding author*

The analytical, experimental and computational methods now used in environmental mineralogy are introduced and illustrated with selected examples. Following a note on the importance of the radiation sources used for diffraction and spectroscopic studies, and a brief reminder of the key role still played by routine methods of mineral characterization (optical microscopy, X-ray powder diffraction [XRD], electron microbeam methods), more specialist techniques for the characterization of bulk solids (including nanoparticles) are considered. These include synchrotron-based advanced XRD and X-ray scattering methods, X-ray absorption spectroscopies, X-ray microprobe techniques, and infrared and Raman spectroscopies. Also considered are transmission electron microscopy, nuclear and magnetic spectroscopies, and particle induced X-ray emission. Following this, attention is focused on the characterization of mineral surfaces and interfaces using low-energy electron diffraction, X-ray reflectivity, X-ray absorption and X-ray standing wave methods, X-ray photoelectron and Auger electron spectroscopies, secondary ion mass spectrometry, environmental scanning electron microscopy, scanning tunnelling and atomic force microscopies and laser confocal microscopy. Computer modelling of solids, surfaces and solution species is discussed briefly before experimental approaches to determining reaction kinetics are considered, followed by experimental studies of the phenomena of adsorption, of evolution of the mineral surface during processes such as silicate weathering, and of how combinations of methods can be used to study the geochemical cycling of arsenic, olivine breakdown, or the comparative reactivities of silicate, carbonate, sulfide and oxide minerals.

1. Introduction

Environmental mineralogy necessarily begins with the characterization of minerals and associated phases in the key environmental systems. Such characterization involves the identification of minerals, and determination of quantities of the phases, together with compositional and textural information. It may extend to more detailed studies of, for example, very small ('nano-') particles, microtextures, or mineral surface chemistry. Beyond such studies of the natural materials, whether 'pristine' or 'contaminated' through some human activity, it is possible to undertake experiments in the laboratory using minerals or synthetic analogues in order to elucidate some processes of environmental importance. In many low-temperature systems, the minerals occur as both important reactants and as reaction products. The advantages of laboratory experiments

© Copyright 2013 the European Mineralogical Union and the Mineralogical Society of Great Britain & Ireland
DOI: 10.1180/EMU-notes.13.2

are that the system being investigated can be controlled closely and monitored; it can also be a model system, selected so as to focus on a particular process, at the same time as eliminating irrelevant complications and ambiguous interpretations. Materials and processes involved in both natural systems and laboratory experiments may also be investigated using computer modelling at scales ranging from the molecular to macroscopic, or even global. Computational studies allow us to gain both a greater understanding of the system(s) concerned and to develop models which have the power to predict as well as to interpret behaviour.

The objective of this chapter is to introduce the analytical (broadly defined) methods of importance in environmental mineralogy, with emphasis on the specialist methods now available to study fine particulates and mineral surfaces, along with their interacting fluids and sorbates. Also, experimental approaches to environmental mineralogy problems together with the essential theory are outlined, certain major mineral systems are considered, and computer modelling at molecular to macroscopic scales are discussed briefly. Specific examples are used throughout the text to draw attention to the strengths and weaknesses of a particular method or approach. In no sense is this account intended to be comprehensive; rather it is illustrative of this rapidly growing field. We necessarily are selective in our choice of examples, in many cases those chosen are from our own investigations, for reasons of familiarity, but the readers should be aware that there are many other excellent studies to be found by researching the literature mentioned in the Appendix to Chapter 1 (Vaughan, 2013). In particular, we wish to emphasize the importance of 'atomic' or 'molecular'-scale methods of analysis in the study of environmental systems. In this sense, environmental mineralogy is a branch of the emergent subject now referred to as 'molecular environmental science'.

1.1. The basis of analytical methods; the importance of radiation sources

With just a few notable exceptions, the methods used to analyse, or otherwise characterize, the materials of importance in environmental mineralogy involve studying their interactions with some form of electromagnetic radiation (visible light, ultraviolet or infrared photons; X-rays; γ-rays) or with particles (ions, protons, molecules). That interaction may involve diffraction, absorption, reflection, fluorescence or scattering of what is usually a beam of radiation. Modern instruments commonly enable such beams to be focused down to a small (micron or even sub-micron) spot and for scanning or 'rastering' of the beam to enable mapping of compositional or other variations in the sample in two, or even three dimensions. Major advances in the development of radiation sources have taken place in recent decades. Most notable have been those advances associated with 'synchrotron radiation'. A conventional laboratory light source uses a simple lamp involving a heated filament; an X-ray source also involves heating a filament in a vacuum, using the emitted electrons to impact on a metal (Cu, Co, *etc.*) target to produce characteristic X-rays. A synchrotron generates electromagnetic radiation in a completely different manner from a laboratory source. Here, a packet of electrons is accelerated around a large diameter (on the order of $20-100$ m) storage ring at relativistic speeds. Simultaneous application of magnetic fields and electric fields keeps the packet on a tightly controlled path as it cycles (hence the name – synchrotron, alluding to the synchronous application of the two fields). Energy is emitted tangentially from particles moving in a circle and, depending on the size and specifications of the storage ring, this can range from the infrared to high energy ('hard') X-rays. One advantage of synchrotron radiation is its far greater intensity ('brightness'), hence providing enough photons to enable focusing down to a microbeam for small spot analysis or mapping. Another advantage is the fact that it is 'white' radiation, *i.e.* emitted over a wide range of energies which can be either scanned, or else a specific energy selected ('tuned') using a monochromator. The latest generation of synchrotron light sources is facilitating analysis at even greater spatial resolution, and at ultradilute concentrations, key developments for the study of many environmental samples. An excellent overview of the applications of synchrotron radiation techniques in environmental science is provided by Fenter *et al.* (2002).

As discussed in more detail below, the development of more conventional laboratory instruments over recent decades has also had a dramatic impact on work in environmental mineralogy. These developments have included improved radiation sources such as field emission guns for electron microscopes, new types of ion beams such as C_{60} beams for mass spectrometry, and new sample chambers facilitating the analysis and imaging of samples which are hydrated or contain delicate biomaterials. Whereas the advances centred around synchrotron radiation have had a very great impact, the other development of great importance to the field has involved instrumentation at the very small scale. Scanning probe microscopy techniques (STM, AFM, see section 3.7 below) involving 'mechanical' scanning of a tip across a solid surface and registering either repulsive and attractive forces, or changes in an electric current when a voltage is applied between tip and sample, have revolutionized surface science. When conducted in vacuum, these methods enable imaging at atomic resolution, but can also be conducted in air, or in the presence of a fluid phase.

1.2. Routine methods of mineral characterization

Before turning to the specialist methods of analysis referred to above, it is worth emphasizing the continuing importance of the established, sometimes very simple methods long employed to study minerals. Indeed, without such basic characterization, the results from more specialized techniques are often of little value. In particular:

(1)   Optical microscopy, particularly using the polarizing microscope to examine thin and/or polished sections in transmitted or reflected light, is often the simplest means of identifying the mineral species present in a sample. It is also extremely sensitive in that a species occurring at a parts per thousand, or even parts per million level as a discrete phase can be identified. Most important of all, optical microscopy reveals textural relationships between phases and inter-relationships that provide information on the processes that have been operating and the sequence of formation (paragenesis) of phases. Comprehensive introductions to transmitted-light studies are provided by such texts as Kerr (1977) or Dyar & Gunter (2008) and to reflected-light studies by Craig & Vaughan (1994).

(2)   X-ray (powder) diffraction (XRD) can provide definitive information for the identification of crystalline phases present at levels $\geq 5\%$ in a sample. The theory and applications of the basic XRD techniques are described in numerous texts and papers (*e.g.* Zussman, 1977). Because of the importance of new developments in XRD methods and their applications to problems in addition to simple phase identification (such as particle-size determination), they are discussed in more detail in a later section of this chapter.

(3)   Electron microbeam methods, including electron probe microanalysis (EPMA), and (analytical) scanning electron microscopy (ASEM), can provide the detailed compositional, morphological, and textural information on the micro-scale to complement the data from 1 and 2 above. The EPMA can provide a full quantitative chemical analysis for a spot on a flat polished sample surface only, ~ 1 μm in diameter (see Reed, 1997, for review). Modern electron probe techniques have extended the capabilities of the instrument both in terms of rapid acquisition of large amounts of data and mapping of element distributions, and by making analysis of light elements (O, C, *etc.*) routinely possible, and analysis of trace elements possible under certain circumstances (*e.g.* Robinson *et al.*, 1998). The SEM is the most widely used microbeam instrument which can give high-resolution morphological information on powders, granular samples, *etc.* and when combined with Energy Dispersive Spectrometry (EDS), provide semi-quantitative chemical analysis. Again the important new developments in SEM studies, particularly those relevant to sensitive environmental samples, lead us to further discussion of this technique later in the chapter (see section 3.6). The other very important electron microbeam method is Transmission Electron Microscopy (TEM). Because of its importance for studying environmental materials, particularly ultrafine ('nano') particles, it is discussed in detail below.

In studying natural samples of materials such as soils, sediments and aerosols, the above techniques will be important, and may also be important in studying the materials used in laboratory experiments. However, as already noted, particular challenges arise in working with many of the key environmental systems because of the importance of fine particle solids, surface coatings and sorbates, and the often critical role played by the mineral surface in contact with a fluid phase. Such studies require additional, more specialist methods with an emphasis on the atomic or molecular scale, as discussed below.

2. Characterizing bulk solids and particles (including nanoparticles)

2.1. X-ray diffraction and X-ray scattering

2.1.1. Diffraction theory

Diffraction is the coherent scattering of waves from a three dimensional object. Constructive interference of scattered waves can be observed at certain specific angles which correspond to positions where the path-length difference between waves reflected off specific parallel planes of atoms are equal to an integer multiple of the wavelength of the incident radiation.

The general condition is familiarly expressed as Bragg's law:

$$n\lambda = 2d\sin\theta$$

(where n is an integer, λ is the X-ray wavelength, d is the interplanar spacing and θ is the glancing angle of incidence). There are physical limits on the d spacings that can be resolved *via* diffraction experiments. At extremely low angles, data collection is limited by the intensity of the straight-through incident beam. Even the most narrow incident beam and most tightly slitted-down detector will eventually allow straight-through beam at some small angle. Similarly, when θ is equal to $90°$ the detector would have to be coaxial with the incident beam and this is clearly not physically possible; therefore the minimum d spacing is $\lambda/2 = d$. In practice, with most laboratory systems, d spacings between 25 and 1 Å can be determined routinely.

As noted above, incident beams in laboratory sources are usually generated by heating a filament in a vacuum, which then emits electrons that impact on a metal target. This method gives intense radiation from the target at characteristic energies, and also causes continuum or *brehmstrahlung* (braking) radiation to be emitted. With a Cu target, the presence of iron in the sample being studied, common in natural samples, will also produce significant amounts of Fe-fluorescence. This increases background noise in the detector at all angles as fluorescence is produced isotropically. Fluorescence can be suppressed by using a secondary monochromator to only pass coherently scattered photons of a selected λ. The secondary monochromator will typically be attached to the front of the detector and scanned along with it. Use of characteristic radiation from a metal source also has the added complexity of the production of other characteristic X-rays besides the intense intended line. The standard method for suppressing this is to use an element with greater atomic number as a filter. By carefully choosing an element with a suitable absorption edge the undesired emission can be suppressed. This will also tend to cut down the *brehmstrahlung* but will not entirely eliminate it.

Another method is to use a primary monochromator to fix the incident radiation at a single energy. This essentially uses Bragg's law to filter the source radiation as follows. If the incident radiation has a range of components throughout the wavelength (energy) spectrum, then by using a fixed-crystal monochromator we can select a single wavelength by locating a slit on the monochromator at the angle corresponding to the desired λ. This has the advantage of removing all unwanted characteristic energies (even in some cases separating $K\alpha_1$ from $K\alpha_2$) as well as suppressing background from the *brehmstrahlung*. However, absorption in the monochromator can reduce the intensity of the desired radiation and so count times may need to be lengthened accordingly. A curved monochromator can compensate for this by both filtering and focusing the beam, thus maximizing the laboratory source in terms of peak resolution and count times. The parallel beam resulting from this can also be used to analyse non-standard samples; that is, it allows diffraction patterns to be acquired from samples that are not entirely on the focusing circle. This means that intact samples that are neither powders nor a single crystal can be analysed, and it opens up XRD to the non-destructive analysis of precious samples (see Chiari, 2013, this volume, for further details and applications).

The parallel beam can also be used to perform depth-resolved XRD. For small angles, the depth penetration of X-rays is a function of incident angle. The higher the angle, the deeper the penetration. Measuring XRD patterns at different incident angles can thus allow distinct mineralogical surface layers to be distinguished as a function of depth. Surface reflectivity is a method closely related to glancing incidence diffraction and will be discussed later in this chapter. The intensity produced by the focusing mirror is required to be able to perform reflectivity analysis with a standard laboratory source, and reflectivity measurements can be optimized by the use of synchrotron radiation (see below).

Diffuse or incoherent scatter will also occur because of the random rough surfaces interacting with the beam. The rougher the surface the more intense the diffuse scatter becomes. Below we discuss how this can be used in well constrained experiments to produce useful information. However, in most cases, the diffusely scattered portion of the beam is undesired noise. Therefore secondary slits, including secondary Soller slits, are typically used to remove this diffuse portion of the scattered beam. Whether the data are collected as a function of angle or a function of energy, the peaks correspond to specific lattice plane distances in the crystalline phases present in a sample. For most routine work, the 2θ angle of the detected peak and the intensity of the peak are both recorded.

In environmental mineralogy, XRD is particularly useful in providing information on phase identification, the nature and kinetics of phase transformations, the quantification of phase percentages, and on particle-size determination. As these are often critical pieces of information, their acquisition will now be discussed further.

2.1.2. Phase identification

Phase identification is the most common use of XRD; this application needs as full a range of d spacings as possible. Measured d spacings and intensities are then compared with a library of available standard diffraction patterns using a search/match algorithm. Often an internal standard such as quartz or halite will be added to the sample to ensure that the measured 2θ values are converted accurately to d spacings. Furthermore, as there are $>20,000$ patterns in the library of powder patterns, some intelligent choices should be made when implementing the search/match program to reduce the number of comparisons made and increase the reliability of the results. We should also mention that automated sample changers are now available which can be highly advantageous for applications that require the analysis of large sample sets, *e.g.* air-filter samples, canal sludges, contaminated land assays. Of course, only crystalline materials can be identified using XRD.

2.1.3. Time-resolved (kinetic) studies

As part of phase identification, applications which involve time resolution are particularly important in Earth and Environmental Sciences. As discussed below, XRD using energy dispersive spectrometry (EDS) methods is ideal for kinetic studies, but that requires access to a synchrotron. For laboratory sources, it is now cost effective to use a position-sensitive detector (PSD) to perform kinetic experiments. Instead of a standard narrow slit and scintillation counter on the detector side of the mechanism, the PSD has a thin wire detector, which covers a fairly long range of 2θ arc, on the order of 8 to 20° of arc. Once calibrated, the detector can convert the detection of photons at points along this length to intensity as a function of 2θ angle; this means the detector can collect a number of peaks simultaneously. The resolution of a PSD is now comparable to many standard systems, angular resolution as good as 0.005° can be obtained. Therefore, a PSD is an excellent tool for studying phase changes where the kinetics or the reaction path needs to be determined.

In the context of kinetic studies, a major advantage is obtained at a synchrotron for diffraction experiments. We generally think of fixing λ and recording d spacing as a unique θ (or 2θ) value. But with a white beam (*i.e.* a wide range of λ), diffraction peaks will occur in a fixed position detector (such as an Energy Dispersive Spectrometer, EDS) as an array of peaks at a range of energies (λ). Because an entire spectrum can be obtained rapidly with this technique, it is ideal for kinetic studies of mineral reactions. Figure 1 shows an example of this from studies of the very rapid precipitation of copper sulfide from aqueous solution. As discussed in detail below, when combined with X-ray absorption spectroscopy (XAS) studies of the same system, it is a powerful way of investigating the structural evolution of a nanoparticulate precipitate.

Another example of the use of energy dispersive XRD for kinetic studies, and one of great relevance to environmental mineralogy, is the study by Vu *et al.* (2008) of the

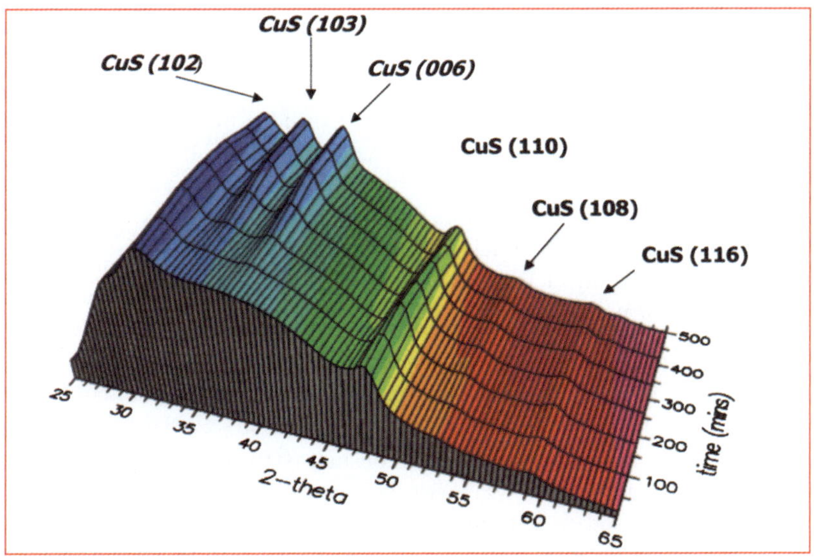

Fig. 1. Energy dispersive, time-resolved XRD trace showing the structural evolution of a copper sulfide precipitated from aqueous solution (unpublished data of Pattrick, Vaughan, Bell and Charnock).

transformation of ferrihydrite to hematite. This study shows that hematite forms from ferrihydrite *via* a two-stage crystallization process with goethite as an intermediate phase. The data were used to support a transformation mechanism involving an aqueous-aided, two-dimensional process.

2.1.4. Quantitative phase analysis

A more difficult analytical problem is that of using XRD for quantitative analysis of the phase percentages in an unknown sample. Essentially this requires using intensities of peaks to calculate phase percentages. However, intensities can be affected by preferred orientation (particularly in the environmentally important phyllosilicate minerals), major element substitution, cation disorder, defect density, and particle-size distribution. Therefore, most studies that attempt to determine the phase percentages from XRD do so by comparison to carefully calibrated standards. Because the range in phase percentages can be quite broad for natural samples, this necessitates the production of a number of standards. Methods for standard-less phase-percentage analysis are available; however, we still recommend employing standards when using this approach, if for no other reason than that this will allow the researcher to place realistic error bars on the results.

2.1.5. Particle-size determination

Many low-temperature reactions are surface controlled and therefore the absolute rate depends critically on surface area. Direct measurement of surface area can be accomplished *via* a gas adsorption method known as BET analysis (after Brunauer *et al.*, 1938) but this is often quite a difficult measurement and requires a fairly large sample size (10–20 g). Furthermore, BET gas adsorption analysis is an *ex situ* technique and, therefore, cannot be used to follow surface-area changes during the course of a reaction. X-ray diffraction can, under certain circumstances, produce information about the grain-size distribution in a sample. Consider the simplest case, that of a uniform powder with all grains of the same dimensions. If the grain diameter is <200 μm, then the XRD peaks will be broader than they would be for a coarser-grained sample. As the grain size of this uniform powder decreases, the peaks continue to broaden. Therefore, for extremely fine powders, the average particle length is an inverse function of the diffraction peak widths. This relationship is quantified in the Scherrer equation for particle size analysis:

$$C_s = \frac{K\lambda}{S\cos\theta}$$

where C_s is average particle length, K is a constant (0.9), λ is wavelength and S is the Full Width at Half Maximum (FWHM) of the peak. A detailed analysis of a number of peak widths can be used to determine the grain-size distribution of a non-uniform unknown. This can then be used to calculate geometric surface areas of a sample but, more importantly, can be used *in situ* to follow grain growth or reduction during reactions.

2.1.6. Other important applications of XRD

There are several other important applications of XRD methods. The first and historically by far the most important of all applications of XRD is the determination of the crystal structure of a mineral or other crystalline solid where that is still unknown. It is beyond the scope of this chapter to give an explanation of the methods used to extract structure from XRD patterns. Excellent introductory references include Zoltai & Stout (1984), and Bloss (1971). Again it is important to note that synchrotron methods offer crucial advances in this area. Single-crystal diffraction studies of the structure of very small crystals (volumes only $50-160$ μm^3) are now possible (see, for example, Burns *et al.*, 2000). It is also possible to use data collected from particulate ('powder') samples to determine a crystal structure (as for the environmentally important mineral mackinawite; see Lennie *et al.*, 1995). Once a reasonable structure is obtained, the cell parameters can be refined by least-squares fitting. The high-angle data have the smallest uncertainties and so, in order to produce the best quality fit to a pattern, the high-angle peaks should be recorded with the best possible accuracy. Common algorithms for least-squares refinement include the Rietveld method (see Young, 1993).

Other areas of application concern extracting other crystal chemical information, such as compositional data where, for example, the mineral is part of a solid solution and this can be related to the XRD pattern or cell parameters derived from it. A rather more specialist area involves strain analysis, where measurement of the shifts in *d* spacing can be correlated with the strain of a deformed crystalline material.

2.1.7. X-ray scattering

X-ray scattering methods, wide-angle X-ray scattering (WAXS) and, particularly, small-angle X-ray scattering (SAXS) are important techniques for the study of very fine-particle solids. These can be used for determining particle sizes, overall shapes, and even structural information for mineral fine particles in the nanometer size range and, usually, suspended in aqueous solution. As emphasized elsewhere in this volume, one of the major developments in environmental mineralogy during the past ten years has been the explosion in research effort expended in understanding what are now termed 'nanoparticles'. An excellent and succinct review of this field is provided by Hochella *et al.* (2008) and a more extensive account of earlier work provided by Banfield & Navrotsky (2001).

Although scattering experiments can be performed with a laboratory X-ray source, most work is now done using the much more intense flux provided at a synchrotron. In the experiment, a monochromatic beam of X-rays is focused on the sample and, whereas most X-rays simply pass through the sample, a small proportion are elastically scattered and the scattering pattern contains information on the structure of the sample. In SAXS, the scattered radiation is recorded at very low angles (typically 0.1 to 10°); the WAXS regime is at higher angles. Information that can be extracted from WAXS data includes that concerned with intermediate range structural order. For example, Gilbert *et al.* (2004) were able to quantify intermediate range order in 3.4 nm diameter ZnS nanoparticles showing that structural coherence is lost over distances >2 nm. In another study, Michel *et al.* (2007) were able to probe the 'crystal' structure of nanocrystalline ('amorphous') ferrihydrite, an iron oxyhydroxide mineral of great environmental importance, and where conventional X-ray diffraction methods are not effective in structural determination.

The SAXS technique has been used in particular to study nucleation and growth phenomena, as when minerals are precipitated from solution. It has been applied to ferric oxyhydroxides, clay minerals (Carrado *et al.*, 2000), the calcium silicates involved in cement technology (Shaw *et al.*, 2002) and the nucleation and growth of silica nanoparticles (Tobler *et al.*, 2009). A general reference on SAXS is the book by Glatter & Kratky (1982).

2.2. Transmission electron microscopy

Transmission Electron Microscopy (TEM) is another long-established technique where instrumentation and sample-preparation developments have provided important new advances. The technique relies on the interpretation of images and electron diffraction patterns obtained when a focused beam of electrons passes through a very thin specimen (usually $\sim$50 nm thick). Resolution on the order of a few angstroms can be obtained using modern instruments, so that surface and interface regions, or the cores and rims of 'nanoscale' particles (see Fig. 2), can be studied. Electron diffraction patterns from very small areas (so-called Selected Area Electron Diffraction, SAED) can provide structural information for very small volumes of crystalline mineral phases, and the combination with energy dispersive spectrometry (EDS) can also provide semi-quantitative compositional data for those very small regions (McLaren, 1991; White, 1985). Electron Energy Loss Spectroscopy (EELS) is another TEM analytical technique, one that is particularly sensitive for lower-atomic-number elements (as well as some metals). In

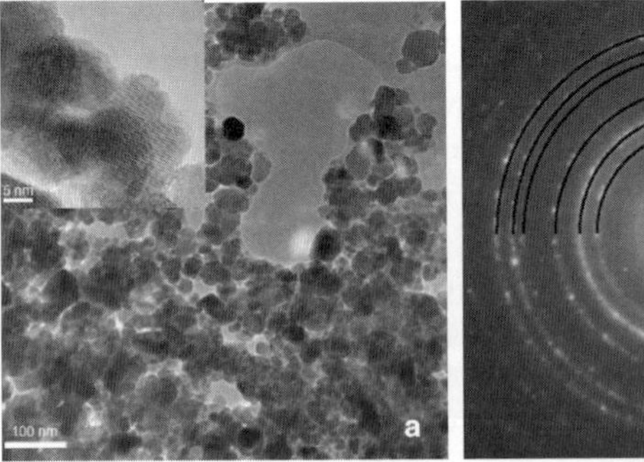
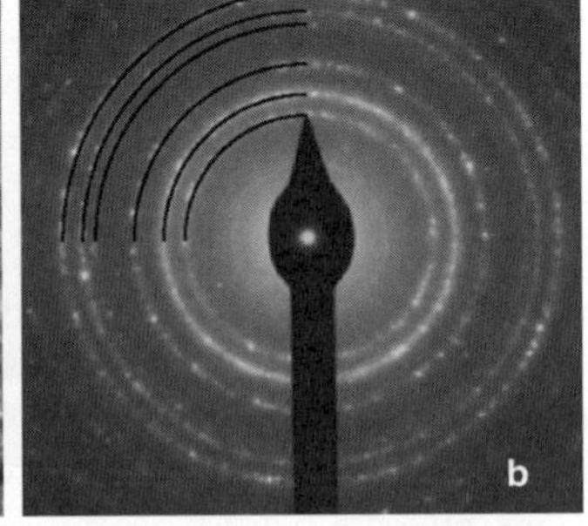

Fig. 2. TEM data for Fe oxide nanoparticles produced by bacterial reduction (with *Geobacter sulfurreducens*) of inorganically synthesized feroxyhyte: (**a**) image of the nanoparticles showing grain sizes and morphologies (note the inset at higher magnification showing differences between rims and cores); (**b**) selected area electron diffraction (SAED) pattern for the sample showing that it has been transformed to magnetite (from Cutting *et al.,* 2009, reprinted from *Geochimica et Cosmochimica Acta,* **73,** Mineralogical and morphological constraints on the reduction of Fe(III) minerals by *Geobacter sulfurreducens*, pp. 4004–4022, © 2009 with permission from Elsevier).

EELS, a proportion of the electrons transmitted through the specimen are inelastically scattered because they have entered the inter-atomic domains of the specimen and have lost energy. The energy loss is characteristic of the atom with which these inelastic electrons have interacted; the inelastic electrons now carry the 'signature' of the atom. So-called 'electron-filtering' TEM instruments have the ability selectively to filter the inelastic electrons of the transmitted beam, so that angstrom-resolution images of the distribution of elements can be produced; for example, the iron and oxygen distribution can be seen in extremely fine-grained hematite. This is called Electron Spectroscopic Imaging (ESI). Important advances have been made in sample preparation and handling for TEM studies by using low temperatures (down to $-196°C$) to preserve microbial cells and other biomaterials and, then, by keeping the sample at these temperatures on a 'cryo-stage' in the TEM. The developments related to studies of biomaterials are discussed in detail by Beveridge *et al.* (2007).

2.3. X-ray absorption spectroscopy (XAS)

The availability of synchrotron radiation sources providing X-rays beams several orders of magnitude more intense than those from the standard X-ray tube has made XAS an important tool for the study of materials. The absorption of an X-ray photon of energy E by a sample of thickness x is given by:

$$I_t(E) = I_0(E)e^{-\mu_{total}(E)x}$$

where $I_t(E)$ and $I_0(E)$ are the transmitted and incident photon intensities, and $\mu_{total}(E)$ is the total absorption coefficient of the sample. If the X-ray photon energy is tuned to the binding energy of some core level of an atom in the material then an abrupt increase in the absorption coefficient occurs producing an absorption edge. A spectrum of the absorption edge shows little or no fine structure in the case of an isolated atom (see Fig. 3a), but for atoms in a molecule whether in a gas, liquid or solid, or in a crystal, the variations of absorption coefficient around and above the absorption edge result in complex fine structure. The former is studied in X-ray absorption near-edge structure (XANES) spectroscopy, and the latter in extended X-ray absorption fine structure (EXAFS) spectroscopy. The energy regions studied in each of these techniques are illustrated in Figure 3b. Near or below the edge absorption peaks often appear due to the excitation of core electrons to bound states (*e.g.* 1s $\rightarrow$ nd, $(n + 1)$s, or $(n + 1)$p orbitals for the K edge). Such processes can provide useful information on chemical bonding and electronic configuration.

The EXAFS is, in fact, a final-state interference effect involving scattering of the outgoing photoelectron from the neighbouring atoms, as simply illustrated in Figure 3c. Thus, whereas in a mono-atomic gas, the photoelectron ejected by the absorption of an X-ray photon will travel outwards unimpeded and the X-ray absorption curve will decay smoothly, in the presence of neighbouring atoms this outgoing photoelectron will be backscattered, producing an incoming wave that can interfere either constructively or destructively with the outgoing wave near the origin. The result is oscillatory

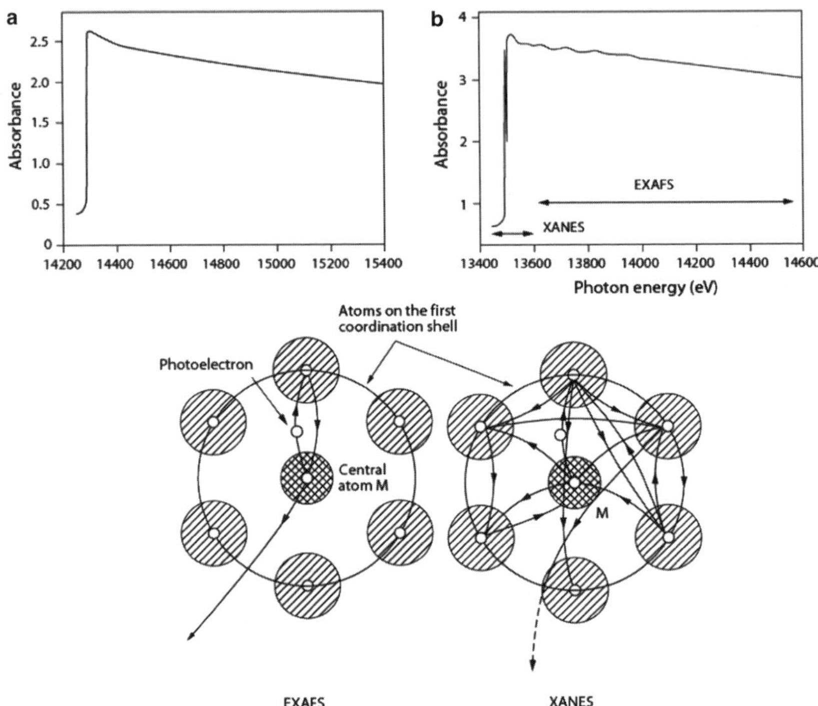

Fig. 3. X-ray absorption spectroscopy – principles of XANES and EXAFS: (**a**) XAS of an isolated atom; (**b**) XAS of a polyatomic system showing XANES and EXAFS; and (**c**) electron scattering processes leading to XANES and EXAFS.

behaviour of the absorption, and the amplitude and frequency of this modulation depend on the type and bonding of the neighbouring atoms and their distances from the absorber. The distinction between XANES and EXAFS can be simply pictured as shown in Figure 3c. The XANES are produced by multiple scattering resonances among neighbouring atoms.

The XANES and EXAFS spectroscopies are a particularly powerful means of investigating materials of interest in environmental mineralogy. This is because they can provide information on a particular element of interest and its chemical and local structural state, even in materials lacking long-range order such as a fluid, nanoparticle precipitate or, as discussed below in section 3.5, sorbate on a surface. Most elements can be studied and in concentrations ranging from parts per million to the pure element. The high intensity of synchrotron radiation means that very small samples (μg) or samples containing the element of interest at very low concentrations (ppm or even less) can be studied under favourable circumstances. Most experiments do not require the sample to be *in vacuo* and, where necessary, samples may be studied under a wide range of temperature or pressure conditions, or in controlled atmospheres. Because XAS is an element-specific bulk method providing information about the

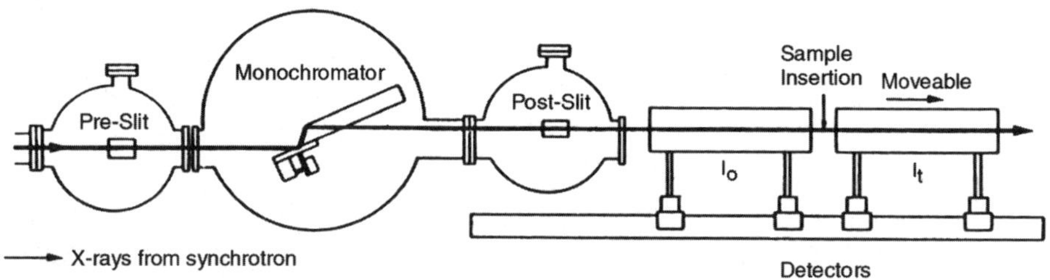

Fig. 4. Experimental system used for measuring EXAFS and XANES in transmission mode.

average local structural and compositional environment of the absorbing atom (generally the two or three closest shells of atoms can be probed extending to a radius of $\sim$6 Å), it can be used to study compositionally complex materials. For many systems, an EXAFS analysis can yield average distances between the absorbing atom and its near-neighbour atoms accurate to $\pm$0.02 Å and average co-ordination numbers to $\pm$10–20%. Analysis of the XANES spectrum can provide information about the oxidation state and local environment of the absorbing atom, most commonly through comparisons with model compounds.

Most XAS experiments are conceptually simple and involve the kind of set-up illustrated in Figure 4, in which the X-rays from the synchrotron are collimated and a crystal monochromator used to "select" and scan through a particular range of energies. The X-ray intensities before and after interaction with the sample are simultaneously recorded as incident (I_0) and transmitted (I_t) intensity values as the X-ray energies (E) are scanned through an appropriate range, from below to above the absorption edge. Typically, this is plotted as $\ln(I_0/I_t)$ *vs.* E (in eV). The relationship between these intensities and the linear absorption coefficient μ of a sample of thickness x (in cm) is $\ln(I_0/I_t) = \mu x$. As an alternative to directly measuring the transmittance of X-rays through the sample, the secondary process of X-ray fluoresence, I_f, can be measured and ratioed with I_0; this is proportional to μ for dilute samples. A particular advantage of measuring this so-called fluorescence yield is that it is more sensitive to low concentrations of elements in highly absorbing matrices. The EXAFS (and XANES) spectra obtained from such experiments (examples of which are shown in Figs 5 and 6) are subject to well established methods of analysis using packages of programs available at synchrotron laboratories. Detailed discussion of such analysis is inappropriate here; suffice to say that after following procedures concerned with spectral calibration and background subtraction, most such packages involve 'fitting' of the experimental data by a process of comparison with spectra generated using theoretical data for the absorber atom in question surrounded by shells of atoms at various distances. Particularly helpful is the examination of the Fourier transform of the EXAFS spectrum which yields a radial structure function (see Fig. 6 below). The magnitude of this function can be plotted against radial distance from the absorber atom (R in Å); peaks in such a plot correspond to shells of atoms around the central absorber (Fig. 6 below). However, it is important to note that it is the EXAFS spectrum that is fitted, not the Fourier transform.

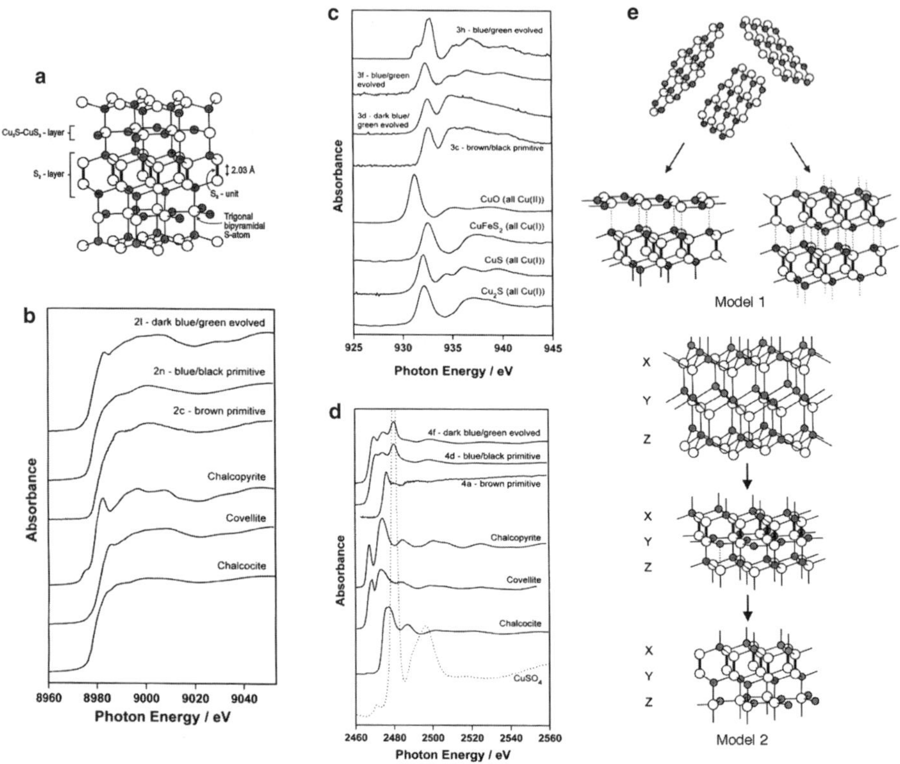

Fig. 5. XAS studies of CuS precipitates (after Pattrick *et al.*, 1997, reprinted from *Geochimica et Cosmochimica Acta*, **61**, The structure of amorphous copper sulfide precipites: An X-ray absorption study, pp. 2023–2036, © 1997 with permission from Elsevier): (**a**) the covellite (CuS) crystal structure; (**b**) Cu *K* edge XANES spectra of CuS precipitates and various model compounds; (**c**) Cu *L* edge XANES spectra of CuS precipitates and various model compounds; (**d**) S *K* edge XANES spectra of CuS precipitates and various model compounds; and (**e**) models for the structural evolution of CuS precipitates.

The EXAFS (and XANES) spectroscopies are particularly powerful methods of studying a range of environmental problems. Because these methods can be applied to liquids and to non-crystalline solids, they have proved important in studying: (1) the chemical form (speciation) of certain elements (notably metals) at low concentrations in aqueous solution; (2) in characterizing the local structure and structural evolution of very fine particle precipitates as a function of time or changing physico-chemical conditions, and (3) in determining the local environments of atoms taken-up from aqueous solution by interaction with the surfaces of minerals, including fine precipitates (see section 3.5). The XAS methods are probably the most powerful way of probing the local environment of a particular atom in a natural (or model) environmental system. A few examples will serve to illustrate the power of XAS methods.

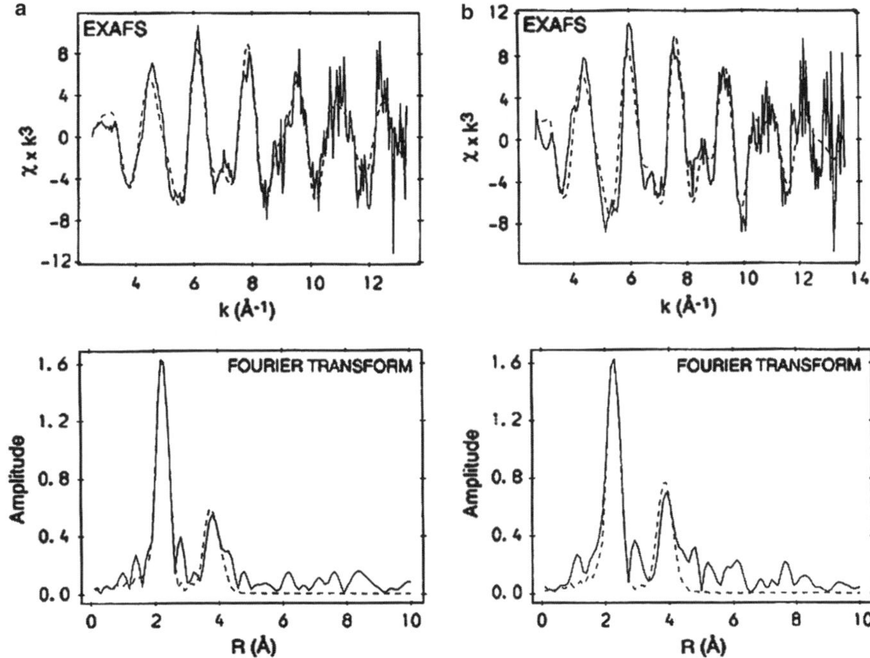

Fig. 6. EXAFS spectra (above) and associated Fourier transforms (below) for anoxic black mud samples from the Afon Goch estuary, Anglesey, Wales (after Parkman *et al.*, 1996): (**a**) experimental (solid line) and best fit theoretical (dashed line) spectra and Fourier transform for Cu *K* edge; (**b**) experimental (solid line) and best fit theoretical (dashed line) spectra and Fourier transform for Zn *K* edge.

The very fine particle (commonly 'nanoparticle') precipitates formed when there are changes in pH or in the activities of other components in natural fluids have been little studied until recent years, not least because of the difficulties involved in their characterization. Examples include the hydrated oxides of iron formed on mixing of surface or groundwaters with consequent changes in pH or redox conditions, and the sulfides of iron, copper and other metals formed just beneath the sediment surface in near-shore areas where the activities of sulfate-reducing bacteria provide a source of sulfide ions to react with dissolved metals. Pattrick *et al.* (1997) have studied the precipitates formed on reaction between copper and sulfur in aqueous solution and their transformation to the stable monosulfide mineral, covellite. The energy dispersive XRD data collected as a function of time from first formation have been shown above (see section 2.1.3 and Figure 1). The Cu *K*, Cu *L* and S *K* edge X-ray absorption near-edge structure (XANES) and extended X-ray absorption fine structure (EXAFS) spectra were also obtained from precipitates that had been 'frozen' in liquid nitrogen within a few seconds of formation to prevent further reaction, or been allowed to age for various periods of time, eventually transforming (with colour changes from brown → blue/ black → blue/green) to the quite complex layer structure of covellite (see Fig. 5a).

Covellite contains both tetrahedrally and trigonally co-ordinated copper, and both single sulfur atoms and S—S groups. As seen from Figure 5b, the Cu K edge XANES spectra are very different for the first-formed precipitates ('brown primitive') and those precipitates allowed to age for some tens of hours at 25°C ('blue/green evolved'). The latter are very similar to crystalline covellite. The Cu L edge spectra and S K edge spectra (see Fig. 5c,d) were used to probe the oxidation state of copper, which is Cu(I) in all the precipitates, and the character of the sulfur which shows evidence for S—S bonds in all cases. Based upon a detailed analysis of these spectra with additional data from XPS studies, the authors were able to suggest two possible models for the structural evolution of CuS precipitates (see Fig. 5e); one involves planar Cu_3S–CuS_3 layers coming together to form the primitive structure, the other involves a wurtzite structure precursor which would undergo a translation of atoms to form the primitive structure (which would undergo further translation and rotation to form the covellite structure).

The CuS example illustrates both the power of X-ray absorption spectroscopy (XAS) as a local structure probe and the complexity of the transformations that may occur in such systems. Other natural sulfide precipitates have been investigated in this way, notably the Mn–S, Hg–S, and Fe–S systems (Moyes *et al.*, 2000; Bell *et al.*, 2010; Butler *et al.*, in prep.), and progress has been made in studying the most important of them, namely the Fe–S system using both XAS methods (Lennie & Vaughan, 1996; Butler *et al.*, in prep.) and synchrotron XRD methods (Benning & Barnes, 1998). Here, the first-formed sulfide (after <1 s of reaction) has been shown to have the local structure of tetragonal FeS, mackinawite, with subsequent transformation on ageing to pyrrhotite when oxygen is rigorously excluded or to greigite when some oxygen is present. The XAS method is also a powerful technique for investigating natural samples containing fine-particle metal sulfides. For example, the anoxic black muds in the estuary of the Afon Goch, downstream from the former mining area of Parys Mountain (Anglesey, Wales, UK) are heavily contaminated with copper and zinc, as revealed by bulk chemical analysis. Analysis by EXAFS at the Cu K edge and Zn K edge as seen from the spectra and their Fourier transforms (Fig. 6), shows that both Cu and Zn are present as discrete sulfides. In the case of Zn, the best fit to the EXAFS spectrum is achieved with four S atoms at 2.31 Å in the first shell, and 12 Zn atoms in the second shell at 3.80 Å, as in the ZnS mineral, sphalerite. On the other hand, the copper also is best fitted with four S atoms at 2.28 Å and a second shell of Cu or Fe atoms at 3.70 Å, indicating that it is present not as a binary copper sulfide but as a copper-iron sulfide, probably chalcopyrite. The XANES spectrum of the anoxic black mud also very closely resembles chalcopyrite and is very different from binary copper sulfides such as covellite (Parkman *et al.*, 1996).

The XAS methods also provide some of the most powerful ways of studying reactions at mineral surfaces, particularly sorption phenomena where the local environment around a sorbate can be investigated in detail. This aspect of XAS applications in environmental mineralogy is discussed in detail in a section (3.5) which follows.

2.4. Advanced microanalysis and mapping (X-ray microprobe, spectromicroscopy, and microtomography, and particle-induced X-ray emission)

Over recent years, major advances have been achieved in the analysis of ever smaller volumes of material at ever lower concentrations of the components of interest. This has involved the focusing of incident radiation beams down to very small spot sizes. The capability to scan (or 'raster') such beams has allowed for the mapping (in two or three dimensions) of samples at micron, sub-micron, or even nanometre scales. For the most part, these methods have only become possible through the use of synchrotron radiation (although the PIXE technique, discussed below, requires a particle accelerator to generate a beam of protons). These advanced methods of microanalysis and mapping may involve elemental analysis or, in some cases, may provide information on the speciation of the elements present. Some of the techniques are discussed briefly below; although representative of this rapidly developing field, this is by no means a comprehensive list. A much more detailed account is to be found in Fenter *et al.* (2002).

2.4.1. Scanning transmission X-ray microscopy (STXM)

STXM is a method operating at a small number of synchrotron facilities worldwide and which uses soft (low energy; $\sim$100–2000 eV) X-rays or photons in the ultraviolet region which are focused down to a beam of $<$40 nm in diameter. Because of the energies employed, this method is a powerful means of probing the form and distribution of the light elements of importance in organic compounds and biomaterials (C, O, N *etc.*), and therefore of processes of biomineralization or of mineral–microbe interactions. For example, in work by Obst *et al.* (2009), STXM has been used for 3-dimensional imaging of calcium carbonate biomineralization by a planktonic freshwater bacterium (*Synechocouss leopoliensis*). The absorption of X-rays tuned to the energies of the C1s and Ca2p edges enabled the mapping of calcite precipitation on the surface of the organism. The STXM energy range also covers the *L*-edges of the first row transition metals, providing important oxidation state information, and allowing organic and inorganic components of systems to be examined.

2.4.2. (X-ray) photoemission electron microscopy – ((X)PEEM)

Photoemission electron microscopy involves excitation of the sample using (synchrotron generated) X-rays; it can also employ UV light as the source of excitation. The incident radiation is linearly or circularly polarized, and the emitted electrons are imaged in the microscope using electromagnetic lenses like those employed in a transmission electron microscope (Scholl *et al.*, 2002). The technique has particular applications in magnetic domain imaging but, more importantly for environmental applications, when an energy filter is added, the electrons contributing to the image can be selected. Hence, photoemission spectra can be selectively acquired with spatial resolution better than 100 nm and with sub-eV energy resolution and used to generate elemental images for particular chemical states (such as oxidation states). Applications to environmental problems are discussed by de Stasio *et al.* (2001).

2.4.3. X-ray microprobe and microtomography

The X-ray (or XRF) microprobe uses the X-ray fluorescence process (as in X-ray emission spectroscopy – see Fig. 12) to identify and quantify specific elements in a sample. The technique can provide quantitative data on the presence and the amounts of major and trace elements, but does not generally yield information on speciation, oxidation state, coordination or other more subtle aspects of chemistry. This is another method which owes its existence to the availability of synchrotron sources, many of which have a beamline providing higher-energy ('hard') X-rays dedicated to the technique. The advantage of using an XRF facility at a synchrotron rather than in a conventional laboratory is the high X-ray flux and capability of focusing the beam down to a spot size as little as 0.1 μm. The extension of the method to scan and map compositional variations in two dimensions is another advantage of the microfocus beam, and one that can be extended to three dimensions for microtomography. The latter is essentially just an extension of the technology used in medical CAT scans to high spatial resolution. A comprehensive review of X-ray microprobe and microtomography methods is given by Sutton *et al.* (2002).

2.4.4. Particle-induced X-ray emission

Particle-induced X-ray Emission (PIXE) is a technique that is related to electron beam methods of analysis but, instead of an incident electron beam, PIXE uses a focused beam of particles, usually protons, to excite the emission of characteristic X-rays from the sample. A particle accelerator is required to generate the beam, so that PIXE is much less widely available than the electron beam methods. The major advantage of this method, one related to the large mass of the proton compared with the electron, is the much better ratio of peak to background in the acquired data; hence, detection limits for many elements are as low as 1 ppm. PIXE can also be used to map element distributions and has better spatial resolution than methods such as EPMA. Most PIXE systems also have the capability of detecting backscattered protons and this can be used to determine the major element composition at the surface and its variation with depth ('depth profiling'). This related technique, known as Rutherford Back-Scattering (RBS) was one of the first surface-analysis methods developed. Such depth profiling and surface analysis methods are considered in detail later in this chapter. Useful reviews of the PIXE technique are provided by Johansson *et al.* (1995) and by Cabri and Campbell (1998). Examples of the application of these methods are discussed later in this chapter.

2.5. Infrared and Raman spectroscopies

Energy in the infrared (IR) region of the electromagnetic spectrum ($\sim 0.75 - 100$ μm) can be absorbed by material through vibrational processes such as the bending and stretching of the bonds between atoms. This can be observed in gases, liquid or solids. In the simplest case, measurement of the positions and intensities of absorption peaks as a function of energy (wavelength) can provide a 'fingerprint' identification of a particular molecule or crystalline material present in a sample. The IR spectra can be acquired in absorption mode by measuring absorption peaks when the beam passes through the sample (in the case of gaseous, liquid, particulate, or more rarely, single-crystal materials), or reflection mode when it is reflected off the surface of a solid crystal or compressed powder. Surface coatings can also be studied in reflection mode. The detailed analysis of an IR spectrum involves correlating individual peaks with the bending or stretching of particular bonds in the material under investigation. This can provide information on molecular or crystalline structure, or on chemical changes taking place in a system, either with time or with changing conditions. Recent decades have seen the development of equipment enabling imaging of samples in the IR (IR microscopes) to facilitate the mapping of particular components in complex samples. As with virtually all of the analytical methods that use electromagnetic radiation, IR spectroscopy is being revolutionized by the application of synchrotron radiation. The intensity of the IR light from synchrotron sources enables small spot analysis and chemical imaging at resolutions down to ~ 1 μm. Because IR radiation is particularly sensitive to the detection of the bending and stretching of bonds such as O—H, C—H, C—O, N—H and C—N, it is a powerful probe of organic compounds and biomaterials, and of their interactions with minerals. An excellent review of developments in this area is provided by Hirschmugl (2002).

One example of a recent advance in this field is that of using multiple internal reflection Fourier transform infrared (MIR-FTIR) to perform *in situ* studies of organic complexation at mineral surfaces (Morris & Wogelius, 2008). Along with information about complex configuration, this technology can also provide data on surface diffusion rates and the polymerization of residual silica during cation-proton exchange. This development will be discussed further below; it may be a critical new method for studying the reactivity of radioactive waste forms.

Vibrational spectra can also be studied using Raman Spectroscopy. Here, monochromatic light from a suitable source (a powerful laser in modern instruments) is directed through a translucent sample (gas, liquid, glass, crystalline solid or powder) and the light scattered at 90° or 180° is recorded. The light scattering is of two kinds, Rayleigh scattering, where no frequency change occurs, and Raman scattering which involves a change of frequencies arising from vibrational effects. As with IR spectroscopy, Raman spectra can be used to provide 'fingerprint' identification of particular materials, or to provide more detailed information on molecular or crystal structure, elastic or thermodynamic properties. It is another of the techniques which has been refined over the years; for example, so as to study spectral signals from very small areas using the 'Raman microprobe' and even to obtain 3-dimensional images on a micro-scale. An example is shown in Figure 7 where a fossil bacterium in ~ 850 Ma old rocks is shown using traditional optical imaging in thin section and using 3-D Raman spectroscopy (data from Schopf & Kudryavtsev, 2005). A thorough review of the application of Raman in Earth sciences is given by Dubessy *et al.* (2012).

2.6. Nuclear and magnetic spectroscopies (NMR, Mossbäuer, XMCD)

2.6.1. Nuclear magnetic resonance spectroscopy

Nuclear magnetic resonance (NMR) spectroscopy involves measuring the absorption of electromagnetic radiation by the nuclei of certain atoms in the presence of an externally

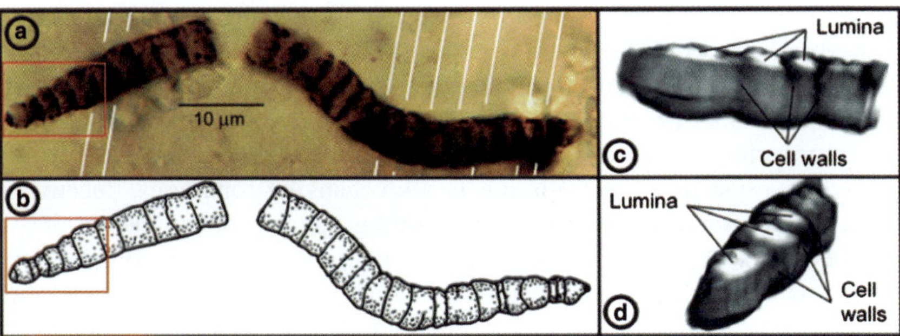

Fig. 7. Fossil cyanobacterium trichome (*Cephalophytarion laticellulosum*) from the Precambrian ($\sim$850 Ma old) Bitter Springs Formation (central Australia): (**a,b**) photomicrograph of a thin section with derived interpretive drawing; (**c,d**) images from 3-D Raman spectroscopy of the region outlined in red in parts **a** and **b** (from Schopf and Kudryavtsev, 2005, reproduced with the permissions of John Wiley and Sons).

applied magnetic field. The simplest type of experiment involves applying a fixed electromagnetic field to the sample, usually in the radiofrequency region, and varying the applied magnetic field until the radiation is most strongly absorbed. Most NMR work conducted by chemists has involved the ^{1}H atom, where the position of the resonance signal with respect to a standard, along with any fine structure and information on width and shape of peaks, can be used to identify protons in different positions in a molecule and hence help to determine the molecular structure. Although solid samples generally give broad and poorly resolved spectra, a variant of the standard NMR experiment can be used to study crystalline solids. Here, the sample is held in a magnetic field and is spun in a specific orientation (the 'magic angle') to give high-resolution data on the nuclei of atoms such as ^{29}Si and ^{27}Al, and to provide insights into the crystal chemistry of silicate and aluminosilicate minerals by pinpointing different environments for these elements in their structures. Such studies of minerals and glasses have been reviewed by Kirkpatrick (1988). Although a significant number of other nuclei of environmental interest are potentially accessible to studies by NMR methods, many have not been studied in any detail. A group of techniques based on the NMR phenomenon involves mapping in three dimensions (tomography) the distribution of a target nucleus and hence atom. This is the basis of the body scanner instruments now widely used in medical diagnosis.

2.6.2. Mössbauer spectroscopy

Mössbauer spectroscopy involves the recoilless emission and resonant absorption of gamma rays by a target nucleus. Although the Mössbauer effect can be observed in over twenty elements, only one element, iron, accounts for the great majority of studies. Of course, iron is an extremely important element in many environmental systems. The absorption by the ^{57}Fe nucleus (2% natural abundance) of a 14.4 keV gamma ray emitted by a ^{57}Co source occurs by the nucleus going from a ground state to an excited state. The position of the absorption peak relative to a standard, and whether it is a single peak, or a doublet or sextet, depends upon factors such as the oxidation state and coordination environment of the iron in the sample, and (for a sextet) if it is magnetically ordered. It is only the iron in the sample that gives rise to the spectrum, although the iron in different crystallographic sites in a single mineral, or in different phases, will all contribute to the overall spectrum and with a peak intensity proportional to its percentage presence in the sample. Because spectral resolution is improved at lower temperatures, and because magnetic properties are temperature dependent, it is often useful to record spectra at liquid nitrogen ($\sim$77 K) or even liquid helium ($\sim$4.2 K) temperatures as well as at room temperature. Like the XAS methods (see below), Mössbauer spectra can be obtained from amorphous and glassy materials as well as crystalline solids (but not from liquids). Mössbauer spectroscopy offers a powerful way in which to determine Fe^{2+}/Fe^{3+} ratios and to determine the distribution of iron between different phases in a system (including the poorly crystalline or amorphous iron oxyhydroxides or sulfides often present in natural sediments and soils). An example is given in Figure 8 (taken from Hery *et al.*, 2010). Good sources of further information are Maddock (1985) and Hawthorne (1988). Although the ^{57}Fe Mössbauer spectra of most iron-bearing minerals were studied in the 1960s and 1970s following the discovery of the Mössbauer effect, the technique has undergone something of a renaissance in recent years because of the importance of iron in many environmental systems and its value as a redox indicator.

2.6.3. X-ray magnetic circular dichroism

X-ray magnetic circular dichroism (XMCD) is the difference spectrum between two X-ray absorption spectra recorded at a synchrotron in circularly polarized light and

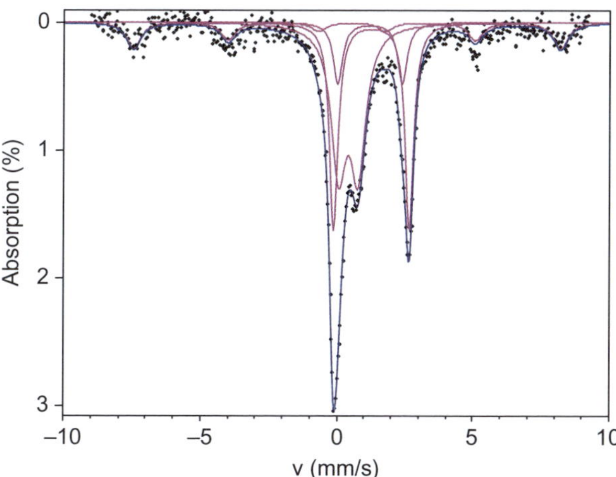

Fig. 8. ^{57}Fe Mössbauer spectrum of sediment from an As-contaminated aquifer in West Bengal (Spectrum produced from authors' data and data from Hery *et al.* (2010). The peaks at extreme velocity (v) positions are part of a sextet of peaks associated with a magnetically ordered iron mineral, probably goethite. The overlapping pairs of peaks arise from Fe^{2+} (pairs with greatest separation) and Fe^{3+} (pairs with smallest separation). The parameters are characteristic of iron in a clay mineral such as clinochlore. These materials were used in 'microcosm' experiments where the role of bacteria in redox reactions leading to release of As was explored.

taken in an alternating magnetic field. The one spectrum is taken with left circularly polarized light and the other with right circularly polarized light. For transition metals such as Fe, Co or Ni, the absorption spectra used are usually measured at the L-edge (where a 2p electron is excited to a 3d state). Because the 3d electrons are the source of the magnetic properties of the element, these spectra give insights into the magnetism of materials. XMCD can only be applied to minerals which are ferromagnetic or ferrimagnetic but this does include some environmentally important phases such as magnetite (and certain other spinel phases), monoclinic pyrrhotite and greigite. The difference spectra can also provide information on the crystal chemistry of magnetic minerals, such as the distribution of cations between different sites in a species such as ferrite spinel. The use of XMCD to determine cation site occupancy in spinel ferrites as an example of a mineralogical application is discussed by Pattrick *et al.* (2002).

3. Characterizing mineral surfaces and interfaces

3.1. Low-energy electron diffraction (LEED) for surface structure

The actual arrangement of atoms at the surface of a mineral (the surface 'monolayer') can be studied using Low-Energy Electron Diffraction (LEED). This method requires the sample to be in UHV and, therefore, dry and clean, and to be a single crystal with a well ordered surface structure. In LEED, a beam of electrons (of typically 20–200 eV energy) is fired at the surface and generates a back-scattered electron diffraction pattern. This pattern of spots, to which only elastically scattered electrons contribute, can be analysed in terms of spot positions and give information on the size, symmetry and orientation of the surface unit cell. In a more sophisticated analysis, the intensities of the spots can be used to provide accurate information on the positions of atoms at the surface of the crystal (or of an ordered overlayer on that surface). A detailed account is provided in the book by Van Hove *et al.* (1986). An example of a LEED 'pattern' is shown in Figure 9 (Fellows *et al.*, 1999).

3.2. X-ray reflectivity, diffuse scattering and X-ray standing wave methods

Synchrotron X-ray techniques are extremely useful for acquiring data from the mineral–fluid interface during reaction. Because of the energy range and high brightness available at a synchrotron, enough signal can be transmitted through a reactant fluid overlayer to perform a variety of measurements without disturbing the system. Roughness of the surface can be determined at the angstrom scale and is a root mean square (r.m.s.) measure of the average peak-to-trough height of features on the surface. Roughening leads to an increase in surface area and to a change in the distribution of reaction sites on a surface. Because a wide range of reactions in low-temperature systems are surface-controlled, it is critical to constrain directly changes in surface area and site distribution. Glancing incidence X-ray measurements of the reflected and diffusely scattered components of the beam are particularly useful for describing the structural adjustments that occur during reaction, even if the surface layer becomes amorphous, and are the only methods available for characterizing buried interfaces. These types of measurements are generally made in reflection mode, which means that the detector is positioned well downstream of the sample and is kept at a fairly low angle ($\theta < \sim 7°$) relative to the sample surface in order to detect various components of the beam scattered from the surface. Low-angle diffraction is also possible in this geometry. By using narrow slits and a long 2θ detector distance, data can be acquired at angles as low as $0.04°$ (θ).

The incident beam can also be used to fluoresce elements present at the mineral surface (van der Hoogenhof & Ryan, 1993; Bowen & Wormington, 1993). With single-crystal experiments, the angle of incidence can be controlled precisely, and as the penetration depth of X-rays is a known function of incident angle, the fluorescence signal can be used as a depth-sensitive probe. This allows extremely low concentrations of adsorbates to be depth profiled. Also, at a synchrotron, the energy of the incident beam can be varied by the use of a monochromator, allowing measurement of XAS spectra giving local co-ordination information for specific elements present in the sample. By performing XAS measurements as a function of incident angle, changes in co-ordination polyhedra or changes in oxidation state can both be monitored as a function of depth in a reacted mineral surface (Farquhar *et al.*, 1999; Wogelius & Fraser, 1996). Combining XAS with scattering measurements means that a full description of the surface structure and chemistry can be obtained.

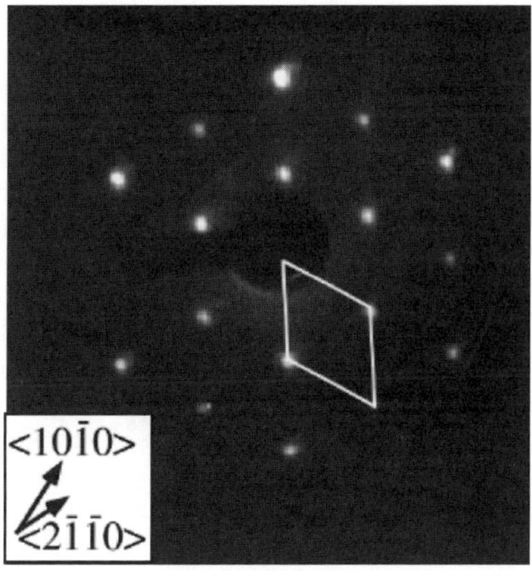

Fig. 9. Low-energy electron diffraction (LEED) pattern from an ilmenite, $FeTiO_3$ (0001) crystal surface clearly showing the hexagonal symmetry of this material. The reciprocal surface unit cell is outlined and the crystallographic directions of the $FeTiO_3$ substrate from Laue diffraction also shown (after Fellows *et al.*, 1999; reproduced with the permission of the Mineralogical Society of America from *American Mineralogist*).

3.2.1. *X-ray reflectivity and diffuse scattering*
Much work has been done on the interpretation of low-angle specular and diffuse X-ray scattering (*e.g.* Andrews & Cowley, 1985; Sinha *et al.*, 1988; Tidswell *et al.*, 1990; Weber & Lengeler, 1992; de Boer *et al.*, 1994; and Pershan, 1994; see also Gibaud & Hazra, 2000, for an overview). Although a comprehensive review is beyond the scope of this work, we will outline the fundamental concepts involved. Computation of the specularly reflected component of the beam is based on the Fresnel equations as applied to a homogeneous multilayer system through the recursion formula derived by Parratt (1954) (Bowen & Wormington, 1993; Wormington, 1995). Both the reflected

and diffuse computations typically use the distorted wave Born approximation (Bowen & Wormington, 1993; Wormington, 1995). The reflectivity coefficient for each layer n can conveniently be calculated (Crabb *et al.*, 1993). If the surface is not perfectly smooth, then the reflected intensity will be reduced for all angles above the critical angle. The reduction in intensity is a direct function of surface roughness. Therefore, for a simple one-phase system, the discrepancy between the reflectivity calculated for a perfectly flat surface and the measured reflectivity allows the surface roughness to be determined, where σ is the root mean square (r.m.s.) roughness (for more details on the theory of scattering see Ogilvy, 1991). For multilayer systems, the reflectivity will depend on the roughness of each of the interfaces (σ_1, ... σ_{n-1}). Furthermore, when the electron densities of layers are different by more than $\sim5\%$, the contrast in the refractive index between layers gives rise to interference fringes in the reflected beam, known as Kiessig fringes. The amplitude of these oscillations is directly proportional to the electron density contrast between the layers (*e.g.* $\rho_1-\rho_2$) and the spacing of the oscillations is inversely proportional to the layer thickness, d_n. Therefore, three key parameters can be determined accurately. These are the r.m.s interfacial roughness, σ_n, the layer thickness, d_n, and the average electron density of the layer ρ_n.

While the specular reflectivity data give the r.m.s. roughness of the surface or surfaces in a system, more can be learned about the in-plane structure of the surface by analysing the diffusely scattered component. In this discussion we will consider the mineral surface as being in the xy plane with the z direction normal to the surface. Diffuse scatter measurements lead directly to an ability to determine the height-height correlation function for an interface (Wormington, 1995; de Boer *et al.*, 1994; Weber & Lengeler, 1993). If we assume that the surface is isotropic in terms of surface features, then the height-height correlation function can be written in one dimension (Bowen & Wormington, 1993). The correlation function is:

$$C(x) = \exp\left(-\left(\frac{x}{\xi}\right)^{2h}\right)$$

where x is a distance in the plane, ξ is the scale over which surface features are correlated, and h is a fractal parameter which varies between 0 and 1 and that describes how jagged ($h \rightarrow 0$) or smooth ($h \rightarrow 1$) is the surface. This augments the reflectivity data. The reflectivity data give an excellent picture of the surface in terms of the layer structure in the z direction and give an extremely accurate measurement of the average roughness of the layers. However, the diffuse data resolve the in-plane spacing of surface features and enable us to assess whether the surface topography is smooth valleys and hills (*e.g.* atomic buckling due to relaxation) or jagged, stark cliffs (*e.g.* stepped faces).

This type of fractal surface model has been used successfully in a number of applications (Andrews & Cowley, 1985; Sinha *et al.*, 1988; Weber & Lengeler, 1992). In the study by Weber & Lengeler (1992), the diffuse scatter results were compared with atomic force microscopy (AFM) profiles (see discussion of this technique below).

The two techniques agreed well, except that the AFM roughness values tend to be biased to slightly smaller values than the X-ray data. This is because the width of the AFM tip places a lower limit of resolution on the profiles which is larger than the X-ray resolution limit. Their results also showed that the interpretation of X-ray scattering data provides a mathematical surface model that is consistent with results obtained by other methods and that X-ray techniques can provide important complementary or corroborative information. An advantage that the X-ray technique has over AFM is that it can be used to probe buried interfaces, such as the topography that might develop at the base of a leached layer on a feldspar crystal.

3.2.2. X-ray standing wave studies

X-ray standing waves (XSW) are created by the interference between incident and reflected X-rays. Short-period standing waves involve Bragg reflection from the surface of a 'perfect' crystal and these can be used to probe the locations of ions within crystals, sorbed ions on mineral surfaces, or ions at a mineral-fluid interface. Long-period standing waves can be created at grazing incidence. Studies of the latter have been used to investigate the 'electrical double layer' in the fluid at a solid/fluid interface. An example is a study of the distribution of Pb^{2+} and Se^{6+} ions in the double layer at metal oxide/aqueous solution interfaces (Trainor *et al.*, 2002). The short-period XSW method has been used, for example, to investigate the positions of sorbed ions (Mn^{2+}, Co^{2+}, Ni^{2+}, Zn^{2+}, As^{3+}, Se^{4+}, Sr^{2+}, Pb^{2+}, U^{6+}) in calcite or at the calcite surface. A detailed review of this relatively complex topic is given by Bedzyk & Cheng (2002).

In recent years, there has also been major progress in using specular reflectivity as the basis for completing X-ray standing-wave measurements. One example of this approach is in work by Yoon *et al.* (2005) where reflectivity was used to measure the thickness of polymer films on metal oxide substrates. The XSW measurements were then completed to determine whether Pb(II) and As(V) were bound directly at the oxide surface or were partitioned into the organic film. XSW uses the fact that the interference between the incident and reflected beams from a flat surface produces a standing wave, with a period (D) that depends on the angle between the incident and reflected beams (θ) and also on the wavelength (λ) of the X-rays used:

$$D = \lambda/(2\sin\theta)$$

By keeping the incident angle small, the angle θ can be optimized and long-period standing waves generated. Therefore, if a fluorescence yield for a given element is measured as a function of changes in incident angle, a spectrum will result with maxima corresponding to angles where the standing wave probe sweeps through the position above the surface where the element is located. Modelling of these spectra, and comparison to standards where the elemental profile is known, then results in an ability to locate precisely dilute contaminants within a biofilm or other surface film with thicknesses greater than those discernible with standard diffraction techniques.

3.3. The chemical composition of surfaces: X-ray Photoelectron Spectroscopy, Auger Electron Spectroscopy and related techniques

3.3.1. X-ray photoelectron spectroscopy

In X-ray photoelectron (also termed 'photoemission') spectroscopy (XPS; see Fig. 11), a monochromatic beam of soft X-rays strikes a surface, ejecting photoelectrons from valence and core levels of surface and near-surface atoms. The kinetic energy (E_k) of the ejected photoelectrons is measured in an electrostatic energy analyser. Photo-electrons which have not lost energy on the way out of the sample have characteristic energies given by the relationship:

$$E = h\nu - E_B - \phi_{sp}$$

where $h\nu$ is the energy of the incident X-ray photon, E_B is the binding energy by which the electron is held in the orbital of the parent atom and $\emptyset_{sp}$ is a machine constant (the spectrometer 'work function'). The value of E_B is characteristic for particular orbitals and, hence, the element concerned; in certain cases, it may also provide information on oxidation state and other aspects of speciation (*e.g.* distinguishing S occurring as elemental sulfur *vs.* sulfide or sulfate). Because electrons only escape from the surface of a sample (depending on exact experimental and sample conditions) infor-mation may come from just a few Å to tens of Å beneath the surface. The XPS technique offers a powerful method for the analysis of surfaces, although the experiment does have to be conducted under a moderate to high vacuum ($\geq 10^{-5}$ mbar).

X-ray photoelectron spectra are normally presented as a plot of electron binding energy *vs.* intensity (Fig. 10a), with intensity providing a quantitative means of deter-mining the concentration of a particular element or species in the surface region. In the several decades since its first development (see Siegbahn *et al.*, 1967), the XPS tech-nique, which was known in the early years as "ESCA" (Electron Spectroscopy for Chemical Analysis), has become the most widely employed method for surface chemi-cal analysis. Although the earlier machines offered only low spatial resolution, it is now possible to achieve analysis of areas of only a few square microns and this has led to the development of scanning XPS instruments capable of mapping surface chemistry. There have also been advances in the ability to resolve energy spectra, enabling the subtleties of surface compositions to be explored. The only elements that cannot be detected with this technique are H and He (due to the low probabilities of photoelectron ejection from these atoms).

A good example of an XPS study is the oxidation and breakdown of chalcopyrite ($CuFeS_2$) in the atmosphere and in aqueous solutions, both the acid solutions typical of certain environmental systems, and the alkali solutions used in mineral processing (flotation) plants (Kelsall *et al.*, 1992; Yin *et al.*, 1995). The photoelectron spectrum shown in Figure 10a is typical of a broad-scan spectrum of a chalcopyrite surface that has been subject to the first stage of oxidation in an (pH 9.2) aqueous solution. In these experiments, oxidation was controlled electrochemically and the oxidized chalco-pyrite surface kept in an inert atmosphere whilst being transferred from the solution to the spectrometer chamber. The spectrum in Figure 10a shows peaks at binding energies

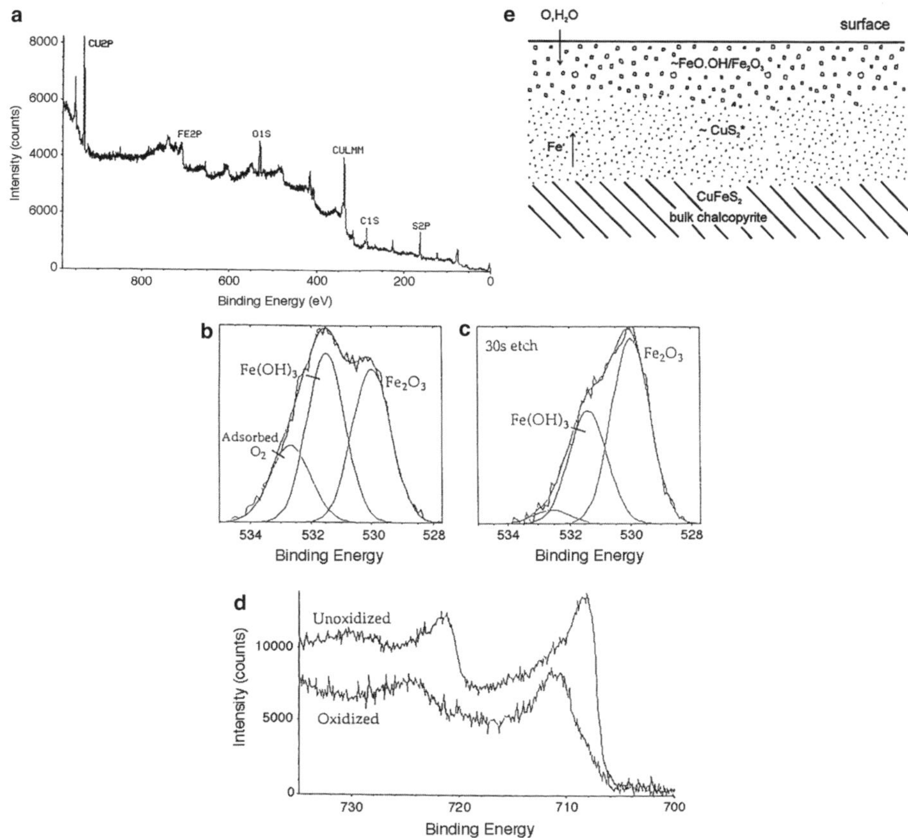

Fig. 10. XPS spectra for oxidized chalcopyrite ($CuFeS_2$) (from Kelsall *et al.*, 1992 [reproduced with the permission of the Electrochemical Society]; Yin *et al.*, 1995 [reprinted from *Geochimica et Cosmochimca Acta*, **59**, 1091–1100, with the permission of Elsevier]): (**a**) broad scan spectrum of oxidized chalcopyrite; (**b**) oxygen 1s spectrum under the same conditions; (**c**) oxygen 1s spectrum after argon ion etching for 30 s; (**d**) iron 2p spectrum of unoxidized and oxidized chalcopyrite; (**e**) model showing the development of oxidized layers on the chalcopyrite surface.

which can be identified with particular electrons in the system (S 2p, Fe 2p, Cu 2p) and with the oxygen associated with the oxidized surface layer (O 1s). Also seen is a copper Auger peak (Cu *LMM*, see below) and a peak from some carbon impurity (C 1s) in the system. Peak intensities give a relative measure of the concentrations of these species in the surface layer; more importantly, closer examination of the O 1s peak (Fig. 10b) reveals fine structure that can be interpreted in terms of three oxygen-containing components, Fe_2O_3, $Fe(OH)_3$ and adsorbed molecular oxygen. Argon ion etching (Fig. 10c, and see below) removes most of the molecular oxygen and more of the $Fe(OH)_3$. Comparison between the Cu 2p and S 2p peaks for oxidized and fresh chalcopyrite surfaces reveals little or no differences between them, but the Fe 2p spectra (see Fig. 10d) show significant differences. Specifically, the peak positions for the oxidized chalcopyrite samples are commensurate with Fe^{3+} in oxide or hydrated oxide phases.

These investigations of chalcopyrite oxidation have led to a model for the development of oxidation products which can be shown in 'cartoon' form (as in Fig. 10e) and has a $Fe_2O_3/Fe(OH)_3$ surface layer underlain by a metastable copper sulfide layer of $\sim CuS_2$ stoichiometry and then the bulk chalcopyrite. The model suggests that diffusion of metal ions to the surface is an important control on oxidation rates in these phases, an idea supported by evidence that metal:sulfur stoichiometry, which would influence diffusion, in turn influences oxidation rates (Vaughan et al., 1995).

As for most X-ray based techniques, XPS can be performed using a synchrotron radiation (SR) source and this (SR-XPS) offers certain advantages over laboratory-based XPS. One advantage is the much greater intensity ('flux') of the radiation which improves the signal-to-noise ratio of the acquired spectra, enabling species to be analysed at significantly lower concentrations. An even more important advantage is that synchrotron radiation is tunable, and not limited to the energies used in laboratory spectrometers (such as Mg Kα or Al Kα). Thus, for surface analysis, it is possible to select an incident X-ray energy that is 50 to 100 eV above the binding energy of the core-level electron of the element of interest, resulting in escape depths of the photoelectron of only a few Å units and, therefore, great surface sensitivity (Lindau & Spicer, 1980). A good example of the use of SR-XPS in mineral sciences is in the study of oxidation of metal sulfides such as pyrite (see Schaufuss et al., 1998).

Reviews of conventional XPS from a mineralogical viewpoint are given by Hochella (1988) and Perry (1990) with briefer accounts in Tossell & Vaughan (1992) and Hochella (1995). Seyama & Soma (2003) discuss applications to environmental and geochemical surface processes, and a lengthy review of applications to clay science is provided by Seyama et al. (2006).

3.3.2. Auger electron spectroscopy

Auger electron spectroscopy (AES) is another technique employed in the chemical analysis of solid surfaces that involves measuring the kinetic energies of electrons ejected from the surface. Indeed, the practical similarities between XPS and AES methods means that they are often conducted on the same instrument (see also Fig. 11a). Auger electrons arise from an 'auto-ionization' process within an excited atom as illustrated in Figure 11. An incident X-ray photon (or electron, as the effect may be stimulated by either X-rays or electrons) causes the ejection (photo-emission) of a core electron. The vacancy ('hole') hereby created in the core level may then be filled by an electron relaxing from a less tightly bound energy level, and the quantum of energy equal to the difference in binding energy between the originally ejected electron and its replacement then becomes available. This may either be emitted as an X-ray photon (X-ray fluorescence) or used to eject another electron, the Auger electron, the kinetic energy of which is approximately:

$$E_{Auger} = E_B - E_C - E_D$$

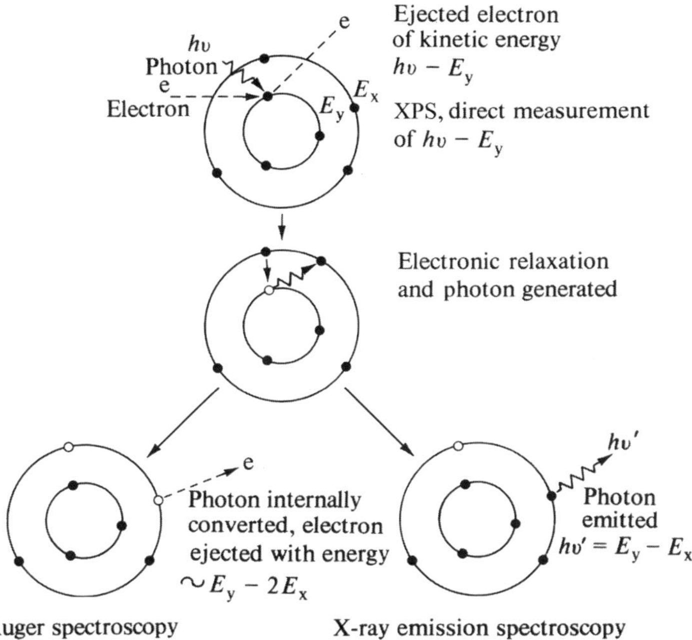

Fig. 11. Diagram showing the principles of X-ray Photoelectron Spectroscopy (XPS), Auger Electron Spectroscopy (AES), and X-ray Emission Spectroscopy which is also the principle of the electron microprobe and of Energy Dispersive (X-ray) Spectroscopy (EDS) analysis used in electron microscopes.

where E_B is the binding energy of the original photoelectron (as in the previous equation), E_C is the binding energy of the electron involved in the process of relaxation and generation of an X-ray photon, and E_D is the binding energy of the level from which the Auger electron comes when in the (doubly ionized) final state. The kinetic energy of the Auger electron, in contrast to the photoelectron, is independent of the energy of the incident radiation and is characteristic of the binding energies of the electrons within the atom concerned. Auger electrons can, therefore, be used for elemental analysis. All elements with three or more electrons (*i.e.* those beyond H and He) give rise to a characteristic Auger spectrum, but it is for elements of low atomic number that Auger processes dominate. As with XPS, peak intensities are proportional to concentrations and AES is similar in surface sensitivity to XPS. It therefore provides a complimentary technique to XPS for the chemical analysis of surfaces, particularly for the study of altered surfaces, coatings and sorbates. In both methods, elemental composition as a function of depth into the sample can be studied by combining spectroscopic measurements with argon ion bombardment. In this technique, known as 'depth profiling', Ar ions etch or 'sputter' away layers of material at a rate that is typically around several monoatomic layers per minute. Such methods have commonly been used for the analysis of thin films, where sputter rate can be calibrated using films of known thickness.

From determinations of sputter rates for materials analogous to specific mineral phases, depth profiling can be done for geological materials in an approximate way.

Electron stimulated Auger spectroscopy has the advantage over X-ray stimulated AES (and photoelectron spectroscopy) in that high energy electron beams can be focused down to spot sizes of a few hundred angstrom units and provide surface analytical data with relatively high spatial resolution. Such beams can also be scanned rapidly (rastered) across a surface so as to build up an image showing the lateral variation in elemental concentration (a technique known as Scanning Auger Microscopy, SAM). An example of such an Auger 'map' of a mineral surface is shown in Figure 12. Here, following on from the above discussion concerning the mechanism of oxidation of chalcopyrite in aqueous solution, are shown SAM maps of the oxidized chalcopyrite surface. The distributions of copper, oxygen, sulfur and iron clearly show the way in which areas of high copper correlate with high sulfur, and high iron with high oxygen as would be expected

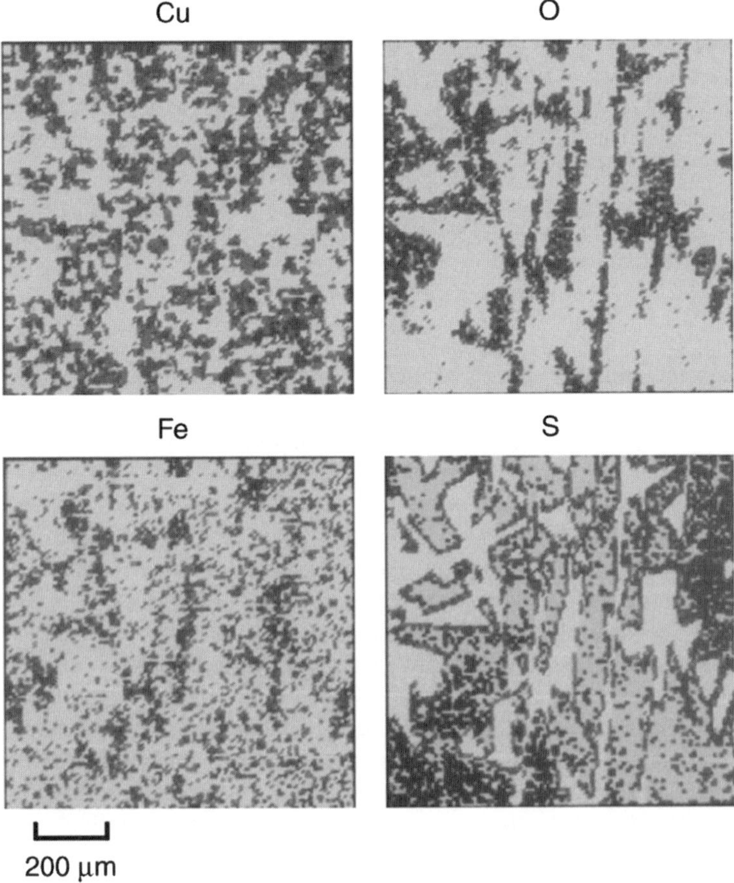

Fig. 12. Scanning Auger Microscopy (SAM) maps of an oxidized chalcopyrite surface.

(Fig. 10e). The major factor controlling the degree of oxidation across the surface appears to be the presence of polishing scratches and other surface imperfections.

3.3.3. Other X-ray emission and photoelectron techniques

X-ray emission spectroscopy (XES; see Fig. 11) is the basis of the most important of the widely used methods of micro-analysis of relevance to environmental mineralogy, the electron probe (EPMA) and the energy dispersive spectroscopic (EDS) analysis conducted in conjunction with electron microscopy. In analysis for a chosen element, a major X-ray emission peak is selected and counts recorded at peak maximum for comparison with a standard to determine its concentration. However, it is also possible to scan through a range of energies and record an X-ray emission spectrum, particularly in the (lower-energy) valence band region of the material. Such spectra can be used to 'dissect' the chemical bonding in the material studied. Some work of this type was conducted using the electron probe but, once again, synchrotron radiation is re-invigorating this technique. Detailed studies of the chemical bonding in small molecules attached to a metal or mineral surface can be undertaken using angle-resolved XES. An example is a study of the bonding of simple carboxylic acids on the (110) surface of Cu by Karis *et al.* (2000).

X-ray photoelectron diffraction (XPD) is another frontline synchrotron method with potential for detailed fundamental studies of mineral surfaces. Here, surface structure is determined (under UHV conditions) by the masking of photoelectron scattering paths by atoms in surface layers as photoelectrons escape from the surface of the solid. It has been applied, for example, in studies of the hematite (0001) surface (Thevuthasan *et al.*, 1999).

3.4. Secondary ion mass spectrometry

Secondary Ion Mass Spectrometry (SIMS) is another 'workhorse' technique employed in surface analysis. Mass spectrometry, the identification and analysis of quantities and ratios of the isotopes of a specific element (or of molecular entities) by discrimination on the basis of atomic or molecular mass, has numerous applications in the natural sciences. SIMS involves bombarding the surface of a sample with a beam of ions, causing the ejection of secondary ions from surface layers. These secondary ions are taken into a spectrometer where determination of their masses enables identification of the atomic or molecular species present and, from the intensities of peaks in the mass spectrum, their percentage contributions. In this way, the chemical composition of the surface can be determined. Because the bombardment of the surface removes material from deeper layers as the process continues, the SIMS technique can provide a profile of compositional changes as a function of depth beneath the surface (generally on scales from nm to μm). As with other microbeam methods, rastering of the beam can be employed to image areas of a surface. In the past, a disadvantage of SIMS has been that the incident beam may cause changes in the chemistry of the sample, particularly if the sample comprises organics or biomaterials, or may disrupt the surface so as to compromise any attempts at 'depth profiling'. However, recent developments involving new types of

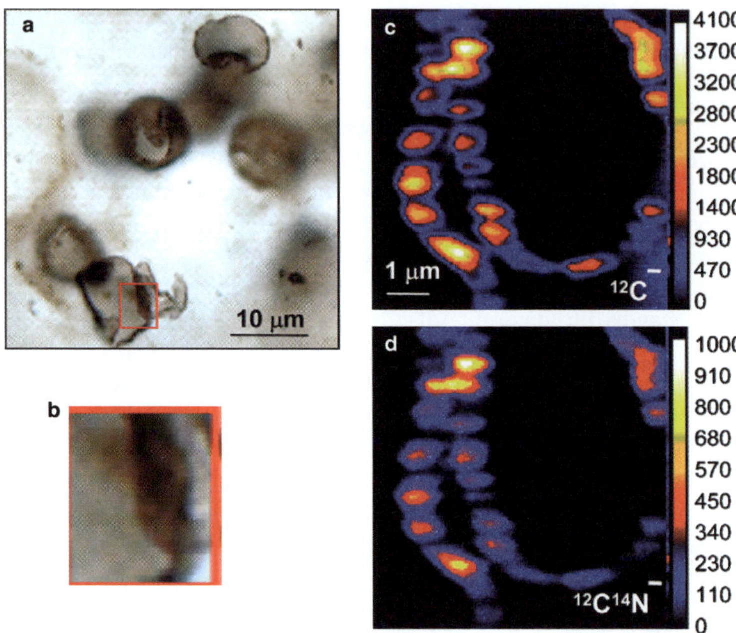

Fig. 13. NanoSIMS of Late Precambrian microfossils from the Bitter Springs Formation of Australia: (**a,b**) transmitted light optical photomicrographs of a polished thin section of chert; **b** is the area within the rectangle marked on **a** where two cells are in contact; (**c,d,e**) NanoSIMS elements maps of ^{12}C, $^{12}C^{14}N$ and ^{32}S, respectively, of the cell contact imaged in **b**. The maps show a one-to-one correspondence between these isotopes. This correspondence, along with the globular aligned character, are strong indicators of biogenicity (after Oehler *et al.*, 2006 [the publisher for this copyrighted material is Mary Ann Liebert, Inc.], 2009 [reprinted from *Precambrian Geology*, **173**, NANOSIMS: insights into biogenicity and syngenicity of Archean carbonaceous structures, pp. 70–78, © 2009 with permission from Elsevier]).

incident beam to bombard the sample (*e.g.* a beam of C_{60} molecules) are revolutionizing this technique, enabling delicate samples to be analysed at nanometer resolution. An example is shown in Figure 13 (from Oehler *et al.*, 2006, 2009). Further details can be found in Macrae (1995) or McMahon & Cabri (1998) and references therein, whereas the articles by Lyon & Henkel (2010) and by Sinha & Hoppe (2010) provide recent reviews of the techniques and their applications in the geosciences.

3.5. X-ray Absorption Spectroscopy and the local structure of molecular clusters at surfaces and interfaces

Another very powerful application of XAS methods is in the study of sorption phenomena. Interactions between mineral phases and species in solution (such as metal ions or metal complexes) can lead to the uptake of such species. As further discussed in later sections of this chapter, such uptake may be associated with precipitation of a new solid on the substrate mineral, replacement of the substrate, exchange of ions between the substrate and the solution, or a true sorption process (either physisorption involving formation of an outer sphere complex, or chemisorption involving an inner sphere complex). X-ray spectroscopy offers one of the very few methods for distinguishing between these different uptake processes. There is now a large number of such studies and much of the earlier work has been reviewed by Brown (1990), Brown *et al.* (1995) and Brown & Sturchio (2002). An example from our own work, that involving cadmium uptake on fine-particle FeOOH (goethite and lepidocrocite) and the sulfides mackinawite (FeS) and pyrite (FeS_2) serves to illustrate the approach.

Parkman *et al.* (1999) measured the uptake of Cd^{2+} ions from aqueous solution by interaction with the surfaces of carefully synthesized fine-particle goethite, lepidocrocite, mackinawite and pyrite. Bulk measurements of solution concentrations of cadmium (which were initially in the approximate range 2–500 ppm) after exposure to minerals showed an initial 100% uptake on the hydroxide surface, falling off to ∼40–50% uptake at the higher initial concentrations. Similar behaviour was exhibited by pyrite, whereas the mackinawite continued to take up 100% from solution even at the highest concentrations. The local environment of the metal taken up at the mineral surface was then defined using EXAFS and XANES.

In the case of Cd on goethite as illustrated in Figure 14, the best fits are for sixfold co-ordination to oxygens at 2.26 Å but with clear evidence for a second shell of Fe atoms at 3.75 Å and, hence, for binding of the Cd at the goethite surface by an inner-sphere complex formation mechanism. Similar results were obtained for Cd on lepidocrocite but with a second shell of Fe atoms at the shorter distance of 3.31 Å. The spectra for Cd interaction with mackinawite showing four S atoms bonded to the Cd at 2.52 Å are consistent with formation of a new phase on the surface, either by precipitation or replacement reactions. This is probably a CdS species, as comparison between the Cd *K*-edge XANES for this system and a pure CdS sample further confirms. These spectra (shown in Fig. 14) are very similar and distinctly different to those for Cd on goethite or in aqueous solution. The behaviour of Cd on pyrite is unexpected and inconsistent with thermodynamic calculations; the Cd appears to be in sixfold co-ordination with oxygen (at 2.28 Å) which could indicate uptake as an outer-sphere complex or reaction with locally oxidized areas of the pyrite surface (although every effort was made to exclude oxygen from the surface in the experiments).

In addition to the core techniques in X-ray spectroscopy of EXAFS and XANES, more specialist methods of studying interactions at surfaces include Surface Extended X-ray Absorption Fine Structure Spectroscopy (SEXAFS) and Reflected Extended X-ray Absorption Fine Structure Spectroscopy (ReflEXAFS). SEXAFS involves studying the sample in a high vacuum and employing low-energy X-rays in order to study low-atomic-number elements such as oxygen. Clearly the limitations already noted in regard to working in high vacuum apply to the kinds of samples which may be studied using SEXAFS. On the other hand ReflEXAFS, which involves an experimental arrangement in which the synchrotron X-ray beam strikes the surface of a very flat sample at a shallow angle ('grazing incidence') can be used on a sample in air, or even with a thin layer of fluid on the surface (see Fig 15). Such an arrangement limits the depth of penetration of the X-rays into the sample surface and by varying the angle of incidence, that depth may be varied. In this way, information on the local chemical environment of an element in a surface coating may be obtained.

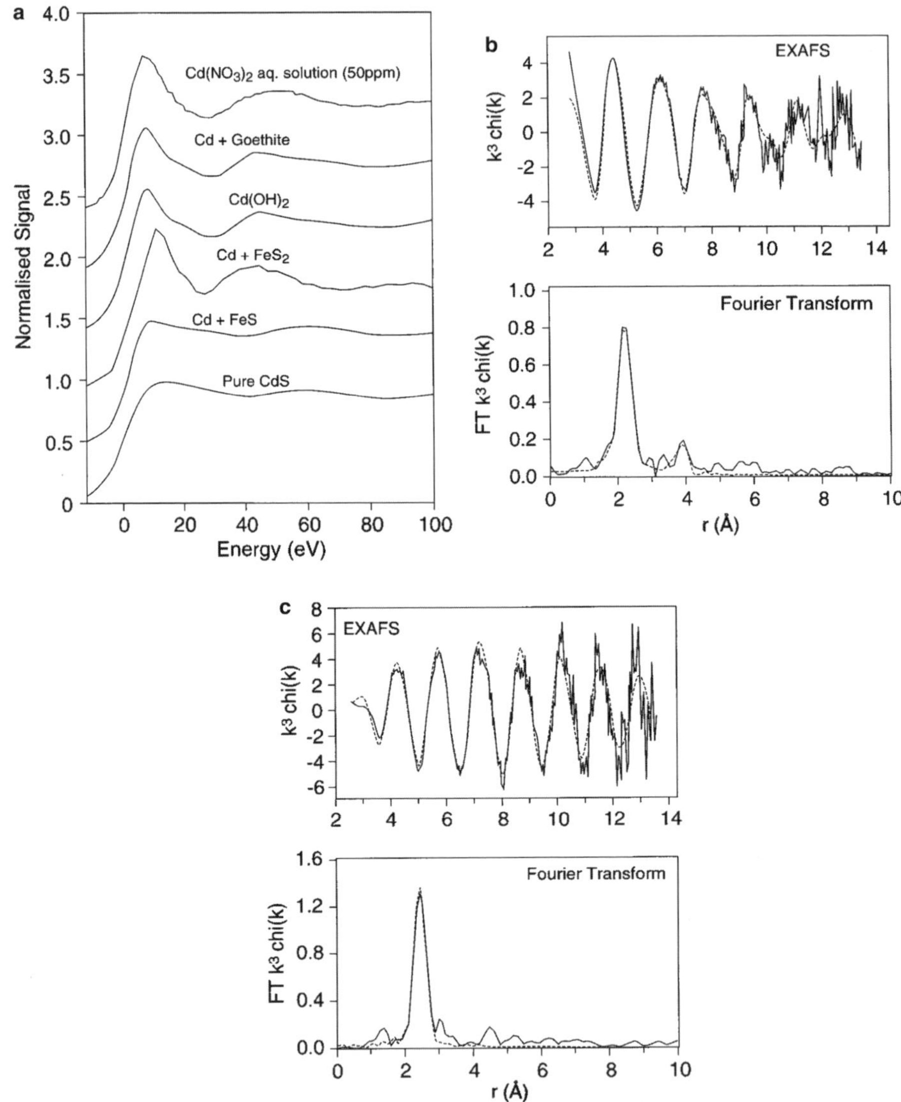

Fig. 14. XAS studies of the uptake of Cd onto the surfaces of goethite and mackinawite (from Parkman *et al.*, 1999; reproduced with the permission of the Mineralogical Society of America from *American Mineralogist*): (a) Cd *K* edge XANES for model compounds and Cd on goethite and mackinawite (FeS); (b) EXAFS and Fourier transform (solid line: experimental; dashed line: best fit theoretical) for Cd^{2+} on goethite; (c) EXAFS and Fourier transform for Cd^{2+} on mackinawite (solid/dashed lines as in **b**).

An example of a ReflEXAFS study is that of Cu(II) sorption from aqueous solution onto the (0001) surface of muscovite mica (Farquhar *et al.*, 1996). As in the other sorption studies discussed above, measurements were made of the bulk uptake of the copper

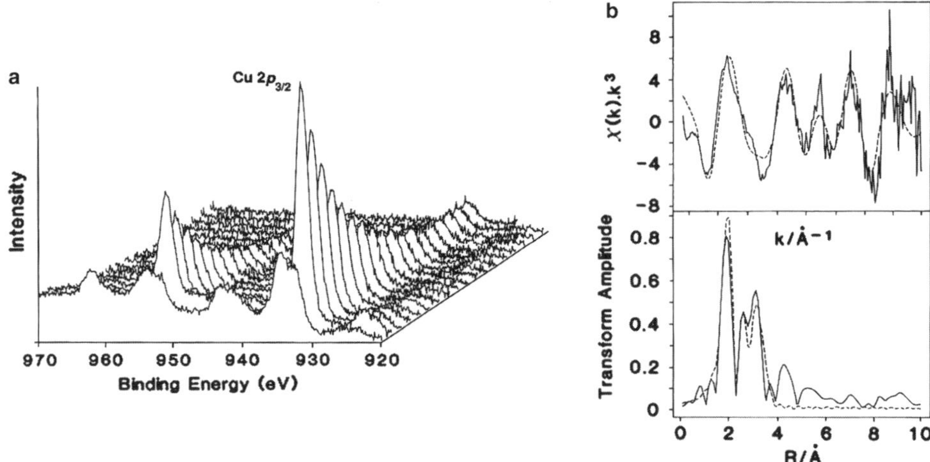

Fig. 15. XAS and XPS studies of Cu uptake by the surface of muscovite mica (from Farquhar *et al.*, 1996 [reproduced from *Journal of Colloid and Interface Science*, **177**, pp. 561–567, with the permission of Elsevier]): (**a**) XPS depth profile of Cu-reacted mica (Cu 2p peaks) collected whilst sputtering the surface with argon ions; (**b**) ReflEXAFS spectra (Cu *K* edge) and Fourier transform of Cu-reacted mica (experimental data: solid line; theoretical best fit: dashed line).

from solution and showed removal of essentially all of the metal at low concentrations, with decreasing percentages removed at higher concentrations. Studies of the reacted mica surface using XPS suggested that the Cu is in a similar bonding environment to that in copper hydroxide, and a depth profile (see Fig. 15) of the reacted surface suggested that there had been no diffusion of the Cu into the mica surface. The ReflEX-AFS studies of the reacted surface showed three shells of atoms around the copper, at 1.97, 2.65 and 3.25 Å which could be best fit to four oxygen atoms, a close copper, and four aluminium or silicon atoms, respectively (see Fig. 15). This provides direct evidence for Cu being bound to the surface at aluminate or silicate groups, and suggests that the Cu species form small multinuclear clusters or ribbons of copper hydroxide-like character where short Cu—Cu distances occur between copper atoms bonded by two bridging oxygen atoms.

Although it relates more indirectly to environmental mineralogy, another important area of application of EXAFS and XANES techniques is in the study of speciation in aqueous solution, particularly of metals or metalloids at low concentrations. For example, EXAFS has been used to demonstrate polynuclear metal–sulfur solution complex formation at 25°C in the Cu—S system (Helz *et al.*, 1993) or that arsenic complexes in sulfide solutions are mononuclear (Helz *et al.*, 1995).

Further details regarding the background to XAS methods and their application to mineralogical and environmental problems are provided in Teo (1986), Brown *et al.* (1988), Greaves (1995), Brown *et al.* (1995) and Brown & Sturchio (2002). In addition, Denecke (2006) has produced a comprehensive review of the use of EXAFS in the study of actinide speciation, a critical area of research as regards nuclear safety. Good examples of the power of XAS methods in addressing environmental problems include work on the redox cycling of another element of great importance for nuclear safety, neptunium. Law *et al.* (2010) used XAS to show that enhanced removal of dissolved Np(V) to sediments is associated with a microbially mediated reduction process, with a poorly soluble Np(IV) species being associated with the solids. Another novel recent application of XAS to a serious environmental problem concerns mercury speciation in complex minewastes and associated soils. Here, although XAS methods cannot detect elemental (liquid) Hg, slow cooling of samples to below 234K produces a transformation to a crystalline α-Hg(0) which can be detected. This has been used to show that ~20% of the mercury in some wastes associated with mercury mining in California occurs as the elemental liquid form (see Jew *et al.*, 2011).

3.6. Environmental scanning electron microscopy (ESEM)

Conventional SEM analysis provides one of the cornerstones of mineralogical techniques by allowing investigators to image surface features with micron or even submicron resolution. When fitted with an energy dispersive spectrometry (EDS) system, an SEM can simultaneously produce chemical data as well as textural and morphological information from microscopic particles. The SEM produces a beam of electrons by heating a filament in a vacuum tube. This beam is accelerated and focused and impacts the sample within an evacuated column. Accelerating voltages range from 10 to 40 kV, and this means that the incident electrons have sufficient energy to eject core electrons from atoms in the targeted sample. Some of the incident electrons do not cause electrons to be ejected but are instead backscattered in the direction of the incident beam. In addition, a finite number of secondary electrons are emitted from the sample *via* the photoelectron and Auger processes (see section 3.3).

Thus, three types of signals can be monitored. Backscattered electrons will have an energy and probability of backscatter that is directly proportional to the average atomic number (Z) of the sample target. By rastering the beam over the sample surface, and using a position-sensitive detector set to energies just below the incident electron energy, the SEM can produce a backscattered image of the sample where bright areas correspond to high-Z regions and dark areas correspond to low-Z regions. In contrast, the secondary electrons have an extremely short mean free path within the solid (see section 3.3) and are only detected if they are produced within the upper 30–50 Å of the sample surface. This short path length also means that topographic features can result in 'shadowing' effects. The detection of secondary electrons will depend strongly on the topography of the sample surface and, hence, by collecting data in the lower portion of the emitted electron range, topographic information can be obtained. This is critical for using crystal habit to identify microcrystalline phases.

The third type of signal is the characteristic X-ray emission. If an EDS detector is used to collect a spectrum of the characteristic X-rays emitted from the sample, a spatially resolved chemical analysis can be acquired. Because most samples analysed *via* SEM have irregular surfaces, fully quantitative analysis is not possible. However, with careful use of calibration standards and avoidance of topographically induced analytical interferences, SEM-EDS systems can produce accurate analytical results. Note that in the case of small particles or reacted surfaces, the electron beam penetrates at least 10 μm into the target (at 15 kV accelerating voltage, up to 20 μm at 40 kV) and characteristic X-rays, particularly for high-Z elements, can be emitted from the base of that excitation volume. Thus the analysis of a 5 μm diameter sphere on a surface will be an analysis of both the sphere itself and a significant volume of the substrate. The beam also spreads out when it strikes the sample, giving an interaction volume that has been described as onion-like in shape. This limits the horizontal resolution to ~5 μm; a value similar to the vertical resolution discussed above.

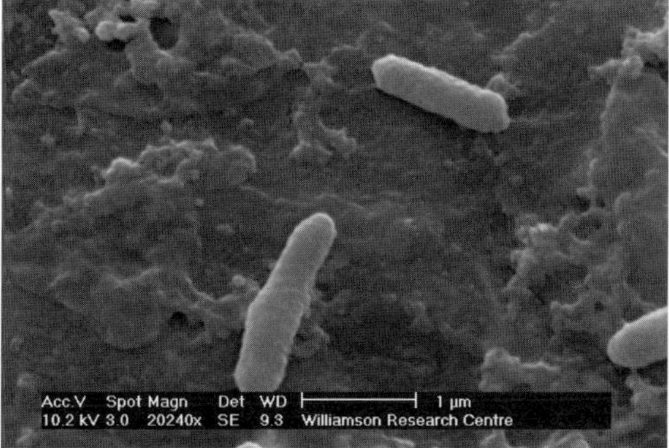

Fig. 16. Environmental Scanning Electron Microscope ('ESEM') image showing cells of *Geobacter sulfurreducens* growing on an Fe oxide substrate (from Wilkins *et al.,* 2007, reprinted by permission of the publisher (Taylor & Francis Ltd, http://www.tandf.co.uk/journals)).

A major disadvantage of the conventional SEM in studying environmental samples is the need to have the sample in a high vacuum. Under these conditions, delicate structures that contain H_2O become desiccated and lose integrity, and with the heating caused by the incident beam, they can be partly volatilized. This makes it virtually impossible to image delicate surface features such as biofilms, or the actual microbes that can be associated with mineral surfaces. Two important technical developments have largely overcome these problems. The first is the cryogenic sample stage, whereby a refrigerant such as liquid nitrogen is used to keep the sample at ultra-low temperatures and to 'freeze-in' any water or hydrous components during measurement. The second has involved developing systems of sample containment and measurement that do not require the sample to be in high vacuum, so that a 'wet' sample can be studied without becoming dehydrated. An image obtained using such an Environmental SEM (or 'ESEM') and showing bacteria on a mineral surface is reproduced here as Figure 16.

3.7. Scanning tunnelling and atomic force microscopy and the topography, atomic and electronic structures of mineral surfaces

It is no exaggeration to say that the study of the surfaces of solids was revolutionized by the development in the 1980s of, firstly, the Scanning Tunnelling Microscope (STM) and then the Atomic Force Microscope (AFM). Using a STM it is possible to image the surfaces of conducting and semiconducting materials such as metals, most metal sulfide and many metal oxide minerals, down to the atomic level. The AFM offers the same atomic resolution capabilities as the STM but also for insulating materials and is, therefore, potentially applicable to all minerals and a very wide range of other materials. A great advantage of both methods is that they can be used to study surfaces in contact with the air, or even a fluid phase, as well as in the pristine ultra high vacuum (UHV) environments required by many of the other methods used to study surfaces.

In the STM, a sharp metallic tip is brought to within a few nanometres of a conducting surface and a small potential difference (usually in the mV range) applied between the tip and the sample. If the tip is biased positively relative to the sample, electrons will flow from the sample to the tip in a process known as 'electron tunnelling' which produces a small, yet measurable, current. The magnitude of this current is exponentially dependent on the tip–surface separation; the larger the distance between the tip and the surface, the smaller the current. This can be expressed by:

$$I_T = \exp\left[\left(\frac{-4s\pi}{h}\right)(2m\varphi)^{\frac{1}{2}}\right]$$

where I_T is the tunnelling current, s is the sample–tip separation, h is Planck's constant, m is the electron rest mass and ϕ is the tunnelling barrier height (a term proportional to the 'electronic work functions' of the sample and the tip, *i.e.* minimum energy required to remove an electron from sample and tip, respectively). Thus, by measuring the magnitude of the tunnelling current as the tip is moved across the surface, a topographic image of the surface can be obtained. In practice, this can be achieved by mounting a tip (which in the simplest case may be formed by cutting a Pt/Ir wire with a sharp pair of scissors) on a piezoelectric tube scanner. Because the piezoelectric material employed has the useful property of expanding or contracting when a voltage is applied across it, and typically by $\sim$1 Å per millivolt, the tip can be manoeuvred with ultra-high precision (to within 0.1 Å). The experimental system, as illustrated in Figure 17, gives the possibility of such high precision movement in x, y, and z directions and hence of producing a 3-dimensional image of the surface at angstrom resolution.

It should be emphasized that topographic information on scales from micrometers to nanometres can be obtained using the STM in one or other of the two main modes of operation. In 'constant height' mode, the tip is scanned in the xy plane of the surface whilst remaining stationary in the z direction. Variations in tunnelling current (I_T) as a function of position in the surface plane are then used to produce an image of the surface topography. In 'constant current' mode, the value of I_T is held at a constant value by movement of the tip in the z direction *via* a feedback mechanism, and the plot of the z-piezo voltage *vs.* lateral position then yields the image. Typically, a modern STM instrument can achieve a resolution of $\sim$1 Å in the plane of the surface and $<$0.1 Å perpendicular to the surface.

A fundamental question that can be addressed using a scanning probe method such as STM is that of the atomic structure of a mineral surface, *i.e.* whether the arrangement of atoms at the surface differs from a simple truncation of the bulk structure. To make progress in this most fundamental area it is necessary to conduct careful studies of very clean surfaces in ultra-high vacuum. An example of such work is provided by the study of the magnetite (Fe_3O_4) (111) surface, as summarized by Cutting *et al.* (2006). As shown in Figure 18, atomic-resolution STM images of this dominant growth and fracture surface show three distinct types of surface (labelled A, A$'$ and B); these are interpreted as different terminations of the magnetite arising from different level 'slices' through the bulk structure (as also shown in Fig. 18). Such direct observations of 'molecular-scale' surface structure is only possible with scanning probe microscope methods.

Another advantage of conducting experiments using a scanning

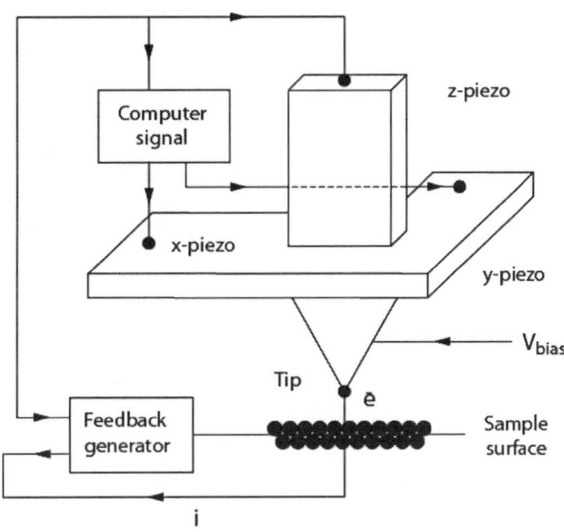

Fig. 17. Schematic diagram showing arrangements for an STM experiment.

probe method in UHV is that small amounts of a reactant can be introduced into the analysis chamber and their interactions with the surface observed directly. In this way the reactivity of specific sites on the surface can be probed. For example, building upon the work noted above, Cutting *et al.* (2006) have studied the interaction of the magnetite (111) surface with three contrasting organic molecules (formic acid, pyridine and carbon tetrachloride). In Figure 19 an image of the so-called A$'$ surface after exposure to a small amount of pyridine is shown. Single molecules of pyridine give rise to the bright features in the image (C_5H_5N), and the topography determined by the cross section labelled $\alpha-\beta$ is consistent with interaction of the N atoms of pyridine with tetrahedral site Fe atoms. Another illustration is provided by the study of the (001) surface of monoclinic pyrrhotite (Fe_7S_8) by Becker *et al.* (1997a). The image of this surface shown in Figure 20 was taken after exposure to a small amount of oxygen. The terraces, which by comparison between experimental and computational evidence were shown to have layers of sulfur atoms at their surfaces, are related by steps that are integer multiples of 2.9 Å which is the distance between consecutive S atom layers in the bulk structure. It is very clear from this image that the reaction with oxygen occurs at step edges and, particularly, at the corners of such edges. Here, it can be suggested, is where the oxygen has access to iron atoms (or electrons with a greater 'Fe character').

The nature of the STM experiment is really such that what are being probed at the highest levels of resolution are the electronic states at the surface. This aspect can be further explored by experiments using an STM in which the tip is held stationary directly over an 'atom' of interest, whilst the bias voltage between the tip and sample is ramped and the changes in tunnelling current are recorded. The data obtained are an electron tunnelling spectrum (hence this is properly termed Electron Tunnelling Spectroscopy, ETS) and can be used to map out the density and energy of electron states in the

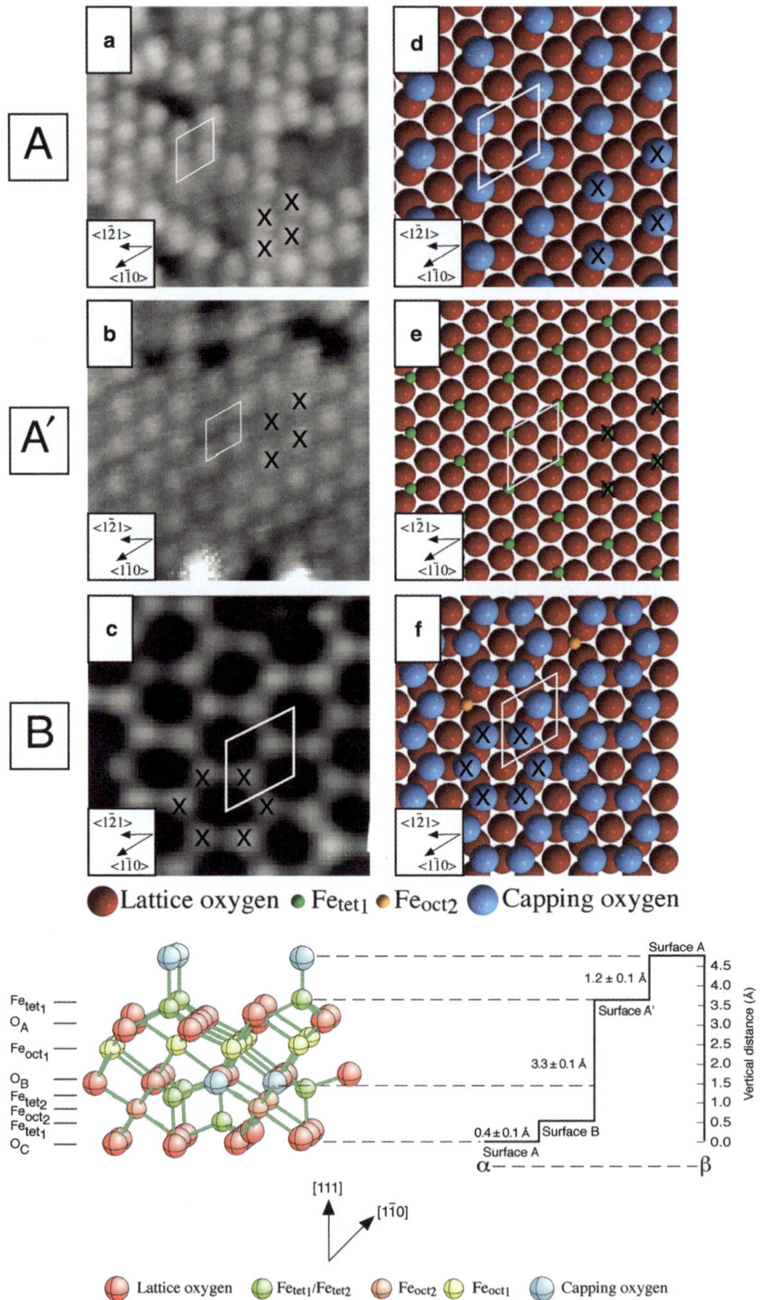

Fig. 18. Scanning Tunnelling Microscope (STM) images of the (111) surface of magnetite obtained under UHV conditions: (**a**) 100 Å × 100 Å image of the A surface; (**b**) 120 Å × 120 Å image of the A′ surface; (**c**) 16 Å × 16 Å image of the B surface, with corresponding models of the surfaces (**d, e, f**, respectively)

atom concerned. However, the potential to employ ETS to characterize and identify individual atoms on mineral surfaces remains to be fully explored.

The major limitation of STM for studying mineral systems, namely that it cannot be used for insulating materials, does not apply to AFM. The principles involved in this technique, as illustrated in Figure 21a, again involve using a piezoelectric scanner to move a tip across the surface of the sample. However, in this case the tip (typically made from silicon nitride and with a diameter of between 1 and 20 nm) is mounted on a cantilever of force constant between $\sim$0.001 and 0.2 Nm^{-1}. When this tip is brought just into contact with the surface, it experiences a very small force (of the order of nanonewtons) as a result of interaction with the surface atoms. In this kind of operation, known as 'contact mode' AFM, the tip is scanned across the surface at a tip-sample separation corresponding to a chemical bond length of the tip-sample combination. As it is scanned across the surface, the tip will be subject to varying attractive and repulsive forces of the kind associated with van der Waals bonds, and the movements of the tip will be registered by deflection of the cantilever. In a typical AFM instrument, as illustrated in Figure 21b, deflection of the cantilever is monitored by reflecting a laser beam from the back of the cantilever on to a segmented photodetector. Another kind of scanning, known as 'non-contact mode' may also be used, particularly for delicate samples that could be damaged by imaging in contact mode. Here the tip is not in contact with the surface and the forces are electrostatic in origin and even smaller than in contact-mode operation. In order to enhance the sensitivity of measurement, the tip is forced to vibrate close to its resonance frequency, and so this method is often called 'tapping mode' AFM. Variations in sample-tip forces will alter the resonant frequency of the tip and this frequency shift can be used to determine the magnitude of the forces involved.

Because AFM imaging can be performed with the mineral surface in contact with a fluid phase, some of the most elegant applications have involved studies of the mechanisms of growth or dissolution of minerals and the factors influencing such processes. Thus, in a series of landmark papers, Dove and De Yoreo and co-workers have explored many aspects of the precipitation and growth of calcite. In earlier work, calcite precipitation and growth were studied directly in real time through AFM measurements of calcite crystal surfaces exposed to CaCO$_3$ solutions of known composition and saturation state (Dove *et al.*, 1992; Dove & Hochella, 1993). These studies led to AFM experiments where observations of microscopic surface processes on {10$\bar{1}$4} faces (see Fig. 22) were linked to classical growth theory (Teng *et al.*, 2000). As well as pure calcite, the systems studied have been used to explore the poisoning of crystal growth by a phosphate inhibitor, and the effects of strontium, magnesium and, in particular, of organic molecules on calcite growth processes (see, for example,

---

Fig. 18. (*Continued*) which are interpreted as arising from different level slices through the bulk structure of magnetite, as illustrated in the lower part of the figure (from Cutting *et al.*, 2006 reprinted from *Geochimica et Cosmochimica Acta*, **70**, Molecular scale investigations of the reactivity of magnetite with formic acid, pyridine, and carbon tetrachloride, pp. 3593–3612, © 2006 with permission from Elsevier).

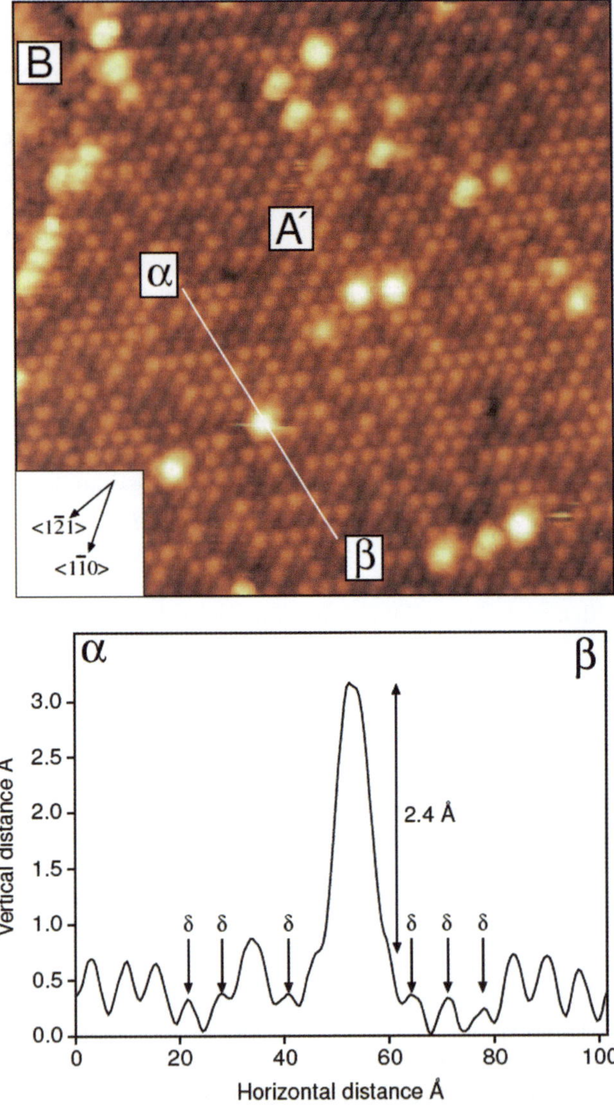

Fig. 19. Scanning Tunnelling Microscope (STM) image (200 Å × 200 Å) of the (111) surface of magnetite obtained under UHV conditions. The 'spheres' imaged in the background (at 6 Å separation) are individual tetrahedral site Fe atoms at the mineral surface. The sample has been exposed to a small amount of the organic molecule pyridine (C_5H_5N); the bright spots are interpreted as individual pyridine molecules bonded to the surface *via* interaction of N with the tetrahedral Fe atoms. The cross section (α–β) shows surface topography and the height of the sorbed pyridine molecule matches this interpretation (from Cutting *et al.*, 2006; reprinted from *Geochimica et Cosmochimica Acta*, **70**, Molecular scale investigations of the reactivity of magnetite with formic acid, pyridine, and carbon tetrachloride, pp. 3593–3612, © 2006 with permission from Elsevier).

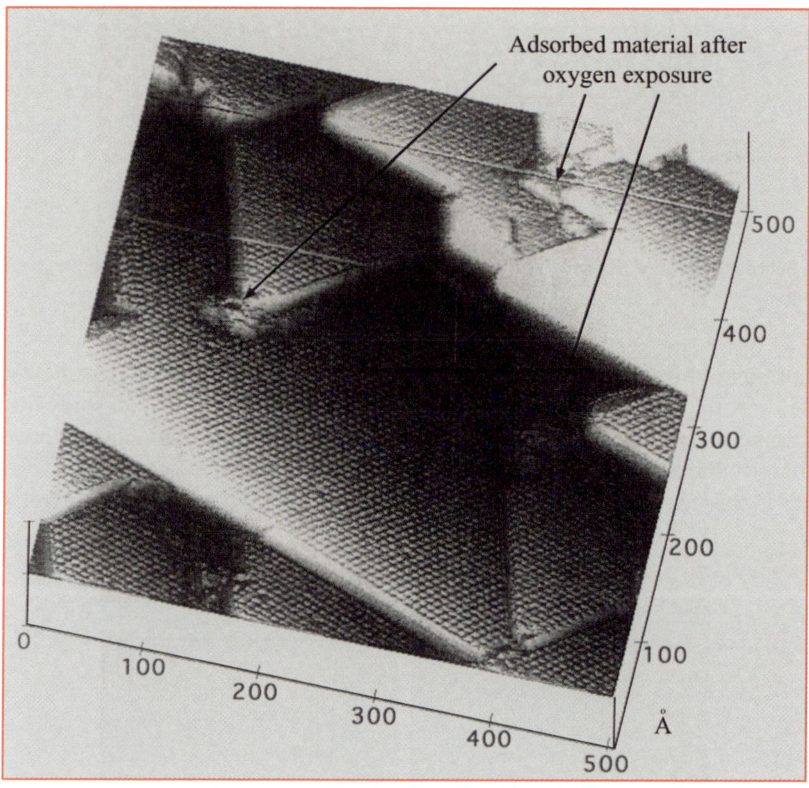

Fig. 20. Experimental STM image of a monoclinic pyrrhotite (001) surface after exposure to oxygen (6000 Langmuirs). The terraces are unaltered by oxygen exposure whereas adsorption features are seen at the corners of steps (from Becker *et al.*, 1997a reprinted from *Surface Science*, **389**, The atomic and electronic structure of the (001) surface of monoclinic pyrrhotite (Fe$_7$S$_8$) as studied using STM, LEED and quantum mechanical calculations, pp. 66–87, © 1997 with permission from Elsevier).

De Yoreo & Dove, 2004). The work involving complex organic or biomolecules has offered critical insights into the complex processes of biomineralization.

Both AFM and STM can, therefore, provide information on surface topography from the micrometre to the angstrom scale and can operate with the surface in contact with air or a fluid phase as well as *in vacuo*. These related techniques form part of what is now a family of Scanning Probe Microscopy (SPM) methods that uniquely can provide high-resolution 3-dimensional information on surfaces in a wide range of environments. Other notable SPM techniques include Magnetic Force Microscopy (MFM) in which a magnetized tip is used to map magnetic properties of the sample surface, and Electro-chemical STM (ECSTM) in which the redox conditions at a solid-fluid interface can be controlled and monitored whilst collecting STM data. Jones *et al.* (2003) also used com-bined scanning electrochemical-atomic force microscopy to study, *in situ*, the initial stages of proton-assisted calcite dissolution in aqueous solution. Another novel

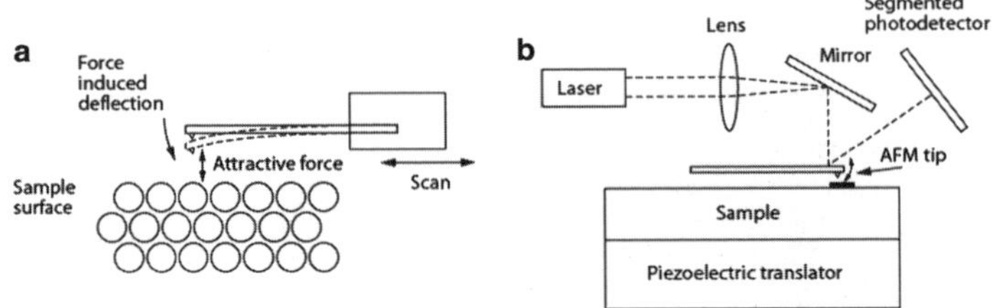

Fig. 21. Schematic diagram showing an AFM experiment: (**a**) the principle involved; (**b**) the experimental configuration.

application based on AFM and termed 'biological force microscopy' involves loading an AFM tip with a bacterium. Lower *et al.* (2001) used this technique to study the infinitesimal forces (measured in piconewtons) that characterize the interactions between the dissimilatory metal-reducing bacterium *Shewanella oneidensis* and the surface of goethite. It was found that the affinity between the bacterium and the mineral surface

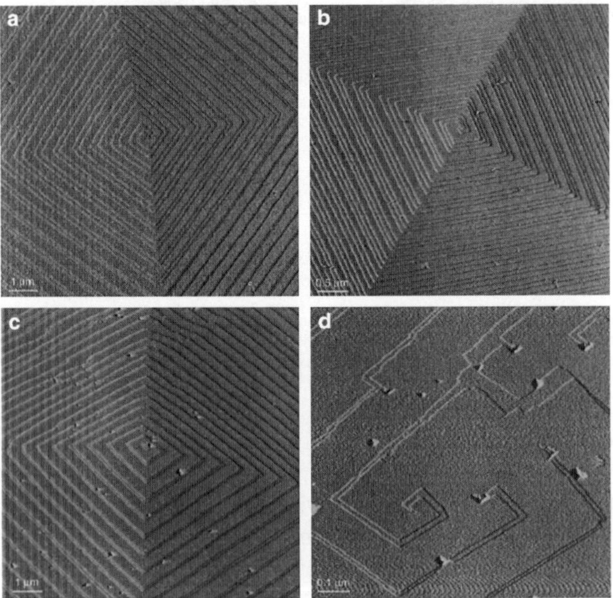

Fig. 22. AFM images of growth hillocks on $\{10\bar{1}4\}$ faces of calcite. (**a**) Growth generated by a dislocation with a burgers vector of two results in a double spiral. (**b**) Growth generated by a dislocation with a Burgers vector of three results in a triple spiral. (**c**) Growth initiated at an obvious surface imperfection to yield a growth hillock with a growth unit of two monomolecular layers. (**d**) Close-up view (1 µm × 1 µm) of growth initiated by a dislocation with a Burgers vector of two but complicated by obvious surface imperfections to result in a multi-sourced, multiple hillock (from Teng *et al.*, 2000 reprinted from *Geochimica et Cosmochimica Acta*, **64**, Kinetics of calcite growth: surface processes and relationships to macroscopic rate laws, pp. 2255–2266, © 2000 with permission from Elsevier).

increases by two to five times under the anaerobic conditions in which electron transfer from bacterium to mineral is expected. Another key feature of the SPM methods is that images can be collected in times as short as 10 s and serve as 'snapshots' of kinetic processes occurring on surfaces over time scales of this order or longer. Thus, dissolution or growth of minerals, and mineral–fluid reactions truly can be observed in 'real time'. There is now an extensive literature on scanning probe methods, and articles concerned with applications to minerals have appeared in the major mineralogical and geochemical periodicals. Reviews from a mineralogical viewpoint are given, for example, by Hochella (1995) and in the volume edited by Brenker & Jordan (2010).

3.8. Laser confocal microscopy and the imaging of biofilms

Recent decades have seen a rapid growth in studies of the interactions between minerals and microbial organisms. In this context, there will always be a role for both transmitted and reflected-light optical microscopy (practical maximum resolution ~ 1 μm) in studying mineral surfaces and interfaces and the associated microbial materials, in particular the extracellular materials associated with bacteria and known as biofilms. Such biomaterials can be further characterized in terms of their components using fluorescent markers, and when these fluorescent markers are used with Confocal Scanning Laser Microscopes (CSLM) an even wider world of microscopy opens up. In these instruments, the use of a laser provides a coherent beam of light for better resolving power and focusing accuracy, such that high-resolution optical slices of a sample can be obtained and, with suitable computing power and optical reconstruction software, three-dimensional imaging can be achieved (Lawrence & Neu, 2006). An example is shown in Figure 23, taken from the work of Brydie *et al.* (2005, 2009). Here, CSLM has been used to image a biofilm grown in a simulated fracture between two quartz glass surfaces. With fluorescent probes it is possible to investigate such biofilms for some of the geochemical parameters (*e.g.* pH) that control mineralization processes (see Hunter & Beveridge, 2005). Furthermore, advances in multiple-photon excitation CSLMs can, through light activation of highly specific fluorescent ligands, pinpoint the distances between such small cellular components as molecules. Recent advances in optical methods are now overcoming the diffraction-imposed limits to resolution of focusing light microscopes (180 nm in the focal plane and 500 nm along the optic axis). As outlined by Hell (2003), techniques such as stimulated emission depletion microscopy (STED) and photoactivated localization microscopy (PALM) offer nanoscale imaging with focused light. These techniques have been used to image intracellular fluorescent proteins at nanometer resolution (Betzig *et al.*, 2006) and even to video record fast physiological phenomena at the nanoscale (Westphal *et al.*, 2008).

4. Computer modelling of solids, surfaces, solution species and their reactions

One of the most important developments over the past 50 years has been the increasing ability of scientists to use computer modelling to interpret and predict the structures and

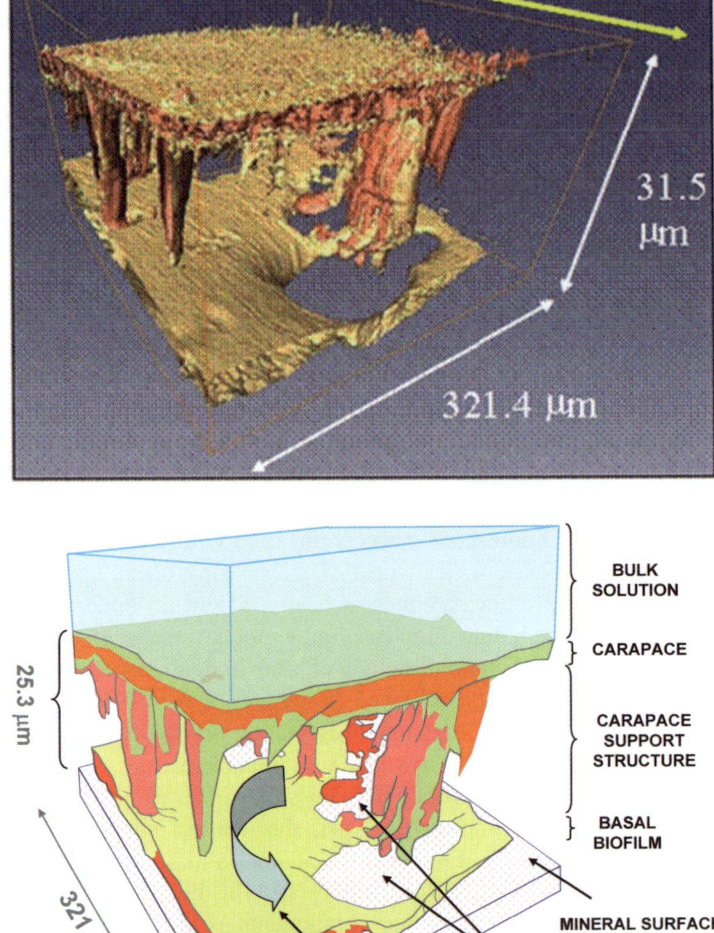

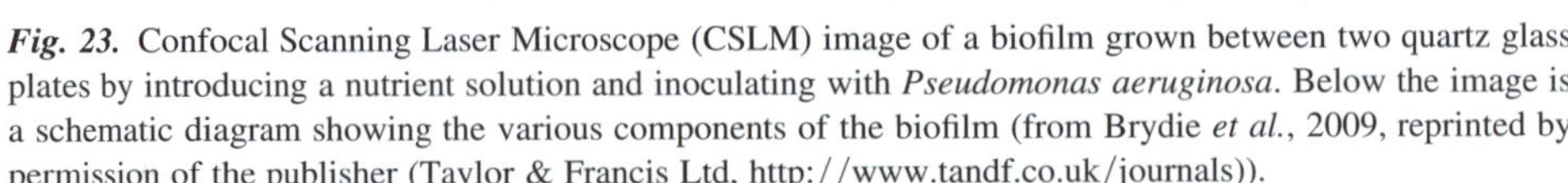

Fig. 23. Confocal Scanning Laser Microscope (CSLM) image of a biofilm grown between two quartz glass plates by introducing a nutrient solution and inoculating with *Pseudomonas aeruginosa*. Below the image is a schematic diagram showing the various components of the biofilm (from Brydie *et al.*, 2009, reprinted by permission of the publisher (Taylor & Francis Ltd, http://www.tandf.co.uk/journals)).

properties of materials. This has resulted from both advances in theory and the advent of high-performance computers. The most fundamental approaches to such modelling are those based on quantum mechanics. Overviews of applications in the Earth and mineral sciences are given by Tossell & Vaughan (1992) and by Cygan & Kubicki (2001), whereas a more general introduction is given by Levine (1983). Any particular approach involves the choice of a model (such as a cluster of atoms in a fluid, a unit cell of a crystal, or a slab representing the surface of a crystal) and of a method. Most methods are concerned with attempting to calculate the energies and 'distribution' of the electrons in a molecule or crystal by solving the Schrödinger equation under particular conditions and with varying degrees of approximation. Such quantum mechanical methods range from the more rigorous, first-principles (so-called *ab initio*) methods to more approximate methods. The quantum mechanical methods give information about electron energies and hence electronic structure, and can help in understanding and predicting properties dependent upon electronic structure (electrical, magnetic and various spectroscopic properties) as well as energetics and molecular or crystal structure. Other methods include so called 'atomistic simulations', based on an ionic model, which employ interatomic potential functions and can be an effective means of calculating minimum-energy structures and properties such as elastic and dielectric constants. Such interatomic potentials, combined with Newtonian mechanics, are also the basis for the computational methods known as molecular dynamics. As briefly discussed below, these enable a range of bulk thermodynamic properties to be calculated.

Broadly, quantum mechanical methods can be divided into those which use a linear combination of atomic orbitals (LCAO) and those which are based around density functional theory (DFT). In both cases the Born−Oppenheimer approximation, whereby the electrons are taken to be in the lowest energy or 'ground state', is assumed. The energy of the system is found by solving the Schrödinger equation either for all the electrons in the system or for only the valence electrons (*i.e.* those electrons involved in bonding). In the latter case, it is assumed that only interactions between valence electrons are important, and that the interactions of the inner electrons and the nucleus can be effectively modelled by a 'pseudopotential'. In its simplest form, the electronic energy can be defined as:

$$E_{\text{Tot}} = E_{\text{k}} + E_{\text{H}} + E_{\text{c}}$$

where E_{k} is the kinetic energy of the electrons, E_{H} is the attractive energy between electrons and nucleus, and E_{c} is the electron correlation energy due to electron−electron interactions. The main difference between the LCAO and DFT methods lies in the way in which electron−electron interactions (E_{c}) are described. The LCAO methods usually ignore electron correlation effects whereas DFT specifically includes electron interactions and calculates the energy in terms of the electron density distribution.

A proper treatment of such computational methods and their applications to mineralogical problems would require a whole book in itself. However, the power of these methods can be illustrated with just a few examples. Thus, higher-level (quantum mechanical) calculations have been used by Becker & Hochella (1996) and Becker *et al.*

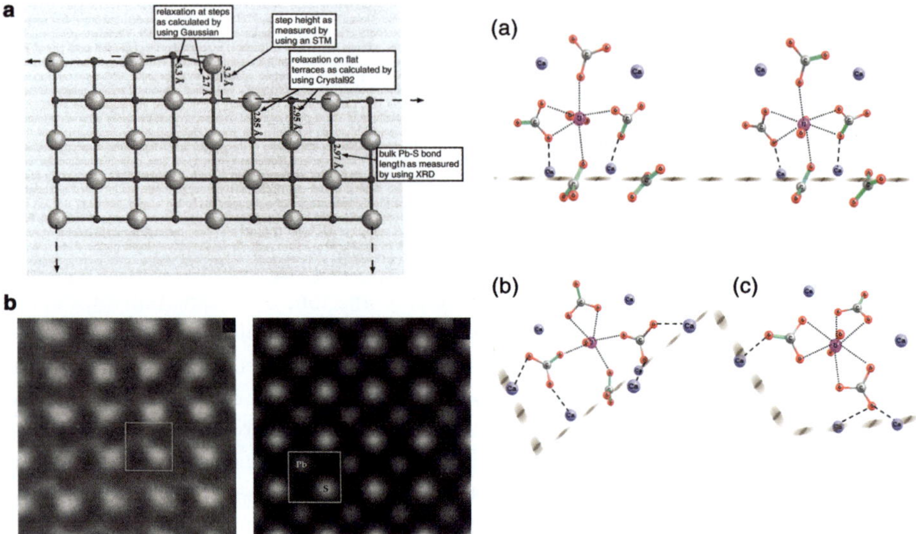

Fig. 24. Left hand side: (**a**) calculated and experimental vertical relaxation at the {001} surface of galena calculated using molecular cluster and periodic lattice methods (after Becker & Hochella, 1996). (**b**) Experimental (left) and calculated (right) STM images for the {001} cleavage surface of galena (after Becker & Hochella, 1996). Right hand side: results of calculations showing an adsorbed [Ca$_2$UO$_2$(CO$_3$)$_3$] complex on different calcite surfaces: (**a**) flat {10$\bar{1}$4} surface; (**b**) acute step {3$\bar{1}$48} surface; (**c**) obtuse step {3$\bar{1}$$\bar{2}$15} face (after Burton *et al.*, 2012).

(1997b) to study the surface of galena (PbS). As illustrated in Figure 24a, relaxation at the {001} surface of galena was calculated using both molecular cluster (*ab initio* LCAO) methods and periodic lattice (DFT) methods. Such calculations could also be used to generate STM images for comparison with experimental data (Fig. 24b) and thereby eliminate the ambiguity over whether bright features in the images arise for Pb or S atoms. In a related study by Wright *et al.* (1999a, b) the interaction and dissociation of water on the surface of galena was studied using both approximate and *ab initio* methods. These studies show that whereas water is a stable species on a perfect {001} surface and dissociation does not occur, at a site where a structural step occurs, the reaction:

$$PbS + H_2O \longrightarrow Pb(OH)^+ + HS^-$$

is exothermic, with a sufficiently low barrier that facile reaction occurs at ambient temperature. These methods are also a powerful way for studying the aqueous complexes of metals; the forms in which metals in solution may co-exist with the relevant mineral species. For example, the various bisulfide complexes of zinc in aqueous solution were studied at *ab initio* level by Tossell & Vaughan (1993). The equilibrium structures of the complexes Zn(SH)$_3$OH^{2-} and ZnS(SH)$_2$(OH$_2$)$^{2-}$ were determined from the calculations which show that the first of these two isomers is more stable by $\sim$50 kJ/mol. Such species are very difficult to study experimentally because of their low solubilities.

Using such approaches, many systems can now be studied using computational methods. In the methods due to Car and Parinello, the *ab initio* DFT approach is combined with molecular dynamics, thereby permitting kinetic processes to be modelled, as in the work of Refson *et al.* (1995) on the MgO surface. Molecular Dynamics (MD) calculations are more accessible to the non-specialist and require less computational muscle than quantum mechanical methods. Here, Newton's equations for the velocity of a particle and for the force on a particle are applied to atoms in order to give a set of atomic trajectories for a system. Obviously, the more atoms involved the more computational resources are required. The forces on each atom will be a function of the attraction and repulsion the atom feels for neighbouring atoms in the system. Molecular dynamics methods use interatomic potentials that are chosen *a priori*. An illustrative example of an equation used to calculate an interatomic potential for atoms i and j is:

$$V = \frac{Z_i Z_j e^2}{r_{ij}} + A_{ij} \exp\left(\frac{-r_{ij}}{\rho_{ij}}\right) + C_{ij} r_{ij}^{-6}$$

where V is potential energy, Z_n is ionic charge on atom n, r_{ij} is distance between the two atoms, A_{ij} and ρ_{ij} are repulsion parameters and C_{ij} is a van der Waals' term (Lasaga, 1990). As mentioned, *ab initio* and MD methods can be combined such that *ab initio* methods are used to calculate interatomic potentials and then the MD simulation is used to calculate trajectories. Velocities and forces, of course, can then be translated *via* statistical mechanics into bulk-thermodynamic properties of a system. This is why MD techniques are so useful; because they can allow equations of state to be calculated for temperatures and pressures where it is difficult to complete experiments. Comparison of thermodynamic quantities calculated *via* MD to known experimental thermodynamic values is a critical way of benchmarking the approach and indicating whether the potentials used are appropriate for the problem under study. One example of MD simulation is the work of Burton *et al.* (2012) in simulating the behaviour of U(VI) in contact with different calcite surfaces (see Fig. 24c,d). The calcium-uranyl-carbonate [Ca$_2$UO$_2$(CO$_3$)$_3$] species was shown to display both inner- and outer-sphere adsorption to the flat [10$\bar{1}$4] and the stepped [3$\bar{1}$48] and [3$\bar{2}\bar{2}$6] planes of calcite. Free energy calculations were used to simulate adsorption paths of the same uranyl species on different calcite surfaces; outer sphere adsorption was found to dominate over inner sphere adsorption. The uranyl complex was found to adsorb preferentially on the acute stepped [3$\bar{1}$48] face of calcite (Fig. 24d) in agreement with experiment.

5. Experimental approaches to environmental mineralogy problems

Reactions with silicates are sluggish, especially at the low temperatures of most environmental systems. Reactions with carbonates and phosphates can be orders of magnitude faster than those involving silicates, but nonetheless under most conditions the reactivity of minerals with aqueous fluids is slow enough that the reactions are surface controlled and not controlled by transport in the fluid phase. For this reason, in order to predict how natural aquatic systems involving minerals will evolve we need a basic understanding of the surface reaction rates and reaction mechanisms. This has involved an enormous experimental effort over the last 40 years and as a result we now have a much better idea of the manner and rate with which mineral surfaces react. The rate laws extracted from these experiments can be used in applications ranging from the modelling of the evolution of a radioactive waste repository host lithology (see Chapter 9; Curtis & Morris, 2013) to calculating the constraints that silicate weathering reactions place on the mean surface temperature of the Earth over geological time (Brady, 1991).

5.1. Kinetics of dissolution and growth

5.1.1. Batch reactors

Measurement of kinetic-rate constants and the order of reaction are critical in being able to understand geochemical systems. One of the simplest ways to determine reaction parameters is by setting up simple sealed batch-reaction vessels. Essentially this involves sealing a reactant mineral with reactant solution in a watertight container, controlling temperature throughout the course of reaction (usually by thermostated water bath), and taking fluid samples as a function of time during the course of the reaction. Changes in fluid composition are then plotted as a function of time and element release/consumption rates extracted.

Batch methods are extremely useful for initial assessments of key rate dependencies in new systems and also for completing a large number of experiments in a short time. The low-maintenance aspect of this approach also makes it useful for allowing experiments to run for extremely long periods of time. There are major drawbacks with this approach, however. Perhaps the biggest problem is that unless pH buffers are used in the reactant fluid, the pH of the reaction will tend to drift. Since most reactions depend strongly on pH this introduces a serious problem in data interpretation. Use of pH buffers mitigates this problem, but the organic acids present in most standard pH buffers used in the near-neutral pH range tend to also interfere with the reaction of interest.

Many of these acids act as chelate complexes and increase rates significantly. Simple aqueous complexation with the buffer ligands can confuse the experimental conclusions. Flow-through systems are more complicated and tend to require more day-to-day care but they are the best way to avoid the buffer/pH drift problem. Batch reactions are also plagued by the fact that as the concentration of solutes builds up in solution, solubility limits may be approached. The decrease in reaction affinity also will affect the rate and must be accounted for. In addition, unwanted precipitation reactions may occur which will completely alter the fluid chemistry and may severely affect the surface area and surface chemistry of the reactant solids.

Figure 25 shows an example of the type of data that can be extracted from batch experiments. In this figure the rates of B and Si release from a simulated nuclear

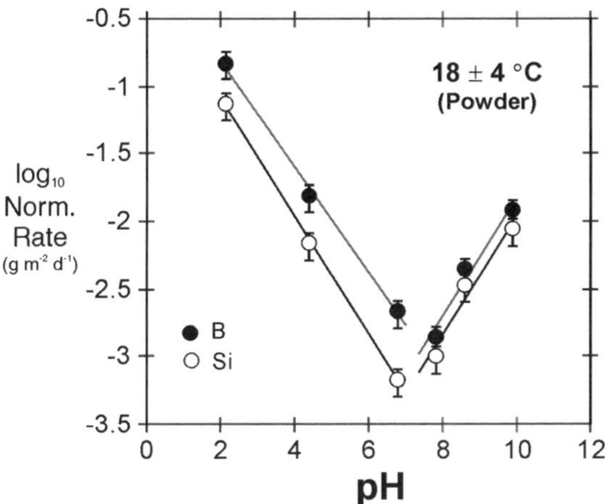

Fig. 25. Rates of B and Si release from a simulated nuclear waste glass in batch dissolution experiments at room temperature (18 ± 4°C) *vs.* the buffer solution pH. The rate data are normalized to the molar composition of the glass. (Reprinted from *Applied Geochemistry*, **15**, from a paper by Abraitis *et al.*: The kinetics and mechanisms of simulated British Magnox waste glass dissolution as a function of pH, silicic acid activity and time in low temperature aqueous systems, pp. 1399–1416. Copyright (2000), with permission from Elsevier.)

waste glass in batch dissolution experiments at room temperature (18 ± 4°C) are presented as a function of buffer solution pH (Abraitis *et al.*, 2000). The rate data are normalized to the molar composition of the glass. As discussed in Chapter 9 (Curtis & Morris, 2013) the reactivity of the primary waste-form is one of the key variables to be accounted for in modelling the long-term evolution of radioactive waste repositories.

5.1.2. *Flow-through reactors*

Flow-through reactors may be either single-pass or recirculating systems. Single-pass systems hold a reactant in a chamber and force fluid through the chamber slowly. The continuous replacement of fluid in the chamber keeps the fluid/solid ratio high and typically this means that the buffer/pH drift problem is circumvented; essentially the fluid composition is buffered by volume. Filters are often used on the effluent side of the chamber to ensure that solids are not 'rafted' out of the reaction vessel. Fluid composition is monitored in the effluent. For constant influent composition the effluent should reach a steady-state composition. Composition of the effluent will be governed by the kinetics of the reaction and by the flow-rate of fluid through the reaction vessel. For slow reactions, the flow rate can be kept at a minimum so that solute concentrations are above detection limits.

A modification of the flow-through system involves adding a recirculating portion into the apparatus. This is most commonly done as a fluidized bed system (Chou & Wollast, 1985). Here, three pumps are used: one to add influent, one to recirculate a portion of the fluid, and the final one to extract effluent. Pumping rates of the first and third are kept equal so that the volume of the system is constant. The second pump is usually faster than the others and simply acts as a stirred reservoir of fluid. The solid sample is kept in a cone-shaped reaction vessel (narrow end pointed down). The particles are entrained by flow, but as the fluid rises to the larger-diameter portion of the vessel, of course, the fluid velocity drops; therefore the grains settle back out and fall towards the bottom of the vessel. This keeps the solid permanently agitated so that no concentration gradients build up in the reaction vessel, and makes it unnecessary to filter the fluid as it leaves the chamber. A portion of the fluid is pumped off at the top of the vessel as effluent for analysis, and a portion is returned to the bottom to be pumped into the chamber again. The equation relating the rate to the experimental flow-through system has the simple form:

$$\text{Rate} = \frac{c \cdot p}{s}$$

where c is solute concentration difference between influent and effluent, p is pumping rate, and s is solid surface area.

5.2. Reaction kinetics

5.2.1. *General form of rate laws and master variables*

Statistical mechanics, surface science, and thermodynamics are all related through kinetics. A sound interpretation of kinetic data or an application of a rate law to a field problem must rely on an understanding of modern kinetic theory. Here, we review some of the basic concepts and provide an introduction to advanced kinetic theory. This will serve as a foundation for a later section which shows how to extract rate laws from experiment.

Reactions can be broadly classified as either 'homogeneous' or 'heterogeneous'. A homogeneous reaction is a reaction that occurs in a single phase. Two examples of homogeneous reactions are: (1) an aqueous speciation change and (2) radioactive decay. A heterogeneous reaction is a reaction that involves two or more phases. Examples of heterogeneous reactions include dissolution, precipitation, and diffusion across a phase boundary.

A typical (heterogeneous) dissolution reaction for periclase (MgO) can be written as:

$$\text{MgO (periclase)} + 2\text{H}^+ \longleftrightarrow \text{Mg}^{2+} + \text{H}_2\text{O}.$$

If far from equilibrium, the rate equation for this is (Wogelius *et al.*, 1995):

$$\frac{d[\text{Mg}^{2+}]}{dt} = \text{rate} = k[\text{H}^+]^{0.4}$$

This involves two phases, mineral plus solution. This equation signifies that the change in Mg^{2+} concentration over time will be equal to the dissolution rate of the solid. The dissolution rate will in turn be a function of a 'rate constant', k, and also will depend on the concentration of one of the reactants; in this case the acidity of the solution. It is important to note that there is an exponential dependence on the proton concentration and that the exponent does not match the stoichiometry of the reaction.

This exponential dependence for periclase dissolution brings us to the concept of 'reaction order'. Consider the general reaction:

$$a\mathrm{A} + b\mathrm{B} \longleftrightarrow c\mathrm{C} + d\mathrm{D}.$$

The rate for this reaction can be written as:

$$\mathrm{rate} = \mathrm{k}[\mathrm{A}]^{\alpha}[\mathrm{B}]^{\beta}$$

The rate constant is again represented by k, while the coefficients on the concentration terms define 'reaction order' with respect to a given reactant. Again, the stoichiometry does not define the reaction order; this is usually true and underlines that determination of reaction order is an experimental problem. Reaction order is simply the value of the exponent on the concentration (or activity) term in the rate law. Overall reaction order is the sum of the exponents. In this case, the reaction is order α in A, order β in B, and order $\alpha + \beta$ overall. Periclase dissolution is 0.4 order in H^{+}, and 0.4 order overall.

Most discussions of kinetic theory assume that the system can be considered as far-from-equilibrium. However, this far-from-equilibrium approximation is not suitable for environmental systems; typically a far-from-equilibrium situation can only be sustained for short periods of time. As we are often concerned with systems such as lakes, rivers, and aquifers that have had long periods of time to react, we must be able to consider the effect on reaction kinetics caused by approaching equilibrium.

In the same way that the force on a spring is related to displacement, the rate of a reaction can be related to the amount of displacement from equilibrium.

$$\text{For a spring: } F = -\mathrm{k}x$$

or

Force = spring constant * displacement.

For a chemical reaction: Rate α $\mathrm{k}C$ (G),

or the rate is proportional to a rate constant times concentration term times thermodynamic term (which is often not written explicitly but depends on displacement from equilibrium). This thermodynamic term is governed by the free-energy displacement from equilibrium (ΔG_{R}). As this term approaches zero, the overall reaction rate must obviously approach zero. Therefore it is useful to include this term explicitly in the rate law. Theory and experiment agree that there is an exponential relationship between rate and ΔG_{R}, so that the full rate law may be written as:

$$\mathrm{rate} = \mathrm{k}C\left(1 - e^{\dfrac{\Delta G_{\mathrm{R}}}{\mathrm{R}T}}\right)$$

When ΔG_{R} is at its most negative value, the exponential is insignificant and the term in parentheses is essentially equal to 1. Thus, this returns the far-from-equilibrium assumption that the thermodynamics can be ignored. However, as ΔG_R approaches zero, the exponential term approaches 1 and, thus, the term in parentheses approaches zero, which brings the overall rate to zero as is required. In the geochemical kinetics literature (Helgeson, 1968), the negative of the free energy change of reaction $(-\Delta G_R)$ has been given a unique name, reaction affinity (A). There are other ways to include displacement from equilibrium in a kinetic law; however, this is the most common method and serves to illustrate one of the ways in which thermodynamics and kinetics are related.

Thus far we have shown that rates depend on an intrinsic rate constant (k) that must be measured, on displacement from equilibrium (ΔG_R or A), and on a reactant concentration term. Obviously, the reactant participates in the formation of the products. We are often interested in the precise atomic configuration that allows the reaction to proceed; this can often lead us to make predictions about how certain variables may affect the rate. Techniques that are sensitive to short-range order, such as XAS, therefore can be extremely useful in studying reaction mechanisms. Concentration dependencies and atomic configurations can often provide insight into exactly how a reaction occurs. As many reactions of interest are surface controlled, we are required to gather as much information about surface structure, topography, and chemistry as possible.

However, as part of this discussion of kinetics, it is crucial that we also introduce the concept of catalysis because, often, for reasons that we will make clear, the most important reactant does not actually appear in the chemical equation. This is often disregarded when mineral kinetics are discussed, but we feel that these ideas are particularly important when dealing with systems that involve mineral surfaces, organic compounds, and biotic processes.

First, we discuss a simple textbook example as given by Levine (1983). Consider the reaction:

$$2H_2O_2 \longleftrightarrow 2H_2O + O_2.$$

The rate law for this reaction can be written as:

$$\text{rate} = k[H_2O_2][I^-].$$

I^- does not appear in the reaction, but its concentration affects the rate. This is because, in this reaction, iodide is a 'catalyst'. How can a species affect a rate if it doesn't appear in the reaction? The answer is that the overall reaction sometimes does not include all of the steps in a reaction, and the catalyst affects one of these steps.

$$
\begin{aligned}
\text{(i)} \quad & H_2O_2 + I^- \longleftrightarrow H_2O + IO^- \\
\text{(ii)} \quad & \underline{IO^- + H_2O_2 \longleftrightarrow H_2O + O_2 + I^-} \\
& 2H_2O_2 \longleftrightarrow 2H_2O + O_2
\end{aligned}
$$

IO^- is a reaction intermediate that appears in step I but is reacted away in step II such that the overall reaction can ignore it and still show the correct stoichiometry. The definition of a catalyst is 'a substance that increases the rate of reaction but that can be recovered unchanged at the end of a reaction.' A catalyst does not change the equilibrium position of a reaction; that is, ΔG of the reaction stays the same. However the activation energy of the reaction is lowered, which increases the reaction rate. Catalysis can be heterogeneous or homogeneous and it can have a tremendous affect on rates.

Consider the peroxide reaction:

$$2H_2O_2 \longleftrightarrow 2H_2O + O_2$$

The activation energies of this reaction depend strongly on catalysis (Levine, 1983):
18 kcal/mol uncatalysed,
14 kcal/mol I^- catalysed (homogeneous),
12 kcal/mol Pt catalysed (heterogeneous on surface), and
6 kcal/mol enzyme catalysed.
Decreasing the activation energy from 18 to 6 kcal/mol increases the rate by approximately a factor of 10^9, an enormous increase.

A reaction that is more relevant to environmental mineralogy concerns the important role that clay mineral interlayer surfaces and edge sites can play in speeding up the rates of organic reactions. Advantage is taken of this property of clays in developing strategies to treat harmful organic pollutants, and is also critical for the generation of valuable hydrocarbon (fossil fuels) resources over geological time. Perhaps the first step on the road to generating long-chain hydrocarbons involves the decarboxylation of fatty acids. Laboratory experiments have shown that the yield from decarboxylation experiments is clearly enhanced by the presence of clay minerals. Recent computational work has shed light on how this happens (see Geatches *et al.*, 2011). Simple fatty acids can be attracted electrostatically to positions within the structure that have charge imbalance due to the non-isovalent substitution $Al^{+3} \leftrightarrow Si^{+4}$. Bound above such a site (which is charge neutralized by interlayer Na^+), the fatty acid can react to become a reaction intermediate (also described as a transition-state complex) consisting of an alcohol plus CO, both surface bound. Further reaction decreases the overall energy of the reactant-product system by between -0.14 and -0.77 eV to produce an alkane (ethane, as shown below) plus CO_2. The clay may be recovered unchanged. The following two-step reaction mechanism followed by the overall reaction is given below:

(i) $C_2H_5COOH + Na\text{-}Al_5Si_7O_{20} (OH)_4 \longleftrightarrow >[\ Na\text{-}Al_5Si_7O_{20} (OH)_4$
     $- CH_3CH_2OH] + CO$
(ii)

$$>[Na\text{-}Al_5Si_7O_20(OH)_4 - CH_3CH_2OH] + CO \leftrightarrow Na\text{-}Al_5Si_7O_{20}(OH)_4 + C_2H_6 + CO_2$$
$$\overline{C_2H_5COOH \rightarrow C_2H_6 + CO_2}$$

The optimized density functional computation-based structures for the initial, transition, and final states of this fundamentally important reaction are shown in Figure 26. (In the above reaction the $>$ indicates a surface species.)

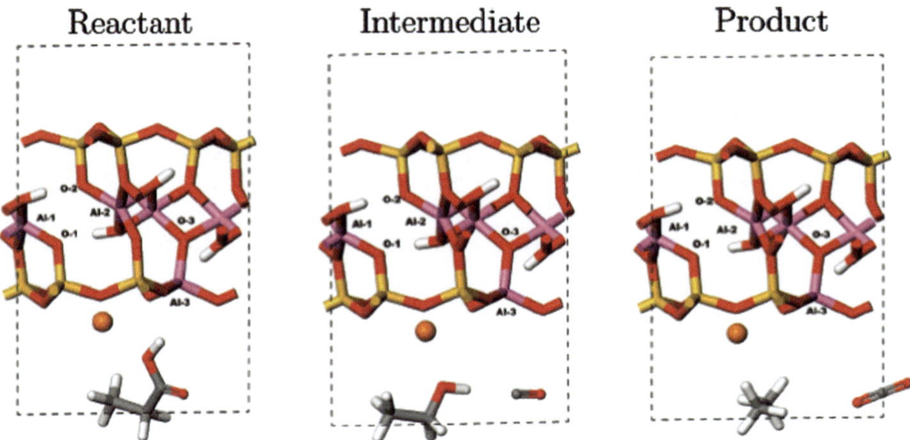

Fig. 26. The initial-, transition- and final-state structures of species involved in the reaction between a fatty acid and clay mineral as shown by computation using Density Functional Theory. Adapted with permission from Geatches *et al.* (2011). Copyright (2011) American Chemical Society.

How does catalysis work? Essentially catalysis works because the ΔH of chemisorption is quite a large negative number. We will discuss chemisorption and physisorption in detail in a later section, but the term essentially refers to the formation of a chemical bond between a surface and an adatom (surface adsorbed atom). Physisorption does not involve bond formation.

The ΔH of 'physisorption' is small:

$$\Delta H_{\text{phys}} \approx -2 \, to -6 \, \text{kcal mol}^{-1},$$

while the ΔH of 'chemisorption' is a large negative number:

$$\Delta H_{\text{chem}} \approx -20 \, \text{kcal mol}^{-1}.$$

This is highly exothermic and tends to reduce the activation energy (E_{a}) of surface reactions. This leads to catalysis in the following way:

$$E_{\text{a}} = E_{\text{surf.rxn.}} + \Delta H_{\text{ads,}} \text{ and if } \Delta H_{\text{ads}} = \Delta H_{\text{chem}} = -x$$

Then:

$$E_{\text{a}} = E_{\text{surf.rxn.}} - x.$$

Thus, the activation energy is reduced by the ΔH of chemisorption and the reaction will proceed more rapidly.

For solid catalysts to work, at least one reactant must be chemisorbed; in fact ΔH_{ads} is the key to catalysis. Any process that increases grain size must decrease surface area and thereby decrease catalytic effect. Strongly adsorbed poisons decrease catalysis at far below monolayer coverage, and therefore we deduce that catalysis only occurs at specific, preferred sites. Because the distribution of sites varies with crystallographic orientation it follows that catalysis will not be the same on any arbitrary crystallographic plane of a given mineral.

How do these kinetic concepts relate to environmental mineralogy? It was argued that silicate mineral dissolution (precipitation) can be adequately modelled as a zero-order reaction of the type:

$$\frac{dM}{dt} = -k,$$

where k is the rate constant, t is time, and dM is the change in moles of reactant. It was further postulated that the activation energies are similar for all silicate phases (Wood & Walther, 1983). Similar activation energies for all phases would mean that the slopes on Arrhenius plots (discussed below) would all be equivalent. Thus, if this were true we could use one rate constant for all silicates and disregard any concentration dependencies. For a typical 2-mm diameter grain of any silicate, the time scales for dissolution would be of the order:

Temperature	Time to dissolve
700°C	70 years
500°C	330 years
100°C	100,000 years

But we now know that activation energies and rate constants are not all similar, so that up to at least 250°C, the details of the kinetic rate constants are critical. And, most reactions are not zero order, including silicate, carbonate, phosphate, and oxide mineral growth and dissolution along with adsorption reactions and electron-transfer (redox) reactions. These tend to be surface-controlled and rates are strong functions of such master variables as pH, temperature, Eh, P_{CO_2}, ionic strength, surface area, organic ligand type and concentration. Note also that the assessed performance lifetime of a radioactive waste repository is usually 10^5 to 10^6 years and so there are regulatory reasons why these reactions are critical to understand. So it is clear that kinetic studies need to delineate not only the far from equilibrium rate constants and activation energies for mineral reactions, but must also determine the reaction dependencies and reaction orders for a number of reactants before we can successfully model natural systems.

5.3. Interpreting kinetic data

Now we turn to a discussion of how to design simple kinetic experiments and how to interpret the data in order to produce a general rate equation.

5.3.1. Surface vs. transport control

One of the most basic questions that needs to be answered is whether the process under study is surface- or transport-controlled. Essentially, surface-controlled reactions are so slow that the solution has no trouble removing dissolved ions from the vicinity of the surface before new ions detach. Transport control means that ions detach so quickly that a significant 'concentration gradient' builds up between the surface and the bulk fluid; thus slowing the reaction.

Rates will then tend to increase (to a limit) as fluid motion increases (which is why you stir the sugar in your tea or coffee).

How do you tell the difference between surface control and transport control? Approaches include the following:

(1)　Change the fluid circulation rate; if the apparent rate increases, then the reaction is 'transport-controlled'.

(2)　Observe the surface morphology; if it shows a crystal-structure-controlled morphology then the reaction is 'surface-controlled'.

(3)　Calculate the diffusion rate in the fluid compared to the reaction rate; if the diffusion rate is rapid compared to the reaction rate then it is probably a 'surface-controlled' reaction.

(4)　Check the magnitude of the activation energy.

$$\text{Surface control:} \quad E_a \geq 8\,\frac{\text{kcal}}{\text{mol}}\left[= 34\,\frac{\text{kJ}}{\text{mol}}\right]$$

$$\text{Transport control:} \quad E_a < 5\,\frac{\text{kcal}}{\text{mol}}\left[= 20\,\frac{\text{kJ}}{\text{mol}}\right]$$

As shown, surface-controlled reactions have high activation energies and hence a strong temperature dependence. Reactions with intermediate activation energies will typically display mixed kinetics.

5.4. Determining reaction rate laws

5.4.1. Initial rate method

One can use this technique if the rate is slow, or if all reactant concentrations except one can be held fixed. You can then determine the order of reaction for various reactants sequentially by using this method. For example, consider a reaction involving three reactants:

$$A + B + C \longrightarrow \text{Products}$$

The initial rate law will be:

$$\frac{d[P]}{dt} = k[A]^{n_a}[B]^{n_b}[C]^{n_c}.$$

If you hold the concentration of B and C constant, then the rate will simply be:

$$\frac{d[P]}{dt} = k'[A]^{n_a},$$

where $k' = k[B]^{n_b}[C]^{n_c}$.

If you then complete a set of experiments each with a 'different initial' concentration of reactant A you can determine n_a by determining the initial rate for each concentration of A. Once n_a is known, you can then keep A and C constant and determine n_b. By completing experiments for all three of the reactants you can finally determine k.

An example is that of pyrite formation from FeS (Rickard, 1975):

$$FeS + S^0 \longleftrightarrow FeS_2$$

$$\frac{d[FeS_2]}{dt} = k[A_{FeS}]^{n_1}[A_s]^{n_2}[P_{H_2S}]^{n_3}[H^+]^{n_4}$$

Experiments were completed with constant reactant solid surface area at fixed pH. Initial rates were measured as a function of P_{H_2S} in order to determine n_3. Subsequent experiments determined the other reaction orders.

5.4.2. *Method of isolation*

In this method, rather than performing a number of experiments with different initial concentrations of one of the reactants, one of the reactants is added to the system at an extremely low concentration and allowed to vary during the course of the reaction.

$$A + B + C \longrightarrow Products$$

To implement this, start with $[A]_0 \ll [B]_0$ and $[A]_0 \ll [C]_0$.

Then, no matter how far the reaction proceeds the concentrations of [B] and [C] are essentially constant. Therefore:

$$\frac{d[P]}{dt} = [A]^{n_a}(k[B]^{n_b}[C]^{n_c})$$

$$\frac{d[P]}{dt} = k''[A]^{\alpha}$$

As reactant A is consumed the rate will change and both k'' and α will be resolvable, but α will be a pseudo-order and may not be exactly equal to n_a (here [A] changes during the experiment, unlike the initial rate method).

5.5. Determining activation energy

This is quite a simple quantity to evaluate. Rate constants are a strong function of temperature, and the strength of this dependence is controlled by the activation energy. Therefore, to determine the activation energy, experiments should be completed at several different temperatures. The Arrhenius equation gives the rate constant relationship to temperature explicitly:

$$k = Ae^{\frac{-E_a}{RT}}$$

(Arrhenius, 1889). Where k is the rate constant, E_a is activation energy, R is the gas constant, T is temperature (in Kelvin), and A is the pre-exponential or Arrhenius constant. Converting to natural logarithms gives two equations at two different temperatures:

$$\ln k_1 = \ln A - \frac{E_a}{RT_1}$$

$$\ln k_2 = \ln A - \frac{E_a}{RT_2}$$

Combining these two equations gives the result:

$$\ln\frac{k_1}{k_2} = \frac{E_a}{R}\left(\frac{1}{T_2} - \frac{1}{T_1}\right)$$

Thus, E_a can be extracted from experiments at different temperatures. If a number of experiments are completed, the data can be plotted as $\ln k$ *vs.* $1/T$ and the slope will be equal to $-E_a/R$. For many geochemical reactions, k doubles or triples with each 10°C increase in T, and A is between 10^8 and 10^{12} for bimolecular reactions, between 10^{12} and 10^{15} for unimolecular reactions.

5.6. Rate-law integration

Concentration as a function of time can be determined by integrating the rate law.

For a zero-order reaction, the integration is as follows.

$$a\text{A} \longrightarrow \text{Products},$$

$$\text{rate} = \frac{-d[\text{A}]}{dt} = k_a,$$

$$-d[\text{A}] = k_a dt$$

$$-\int_{[\text{A}]_o}^{[\text{A}]_t} d[\text{A}] = \int_0^t k_a dt.$$

The integrated form is:

$$[\text{A}]_0 - [\text{A}]_t = k_a t,$$

or

$$[A]_t = [A]_0 - k_a t.$$

If a plot of [A] *vs. t* is made, then the slope equals $-k_a$ and the intercept equals $[A]_0$. 'First-order reactions' are a special case.

$$\frac{d[A]}{dt} = -k_a[A]$$

$$\int_{[A]_0}^{[A]_t} \frac{d[A]}{[A]} = -\int_0^t k_a dt$$

But we take advantage of the fact that: $\int \frac{dx}{x} = \ln x$.
Thus:

$$\ln[A]_t - \ln[A]_0 = -k_a t,$$

$$\ln \frac{[A]_t}{[A]_0} = -k_a t.$$

If we plot ln [A] *vs. t*, then the slope equals $-k_a$, and the intercept is $\ln[A]_0$.

For *'second-order reactions'*, rather than do the full derivation we simply present the result:

$$\frac{1}{[A]_t} = k_a t + \frac{1}{[A]_0}$$

Similar to the examples above, on a plot of $1/[A]$ *vs. t*, the slope will be equal to k_a and the intercept will be equal to $1/[A]_0$. In general, for all except $N = 1$:

$$[A]_t^{1-N} = (N-1)k_a t + [A]_0^{1-N}$$

Occasionally, the form is too difficult to integrate, but this can be circumvented by performing 'numerical' integration.

For illustration we consider a first-order example. If we assume that:

$$\frac{d[A]}{dt} \approx \frac{\Delta A}{\Delta t},$$

then: $\Delta[A] = - k_a[A]\Delta t$.

So, it follows that as time steps forward from 0 to 1 to 2 *etc.*,

$$A_1 - A_0 = -k_a A_0(t_1 - t_0)$$

or,

$$A_1 = A_0 - k_a A_0 (t_1 - t_0),$$
$$A_2 = A_1 - k_a A_1 (t_2 - t_1),$$
$$A_3 = A_2 - k_a A_2 (t_3 - t_2),$$

and in general:

$$A_j = A_{j-1} - k_a A_{j-1} (\Delta t).$$

This is the basis of how numerical models of kinetic systems step forward in time to calculate new concentrations as the system evolves towards equilibrium.

5.7. Reaction progress variable

The 'reaction progress variable', often written as the symbol ξ, represents the thermodynamic position of the system indexed to time or to the fractional completion of reaction. This can be used to describe the reaction of a single phase, but it is a particularly useful concept when considering a complicated system which involves more than one reaction occurring in parallel. Say, for example, there are two phases dissolving and one phase precipitating. There will be an affinity to react associated with each of these phases but there will also be an equilibrium point associated with the system as a whole. The state of the system can then conveniently be described with reference to overall equilibrium by using the reaction progress variable.

5.8. Transition state

Our understanding of kinetics has been greatly enhanced by the work of Henry Eyring (1938) and his ideas concerning transition state theory. This has been discussed in detail in a mineralogical context by several authors including Lasaga (1981), and Oelkers *et al.* (1994). Essentially, transition state theory posits that in order for a chemical reaction to proceed, an activated, transitory configuration of atoms must occur. Reactants become products only by passing through this activated state. Activation energy represents the energy required to produce this 'activated' complex. The balance between the number of these complexes that break down into products and the number that relax back into the reactant configuration in a unit of time determines the rate constant. We introduce this complexity because it serves to illustrate how the concentration term in a typical rate law can be interpreted. The concentration dependence of a reaction essentially indicates which chemical moieties participate in the formation of this activated complex. Hence a catalytic surface site, proton, or other participant can reasonably be included in the full transition state theory rate law. In the case of surface-controlled reactions, surface area can be included and, as we have seen, a displacement from equilibrium term is also important. Thus, the full transition state theory rate law can be written as:

$$\frac{d\xi}{dt} = S_a \left(\sum_{i=1}^{m} k_i \prod_{j=1}^{n} a_j \right) \left(1 - e^{\frac{-A}{RT}} \right)$$

The three terms on the right hand side emphasize the importance of surface science, kinetics, and thermodynamics in understanding mineral reactivity. In addition, we can break down the rate constant itself. The following equation shows the different components that go into the rate constant (Lasaga, 1981):

$$k = \frac{kT}{\hbar} \frac{\gamma_A \gamma_{BC}}{\gamma^{\#}} e^{\frac{\Delta S^{\#}}{R}} e^{\frac{-\Delta H^{\#}}{RT}}$$

where:

$\frac{kT}{\hbar}$ is the fundamental frequency of the reaction resulting from statistical mechanics, $\frac{\gamma_A \gamma_{BC}}{\gamma^{\#}}$ represents the activity coefficient ratios for surface species participating in a surface-controlled reaction, and

$e^{\frac{\Delta S^{\#}}{R}}$ and $e^{\frac{-\Delta H^{\#}}{RT}}$ are two exponential terms representing the overall energy change for formation of the activated complex and which result from thermodynamics. Thus, mineral reaction kinetics clearly draws on ideas from statistical mechanics, surface science, and thermodynamics.

5.9. Adsorption

Before addressing this critically important topic, the authors would like to make it clear that much of the following discussion is based on two excellent chapters in MSA *Reviews in Mineralogy* volume **23** (Davis & Kent, 1990; Sposito, 1990). Because of space concerns, we will limit this section to some basic concepts. The interested reader is encouraged to refer to those chapters for a full development of this topic.

Related to the detachment and attachment of atoms from surfaces caused by dissolution and precipitation is the concept of surface adsorption. Here we refer to the uptake of a foreign atom at a surface from solution. Two broad classes of reaction are identified. The first is termed 'outer-sphere complexation', 'physisorption', or 'non-specific binding'. All of these terms allude to the fact that in this type of surface uptake no identifiable chemical bond is formed.

'Physisorption' is caused by simple Coulombic interaction which can be calculated according to the equation:

$$F = k \frac{q_1 q_2}{r^2} = \frac{K q_1 q_2}{\varepsilon r^2}$$

F is force, q_i refers to the charges on the particles, r is distance, and K is a proportionality constant and ε is a solvent dielectric constant. As we have already seen, the ΔH of physisorption is small.

'Chemisorption' is also referred to as 'inner-sphere complexation' or 'specific site binding'. The ΔG of chemisorption is quite large, and it is because of this that even in quite dilute systems, appreciable quantities of sorbates may be found at the mineral-fluid interface. Essentially the process of chemisorption reduces the free energy of the surface and hence minimizes the total potential energy of the system. There are three main approaches to the modelling of adsorption (Davis & Kent, 1990): (1) the empirical approach to the physisorption of gases; (2) the empirical approach to adsorption of solutes onto minerals from aqueous solution; this essentially assumes that chemisorption can be represented by physisorption; and (3) Surface Complexation Models (SCM), which explicitly attempt to account for chemistry.

5.9.1. Empirical approach to physisorption of gases

The fundamental treatment is the Langmuir adsorption isotherm which assumes that the adsorbate is present up to a monolayer, that there are no sorbate–sorbate interactions, and that the surface is uniform. Refinements on this include the Freundlich isotherm, and the extremely useful BET adsorption equation (Brunauer *et al.*, 1938) which allows direct measurement of the surface area of fine-grained materials. Attempts to model uptake of solutes onto mineral surfaces grow out of these early methodologies.

5.9.2. Empirical models for adsorbate-uptake reactions at the mineral/fluid interface

The first is the simple K_d approach which is often used but is extremely limited in terms of predictive power. K_d is the ratio between the surface density of the sorbate (Γ_j) and the concentration of the sorbate in solution (C_j).

$$K_d = \frac{\Gamma_j}{C_j}$$

The next level of complexity is the Langmuir approach, where the gas adsorption equation is applied to an aqueous sorption equilibrium. All quantities are analogous to the Langmuir gas adsorption equation given above. We may write the equation for adsorption as:

$$>S + J_{aq} \longleftrightarrow >SJ$$

This is a shorthand method for writing a surface reaction, where the ($>$) again indicates a surface species. A surface site $>S$ reacts with a solute, J_{aq}, to produce a new surface sorbate species $>SJ$. The equilibrium constant describing this reaction is a ratio between the surface sorption density of sorbate J (Γ_j) and the product of the aqueous concentration of J (C_j) and the surface site density of unreacted sites $>S$ (Γ_s):

$$K_L = \frac{\Gamma_j}{C_j \Gamma_s}$$

If g_m is the maximum that can be adsorbed, then:

$$\Gamma_j = \frac{g_m K_L C_j}{1 + K_L C_j}.$$

The above equation closely resembles the BET equation for gas adsorption. To linearize:

$$\frac{C_j}{\Gamma_j} = \frac{1}{K_L g_m} + \frac{C_j}{g_m}$$

Adapting the Freundlich isotherm to aqueous adsorption gives:

$$\Gamma_j = K_F C_j{}^a$$

Now, we turn to a higher level of complexity, where an attempt is made to deal with the chemistry of the reaction.

5.9.3. Surface complexation models

This approach takes the speciation modelling of fluids and applies it to chemical species on surfaces. Davis & Kent (1990) identify four key concepts: (1) the surface is composed of specific functional groups; (2) equilibria can be described by mass action laws (corrected for electrostatics); (3) surface charge and electrical potential are consequences of the chemical reactions of surface groups; and (4) apparent binding constants for mass law equations are empirical parameters related to thermodynamic constants *via* the activity coefficients of the surface species.

This type of modelling can get quite complicated, and therefore some concepts need to be made clear from the outset. First of all, much of the interaction between solutes and surfaces occurs because these are charged entities. Charge on a surface can be described in three ways (Davis & Kent, 1990): (1) permanent (or structural); (2) co-ordinative (mostly due to H^+ and OH^- exchange); and (3) dissociated (due to the diffuse swarm of ions in the electrolyte).

Attempts to model this are called Electrical Double Layer (EDL) or Diffuse Double Layer (DDL) theory. Initial attempts were made independently by Gouy in 1910 and Chapman in 1913. The resulting equation is:

$$\frac{C}{C_0} = \exp\left[\frac{-ze\psi}{kT}\right]$$

where C is concentration at a point, C_0 is bulk concentration, e is unit of electron charge, k is Boltzmann's constant, T is temperature, z is formal charge, and ψ is the electrical potential at a point. All subsequent equations and models are based on this simple form.

The Gouy-Chapman model is simply a solid plus a diffuse layer. This model produced a fairly poor fit to experiments, so it was refined by Stern in 1924 to include a "compact" layer of inner-sphere adsorbates. The assumption was that ψ decays linearly between the surface and the Stern layer (the Stern layer is also known as the Helmholtz layer). The amount of adsorbate in the Stern layer depends on four things: (1) composition of the surface; (2) sorbate ion; (3) solution composition; and (4) temperature. In the soil-chemistry literature, a general series for the ease of removal of cations from soil mineral surfaces is called the 'Lyotropic Series'. Starting from the easiest to remove (lithium), the trend includes the major alkali and alkaline-earth cations:

$$Li^+ > Na^+ > K^+ > Rb^+ \approx Mg^{2+} > Ca^{2+} > Sr^{2+}$$

This is a practical illustration that simply changing the size of the sorbate ion can significantly change sorption behaviour. The trend reflects the fact that small, highly charged ions hold their waters of hydration tightly and hence are usually weakly bound on clay mineral surfaces.

If we consider a binary salt surface in contact with a fluid that contains some of this salt dissolved in it, and if we ignore the adsorption of H^+ or OH^- species from solution, at a certain solution cation concentration, the surface charge will be zero because the positive and negative charges will balance each other. This point of balanced charge is called the point of zero charge (pzc). Thus, for NaCl:

$$\psi_0 = \frac{2.3RT}{F}(pNa_{pzc} - pNa)$$

Thus, this modified Nernst equation gives surface potential (ψ) directly from solution concentration, where F is the Faraday constant and pNa is the negative $\log_{10}$ of the sodium concentration.

For oxides this is complicated by the uptake of protons from solution:

$$>FeOH^0 \longleftrightarrow \; >FeO^- + H^+ (base)$$

$$>FeOH^0 + H^+ \longleftrightarrow \; >FeOH_2^+ (acid)$$

This adsorption process creates 'proton surface charge':

$$\sigma_H = \Gamma_{H^+} - \Gamma_{OH^-}$$

Surface-charge density is usually written as σ, in units of $C\ m^{-2}$.
Proton surface charge is routinely determined by acid-base titration.
Sometimes the literature will show pH_{pzc} as equivalent to:

$$\sigma_H = 0.$$

But surface-charge density includes protons and other species, $\sigma_0 = \sigma_H + \sigma_{cc}$, and so it is more correct to refer to the point where $\sigma_H = 0$ as the point of zero net proton charge, or pH_{PZNPC}.

For adsorption of protons we have the following:

$$\{>OH_2^+\} \longleftrightarrow \{>OH\} + H_s^+,$$

where $\{>\}$ indicates a surface complex.

Just as for any chemical reaction we may write a mass action law to describe this protonation reaction (see Davis *et al.*, 1978, for further details):

$$K_{a1}^{int} = \frac{[>OH][H_s^+]}{[>OH_2^+]}$$

K_{a1}^{int} represents an 'intrinsic' equilibrium constant for this protonation reaction. By intrinsic we mean that the equation is only strictly valid if 'surface' concentrations (or activities) are used. Hence the 's' subscript on the proton concentration refers to the concentration directly at the surface. Of course, this is a difficult quantity to measure and so when bulk titration data are used in calculations, the resulting constants are referred to as 'apparent' constants. These apparent values will be a function of solution ionic strength (Davis *et al.*, 1978).

The concentrations of protons at the surface can be calculated using the Boltzmann distribution:

$$[H_s^+] = [H_b^+] \exp(-e\Psi_0/kT)$$

where 'b' refers to the bulk concentration, e is the unit electron charge, Ψ_0 is surface potential, k is Boltzmann's constant and T is temperature. Therefore we can rewrite the equilibrium law such that only the ratio of singly protonated to doubly protonated species are present as surface terms:

$$K_{a1}^{int} = \frac{[>OH]}{[>OH_2^+]}[H_b^+] \exp(-e\Psi_0/kT)$$

A similar approach could be used for an adsorbed metal cation. Consider:

$$\{>O^-\} + Na^+ \longleftrightarrow \{>ONa\}.$$

Using the Boltzmann distribution relating bulk to surface sodium concentration and solving for the surface sodium concentration leads to:

$$[>ONa] = K_{Na^+}^{int}[>O^-][Na_b^+] \exp(-e\Psi_0/kT).$$

Rigorous implementation of these equations requires 'activities' and not concentrations to be used. Developing activity coefficients for surface species essentially means that surface complexation modelling can be thought of as an extension of

Debye–Hückel aqueous speciation theory to surface speciation with included corrections for the electrostatic interaction between the charged surface and the solute. There has been a tremendous amount of effort, both experimental and theoretical, in dealing with the problem of surface complexation modelling. Some of the key references include: Davis *et al.* (1978), Davis & Kent (1990), Sposito (1983), Yates (1975), Hiemstra *et al.* (1999), Stumm (1987), Sverjensky & Sahai (1996), Westall (1982), Hiemstra and van Riemsdijk (1999) and references contained therein.

Applications of surface complexation modelling to environmental problems are myriad. Problems studied include: (1) mobility of toxic metals and organic contaminants in hazardous waste and radioactive waste repositories; (2) cycling of key trace metals such as Sr, Zn and V in pristine aquatic environments; (3) Fe, W, Ni, Cu, Zn, Pb, and As in acid mine-drainage affected rivers; (4) regulation of the solute flux from the water column to the sediments in lakes, estuaries, and coastal marine settings; (5) controls on toxic element bioavailability to marine and terrestrial organisms; and (6) influence on the final composition of solids during non-equilibrium precipitation under a range of geochemical conditions.

5.10. Evolution of the surface; X-ray, FTIR and *in situ* SPM methods

5.10.1. Synchrotron experiments

As discussed in the section above on scattering techniques, synchrotron radiation allows us to probe surface evolution *in situ*, that is as the mineral surface reacts in contact with an aqueous solution. Purpose-built *in situ* reaction cells are used for these experiments. Measurements can be completed both *in situ* during reaction and *ex situ* with fluid removed from the cell. The reaction cell is usually an inert material (teflon) with X-ray transparent windows. Careful cell design may allow the reflected portion of the beam to be monitored while simultaneously the fluorescent signal can be monitored with an overhead detector. Figure 27 gives the details of a typical experimental geometry and of experimental control. Fluid flow through the cell may be accomplished by influent and effluent feedholes in the cell through which a peristaltic pump may be used to force fluid.

Reflectivity, diffuse scatter, X-ray Absorption Spectroscopy (XAS), and Total External Reflection X-ray Fluorescence (TXRF) measurements may all be completed in glancing incidence geometry at a synchrotron (*ex situ* reflectivity and diffuse scatter may be completed with a conventional source). The high brightness of a synchrotron source is required for the *in situ* experiments for two main reasons. Firstly, the significant absorption of both the incident and reflected beam by the fluid phase during *in situ* measurements degrades the signal-to-noise ratio of the reflected beam considerably. By using a high brightness source, reflectivity profiles can be measured at higher incident angles, thus giving much better constrained data. Secondly, even with fluid absent, the diffuse portion of the scattered beam is of such low intensity that a synchrotron is required. This is especially true for low roughness surfaces. A high brightness source has other advantages, in particular the much more rapid data acquisition time.

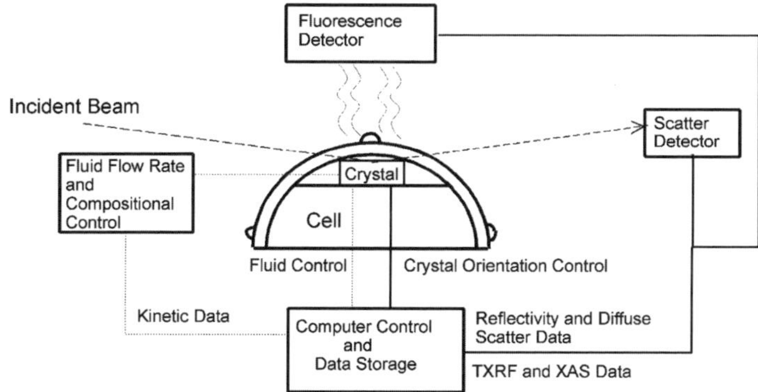

Fig. 27. Schematic diagram of an *in situ* X-ray cell for use in synchrotron experiments; cross-section showing positioning of detectors and control systems.

5.10.2. X-ray reflectivity and silicate weathering

One of the critical problems in mineral-surface chemistry is that of non-stoichiometric dissolution. Preferential loss of alkali and alkaline-earth cations from silicates is well documented by bulk techniques. This phenomenon greatly complicates our ability to model the dissolution reaction as well as other mineral-fluid interactions. Leaching creates a thin reacted film on the surface which has a completely different chemistry and structure from the bulk. Reactive functional groups on organic acids in solution can greatly increase the loss of cations and speed up overall dissolution rates by forming surface chelates which strip exposed metals from the surface. Because the cations lost are fairly large, creation of a reacted layer may in some cases open up the structure to solvent attack, such that we must consider a reactive volume and not simply a reactive surface plane. At the base of the cation-depleted layer is a second interface which cannot be analysed with surface scanning methods such as AFM or SEM. In addition, in order to maintain charge balance in this reacted layer the remaining network-forming silica tetrahedra may polymerize and, thereby, change the rates of reaction. Direct observation of these processes is challenging; however, by combining X-ray reflectivity with MIR-FTIR it has been possible to characterize *in situ* the formation of bidentate surface complexes between dicarboxylic acids and alkaline earth cations, the rapid surface loss of the cation through non-stoichiometric chelated dissolution, and the polymerization of residual silica. These observations have been corroborated by using other methods such as XPS and by the analysis of the product fluids.

Figure 28 shows the reflectivity profile for an olivine glass in the vicinity of the critical angle. An unreacted surface is compared with: (1) a surface reacted with only deionized water, and (2) a surface reacted with a solution containing phthalic acid. The critical angle for the sample reacted with phthalic acid is shifted significantly further to lower incident angle, consistent with a significantly greater decrease in electron density relative to the sample reacted in water. This is consistent with fluid analyses which indicate that magnesium loss from the surface is chelated by phthalate. In an

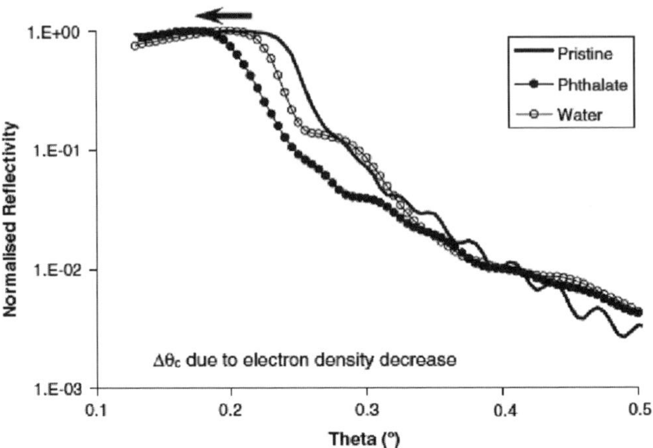

Fig. 28. X-ray reflectivity comparison of a pristine forsteritic glass surface with two reacted surfaces; one reacted for 2.8 h in a solution of 1.7×10^{-4} mol/l phthalate at pH 4 and a second reacted for 2.8 h in deionized water at pH 5.8. A shift in the critical angle (θ_c) for the reacted surfaces to lower angles is indicated by the thick black arrow, indicating an electron density decrease in both near-surface regions. A much more drastic θ_c decrease occurs after reaction with phthalate, however, showing that the removal of Mg from the near-surface region is much more efficient with phthalate present (reprinted from *Geochimica et Cosmochimica Acta,* **72**, Morris & Wogelius, 2008. Copyright (2008), with permission from Elsevier.)

attempt to understand the details of this surface complex, *in situ* MIR-FTIR experiments were completed. For phthalic acid pK1 = 3 and pK2 = 4.6; therefore at pH 4, the aqueous species of phthalic acid is singly protonated while at pH 6 the molecule is completely deprotonated. In Figure 29, the FTIR spectra of dissolved phthalic acid at pH 4 and 6 (top and middle spectra, respectively) are compared, and this shows how the low-amplitude peaks at 1711 cm^{-1} and also those between 1258 and 1296 cm^{-1} effectively disappear as solution pH changes from 4 to 6. The 1711 cm^{-1} peak is due to C$=$O stretching and the lower wavenumber peaks are due to C$-$O$-$H vibrations. All are lost because the protons are removed and this shifts the carboxyl group to a state of resonance because now both C$-$O bonds are equivalent. The lowermost spectrum is an *in situ* measurement for phthalate at the surface of olivine glass at pH 4. Note that the molecule is deprotonated although, at this pH, it should retain one proton; this indicates that it most likely has exchanged the proton for a surface-exposed magnesium. Note also that the absorption bands at 1558 cm^{-1} and 1400 cm^{-1} (carboxyl v_{as} and v_s, respectively) shift significantly relative to the uncomplexed solution species. This implies that a bidentate surface complex has formed, as depicted in Figure 29. Examination of another part of the IR spectrum reveals that the location of the peak corresponding to the Si$-$O stretch also changes during the reaction. Figure 30 shows how the unreacted, minimally polymerized starting glass has an Si$-$O stretch below 1000 cm^{-1}, and how the sample reacted with deionized water simply displays a minimum at the same wavenumber after reaction. This is simply due to a loss of silica and corresponding decrease

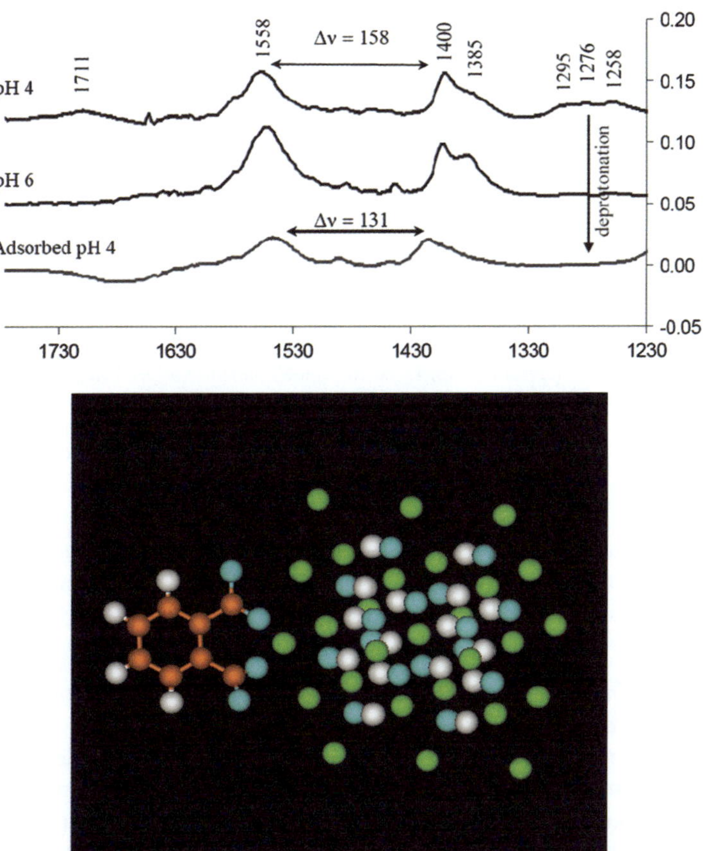

Fig. 29. Upper panel: MIR-FTIR solution species spectra of singly protonated (pH 4) and deprotonated phthalate (pH 6) compared to a spectrum of inner-sphere complexed phthalate adsorbed onto a forsteritic glass thin film at pH 4 after 29 min. The disappearance of the triplet peaks at $1295-1258\,cm^{-1}$ indicates that the adsorbate is deprotonated even though the solution pH is below the pKa2 of 4.6 for phthalate. Shifts in the carboxyl peaks resulting in a decrease in Δv as shown by the arrows suggest that the phthalate forms a bidentate complex with surface Mg atoms. Lower panel: Model of a bidentate phthalate-Mg surface complex on brucite (red = C, blue = O, grey = H, green = Mg). (reprinted from *Geochimica et Cosmochimica Acta*, **72**, Morris & Wogelius, 2008. Copyright (2008), with permission from Elsevier.)

in absorption. However, the sample reacted in phthalic acid displays both a significantly larger amount of silica loss as well as the appearance of a distinct peak at $\sim 1200\,cm^{-1}$. This is consistent with polymerization as observed during annealing studies with amorphous SiO_2 by Xu and Goodman (1992).

This study demonstrates the power and precision of spectroscopic techniques for following the progress of leached-layer formation *in situ*. This technique will be applicable to a variety of surface-alteration processes such as oxidation/reduction reactions, cation-proton exchanges and adsorption/desorption reactions. Direct quantification of

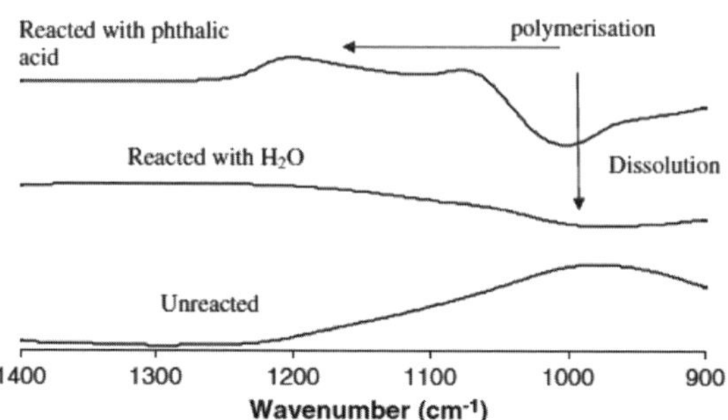

Fig. 30. *In situ* FTIR spectra of a forsteritic glass thin film before reaction (bottom), after 1.5 h reaction with water (middle), and after 1.5 h reaction with 1.7×10^{-4} mol/l phthalate at pH 4 (top). The negative peak at $1000\,cm^{-1}$ in the middle spectrum indicates silica loss due to dissolution of the film. After reaction with phthalate, a peak grows at $1200\,cm^{-1}$ due to polymerization of residual silica. (Reprinted from *Geochimica et Cosmochimica Acta,* **72**, Morris & Wogelius, 2008. Copyright (2008), with permission from Elsevier.)

the rates of leaching and roughening reactions are now possible in the third dimension, in a non-destructive manner and *in situ*. This will contribute to our fundamental understanding of mineral–fluid interface reactions in many important areas including biofilm formation, sulfide oxidation, scale formation and repolymerization.

5.10.3. *Particle Induced X-ray Emission, Rutherford Back-Scattering and X-ray Absorption Spectroscopy in Mass Transfer Studies*

The PIXE, RBS and XAS techniques have been applied to better understand aspects of the mineral chemistry and geochemistry of arsenic. Arsenic is a well known toxin and its poisoning of drinking water by release from sediments in aquifers, particularly in SE Asia, is one of the biggest environmental problems of recent years (for a brief overview, see Vaughan, 2006). Because arsenic is relatively abundant in the Earth's crust, $\approx 10^{-6}$ mass fraction as compared to Ag or Cd at $\sim 10^{-7}$, natural processes which concentrate arsenic can easily build up levels that threaten components of the biosphere. Certain aspects of the global cycling of arsenic are poorly understood. Although arsenic has generally been considered as a 'fellow traveller' with sulfur in sulfide minerals, due to the similar chemistries shared by reduced As and S species, recent work has shed light on how arsenic is incorporated under oxidizing conditions into non-sulfide minerals.

In this context, Jamtveit *et al.* (1993) studied growth-zoned garnets from skarn deposits in southern Norway. Electron microprobe analyses revealed that the core of the garnets was calcium and aluminium-rich; later growth stages of garnet showed much more Fe-rich compositions (andradite: $Ca_3Fe(III)_2Si_3O_{12}$). These later growth zones were analysed by PIXE. Trace-element concentrations were, without question, controlled by the various growth stages suggesting that the zoning resulted from open-system fluid flow. These garnets hosted the highest arsenic concentrations ever measured in silicate minerals, levels in certain zones reaching over 2500 ppm. Initially it was assumed that As was incorporated as As(III) as a substitute for Fe(III) in andraditic parts of the garnet, and that it had substituted into the octahedral sites usually occupied by Fe^{3+} or Al^{3+}. However, XAS analyses revealed from the arsenic XANES that it occurs as As^{5+} surrounded by oxygen atoms in its inner coordination sphere. Furthermore, contrary to previous suggestions that arsenic was incorporated as As^{3+} in the Al/Fe^{3+} octahedral site (Clark *et al.*, 1993), arsenic occurs predominantly as As^{5+} in the tetrahedral Si^{4+} site. It was concluded from this work that under oxidizing conditions arsenic can be partitioned preferentially into low-density silicate phases and this may serve to keep arsenic at relatively high levels in certain parts of the crust. High arsenic levels have been documented in phyllosilicates produced near subduction zones (Hattori *et al.*, 2005), in skarn garnets from regions of focused fluid expulsion from the deep crust (Connemara, Ireland) and also associated with smectites formed in a geothermal environment (Pascua *et al.*, 2005). So, the chemical concentration anomaly revealed by PIXE analysis led to X-ray absorption spectroscopy results which revealed a potential pathway for concentrating arsenic in certain parts of the crust. Release by weathering of these primary phases then begins a shallow crustal cycle whereby this arsenic can be further concentrated in sediments. Here, as suggested by XAS studies (see, for example, Farquhar *et al.*, 2002), it adsorbs onto iron-(hydr) oxides and is sequestered during rapid burial, as postulated in the deltaic environments of SE Asia.

PIXE analysis has also been used at the final stage of the As cycle to study anomalous concentrations of arsenic in human fingernails (Gault *et al.* 2008). Point analyses and maps of arsenic distributions were completed in a fingernail sampled from an individual inhabiting an at-risk region. Arsenic concentrations in the nail averaged ~5 ppm, ranging from 1 to 27 ppm (the high-point analysis is shown in Fig. 31a). High As correlated with high Fe in a small elliptical region of the nail which could be an indicator of a pulse of As into the individual due to sporadic exposure, or the anomaly could be caused by surface contamination. One of the advantages of PIXE is that RBS spectra can be obtained simultaneously, and this can provide powerful supporting information. The RBS data (Fig. 31b) suggests that the iron is evenly distributed throughout the depth analysed and that the stoichiometry remains consistent with that of normal hard keratin. Although the arsenic concentration is too low to make a significant contribution to the RBS fit, the iron data and regular nail stoichiometry both indicate that the observed high iron region is not caused by contamination of the surface but is pervasive below the interface. PIXE analysis of human nails could, indeed, indicate intermittent exposure, and imply a temporal complexity which should be accounted for in understanding exposure patterns.

5.10.4. Synchrotron XRF combined with EXAFS

Recently, advances in scanning technologies have meant that X-ray fluorescence imaging and quantitative analysis is growing as a methodology. Synchrotron rapid scanning XRF (SRS-XRF) now allows large objects of geological interest to be imaged quickly and with excellent resolution (Bergmann *et al.*, 2010). For smaller objects, micro-XRF has also been developed to give images with micron resolution, comparable to that available with electron optics (Wogelius *et al.*, 2011). Key advantages come with using a synchrotron beam for XRF analysis; high flux reduces analysis time and can lead

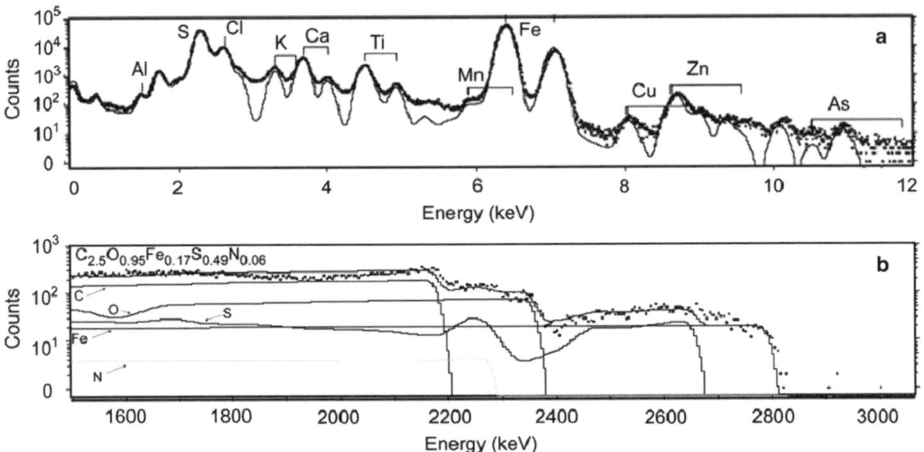

Fig. 31. (**a**) PIXE spectrum of a human fingernail sample (Kandal province, Cambodia) taken from a location with anomalously high arsenic and iron concentrations (27 and 66,400 µg/g, respectively). Dots represent the data; the line is a fit. (**b**) RBS spectrum from the same location (symbols are as in Fig. 31a). Stoichiometry at this point is equivalent to normal keratin but with elevated Fe levels, indicating that the high metal concentrations are within the keratinous tissue and are not due to contamination by mineral dust particles (see Gault *et al.*, 2008, reprinted from *Science of the Total Environment*, **393**, pp. 168–176, © 2008, with permission from Elsevier).).

to extremely good concentration sensitivity. The flexible nature of many beamlines also allows the investigator to change experimental parameters in order to optimize a single measurement, and the use of a synchrotron can mean that supplementary analyses can be completed on the sample in order to constrain the data in ways that are just not possible with conventional methods. Due to the fact that incident beam energy can be controlled precisely by a monochromator on most beamlines, XAS spectra can be collected from points within a specimen along with XRF or XRD measurements. Or, XAS spectra can be used to select the incident beam energy to be used during XRF imaging in order to highlight specific features. A striking example of this can be found in Edwards *et al.* (2011). Here, X-ray and FTIR analyses of organic material from the Green River Formation, an oil-shale lithology with enormous quantities of trapped hydrocarbons, were undertaken. Fossilized remains were targeted for analysis in order to constrain the breakdown pathways of various organic molecules. Figure 32a shows a photograph of one of the bedding plane fossils. Because reptile skin has high concentrations of sulfur (~7% by weight), maps of sulfur were expected to reveal the chemical remains of the fossil organism. Therefore, a standard method was used to map total sulfur on this bedding plane, as shown in Figure 32b. In this image, it is difficult to resolve the fossil; this is because the bedding plane contains relatively large amounts of an evaporite sulfate mineral (probably thenardite as suggested by glancing incidence XRD analysis). However, by completing XANES analysis at several points on the fossil skin, it was shown that organic sulfur compounds were present; they were simply

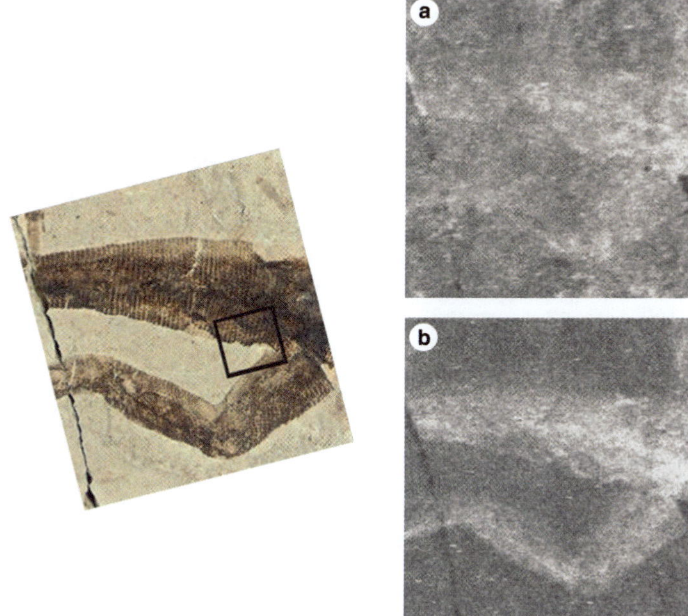

Fig. 32. Left panel: Optical photograph of exceptionally well preserved reptile skin from the Eocene Green River Formation, USA (∼50 million years old). (**a**) SRS-XRF map of total sulfur distribution on the specimen surface (incident beam energy: 2481.5 eV, above the sulfate edge energy). (**b**) Second SRS-XRF map at lower incident beam energy performed to distinguish the organic sulfur distribution. This map shows that nearly all the organic sulfur is concentrated within the fossilized skin, as would be predicted within the breakdown products of keratin. Sulfur species outside the fossil on the bedding plane are dominantly inorganic sulfate (incident beam energy 2479.9 eV, below the sulfate edge energy). (Reproduced from Edwards *et al.*, 2011, with the permission of the Royal Society.) Left panel ∼2 cm wide, boxed area ∼5 mm wide.

swamped by inorganic sulfate. The sample was scanned again, this time with the incident beam energy decreased to a point below the critical excitation energy for inorganic sulfate. The beam was still energetic enough to excite the organic species. Figure 32c shows that, in this case, the sulfur distribution is indeed controlled by the organic material derived from the fossil. Here, for the first time, the role of mineral surfaces and trace metals in the breakdown or preservation of organic material is being explored with spatial resolution.

5.10.5. Diffuse X-ray scatter and olivine breakdown

One of the most basic parameters in mineral reaction kinetics is surface area. This is usually measured by BET gas adsorption isotherm (Brunauer *et al.*, 1938) from a reactant powder. BET measurements need large surface areas to work and are difficult, if not impossible, to apply to single crystals. This means that it is difficult to compare rates measured from powders with those measured using single crystals, although experiments with single crystals give us crystallographically resolved information which proves useful in constraining reaction mechanisms. While the specular reflectivity data give the r.m.s. roughness of the surface or surfaces in a system, more can be learned about the in-plane structure of the surface by analysing the diffuse scatter component.

Measurements of specular reflectivity and of diffuse scatter have been made on reacted single crystals of the mineral olivine ($Mg_{1.8}Fe_{0.2}SiO_4$) using synchrotron X-ray radiation (Wogelius *et al.*, 1999). Olivine, due to the non-linkage of the silica tetrahedra in the structure, is one of the fastest dissolving silicate minerals. Figure 33 shows a comparison of the data from the specular reflectivity scans taken for the samples reacted for 10^4 and 10^6 seconds. The change in the quality of the surface during reaction at pH 4 is dramatic. Specular reflectivity measurements on samples reacted for 10^4 s (2.78 h) and 10^6 s (11.57 days) give surface roughness values, σ, of 8.5 ± 1.0 Å and 19.0 ± 1.2 Å, respectively. The initial roughness of the starting material was $\sim$7 Å. Surface roughening is, therefore, only slightly over the time scale of 10^4 seconds. However, surface roughness more than doubles over 10^6 s of reaction. Diffuse scattering measurements also showed marked changes between the two samples. Figure 34 shows a comparison of the diffuse scatter for the 10^4 and 10^6 s reaction times.

The surface roughness values determined from the specular reflectivity data were used to constrain simulations of the diffuse scatter data. Correlation lengths (ξ) and fractal parameters (h) were varied until the simulations agreed with the data. Corrections for incident beam intensity and geometry were applied. The correlation lengths (a measure of the average distance between surface features) and fractal parameters

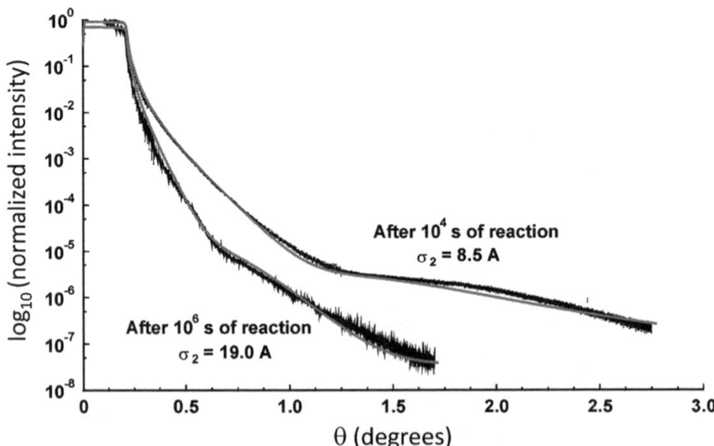

Fig. 33. Specular reflectivity scans for olivine samples reacted for 10^4 s and 10^6 s (both scans have each data point plotted as a short vertical line with line height equal to error) with pH 4 nitric acid at 25°C. Data are presented as log (normalized intensity) as a function of incident angle (°). Also shown are the simulations of the data (grey curves) calculated as discussed in the text. Increase in surface roughness during reaction from 8.5 Å to 19.0 Å causes the drastic reduction in the reflected intensity as a function of incident angle in the longer reacted sample compared to that reacted for a shorter time. The critical angle is evident at $\sim$0.21° (from Wogelius *et al.*, 1999, reproduced with kind permission from Springer Science & Business Media B.V.).

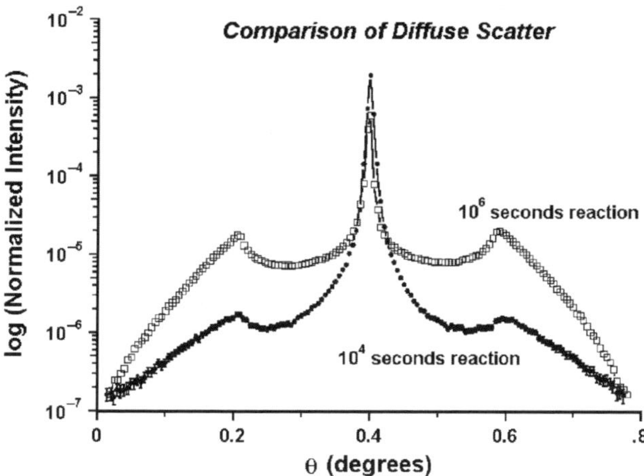

Fig. 34. Comparison of the diffuse scatter component as measured by rocking curve scans for the 10^4 s (solid circles with error bars) and 10^6 s samples (open squares, symbol size equals error). The detector angle was held fixed at 0.8°. The central peak is the specular reflectivity peak while the symmetric features with lower intensity peaks are the Yoneda wings. Increase in surface roughness during reaction decreases the intensity of the specular peak but increases the intensity of the Yoneda wings. Locations of the Yoneda peaks correspond to the critical angle for the incident (0.21°) and the scattered (0.8°−0.21° = 0.59°) beam (from Wogelius *et al.*, 1999, reproduced with kind permission from Springer Science & Business Media B.V.).

(quantity describing the jaggedness of the surface) both changed during reaction. ξ decreased significantly from essentially infinity (16 μm $\pm$ 6) to $\sim$2750 Å ($\pm$260), while *h* decreased slightly from 0.23 ($\pm$0.02) to 0.195 ($\pm$0.03). This is consistent with the specular reflectivity information and indicates that during reaction, as the height of surface imperfections doubled, the average distances between them decreased by over two orders of magnitude and the profile of the surface became more jagged. From the data, minimum rates of roughening and minimum rates of decrease in correlation length can be estimated for the initial period of dissolution, with $d\sigma/dt \approx 1.06 \times 10^{-5}$ Å s^{-1} and $d\xi/dt \approx -0.157$Å s^{-1}. The rate of roughening is $\sim$1/30th of the rate of material removal from the surface. Below, we show how the diffuse scatter data can be used to model the change in surface area.

5.10.6. SPM experiments

As regards experimental studies of silicate dissolution at the molecular scale, Rufe & Hochella (1999) have used AFM to observe directly the dissolution of the basal surface of a mica crystal. As seen in Figure 35, a pit forms, the size of which can be measured accurately so as to determine the dissolution rate for this surface. The study shows that whereas there is active dissolution of material from the sides of this pit, some material is also removed *via* diffusion through the layers. A more extensive study, also involving these authors (Bickmore *et al.*, 2001) employed *in situ* AFM methods to explore the dissolution in acid solution of two other phyllosilicates, hectorite

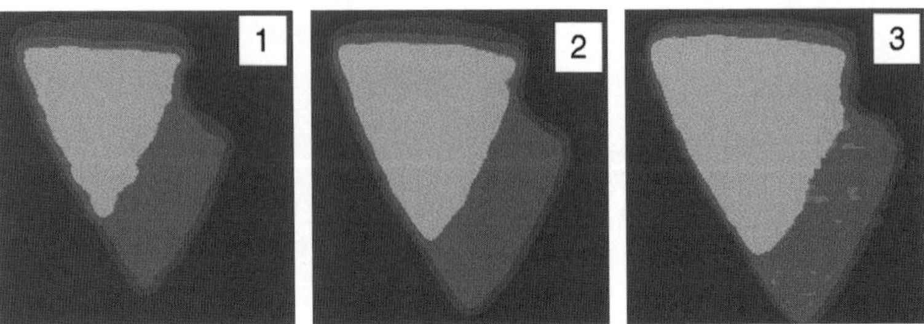

Fig. 35. AFM images of an evolving dissolution pit on the {0001} plane of a mica. The darkest layer is the topographic layer; each successive layer is 10 Å lower: (1) original pit after HF etching; (2) same pit after 39 h in pH 2 HCl; and (3) same pit after 63 h in pH 2 HCl. All images are 625 nm wide (after Rufe & Hochella, 1999).

and nontronite. Crystallites were found to dissolve inward from the edges and basal surfaces appeared to be unreactive over the timescale of the experiment. In this study, the stabilities of different faces were evaluated and explained using periodic bond chain theory to predict the topology of the surface functional groups on the edge faces of the minerals. Of course, the dissolution rates of a much wider range of minerals have been studied using SPM techniques and following the types of approaches outlined by Dove & Platt (1996) with reference to calcite, barite and celestite.

6. Comparative reactivity of silicates, carbonates, sulfides and oxides: Experiments *vs.* the natural environment

Above, we have attempted to summarize the analytical methods most commonly used in environmental mineralogy, along with the experimental approaches most often used to extract kinetic and sorption data from low-temperature systems. In this final section, we will discuss some important examples, both from experiments and from the natural environment, regarding the reactivities of minerals.

6.1. Structural controls on dissolution

The first systematic study of weathering rates of silicate minerals was completed by Goldich (1938), >70 years ago. He studied igneous terrains in Minnesota and noticed that olivine and anorthitic plagioclase tended to disappear from the weathering residue first, and that potassium feldspar and quartz tended to be the most stable of the primary phases. This essentially mimicked Bowen's reaction series, with the highest-temperature phases being the least stable and the lowest-temperature phases being the most stable under Earth-surface conditions. Of course, this seems obvious when one considers that quartz beach sands are the rule and olivine beach sands are extremely rare, but it provided a framework for understanding the sequence of silicate reactivity. Thus, the stability sequence of the common rock-forming igneous minerals is known as Goldich's weathering series and is illustrated in Figure 36.

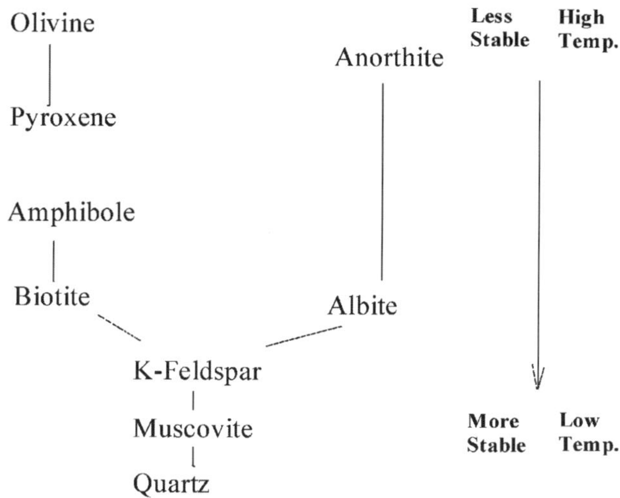

Fig. 36. Derived from Bowen's reaction series for crystallization from a melt, Goldich's (1938) stability series shows that igneous minerals which crystallize at low temperatures are slower to dissolve and therefore more stable in crustal weathering processes than those which crystallized at high temperatures.

This series reflects the structure of the silicates and suggests that the bonding within the mineral controls the breakdown mechanism. The silica tetrahedron (SiO_4) is the fundamental building block of all silicate minerals. From top to bottom of both the continuous (or sialic) and the discontinuous (or mafic) sides of Goldich's series, the degree of linkage of the silica tetrahedra in the structure increases. Olivine has no Si—O—Si bonds; all of the silica tetrahedra are isolated from each other and therefore olivine is called a nesosilicate (island silicate). Similarly, even though anorthite is a tectosilicate (framework silicate) with all tetrahedral sites linked to each other, anorthite has half of its tetrahedral sites occupied by Al, and hence again there are no Si–O–Si bonds in the structure (unless Si-Al disordering occurs). Quartz, by comparison, consists of only silica tetrahedra and all bonds in the structure are Si—O—Si bonds. The other phases in Figure 36 lie somewhere between these end-members in terms of degree of polymerization of silica. Therefore the reactivity of the silicates can be understood in terms of the atomic-scale bond strengths of the mineral in question. Because the reaction rates of silicates are extremely slow and because secondary silicate phases have extremely low solubilities, it was difficult to accurately quantify the absolute reaction rates of silicates until the late 1960s. Advances in fluid analytical techniques made it possible to measure accurately solutes from dissolution experiments at levels well below saturation limits for secondary phases. Figure 36 summarizes the dissolution rates of the common igneous rock-forming minerals and shows the clear structural control on reactivity.

Brady & Walther (1989) attempted to use the electrostatic site potential of oxygen in the mineral as an index for reaction rates. The resulting relationship is given in Figure 37 and shows that, indeed, the rate of reaction is a function of the average

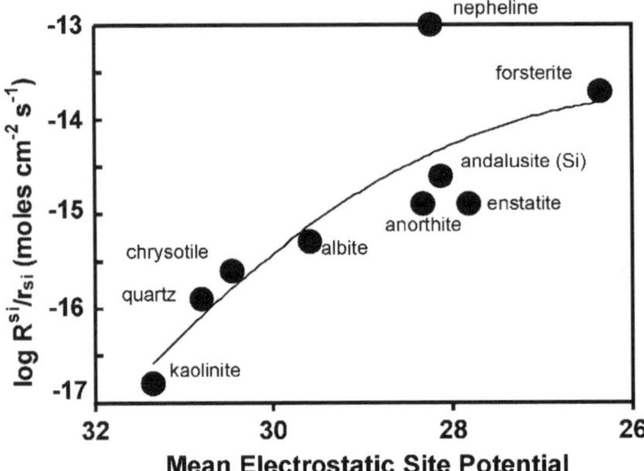

Fig. 37. 25°C pH 8 dissolution rates for silicates plotted against site potentials (volts) of oxygen bonds connecting detaching silicon atoms to the respective mineral surfaces. The rates are normalized: R^{si} is the release rate of silica, r_{si} is the stoichiometric ratio of moles of silica to total non-alkali oxides in the mineral formula unit (from Brady & Walther, 1989 reprinted from *Geochimica et Cosmochimica Acta*, **53**, pp. 2823–2830, © 1989 with permission from Elsevier).

cation–oxygen bond strength. However, this graph is intended to be used at near-neutral pH and at low concentration of solutes and hence may not predict relative rates under conditions prevalent in natural systems. Another predictive approach has been attempted by Casey in a number of contributions (*e.g.* Casey *et al.*, 1988) where he has accounted for the mixed oxide character of silicates. Essentially this involves dealing with silicates as being formed from silica plus another metal oxide; *e.g.* forsterite is silica plus magnesium oxide, fayalite is silica plus iron oxide, *etc.* In this way the reactivity of silicates can be thought about as the sum of the reactivity of one or more constituent oxides. This can most

broadly be seen as an explanation for the characteristic U-shaped log rate *vs.* pH curve for silicate dissolution as shown on Figure 38 for olivine. At low pH, the extensive protonation of surface oxygens (and perhaps penetrative cation-proton exchange) polarizes the structural Mg—O bonds and promotes Mg detachment. At high pH, the nearly complete deprotonation of surface oxygen atoms electrostatically destabilizes the silica tetrahedra and promotes the detachment of silica. Thus the rate has maxima at both low and high pH with a rate law minimum at neutral pH. This also explains why the single oxide silica has a dissolution rate that is pH independent at low pH (Fig. 39 compares the behaviour of quartz to that of albite). Sverjensky & Sahai (1996) and Hiemstra *et al.* (1989) have both developed models to deal with the multisite nature of protonation onto silicate phases.

The concept of thinking about the reactivity of mineral phases as mixed phases is a useful framework for understanding relative rates, not only for silicates but also for carbonates. Figure 40 shows the rates of reaction for a number of simple carbonates as measured by Chou & Wollast (1989). It is clear that the rate of reaction is controlled by ligand exchange reaction for the cation in the structure, with cation radius a good indicator of reactivity for all phases except magnesite. As for silicates, the release of the cation is a ligand exchange reaction, typically of structural oxygen (oxygen bonded to silicon or in this case carbon) for the oxygens of the aqueous solvent. The anomalous behaviour of magnesite (and dolomite) is due to the extremely high affinity of Mg for carbonate and underlines the fact that mechanisms of reactions must be considered along with initial and final conditions.

6.2. Solute and solvent effects on adsorption reactions

Solution composition has a tremendous influence on the different ways in which the solid phase interacts with the aqueous phase. Perhaps the most general affect is the way in which simply by increasing the ionic strength of the fluid, reaction rates are significantly changed. This is due to the fact that increasing the ionic strength of the fluid collapses the electrical double layer at the solid-fluid interface. This changes the adsorption behaviour of solutes as well as the structure of the interface. In turn, this can appreciably change the kinetics of dissolution reactions.

In contrast, certain solutes can form inner-sphere complexes which drastically change surface reactivity. These can either increase or decrease rates. In the former case, strong bonds can be formed between bidentate organic ligands and cations exposed at mineral surfaces. These ligands can then de-stabilize the solid

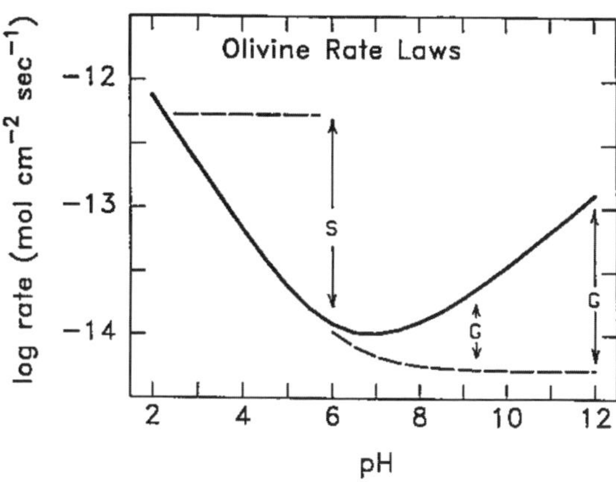

Fig. 38. Application of rate laws to natural systems. The bold solid line is forsteritic olivine rate law at 25°C without chelation or inhibition effects. The dashed line to the left is the rate law postulated for activities of organic species in natural aquatic environments. The dashed line to the right is the rate law postulated for systems in equilibrium with the natural atmospheric p_{CO_2}. The symbol S denotes surface-water pH measured on troctolites from the Duluth complex, an ultramafic terrain. The two G's denote groundwater pH values typical of ultramafic emplacements. Here, pH 9, also measured in the Duluth complex troctolites, corresponds to Barnes & O'Neil's (1969) Mg-HCO$_3^-$-type waters; pH 12 corresponds their Ca-OH$^-$-type waters. The distance between arrowheads shows the discrepancy between the fundamental rate law and dissolution rates likely to be observed in natural systems (from Wogelius & Walther (1991), reprinted from *Geochimica et Cosmochimica Acta*, **55**, pp. 943–955, © 1991, with permission from Elsevier).

structure by forming chelate complexes which preferentially detach and increase reaction rates. Furrer & Stumm (1986) have shown how the number of members in the ring of a bidentate chelate can be a good indicator of the strength of the metal–ligand bond and hence an indicator of increased rate. Conversely, some solutes can poison reactions by binding at surface sites to form stable, unreactive complexes. Perhaps the best known example is that of phosphate complexation on carbonate minerals. This is among the first examples of surface poisoning, as already discussed (see Fig. 22). Figure 38 shows how for olivine both chelation and poisoning (or inhibition) needs to be understood before reasonable predictions can be made about the reactivity of a phase.

6.3. Electron transfer reactions and surface evolution

Reactivities of mineral phases are often complicated by electron transfer reactions, especially in the case of iron since it is such an important component of both organic and inorganic reaction paths. In a rigorous thermodynamic treatment of a system, Fe^{2+} and Fe^{3+} must be treated as two completely different components and, therefore, the phase rule gains a degree of freedom for each possible oxidation state of the element. Thus the reaction paths for iron sulfides are complicated because they involve not only the loss of an electron from the Fe in the structure, but also the change in oxidation state of sulfur from -2 to $+6$, a total loss of 17 electrons from a molecule of FeS_2, for example, as it reacts to iron oxide and sulfate. Rimstidt & Vaughan (2003) have proposed a series of six surface reactions describing pyrite oxidation in aqueous solution, defining the reaction as an electrochemical process with three distinct steps: a cathodic reaction, electron transport, and anodic reaction. It is argued that the cathodic reaction is the rate-determining step and that the pyrite oxidation rate depends on the concentration of O_2 or another oxidant such as Fe^{3+}. A fuller account of the surface reactivity of sulfide minerals is given by Rosso & Vaughan (2006).

Another example of such electron transfer reactions is provided by the manganese silicate mineral rhodonite ($MnSiO_3$); here, a study was conducted which combined X-ray reflectivity with diffuse scatter and X-ray absorption spectroscopy in order to fully determine the changes in surface structure normal to the surface, in plane topography, and oxidation state as a function of depth below the solid/solution interface (Farquhar *et al.*, 2003). A rhodonite surface was prepared and showed *via* X-ray reflectivity that during the initial stage of aqueous oxidation, a well organized thin film was produced on the mineral surface; this is shown in Figure 41 as the development of Kiessig fringes in the reflectivity profile. These oscillations disappear in the next stage of reaction, indicating that the coherent nature of this reacted thin film was lost after additional reaction. In one of

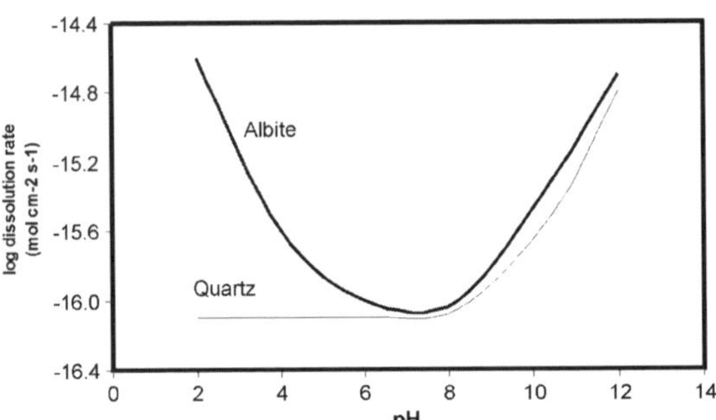

Fig. 39. Schematic comparison of the reactivity of quartz (SiO_2, thin line; data from Brady & Walther, 1989) and of albite ($NaAlSi_3O_8$, thick line; data from Chou & Wollast, 1985). Quartz dissolution is pH independent in the acidic and near-neutral pH regions, while albite dissolution is strongly pH dependent in acid. The albite curve is typical of most silicates.

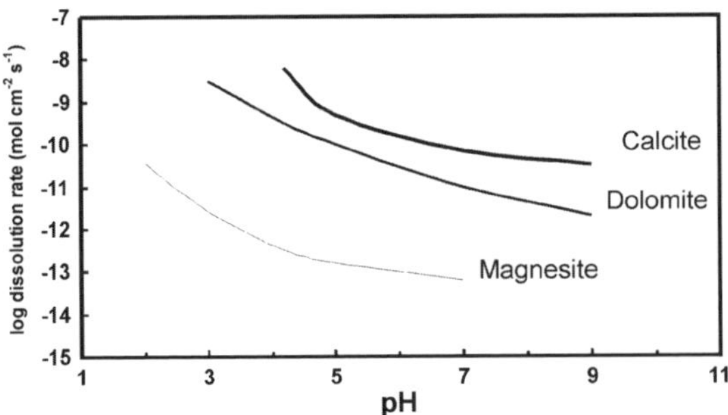

Fig. 40. Dissolution rate of selected carbonate minerals as a function of pH (based on data in Chou & Wollast, 1989). Note the dramatic decrease in reactivity as the quantity of Mg in the structure increases.

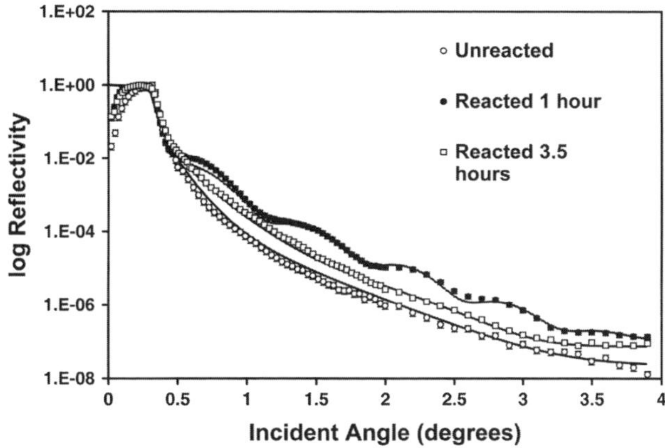

Fig. 41. Specular reflectivity profiles of a rhodonite ($MnSiO_3$) surface. Fits to each profile are shown as the solid curves. The unreacted surface (open circles) has a starting roughness of 15.5 Å. After the first 1-h reaction step (filled circles), a thin film of 75 Å has developed which causes the oscillations in the profile. The film is of lower density than bulk rhodonite and the primary surface roughness has decreased to 4.5 Å. After further reaction time (open squares) much of the film dissolves away and the oscillations become attenuated (see Farquhar *et al.*, 2003, reproduced with permission from the Mineralogical Society of Great Britain & Ireland).

the first examples of XANES depth profiling, Figure 42 presents a set of XANES scans taken as a function of incident angle from the reacted surface. The shift in edge position is a function of oxidation state, with high-energy edges due to higher oxidation states. The decrease in edge position as a function of incident energy shows that the surface Mn is highly oxidized, with the Mn(II)/Mn(IV) ratio increasing as a function of depth. Combining the X-ray reflectivity data with the oxidation state information allowed a full picture of the pathway that the rhodonite surface follows during coupled dissolution and oxidation. This was only possible by using a flexible synchrotron beamline which allowed both types of scans to be collected from the same sample during the same experimental run.

6.4. Surface area and scaling laws

The data presented above (in section 5.10.5) clearly show that the in-plane olivine surface structure changes in a dramatic but measurable way during reaction with mildly acidic solutions at 25°C over time periods on the order of 10^6 s. Most published kinetics experiments with powdered samples show an initial 'surface adjustment period' (*e.g.* Chou & Wollast, 1985; Mast & Drever, 1987; Wogelius & Walther, 1991, 1992). Precisely how the surface adjusts is not known. Carroll-Webb & Walther (1988) attempted to measure changes in surface area with reacted aluminium oxide and aluminosilicate (kaolinite) by applying BET-type measurements to powders after reaction and comparing these areas with data from the starting material. Surface areas were the same within the error of the measurement. However, the development of etch pits during surface-controlled dissolution is well documented, and X-ray reflectivity has recently shown that the terraces between pits also roughen during dissolution (Chiarello *et al.*,

R.A. Wogelius & D.J. Vaughan

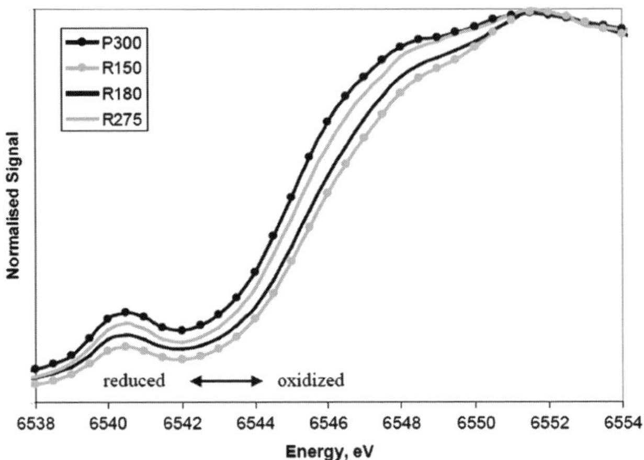

Fig. 42. XANES spectra from a rhodonite sample obtained using glancing incidence. P300 refers to the pristine crystal at 300 mdeg incident angle which probes the surface down to 140 Å. R150, R180 and R275 are all spectra taken after an initial reaction step at progressively higher incident angles corresponding to progressively greater X-ray penetration depth. The angles are 150, 180 and 275 mdeg, respectively, corresponding to penetration depths of 32.6, 35.1 and 65.3 Å. Note that the highest angle measurement (R275) most closely resembles the pristine sample, indicating that the amount of Mn oxidation is greatest near the surface and decreases away from the mineral surface (Farquhar *et al.*, 2003, reproduced with permission from the Mineralogical Society of Great Britain & Ireland).

1993). So an increase in surface area during dissolution must reasonably be expected; Carroll-Webb & Walther's (1988) result simply means that this increase may not be resolvable by BET.

We can use r.m.s. roughness values and correlation lengths to constrain the surface-area changes on the olivine surface in two different ways. The first calculation is a simple Euclidean-geometry type of estimate of surface area change.

(1) The reactive surface is taken as a flat reference plane which is equivalent to the exposed surface area of the mineral wafer used in these experiments. The starting surface area is S_0, in this case 5 mm $\times$ 5 mm or 25 mm^2. If we assume that the surface is isotropic, then in the plane there will be S_0/ξ^2 features on the surface. We further assume for the sake of simplicity that each feature is a depression bounded by four walls of the size $\sigma^*\xi$. Therefore the additional surface area (S_a) will be equal to:

$$\frac{4\sigma\, S_0}{\xi} = S_a$$

The total surface area (S_t) will then be: $S_t = S_a + S_0$. Application of this simple model to the 10^4 s data gives a 0.02% increase in surface area. For the 10^6 s sample however, the much smaller ξ value results in a much larger calculated value, with a roughly 3% increase in surface area. Neither of these values would be resolvable with standard BET techniques on powdered samples; however, the 3% increase for the 10^6 s sample is not negligible.

(2)   This second calculation takes advantage of the fact that the diffuse scatter has allowed us to estimate the fractal dimension of the surface. We assume that the surface is isotropic in the xy plane. The fractal dimension (D) of a line is calculated from the relationship: $D = 2 - h$. Therefore, a profile line across the surface that was reacted for 10^4 s would have a fractal dimension of 1.77 ($=2-0.23$). The 10^6 s sample would have a profile line with a fractal dimension of 1.805 ($=2-0.195$). Thus the length of the line that describes that profile has increased, in this case from $5^{1.77}$ mm to $5^{1.805}$ mm. By assuming that the xy plane is isotropic, we can extrapolate from a line to a plane. In this case, the fractal dimension of the surface plane will be:

$$D = 3 - h$$

(after Turcotte, 1992), and for a square plane:

$$S_t = L^D$$

where L is the length of a side of the plane. This allows us to estimate the total surface area after 10^4 s of reaction, which gives $S_t = 8.63 \times 10^{-5}$ m^2. After 10^6 s of reaction the total surface area is: $S_t = 9.13 \times 10^{-5}$ m^2. This corresponds to a 5.8% increase in surface area. The absolute values of the surface areas for the two reaction times are nearly a factor of four larger than those calculated with a Euclidean model. However, the 'change' in surface area calculated by the two methods only differs by a factor of two. Therefore, by either calculation, there has been a significant increase in surface area over the course of the reaction. However, even a 5.8% increase in surface area would be difficult to measure accurately *via* BET adsorption isotherm, and would be nearly impossible to apply to an individual face of an olivine single crystal.

This time-resolved change in surface area is of particular importance in understanding the reactions of silicates with aqueous solutions because, under most shallow-depth geochemical processes, the reactions between silicates and fluids are surface controlled. Therefore, a surface-area term appears explicitly in the rate law describing the reaction kinetics (Aagaard & Helgeson, 1982). As shown, even though the surface roughness value may only increase by a factor of two during the initial stages of dissolution, the decrease in ξ as a function of time may make major changes in the type of surface presented to the fluid for reaction. This essentially means that the retreat of a surface during dissolution, although not entirely uniform, will not sustain the development of a large number of features with long peak to trough distances: increases in surface area from the deepening of troughs will be in part balanced by the removal of material from peaks. However, the number of features on a surface may increase drastically, as measured by a decreasing correlation length, and this is probably the key factor in increasing the surface area. Note that the diffuse scatter is produced from the entire surface, and it represents an average representation of the surface as a whole. In order to produce a comparable statistic by AFM profilometry at the same resolution, many scans would have to be taken and averaged.

Increase in surface area is critical because silicates dissolve in acidic fluids by protonation reactions on oxygen atoms exposed at the mineral surface. In the case of olivine, protonation of surface oxygen atoms coordinated to near-surface magnesium promotes cation detachment (Casey & Westrich, 1992). Because the silica tetrahedra are not polymerized, removal of Mg^{2+} leaves the individual tetrahedra free to go into solution with no further bond-breaking required. Therefore, an increase in the amount of exposed under-coordinated Mg atoms should make dissolution more energetically favourable. Production of unit-cell scale roughness at the surface, and the decrease in distance between surface topographic features as described here, not only increases the surface area but also changes the types of sites available for solvent attack, which obviously changes the energetic distribution of sites available for reaction. The types of surface sites available play a critical role in determining how other Mg-bearing phases react with fluids (Refson *et al.*, 1995; Wogelius *et al.*, 1995).

6.5. Computer modelling from molecular to macroscopic scales; reaction path modelling

Reaction path modelling is the use of a thermodynamic and kinetic model to calculate the progress of a reaction from a point of disequilibrium to another point in time which is usually at, or closer, to equilibrium. As an introduction to the topic, we develop the idea of reaction path by noting that activity diagrams are extremely useful for plotting the reaction path, just as contoured ternary diagrams show reaction paths in igneous petrology.

Consider albite dissolution:

$$NaAlSi_3O_8 + 4H^+ \longleftrightarrow Al^{3+} + Na^+ + 3\ SiO_2{}^{aq} + 2\ H_2O$$

$$K_a = \frac{a^3_{SiO_2}\, a_{Al^{3+}}\, a_{Na^+}}{a^4_{H^+}}$$

Plotting the evolution of the product fluid on an activity diagram of aNa *vs.* aSiO$_2$ would allow one to see if this reaction would reach equilibrium in tandem with gibbsite, kaolinite or quartz? Other questions could be answered such as the following: "How long will the fluid be in the gibbsite stability field?" Of course, albite and gibbsite are never in equilibrium; other phase fields divide them. But what is the trajectory for albite dissolution? If pH is fixed, on a linear scale the slope would be 1/3. If pH floats with the reaction, then the slope would be $(4 + 1)/3$. These will *not* be the slopes on a logarithmic plot, curvature will be displayed, although the general trend of the curves will reflect the stoichiometry. So, graphically we can follow a reaction path, but for many problems we want to calculate this in a time-resolved way.

Modelling is completed in essentially two steps: (1) fluid (and surface) speciation which is a static thermodynamic calculation, and (2) time-stepping using a rate law which is a kinetic calculation. Below we show how this proceeds.

(1) Speciate the fluid with the Newton–Raphson Method. This essentially takes advantage of the fact that, no matter how complicated a solution is, as long as you know pH and the total concentration of each element, you can solve the system of equations describing speciation. Each new solute brings one unknown, but by measuring its concentration the system can be solved. Take a simple solution with just aluminium in it. There will be five species:

$$Al^{3+}, \; Al(OH)^{2+}, \; Al(OH)_2{}^{+}, \; Al(OH)_3{}^{0}, \; Al(OH)_4^{-}$$

This will give four speciation equations with corresponding mass action laws. For example:

$$Al^{3+} + H_2O \leftrightarrow Al(OH)^{2+} + H^{+}$$

$$K_1 = \frac{a_{H^{+}} a_{Al(OH)^{2+}}}{a_{Al^{3+}}} \quad etc.$$

There will also be one mass balance equation for each solute:

$$Al_t = Al^{3+} + Al(OH)^{2+} + Al(OH)_2{}^{+} + Al(OH)_3{}^{0} + Al(OH)_4^{-}.$$

This results in five equations with seven unknowns (activity of each species, total dissolved Al and pH). Measure pH and Al_t and the result will be five equations and five unknowns, which makes the system solvable – in theory. But solving all of these equations simultaneously can be tedious. So, we use an iterative technique based on Newton's method.

$$0 = Al_t - (Al^{3+} + Al(OH)^{2+} + Al(OH)_2{}^{+} + Al(OH)_3{}^{0} + Al(OH)_4^{-})$$

Let $x = Al^{3+}$, then:

$$f(x) = Al_t - (Al^{3+} + Al(OH)^{2+} + Al(OH)_2{}^{+} + Al(OH)_3{}^{0} + Al(OH)_4^{-}$$

$f(x) = 0$ is value of Al^{3+} that solves the equation. Take a guess at x_1, and you then can compute $f(x_1)$ and $f'(x_1)$, which is just the tangent at $f(x_1)$.
Your next guess might be:

$$x_2 = x_1 - \frac{f(x_1)}{f'(x_1)}.$$

This keeps getting you closer to zero (hopefully the zero you intended) and the solution to the equation. The calculation is iterated until the difference between iterations falls below a set tolerance level. When used for a system of equations this is called Newton–Raphson iteration.

(2) Time stepping *via* the kinetic rate law.

We use the fact that, as we approach equilibrium, the rate must decrease.

$$R = k\left(1 - e^{\frac{-A}{RT}}\right)$$

For simplicity we use a zero-order reaction. If the affinity is large, the exponential approaches zero and $R = k$. If the affinity is small, say close to zero, then $e^0 = 1$ and the rate is zero.

The derivative of this equation can be used to 'predict the future'. You don't know the actual change in the function over the interval, but you can predict what it might be based on a change in Affinity.

For example, at time $t = 7$ you might use speciation and the rate law to predict the rate, R_8, at $t = 8$:

$$R_8{}^P = R_7 + \Delta t^* \frac{dr}{dt}$$

(The slope between points 7 and 8 would be this dr/dt.)
Simply put:

$$R_{t+1}{}^P = R_t - \Delta R_{(t+1,t)}$$

But we are trying to approximate an instantaneous slope (the derivative) while Δt is an incremental. Your estimated slope will be based on t_7 and will always be (slightly) wrong, so we need to go back and correct this estimate. We do this by calculating a slope at the new time step and averaging slopes.

$$R_{t+1}{}^{Corrected} = R_t - \Delta t \left[\frac{\left(\frac{dr}{dt}\right)_{t+1} + \left(\frac{dr}{dt}\right)_t}{2} \right]$$

The second slope was calculated at the new time step from the re-speciated and perturbed system (new affinities). This therefore allows us to correct our initial estimate.

Check that $R_{t+1}^c - R_{t+1}^P = \delta$

If $\delta <$ a preset tolerance limit, go on to $t + 2$

If $\delta >$ tolerance, go back to the start and recalculate.

This is the 'Predictor-Corrector' method for time-stepping.

The speciation/time-stepping approach was first used with success by Wolery (1983) in the code EQ3/6 but this is the fundamental methodology used by nearly all reaction-path models. In order to function, the code needs an extensive, consistent thermodynamic data base. The next level of complexity would involve coupling detailed chemistry with system hydrodynamics.

7. Concluding remarks

Great progress has been made in the past 30 years both in the development of experimental methods to study mineral-containing systems of environmental relevance, and in the development of theoretical approaches to modelling such systems. Indeed, in the decade since the first edition of this book was published, there has been an explosion of interest and activity in the field of environmental mineralogy. The rate of this progress looks set to accelerate even more, with further methodological developments and much wider applications to 'real world' problems.

Acknowledgements

The authors acknowledge the financial support for research described here, much received over the years from the NERC and EPSRC. They also thank Paul Wincott, John Charnock, Paul Lythgoe, Alistair Bewsher and John Waters for technical support, and colleagues at Manchester, particularly, Jon Lloyd, Richard Pattrick and Dave Polya, for stimulating discussions.

References

Aagaard, P. & Helgeson, H.C. (1982) Thermodynamic and kinetic constraints on reaction rates among minerals and aqueous solutions. I. Theoretical considerations. *American Journal of Science*, **282**, 237–285.

Abraitis, P.K., Livens, F.R., Monteith, J.E., Small, J.S., Trivedi, D.P., Vaughan, D.J. & Wogelius, R.A. (2000) The kinetics and mechanisms of simulated British Magnox waste glass dissolution as a function of pH, silicic acid activity and time in low temperature aqueous systems. *Applied Geochemistry*, **15**, 1399–1416.

Andrews, S.R. & Cowley, R.A. (1985) Scattering of X-rays from crystal surfaces. *Journal of Physics C, Solid State Physics*, **18**, 6427–6439.

Arrhenius, S. (1889) On the reaction rate of the inversion of non-refined sugar upon souring. *Zeitschrift für Physikalische-Chemie*, **4**, 226–248.

Banfield, J.F. & Navrotsky, A. (editors) (2001) *Nanoparticles and the Environment*. Reviews in Mineralogy & Geochemistry, **44**. Mineralogical Society of America, Washington, D.C., 349 pp.

Barnes, I. & O'Neil, J.R. (1969) The relationship between fluids in some fresh Alpine-type ultramafics and possible modern serpentinization, Western United States. *Geological Society of America Bulletin*, **80**, 1947–1960.

Becker, U. & Hochella, M.F., Jr. (1996) The calculation of STM images, STS spectra, and XPS peak shifts for galena: new tools for understanding mineral surface chemistry. *Geochimica et Cosmochimica Acta*, **60**, 2413–2426.

Becker, U., Munz, A., Lennie, A.R., Thornton, G. & Vaughan, D.J. (1997a) The atomic and electronic structure of the (001) surface of monoclinic pyrrhotite (Fe_7S_8) as studied using STM, LEED and quantum mechanical calculations. *Surface Science*, **389**, 66–87.

Becker, U., Greatbanks, S.P., Rosso, K., Hillier, I.H. & Vaughan, D.J. (1997b) An embedding approach for the calculation of STM images: method and application to galena (PbS). *Journal of Chemical Physics*, **107**, 7537–7542.

Bedzyk, M.J. & Cheng, L. (2002) X-ray standing wave studies of minerals and mineral surfaces: principles and applications. In: *Applications of Synchrotron Radiation in Low Temperature Geochemistry and Environmental Science* (P.A. Fenter, M.L. Rivers, N.C. Sturchio & S. R. Sutton, editors). Reviews in Mineralogy, **49**. Mineralogical Society of America, Washington, D.C., pp. 221–266.

Bell, A.M.T., Pattrick, R.A.D. & Vaughan, D.J. (2010) Structural evolution of aqueous mercury sulphide precipitates: an energy dispersive X-ray diffraction study. *Mineralogical Magazine*, **74**, 85–96.

Benning, L.G. & Barnes, H.L. (1998) *In situ* determination of the stability of iron monosulphides and kinetics of pyrite formation. *Mineralogical Magazine*, **62**, 151–152.

Bergmann, U., Morton, R.W., Manning, P.L., Sellers, W.I., Farrar, S., Huntley, K.G., Wogelius, R.A. & Larson, P. (2010) *Archaeopteryx* feathers and bone chemistry fully revealed via synchrotron imaging. *Proceedings of the National Academy of Sciences*, **107**, 9060–9065.

Betzig, E., Patterson, G.H., Sougrat, R., Linwasser, O.W., Olenych, S., Bonifacino, J.S., Davidson, M.W., Lippincott-Schwarz, J. & Hess, H.F. (2006) Imaging intracellular fluorescent proteins at nanometer resolution. *Science*, **313**, 1642–1645.

Beveridge, T.J., Moyles, D. & Harris, B. (2007) Electron microscopy. In: *Methods for General and Molecular Microbiology* (C.A. Reddy, T.J. Beveridge, J.A. Breznak, G.A. Marzluf, T.M. Schmidt and L.R. Snyder, editors). ASM Press, Washington D.C., pp. 19–33.

Bickmore, B.R., Bosbach, D., Hochella, M.F., Jr., Charlet, L. & Rufe, E. (2001) In situ atomic force microscopy study of hectorite and nontronite dissolution: implications for phyllosilicate edge surface structures and dissolution mechanisms. *American Mineralogist*, **86**, 411–423.

Bloss, F.D. (1971) *Crystallography and Crystal Chemistry*. Holt, Reinhart and Winston, New York, 545 pp.

Bowen, D.K. & Wormington, M. (1993) Characterization of materials by grazing-incidence X-ray scattering. *Advances in X-ray Analysis*, **36**, 171–184.

Brady, P.V. (1991) The effect of silicate weathering on global temperature and atmospheric CO_2. *Journal of Geophysical Research*, **96B**, 18101–18106.

Brady, P.V. & Walther, J.V. (1989) Controls on silicate dissolution rates in neutral and basic solutions at 25°C. *Geochimica et Cosmochimica Acta*, **53**, 2823–2830.

Brenker, F. & Jordan, G. (editors) (2010) *Nanoscopic Approaches in Earth and Planetary Sciences*. EMU Notes in Mineralogy, **10**, European Mineralogical Union and the Mineralogical Society of Great Britain & Ireland, 420 pp.

Brown, G.E., Jr. (1990) Spectroscopic studies of chemisorption reaction mechanisms at oxide-water interfaces. In: *Mineral–Water Interface Geochemistry* (M.F. Hochella Jr. & A.F. White, editors). Reviews in Mineralogy and Geochemistry, **23**. Mineralogical Society of America, Washington, D.C., pp. 309–364.

Brown, G.E., Jr. & Sturchio, N.C. (2002) An overview of synchrotron radiation applications to low temperature geochemistry and environmental science. In: *Applications of Synchrotron Radiation in Low Temperature Geochemistry and Environmental Science* (P.A. Fenter, M.L. Rivers, N.C. Sturchio & S.R. Sutton, editors). Reviews in Mineralogy and Geochemistry, **49**. Mineralogical Society of America, Washington, D.C., pp. 1–116.

Brown, G.E., Jr, Calas, G., Waychunas, G.A. & Petiau, J. (1988) X-ray absorption spectroscopy and its applications in mineralogy and geochemistry. In: *Spectroscopic Methods in Mineralogy and Geology* (F.C. Hawthorne, editor). Reviews in Mineralogy and Geochemistry, **18**. Mineralogical Society of America, Washington D.C., pp. 431–512.

Brown, G.E., Jr., Parks, G.A. & O'Day, P.A. (1995) Sorption at mineral-water interfaces: macroscopic and microscopic perspectives. Pp. 129–184 in: *Mineral Surfaces* (D.J. Vaughan & R.A.D. Pattrick, editors). Mineralogical Society Series, **5**, Chapman & Hall, London.

Brunauer, S., Emmet, P.S. & Teller, E. (1938) Adsorption of gases in multimolecular layers. *Journal of the American Chemical Society*, **60**, 309–319.

Brydie, J.R., Wogelius, R.A., Merrifield, C.M., Boult, Gilbert, P., Allison, D. & Vaughan, D.J. (2005) The u2M project on quantifying the effects of biofilm growth on hydraulic properties of natural porpous media and on sorption equilibria: an overview. In: *Understanding the Micro to Macro Behaviour of Rock-Fluid Systems* (R.P. Shaw, editor). Special Publication, **249**, Geological Society of London, pp. 131–144.

Brydie, J.R., Wogelius, R.A., Boult, S., Merrifield, C.M. & Vaughan, D.J. (2009) Model system studies of the influence of bacterial biofilm formation on mineral surface reactivity. *Biofouling*, **25**, 463–472.

Burns, P.C., Pluth, J.J., Smith, J.V., Eng, P., Steele, I. & Housley, R.M. (2000) Quetzalcoatlite: New octahedral-tetrahedral structure from 5-micrometer crystal at the Advanced Photon Source-GSE-CARS facility. *American Mineralogist,* **85**, 604–607.

Burton, N.A., Doudou, S., Vaughan, D.J. & Livens, F.R. & (2012) Atomistic simulations of uranyl (VI) adsorption on calcite and stepped calcite mineral surfaces. *Environmental Science and Technology,* (DOI: http://dx.doi.org/10.1021/es300034k).

Cabri, L.J. & Campbell, J.L. (1998) The proton microprobe in ore mineralogy (micro-PIXE technique). In: *Modern Approaches to Ore and Environmental Mineralogy* (L.J. Cabri & D.J. Vaughan, editors). Mineralogical Association of Canada, Ottawa, Canada, pp. 181–198.

Carrado, K.A., Xu, L. Gregory, D., Song, K., Seifert, S. & Botto, R.E. (2000) Crystallization of a layered silicate clay as monitored by small-angle X-ray scattering and NMR. *Chemical Materials*, **12**, 3052–3059.

Carroll-Webb, S.A. & Walther, J.V. (1988) A surface complex reaction model for the pH-dependence of corundum and kaolinite dissolution rates. *Geochimica et Cosmochimica Acta*, **52**, 2609–2623.

Casey, W.H. & Westrich, H.R. (1992) Control of dissolution rates of orthosilicate minerals by divalent metal-oxygen bonds. *Nature*, **355**, 157–159.

Casey, W.H., Westrich, H.R. & Arnold, G.W. (1988) Surface chemistry of labradorite feldspar reacted with aqueous solutions at pH 2, 3 and 12. *Geochimica et Cosmochimica Acta*, **52**, 2795–2807.

Chiarello, R.P., Wogelius, R.A. & Sturchio, N.C. (1993) *In situ* synchrotron X-ray reflectivity measurements at the calcitewater interface. *Geochimica et Cosmochimica Acta*, **57**, 4103–4110.

Chiari, G. (2013) Mineralogy and cultural heritage. In: *Environmental Mineralogy II* (D.J. Vaughan & R.A. Wogelius, editors), EMU Notes in Mineralogy, **13**. European Mineralogical Union and the Mineralogical Society, London, pp. 405–440.

Chou, L. & Wollast, R. (1985) Steady-state kinetics and dissolution mechanisms of albite. *American Journal of Science*, **285**, 963–993.

Chou, L. & Wollast, R. (1989) Comparative study of the dissolution kinetics and mechanisms of carbonates in aqueous solutions. *Chemical Geology*, **78,** 269–282.

Clark, R.J.H., Hester, R.E., Grime, G.W., Watt, F., Wogelius, R.A. & Jamtveit, B. (1993) Processing micro-PIXE linescan data – studies of arsenic zoning in skarn garnets. *Nuclear Instruments and Methods in Physics Research,* **B77**, 410–414.

Crabb, T.A., Gibson, P.N. & Roberts, K.J. (1993) REX – a least-squares fitting program for the simulation and analysis of X-ray reflectivity data. *Computational Physics Communications*, **77**, 441–449.

Craig, J.R. & Vaughan, D.J. (1994) *Ore Microscopy and Ore Petrography*. 2nd edition. Wiley-Interscience, New York, 434 pp.

Curtis, C.D. & Morris, K. (2013) Mineralogy in long-term nuclear waste management. In: *Environmental Mineralogy II* (D.J. Vaughan & R.A. Wogelius, editors). EMU Notes in Mineralogy, **13**. European Mineralogical Union and the Mineralogical Society, London, pp. 383–404.

Cutting, R.H., Muryn, CA., Thornton, G. & Vaughan, D.J. (2006) Molecular scale investigations of the reactivity of magnetite with formic acid, pyridine, and carbon tetrachloride. *Geochimica et Cosmochimica Acta*, **70**, 3593–3612.

Cutting, R.H., Coker, V.S., Fellowes, J.W., Lloyd, J.R. & Vaughan, D.J. (2009) Mineralogical and morphological constraints on the reduction of Fe(III) minerals by *Geobacter sulfurreducens*. *Geochimica et Cosmochimica Acta*, **73**, 4004–4022.

Cygan, R.T. & Kubicki, J.D. (2001) *Molecular Modelling Theory: Applications in the Geosciences*. Reviews in Mineralogy and Geochemistry, **42**, Mineralogical Society of America, Washington, D.C., 531 pp.

Davis, J.A. & Kent, D.B. (1990) Surface complexation modeling in aqueous geochemistry. In: *Mineral–Water Interface Geochemistry* (M.F. Hochella Jr. & A.F. White, editors). Reviews in Mineralogy and Geochemistry, **23**, Mineralogical Society of America, Washington, D.C., pp. 177–260.

Davis, J.A., James, R.O. & Leckie, J.O. (1978) Surface ionization and complexation at the oxide/water interface. I. Computation of electrical double layer properties in simple electrolytes. *Journal of Colloid and Interface Science*, **63**, 480–499.

de Boer, D.K.G., Leenaers, A.J.G. & van der Hoogenhof, W.W. (1994) Influence of roughness profile on reflectivity and angle-dependent X-ray fluorescence. *Journal of Physics III, 4*, 1559–1564.

Denecke, M.A. (2006) Actinide speciation using X-ray absorption fine structure spectroscopy. *Coordination Chemistry Reviews*, **250**, 730–754.

de Stasio, G., Gilbert, B., Frazer, B.H., Nealson, K.H., Conrad, P.G., Livi, V., Labrenz, M. & Banfield, J.F. (2001) The multidisciplinarity of spectro-microscopy: from geomicrobiology to archeology. *Journal of Electron Spectroscopy and Related Phenomena*, **114**, 997–1003.

De Yoreo, J.J. & Dove, P.M. (2004) Shaping crystals with biomolecules. *Science*, **306**, 1301–1302.

Dove, P.M. & Hochella, M.F., Jr. (1993) Calcite precipitation mechanisms and inhibition by orthophosphate: In situ observations by scanning force microscopy *Geochimica et Cosmochimica Acta*, **57**, 705–14.

Dove, P.M. & Platt, F.M. (1996) Compatible real-time rates of mineral dissolution by Atomic Force Microscopy (AFM). *Chemical Geology*, **127**, 331–338.

Dove, P.M., Hochella, M.F., Jr. & Reeder, R.J. (1992) In situ investigation of near-equilibrium calcite precipitation by atomic force microscopy. In: *Water–Rock Interaction – Proceedings of the 7th International Symposium, WRI-7* (Y.K. Kharaka and A.S. Maest, editors). Balkema, Rotterdam, pp. 141–144.

Dubessy, J., M.-C. Caumon & F. Rull (editors) (2012) *Applications of Raman Spectroscopy to Earth Sciences and Cultural Heritage*. EMU Notes in Mineralogy, **12**. European Mineralogical Union and the Mineralogical Society, 498 pp.

Dyar, M.D. & Gunter, M.E. (2008) *Mineralogy and Optical Mineralogy*. Mineralogical Society of America, Washington D.C., 690 pp.

Edwards, N.P., Barden, B.E., van Dongen, B.E., Manning, P.L., Bergman, U., Sellers, W.I. & Wogelius, R.A. (2011) Infra-red mapping resolves soft-tissue preservation in 50 Million year old reptile skin. *Proceedings of the Royal Society, Series B*, **278**, 3209–3218.

Eyring, H. (1935) The activated complex and the absolute rate of chemical reactions. *Chemical Reviews*, **17**, 65–82.

Farquhar, M., Charnock, J.M., England, K. & Vaughan, D.J. (1996) Adsorption of Cu (II) on the 001 plane of mica: a REFLEXAFS and XPS study. *Journal of Colloid and Interface Science*, **177**, 561–567.

Farquhar, M.L., Wogelius, R.A. & Tang, C.C. (1999) *In situ* synchrotron X-ray reflectivity study of the oligoclase feldspar mineral-fluid interface. *Geochimica et Cosmochimica Acta*, **63**, 1587–1594.

Farquhar, M.L., Charnock, J.M., Livens, F.R. & Vaughan, D.J. (2002) Mechanisms of arsenic uptake from aqueous solution by interaction with goethite, lepidocrocite, mackinawite and pyrite: an X-ray absorption spectroscopy study. *Environmental Science & Technology*, **36**, 1757–1762.

Farquhar, M.L., Wogelius, R.A., Charnock, J.M., Wincott, P., Tang, C.C., Newville, M., Eng, P.J. & Trainor, T.P. (2003) Surface oxidation of rhodonite: structural and chemical study by surface scattering and glancing incidence XAS techniques. *Mineralogical Magazine*, **67**, 1205–1219.

Fellows, R.A., Lennie, A.R., Munz, A.W., Vaughan, D.J. & Thornton, G. (1999) Structures of $FeTiO_3$ (0001) surfaces observed by scanning tunneling microscopy. *American Mineralogist*, **84**, 1384–1391.

Fenter, P.A., Rivers, M.L., Sturchio, N.C. & Sutton, S.R. (editors) (2002) *Applications of Synchrotron Radiation in Low-temperature Geochemistry and Environmental Science*. Reviews in Mineralogy and Geochemistry, **49**, Mineralogical Society of America, Washington D.C., 579 pp.

Furrer, G. & Stumm, W. (1986) The coordination chemistry of weathering: I. Dissolution kinetics of δ-Al_2O_3 and BeO. *Geochimica et Cosmochimica Acta*, **50**, 1847–1860.

Gault, A.G., Rowland, H.A.L., Charnock, J.M., Wogelius, R.A., Gomez-Morilla, I., Vong, S., Leng, M., Samreth, S., Sampson, M.L. & Polya, D.A. (2008) Arsenic in hair and nails of individuals exposed to arsenic-rich groundwaters in Kandal Province, Cambodia. *Science of the Total Environment*, **393**, 168–176.

Geatches, D.L., Greenwell, H.C. & Clark, S.J. (2011) Ab initio transition state searching in complex systems: fatty acid decarboxylation in minerals. *Journal of Physical Chemistry A*, **115**, 2658–2667.

Gibaud, A. & Hazra, S. (2000) X-ray reflectivity and diffuse scattering. *Current Science*, **78**, 1467–1477.

Gilbert, B., Huang, F., Zhang, H., Waychunas, G. & Banfield, J. (2004) Nanoparticles: strained and stiff. *Science*, **305**, 651–654.

Glatter, O. & Kratky, O. (1982) *Small Angle X-ray Scattering*. Academic Press, New York.

Goldich, S.S. (1938) A study in rock weathering. *Journal of Geology*, **46**, 17–58.

Greaves, G.N. (1995) New X-ray techniques and approaches to surface mineralogy. In: *Mineral Surfaces* (D.J. Vaughan, D.J. & R.A.D. Pattrick, editors). Mineralogical Society Series, **5**, Chapman & Hall, London, pp. 87–128.

Hattori, K., Takahashi, Y., Guillot, S. & Johanson, B. (2005) Occurrence of arsenic (V) in forearc mantle serpentinites based on X-ray absorption spectroscopy study. *Geochimica et Cosmochimica Acta*, **69**, 5585–5596.

Hawthorne, F.C. (1988) Mössbauer spectroscopy. In: *Spectroscopic Methods in Mineralogy and Geology* (F.C. Hawthorne, editor). Reviews in Mineralogy and Geochemistry, **18**, Mineralogical Society of America, Washington D.C., pp. 255–340.

Helgeson, H.C. (1968) Evaluation of irreversible reactions in geochemical processes involving minerals and aqueous solutions: I. Thermodynamic relations. *Geochimica et Cosmochimica Acta*, **32**, 853–877.

Hiemstra, T., van Riemsdijk, W.H. & Bolt, G.H. (1989) Multisite proton adsorption modeling at the solid/solution interface of (hydr)oxides: a new approach I. Model description and evaluation of reaction constants. *Journal of Colloid and Interface Science*, **133**, 91–104.

Hell, S.W. (2003) Toward fluorescence nanoscopy. *Nature Biotechnology*, **21**, 1347–1355.

Helz, G.R., Charnock, J.M., Vaughan, D.J. & Garner, C.D. (1993) Multinuclearity of aqueous copper and zinc bisulfide complexes: an EXAFS investigation. *Geochimica et Cosmochimica Acta*, **57**, 15–25.

Helz, G.R., Tossell, J.A., Charnock, J.M., Pattrick, R.A.D., Garner, C.D. & Vaughan, D.J. (1995) Oligomerization in As(III) sulfide solutions: theoretical constraints and spectroscopic evidence. *Geochimica et Cosmochimica Acta*, **59**, 4591–4606.

Hery, M., van Dongen, B.E., Gill, F., Mondal, D., Vaughan, D.J., Pancost, R.D., Polya, D.A & Lloyd, J.R. (2010) Arsenic release and attenuation in low organic carbon aquifer sediments from West Bengal. *Geobiology*, **8**, 155–168.

Hiemstra, T. & van Riemsdijk, W.H. (1999) Surface structural ion adsorption modeling of competitive binding of oxyanions by metal (hydr)oxides. *Journal of Colloid and Interface Science*, **210**, 182–193.

Hirschmugl, C.J. (2002) Applications of storage ring infrared spectromicroscopy and reflection-absorption spectroscopy to geochemistry and environmental science. In: *Applications of Synchrotron* Radiation in Low-temperature Geochemistry and Environmental Science (P.A. Fenter, M.L. Rivers, N.C. Sturchio. & S. R.Sutton, editors). Reviews in Mineralogy and Geochemistry, **49**, Mineralogical Society of America, Washington D.C., pp. Pp. 317–340.

Hochella, M.F., Jr. (1988) Auger electron and X-ray photoelectron spectroscopies. In: *Spectroscopic Methods in Mineralogy and Geology* (F.C. Hawthorne, editor). Reviews in Mineralogy and Geochemistry, **18**, Mineralogical Society of America, Washington D.C., pp. 573–637.

Hochella, M.F., Jr. (1995) Mineral surfaces: their characterization and their chemical, physical and reactive nature. Pp. 17–60 in: *Mineral Surfaces* (D.J. Vaughan & R.A.D. Pattrick, editors). Mineralogical Society Series, **5**, Chapman & Hall, London.

Hochella, M.F., Jr., Lower, S.K., Maurice, P.A. Penn, R.L., Sahai, N., Sparks, D.L. & Twining, B.S. (2008) Nanominerals, mineral nanoparticles, and Earth systems. *Science*, **319**, 1631–1635.

Hunter, R.C. & Beveridge, T.J. (2005) Application of a pH-sensitive fluoroprobe (C-SNARF-4) for pH microenvironmental analysis in *Pseudomonas aeruginosa* biofilms. *Applied and Environmental Microbiology*, **71**, 2501–2510.

Jamtveit, B., Wogelius, R.A. & Fraser, D.G. (1993) Zonation patterns of skarn garnets: stratigraphic records of hydrothermal systems. *Geology*, **21**, 113–116.

Jew, A.D., Kim, C.S., Rytuba, J.J., Gustin, M.S. & Brown, G.E.m Jr. (2011) New technique for quantification of elemental Hg in mine wastes and its implications for mercury evasion into the atmosphere. *Environmental Science & Technology*, **45**, 412–417.

Johansson, S.A.E., Campbell, J.L. & Malmqvist, K.G. (1995) *Particle-induced X-ray Emission Spectrometry*. Chemical analysis Series, **133**, J. Wiley & Sons, New York, 434 pp.

Jones, C.E., Unwin, P.R. & Macpherson, J.V. (2003) In situ observation of the surface processes involved in dissolution from the cleavage surface of calcite in aqueous solution using combined scanning electrochemical-atomic force microscopy (SECM-AFM). *CHEMPHYSCHEM*, **4**, 139–146.

Karis, O., Hasselstrom, J., Wassdahl, N., Weinelt, M., Nillson, A., Nyberg, M., Petterson, L.G.M., Stohr, J. & Samant, M.G. (2000) *Journal of Chemical Physics*, **112**, 8146–8155.

Kelsall, G.H., England, K., Vaughan D.J. & Yin, Q. (1992) Electrochemical oxidation of chalcopyrite (CuFeS$_2$) in alkaline solutions. In: *Proceedings of the Third International Symposium on Electrochemistry in Mineral and Metal Processing* (R. Woods & P.E. Richardson, editors) *Electrochemical Society Proceedings*, **92**, Pennington, New Jersey, USA, pp. 318–341.

Kerr, P.F. (1977) *Optical Mineralogy* (4th edition). McGraw-Hill, New York, 492 pp.

Kirkpatrick, R.J. (1988) MAS-NMR spectroscopy of minerals and glasses. In: *Spectroscopic Methods in Mineralogy and Geology* (F.C. Hawthorne, editor). Reviews in Mineralogy and Geochemistry, **18**, Mineralogical Society of America, Washington D.C., pp. 341–404.

Lasaga, A.C. (1981) Transition state theory. In: *Kinetics of Geochemical Processes* (A.C. Lasaga & R.J. Kirkpatrick, editors). Reviews in Mineralogy and Geochemistry, **8**, Mineralogical Society of America, Washington D.C., pp. 135–170.

Lasaga, A.C. (1990) Atomic treatment of mineral-water surface reactions. In: *Mineral–Water Interface Geochemistry* (M.F. Hochella Jr. & A.F. White, editors). Reviews in Mineralogy and Geochemistry, **23**, Mineralogical Society of America, Washington, D.C., pp. 17–86.

Law, G.T., Geissler, A., Lloyd, J.R., Livens, F.R., Boothman, C., Begg, J.D.C., Denecke, M., Rothe, J., Dardenne, K., Burke, I.T., Charnock, J. & Morris, K. (2010) Geomicrobiological redox cycling of the transuranic element neptunium. *Environmental Science & Technology*, **44**, 8924–8929.

Lawrence, J.R. & Neu, T.R. (2006) Laser scanning microscopy. In: *Methods for General and Molecular Microbiology* (C.A. Reddy, T.J. Beveridge, Breznak, J.A., Marlzuf, G.A., Schmidt, T.M. & Synder, L.R., editors). ASM Press, Washington, D.C., pp. 34–53.

Lennie, A.R. & Vaughan, D.J. (1996) Spectroscopic studies of iron sulfide formation and phase relations at low temperatures. In: *Mineral Spectroscopy: A tribute to Roger G. Burns* (M.D. Dyar, C. McCammon & M.W. Schaefer, editors). Geochemical Society Special Publication, **5**, Geochemical Society, Houston, Texas, USA, pp. 117–131.

Lennie, A.R., Redfern, S.A.T., Schofield, P.F. & Vaughan, D.J. (1995) Synthesis and Rietveld crystal structure refinement of mackinawite, tetragonal FeS. *Mineralogical Magazine,* **59**, 677–683.

Levine, I.N. (1983) *Physical Chemistry* (2nd edition). McGraw-Hill, New York, 890 pp.

Lindau, I. & Spicer, W.E. (1980) Photoemission as a tool to study solids and surfaces. In: *Synchrotron Radiation Research* (H. Winick & S. Doniach, editors). Plenum Press, New York, pp. 159–221.

Lower, S.K., Hochella, M.F. & Beveridge, T.J. (2001) Bacterial recognition of mineral surfaces: nanoscale interactions between *Shewanella* and α-FeOOH. *Science*, **292**, 1360–1363.

Lyon, I. & Henkel, T. (2010) Secondary ion mass spectrometry – less conventional applications: TOF-SIMS, molecules and surfaces. In: *Nanoscopic Approaches in Earth and Planetary Sciences* (F. Brenker & G.. Jordan, G., editors.). *EMU Notes in Mineralogy*, **8**, European Mineralogical Unioin and the Mineralogical Society, London, pp. 111–136.

Macrae, D. (1995) Secondary ion mass spectrometry and geology. *The Canadian Mineralogist,* **33**, 219–236.

Maddock, A.G. (1985) Mössbauer spectroscopy in mineral chemistry. Pp. 141–208 in: *Chemical Bonding and Spectroscopy in Mineral Chemistry* (F. J. Berry & D.J. Vaughan, editors). Chapman & Hall, London.

Mast, M.A. & Drever, J.I. (1987) The effect of oxalate on the dissolution rates of oligoclase and tremolite. *Geochimica et Cosmochimica Acta*, **51**, 2559–2568.

McLaren, A.C. (1991) *Transmission Electron Microscopy of Minerals and Rocks*. Cambridge University Press, Cambridge, UK, 387 pp.

McMahon, G. & Cabri, L.J. (1998) The SIMS technique in ore mineralogy. Pp. 153–180 in: *Modern Approaches to Ore and Environmental Mineralogy* (L.J. Cabri & D.J. Vaughan, editors). MAC Short Course Series, **27**, Mineralogical Association of Canada, Ottawa, Canada.

Michel, F.M., Ehm, L., Antao, S.M., Lee, P.L., Chupas, J., Liu, G., Strongin, D.R., Schoonen, M.A.A., Phillips, B.L. & Parise, J.B. (2007) The structure of ferrihydrite, a nanocrystalline material. *Science*, **316**, 1726–1729.

Morris, P.M. & Wogelius, R.A. (2008) Probing solid-fluid interfaces via in situ Multiple Internal Reflection FTIR and X-ray scattering techniques: A kinetic study of the reactivity of phthalic acid on an olivine glass surface. *Geochimica et Cosmochimica Acta*, **72**, 1970–1985.

Moyes, L., Livens, F.R., Pattrick, R.A.D., Vaughan, D.J. & Charnock, J.M. (2000) Structural development of amorphous transition metal sulphides. *Geoscience 2000*, Manchester (abstract).

Obst, M., Wang, J. & Hitchcock, A.P. (2009) 3-d chemical imaging with STXM tomography. *Journal of Physics Conference Series*, **186**, 012045.

Oehler, D.Z., Robert, F., Mostefaoui, S., Meibom, A., Selo, M. & McKay, D.S. (2006) Chemical mapping of Proterozoic organic matter at sub-micron spatial resolution. *Astrobiology*, **6**, 838–850.

Oehler, D.Z., Robert, F., Walter, M.R., Sugitani, K., Allwood, A., Meibom, A., Mostefaoui, S. Selo, M., Thomen, A. & Gibson, E.K. (2009) NANOSIMS: insights into biogenicity and syngenicity of Archaean carbonaceous structures. *Precambrian Research*, **173**, 70–78.

Oelkers, E.H., Schott, J. & Devidal, J.L. (1994) The effect of aluminium pH, and chemical affinity on the rates of aluminosilicate dissolution reactions. *Geochimica et Cosmochimica Acta*, **58**, 2011–2024.

Ogilvy, J.A. (1991) *Theory of Wave Scattering from Random Rough Surfaces.* Adam Hilger, Bristol, UK, 277 pp.

Parkman, R.H., Curtis, C.D., Charnock, J.M. & Vaughan, D.J. (1996) Metal fixation and mobilisation in the sediments of the Afon Goch Estuary, Dulas Bay, Anglesey. *Applied Geochemistry*, **11**, 203–210.

Parkman, R.H., Charnock J.M., Bryan, N.D., Livens, F.R. & Vaughan, D.J. (1999) Reactions of copper and cadmium ions in aqueous solutions with goethite, lepidocrocite, mackinawite and pyrite. *American. Mineralogist*, **84**, 407–419.

Parratt, L.G. (1954) Surface studies of solids by total reflection of X-rays. *Physical Review*, **95**, 359–369.

Pascua, C., Charnock, J., Polya, D.A., Sato, T., Yokoyama, S. & Minato, M. (2005) Arsenic-bearing smectite from the geothermal environment. *Mineralogical Magazine*, **69**, 897–906.

Pattrick, R.A.D., Mosselmans, J.F.W., Charnock, J.M., England, K. & Vaughan, D.J. (1997) The structure of amorphous copper sulfide precipitates: an X-ray absorption study. *Geochimica et Cosmochimica Acta*, **61**, 2023–2036.

Pattrick, R.A.D., van der Laan, G., Henderson, C.M.B., Kuiper, P., Dudzik, E. & Vaughan, D.J. (2002) Cation site occupancy in spinel ferrites studied by X-ray magnetic circular dichroism: developing a method for mineralogists. *European Journal of Mineralogy*, **14**, 1095–1102.

Perry, D.L. (editor) (1990) *Instrumental Surface Analysis of Geological Materials.* VCH Publishers, New York, 373 pp.

Pershan, P.S. (1994) The effects of surface profile and interface correlation on X-ray reflectivity from fluid interfaces. *Journal of Physics: Condensed. Matter*, **6**, A37–A50.

Reed, S.J.B. (1997) *Electron Microprobe Analysis* (2nd edition). Cambridge University Press, Cambridge, UK, 344 pp.

Refson, K., Wogelius, R.A. & Fraser, D.G. (1995) Water chemisorption and reconstruction of the MgO surface. *Physical. Review B, Condensed Matter*, **52**, 10823–10826.

Rickard, D.T. (1975) Kinetics and mechanism of pyrite formation at low temperatures. *American Journal of Science*, **275**, 636–652.

Rimstidt, J.D. & Vaughan, D.J. (2003) Pyrite oxidation: a state-of-the-art assessment of the reaction mechanism. *Geochimica et Cosmochimica Acta*, **67**, 873–880.

Robinson, B.W., Ware, N.G. & Smith, D.G.W. (1998) Modern electron microprobe trace- elements analysis in mineralogy. Pp. 153–180 in: *Modern Approaches to Ore and Environmental Mineralogy* (L.J. Cabri & D.J. Vaughan, editors). MAC Short Course Series, **27**. Mineralogical Association of Canada, Ottawa, Canada.

Rosso, K.M. & Vaughan, D.J. (2006) Reactivity of sulfide mineral surfaces. In: *Sulfide Mineralogy and Geochemistry* (D.J. Vaughan, editor). Reviews in Mineralogy and Geochemistry, **61**. Mineralogical Society of America, Washington, D.C., pp. 557–608.

Rufe, E. & Hochella, M.F., Jr. (1999) A quantitative assessment of reactive surface area in silicate dissolution. *Science*, **985**, 874–876.

Schaufuss, A.G., Nesbitt, H.W., Kartio, I., Laajalhto, K., Bancroft, G.M. & Szargan, R. (1998) Reactivity of surface chemical states on fractured pyrite. *Surface Science*, **411**, 321–328.

Scholl, A., Ohldag, H., Nolting, F., Stohr, J. & Padmore, H.A. (2002) X-ray photoemission electron microscopy, a tool for investigation of complex magnetic structures. *Reviews of Scientific Instruments*, **73**, 1362–1366.

Schopf, J.W. & Kudryavtsev, A.B. (2005) Three-dimensional Raman imagery of Precambrian microscopic organisms. *Geobiology*, **3**, 1–12.

Seyama, H. & Soma, M. (2003) Surface analytical studies on environmental and geochemical surface processes. *Analytical Sciences*, **19**, 487–497.

Seyama, H., Soma, M. & Theng, K.G. (2006) X-ray photoelectron spectroscopy. Pp. 865–878 in: *Handbook of Clay Science* (F. Bergaya, B.K.G. Theng & G. Lagaly, editors). Elsevier, Amsterdam.

Shaw, S., Henderson, C.M.B. & Clark, S.M. (2002) *In situ* synchrotron study of the kinetics, thermodynamics and reaction mechanisms of the hydrothermal crystallization of gyrolite ($Ca_{16}Si_{24}O_{60}(OH)_8.14H_2O$). *American Mineralogist*, **87**, 533–541.

Siegbahn, K., Nordling, G.N., Fahlman, A., Nordberg, H., Hamrin, K., Hedman, J., Johansson, G., Bergmark, T., Karlsson, S.E., Lindgren, J. & Lindberg, B. (1967) ESCA: Atomic molecular and solid state structure studied by means of electron spectroscopy. *Nova Acta Regiae Soc Scientia Upssala*, Ser 4, **20**, Almqvist & Wiksells, Uppsala, 283 pp.

Sinha, B.W. & Hoppe, P. (2010) Ion microprobe analysis: basic principles, state-of-the-art instruments and recent applications with emphasis on the geosciences. In: *Nanoscopic Approaches in Earth and Planetary Sciences* (F. Brenker & G. Jordan, editors). EMU Notes in Mineralogy, **8**, European Mineralogical Union and the Mineralogical Society, London, pp. 137–168.

Sinha, S.K., Sirota, E.B., Garof, S. & Stanley, H.B. (1988) X-ray and neutron scattering from rough surfaces. *Physical Review B, Condensed Matter*, **38**, 2297–2311.

Sposito, G. (1983) On the surface complexation model of the oxide-aqueous solution interface. *Journal of Colloid and Interface Science*, **91**, 329–340.

Sposito, G. (1990) Molecular models of ion adsorption on mineral surfaces. In: *Mineral-Water Interface Geochemistry* (M.F. Hochella Jr. & A.F. White, editors). Reviews in Mineralogy and Geochemistry, **23**. Mineralogical Society of America, Washington, D.C., pp. 261–279.

Stumm, W. (editor) (1987) *Aquatic Surface Chemistry: Chemical Processes at the Particle–Water Interface*. Wiley, New York, 520 pp.

Sutton, S.R., Bertsch, P.M., Newville, M., Rivers, M., Lanzirotti, A. & Eng, P. (2002) Microfluorescence and microtomography analyses of heterogeneous earth and environmental materials. In: *Applications of Synchrotron Radiation in Low Temperature Geochemistry and Environmental Science* (P.A. Fenter, M.L. Rivers, N.C. Sturchio & S.R. Sutton, editors). Reviews in Mineralogy and Geochemistry, **49**. Mineralogical Society of America, Washington, D.C., pp. 429–484.

Sverjensky, D.A. & Sahai, N. (1996) Theoretical prediction of single-site surface-protonation equilibrium constants for oxides and silicates in water. *Geochimica et Cosmochimica Acta*, **60**, 3773–3797.

Teng, H.H., Dove, P.M. & De Yoreo, J.J. (2000) Kinetics of calcite growth: surface processes and relationships to macroscopic rate laws. *Geochimica et Cosmochimica Acta*, **64**, 2255–2266.

Teo, B.K. (1986) *EXAFS: Basic Principles and Data Analysis*. Springer-Verlag, Berlin, 349 pp.

Thevuthasan, S., Kim, Y.J., Yi, S.I., Chambers, S.A., Morais, J., Deneke, R., Fadley, C.S., Liu, P., Kendelewicz, T. & Brown, G.E. Jr. (1999) Surface structure of MBE grown α-$Fe_2O_3(0001)$ by intermediate energy X-ray photoelectron diffraction. *Surface Science*, **425**, 276–286.

Tidswell, I.M., Ocko, B.M., Pershan, P.S., Wasserman, S.R., Whitesides, G.M. & Axe, J.D. (1990) X-ray specular reflection studies of silicon coated by organic monolayers (alkylsiloxanes). *Physical Review B, Condensed Matter*, **41**, 1111–1128.

Tobler, D.J., Shaw, S. & Benning, L.G. (2009) Quantification of initial steps of nucleation and growth of silica nanoparticles: an *in-situ* SAXS and DLS study. *Geochimica et Cosmochimca Acta*, **73**, 5377–5393.

Tossell, J.A. & Vaughan, D.J. (1992) *Theoretical Geochemistry: Applications of Quantum Mechanics in the Earth and Mineral Sciences.* Oxford University Press, Oxford, UK, 528 pp.

Tossell, J.A. & Vaughan, D.J. (1993) Bisulfide complexes of zinc and cadmium in aqueous solution: calculation of structure, stability, vibrational and NMR spectra, and of speciation on sulfide mineral surfaces. *Geochimica et Cosmochimica Acta,* **57**, 1935–1945.

Trainor, T.P., Templeton, A.S., Brown, G.E. Jr. & Parks, G.A. (2002) Application of the long-period X-ray standing wave technique to the analysis of surface reactivity: Pb(II) sorption at α-Al$_2$O$_3$/aqueous solution interfaces in the presence and absence of Se(VI). *Langmuir,* **18**, 5782–5791.

Turcotte, D.L. (1992) *Fractals and Chaos in Geology and Geophysics.* Cambridge University Press, Cambridge, UK, 221 pp.

van der Hoogenhof, W.W. & Ryan, T.W. (1993) Structural characterization of Au/Co multi-layers by X-ray diffraction, X-ray reflectivity, and glancing incidence X-ray fluorescence. *Journal of Magnetism and Magnetic Materials,* **121**, 88–93.

Van Hove, M.A., Weinberg, W.H. & Chan, C-M. (1986) *Low Energy Electron Diffraction.* Springer Verlag, Berlin.

Vaughan, D.J. (2006) Arsenic. *Elements,* **2**, 71–75.

Vaughan, D.J. (2013) Introduction: the nature and scope of environmental mineralogy. In: *Environmental Mineralogy II* (D.J. Vaughan & R.A. Wogelius, editors), EMU Notes in Mineralogy, **13**, European Mineralogical Union and the Mineralogical Society, London, pp. 1–4.

Vaughan, D.J., England, K.E.R, Kelsall, G.H. & Yin, Q. (1995) Electrochemical oxidation of chalcopyrite (CuFeS$_2$) and the related metal-enriched derivatives Cu$_4$Fe$_5$S$_8$, Cu$_9$Fe$_9$S$_{16}$, and Cu$_9$Fe$_8$S$_{16}$. *American Mineralogist,* **80**, 725–731.

Vu, H.P., Shaw, S. & Benning, L.G. (2008) Transformation of ferrihydrite to hematite: an in situ investigation on the kinetics and mechanisms. *Mineralogical Magazine,* **72**, 217–220.

Weber, W. & Lengeler, B. (1992) Diffuse scattering of hard X-rays from rough surfaces. *Physical Review B, Condensed Matter,* **46**, 7953–7956.

Westall, J.C. (1982) *FITEQL, a computer program for determination of chemical equilibrium constants from experimental data, version 2.0.* Report 82–02, Dept. of Chemistry, Oregon State University, Corvallis, Oregon, USA.

Westphal, V., Rizzoli, S.O. Lauterbach, M.A., Kamin, D., Jahn, R. & Hell, S.W. (2008) Video-rate far-field optical nanoscopy dissects synaptic vesicle movement. *Science,* **320**, 246–249.

White, J.C. (editor) (1985) *Application of Electron Microscopy in the Earth Sciences.* MAC Short Course Handbook, **11**. Mineralogical Association of Canada, Ottawa, 213 pp.

Wilkins, M.J., Wincott, P.L., Vaughan, D.J., Livens, F.R. & Lloyd, J.R. (2007) Growth of *Geobacter sulfurreducens* on poorly crystalline Fe(III) oxyhydroxide coatings; an integrated study combining light and electron microscopy with X-ray photoelectron spectroscopy. *Geomicrobiology Journal,* **24**, 199–204.

Wogelius, R.A. & Fraser, D.G. (1996) Surface oxidation and hydroxylation of olivine produced by reaction with aqueous solutions: an ex situ XAS (REFLEXAFS) and ERDA study. *Journal of Conference Abstracts,* **1**(1) (*Sixth V.M. Goldschmidt Conference, Heidelberg*) p. 684.

Wogelius, R.A. & Walther, J.V. (1991) Olivine dissolution: effects of pH, carbon dioxide, and organic acids. *Geochimica et Cosmochimica Acta,* **55**, 943–955.

Wogelius, R.A. & Walther, J.V. (1992) Olivine dissolution at near surface conditions. *Chemical Geology,* **97**, 101–112.

Wogelius, R.A., Refson, K., Fraser, D.G., Grime, G.W. & Goff, J.P. (1995) Periclase surface hydroxylation during dissolution. *Geochimica et Cosmochimica Acta,* **59**, 1875–1881.

Wogelius, R.A., Farquhar, M.L., Fraser, D.G. & Tang, C.C. (1999) Structural evolution of the mineral surface during dissolution probed with synchrotron X-ray techniques. In: *Growth, Dissolution, and Pattern Formation in Geosystems* (B. Jamtveit & P. Meakin, P., editors). Kluwer, Dordrecht, The Netherlands, pp. 269–289.

Wogelius, R.A., Manning, P.L., Larson, P.L., Barden, H., Edwards, N.P., Webb, S.M., Sellers, W.I., Taylor, K.G., Dodson, P., You, H., Da-qing, L. & Bergmann, U. (2011) Trace metals as biomarkers for eumelanin pigment in the fossil record. *Science,* **333**, 1622–1626.

Wolery, T.J. (1983) *EQ3NR, a computer program for geochemical aqueous speciation −solubility calculations: user's guide and documentation.* UCRL-53414. Livermore (Cal.) Lawrence Livermore National Laboratory, California, USA.

Wood, B.J. & Walther, J.V. (1983) Rates of hydrothermal reactions. *Science,* **222**, 413–415.

Wormington, M. (1995) *GIXS: Grazing Incidence X-ray Scattering Software Manual.* Bede Scientific Instruments, Durham, UK.

Wright, K., Hillier, I.H., Vaughan, D.J. & Vincent, M.A. (1999a) Cluster models of the dissociation of water on the surface of galena (PbS). *Chemical Physics Letters,* **299**, 527–531.

Wright, K., Hillier, I.H., Vincent, M.A. & Kresse, G. (1999b) Dissociation of water on the surface of galena (PbS) a comparison of periodic and cluster models. *Journal of Chemical Physics,* **111**, 6942–6946

Xu, X. and Goodman, D.W. (1992) New approach to the preparation of ultrathin silicon dioxide films at low temperatures. *Applied Physics Letters,* **61**, 774–776.

Yates, D.E. (1975) *The Structure of the Oxide/Aqueous Electrolyte Interface.* PhD dissertation, University of Melbourne, Australia.

Yin, Q., Kelsall, G.H., Vaughan, D.J. & England, K.E.R. (1995) Atmospheric and electrochemical oxidation of the surface of chalcopyrite (CuFeS$_2$). *Geochimica et Cosmochimica Acta,* **59**, 1091–1100.

Yoon, T.H., Trainor, T.P., Eng, P.J., Bargar, J.R. & Brown, G.E., Jr. (2005) Metal ion partitioning at thin organic film–mineral interfaces: Resonance-enhanced long-period X-ray standing wave (XSW) study on the partitioning of Pb(II) and As(V) ions at mineral-PAA film interfaces. *Langmuir,* **21**, 4503–4511.

Young, R.A. (1993) *The Rietveld Method.* Oxford University Press, Oxford, UK.

Zoltai, T. & Stout, J.H. (1984) *Mineralogy: Concepts and Principles.* Burgess, Minneapolis, Minnesota, USA, 505 pp.

Zussman, J. (1977) X-ray diffraction. In: *Physical Methods in Determinative Mineralogy,* 2nd edition (J. Zussman, editor). Academic Press, New York, pp. 391–473.

Minerals and soil development

Daᴠɪᴅ A.C. MANNING

School of Civil Engineering and Geosciences, Newcastle University,
Newcastle upon Tyne NE1 7RU, UK.

Minerals are a dynamic component of the soil system, interacting with biological and abiological systems and processes at the interface between the geosphere and the biosphere. They supply the majority of the key nutrients required for life, and through weathering reactions they influence atmospheric chemistry. They act as sinks for contaminants, ranging from metals to carbon dioxide.

This chapter provides an overview of key mineral reactions within soils, and shows how these influence the mineralogical composition of soils in different environments. It emphasizes the importance of thermodynamic and kinetic factors as controls on soil mineralogy, and explores how these influence the supply of nutrients to plants. The role of soils as a sink for atmospheric CO_2 is discussed in the context of the formation of pedogenic carbonate minerals.

1. Introduction and context

Soils lie at the interface between the atmosphere and the geosphere. For as long as they have existed throughout geological time, continental areas have acted as the locations where rocks and the atmosphere interact. Although rock and mineral compositions have remained the same, atmospheric compositions have changed, and so the process and products of weathering have also changed. Attention has been paid to ancient soils (palaeosols) and rock weathering to understand the evolution of the atmosphere (Berner, 1992; Lenton, 2001). The nature of iron within palaeosols has been used in attempts to determine the oxygen content of the atmosphere throughout geological time (*e.g.* Rye & Holland, 1998). The occurrence of carbonate minerals, formed as a consequence of weathering, has been used to determine how atmospheric CO_2 contents have changed from the Precambrian (*e.g.* Sheldon, 2006) to the present (*e.g.* Liu, 2011; Ryzkov *et al.*, 2008).

As well as recording reactions between rock minerals and the atmosphere, soils also record the effects of the colonization of the land by plants, starting from the very first fungal associations with liverworts in the middle Ordovician (470 Ma; Humphreys *et al.*, 2010) to the present day. Palaeosols are an important part of the fossil record, as an indicator of palaeogeography and past climates (Cerling, 1984; Retallack, 2009). From the first colonization of the land by plants, soils have evolved as both the plant and animal kingdoms have evolved. Soils act as a place in which organisms live, and as a place where plants grow. They support terrestrial life in many ways.

© Copyright 2013 the European Mineralogical Union and the Mineralogical Society of Great Britain & Ireland
DOI: 10.1180/EMU-notes.13.3

And when it rains, soils act as the first step in the coupled hydrological-hydrogeological cycle, a first stage of filtration and treatment of rain water, then surface waters.

As human beings, we utterly depend on soils for our existence, and the development of agriculture over the last 7000 years has had a major and continuous influence on the composition of the atmosphere (Ruddiman, 2003). With industrialization, we have added natural and artificial materials to soils to enhance fertility and compensate for what we remove to consume in the form of crops. We build on soils – our houses, offices, factories *etc.*, but also our roads and railways. We struggle with a legacy of soils that have been contaminated by industry or war, and that have become unsafe for life. At the start of the 21st Century, we agree that soils should be protected, but we are a long way from sharing a common vision concerning how their care should be regulated.

Figure 1 shows a simplified summary of the carbon cycle. In this figure, the dominant exchange is between the atmosphere and the coupled plant-soil system. It is estimated that $\sim$150 GT (gigatonnes)/year of C is removed from the atmosphere by plants and soils combined, and an equivalent amount is returned as a consequence of respiration and other processes. The system is effectively in equilibrium. Importantly, an amount of carbon equivalent to all that held in the atmosphere passes through the coupled plant-soil system in just 5 years. This is potentially a very responsive system.

The ability of soils to remove carbon from the atmosphere has obvious implications for climate change, given that atmospheric CO_2 levels correlate with increasing global temperatures (Petit *et al.*, 1999). With care, it is possible to demonstrate that atmospheric CO_2 levels have been rising since the start of agriculture (Figure 2; Ruddiman, 2003). The rapid increase in atmospheric CO_2 associated with the combustion of fossil fuels is the most recent phase in the long human history of bioturbation of carbon in soil and geological systems. Given that soils are evidently active participants in the global carbon cycle, it makes sense to encourage soil processes that remove CO_2 from the atmosphere.

While attention commonly focuses on the carbon cycle, there are analogous cycles concerning nitrogen and other chemical species. The impact on global warming of emissions from soils of N oxides is highly significant. However, with the exception of ammonium exchange on clays, minerals play much less of a part in the nitrogen cycle compared with the carbon cycle.

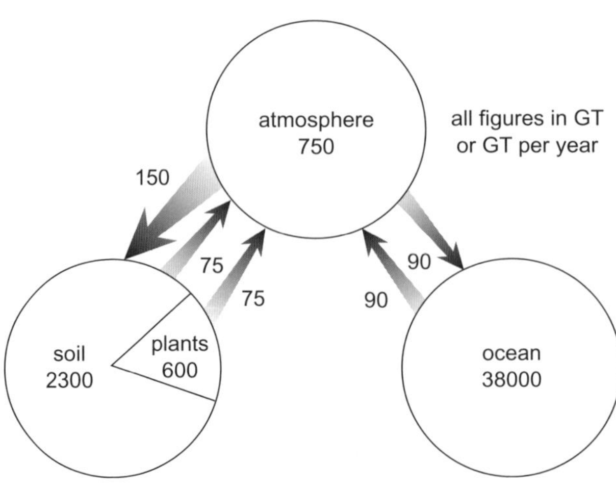

Fig. 1. Summary of the global carbon cycle.

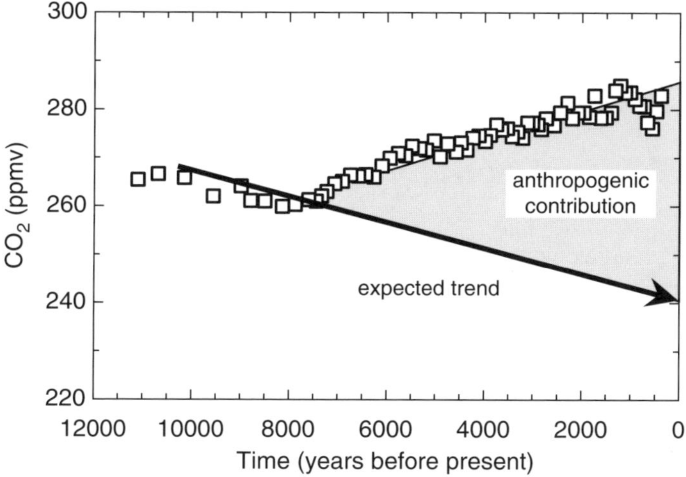

Fig. 2. Deviation from the expected interglacial trend in atmospheric CO_2 over the last 12,000 years (based on Ruddiman, 2003; CO_2 data expressed as parts per million by volume (ppmv) are from the Taylor Dome ice core, Indermühle *et al.*, 1999; www.ncdc.noaa.gov/paleo/taylor/taylor.html).

Plant cultivation in agriculture involves removal of nutrients from soils, and of these all but N (and C) are supplied ultimately by minerals – either directly in pre- and post-industrial (*e.g.* organic) agriculture, or as artificial fertilizers manufactured from mineral raw materials. Globally, the scale of nutrient offtake (mineral nutrients within crop and plant tissue that is harvested) is enormous (Sheldrick *et al.*, 2002) and unbalanced from country to country. In Europe and North America, nutrient offtake is generally balanced by the application of fertilizer, sometimes to excess. Figures from the Food and Agriculture Organization of the United Nations (FAO, 2008) show, in marked contrast, that Africa, with 15% of the world's population, uses 1.5% of global fertilizer production. Given that trade in food is global, consumption and application of minerals and mineral-derived products as fertilizers needs to become global to compensate for offtake.

In the context of water supply, soils play a vital role in influencing the quality and rate of delivery of water to the surface and groundwater systems that provide drinking water, or are protected as part of wider management of ecosystem services. At the same time, water management serves to protect soils by reducing erosion and the associated loss of soil resource. In the developed world, management of farmland includes measures such as leaving buffer strips of uncultivated land adjacent to water-courses; the soils in these strips serve a range of functions, including removal of pollutants. In this context, mineralogical reactions such as cation exchange combine with biological processes to minimize the possibility of harm to a water-course. More generally, aquifer recharge involves initial flow through a soil, and reactions in soils influence the quality of water entering groundwater systems.

As well as providing a number of functions *in situ*, soils provide a source of materials that are removed for a variety of purposes. Excavation accompanies construction;

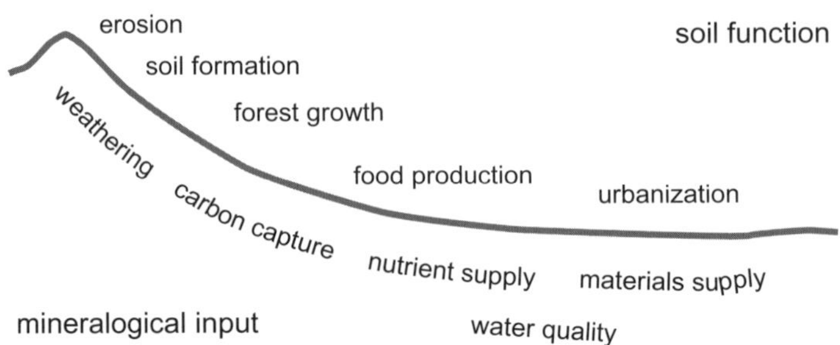

Fig. 3. Schematic summary of soil function in different geographical environments, from mountains to developed lowland areas. Related mineralogical processes include weathering associated with erosion, carbon capture in long-term vegetated areas such as forests, and nutrient supply for growth of food crops. Minerals are also important as components of aquifers, affecting water quality and as a source of materials for the construction of the built environment.

materials taken from one location can be used as fill in another, providing that their geotechnical properties are satisfactory. In the case of peats (increasingly historically), soils can be mined to provide a saleable product, and many alluvial soils grade downwards into mineable deposits of sand or gravel. As well as removing soils, human activities impose burdens on soils through the construction of potentially hazardous accumulations of waste. Mine wastes may be impounded in large dams, and occasionally these fail, spreading waste over nearby soils (*e.g.* the alumina waste incident in Hungary in 2010; Mayes *et al.,* 2011). Landfill sites are constructed to prevent damage to adjacent soils, but this was not always the case. There is a legacy of old landfills, built before there was sufficient awareness of the possible consequences, that have leaked pollutants in the past either through water leaving through the base of the site or gas leaking through a cap.

Figure 3 summarizes very briefly the diverse overarching functions of soils on which we depend, and highlights the mineralogical influences on those functions. The role of mineralogical processes in soil systems will be developed in more detail in this chapter, emphasizing the vital role played by minerals.

2. Minerals in soils

Soils are produced by interaction between the underlying rocks, biological systems and soil solution and gases. They contain residual minerals derived from the geological parent, and minerals newly formed as a consequence of: (1) weathering reactions, and (2) direct precipitation. In general terms, soils contain an immense diversity of minerals, over a range of sizes from nanoparticles through to multi-mineral rock fragments.

Definitive descriptions of the rock-forming minerals are provided by Deer *et al.* (1992), in encyclopaedic publications that have grown with time (www.geolsoc.org.uk/en/Publications/Bookshop/Search/RMFSET.aspx), keeping up to date with developments in mineralogical research. The corresponding description of soil minerals is provided by Dixon & Weed (1989), and this remains a classic text. However, those interested in soil mineralogy need to look more widely to maintain awareness of subsequent and current research.

Within soils, rock-forming silicate minerals occur as residual materials that have undergone varying amounts of chemical and physical weathering. In most temperate soils, **quartz** represents the ultimate stable rock-forming mineral, persisting when all other minerals have been removed by chemical weathering. Thus quartz provides an important part of the textural framework of a soil, but chemically it plays a very limited role compared with other rock-forming minerals. The **feldspars** occur widely within soils developed on feldspar-bearing parent rocks, and can contribute to texture. However, they are chemically unstable, and so contribute to mineral reactions within soils. The ability of feldspars to resist weathering varies according to feldspar type; alkali feldspars are more stable than plagioclase feldspars. At the opposite end of the spectrum, in terms of stability, is **olivine**, which very rarely occurs within soils. Similarly, **pyroxenes** and **amphiboles** are inherently unstable in soils, occurring as residual material in soils developed on appropriate parent rocks (especially basic igneous rocks and their metamorphosed equivalents). Because of their instability in soils and susceptibility to weathering, these ferromagnesian minerals play an important role, in particular as sources of Mg which is a vital plant nutrient. Their reactions also affect bulk soil properties, such as pH.

Layer silicates occur in two important forms in soils. Firstly, **micas** (especially **biotite** and **muscovite**) occur as residual material and, secondly, **clay minerals** occur as newly formed materials. There is a continuum between these two end-member forms. Those layer silicates that contain interlayer cations (micas and most clays) have the potential to participate in cation exchange reactions, and these principally involve the alkalis (Na and K) and alkaline earths (Mg and Ca). Cation exchange in layer silicates also influences ammonium in solution (Garrells & Christ, 1965), having a potential impact on the nitrogen cycle.

The clay minerals form a diverse group within soils. The **kaolinite** group has the simplest chemical composition and structure (Table 1). A 1:1 clay, kaolinite, combines a tetrahedral silica sheet with an octahedral alumina sheet, with no interlayer cations. Kaolinite is a relatively passive component of soils, a common product of mineral weathering. **Illite** is a 2:1 clay, with a layer structure in which two tetrahedral silica sheets sandwich a single alumina sheet. It has interlayer K, and is chemically analogous to muscovite mica. The **smectites** are a family of 2:1 clays, in which the layer structure is the same as illite, but with variation in interlayer cation (Na, Mg, Ca). Like kaolinite, smectites are common products of weathering, and so are common in soil systems under certain conditions. They contrast with both kaolinite and illite in their high reactivity. Cation exchange reactions are common, and smectites are able to adsorb water and organic molecules, giving their characteristic shrink-swell behaviour. The **chlorite** group of clays has a 2:1:1 structure, in which the 2:1 layers, analogous to those found in illites and smectites, are separated by an octahedral sheet, most commonly of Mg coordinated with oxygen. Like the kaolinites, this group of clays has very little cation exchange capacity, and it commonly occurs as a product of weathering of

Table 1. Clay minerals which occur commonly in soil environments (from Cotter-Howells and Paterson, 2000).

Mineral	Type	XRD basal spacing (nm)	Layer charge	Cation exchange capacity (meq/100 g)	Specific surface area (m^2/g)	
					External*	Internal**
Kaolinite	1:1	0.7	~0	1–10	5–35	5–35
Illite	2:1	1.0	~1.0	1–10	50–80	50–80
Vermiculite	2:1	1.4	0.6–0.9	100–200	5–40	750–800
Smectite	2:1	1.2–1.6	0.25–0.6	50–80	30–80	750–800
Chlorite	2:1:1	1.4	0.6–0.9	1–10	10–35	10–35

* as measured by adsorption of nitrogen at 77 K
** as measured by desporption of ethylene glycol at 298 K

ferromagnesian silicate minerals. **Vermiculite** represents a transitional 2:1 clay mineral. Produced by the weathering of both biotite (trioctahedral vermiculite) and muscovite (dioctahedral vermiculite), it is common in soils, providing significant cation exchange capacity.

As products of reactions between rock-forming minerals and aqueous solutions, soil mineral populations commonly contain oxyhydroxide minerals. Iron and aluminium oxyhydroxides predominate, and (especially in the case of iron) as pigments are often responsible for the colour of a soil. **Hematite** (Fe_2O_3) occurs in oxidized arid soils. Other iron oxyhydroxides include **goethite** and **lepidocrocite** (both FeOOH). These fully oxidized iron minerals commonly occur in association with the aluminium hydroxide and oxyhydroxides (**gibbsite**, $Al(OH)_3$, **boehmite**, AlOOH and **diaspore**, AlOOH), forming the dominant mineral assemblages in deeply weathered tropical oxisols.

Less widely reported are occurrences of the nanoparticulate oxyhydroxide aluminosilicates **allophane** and **imogolite**. These typically occur as tubes a few nm in diameter in soils and weathering profiles developed on glassy basic igneous rocks.

Other non-silicate minerals in soils include carbonates, phosphates, oxalates and salts. **Calcite** is a very common pedogenic mineral produced by reaction between Ca released by weathering of silicate minerals and carbonate that ultimately has a biological origin (Cerling, 1984). Ca is also precipitated by reaction with oxalate (commonly of fungal origin) to give the minerals **whewellite** ($Ca(COO)_2.H_2O$) and **weddellite** ($Ca(COO)_2.2H_2O$). These occur within plant tissue as well as in soils, where they predominantly occur associated with plant litter (Manning, 2000). Phosphate minerals in soils are dominated by **apatite** ($Ca_5(PO_4)_3(OH)$, most commonly as a carbonated form. Other phosphates may form by precipitation with cations such as Pb in polluted soils (Hodson *et al.,* 2001). **Gypsum** ($CaSO_4.2H_2O$) occurs widely in arid soils, and **halite** (NaCl) occurs in saline soils where evaporation is high (Dixon & Weed, 1989).

The above list of minerals encountered in soils covers perhaps the majority of the mineral kingdom. It has not included sulfide minerals, and these can of course occur in soils as residual (highly reactive) materials at appropriate locations. Newly-formed iron monosulfides can occur in anaerobic soils where sulfide is stable, and these soils are often black in colour, with a sulfurous smell.

3. Mineral reactions in soils

As a process, mineral weathering controls many aspects of the Earth system, and so has attracted attention for decades. Early observations were empirical, and have stood the test of time. The Goldich weathering series (Goldich, 1938) is shown in Figure 4. Based on observation, it gives the relative stabilities of the rock-forming minerals in soil systems. The validity of the series has subsequently been confirmed by Curtis (1976), who showed that the Gibbs free energy of reaction for appropriate weathering reactions

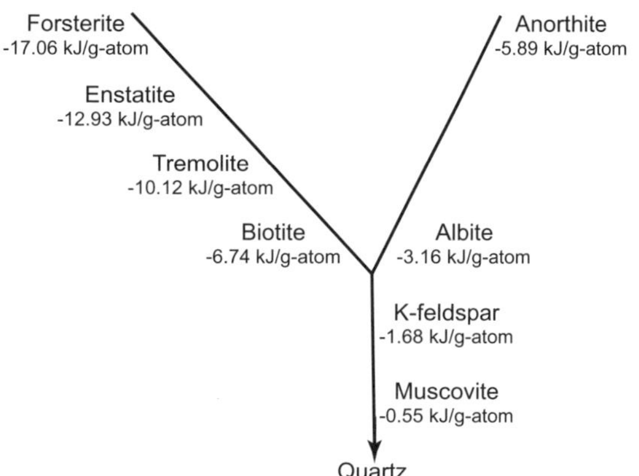

Fig. 4. The Goldich reaction series (Goldich, 1938), with corresponding Gibbs free energy of reaction normalized to the total number of reactant atoms (Curtis, 1976; calculated using thermodynamic data from Robie and Hemingway, 1995).

follows the same sequence. Thus the Goldich weathering series is a consequence of the fundamental thermodynamic properties of individual minerals and their reaction products. Importantly, weathering reactions are irreversible at the Earth's surface, reflecting the instability under Earth-surface conditions of minerals produced at high temperatures and pressures.

Mineral weathering reaction rates have been studied in a large number of laboratory and field experiments, and are reported in substantial reviews (*e.g.* White & Brantley, 1995; Palandri & Kharaka, 2004). It has commonly been found that dissolution-rate experiments in the laboratory give results that differ significantly from those conducted in the field, derived either from soil weathering or catchment studies. This is partly because of inability to control key variables outside the laboratory (*e.g.* reacting surface area, reaction time, *etc.*).

The dissolution rate of aluminosilicate minerals varies as a function of pH, as a consequence of the amphoteric nature of Al which means that the dissolved Al complex varies as a function of pH. In the case of albite, as pH increases, the dissolved Al species become increasingly hydroxyl-rich:

$$NaAlSi_3O_8 + 4H^+ + 4H_2O = Na^+ + Al^{3+} + 3H_4SiO_4$$

$$NaAlSi_3O_8 + 3H^+ + 5H_2O = Na^+ + Al(OH)^{2+} + 3H_4SiO_4$$

$$NaAlSi_3O_8 + 2H^+ + 6H_2O = Na^+ + Al(OH)_2^+ + 3H_4SiO_4$$

This is illustrated in Figure 5, which contrasts the dissolution rates of albite ($NaAlSi_3O_8$) and the olivine-group mineral forsterite (Mg_2SiO_4) which is Al-free, using data collated by Palandri & Kharaka (2004). In both cases, the reaction rate

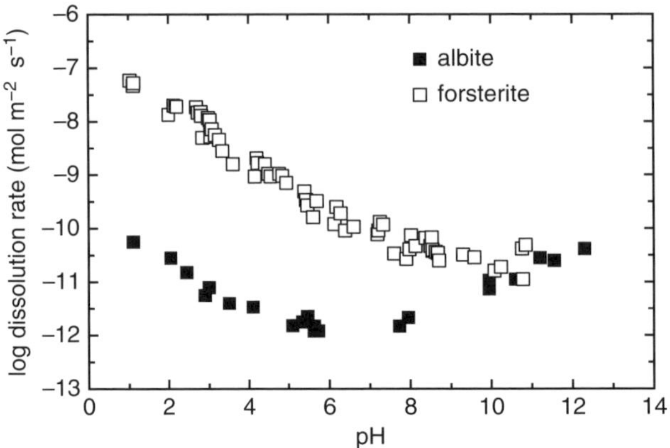

Fig. 5. Dissolution rates for forsterite and albite as a function of pH (data from Palandri and Kharaka, 2004).

increases towards low pH, differing by up to three orders of magnitude. At near neutral pH, the dissolution rates differ by 1–2 orders of magnitude, and under alkaline conditions the dissolution rates converge. In the vast majority of soils, pH lies between 5 and 8, a region with widely contrasting dissolution rates.

The Goldich series (Figure 4) is empirically derived, and reflects mineral stabilities in soils. Figure 6 is a plot of mineral dissolution rates for members of the Goldich series. This shows that there is a regular decrease in reaction rate for the plagioclase series (anorthite–albite), and that variation in dissolution rate for the ferromagnesian minerals is much less systematic. Nevertheless, under pH conditions appropriate for chemically weathered soils, quartz, K-feldspar and muscovite have the slowest dissolution rates, again consistent with their resistance to weathering.

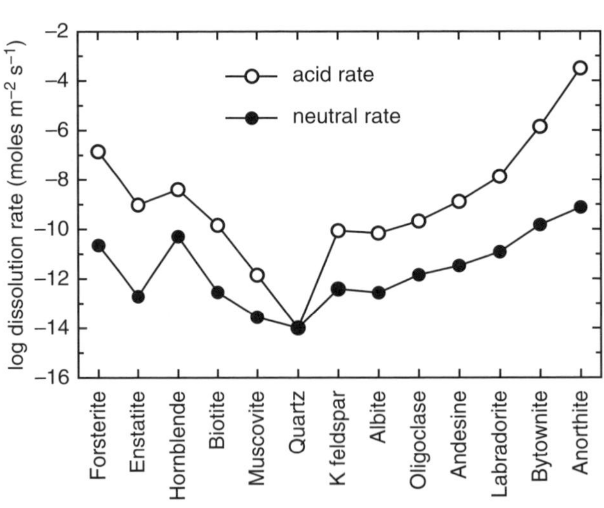

Fig. 6. Dissolution rates under acidic (pH 5) and neutral (pH 7) conditions for minerals in the Goldich series (data from Palandri and Kharaka, 2004).

One reason why reported reaction rates differ arises from the difficulties inherent in quantifying and characterizing the mineral surfaces available for reaction within soil systems. Physical weathering reduces the grain size of minerals and the associated geometrical surface area increases in proportion to the square of the particle diameter. But it is only reasonable to assume that the geometrical surface area applies to quartz and other unreactive mineral species. All other major rock-forming minerals corrode during weathering and, within soil systems, their residual grains can have exceedingly complex surfaces (Figure 7). Despite problems associated with the determination of surface area, the time required to dissolve a hypothetical 1 mm sphere of a mineral is given in Table 2, to illustrate the effect of difference in dissolution rate on persistence.

Geometrically, the clay minerals provide particular challenges that influence their properties and behaviour. The specific surface areas of the dominant clay mineral groups determined

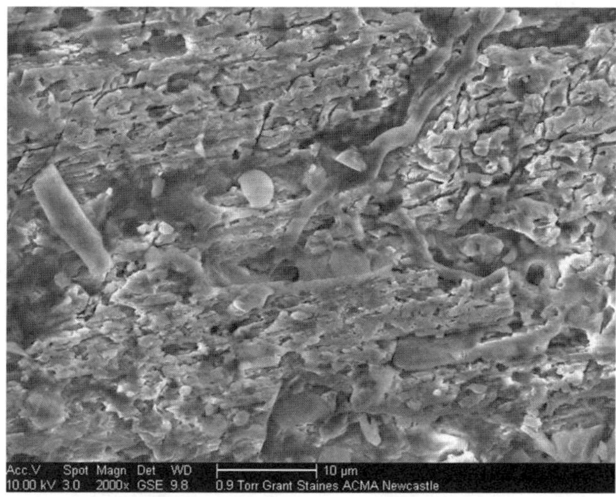

Fig. 7. Scanning electron microscope (SEM) image of fractured and corroded feldspar.

by different methods are given in Table 1. Kaolinite has the lowest specific surface area irrespective of the method used, and smectite the highest. The question then arises, what do the measured surface areas represent? In kaolinite, illite and chlorite, it may be reasonable to assume that the measured surface area represents the external surface of the grain, given the inability of these groups to absorb interlayer species. In contrast, measurement of surface area in smectites and vermiculite gives different results depending on the technique used. Some techniques depend on sorption of organic species, and so inherently measure the interlayer surface, perhaps in preference to the external grain surface.

A further complication with clays is their common occurrence as interlayered mixtures. This phenomenon is widely known from clays within sedimentary rocks,

Table 2. Approximate lifetime in soils (containing a dilute solution, pH 5) of a 1 mm sphere of different rock-forming minerals (based on Drever, 1997, and Lasaga *et al.*, 1994).

Mineral group	Mineral	Lifetime (y)
Quartz	Quartz	34,000,000
Clay	Kaolinite	6,000,000
Mica	Muscovite	2,600,000
K-feldspar	Microcline	921,000
Mica	Biotite	900,000
Na-plagioclase	Albite	575,000
Ca-plagioclase	Bytownite	40,000
Mg pyroxene	Enstatite	10,100
Ca-Mg pyroxene	Diopside	6,800
Mg olivine	Forsterite	2,300
Carbonate	Dolomite	1.6
Carbonate	Calcite	0.1

and also occurs in clays within soils. For example, Psyrillos *et al.* (1999) showed clearly that kaolinite from weathered granite contains interlayered illite. Indeed, the characteristic vermiform appearance of kaolinite may be a consequence of mineralogical impurity, associated with the mixing of alternate layers of kaolinite and illite (Fig. 8).

The textural variation and the highly variable surface character of soil minerals arises as a consequence of a combination of physical and chemical weathering. It underpins the processes by which new minerals are formed in soils. Pre-existing minerals can act as substrates or templates for new mineral growth. They can also act as surfaces on which biological activity can take place (Ritz & Young, 2011). Additional textural variation arises from the biological precipitation of minerals within soils, such as the needle-shaped calcite produced by fungi (Burford *et al.,* 2006).

Minerals formed within soils principally include carbonates, clays and oxyhydroxides. Of these, the carbonates in particular present a record of interaction between plants and soils; carbonate minerals formed in soils are known as pedogenic carbonates. They occur as individual grains, in association with root systems or networks of fungal filaments, as stalactite-like pendants on pebbles within well drained soils, and as coalesced layers (hardpans). Jenny (1980) demonstrated that the depth of occurrence of pedogenic carbonates increases with increasing rainfall, and this means that, in some countries, they may have been overlooked where soil surveys extend no deeper than 1 m. Importantly, the carbon isotope signature of pedogenic carbonates reflects the source of the carbon, and can be related to the vegetation growing on the soil (Cerling, 1984). In this way, transitions in vegetation type from one plant metabolic pathway to another (*e.g.* C3 to C4) can be recognized, and used to interpret the

Fig. 8. Kaolinite-illite intergrowths (Psyrillos *et al.*, 1999). **(a)** SEM image of a kaolinite verm, **(b)** back-scattered electron image of a polished section of vermiform kaolinite, illustrating alternation of illitic (white) and kaolinite (grey) layers on a sub-micron scale).

biological history of the coupled plant-soil system (*e.g.* Landi *et al.,* 2003; Stevenson *et al.,* 2005).

4. Influences on mineral reactions in soils

So far, we have considered minerals largely in isolation from the soil environment. This is complex, involving interaction between minerals, water, and biological systems (Ritz & Young, 2011). Minerals play a dynamic role in the soil system, and participate in reactions that take place over a range of scales. At the smallest scale, mineral grains (especially clays) in the micron size range are similar in size to bacteria and other microbes. At the millimeter scale, mineral grains (especially quartz and rock fragments) provide structural strength to a soil, and enable it to develop porosity and permeability, essential to permit the flow of water and soil gases.

Biological processes occur within soils over a range of scales, from microbial to the scale of plant roots and burrowing animals. Worms, in particular, interact with sediment and soil minerals, which pass through their guts (Fraser *et al.,* 2011; Needham *et al.,* 2004). Grazing animals provide faeces to soil, contributing to a flux of biologically derived material that becomes soil organic matter.

Plants, in particular, can be regarded as an engine for change within a soil (Fig. 9). Photosynthesis drives a complex biological pump that takes CO_2 from the atmosphere, converting it into a range of biological products. We see the solid products which take the form of the plant biomass on which we depend for our existence. What we do not see day-to-day are the consequences of photosynthesis for the soil system. Plant roots (underground biomass) are a dynamic system, constantly changing in geometry, and delivering solutions (root exudates; Jones, 1998) into the soil. Plant root exudates are rich in organic acid anions, with a number of specific functions: they corrode minerals,

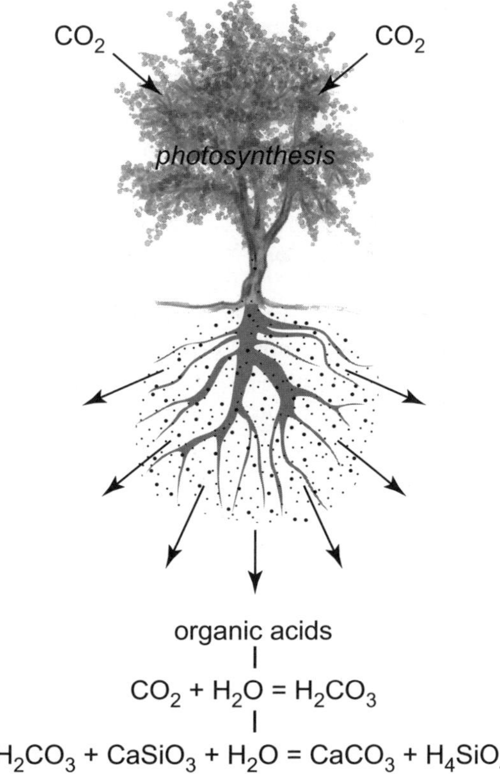

$$CO_2 + H_2O = H_2CO_3$$

$$H_2CO_3 + CaSiO_3 + H_2O = CaCO_3 + H_4SiO_4$$

Fig. 9. Schematic summary of the impact of plant photosynthesis on soil minerals.

enabling the nutrients within minerals to become available to the plant, and they complex elements such as Al that are potentially harmful, reducing their uptake by the plant.

Plant-root acids ultimately decompose to give dissolved CO_2, which, as carbonic acid, enhances dissolution of silicate minerals, permitting precipitation of calcium carbonate minerals if conditions are appropriate.

Dominated by organic acid anions such as citrate and acetate, plant root exudates are inherently unstable within soils. They provide an ideal substrate for soil microbial communities to use in their metabolism. The decomposition of plant root exudates gives CO_2 as the ultimate end-product. Thus soil gases become enriched in CO_2, and a proportion of this enters the soil solution where it forms bicarbonate (under most soil pH conditions).

A further source of mineralogically significant material to soils is *via* plant litter. The debris derived from plants includes leaves and other tissue that contains oxalic acid and calcium oxalates that may be formed while the plant tissue is alive, or as a product of fungal degradation (Manning, 2000). These are of biological origin, and are well documented (Burford *et al.*, 2006). The calcium oxalate minerals whewellite and weddellite have very low solubilities, and it is a paradox that their persistence in soils appears to be very limited. Something consumes calcium oxalate, as its minerals rarely occur other than in close association with organic matter.

Geographical influences on soils are very well known (Jenny, 1941). In the context of soil mineralogy, there are clear differences between newly formed soils dominated by the presence of erosion products (for example, those forming after glacial retreat) and mature soils formed as the product of, in exceptional cases, millions of years of stable exposure at the Earth's surface (for example those of western Australia; Twidale, 1997). Plants have evolved to accommodate these differences in soil mineralogy, and use different strategies to extract mineral-hosted nutrients such as P (Lambers *et al.*, 2008).

In general, systematic variation in the clay mineralogy of soils can be considered in the context of a stability diagram (Fig. 10). In this, the clay mineral assemblage in equilibrium with a specific water composition is defined according to the thermodynamic concentrations (activities) of specific chemical components.

Referring to Figure 10, at low pH and low silica activity, oxyhydroxides such as gibbsite occur. With increasing silica activity, gibbsite reacts to form kaolinite:

$$2Al(OH)_3 + 2H_4SiO_4 = Al_2Si_2O_5(OH)_4 + 5H_2O$$

  gibbsite   *dissolved* *kaolinite*
        silica

With continued increase in silica activity, quartz saturation is reached, and to the right of that line (Fig. 10) quartz will be present in a soil. Kaolinite eventually becomes unstable, and a smectite (represented here by pyrophyllite, $Al_2Si_4O_{10}(OH)_2$) is formed in its place:

$$Al_2Si_2O_5(OH)_4 + 2H_4SiO_4 = Al_2Si_4O_{10}(OH)_2 + 5H_2O$$

  kaolinite   *dissolved* *pyrophyllite*
        silica

Similar reactions can be written to show transformations that produce kaolinite or gibbsite from feldspars or illite. These involve dissolved ionic species, K^+ and H^+, reflecting the importance (a) of pH and (b) of these reactions as controls on soil K:

$$3KAlSi_3O_8 + 12H_2O + 2H^+ = KAl_3Si_3O_{10}(OH)_2 + 6H_4SiO_4 + 2K^+$$
$$K-feldspar \qquad\qquad illite$$

and

$$2KAl_3Si_3O_{10}(OH)_2 + 3H_2O + 2H^+ = 3Al_2Si_2O_5(OH)_4 + 2K^+$$
$$illite \qquad\qquad kaolinite$$

As shown by these reactions and by Figure 10, the mineralogical composition of a soil can be explained in terms of its history of reaction with water. In heavily leached soils, silica and cation activities are low, favouring the development of oxisols in which kaolinite occurs; quartz and gibbsite are mutually exclusive (*i.e.* the phase diagram predicts that they cannot coexist at equilibrium). In waterlogged soils, with little throughflow of water, silica and cation activities can be relatively high, favouring the formation of smectites.

Human influences on the mineralogy of soils are widespread. Most obviously, we add minerals to soils in the form of wastes – from mine spoil to domestic waste. Agriculture systematically involves the addition of minerals to soils through the application of lime and fertilizers including rock dusts (van Straaten, 2007) and slags. Urbanization involves the creation of soils within urban areas that contain a wealth of artificial minerals, and minerals formed or introduced as a consequence of contamination.

5. Mineral reactions with the soil solution

Reactions that take place at mineral surfaces in systems that are relevant to soils have been studied widely, and are described in a number of reviews (Vaughan & Pattrick, 1995). For the purposes of this chapter, the nature of the mineral surface will be considered.

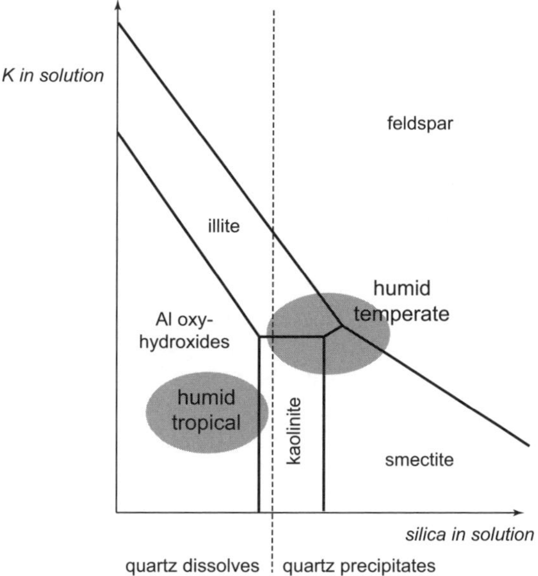

Fig. 10. Phase diagram showing the stability of different soil mineral phases for different soil-solution compositions, highlighting compositions typical of humid-tropical and humid-temperate regions.

Mineral surfaces vary according to the scale of observation and the nature of the mineral. In the simplest case, the physical surface of a poorly reactive material like quartz can be regarded as directly related to the geometrical properties of the grain. Feldspars, often reported in the sand particle-size fraction, typically show the effects of corrosion (Fig. 7). This results in a geometrically complex surface, with abundant niches accessible to microbes, fungal filaments and fine roots. The global importance of feldspar as a mineral in continental crustal rocks and soils, coupled with the evidence from corrosion of its accessibility to biological systems, may arise from its importance as a source of P, typically present in sub-percentage levels in the feldspar framework (Manning, 2008).

At a finer scale, the nature of a mineral surface becomes more difficult to define in geometrical terms. With clay minerals, reference is often made to the internal surface, representing the surfaces on the opposing faces of individual structural sheets. In 2:1 clays, these are composed of the bases of silica tetrahedra, and in 1:1 clays such as kaolinite one surface 'exposes' the base of the silica tetrahedra whereas the other surface exposes the edge of alumina octahedra.

The nature of clay surfaces influences the measurement of surface areas. Gas-adsorption methods measure the surface that is accessible to the sorbing gas (*e.g.* nitrogen). Methods that involve adsorption of chemicals such as ethylene glycol or ethylene glycol monoethyl ether (EGME; Quirk & Murray, 1999) are associated with greater extents of chemical reaction. The measurement of the surface area of a smectite using the EGME method involves expansion of the interlayer site to absorb the organic solvent, and so provides an estimate of the surface area of the clay lamellae, or the internal surface area of the clay. Use of the ethylene glycol desorption method on a non-swelling clay, such as kaolinite, reports the external surface area of the grains (Table 1).

Ion-exchange processes are vital aspects of the function of soil minerals. The 2:1 clays in particular have the potential for cation exchange (Table 1). This particularly affects the alkalis and alkaline earths, and is an important control on ammonium (Garrells & Christ, 1965; Manning & Hutcheon, 2004).

Cation exchange provides one process, in addition to dissolution-precipitation reactions, whereby clay minerals influence the composition of a soil solution. Figure 11 gives an example of this for potassium. Conventionally, soil K is regarded as existing in three pools. Fixed K represents that held within relatively stable minerals, or grains that have yet to be weathered including feldspars and freshly eroded micas. The exchangeable fraction is that held with weathered micas and clays, and the available

Fig. 11. Summary of reactions affecting soil potassium.

fraction is K in solution, readily available for plant uptake. As plants remove K, exchange between the micas/clays and the solution replenishes the solution with K to maintain equilibrium. If K salts are added as fertilizers, the exchange reaction is reversed, replenishing K stocks. In contrast, reaction between the fixed fraction is irreversible and slow.

6. Application of soil mineralogy to global problems

6.1. Food security

Ultimately, minerals provide all the nutrients required by plants except nitrogen and carbon, which are derived from the atmosphere. Human activity involves the mining of minerals at a remarkably small number of locations for the manufacture of chemical fertilizers, which are often blended with nitrogen compounds removed from the atmosphere by the Haber process.

The production of crops involves the removal of mineral nutrients from a soil (Manning, 2012). These enter the plant biomass, and harvesting involves the 'offtake' of nutrients from the soil. Good farming practice involves replenishment of offtake by application of fertilizers to ensure that sufficient nutrients remain in the soil for the next crop.

A number of studies have been carried out to determine the offtake of nutrients on a global scale, and these clarify the extent to which global agricultural production is mining nutrients from a soil. A comprehensive study by Sheldrick *et al.* (2002) has shown that offtake of K and P varies widely from country to country. In the developed world, the latter part of the 20th century involved application of fertilizers in excess of offtake of these nutrients. In marked contrast, offtake of K and P for many African countries greatly exceeded inputs from any source, including locally sourced materials such as crop residues, composted wastes and manures. According to Sheldrick & Lingard (2004), the situation is much more serious for K than for P. In these circumstances, soils are being mined of their nutrient content, and the reactions summarized graphically in Figure 11 are being strongly driven to the right. It is not clear how long this unsustainable situation can continue, and there are grounds for pessimism: the FAO reports that only 15 of 57 African countries import any fertilizer products, and on the basis of these figures it is evident that 1.5% of the world's K fertilizer production is expected to feed 15% of the world's population.

One solution to the difficulties in providing adequate K, bearing in mind the relatively high price and limited availability of the raw materials used to produce K fertilizers, is to consider innovative approaches to nutrient supply based on knowledge of the mineralogical reactions summarized in Figure 11 (Manning, 2010). Plants require K in solution, and the reversible exchange reactions between the soil solution and clays/weathered micas provides a buffer that allows the mineralogical component of the soil to replace K removed from solution by plant roots. 'Fixed' K is released by the weathering of primary silicates, and rates of weathering vary greatly (Table 2). One possible solution to the problem of K supply is to consider using locally sourced low-grade sources of K that weather quickly enough and provide K that would otherwise have to be imported from a remote production site. There is much controversy over the effectiveness of rock weathering as a source of K (Harley & Gilkes, 2000), but given the high price of K ($1000 per tonne at its peak in 2008) conventional fertilizers are increasingly inaccessible to farmers in poorer parts of the world.

Phosphorus provides a contrast. There is a paradox that some authors report impending exhaustion of P from mined sources (Cordell *et al.*, 2009), yet P pollution is a major concern for waste water-treatment to protect surface waters from excess P which can lead to eutrophication. Fortunately, the USGS estimates resources of P to be sufficient for several hundred years, and the reports of exhaustion appear to refer to exhaustion of reserves of P, which typically are only reported with a time horizon relevant to a commercial operation (a few years). Additionally, P recovery from wastewaters is feasible, for example in the form of the mineral struvite ($(NH_4)MgPO_4 \cdot 6H_2O$), which is an excellent fertilizer in its own right (Ponce & De Sa, 2007). There is no need to lose sleep over P, but there is over K.

6.2. Climate change

We are all well aware of the effects of climate change, although some dispute the causes and many debate the details. Irrespective of the arguments, there is no doubt that atmospheric CO_2 concentrations are rising at an alarming rate that corresponds roughly to the rate of consumption of fossil fuels. Minerals have a vital role to play in mitigation, exploiting weathering reactions to remove atmospheric CO_2 in partial compensation for artificial emissions (*e.g.* Schuiling & Krijgsman, 2006).

In soils, the mineral olivine weathers and has the potential to combine with CO_2 in accordance with the following reaction:

$$Mg_2SiO_4 + 2CO_2 + 2H_2O = 2MgCO_3 + H_4SiO_4$$

Evidence for the occurrence specifically in soils of this reaction is limited. However, Schuiling & Krijgsman (2006) suggest that an application of 80 kg of olivine per hectare will weather in a period of 30 years, withdrawing $\sim$35 kg of CO_2 from the atmosphere. Global CO_2 emissions are currently of the order of 30×10^9 T (www.iea.org). To compensate completely for this, whatever carbonation process is envisaged, would require of the order of 70×10^9 T of olivine, which is more than 1000 times current global production (estimated using data from Newman, 2011 for Norway, which produces 60% of the world's olivine).

In urban settings, a similar carbonation reaction has been observed involving calcium silicates (Renforth *et al.*, 2009) originating from cement derived from demolition products within the soil. This process gives rise to soil C contents of 300 tonnes of C per hectare, well in excess of amounts held as organic C in agricultural soils. It is believed that the global annual potential for carbon capture by this process is of the order of 200 million tones of C/year (Renforth *et al.*, 2011). However, as cement is produced by the calcining of limestone, the carbonation of such materials in soils merely replaces the carbon lost from the mineral raw material during manufacture (and does not compensate for emissions from fuels used in the process).

7. Conclusions

Minerals are dynamic components of soil systems. Rock-forming minerals produced by igneous processes at highest temperature, such as olivine, are destroyed by weathering, and this explains the fertility commonly associated with new soils formed on recently erupted basaltic lavas. Other primary minerals such as quartz are stable in soils for extremely long periods. Newly formed minerals, precipitated as other minerals break down, include the clays and carbonate minerals which record the isotopic signature of the source of their carbon, often vegetation.

An understanding of the rates and mechanisms by which minerals react in soil systems underpins two key aspects of land management. First, geological minerals are the source of vital nutrients required for plant growth, and much needs to be done to sustain their supply to agriculture. Secondly, the dynamic nature of the coupled soil-plant system, that processes all of the CO_2 in the atmosphere in 5 years, is a phenomenon that can be exploited to remove CO_2 from the atmosphere in mitigation of anthropogenic conditions. The practical solution of these problems depends on a sound knowledge of mineral reactions in soils.

References

Berner, R.A. (1992) Weathering, plants and the long-term carbon cycle. *Geochimica et Cosmochimica Acta*, **56**, 3225–3231.

Burford, E.P., Hillier, S. & Gadd, G.M. (2006) Biomineralization of fungal hyphae with calcite ($CaCO_3$) and calcium oxalate mono- and dihydrate in carboniferous limestone microcosms. *Geomicrobiology Journal*, **23**, 599–611.

Cerling, T.E. (1984) The stable isotope composition of modern soil carbonate and its relationship to climate. *Earth and Planetary Science Letters*, **71**, 221–240.

Cordell, D., Drangerta, J.-O. & White, S. (2009) The story of phosphorus: Global food security and food for thought. *Global Environmental Change*, **19**, 292–305.

Cotter-Howells, J.D. & Paterson, E. (2000) Minerals and soil development. pp. 91–124 in: *Environmental Mineralogy* (D. J. Vaughan and R. A. Wogelius, editors) EMU Notes in Mineralogy, **2**. European Mineralogical Union, Eötvös University Press, Budapest.

Curtis, C.D. (1976) Stability of minerals in surface weathering reactions: a general thermochemical approach. *Earth Science Processes*, **1**, 63–70.

Deer, W.A., Howie, R.A. & Zussman, J. (1992) *An Introduction to the Rock-Forming Minerals*. Longman, London.

Dixon, J.B. & Weed, S.B. (1989) *Minerals in Soil Environments*. 2nd edition, Soil Science Society of America, Madison, Wisconsin, USA.

Drever, J.I. (1997) *The Geochemistry of Natural Waters, Surface and Groundwater Environments*, 3rd edition. Prentice-Hall, Englewood Cliffs, New Jersey, USA.

FAO (2008) *Current world fertilizer trends and outlook to 2011–12*. (FAO: Rome). www.fao.org/newsroom/en/news/2008/1000792/index.html.

Fraser, A., Lambkin, D.C., Lee, M.R., Schofield, P., Mosselmans, J.F.W. & Hodson, M.E. (2011) Incorporation of lead into calcium carbonate granules secreted by earthworms living in lead-contaminated soils. *Geochimica et Cosmochimica Acta*, **75**, 2544–2556.

Garrels, R.M. & Christ, C.L. (1965) *Solutions, Minerals and Equilibria.* Harper & Row, New York.

Goldich, S.S. (1938) A study in rock weathering. *Journal of Geology*, **46**, 17–58.

Harley, A.D. & Gilkes, R.J. (2000) Factors influencing the release of plant nutrients from silicate rock powders: a geochemical overview. *Nutrient Cycling in Agroecosystems*, **56**, 11–36.

Hodson, M.E., Valsami-Jones, E., Cotter-Howells, J.D., Dubbin, W.E., Kemp, A.J., Thornton, I. & Warren, A. (2001) Effect of bone meal (calcium phosphate) amendments on metal release from contaminated soils – a leaching column study. *Environmental Pollution*, **112**, 233–243.

Humphreys, C.P., Franks, P.J., Rees, M., Bidartondo, M.I., Leake, J.R. & Beerling, D.J. (2010) Mutualistic mycorrhiza-like symbiosis in the most ancient group of land plants. *Nature Communications,* **1**, 103 doi:10.1038/ncomms1105.

Indermühle, A., Stocker, T.F., Joos, F., Fischer, H., Smith, H.J., Wahlen, M., Deck, B., Mastroianni, D., Tschumi, J., Blunier, T., Meyer, R. & Stauffer, B. (1999) Holocene carbon-cycle dynamics based on CO_2 trapped in ice at Taylor Dome, Antarctica. *Nature*, **398**, 121–126.

Jenny, H. (1941) *Factors of Soil Formation.* McGraw-Hill, New York, London.

Jenny, H. (1980) *The Soil Resource.* Springer Verlag, New York.

Jones, D.L. (1998) Organic acids in the rhizosphere – a critical review. *Plant and Soil*, **205**, 25–44.

Lambers, H., Raven, J.A., Shaver, G.R. & Smith, S.E. (2008) Plant nutrient-acquisition strategies change with soil age. *Trends in Ecology and Evolution.* **23**, 95–103.

Landi, A., Mermut, A.R. & Anderson, D.W. (2003) Origin and rate of pedogenic carbonate accumulation in Saskatchewan soils, Canada. *Geoderma*, **117**, 143–156.

Lasaga, A.C., Soler, J.M., Ganor, J., Burch, T.E. & Nagy, K.L. (1994) Chemical weathering rate laws and global geochemical cycles. *Geochimica et Cosmochimica Acta*, **58**, 2361–2386.

Lenton, T.M. (2001) The role of land plants, phosphorus weathering and fire in the rise and regulation of atmospheric oxygen. *Global Change Biology*, **7**, 613–629.

Liu, Z.-H. (2011) Is pedogenic carbonate an important atmospheric CO_2 sink? *Chinese Science Bulletin*, **56**, 1–3.

Manning, D.A.C. (2000) Carbonates and oxalates in sediments and landfill: monitors of death and decay in natural and artificial systems. *Journal of the Geological Society of London*, **157**, 229–238.

Manning, D.A.C. (2008) Phosphate minerals, environmental pollution and sustainable agriculture. *Elements*, **4**, 105–108.

Manning, D.A.C. (2010) Mineral sources of potassium for plant nutrition: a review. *Agronomy for Sustainable Development*, **30**, 281–294.

Manning, D.A.C. (2012) Plant nutrients. In: *Soils and Food Security* (R.E. Hester and R.M. Harrison, editors). Issues in Environmental science and Technology, **35**, Royal Society of Chemistry, Cambridge, UK, pp. 183–197.

Manning, D.A.C. & Hutcheon, I.E. (2004) Distribution and mineralogical controls on ammonium in deep groundwaters. *Applied Geochemistry*, **19**, 1495–1503.

Mayes, W.M., Jarvis, A.P., Burke, I.T., Walton, M., Feigl, V., Klebercz, O. & Gruiz, K. (2011) Dispersal and attenuation of trace contaminants downstream of the Ajka bauxite residue (red mud) depository failure, Hungary. *Environmental Science & Technology*, **45**, 5147–5155.

Needham, S.J., Worden, R.H. & McIlroy, D. (2004) Animal-sediment interactions: the effect of ingestion and excretion by worms on mineralogy. *Biogeosciences*, **1** 113–121.

Newman, H.R. (2011) *The Mineral Industry of Norway.* United States Geological Survey 2009 Minerals Yearbook, United States Geological Survey, myb3-2009-no, 8 pp.

Palandri, J.L. & Kharaka, Y.K. (2004) A compilation of rate parameters of water-mineral interaction kinetics for application to geochemical modeling. *U.S. Geological Survey Open File Report* 2004–1068, Menlo Park, California, USA.

Petit, J.R., Jouzel, J., Raynaud, D., Barkov, N.I., Barnola, J.-M., Basile, I., Bender, M., Chappellaz, J., Davis, M., Delaygue, G., Delmotte, M., Kotlyakov, V.M., Lipenkov, V., Lorius, C., Pepin, L., Ritz, C., Saltzman,

E. & Stievenard, M. (1999) Climate and Atmospheric History of the Last 420,000 Years from the Vostok Ice Core, Antarctica, *Nature*, **399**, 429.

Ponce, R.G. & De Sa M.E.G.L. (2007) Evaluation of struvite as a fertilizer: a comparison with traditional P sources. *Agrochimica*, **51**, 301–308.

Psyrillos, A., Howe, J.H., Manning, D.A.C. & Burley, S.D. (1999) Geological controls on kaolin particle shape and consequences for mineral processing. *Clay Minerals*, **34**, 193–208.

Quirk, J.P. & Murray, R.S. (1999) Appraisal of the ethylene glycol monoethyl ether method for measuring hydratable surface area of clays and soils. *Soil Science Society of America Journal*, **63**, 839–849.

Renforth, P., Manning, D.A.C. & Lopez-Capel, E. (2009) Carbonate precipitation in artificial soils as a sink for atmospheric carbon dioxide. *Applied Geochemistry*, **24**, 1757–1764.

Renforth, P., Washbourne, C.-L., Taylder, J. & Manning, D.A.C. (2011) Silicate production and availability for mineral carbonation. *Environmental Science & Technology*, **45**, 2035–2041.

Retallack, G. (2009) Cambrian, Ordovician and Silurian pedostratigraphy and global events in Australia. *Australian Journal of Earth Sciences*, **56**, 571–586.

Ritz, K. & Young, I.M. (2011) *The Architecture and Biology of Soils: Life in Inner Space.* CABI Publishing, Wallingford, UK.

Robie, R.A. & Hemingway, B.S. (1995) Thermodynamic properties of minerals and related substances at 298.15 K and 1 Bar (10^5 Pascals) pressure and at higher temperatures. *U.S. Geological Survey Bulletin*, **2131**, Washington.

Ruddiman, W.F. (2003) The anthropogenic greenhouse era began thousands of years ago. *Climatic Change*, **61**, 261–293.

Rye, R. & Holland, H.D. (1998) Paleosols and the evolution of atmospheric oxygen: a critical review. *American Journal of Science*, **298**, 621–672.

Ryskov, Y.G., Demkin, V.A., Oleynik, S.A. & Ryskova, E.A. (2008) Dynamics of pedogenic carbonate for the last 5000 years and its role as a buffer reservoir for atmospheric carbon dioxide in soils of Russia. *Global and Planetary Change*, **61**, 63–69.

Schuiling, R.D. & Krijgsman, P. (2006) Enhanced weathering: An effective and cheap tool to sequester CO_2. *Climatic Change*, **74**, 349–354.

Sheldon, N.D. (2006) Precambrian paleosols and atmospheric CO_2 levels. *Precambrian Research*, **147**, 148–155.

Sheldrick, W.F. & Lingard, J. (2004) The use of nutrient audits to determine nutrient balances in Africa. *Food Policy*, **29**, 61–98.

Sheldrick, W.F., Syers, J.K. & Lingard, J. (2002) A conceptual model for conducting nutrient audits at national, regional and global scales. *Nutrient Cycling in Agroecosystems*, **62**, 61–67.

Stevenson, B.A., Kelly, E.F., McDonald, E.V. & Busacca, A.J. (2005) The stable carbon isotope composition of soil organic carbon and pedogenic carbonates along a bioclimatic gradient in the Palouse region, Washington State, USA. *Geoderma*, **124**, 37–47.

Twidale, C.R. (1997) The great age of some Australian landforms: examples of, and possible explanations for, landscape longevity. pp. 13–23 in: *Palaeosurfaces: Recognition, Reconstruction and Palaeoenvironmental Interpretation* (M. Widdowson, editor). Special Publication, **120**, Geological Society, London, pp. 13–23.

van Straaten, P. (2007) *Agrogeology. The use of Rocks for Crops.* Enviroquest Ltd, Cambridge, Ontario, Canada.

Vaughan, D.J. & Pattrick, R.A.D. (1995) *Mineral Surfaces.* Mineralogical Society of Great Britain and Ireland Series, **5**, Chapman & Hall, London.

White, A.F. & Brantley, S.L. (1995) *Chemical Weathering Rates of Silicate Minerals.* Reviews in Mineralogy, **31**. Mineralogical Society of America, Washington, D.C.

Mineralogy of marine sediment systems: a geochemical framework

ANDREW C. APLIN[1] and KEVIN G. TAYLOR[2]

[1]*School of Civil Engineering and Geosciences, Newcastle University,
Newcastle upon Tyne NE1 7RU, UK, e-mail: andrew.aplin@newcastle.ac.uk*
[2]*School of Earth, Atmospheric and Environmental Sciences, University of Manchester,
Oxford Road, Manchester M13 9PL, UK, e-mail: Kevin.taylor@manchester.ac.uk*

Sediments are a major component of the ocean system and the composition of, and chemical processes acting upon, these sediments are controlled by many factors. In this chapter we outline the major types of marine sediment (lithogenous, biogenic, hydrogeneous and authigenic), discuss the sources of these and their chemical and mineral characteristics. The most abundant minerals are (oxyhydr)oxides, clay minerals, sulfides and carbonates, and all of these have been studied extensively, using a range of mineralogical techniques. A major aim of this chapter is to provide a framework by which the mineral transformations that take place upon, or just below, the sea-floor can be understood. There are significant thermodynamic, kinetic and biological controls on these transformations, and the interaction of these plays a major role in element cycling between the oceans and the lithosphere, and trace element-enrichment on the sea-floor and carbon burial and remineralization. Further, we conclude that more research is required on the interactions between minerals and bacteria within marine sediments, and on the role of amorphous oxide minerals delivered from land in early burial reactions and mineral precipitations.

1. Introduction

Oceans cover $\sim$70% of the Earth's surface so that the processes operating within the marine realm and upon ocean floors play a marked role in element and nutrient cycling, biological productivity, ultimately exerting an important influence on global climate. This chapter will focus on the distribution, composition and mineralogy of sediments that are actively accumulating on the sea floor, both in the shallow marine and deep ocean realms. Further, it will describe the geochemical and mineralogical processes operating in these sediments.

Recently deposited sediments comprise a diverse mixture of detrital minerals and amorphous or poorly crystalline materials, biogenic material, organic matter and porewater. The mixture is inherently unstable and progresses towards equilibrium through a series of diagenetic reactions (Garrels, 1986), ultimately generating the muscovite–chlorite–quartz (–Fe oxide–organic carbon) mineralogy typical of low-grade metamorphic phyllites. Diagenesis begins immediately upon sedimentation, initially involving change of the most reactive components: organic matter, oxyhydroxides, dissolved oxidants such as oxygen and sulfate, and poorly crystalline or amorphous

© Copyright 2013 the European Mineralogical Union and the Mineralogical Society of Great Britain & Ireland
DOI: 10.1180/EMU-notes.13.4

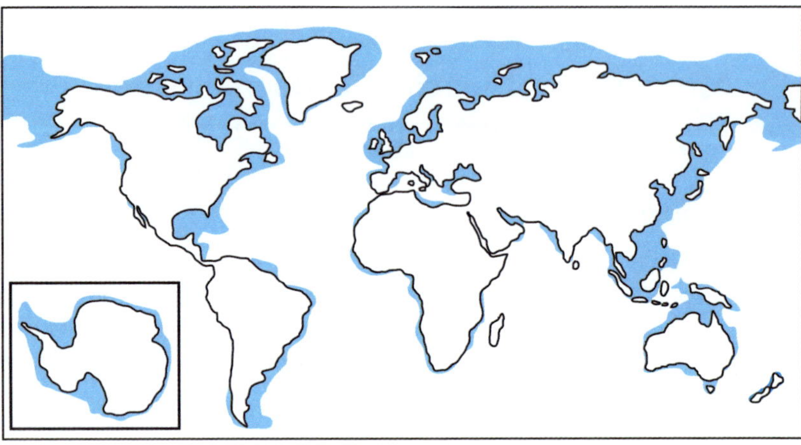

Fig. 1. Distribution of modern continental shelves (dark grey) (modified after Johnson & Baldwin, 1996).

aluminosilicates. Over the past 50 years, the chemistry and mineralogy of recent sediments has been studied extensively. Early mineralogical descriptions have been superseded by combined studies of interstitial water chemistry and mineralogical changes in a wide range of sedimentary settings. These have generated qualitative and increasingly quantitative insights into the nature and rates of early diagenetic reactions and the processes of authigenic mineral formation.

For the purposes of this chapter, the marine realm can be split into two parts: the continental shelf and the deep ocean basin. The continental shelf is commonly defined as sea floor shallower than 200 m adjacent to continental land masses, with the outer margin marked by the continental slope (*e.g.* Whitten & Brooks, 1972). The distribution of modern continental shelves is shown in Figure 1. The deep ocean basins incorporate areas further downslope from the continental slope and include the continental rise, the abyssal plain, ocean trenches and submarine ridges.

2. Marine sediment types

Sediment that accumulates in the shallow marine and ocean realm can be split into four main types: (1) lithogenous (material derived from the breakdown of rocks in terrestrial settings); (2) biogenic (the skeletal remains of organisms); (3) hydrogenous (chemical precipitates); (4) authigenic (formed within the sediment) as described below.

(1)   *Lithogenous* (or *detrital*) material is generated by continental weathering processes and is transported to the ocean by water (and air). Detrital material comprises the vast bulk of lacustrine, estuarine and nearshore sediments and therefore exerts a major control on their physical and, to a lesser extent, their

chemical properties. Although any mineral can be supplied to sediments as part of the detrital load, the most common detrital phases are clay minerals and quartz.

(2) *Biogenic* material is produced in the water column and mainly comprises calcareous and siliceous microfossils, plus organic matter. Rarely of quantitative importance in lacustrine or nearshore marine sediments, this material becomes a significant fraction of marine sediments in places where there is a limited supply of detrital material: some pelagic settings and on broad, shallow ocean shelves starved of terrestrial input.

(3) *Hydrogenous* material is formed inorganically in the water column. The most common and geochemically most important examples of hydrogenous material are iron and manganese oxyhydroxides. These form, for example, where metal-rich hydrothermal fluids mix with cold seawater along mid-ocean ridges, and close to sediment–water interfaces in cases where dissolved Mn or Fe has diffused from sediment pore waters following the reductive dissolution of previously sedimented oxides. Hydrogenous material is only quantitatively important in sediments which are starved of both terrestrial and biogenic input: in pelagic settings far from the continents and underneath waters with very low biological productivity, for example the red clays of the south central Pacific Ocean. Note here the inevitably artificial nature of this simple classification, in that oxyhydroxides, and also organic matter plus some biologically produced inorganic components, are also derived as detrital phases as a result of continental weathering.

(4) *Authigenic* material is formed within the sediment as a result of reactions involving the chemically least stable sediment components. In most cases these components comprise <10% of the supplied sediment; their great importance lies in their reactivity and thus the major influence they exert both on the chemistry of interstitial water and on the minerals (carbonates, sulfides, phosphates and aluminosilicates) which form during early diagenesis.

2.1. Lithogenous sediment

Delivery of land-derived sediment to the oceans can take place *via* three mechanisms: fluvial discharge, wind-blown dispersal of dust, and glacially eroded and discharged material.

2.1.1. Fluvial inputs

The chemical and physical weathering of rocks within terrestrial catchments results in the transport and delivery of sediment to the marine realm. The total global load of sediment delivered in suspension by rivers to oceans globally has recently been estimated as $\sim 12.6 - 24 \times 10^9$ t y^{-1} (Walling, 2006). This global distribution of fluvially derived sediment is distributed heterogeneously (see Fig. 2), with areas undergoing high rates of tectonic uplift and/or the most intense rainfall contributing by far the majority of this material (*e.g.* the Indonesian Archipelago, the Amazon). Rivers draining old, arid and stable land surfaces (*e.g.* Australia) contribute negligible loads. It has been

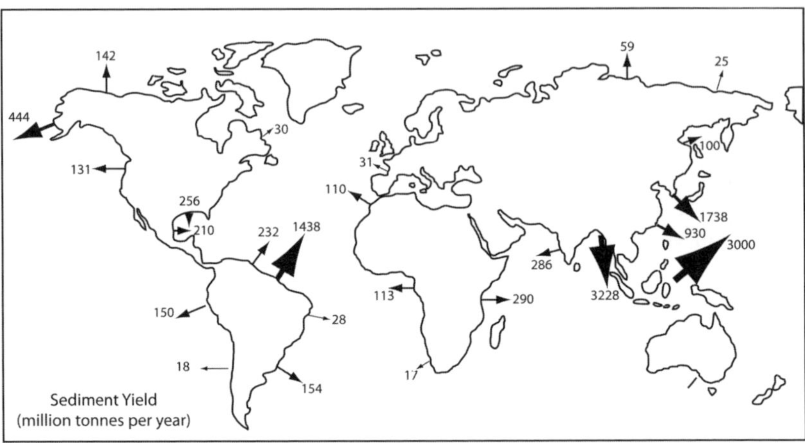

Fig. 2. Volume of riverine silt supplied to continental shelves (adapted from Milliman & Meade, 1993).

recognized, however, that anthropogenic impacts upon catchment basins (*e.g.* land-use changes, deforestation, climate change) have led to an increase in sediment delivery (*e.g.* Walling & Fang, 2003). It should be stressed, however, that the estimation of global sediment loads is fraught with difficulty, due mainly to the poor level of monitoring of many rivers. Furthermore, much of the material delivered by rivers may be trapped in estuaries, rather than being delivered to open marine waters (Martin & Whitfield, 1983).

The minerals that are transported from the land to the oceans in suspended river sediment include resistant primary minerals (quartz, zircon) and secondary minerals derived from soil profiles (*e.g.* clay minerals, iron and manganese oxides and oxyhydroxides, aluminium oxides). Data summarizing the mineralogical composition of suspended particles in the world's main rivers are given by Konta (1985) and Emeis (1985) (see Fig. 3). The composition of the clay minerals, which dominate the mineral assemblages, largely reflects the soils from which they were derived and are thus an integrated record of the chemical and physical weathering processes for those parts of the catchment supplying particles to the river. Illite is ubiquitous and is the most common clay mineral in all rivers except the Niger, with lesser amounts of quartz and feldspars in almost all rivers. The Niger is the most obvious example of a hot and humid weathering regime, dominated by kaolinite, depleted in quartz and feldspars, and containing gibbsite.

The elemental composition of this material varies in response to catchment geology, soil type and climate; global element flux estimates have been made by Martin & Maybeck (1979), Martin & Whitfield (1983) and, more recently, by Viers *et al.* (2009). Increased chemical weathering leaves an aluminium + iron oxide-enriched residue, culminating in the formation of laterites (Persons, 1970). Suspensates in rivers draining humid, tropical climatic zones (like the Niger and Orinoco) are correspondingly enriched in aluminium and depleted in sodium, potassium and calcium

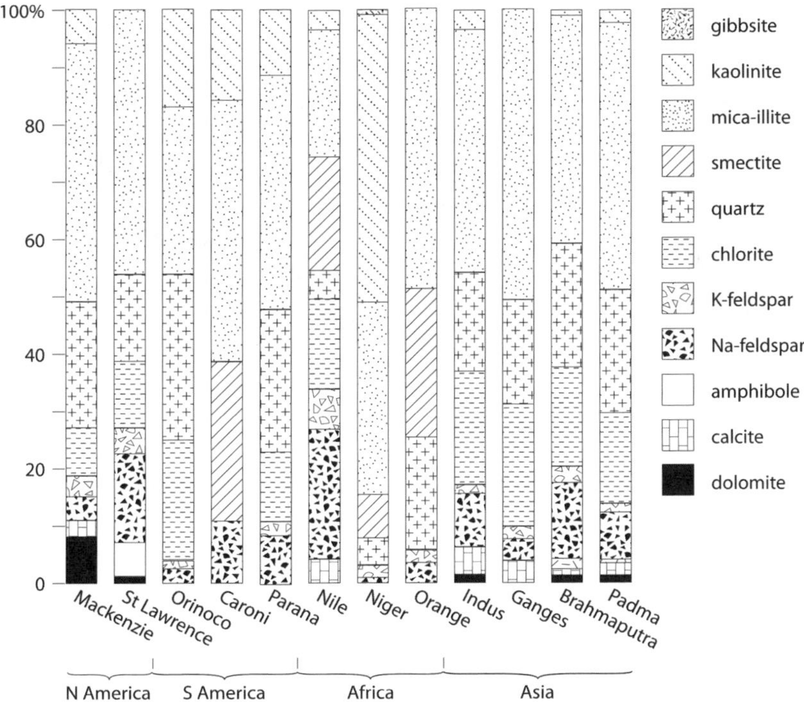

Fig. 3. Average abundance of major minerals in suspended material in some major rivers (after Konta, 1985).

compared to rivers draining cooler and/or drier catchments (Konta, 1985). And, although suspensates in tropical rivers may not show an absolute enrichment in iron, it is likely that the nature of the iron residues will be very different from those in rivers draining cool climatic zones. Labile oxides and oxyhydroxides will be important ferroan phases in the tropical rivers, with a greater proportion of relatively refractory Fe-bearing silicates (like chlorite) in rivers carrying sediment derived from weathering in cool climates. Poulton & Raiswell (2000) and Viers *et al.* (2009) have shown that there is a significant anthropogenic contribution to some element fluxes from land to ocean.

A significant proportion of fluvial suspensates are effectively 'X-ray amorphous'. Figure 4 shows X-ray diffraction (XRD) patterns of suspensates from the Nile and Orinoco rivers (Konta, 1985). Peaks representing crystalline phases are perched on a broad background indicative of poorly crystalline material. Such material is poorly characterized but includes: degraded aluminosilicates; iron, manganese and aluminium oxyhydroxides; opal, and organic matter (Perdue *et al.*, 1976; Sigleo & Helz, 1981; Leppard, 1992; Wilkinson *et al.*, 1999). Its importance lies not only in its apparent abundance, but also in its chemical reactivity. X-ray amorphous material is amongst the first to disappear during mud and mudstone diagenesis (Foscolos & Powell,

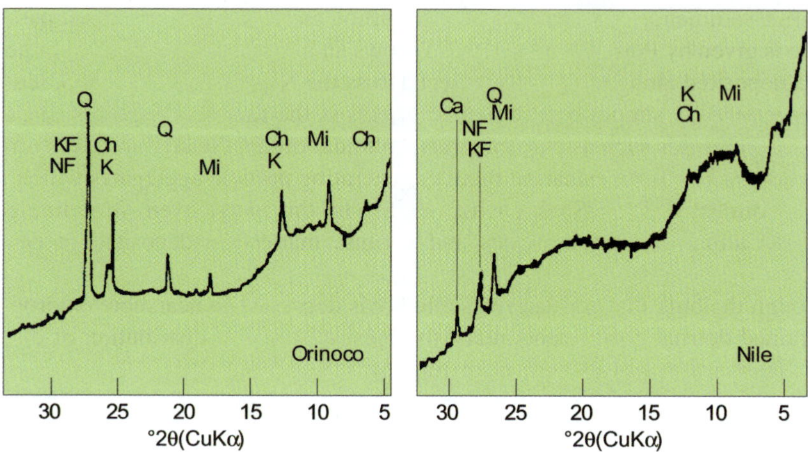

Fig. 4. XRD patterns of suspensates from the Nile and Orinoco Rivers (data from Konta, 1985). Ca: calcite, Ch: chlorite, K: kaolinite, KF: K feldspar, Mi: mica, NF: Na feldspar, Q: quartz. The vertical axis is intensity.

1980) and is almost certainly an important feedstock for the growth of early diagenetic clay minerals.

Suspensates are sedimented when the fluid is no longer capable of carrying particles in suspension. As mineralogy is strongly influenced by grain size, the mineralogy of river suspensates also depends on the ability of the river to carry material of a given size. For example, Gibbs (1977) shows that in the Amazon River, distinct mineral phases occur in specific size fractions (Fig. 5). The relationship between grain size and mineralogy is ultimately reflected in the mineralogical composition of estuarine

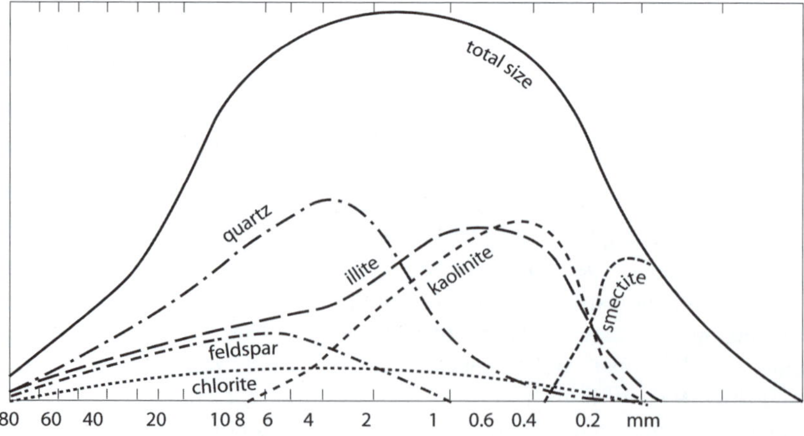

Fig. 5. Size distribution of minerals in suspended material carried by the Amazon (after Gibbs, 1977).

and marine sediments as a result of particle settling along sediment transport paths. One example is given by Porrenga (1966), who shows an increase in smectite:kaolinite ratios in muds deposited along a 120 km transect across the Niger delta. Superimposed on the trend generated by simple hydrodynamic effects is the fact that fine-grained particles with surface charges such as clay minerals, colloidal organic matter and oxyhydroxides tend to flocculate during estuarine mixing, generating particle aggregates which can be rapidly sedimented (*e.g.* Sholokovitz, 1976). In this way, even very fine-grained (<0.1 μm) aluminosilicates, oxides and organic matter are deposited in nearshore sediments.

Although the bulk of river-derived sediment is deposited in nearshore environments, finer-grained detrital components reach the open ocean. The distribution of clay minerals in deep-sea sediments has been reported by Biscaye (1965) and Griffin *et al.* (1968), and reflect significantly the supply routes from the continents (see Figs 6–9). The highest concentrations of kaolinite are found in low-latitude sediments, chlorite is released by physical weathering of rocks in polar regions and, in the absence of significant chemical weathering, is dispersed to the oceans by glaciers and ice rafting. Its concentration in deep-sea sediments declines sharply at ∼50°N and 50°S, corresponding to the iceberg transport limit. Illite (and mixed-layer illite-smectites) is the most common clay mineral in deep-sea sediments and, because it forms under a wide range of weathering regimes, it is not confined to particular latitudinal bands. Its distribution is controlled by: (1) the amount of land surrounding an oceanic area, and (2) dilution by other minerals (Chester, 1990). Of the main clay mineral groups, the smectites are the only ones which are mainly formed by authigenic processes in marine sediments. The highest concentrations of smectite in marine sediments thus occur in the very slowly accumulating sediments of the central and south Pacific Ocean, where there is little dilution by detrital or biogenic inputs (Fig. 9).

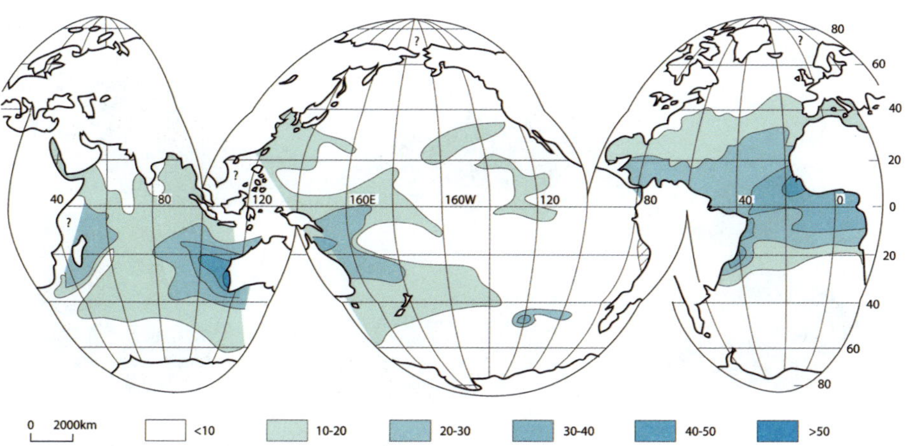

Fig. 6. Kaolinite percentages in the clay fraction of surface sediments in the world ocean (after Windom, 1976).

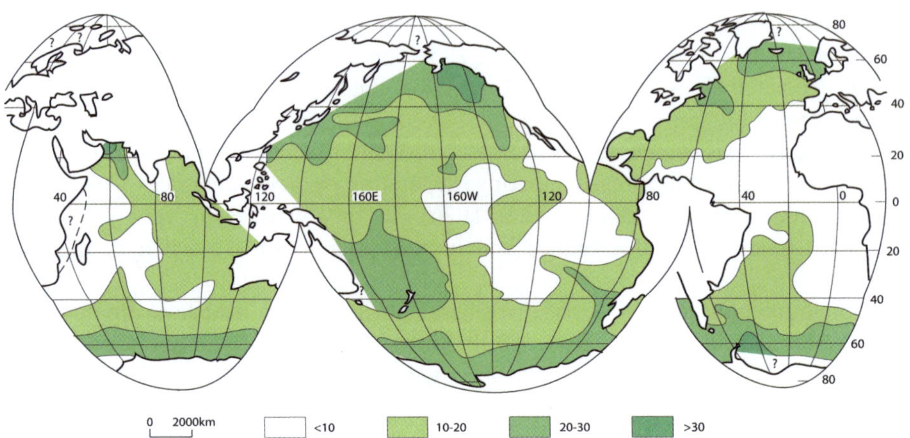

Fig. 7. Chlorite percentages in the clay fraction of surface sediments in the world ocean (after Windom, 1976).

2.1.2. Wind-blown dust

A significant vector for sediment delivery from land to oceans is by wind transport (Fig. 10). The best documented example is the aeolian transfer of dust from North Africa, along the Sahara-Sahel Dust Corridor. This, the world's largest dust source, is a zone lying from 12°N and 28°N, and running 4000 km east to west from Chad to Mauritania (Goudie & Middleton, 2001; Bristow, 2005). The mineralogy and composition of this dust are highly dependent on the source areas, and include soil derived minerals (iron oxyhydroxides, kaolinite), opaline silica (derived from diatom-rich lake deposits) and rare earth element-rich phosphate minerals (Moreno *et al.*, 2006).

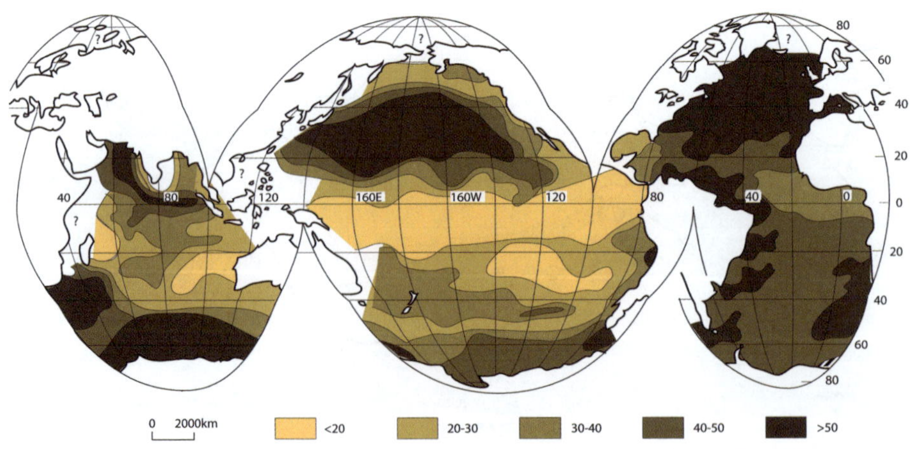

Fig. 8. Illite percentages in the clay fraction of surface sediments in the world ocean (after Windom, 1976).

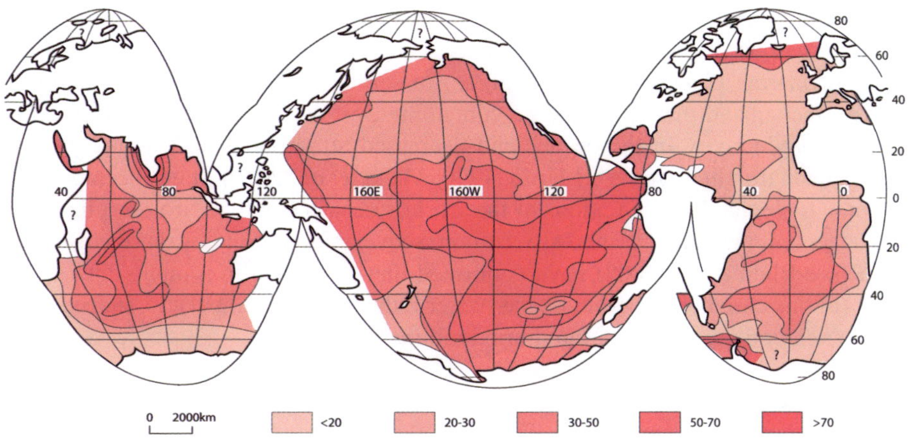

Fig. 9. Smectite percentages in the clay fraction of surface sediments in the world ocean (after Windom, 1976).

Indeed, mineralogical and trace element composition has long been used as a source-discriminator for such dust (Chester *et al.*, 1971; Chiapello *et al.*, 1997). As early as the 1970s it was recognized (Chester & Johnson, 1971) that the mineralogy of dust collected from the coast of Morocco was similar to deep-ocean sediments deposited directly west of the collection site. Of major significance to ocean chemistry and biological productivity is that that wind-blown dust is now believed to be a major supplier of iron to surface waters far from continents, acting as a limiting nutrient for primary productivity (Jickells *et al.*, 2005). Iron in dust is generally in the form of clay minerals and iron (oxyhydr)oxides, which are only very sparingly soluble in seawater. It is believed that the iron is complexed by microbially produced organic compounds termed *siderophores*. These preferentially bind Fe^{3+}, keeping a fraction of the iron in solution thus preventing it from precipitating as a mineral phase, and so rendering it bioavailable (Raiswell, 2011). It has also been suggested that atmospheric processes can transform soil-derived iron (oxyhydr)oxides into more soluble species (Shi *et al.*, 2009). As a result of this delivery of

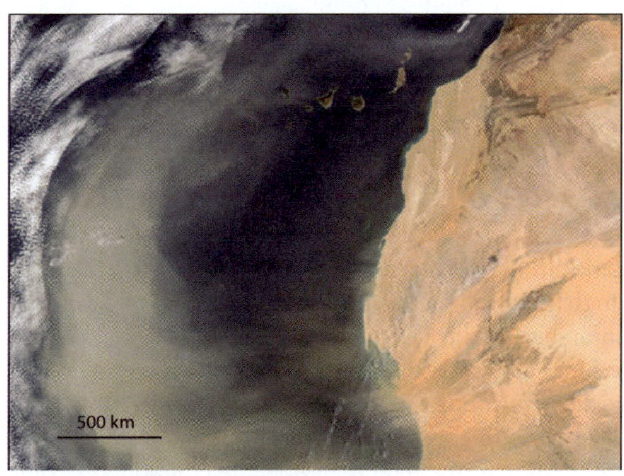

Fig. 10. Saharan dust from space. Image reproduced courtesy of NASA.

iron to deep ocean surface waters, dust delivery plays a major role in controlling the cycling of carbon through its importance as an essential, and limiting, nutrient for the growth of marine plankton (bacteria, algae) in surface ocean waters. For example, it has been recognized that increased Fe-containing dust deposition during glacial times increased productivity in the oceans and decreased atmospheric CO_2 concentrations, the so called "iron hypothesis" (Martin, 1990; Maher et al., 2010). It has been proposed that some biologically poor regions of the world's oceans could be 'iron fertilized' and hence 'geo-engineered' to increase carbon sequestration. Studies in the Southern Ocean (see Blain et al., 2007) have confirmed that long-term iron inputs into surface waters (in this case derived from deeper water rather than dust input) do, indeed, result in greater carbon sequestration into buried sediment.

2.1.3. Glacially derived sediment

There is a significant supply of sediment by glaciers and glacial meltwaters to ocean basins. Due to the high physical erosion in glaciated terrains, there is a ready availability of sediment to be transported by glacial meltwaters as both suspended load and bedload (Gurnell, 1987). Raiswell et al. (2006) estimated the global flux of sediment delivery from both meltwaters and icebergs combined to be 2.9×10^9 kg y^{-1}. As well as providing an important sediment source to the marine realm, it was shown by Raiswell et al. (2006) that the presence of nanoparticulate iron (oxyhydr)oxides in melting icebergs were an important source of iron to the deep ocean surface waters.

3. Biogenic components

Calcareous (foraminifera, coccolithophorids, pteropods) and siliceous (radiolaria, diatoms) tests are the dominant forms of biologically produced mineral matter in most modern sediments, except in broad, sediment-starved, shallow waters where coral-derived and inorganically precipitated calcium carbonate ($CaCO_3$) can be common. The calcareous fossils comprise calcium carbonate, whilst the siliceous fossils are formed of opal (sometimes termed 'amorphous silica' or opal-A). Controls on the abundance of biogenic material in marine sediments has been reviewed by Berger (1976) and Chester (1990); maps showing the distribution of calcium carbonate in marine sediments have been published by Berger et al. (1976); Kolla et al. (1976) and Biscaye et al. (1976), and of opaline silica by Bostrom et al. (1973).

The abundance of biogenic material in recent sediments (Fig. 8) reflects the balance of: (1) its rate of production in biologically productive surface waters; (2) its rate of dissolution both within the water column and at the sediment-water interface, and (3) the extent to which it is diluted by non-biogenic material, especially detrital material. Sediments rich in biogenic material tend, therefore, to occur in pelagic settings where the rate of supply of detrital material is low. In most lacustrine and nearshore settings, the rate of supply of detrital material is sufficiently great that biogenic components comprise only a few weight percent of the sediment. The only nearshore sediments which contain significant amounts of biologically produced material are those forming under warm, biologically productive waters in regions which are starved of detrital sediment (*e.g.* Bahamas; Persian Gulf); these only form a few percent of the world ocean. In these tropical shallow marine settings, $CaCO_3$-rich sediments are derived from both biologically and chemically derived sources. Skeletal carbonate is sourced from a wide range of both framework-building organisms (*e.g.* coral, bryzoa) and benthic and pelagic organisms (*e.g.* molluscs, foraminfera, echinoids). These skeletal-derived sediments may be composed of low-Mg calcite, high-Mg calcite, or aragonite – all essentially forms of calcium carbonate. Chemical precipitates include ooids – concentrically layered aragonite grains – and whitings – fine-grained crystals of calcium carbonate. Recent research has suggested that excreta from fish may also be a significant source of very fine-grained calcium carbonate in shallow marine seas (Perry *et al.*, 2011).

The rate of production of shells in the oceans is a function of biological fertility, which is in turn based on the rate of supply of nutrients to surface waters. Different organisms thrive in waters of differing fertility, such that diatoms dominate in more productive waters, with coccoliths dominant in the less productive, oceanic central gyres. As seawater is undersaturated with respect to opaline silica and, except in surface waters, undersaturated with respect to calcite, skeletal material undergoes dissolution as it sinks through the water column. Since seawater is universally undersaturated with respect to opaline silica, most skeletal opal is dissolved within the water column and recycled back into the production of new skeletons. However, given sufficient productivity a fraction arrives at the sediment-water interface. Away from areas of significant detrital input, the occurrence of opal-rich sediments therefore mirrors the pattern of opal production in surface waters.

Dissolution processes also strongly influence the distribution of skeletal calcite in the pelagic oceans. The situation is more complex than that for opal, because the extent to which seawater is undersaturated with respect to calcite varies significantly, generally increasing with increasing depth (*e.g.* Edmond, 1974; Broecker & Peng, 1982). As the calcium content of seawater is nearly constant, Broecker & Peng (1982) simplified the expression for the degree of saturation to:

$$D = [CO_3^{2-}]_{measured}/[CO_3^{2-}]_{saturation}$$

Broecker & Peng (1982) mapped saturation horizons for each major ocean, below which calcite (or aragonite) should dissolve. The saturation horizon for calcite is $\sim$4.5 km in the Atlantic Ocean, 3.5 km in the Indian Ocean and 3 km in the North Pacific. Saturation horizons for aragonite are much shallower than those for calcite, around 3.5 km in the Atlantic and <1 km in the Indian and Pacific Oceans.

Water depth, therefore, exerts a primary influence on the distribution of both calcite and aragonite in the oceans (Fig. 11). Two key horizons are delineated: the *Calcite Compensation Depth* (CCD), below which sediments contain very little carbonate, and the *lysocline*, above which shells are preserved more or less intact. The CCD occurs at 4.5–5 km over much of the Atlantic, at $\sim$5 km in the tropical Indian Ocean and at 3–5 km in the Pacific. Consonant with its saturation horizons, the Aragonite Compensation Depth is much shallower than the CCD, ranging from 3 km in the western North

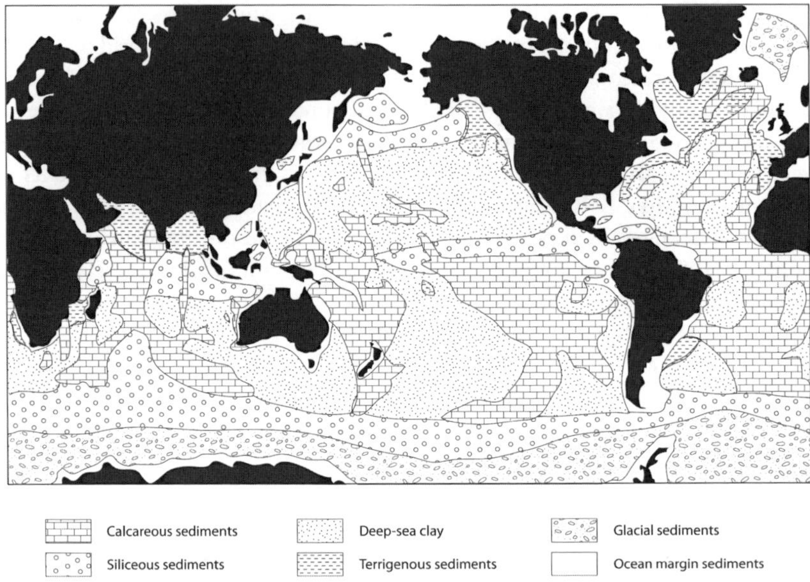

Fig. 11. Distribution of the main types of marine sediments (after Davies & Gorsline, 1976).

Atlantic to 1–2 km in the western tropical Pacific, and a few hundred metres in the tropical North Pacific (Berger, 1976; Chester, 1990).

4. Hydrogenous components

Having differentiated earlier between the hydrogenous and authigenic components of modern sediments, this discussion should properly be restricted to those sediment components that accrete directly from the water column. However, the geochemically most important hydrogenous components in sediments are iron and manganese oxides, which are reductively remobilized in sub-oxic sediment pore waters and then reprecipitated in surficial, oxic sediments. It is clear that sedimentary iron and manganese oxides can accrete from multiple sources and, for clarity, we shall deal with the mineralogy and geochemistry of both authigenic and hydrogenous manganese and iron oxides within this section.

Although iron and manganese oxides are supplied ubiquitously to sediments, they are only quantitatively important sediment phases in slowly accumulating, oxic, pelagic sediments. The best examples of these are the red clays which cover wide areas of the deep Pacific Ocean below the CCD. Oxide-rich sediments also occur along mid-ocean ridges, where iron-rich and manganese-rich hydrothermal solutions debouch to the ocean. Oxides are then distributed by ocean currents to ridge sediments and into adjacent basins; an excellent example of this is the Bauer Deep, to the east of the East Pacific Rise (*e.g.* Bostrom & Petersen, 1969; Dymond, 1981).

Iron and manganese oxides are often intimately associated in marine sediments and are classified together as marine *ferromanganese* deposits (see Glasby, 1977; Cronan, 1980, for extensive reviews). Ferromanganese deposits occur as stains and encrustations on pebbles and rocks, as nodules at the surface of slowly deposited pelagic sediments, as dispersed micronodules and fine particles within sediments, and as massive, hydrothermal precipitates. Manganese, iron and trace metals can therefore be supplied directly from the water column, which may commonly be derived from locallized hydrothermal solutions. They are also cycled within sediments as a result of the diagenetic remobilization of metal-bearing phases such as organic matter and oxides. The mineralogy and geochemistry of the oxide deposits significantly reflect the various modes by which metals are supplied to them. The mineralogy of marine ferromanganese deposits was originally described by Buser & Grutter (1956) and has been reviewed comprehensively by Burns & Burns (1977).

The dominant minerals in marine ferromanganese precipitates are amorphous iron oxyhydroxide and the manganese oxides birnessite and todorokite. Goethite and lepidocrocite are also found occasionally. Birnessite (δ-MnO_2) and todorokite have a double layer structure consisting of ordered sheets of MnO_2 with disordered layers of metal ions such as Mn^{2+}, Cu^{2+}, Ni^{2+} and Fe^{2+}, coordinated with O^{2-}, OH^- and H_2O. Significant concentrations of elements such as Mg, Ba, Cu, Ni and Ca can occur in the birnessite and todorokite crystal structures (*e.g.* Sherman & Peacock, 2010). Ferromanganese deposits which accrete directly from seawater and which are not influenced by hydrothermal activity comprise mainly δ-MnO_2 and FeOOH (Aplin & Cronan, 1985). They have Mn:Fe ratios of around unity and are enriched in Co compared to other trace metals such as Cu, Ni and Zn (Table 1). Cobalt is present as Co^{3+}, and it is proposed that the oxidation of Co^{2+} to Co^{3+} is catalysed by the MnO_2 surface.

The composition of sediment-hosted ferromanganese nodules is influenced strongly by the extent to which metals are accreted directly from the water column and from sediment pore waters. Nodules and dispersed sediment oxyhydroxides which are rich in δ-MnO_2 and FeOOH are chemically similar to encrustations and accrete mainly from

Table 1. Chemical composition of some oceanic Fe-Mn deposits (from Chester, 1990, reproduced with the permission of Springer).

Element	Element abundance in μg/g			
	Hydrogenous crust	**Oxic nodule**	**Sub-oxic nodule**	**Hydrothermal crust**
Mn	222,000	316,500	480,000	550,000
Fe	190,000	44,500	4900	2000
Co	1300	280	35	39
Ni	5500	10,100	4400	180
Cu	1480	4400	2000	50
Zn	750	2500	2200	2020
Mn:Fe	1.2	7.1	98	275
Mineralogy of Mn phase	δ–MnO_2	Todorokite, δ–MnO_2	Todorokite	Birnessite, todorokite
Growth rate (mm/10^6 y)	1–2	10–50	100–200	1000–2000

seawater. Under conditions of oxic diagenesis, nodules become enriched in Cu, Ni and Zn as a result of their release from decaying organic matter. Cu + Ni + Zn abundances of up to 3% are known (*e.g.* Cronan, 1980) in some nodules from the NE Pacific Ocean. These nodules also have Mn:Fe ratios of 2–7 and contain todorokite, implying a degree of reductive remobilization of primary Mn oxides and their subsequent reprecipitation within nodules. Under sub-oxic conditions, *e.g.* in hemipelagic sediments or lacustrine sediments, the remobilization of primary Mn oxides is extensive, yielding nodules with Mn:Fe ratios in excess of 10 and, compared to todorokite-bearing nodules formed under mainly oxic conditions, a depletion in transition metals such as Cu, Ni and Zn (Calvert & Price, 1977; Dymond *et al.*, 1984; Table 1).

The mineralogy and geochemistry of oxides deposited as encrustations on rocks close to hydrothermal vents have been described by, for example, Scott *et al.* (1974), Moore & Vogt (1976), Toth (1980) and Moorby *et al.* (1984). These deposits often have very high Mn:Fe ratios as a result of the incorporation of Fe into sulfide or silicate minerals closer to the exit point of the hydrothermal system. Compared to hydrogenous ferromanganese deposits, which grow at ~ 1 mm$/10^6$ y, hydrothermal deposits accrete rapidly. For example, Toth (1980) calculated mean accretion rates of 100–200 mm$/10^6$ y for Mn-rich crusts from the TAG area of the mid Atlantic Ridge. Despite the fact that they contain abundant todorokite, into which transition metals can readily be incorporated, the rapid growth rate of hydrothermal crusts appears to preclude the uptake of significant amounts of trace metals; compared to hydrogenous crusts, they are depleted in all trace metals (Table 1).

Recent data from deep-sea muds from the central North Pacific (Kato *et al.*, 2011) have shown that significant concentrations of rare-earth elements and yttrium are present in association with iron oxides and clay minerals. Such muds, therefore, represent a potential economic source of these elements.

5. Authigenic clay minerals

Clay mineral diagenesis begins within centimetres of the sediment–water interface and is a process which, in surface sediments, largely involves the formation of clay minerals from reactive components such as biogenic opal, degraded aluminosilicates and amorphous aluminium and iron oxyhydroxides (Foscolos & Powell, 1980; Aplin, 1993). These processes have been summarized by Aplin (1993) and this section borrows significantly from that work.

5.1. Mineralogical and chemical trends

The wide variety of authigenic clay minerals recognized in surface marine and lacustrine sediments are listed in Table 2 (in addition to references in Table 2, see Bischoff, 1972; Elderfield, 1976; Kastner, 1981; Odin & Matter, 1981; Cole & Shaw, 1983; McMurtry *et al.*, 1983; Odin, 1988a, 1990). Most of the minerals listed in Table 2 are described extensively in standard texts (Brindley & Brown, 1984; Newman, 1987; Chamley,

Table 2. Authigenic clay minerals in recent marine and lacustrine sediments (from Aplin, 1993, reproduced with the permission of Springer).

Group	Species	Model environment	Example	Ref
Smectite	Fe(II)/Mg-rich	Si, Mg-rich anoxic hydrothermal	Red Sea	[1]
Smectite	Fe(II)/Fe(III)-rich	Si, Fe-rich anoxic/oxic hydrothermal	Red Sea	[1],[2]
Smectite	Nontronite	Si, Fe-rich oxic hydrothermal	Galapagos Mounds Famous Area, Mid-Atlantic Ridge	[3] [4]
		Si, Fe-rich oxic lacustrine	Lake Chad	[5]
Smectite	Fe(III)-rich	Si, Fe-rich oxic pelagic	Bauer Deep	[6]
Smectite	Fe(III)/Al-rich	Si, Fe, Al-rich oxic pelagic	Central Pacific	[7],[8]
Smectite	Mg-rich	Si, Mg-rich evaporative	Lake Chad Salina Ometepac	[9],[10] [11]
?Smectite	"Glauconitic smectite"	Si-Fe-rich sub-oxic continental shelf edge	Gulf of Guinea	[12]
?Mica	"Glauconitic mica"	Si-Fe-rich sub-oxic continental shelf edge	Gulf of Guinea	[12]
Chlorite	Al-rich	Al-rich anoxic continental shelf	East China Sea	[13]
Kaolinite-serpentine	Odinite	Fe, Al-rich sub-oxic continental shelf	Senegalese continental shelf	[14]
?	Phyllite C	Si, Fe, Al-rich sub-oxic continental shelf	Senegalese continental shelf	[14]
Pyrophyllite-talc	Talc	Si, Mg-rich high temperature hydrothermal	Red Sea	[15]
Palygorskite-sepiolite	Palygorskite, sepiolite	Low/moderate temperature alteration of volcanic debris (?)	Atlantic fracture zones	[16]

References: [1] Cole (1988); [2] Badaut *et al.* (1985); [3] Corliss *et al.* (1978); [4] Hoffert *et al.* (1978); [5] Pédro *et al.* (1978); [6] Cole (1985); [7] Hein *et al.* (1979); [8] Aplin (1993); [9] Tardy *et al.* (1974); [10] Gac *et al.* (1977); [11] Hover *et al.* (1999); [12] Odin (1988b); [13] Mackin & Aller (1984a); [14] Odin & Masse (1988); [15] Zierenberg & Shanks (1983); [16] Bowles *et al.* (1971).

1989). Exceptions are the so-called green marine clays (glauconitic minerals, odinite and phyllite C), which have been the subject of extensive work by Odin and his coworkers (see Odin, 1988a). Odin has defined two marine facies, both of which are characterized by the occurrence of green grains, but green grains which have different mineralogical and chemical compositions. The glaucony facies is characterized by glauconitic minerals whilst the verdine facies is characterized by odinite and phyllite C. Glauconitic minerals exhibit a spectrum of chemical and physical properties between dioctahedral smectitic and micaceous (glauconite *sensu stricto*) end-members and may be thought of in the same way as mixed-layer illite-smectite. Odinite is a recently defined (ferric) iron magnesium, dioctahedral-trioctahedral 1:1 serpentine-like mineral with a 7 Å basal spacing (Odin, 1990). Previously, this mineral was known as "phyllite V" (Odin, 1985), but phyllite V is now thought to be a mixture of odinite, a (ferric) iron chlorite, a 7–14 Å mixed-layer clay and a pyrophyllite-like clay (Odin, 1990). The precise nature of phyllite C is enigmatic but it has properties which are intermediate between those of a smectite and a swelling chlorite (see Odin, 1988a for greater detail).

Until Odin's investigations, berthierine (a 7 Å trioctahedral serpentine mineral) was thought to be the green mineral which gives the verdine facies its name (Porrenga, 1967). However, Odin (1988a and references therein) suggests that berthierine is rarely found in recent sediments.

5.2. Geographical distribution of minerals

It is striking that each group or species of authigenic clay minerals occurs in specific geographical regions of the modern ocean (Table 2; Fig. 12). Odinite and phyllite C are characteristic of sediments accumulating between 10 and 60 m water depth on tropical continental shelves. Glauconitic minerals occur mainly between 60 and 500 m water depth at latitudes between 0 and 60°. Both glauconitic minerals and odinite occur as pellets, or replacing sand-sized minerals, or on hardgrounds implying that both form preferentially in environments that are at least periodically subjected to wave or current activity. Smectite is the major authigenic clay mineral in pelagic sediments, where hydrothermal solutions debouch into the ocean, and in lacustrine sediments (Table 2). Authigenic palygorskite and sepiolite typically occur in evaporative environments.

5.3. Chemistry

Representative chemical analyses of early diagenetic clay minerals are given in Table 3 and are displayed as a series of triangular diagrams in Figure 13. Specific minerals occupy particular regions of the seafloor. One example of this is smectite in sediments from the Pacific Ocean, where there is geographical uniformity of composition with clear species distributions (Tables 2 and 3). Variations in the Al and Fe contents of the smectites are especially striking and, based on the available data, appear to be related to the proximity to centres of hydrothermal activity. On the Galápagos spreading centre, where iron and silica-rich hydrothermal solutions bleed onto the ocean floor, Fe(III) smectite (nontronite) dominates (McMurtry *et al.*, 1983). In the Bauer Deep, where hydrothermally derived Fe oxides from the nearby East Pacific Rise are ponded, smectite is somewhat more aluminous, but is still dominated by ferric iron (Cole, 1985). In the Central North and South Pacific, which are unaffected by hydrothermal activity, smectite contains more Al than Fe (Hein *et al.*, 1979). All the smectites are dioctahedral and contain no ferrous iron.

Trioctahedral smectites are rather rare in recent sediments. Mg-rich varieties occur in areas where the Mg content of ambient waters are enhanced by evaporation (saline lakes and evaporative ocean margins; Tardy *et al.*, 1974; Hover *et al.*, 1999) or hydrothermal activity (Red Sea; Cole, 1988), whilst Fe(II)-rich smectites have been isolated from sediments underlying the anoxic brine pools in the Red Sea (Badaut *et al.*, 1985; Cole, 1988).

Glauconitic minerals can be distinguished from early diagenetic smectites by their slightly lower SiO_2 contents and the occurrence of small amounts of ferrous iron and high concentrations of potassium. Odinite and phyllite C are compositionally distinct from both early diagenetic smectites and glauconitic minerals. They are depleted in SiO_2, enriched in Al_2O_3 and contain very significant amounts of both MgO and FeO (Fig. 10).

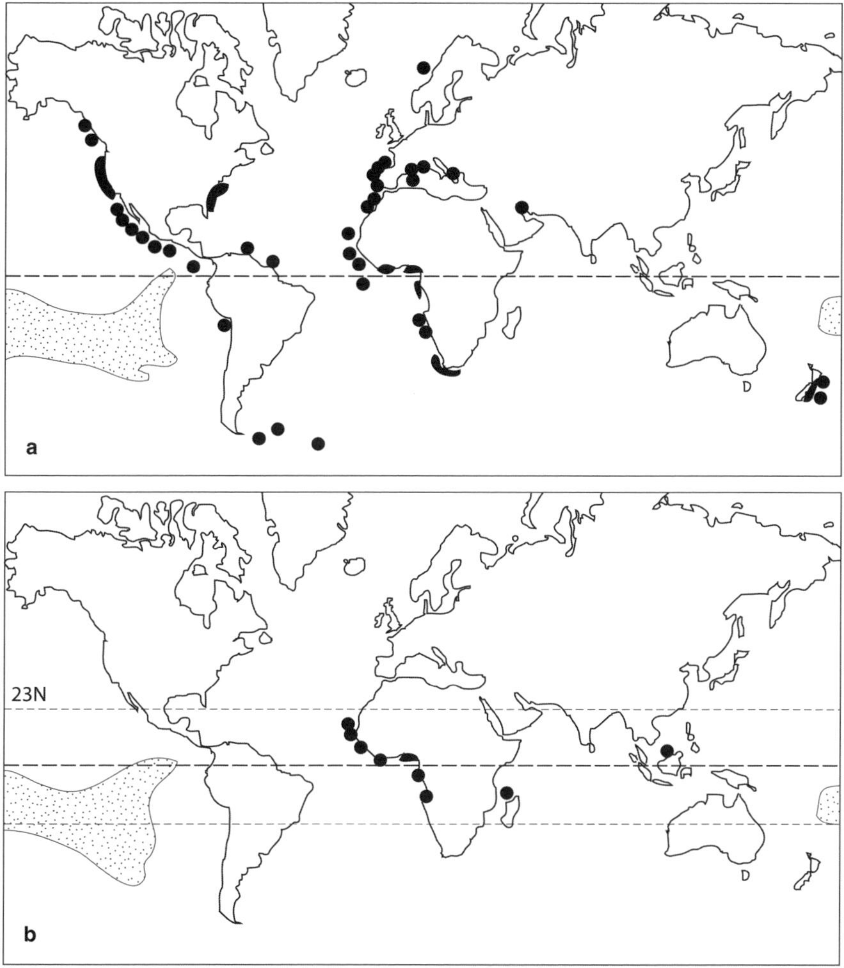

Fig. 12. Distribution of **(a)** authigenic glauconitic minerals (glaucony facies), **(b)** odinite (verdine facies) and smectite (speckled) in marine surface sediments (from Aplin, 1993, reproduced with the permission of Springer).

5.4. Controls on chemistry and mineralogy

Aplin (1993) has argued that a primary control on the type and chemical composition of early diagenetic clay minerals is the composition of unstable, often X-ray amorphous, phases supplied to the sediment. These phases are derived mainly from continental weathering (degraded aluminosilicates, Al-Fe oxyhydroxides) and hydrothermal activity at plate margins (silica, Fe oxyhydroxide), with additional reactants such as opal and organic matter supplied by biological activity within the water column. The proposed links between the formation of particular clay minerals and the relative

Table 3. Major oxide compositions of authigenic clay minerals in recent marine and lacustrine sediments (from Aplin, 1993, reproduced with the permission of Springer).

Group	Species	Example	SiO_2	Al_2O_3	Fe_2O_3	FeO	MgO	K_2O	Na_2O	CaO	Ref
Smectite	Fe(II)/Mg-rich	Red Sea	45.1	4.8	8.0	n.d.	22.9	0.0	3.4	0.8	[1]
Smectite	Fe(II)/Fe(III)-rich	Red Sea	43.2	2.4	33.0	n.d.	1.8	0.2	3.5	0.9	[1]
Smectite	Nontronite	Galapagos Mounds	51.6	0.1	26.5	n.d.	3.2	3.3	0.1	0.3	[2]
Smectite	Fe(III)-rich	Bauer Deep	53.3	5.4	19.2	n.d.	4.8	0.6	1.4	0.3	[3]
Smectite	Al/Fe(III)-rich	Central Pacific	49.7	15.2	10.9	n.d.	4.1	1.5	2.7	0.8	[4]
Smectite	Mg-rich	Lake Chad	57.1	5.9	2.8	n.d.	17.9	0.2	0.9	0.2	[5]
Smectite (?)	"Glauconitic smectite"	Gulf of Guinea	46.9	6.9	21.2	0.9	3.6	3.4	0.3	1.4	[6]
Minca (?)	"Glauconitic mica"	Galician Margin	43.4	6.0	21.3	1.5	3.5	8.2	0.2	0.7	[7]
Kaolinite-serpentine	Odinite	Senagalese shelf	34.3	9.2	21.4	6.5	13.0	0.5	0.2	0	[8]
?	Phyllite C	Senagalese shelf	41.9	9.9	17.4	4.1	11.2	0.5	0	0	[8]
Pyrophyllite-talc	Talc	Red Sea	49.5	3.1	0.2	3.9	25.9	0.1	2.2	0.2	[9]
Palygorskite-sepiolite	Palygorskite	Marianas volcanics	59	11	4	n.d.	9	1	n.d.	1	[10]
Palygorskite-sepiolite	Sepiolite	Mid-Atlantic Ridge	58.6	0.4	0.2	0.05	25.4	n.d.	n.d.	0.3	[11]

n.d. = not determined. Where there are no data from FeO, total iron is given as Fe_2O_3. All data normalized to 85% total oxides.

References: [1] Cole (1988); [2] McMurtry *et al.* (1983); [3] Cole (1985); [4] Aplin (1993); [5] Tardy *et al.* (1974); [6] Odin (1988b); [7] Odin & Lamboy (1988); [8] Odin & Masse (1988); [9] Zierenberg & Shanks (1983); [10] Desprairies (1982); [11] Hathaway & Sachs (1965).

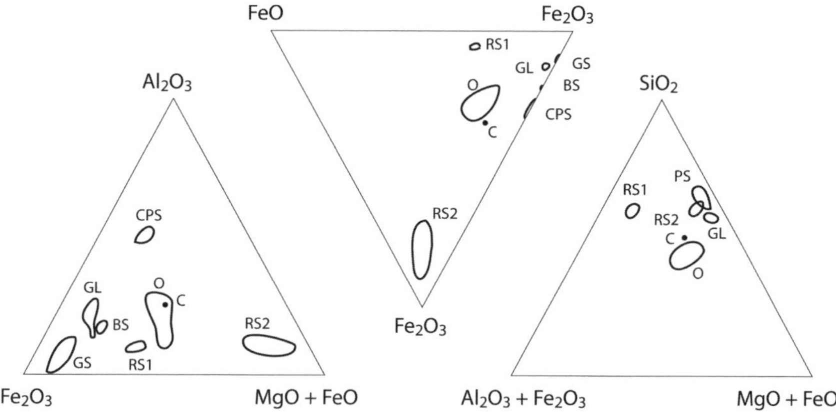

Fig. 13. Triangular diagrams displaying the known range of chemical composition of authigenic smectite, odinite, glauconitic minerals and phyllite C in marine surface sediments. Data from McMurtry *et al.* (1983); Hein *et al.* (1979); Cole (1985, 1988); Odin (1988a); Aplin, (1993). CPS = Central Pacific smectite; GL = glauconitic minerals; BS = Bauer Deep smectite; GS = Galápagos Mounds smectite; O = odinite; C = phyllite C; PS = Galápagos smectite + Bauer smectite + Central Pacific smectite; RS1 = Fe(II)/Mg-rich smectite, Red Sea; RS2 = Mg-rich smectite, Red Sea (from Aplin, 1993, reproduced with the permission of Springer).

rates of supply of the various poorly crystalline and dissolved reactant phases are summarized in Figure 14 (see also Table 2).

Smectites form in environments where reactive silica, either as opal or in hydrothermal solutions, is relatively abundant. The particular species of authigenic smectite depends on the composition of reactants supplied along with the silica. Smectites forming in the Atlantis II Deep in the Red Sea provide a good example of this. There, hot (>200°C), anoxic brines are discharged from the underlying basement and are ponded beneath oxygenated seawater in naturally occurring depressions on the seafloor (Degens & Ross, 1969). The brines contain high concentrations of silica, reduced iron and (often) magnesium. Petrographic, chemical and isotopic data show that distinct species of smectite form as the hydrothermal waters are cooled and oxidized by contact with overlying seawater (Zierenberg & Shanks, 1983; Cole, 1983, 1988; Badaut *et al.*, 1985). Trioctahedral, Mg-rich smectite (Fe(II)-bearing saponite) and talc form in anoxic, high temperature (>150°C) conditions close to hydrothermal vents. Removal of Mg from solution and partial oxidation of Fe(II) to Fe(III) leads to the formation of a mixed trioctahedral-dioctahedral Fe-rich smectite at temperatures between ~80 and 140°C. Finally, when solutions have cooled to ~80°C and iron occurs mainly as Fe(III), dioctahedral nontronite forms. The Red Sea is, in two respects, a rather unusual seafloor hydrothermal system. Firstly, the hydrothermal fluids often contain Mg; secondly, anoxic conditions occur above the sediment–water interface. At mid-ocean ridges, it is more typical for hydrothermal solutions to be depleted in Mg and to debouch directly into oxygenated seawater (Edmond *et al.*, 1982). These solutions are also greatly enriched in iron relative to aluminium (Michard *et al.*, 1984). In these

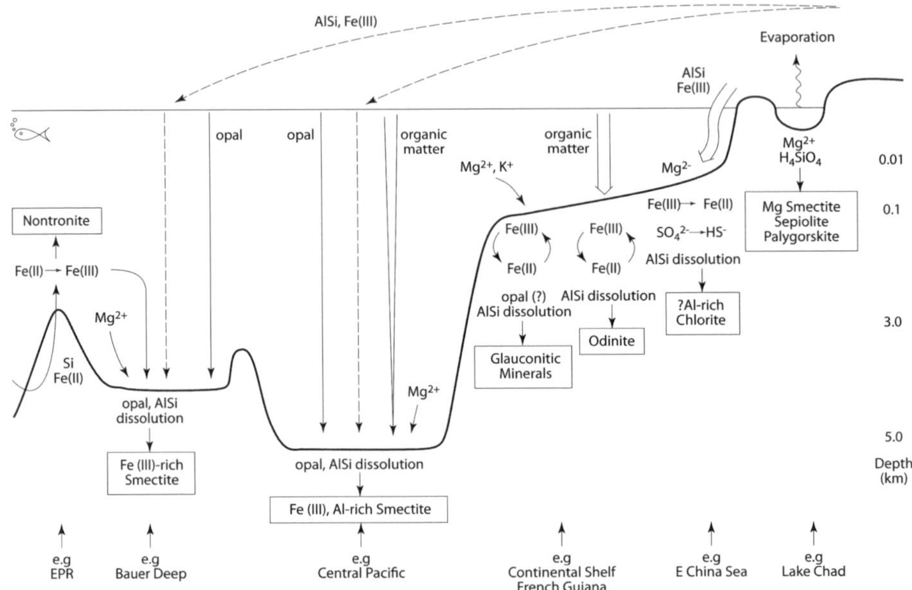

Fig. 14. Proposed links between (1) the supply of reactive materials (opal, degraded aluminosilicates (AlSi), Fe and Al oxyhydroxides, organic matter) and (2) the mineralogy of early diagenetic clays (after Aplin, 1993, reproduced with the permission of Springer).

systems, reduced iron is precipitated mainly as sulfides and Mg smectites are absent. Any iron that is not precipitated as sulfide is rapidly converted to oxyhydroxides and either combines with silica to form nontronite (as in the Galápagos hydrothermal mounds; Corliss *et al.*, 1978), or is dispersed by ocean currents. In the Pacific Ocean, hydrothermally derived Fe oxides occur in sediments accumulating along the East Pacific Rise and adjacent Bauer Deep. These deep-water sediments are oxic and, in addition to hydrothermally derived material, contain abundant opal and lesser amounts of terrestrially derived aluminosilicates. Authigenic Fe(III)-rich smectite is common and contains small but significant amounts of aluminium (Cole, 1985; Table 3). Within pelagic sediments in the Central Pacific, away from the influence of hydrothermal activity, an Al-Fe(III) smectite forms. Like the Bauer Deep sediments, these sediments are both oxic and opaline but contain relatively more terrestrially-derived aluminosilicates and fewer (if any) hydrothermally derived iron oxyhydroxides (Aplin & Cronan, 1985).

Terrestrially derived aluminosilicate debris, Fe and Al oxyhydroxides and organic matter are all supplied rapidly to nearshore sediments. In general, the authigenic clays forming in these sediments (odinite, phyllite C and perhaps, chlorite) are more aluminous than deep-sea smectites, reflecting the relatively greater supply of reactive alumina compared to silica. The formation of particular mineral species seems to be controlled by the chemical composition of the terrestrial input plus the redox status of the sediments (Fig. 12; Table 2). In sediments accumulating rapidly in the East China Sea, Mackin & Aller (1984a) suggested that an aluminous dioctahedral chlorite might be the main authigenic clay mineral. The detrital phases in these sediments are derived from rivers draining the Chinese mainland and are of average chemical maturity (Li *et al.*, 1984). Pore-water concentrations of dissolved Fe are low (Mackin & Aller, 1984a). In these sediments, where the supply of reactive Al is high relative to silica and iron, aluminous chlorite may be the dominant authigenic clay. The same logic can be extended to argue that in sulfidic, nearshore sediments, reduced Fe would be precipitated as iron sulfide with the result that aluminous clay minerals would form preferentially over iron-bearing minerals.

Aluminium and Fe oxyhydroxides are major weathering products in tropical climates and are the apparent feedstock for the formation of Al, Fe-rich odinite and phyllite C on shallow, tropical continental shelves. These minerals define a particular redox niche because they contain both ferrous and ferric iron. In surface sediments, organic matter is oxidized by bacterially catalysed reactions involving the sequential reduction of oxygen (oxic zone), Mn oxides, nitrate and Fe oxides (sub-oxic zone) and sulfate (anoxic zone; Froelich *et al.*, 1979). As they contain both ferrous and ferric iron, odinite and phyllite C must form close to the oxic/sub-oxic redox boundary. Furthermore, specific redox conditions must be maintained for long periods in order for the clay minerals to develop. Such conditions must be maintained by a delicate balance between the supply of organic matter, via biological productivity, and the supply of oxygen by current and wave activity. In many muddy nearshore sediments, the sub-oxic zone is thin and sulfate reduction is the dominant oxidation process. In these sediments Fe is rapidly stripped from pore-waters by reaction with sulfide. Odinite and phyllite C form in sediments in which a significant supply rate of organic matter is balanced by a supply of oxygen from current activity. Both mineral species form in micro-environments (pellets, intragranular cracks) in which there is restricted exchange of solutes with overlying seawater. These may be sub-oxic niches in an otherwise oxic environment. Both compositionally and geographically, glauconitic minerals are intermediate between Fe(III) smectite and odinite (Fig. 10; Tables 2 and 3). Compared to smectite, glauconitic minerals are somewhat depleted in Si and contain small but significant amounts of Fe(II). Compared to odinite, they are enriched in Si and depleted in Fe(II). The formation of glauconitic minerals is favoured in sediments which are (1) marginally or occasionally sub-oxic (moderate supply of organic matter) and (2) deposited away from point sources of Al-rich silicates and oxyhydroxides. Like odinite, glauconitic minerals often form within pellets and microfossil tests, which may be sub-oxic microenvironments in a generally oxic environment.

5.5. Role of aqueous phase

Despite the clear link between the compositions of reactive precursors and authigenic clay minerals, it is highly unlikely that the clays form *via* energetically unfavourable solid-state reactions. The reactions are much more likely to be mediated through the aqueous phase and, if so, this should be reflected in the composition of associated pore-waters. Evaluation of the role of the aqueous phase requires pore-water data for all the major clay-forming components (silica, aluminium, magnesium and iron) from sediments in which authigenic clay minerals have been chemically and mineralogically characterized. Unfortunately, there are no published examples, so that it is impossible to make definitive statements about the equilibrium state of the sediment–pore water system or the precise role of the aqueous phase in the formation of early diagenetic clay minerals. Profiles of dissolved silica have, however, been determined in a wide variety of marine sediments. All the profiles show similar features, with low concentrations in surface sediments increasing to constant values within 1 m of the sediment–water interface (Hurd, 1973; Mackin & Aller, 1984a, 1984b; Fig. 15). Maximum values vary from sediment to sediment but are not related to sediment type or depositional environment. Maximum values are also much greater than the equilibrium solubility of quartz, but much lower than the equilibrium solubility of amorphous silica (Lindsay, 1979). Silica activities are not buffered by quartz or opal, but perhaps by aluminosilicate phases. Unfortunately, there are no aluminium data from the pore-waters of any sediments where authigenic clays have been fully characterized. The most complete set of porewater aluminium data is probably that collected by Mackin & Aller (1984a, 1984b) from rapidly accumulating, nearshore and estuarine sediments. In the East China Sea, Mackin & Aller (1984a) found maximum Al concentrations in overlying seawater, declining downwards into the sediment pore-waters. Silica showed the opposite trend, increasing with depth into the sediment (Fig. 16). Mackin & Aller (1984a) interpreted these trends as indicating control of dissolved silica and aluminium by equilibrium with an authigenic clay mineral. In such rapidly accumulating sediments, authigenic phases are swamped by detrital phases and it is impossible to identify any authigenic phase. Statistical analysis of the pore-water data, based on the assumption of equilibrium between clay mineral and pore water, suggested that the clay might have been an Al-rich dioctahedral chlorite.

It is also possible that Mackin & Aller's (1984a) data reflect a steady state between the dissolution of reactants (degraded aluminosilicates) and the precipitation of products

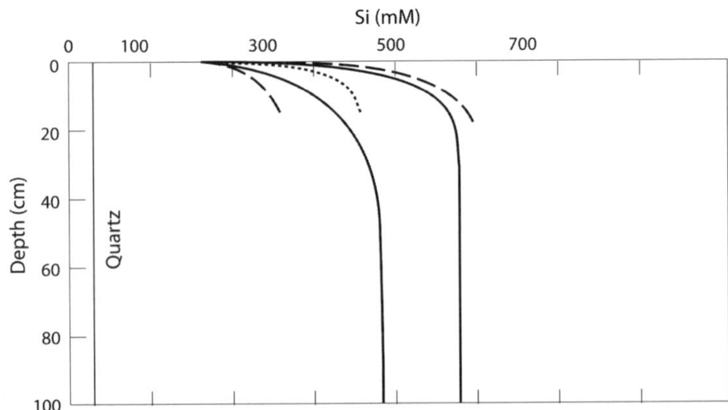

Fig. 15. Profiles of dissolved silica in recent sediments. Data from Hurd (1973); Mackin & Aller (1984a, 1984b). Solid line: pelagic sediments from the Central Pacific; dashed line: shelf muds from the East China Sea; dotted line: estuarine mud from Mud Bay, eastern USA.

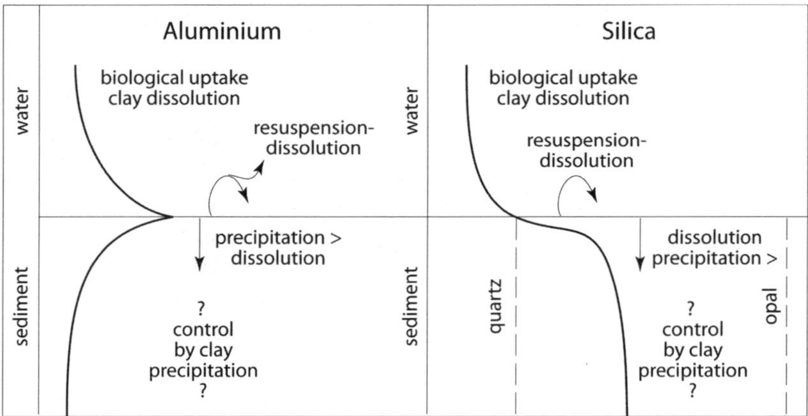

Fig. 16. The cycles of aluminium and silica in recent muds (based on data in Mackin & Aller, 1984a).

(authigenic clay). Enhanced concentrations of aluminium occur in waters overlying both nearshore and pelagic sediments, suggesting that aluminosilicates are being dissolved (Mackin & Aller, 1984a; Orians & Bruland, 1986). Dissolution may continue within the sediment column, but if the rate of precipitation is greater than the rate of dissolution, concentrations of dissolved Al will decline. In contrast, increased levels of dissolved silica in pore-waters suggest that the rate of dissolution of reactive Si-bearing phases (opal, degraded aluminosilicates) is initially greater than the rate of uptake of silica into authigenic clays (Fig. 16).

More recently, Michalopoulos & Aller (2004) documented that biogenic silica deposited in deltaic sediments of the Amazon River provides the silica that leads to the formation of K-Fe-rich aluminosilicates. These workers showed that in the past, standard leach methods previously used to determine the availability of reactive biogenic silica phases in such sediments, have under-represented the true concentrations, and as such their importance was overlooked. Pore water analysis indicates that these authigenic aluminosilicates form rapidly (within 1 y) in the upper 2 m of the sediment. These authigenic aluminosilicates have higher cation-to-Si ratios than detrital clays delivered by the Amazon River and, therefore, represent the reverse weathering process proposed by Mackenzie & Garrels (1966). Such deltaic sediments, which are major sediment deposits in modern oceans, therefore represent a sedimentary sink of Si and cations and probably represent an overlooked part of the oceanic budgets for these elements.

In contrast to the reverse weathering reaction noted in sediments off the Amazon, Wallmann *et al.* (2008) documented the occurrence of significant silicate weathering within the methanogenic zone underlying the sulfate reduction zone in sediments off Sakhalin Island in the Pacific. Here, although the presence of reverse weathering, and associated CO_2 release, was noted in surface sediments, within the methanogenic sediments silicate weathering resulted in an order of magnitude greater consumption of CO_2. These workers concluded that marine silicate weathering within anoxic sediments is as important a CO_2 sink as terrestrial weathering.

6. Early diagenetic redox reactions

6.1. Supply and remineralization of organic matter: the 'zonal scheme'

Organic carbon is deposited in sediments, along with a range of dissolved and solid-phase oxidants: oxygen, nitrate, sulfate, and Fe and Mn oxides. The inherent energy of the system is exploited by a diverse population of heterotrophic microorganisms, which utilize the full range of electron acceptors to oxidize organic carbon. A large literature has explored the occurrence, rates and relative importance of many of these reactions in a wide range of marine and non-marine sediments (*e.g.* Jørgensen, 1982; Bender & Heggie, 1984; Iversen & Jørgensen, 1985; Henrichs & Reeburgh, 1987; Canfield, 1989a; Canfield *et al.*, 1993a, 1993b; Thamdrup *et al.*, 1994), and there is an equally large body of work describing the minerals (carbonates, Fe sulfides, phosphates, silicates) which characteristically form as a result of early diagenetic redox reactions (*e.g.* Berner, 1970, 1984; Goldhaber & Kaplan, 1974; Irwin *et al.*, 1977; Kastner, 1984; Curtis *et al.*, 1986; Canfield & Raiswell, 1991a, 1991b; Jarvis *et al.*, 1994; Taylor & Curtis, 1995; Taylor & Macquaker, 2000, 2011).

In principle, heterotrophic organisms utilize terminal electron acceptors to oxidize organic matter in a sequence which relates to the energy released per mole of carbon oxidized. Oxidants should be utilized in the following sequence: oxygen, nitrate, manganese oxides, iron oxides and sulfate, according to the following reactions (Froelich *et al.*, 1979):

Oxygen:

$$(CH_2O)_{106}(NH_3)16(H_3PO_4) + 138O_2$$
$$\longrightarrow 106CO_2 + 16NO_3^- + H^3PO_4 + 122H_2O + 16H^+$$

Nitrate:

$$(CH_2O)_{106}(NH_3)16(H_3PO_4) + 84.8HNO_3$$
$$\longrightarrow 16NH_3 + 106CO_2 + H_3PO_4 + 148.4H_2O$$

Mn oxide:

$$(CH_2O)_{106}(NH_3)16(H_3PO_4) + 236MnO_2 + 472H^+$$
$$\longrightarrow 236Mn^{2+} + 106CO_2 + 8N_2 + H_3PO_4 + 336H_2O$$

Fe oxide:

$$(CH_2O)_{106}(NH_3)16(H_3PO_4) + 212Fe_2O_3 + 848H^+$$
$$\longrightarrow 424Fe^{2+} + 106CO_2 + 16NH_3 + H_3PO_4 + 530H_2O$$

Sulfate:

$$(CH_2O)_{106}(NH_3)16(H_3PO_4) + 53SO_4^{2-}$$
$$\longrightarrow 53H_2S + 106HCO_3^- + 16NH_3 + H_3PO_4$$

Following sulfate reduction, biogenic methane can be formed by fermentation reactions and by CO_2 reduction:

$$2(CH_2O)_{106}(NH_3)16(H_3PO_4) \longrightarrow 106CO_2 + 106CH_4 + 32NH_3 + 2H_3PO_4$$
$$CO_2 + 4H_2 \longrightarrow CH_4 + 2H_2O$$

The above simple set of reactions undoubtedly provides a very useful framework with which to understand changes in pore-water chemistry and the formation of diagenetic minerals in modern sediments. Furthermore, studies of the underlying microbial processes support its general applicability, and the fact that the sequence is linked to microbial competition for available energy. For example, Lovley & Phillips (1987) have shown that manganese- and iron-reducing bacteria competitively inhibit sulfate reduction when highly reactive Mn and Fe oxides are present; Lovley & Phillips (1988) and Myers & Nealson (1988) have shown that Mn reduction can partly inhibit Fe reduction. However, in many sediments, especially rapidly deposited and reworked nearshore sediments, the zones are not sharply defined (*e.g.* Sørensen & Jørgensen, 1987) and are complicated by the occurrence of chemically distinct microenvironments; for example, within burrows or organic aggregates (*e.g.* Aller, 1977; Canfield & Raiswell, 1991b; Raiswell *et al.*, 1993; Briggs *et al.*, 1996). Especially in nearshore sediments, which comprise the bulk of globally deposited sediment, a more complex series of redox reactions exists which does not, in many cases, define mutually exclusive zones of sediments. Although the overall driving force for the reactions is the initial coexistence of organic carbon and mineral oxidants, secondary reactions involving the interaction of oxidized and reduced manganese, iron, carbon and sulfur species are also important and can add significant complexity to the simple scheme presented above (*e.g.* Canfield *et al.*, 1993a; Thamdrup *et al.*, 1994). The many potential C-S-Fe-Mn-N-O redox reactions are outlined in Figure 17. Wherever the free energy of a reaction is negative, it is likely that the reaction will proceed, chemically and/or catalysed by microbes, in sediments where the reactants are in close physical proximity.

The relative importance of individual organic carbon mineralization pathways within sediments is primarily controlled by the relative rates of supply of carbon and oxidants, with an additional influence exerted by sedimentation rate. The rate at which organic matter is supplied to sediments depends on the primary productivity of overlying surface waters and, because much of the organic matter is recycled within the water column, on water depth. In the oceans, rates of primary production are highest on continental margins and in areas of upwelling. The rate of supply of organic matter to sediments is thus greatest on continental margins and is several orders of magnitude lower in pelagic sediments within the oceans-central gyres.

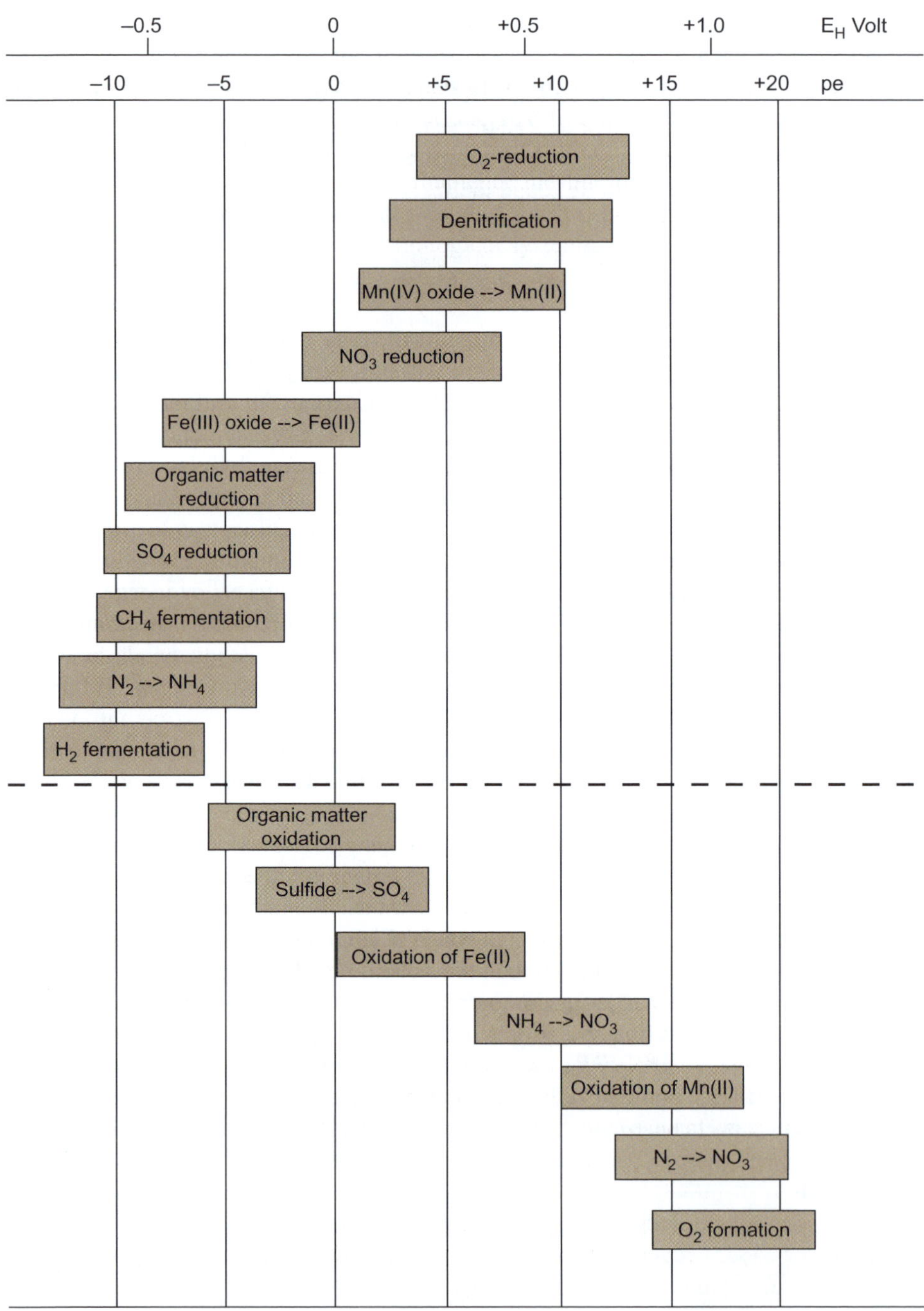

Fig. 17. Reduction and oxidation reactions which may occur in modern sediments (from data in Stumm & Morgan, 1981).

In most (but not all) marine sediments, the two most important carbon mineralization pathways are oxic respiration and sulfate reduction. The rates of both oxic respiration and of sulfate reduction increase with increasing sedimentation rate, reflecting primarily the greater supply of metabolizable organic matter to more rapidly deposited, nearshore sediments (Toth & Lerman, 1977; Berner, 1978; Canfield, 1989a; Fig. 18). Canfield (1989a) compiled data for sulfate reduction rates and oxic respiration rates for a wide range of rates of marine sedimentation and sedimentary environments, including several euxinic sites (sites where there is dissolved sulfide in the water column). Canfield (1989a) concluded that, in nearshore sediments, approximately equal amounts of organic matter were oxidized by sulfate reduction and oxic respiration. As sedimentation rates decrease, oxic respiration becomes progressively more important and, in

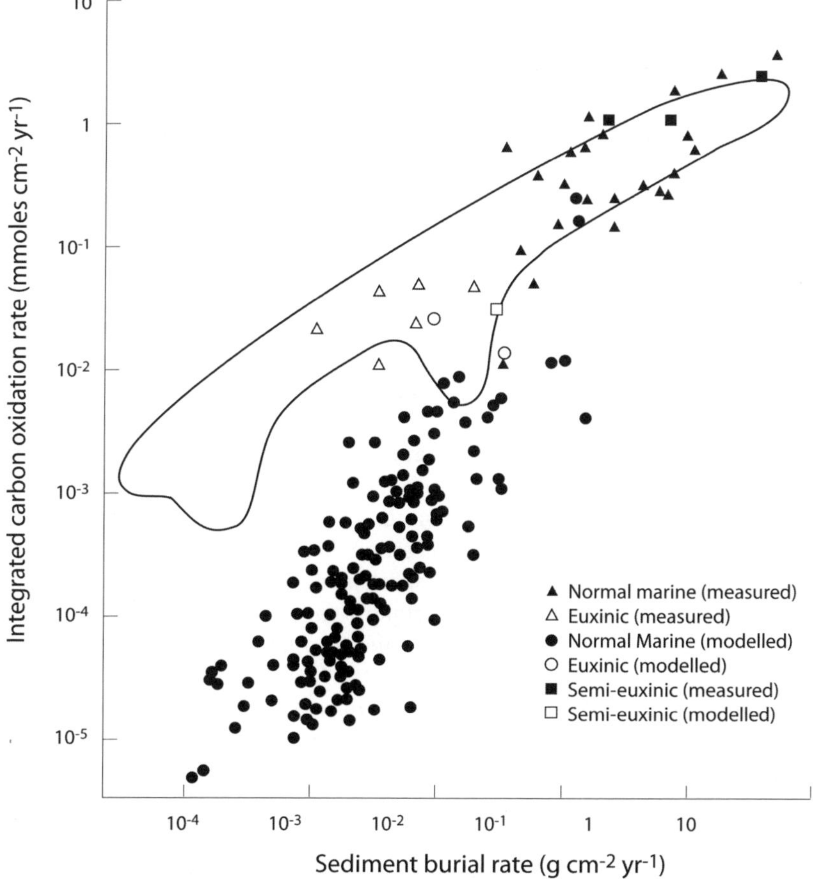

Fig. 18. Depth integrated organic carbon oxidation rates for modern sediments, plotted against sediment burial rate. Rates of oxidation by sulfate reduction are plotted as individual points. Rates of oxidation by oxic respiration occur within the outlined area (reproduced from Canfield, 1989a with the permission of Elsevier).

deep-sea sediments, 100–1000 times more organic carbon is oxidized by oxic respiration than by sulfate reduction (Fig. 18).

The relative importance of oxic respiration is greater in slowly deposited sediments because the rate of supply of metabolizable organic matter to these sediments is less and because oxygen can penetrate, by diffusion or biologically-driven mixing of sediment, to greater depths. Lyle (1983) estimated the thickness of the surface oxic layer in eastern Pacific sediments by noting the depth of a colour change from red-brown (Fe oxide bearing) to grey-green (no Fe oxide). He showed that the thickness of the brown layer decreased from <2 cm on the California/Central American/Peruvian continental margin to >100 cm in low-productivity, pelagic regimes. In extreme cases, dissolved oxygen may penetrate metres into the sediment; *e.g.* Wilson *et al.* (1985) showed that oxygen was present throughout a 2 m core from the NE Atlantic which had an accumulation rate of $\sim$0.4 cm/10^3 y. The classic zonal scheme proposed by Froelich *et al.* (1979) is most clearly defined in hemipelagic, marine sediments, where the main redox reactions involve the dissimilatory oxidation of carbon by oxygen, nitrate, Mn and Fe oxides. Each zone is defined by characteristic changes in the composition of interstitial waters (Fig. 19), reflecting the reactions outlined above. Remobilization of disseminated Mn oxides in the Mn reduction zone can lead to the deposition of an Mn oxide spike at the oxygen/Mn^{2+} interface. With continuing burial, the freshly precipitated Mn oxides will be reduced for a second time, diffusing upwards to, once again, reprecipitate at the O_2/Mn^{2+} interface. Similar effects have been noted in a number of deep-sea sediments close to horizons where shifts in the organic carbon of sediments represent changes in depositional conditions. These occur, for example, close to the glacial-Holocene transition zone and where relatively organic-rich turbidites have been instantaneously emplaced into organic-lean pelagic sediments (Wilson *et al.*, 1986; Wallace *et al.*, 1988; Rosenthal *et al.*, 1995). In these cases, a progressive oxidation front migrates down into the more organic-rich sediments, its depth approximately fixed by the balance of downwards-penetrating oxidant and upwards migrating Mn^{2+} and Fe^{2+}. This specific steady state, reflecting the balance of dissolution, migration and

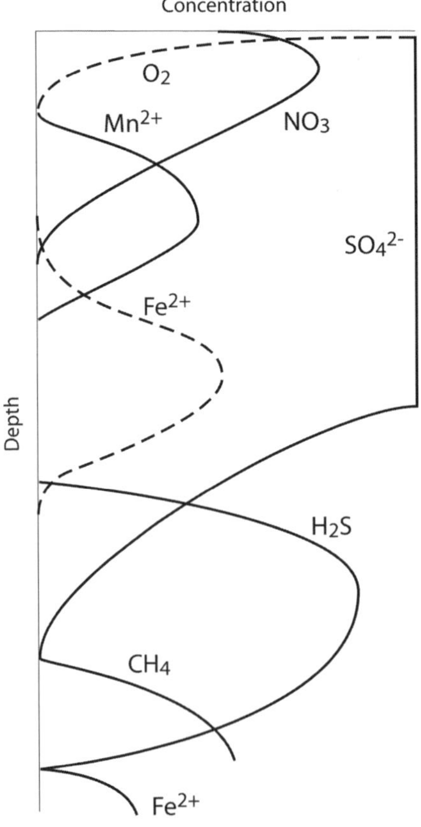

Fig. 19. Schematic representation of the sequence of reduction processes in modern sediments as reflected by the composition of interstitial water.

precipitation of manganese, has been modeled by Froelich *et al.* (1979) and Burdidge & Gieskes (1983). In general, this type of advection-diffusion-reaction model has been used extensively to describe many early diagenetic reactions in modern sediments, not least those involving the remineralization of organic carbon (*e.g.* Berner, 1980; Wang & Van Cappellen, 1996; Van Cappellen & Wang, 1996).

6.2. Carbon remineralization and secondary redox reactions

Most marine sediment is deposited rapidly on continental shelves and margins. In these settings, oxygen does not penetrate more than a few centimetres into the sediment column. Diagenetic zones are highly compressed towards the sediment–water interface and less than half (and as little as a few percent) of organic carbon mineralization occurs *via* oxic respiration (*e.g.* Jørgensen, 1982; Canfield, 1989a; Canfield *et al.*, 1993a, 1993b). Much of the carbon remineralization is driven by sulfate reduction, although in metal oxide-rich sediments, particularly where there is intense recycling of oxides in surface sediments, dissimilatory reduction of Fe and Mn oxides may be a quantitatively important route by which organic carbon is remineralized (Aller *et al.*, 1986; Sørensen & Jørgensen, 1987; Aller, 1990; Canfield *et al.*, 1993a, 1993b; Thamdrup *et al.*, 1994; Aller & Blair, 2006). Nitrate reduction is quantitatively unimportant in most cases.

Compression of the classic diagenetic zones in nearshore sediments results in the juxtaposition of reactive, oxidized and reduced Fe, Mn and S species. Rapid redox reactions which have been identified in recent sediments include the oxidation of H_2S by oxidized Fe and Mn, the oxidation of Fe^{2+} by oxidized Mn and the oxidation of pyrite by oxidized Fe (Pyzik & Sommer, 1981; Postma, 1985; Burdidge & Nealson, 1986; Lovley & Phillips, 1988; Thamdrup *et al.*, 1994):

$$2FeOOH + H_2S \longrightarrow 2Fe^{2+} + S^0 + 4OH^-$$

$$3H_2S + 2FeOOH \longrightarrow S^0 + 2FeS + 4H_2O$$

$$MnO_2 + H_2S \longrightarrow Mn^{2+} + S^0 + 2OH^-$$

$$MnO_2 + 2Fe^{2+} + 2H_2O \longrightarrow Mn^{2+} + 2FeOOH + 2H^+$$

$$FeS_2 + 14FeOOH + 6H_2O \longrightarrow 15Fe^{2+} + 2SO_4^{2-} + 26OH^-$$

$$4Fe^{2+} + O_2 + 6H_2O \longrightarrow 4FeOOH + 8H^+$$

In recent years, techniques have been developed which have allowed a much more precise quantification of these reactions in modern sediments. For example, microsensors and other fine-scale sampling techniques have considerably enhanced the spatial resolution with which pore water profiles can be measured (*e.g.* Revsbech & Jørgensen, 1986; Reimers, 1987; Davison *et al.*, 1991; Krom *et al.*, 1994; Brendel & Luther, 1995; Zhang *et al.*, 1995). Incubation experiments and the use of benthic chambers deployed on the sediment surface, provide data on the rate and spatial distribution of reactions, and on the rate of chemical exchange across the sediment–water interface

(Aller & Mackin, 1989; Jahnke & Christiansen, 1989; Canfield *et al.*, 1993a, 1993b). Some of the best data sets have been collected on Danish coastal marine sediments and serve to illustrate the quantitative importance of redox reactions specifically involving manganese, iron and sulfur and only indirectly involving organic carbon through the production of H_2S by sulfate reduction (Canfield *et al.*, 1993a, 1993b; Thamdrup *et al.*, 1994). Thamdrup *et al.* (1994) showed that below a $1-5$ mm thick oxic zone, a zone of net Mn reduction extended to $1-2$ cm depth, with Fe reduction occurring to a depth of $4-6$ cm. Most of the H_2S produced from sulfate reduction precipitated as Fe sulfides and S^0 by reaction with Fe oxides. However, much of the reduced sulfur was reoxidized; only 15% of the precipitated sulfide was buried permanently, and most of the reoxidation of reduced sulfide occurred within 1 cm of the sediment–water interface. All of the estimated Mn reduction could be accounted for by reoxidation of reduced sulfur and iron, and partial oxidation of H_2S to S^0 and pyrite accounted for about two thirds of the estimated Fe reduction. The remaining iron reduction was coupled to complete oxidation of reduced sulfur or carbon mineralization.

Canfield *et al.* (1993a, 1993b) studied three sites at water depths of 190 m, 380 m and 695 m in the eastern Skagerrak. The style of diagenesis was distinct at each site. At the deepest site, surface sediments were enriched in Mn to 3.5 wt.% and Mn reduction was the only important anaerobic carbon oxidation process in the upper 10 cm of the sediment. At the other two sites, carbon oxidation coupled to Fe reduction was at least as important as was sulfate reduction, and most of the Mn reduction was thought to be related to the oxidation of Fe monosulfides, not dissimilatory reduction. The high rates of oxide reduction were driven by rapid recycling of Mn and Fe. Canfield *et al.* (1993b) calculated that the residence time of Fe and Mn oxides, with respect to reduction, was $70-250$ days, requiring that on average, a molecule of Fe or Mn be oxidized and recycled between 100 and 300 times before its final burial into the sediment. Dissolved Mn was completely adsorbed onto fully oxidized Mn oxides until the oxidation level of the oxides was ~ 3.8, inhibiting diffusive loss of Mn to the overlying water column.

The Danish studies illustrate very well the suite of early diagenetic redox reactions which are ultimately driven by organic carbon but which do not directly involve organic carbon. They also show the dynamic nature of nearshore sediments and the extent to which redox sensitive elements such as Mn and Fe, and presumably any metals associated with the Mn and Fe oxides, are recycled before their eventual burial into the sediment.

Two other systems in which an extensive, interdisciplinary data set has shown the dominance of metal oxide reduction are the mobile mudbelt systems along the Amazon–Guianas shelf, South America, and the Papua shelf (Aller *et al.* 2004; Aller & Blair, 2006). In the former, the mud-dominated sediments are derived from the Amazon, which discharges an estimated 1200×10^6 tonnes of sediment annually; on the Papua shelf on the south coast of Papua New Guinea, rivers annually supply 365×10^6 tonnes of sediment. In both cases, large supplies of sediment result in mud-dominated sediment being deposited as mobile mudbanks that are episodically reworked, transported along shore and redeposited. Extensive mobilization and the reworking of sediment between the seabed and oxygenated overlying water, coupled with the input of tropically-derived iron oxides, results in the dominance of iron reduction during early diagenesis within the sediments. With the exception of local environments where sediments have been stabilized in mangrove swamps, the contribution of sulfate reduction to organic carbon oxidation is considered to be minor. These systems also play a major role in iron cycling and supply of iron to the oceans. The re-oxidation of Fe^{2+} during sediment reworking and re-suspension generates fine particulate iron oxides that can be transported off-shelf into deeper water (see Raiswell, 2011). Iron isotope analysis of iron oxides within both shallow marine and deep water sediment systems suggests that this phenomenon, commonly termed the *benthic iron shuttle,* is a major pathway by which iron is supplied to the oceans (Severmann *et al.,* 2008; Homoky *et al.,* 2009). The dominance of iron reduction results in high concentrations of Fe^{2+} in pore waters which, in the absence of sulfide, can lead to the precipitation of non-sulfide iron mineral phases, including iron-rich clays such as berthierine, and carbonate minerals such as siderite. It has been recognized for some time that the precipitation of iron-rich clays takes place within the Amazon-Guianas shelf sediment (Aller *et al.,* 1986) and similar conditions have been postulated to result in iron-rich clays in other modern-day settings (Ku & Walter, 2003).

Hand in hand with the accumulation of high-quality datasets quantifying early diagenetic reactions involving C, O, N, S, Fe and Mn, computer-based multicomponent reactive transport models have been developed to explain and predict the field observations. One of the most important of the models is that developed by Van Cappellen & Wang (1996) and Wang & Van Cappellen (1996), which is designed to account for reaction couplings among the elements C, O, N, S, Fe and Mn. The model includes descriptions of chemical species and reactions, reaction kinetics and both diffusive and advective transport. Wang and Van Cappellen used Canfield *et al.*'s (1993a, 1993b) data sets to test their model. The model accurately matches Canfield *et al.*'s observations and lends support to their original interpretations. Aerobic respiration accounted for 22–46% of carbon oxidation, with the rest of the oxygen being used to oxidize secondary reduced species, especially reduced Fe and Mn. The model suggested at the two shallower sites, 75 and 97% of the reduction of Mn oxide was coupled to Fe^{2+} oxidation, with most of the Fe oxides being used by bacteria to oxidize organic carbon. At the deepest site, dissimilatory reduction was the main dissolution pathway for Mn oxides. The impressive agreement between data and model indicates the future potential of mixing data and models in order to develop a truly quantitative description of the early diagenetic C-S-O-N-Mn-Fe cycle.

In systems that contain negligible iron and manganese oxides (*e.g.* tropical carbonate-rich sediments), the oxidation of H_2S is not linked to the reduction of metal oxides but to free oxygen which may diffuse from the overlying water column or be mixed into pore-waters by bioturbation:

$$H_2S + 2O_2 \longrightarrow SO_4^{2-} + 2H^+$$

In contrast to metal oxide reduction, this is an acid-producing reaction. Walter & Burton (1990) and Ku *et al.* (1999) showed that sulfate reduction within tropical carbonates on the Florida Platform produces free hydrogen sulfide which is oxidized by oxygen introduced into the sediment by bioturbation, thereby increasing porewater acidity. These researchers documented that the acid porewaters dissolved calcium carbonate grains within the sediment. From these observations they calculated that up to 50% of the original calcium carbonate deposited in the sediment was dissolved, with important consequences for the carbon budget in such systems. Recently, Perry & Taylor (2006) & Taylor *et al.* (2007) showed that in modern tropical carbonate sediments receiving a supply of terrestrial detrital iron, the presence of iron-inhibited free hydrogen sulfide generation and, as such, no carbonate grain dissolution took place. This implies that increasing input of land-derived iron minerals, possibly as a result of changes in land use and climate, is likely to affect carbon budgets in tropical sediments.

7. Pyrite formation

The most obvious consequence of the early diagenetic reduction of sulfate and ferric iron is the formation of pyrite, which is the most common sulfide mineral in clay-rich marine sediments and typically comprises 0.5–2 wt.% of nearshore sediments, and up to 5–10 wt.% in some organic-rich sediments. The main facets of pyrite formation have been described by Berner (1970, 1984) and the subject has been well reviewed by Canfield & Raiswell (1991b) and Rickard and Luther (2007), who provide excellent descriptions of how variations in the balance of iron and sulfate reduction affect pore-water geochemistry and give rise to the various textural forms of sedimentary pyrite. Pyrite does not generally form directly from the reaction of sulfide and Fe oxides but *via* intermediate iron sulfides, of which mackinawite and greigite are the most important. One set of reactions describing pyrite formation is (Berner, 1970, 1984; Fig. 20):

$$18CH_2O + 9SO_4^{2-} \longrightarrow 18HCO_3^- + 9H_2S$$

$$6FeOOH + 9H_2S \longrightarrow 6FeS + 3S^0 + 12H_2O$$

$$3FeS + S^0 \longrightarrow Fe_3S_4$$

$$Fe_3S_4 + 2S^0 \longrightarrow 3FeS_2$$

Overall: $18CH_2O + 9SO_4^{2-} + 6FeOOH \longrightarrow 18HCO_3^- + 3FeS + 3FeS_2 + 12H_2O$

In this sequence, sulfide reacts with Fe oxyhydroxides to form an iron monosulfide (Goldhaber & Kaplan, 1974; Lennie & Vaughan, 1996). Elemental sulfur is also produced which, probably as coexisting dissolved polysulfide, oxidizes monosulfide to greigite and then to pyrite (Berner, 1970; Sweeney & Kaplan, 1973; Rickard, 1975; Luther, 1991). Greigite is a cubic (spinel structure) iron sulfide with the composition Fe_3S_4 and contains mixed valence iron: $Fe^{3+}(Fe^{3+}, Fe^{2+})(4S^{2-})$ (Vaughan & Ridout, 1971). As the ratio of pyrite to monosulfide in modern and ancient marine sediments is typically >10, the reaction sequence given above cannot be a complete description of pyrite formation. Specifically, it does not produce sufficient oxidant (elemental sulfur or polysulfides) to completely convert FeS to FeS_2. In some cases, *e.g.* where bioturbation draws

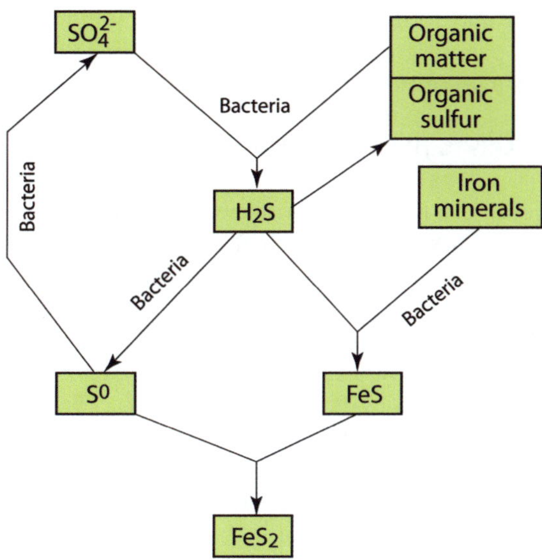

Fig. 20. Schematic representation of the early diagenetic sulfur cycle, stressing the general route by which pyrite forms.

molecular oxygen into the sulfate reduction zone, additional elemental sulfur and poly-sulfides could be derived from the oxidation of H_2S by molecular oxygen. Rickard (1997) and Rickard & Luther (1997) have shown that H_2S can oxidize FeS to FeS_2:

$$FeS + H_2S \longrightarrow FeS_2 + H_2$$

This reaction might explain why FeS/FeS_2 ratios are often (but not always) greater in lacustrine and brackish water sediments compared to fully marine sediments (Berner *et al.*, 1979; Davison *et al.*, 1985; Boesen & Postma, 1988). In sulfate-depleted lacustrine waters, the importance of iron reduction will be relatively greater than that of sulfate reduction and insufficient H_2S may be produced to allow complete conversion of FeS to FeS_2.

Except in euxinic settings (bottom waters containing free sulfide), all sedimentary pyrite is diagenetic. Ultimately, the extent to which detrital Fe minerals may be converted to pyrite is controlled by the relative rates at which: (1) sulfate is reduced to form sulfide, and (2) detrital Fe minerals are supplied to the sediment. However, two other factors are highly influential: (1) the rates at which the wide range of detrital iron minerals are sulfidized compared to the timespan over which sulfate reduction occurs (Canfield *et al.*, 1992), and (2) the extent to which, rather than being preserved, iron sulfides are reoxidized in surface sediments by Fe oxides or molecular oxygen (*e.g.* Thamdrup *et al.*, 1994). The proportion of iron occurring as pyrite in a sediment is known as the "Degree of Pyritisation" (DOP_T) (Berner, 1970):

$$DOP_T = \text{Pyrite Fe}/\text{Total Fe}$$

As iron monosulfide may form a significant fraction of sulfidized iron in modern sediments, an analogous parameter, the Degree of Sulfidation (the proportion of sulfur occurring as pyrite in a sediment; DOS_T) is also defined as:

$$DOS_T = [Pyrite\ Fe + Monosulfide\ Fe]/Total\ Fe$$

DOP_T rarely exceeds 0.6 in modern or ancient sediments, even in highly organic-rich sediments or sediments with sulfide-rich interstitial waters (*e.g.* Berner, 1970; Raiswell *et al.*, 1988; Mossmann *et al.*, 1991; Macquaker *et al.*, 1997). Sulfidation kinetics of detrital Fe phases thus play a crucial role in defining the real potential for early diagenetic pyrite formation. Table 4 gives the half-lives of a range of typical detrital Fe minerals, with respect to their sulfidization (Canfield *et al.*, 1992; Raiswell & Canfield, 1996). This work was extended by Poulton *et al.* (2004). The minerals fall into three groups: highly reactive minerals (ferrihydrite, lepidocrocite, goethite, hematite) with half-lives of less than a year; moderately reactive minerals (magnetite, 'reactive' silicates) with half-lives of $\sim 10^2$ years, and poorly reactive minerals (sheet and framework silicates, ilmenite) with half-lives of $>2 \times 10^6$ years. Only rarely, for example on the Peru–Chile Margin, does sulfide persist in pore-waters for millions of years, in which case some sheet silicates become sulfidized (Raiswell & Canfield, 1996). The minerals that are potentially pyritized under the most extreme diagenetic conditions are all readily soluble in concentrated HCl so that a more meaningful quantification of DOP is:

$$DOP_H = Pyrite\ Fe/[Pyrite\ Fe + HCl\ soluble\ Fe] \quad (Berner,\ 1970).$$

In most nearshore and rapidly deposited sediments, sulfate reduction persists for around $10^2 - 10^4$ years so that pyrite formation is commonly limited by the supply of iron oxides and oxyhydroxides. For example, Canfield (1989b) showed that $\sim 80\%$ of the pyrite formed in the FOAM (Friends of Anoxic Mud) field-sampling site in Long Island Sound was contributed by iron oxides. By assuming that pyrite formation is limited by the availability of reactive Fe, Canfield (1988) calculated the *maximum* burial rates of pyrite for a range of sediments and compared them to measured or

Table 4. Rate constants and half-lives of sedimentary iron minerals with respect to their sulfidation (data from Canfield *et al.* 1992; and Raiswell & Canfield, 1996, Reprinted from *Geochemica et Cosmochemica*, **60**, R. Raisewell and D.E. Canfield, Rates of reaction between silicate iron and dissolved sulfide in Peru Margin sediments, pp. 2777–2787, with permission from Elsevier).

Iron mineral	Rate constant (y^{-1})	Half-life
Ferrihydrite	2200	2.8 h
Lepidocrocite	>85	<3 days
Goethite	22	11.5 days
Hematite	12	31 days
Magnetite (uncoated)	6.6×10^{-3}	105 y
'Reactive' silicates	3.0×10^{-3}	230 y
Sheet silicates	0.3×10^{-6}	2.4×10^6 y
Ilmenite, garnet, augite, amphibole	$<0.3 \times 10^{-6}$	2.4×10^6 y

estimated sulfate reduction rates. His data show that at low sedimentation rates typical of the deep sea, pyrite formation is limited by the low production rate of sulfide. At higher sedimentation rates, and in euxinic sediments, sulfide production rates equal or exceed the rates of supply of reactive Fe, so that complete pyritization of reactive Fe is possible. In a suite of rapidly deposited deltaic sediments, however, sulfur burial rates are lower than those predicted for complete pyritization of Fe oxide, despite rates of sulfate reduction high enough to drive complete sulfidation. In this case, Fe oxides appear to be preserved and it is likely that considerable oxidation of sulfide has occurred as a result of bioturbation and physical resuspension of sediment. This is likely to be a widespread phenomenon; *e.g.* Thamdrup *et al.* (1994) found in coastal sediment from Denmark that only 15% of precipitated sulfide was buried permanently, and that most of the reoxidation of solid phase sulfide occurred in the top centimetre of the sediment, by reaction with O_2 and Fe oxides.

The rapid reaction of sedimentary Fe oxides with sulfide to form iron sulfide keeps dissolved sulfide in pore-waters at low levels, even when rates of sulfate reduction are high (Canfield, 1989b). Figure 21 reproduces Canfield's (1989b) data for dissolved Fe, dissolved sulfide, sulfate reduction rate and Fe mineralogy for three sediments from Long Island Sound, the Mississippi Delta and the Santa Barbara Basin. Only when easily extractable iron oxides are depleted do significant levels of dissolved sulfide accumulate. In surface sediments, where oxides persist and where rates of sulfate reduction are high, elevated concentrations of dissolved iron occur. The framework provided by the pore-water chemistry allows a clearer understanding of the mechanisms by which morphologically different types of pyrite form (Canfield & Raiswell, 1991b; Briggs *et al.*, 1996). In Fe-rich and sulfide-poor pore-waters, iron migrates to the site of sulfide formation, most obviously where there are concentrations of organic matter. Biologically recognizable organic matter can be pyritized and preserved in this way. In Fe-poor, sulfide-rich pore-waters, sulfide migrates to relatively unreactive, detrital Fe minerals such as magnetite or biotite, in which case sulfide coatings form.

Pyrite produced by the above reactions can vary in morphology and can take the form of framboids (spherical aggregates of micrometer-sized pyrite crystals; Fig. 22) and micrometer-sized euhedral crystals. Such differences in morphology have been linked to the geochemical environment of formation (see Taylor & Macquaker, 2000) whereby framboidal iron monosulfides form in spatial association with organic matter and euhedral pyrite forms in spatial association with iron oxide minerals. In euxinic seawater (*i.e.* water in which free oxygen is absent and free sulfide is present), pyrite framboids can form directly in the water column and settle onto the seafloor. Framboids formed by this process are smaller in size than those formed within sediments under oxygenated water, raising their possible use as a palaeoenvironmental indicator for ancient shallow marine sediments (Wilkin *et al.*, 1996). Recently, widespread marcasite (FeS_2) has been reported in organic-rich mudstones (Schieber, 2007). Marcasite precipitates at low pH, and its presence has been interpreted as indicating the penetration of oxygen into previously anoxic sediments and the consequent oxidation of previously precipitated pyrite, leading to acidic conditions.

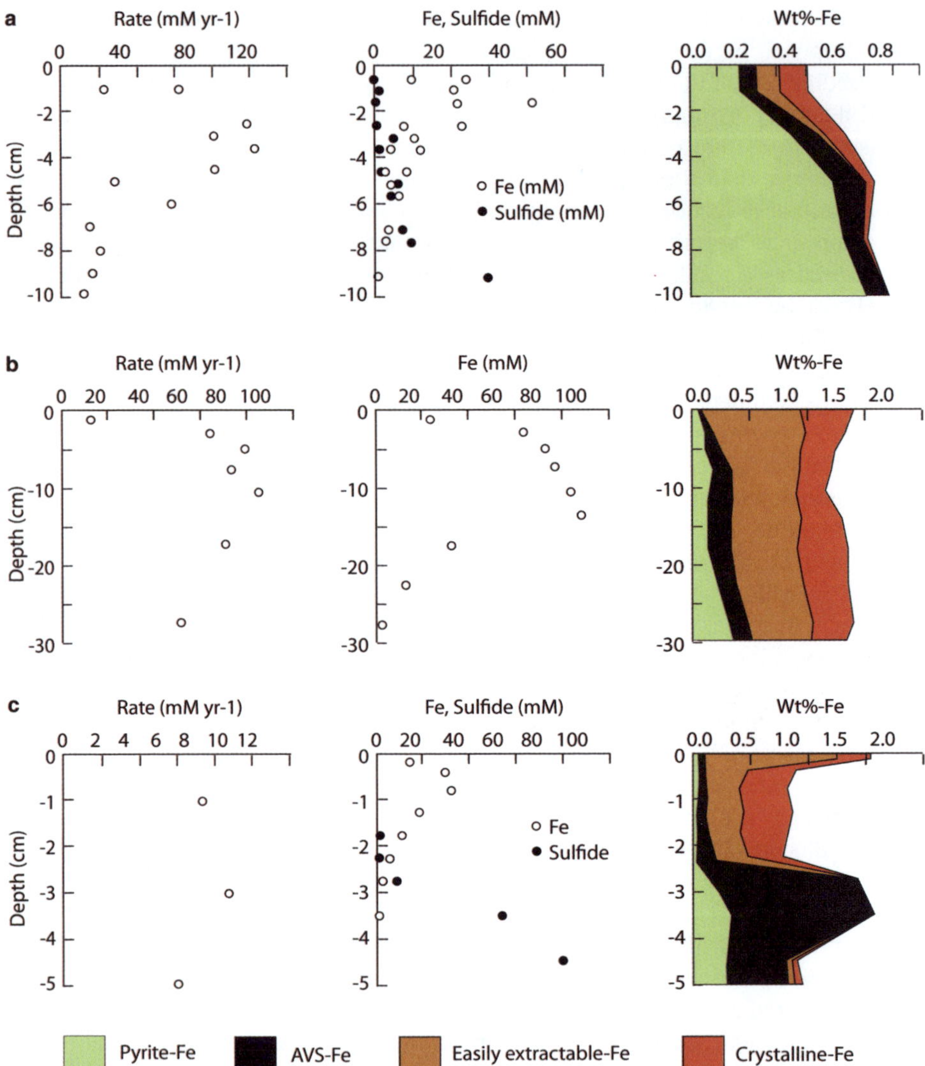

Fig. 21. Sulfate reduction rates, dissolved iron and sulfide concentrations and the speciation of solid phase iron minerals in three sediments from **(a)** Long Island Sound (FOAM), **(b)** the Mississippi Delta (Station 18) and **(c)** the Santa Barbara Basin. AVS-Fe is acid volatile sulfide; easily extractable Fe is iron removed by ammonium oxalate (*e.g.* lepidocrocite, ferrihydrite); crystalline Fe is iron removed by dithionite (*e.g.* goethite, hematite). From Canfield & Raiswell (1991a, reproduced with the permission of Springer).

Although the bacterial reduction of Fe^{3+} has generally been considered to be restricted to the 'reactive' pools of iron oxides described above, direct reduction of structurally bound Fe^{3+} in clay minerals has recently been conjectured by Vorhies and Gaines (2009) in some ancient sediments, based upon mineralogical and textural

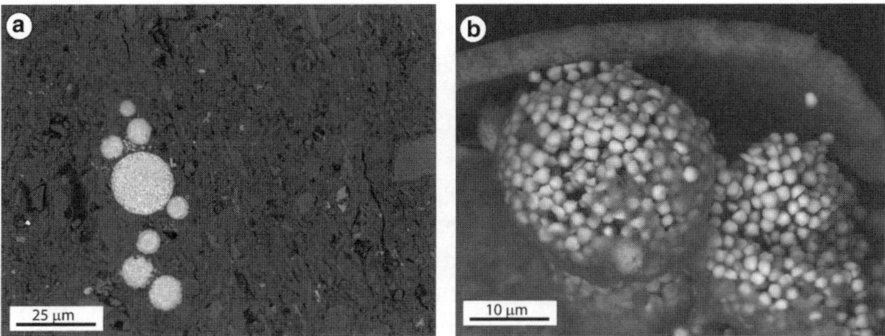

Fig. 22. Images of early diagenetic framboids from (**a**) marine mudstone and (**b**) marine carbonate sediment.

data. Such a pool of bioavailable iron may, therefore, be important in sediments in some cases, although Macquaker *et al.* (1997) reported significant pools of unreduced Fe^{3+} in some deeply buried mudstones.

8. Formation of carbonates

Except for the deepest waters, seawater is supersaturated with respect to calcite and, more so, with respect to dolomite. During early diagenesis, however, the saturation state of interstitial waters with respect to carbonates can change in response to the suite of primary and secondary redox reactions related to the oxidation of organic carbon. Many of the potential reactions have been given above and demonstrate the involvement of a range of protolytic species, which are related to each other by association and dissociation reactions involving protons. These species include carbonate, sulfide, ammonia, iron and manganese. The saturation state of interstitial water with respect to carbonates thus depends on the relative importance of many potential redox and adsorption reactions, and the resulting acid-base balance of the system.

Models of varying complexity have been developed to examine the effects of a suite of early diagenetic reactions on the degree of carbonate saturation (*e.g.* Ben Yaakov, 1973; Gardner, 1973; Boudreau, 1987; Boudreau & Canfield, 1988; Canfield & Raiswell, 1991b; Van Cappellen & Wang, 1996). Canfield & Raiswell's (1991b) model starts with a parcel of oxygen-saturated seawater which undergoes a series of sequential carbon oxidation reactions involving oxygen, nitrate and sulfate, followed by methane oxidation by sulfate and finally methane production. The model is simplified not only in its assumption of sequential carbon oxidation reactions but also in that it assumes a closed system and does not include any solute transport by advection or diffusion. Nor are Fe and Mn oxides involved in the model, although the reaction of sulfide with ferric iron to form pyrite is examined.

Despite these simplifications, Canfield & Raiswell's (1991b) model does match field observations in several well studied modern sediments, and also illustrates some interesting facets of the early diagenetic carbonate system. Two figures from their paper are reproduced here as Figure 23. These show that undersaturation with respect to calcite occurs under rather specific circumstances, most notably: (1) in the lower part of the oxic zone and the denitrification zone, and (2) in the lower part of the sulfate reduction zone, in circumstances where sulfide is not precipitated by reaction with Fe minerals. Where sulfide reacts with Fe minerals, calcite supersaturation is maintained until almost all sulfate has been reduced. These results suggest that significant carbonate dissolution is only likely in oxic sediments, and in circumstances where sulfate reduction is not accompanied by the precipitation of sulfide, for example in highly bioturbated, near-shore sediments where sulfide and alkalinity are rapidly mixed out of the sediment by the activity of macroorganisms (*e.g.* Aller, 1982; Berner & Westrich, 1985). In other circumstances, calcite will be preserved or may be precipitated as a result of the addition of alkalinity to sediment pore waters. Canfield & Raiswell's (1991b) model shows a striking difference between the carbonate saturation states of pore waters in which sulfide resulting from sulfate reduction is precipitated as pyrite and pore waters in which sulfide is not precipitated (Fig. 23). This is an indication of the potentially important role that reactions involving Fe (and Mn) minerals can play in the acid-base balance

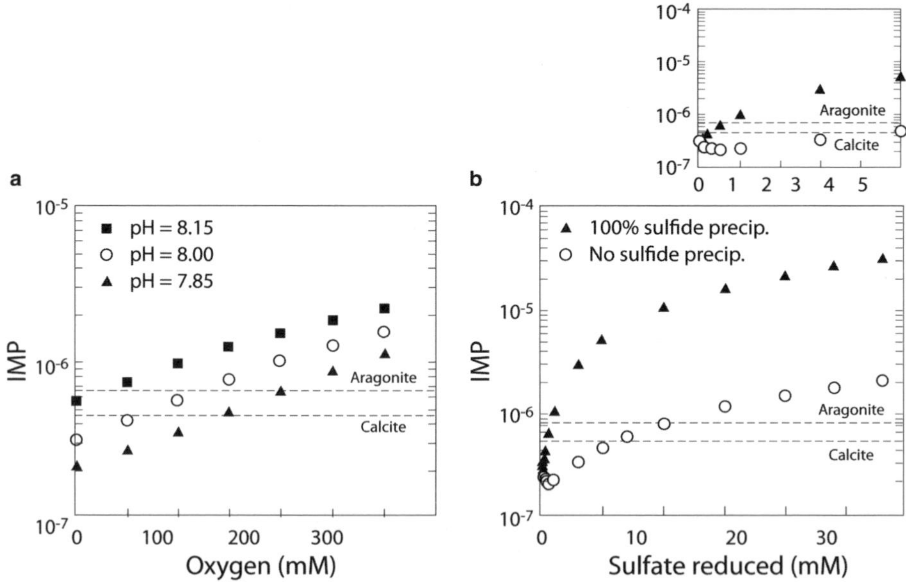

Fig. 23. Carbonate saturation (expressed as Ion Molarity Product, IMP) as a function of (a) progressive oxygen utilization and (b) progressive sulfate reduction. The solubility lines for calcite and aragonite are also shown. In the oxygen diagram, differing starting pH values are used. In the sulfate diagram, two models are shown, one in which all of the produced sulfide accumulates in solution and one in which all of the sulfide is precipitated. See text for further details (from Canfield & Raiswell, 1991b, reproduced with the permission of Springer).

of modern sediments and, thus, the carbonate saturation state of the pore waters. Dissimilatory Fe reduction is a strongly proton-consuming reaction which will lead to an increase in the saturation state of carbonate minerals. Depending on the Fe/Ca ratio of pore waters, this could lead to the precipitation of ferroan calcite or siderite. Siderite is more likely in freshwater sediments, where the activity of calcium will be lower and where there are lower rates of production of sulfide from sulfate reduction (resulting in iron sulfide precipitation) (Postma, 1981, 1982). In marine sediments, siderite is uncommon, probably because of the common juxtaposition of sulfate and iron reduction zones in many sediments and the resulting low activity of dissolved iron.

As described previously, Van Cappellen & Wang (1996) developed a more complex, open-system early diagenetic model which allowed for transport of species and, also, a fuller suite of redox reactions than that of Canfield & Raiswell (1991b). Testing of the model on data produced by Canfield *et al.* (1993a, 1993b) for a nearshore Danish marine sediment illustrates the importance of reactions involving Fe and Mn to pH and, thus, carbonate saturation state. In this sediment, $\sim$70% of Fe reduction was estimated to be dissimilatory, and 30% by the oxidation of H_2S. The modelling showed that $FeCO_3$ (or a mixed Fe-Ca carbonate) was supersaturated very close to the sediment–water interface and continued to precipitate over the top 4–5 cm of the sediment. The amounts are, however, small and would be hard to detect using conventional techniques such as XRD: $\sim$25 μmoles $FeCO_3$ per cm^3 of sediment, or $\sim$0.2 wt.%.

The importance of Fe and Mn cycling to the pH and carbonate saturation state of sediment pore water is illustrated in Figure 24 (Van Cappellen & Wang, 1996). In this nearshore sediment, oxygen penetrates a few millimetres into the sediment, below which the concentrations of dissolved Fe and Mn increase; pH shows substantial variations over the top few centimetres of the sediment. The calculated pH and alkalinity decrease in the aerobic layer of sediment but both parameters increase in the zone of intense, metal hydroxide reduction. Model simulations (Fig. 24) show how Fe/Mn cycling is a key control on the variations in pH. In the absence of iron and manganese, pH decreases in the top centimetre of sediment as a result of dissimilatory and non-dissimilatory oxygen reduction and does not increase at greater depth. By allowing Fe and Mn cycling to occur, pH increases as a result of rapid Fe/Mn oxide reduction, resulting in increasing levels of carbonate supersaturation. The precipitation of authigenic carbonate and sulfide below the main zone of Mn and Fe reduction leads to a decrease in pH.

Although seawater is more supersaturated with respect to dolomite ($CaMg(CO_3)_2$) than calcite, calcite is by far the more common carbonate mineral in marine environments, other than those in which Mg/Ca ratios have been altered by evaporation and the precipitation of anhydrite or gypsum. Although traces of dolomite occur in a variety of deep sea sediments (Bonatti, 1966), significant amounts of authigenic dolomite are restricted to sediments which are relatively enriched in organic carbon (Garrison *et al.*, 1984; Baker & Burns, 1985; Compton, 1988; Middelburg *et al.*, 1990). Dolomite precipitation appears to be kinetically hindered (Sibley *et al.*, 1987; Land, 1998; Arvidson & Mackenzie, 1999), although subsequent growth can be fast. For example, Middelburg *et al.* (1990) estimated dolomite growth rates of the order of a

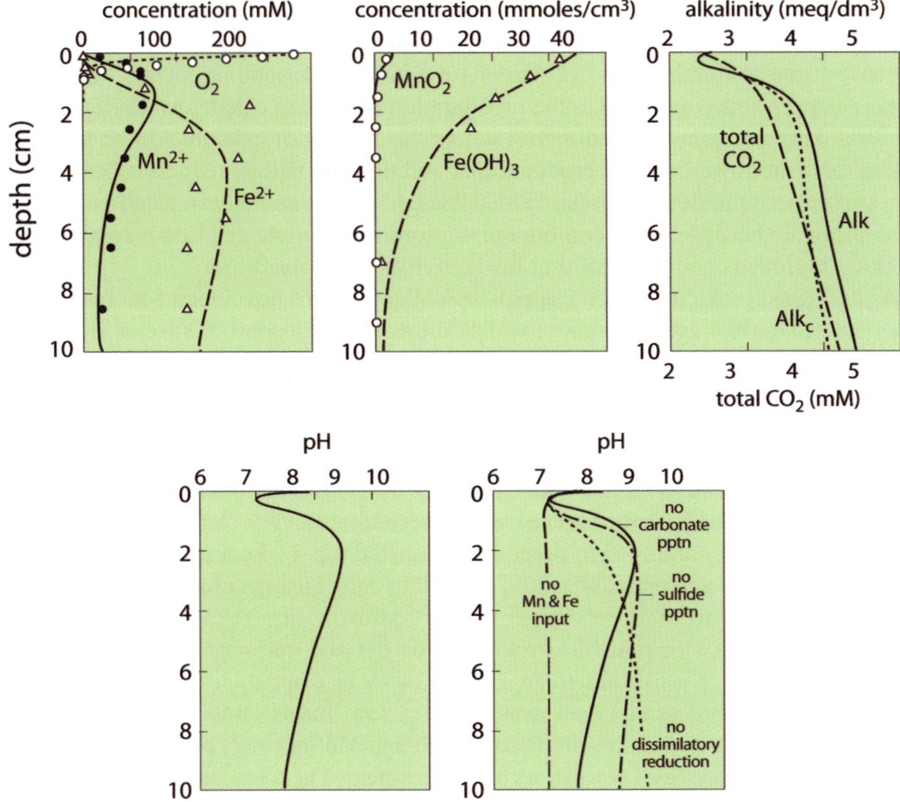

Fig. 24. Measured and calculated distributions of some solid-phase and dissolved components from a coastal marine sediment off Denmark. The figures are from Van Cappellen & Wang (1996 and reproduced here with the permission of Yale University from *American Journal of Science*). The modelled distributions are from a steady state, reactive transport model which allows for transport of species and involves a wide range of primary and secondary redox reactions. More details are given in the text. **(a)** Measured and calculated distributions of oxygen, iron and manganese; **(b)** measured and calculated distributions of reactive manganese and iron oxides; **(c)** calculated profiles of total dissolved inorganic carbon, total alkalinity and carbonate alkalinity; **(d)** calculated profiles of pH, for which there is a minimum at the locus of intense cycling of Fe and Mn; **(e)** calculated pH profiles for several scenarios in which Fe and Mn are cycled in different ways. In one case, there is no input of Fe and Mn, and pH remains low. In cases where Fe and Mn oxides are supplied to the sediment, pH increases below the zone of intense Fe/Mn recycling. The figure demonstrates the importance of Fe/Mn cycling as a control on pH and thus on the saturation state of the interstitial waters with respect to carbonate (and other) minerals.

few centimetres per thousand years in organic-rich sediments from Kau Bay, Indonesia. Conditions favouring dolomite formation are organic-rich sediments in which alkalinity levels are increased substantially as a result of sulfate reduction and/or anaerobic methane oxidation. As they are associated with organic-rich sediments, these dolomites are often referred to as "organogenic". With increasing alkalinity in organic-rich sediments, levels of dolomite supersaturation rise to 100–1000 (Compton, 1988; Middelburg *et al.*, 1990), at which level dolomite nucleates and develops quickly. In some cases stoichiometric dolomite forms (Middelburg *et al.*, 1990), whereas in others a less stable, calcium-rich protodolomite may form first (Compton, 1988). The carbon isotopic composition of both recent and ancient dolomites support a model of formation at the base of the sulfate reduction zone. Values are often more negative than $-10‰$, indicative of a substantial input of carbon derived from the oxidation of organic matter, and can be high as $+15‰$, indicative of precipitation in the methanogenic zone where the reduction of dissolved carbonate leads to heavier isotope values (*e.g.* Claypool & Kaplan, 1974). The magnesium required for dolomite formation is derived by diffusion from overlying seawater and may be a limiting factor in its formation. Pore-water data from Kau Bay suggests that calcium may also be derived from seawater; in other cases, the calcium could also be derived from the dissolution of biogenic calcite (Baker & Burns, 1985).

In both modern and ancient settings, dolomite (and often other carbonate minerals) form as concretions or as laterally extensive bands, most typically with vertical dimensions of a (few) decimetre(s). The pore water data presented by Middelburg *et al.* (1990) shows that once carbonates such as dolomite have nucleated, diffusion of key ions to the site of initial precipitation will favour the development of bands or concretions, especially if the zone of excess alkalinity is stabilized by a hiatus in sedimentation. Various diffusive and surface reaction models have been developed to estimate the rate of concretion growth (*e.g.* Berner, 1968; Raiswell, 1988; Coleman & Raiswell, 1993; Raiswell *et al.*, 1993). What is less clear is what controls the site of initial precipitation; nucleation on pre-existing carbonate or the local development of critically high levels of supersaturation have both been suggested (see Canfield & Raiswell, 1991b), as has the removal of kinetic barriers imposed by the occurrence of proposed inhibitor anions, in particular sulfate (Baker & Kastner, 1981).

9. Formation of carbonate fluorapatite

Reactive phosphorus is supplied to sediments primarily in association with organic matter and Fe oxyhydroxides. In oceanic settings, the burial of phosphorus is a major control on marine primary productivity (Van Capellen & Ingall, 1996). Indeed, the burial of P in marine sediments may account for up to 73% of the present-day burial flux of P to the oceans (Benitez-Nelson, 2000). A combination of pore-water analyses and the analysis of solid-phase P-bearing phases using sequential leach techniques have shown that a significant fraction of the reactive P is converted into francolite (carbonate fluorapatite: CFA) in a very wide range of marine sediments (*e.g.* Jahnke *et al.*, 1983; Froelich *et al.*, 1988; Ruttenberg & Berner, 1993; Schuffert *et al.*, 1994, 1998; Reimers *et al.*, 1996; Filippelli & Delaney, 1996; Slomp *et al.*, 1996; Kim *et al.*, 1999). Although the composition of CFA varies with depositional environment and diagenesis (*e.g.* McArthur, 1985), its approximate formula is $Ca_{10}(PO_4)_{5.74}(CO_3)_{0.26}F_{2.26}$ (Froelich *et al.*, 1988; Kim *et al.*, 1999).

In modern sediments, CFA occurs as two main morphologies: as disseminated crystals, often hard to identify and thus inferred from changes in pore-water chemistry, and as 1–5 cm thick bands, most commonly in sediments which are accumulating relatively slowly and which receive a high flux of P-bearing organic matter. The latter sediments often occur under high productivity upwelling zones such as those along the western coasts of North, Central and South America (*e.g.* Froelich *et al.*, 1988; Schuffert *et al.*, 1994; Kim *et al.*, 1999). In high-productivity regions where sediment deposition is inhibited, for example by current activity, massive phosphorite deposits can result (see, for example, Bentor, 1980; Jarvis *et al.*, 1994).

Although the details of CFA formation are still debated, particularly with respect to the nature of the early diagenetic P cycle, the basic mechanisms are known. Pore-water data show clearly that fluoride is supplied by diffusion from overlying seawater (*e.g.* Froelich *et al.*, 1983; Schuffert *et al.*, 1994). The precise origin of the P is less clear but involves primarily release from organic matter and Fe oxyhydroxides (*e.g.* Ruttenberg & Berner, 1993; Reimers *et al.*, 1996; Slomp *et al.*, 1996; Schuffert *et al.*, 1998; Kim *et al*, 1999). Based on the Redfield ratio (ratio of C:N:P in organic matter), it is possible to estimate the relative amounts of pore water P, NH_4^+ and alkalinity resulting from the oxidative degradation of organic matter. Relative levels of P, which are lower than those of NH_4^+ or alkalinity, suggest the loss of P through the precipitation of mineral phases such as CFA, as seen, for example, in rapidly accumulating continental sediments from Long Island Sound (Ruttenberg & Berner, 1993) and below 10–15 cm depth in organic-rich sediments off western Mexico, where CFA bands are forming (Schuffert *et al.*, 1994). Greater-than-anticipated concentrations of dissolved P indicate a source other than decaying organic matter. In some sediments, high levels of dissolved P occur where reductive dissolution of Fe oxides is prevalent (*e.g.* Slomp *et al.*, 1996; Reimers *et al.*, 1996), indicating a major release of adsorbed P from the oxides.

In some instances, apparent decoupling of the early diagenetic P and F cycles suggests that CFA does not precipitate directly but forms as a two (or more) stage process which can be summarized as the formation of a metastable Ca-P phase followed by its conversion to CFA:

$$3Ca^{2+} + 2PO_4^{3-} \longrightarrow Ca_3(PO_4)_2$$

$$3Ca_3(PO_4)_2 + Ca_{2+} + xCO_3^{2-} + (2+x)F^- \longrightarrow Ca_{10}(PO_4)_{6-x}(CO_3)_xF_{2+x},$$

where x is ~0.26 (Gulbrandsen *et al.*, 1984; Froelich *et al.*, 1988; Kim *et al.*, 1999).

Recent studies of bacterial processes within organic-rich sediments in upwelling zones has shown that sulfide-oxidising bacteria may play a major role in CFA precipitation and phosphorus sequestration into sediments (Schulz & Goldhammer *et al.*, 2010). The sulfide-oxidizing bacteria have been shown to accumulate P in their cells and catalyse the almost instantaneous conversion of phosphorus to CFA.

10. Concluding remarks

Over the past 50 years, increasingly detailed studies of interstitial water chemistry, allied to the characterization of solid phases, have provided a solid framework for understanding the early diagenetic processes which modify the initial composition of modern marine sediments. Many of the reactions and processes highlighted in this chapter are known to be, or are very likely to be, microbially mediated. In the last decade there have been major advances in determining and, in some cases, isolating the microbial populations present in marine sediments (*e.g.* Bottcher *et al.*, 2000; Todorov *et al.*, 2000; Geets *et al.*, 2006; Hamdan *et al.*, 2011) and in the role that isolated and pure-culture bacteria can play in sediment diagenesis and mineral precipitation (*e.g.* Liu *et al.*, 2001; Roden, 2003, 2004). However, it has proved difficult to determine the exact role that individual microbial species play in sediments, given the commonly diverse populations which often operate as consortia in sediments. Future research will hopefully be able to use novel genetic and biomarker techniques to probe this further.

One area in which there is still much to be learnt is in understanding the role that amorphous and poorly crystallized aluminium and silicon oxides and aluminosilicates play in early diagenesis and element cycles. These are known to be important inputs to marine sediments but their characteristics and mineral changes during diagenesis are still poorly known. The spectroscopic techniques to help in this research (*e.g.* Raman, FT-IR, synchrotron radiation methods) have developed significantly in the last decade (Wogelius & Vaughan, 2013, this volume) and need to be applied more widely to modern marine sediments.

References

Aller, R.C. (1977) *The influence of macrobenthos on chemical diagenesis of marine sediments.* Ph.D. dissertation, Yale University, New Haven (Conn.), U.S.A., 600 pp.

Aller, R.C. (1982) Carbonate dissolution in near-shore muds: The role of physical and biological re-working. *Journal of Geology*, **90**, 79–95.

Aller, R.C. (1990) Bioturbation and manganese cycling in hemipelagic sediments. *Philosophical Transactions of the Royal Society* (London), **A331**, 51–68.

Aller, R.C. & Mackin, J.E. (1989) Open-incubation, diffusion methods for measuring solute reaction rates in sediments. *Journal of Marine Research*, **47**, 411–440.

Aller, R.C., Mackin, J.E. & Cox, R.T., Jr. (1986) Diagenesis of Fe and S in Amazon Inner Shelf muds: Apparent domination of Fe reduction and implications for the genesis of ironstones. *Continental Shelf Research*, **6**, 263–289.

Aller, R.C. & Blair, N.E. (2006) Carbon remineralization in the Amazon–Guianas tropical mobile mudbelt: A sedimentary incinerator. *Continental Shelf Research*, **26**, 2241–2259.

Aller, R.C., Hannides, A., Heilbrun, C., Panzeca, C. (2004) Coupling of early diagenetic processes and sedimentary dynamics in tropical shelf environments: the Gulf of Papua deltaic complex. *Continental Shelf Research*, **24**, 2455–2486.

Aplin, A.C. (1993) The composition of authigenic clay minerals in recent sediments: links to the supply of unstable reactants. In: *Geochemistry of Clay-Pore Fluid Interactions* (D.A.C. Manning, P.L. Hall & C.R. Hughes, editors). Mineralogical Society Series, **4**, Chapman & Hall, London, pp. 81–106.

          A.C. Aplin and K.G. Taylor

Aplin, A.C. & Cronan, D.S. (1985) Ferromanganese deposits from the central Pacific Ocean: I. Encrustations from the Line Islands Archipelago. *Geochimica et Cosmochimica Acta*, **40**, 427–436.

Arvidson, R.S. & Mackenzie, F.T. (1999) The dolomite problem: Control of precipitation kinetics by temperature and saturation state. *American Journal of Science*, **299**, 257–288.

Badaut, D., Besson, G., Decarreau, A. & Rautureau, R. (1985) Occurrence of a ferrous, trioctahedral smectite in recent sediments of Atlantis II Deep, Red Sea. *Clay Minerals*, **20**, 389–404.

Baker, P.A. & Burns, S.J. (1985) Occurrence and formation of dolomite in organic-rich continental margin sediments. *American Association of Petroleum Geologists Bulletin*, **69**,1917–1930.

Baker, P.A. & Kastner, M. (1981) Constraints on the formation of sedimentary dolomite. *Science*, **213**, 214–216.

Ben Yaakov, S. (1973) pH buffering of pore water of recent anoxic marine sediments. *Limnology Oceanography*, **18**, 86–94.

Bender, M.L. & Heggie, D.T. (1984) Fate of organic carbon reaching the deep sea floor: a status report. *Geochimica et Cosmochimica Acta*, **48**, 977–986.

Benitez-Nelson, C.R. (2000) The biogeochemical cycling of phosphorus in marine systems. *Earth Science Reviews*, **51**, 109–135.

Bentor, Y.K. (1980) Phosphorites – the unsolved problems. In: *Marine Phosphorites – Geochemistry, Occurrence, Genesis: A symposium held at the Xth International Congress on Sedimentology in Jerusalem, Israel, 9–14 July, 1978.* (Y.K. Bentor, editor). Society of Economic Paleontologists and Mineralogists Special Publication, **29**, pp. 3–18.

Berger, W.H. (1976) Biogenous deep sea sediments: production, preservation and interpretation. In: *Chemical Oceanography*, **5** (J.P. Riley & R. Chester, editors). Academic Press, London, pp. 265–388.

Berger, W.H., Adelseck, C.G. & Mayer, L.A. (1976) Distribution of carbonate in surface sediments of the Pacific Ocean. *Journal of Geophysics Research*, **81**, 2617–2627.

Berner, R.A. (1968) Rate of concretion growth. *Geochimica et Cosmochimica Acta*, **32**, 477–484.

Berner, R.A. (1970) Sedimentary pyrite formation. *American Journal of Science*, **268**, 1–23.

Berner, R.A. (1978) Sulfate reduction and the rate of deposition of marine sediments. *Earth and Planetary Science Letters*, **37**, 492–498.

Berner, R.A. (1980) *Early Diagenesis: A Theoretical Approach.* Princeton University Press, Princeton, New Jersey, USA, 241 pp.

Berner, R.A. (1984) Sedimentary pyrite formation: An update. *Geochimica et Cosmochimica Acta*, **48**, 605–615.

Berner, R.A. & Westrich, J.T. (1985) Bioturbation and the early diagenesis of carbon and sulfur. *American Journal of Science*, **285**,193–206.

Berner, R.A., Baldwin, T. & Holdren, G.R., Jr. (1979) Authigenic iron sulfides as paleosalinity indicators. *Journal of Sedimentary Petrology*, **49**,1345–1350.

Biscaye, P.E. (1965) Mineralogy and sedimentation of recent deep-sea clay in the Atlantic Ocean and adjacent seas and oceans. *Geological Society of America Bulletin*, **76**, 803–832.

Biscaye, P.E., Kolla, V. & Turekian, K.K. (1976) Distribution of calcium carbonate in surface sediments of the Atlantic Ocean. *Journal of Geophysical Research*, **81**, 2595–2603.

Bischoff, J.L. (1972) A ferroan nontronite from the Red Sea geothermal system. *Clays and Clay Minerals*, **20**, 217–223.

Blain, S., Quéguiner, B., Armand, L. *et al.* (2007) Effect of natural iron fertilization on carbon sequestration in the Southern Ocean. *Nature*, **446**, 1070–1074.

Boesen, C. & Postma, D. (1988) Pyrite formation in anoxic environments of the Baltic Sea. *American Journal of Science*, **288**, 575–603.

Bonatti, E. (1966) Deep-sea authigenic calcite and dolomite. *Science*, **153**, 534–537.

Bostrom, K. & Petersen, M.N.A. (1969) The origin of aluminium-poor ferromanganoan sediments in areas of high heat flow on the East Pacific Rise. *Marine Geology*, **7**, 427–447.

Bostrom, K., Kraemer, T. & Gartner, S. (1973) Provenance and accumulation rates of opaline silica, Al, Ti, Fe, Mn, Cu, Ni and Co in Pacific pelagic sediments. *Chemical Geology*, **11**, 123–148.

Bottcher, M.E., Hespenheide, B., Llobet-Brossa, E., Beardsley, C., Larsen, O., Schramm, A., Wieland, A., Bottcher, G., Berninger, U-G. & Amann, R. (2000) The biogeochemistry, stable isotope geochemistry, and microbial community structure of a temperate intertidal mudflat: an integrated study. *Continental Shelf Research*, **20**, 1749–1769.

Boudreau, B.P. (1987) A steady-state diagenetic model for dissolved carbonate species and pH in the pore waters of oxic and suboxic sediments. *Geochimica et Cosmochimica Acta*, **51**, 1985–1996.

Boudreau, B.P. & Canfield, D.E. (1988) A provisional model for pH in anoxic porewaters: Applications to the FOAM site. *Journal of Marine Research*, **46**, 429–455.

Bowles, F.A., Angino, E.A., Hosterman, J.W. & Galle, O.K. (1971) Precipitation of deep-sea palygorskite and sepiolite. *Earth and Planetary Science Letters,* **11**, 324–332.

Brendel, P.J. & Luther, G.W. (1995) Development of a gold amalgam voltammetric microelectrode for the determination of dissolved Fe, Mn, O2 and S(-II) in porewaters of marine and fresh-water sediments. *Environmental Science and Technology*, **29**, 751–761.

Briggs, D.E.G., Raiswell, R., Bottrell, S.H., Hatfield, D. & Bartels, C. (1996) Controls on the pyritization of exceptionally preserved fossils: An analysis of the lower Devonian Hunsrück slate of Germany. *American Journal of Science*, **296**, 633–663.

Brindley, G.W. & Brown, G. (eds.) (1984) Crystal structures of clay minerals and their X-ray identification. Monograph, **5**, Mineralogical Society (London), 495 pp.

Bristow, C. (2005) The dustiest place on Earth. *Nature*, **434**, 816–819.

Broecker, W.S. & Peng, T.-H. (1982) *Tracers in the sea*. Eldigio Press, Palisades, New York. 690 pp.

Burdidge, D.J. & Gieskes, J.M. (1983) A pore water/solid phase diagenetic model for manganese in marine sediments. *American Journal of Science*, **283**, 29–47.

Burdidge, D.J. & Nealson, K.H. (1986) Chemical studies of sulfide-mediated manganese reduction. *Geomicrobiology Journal*, **4**, 361–397.

Burns, R.G. & Burns, V.M. (1977) Mineralogy. In: *Marine Manganese Deposits* (G.P. Glasby, editor). Elsevier, Amsterdam, pp. 185–248.

Buser, W. & Grutter, A. (1956) Über die Natur der Mangankollen. *Schweizer Mineralogische und Petrographische Mitteilungen*, **36**, 49–62.

Calvert, S.E. & Price, N.B. (1977) Shallow water, continental margin and lacustrine nodules: distribution and geochemistry. In: *Marine Manganese Deposits* (G.P. Glasby, editor). Elsevier, Amsterdam, pp. 45–86.

Canfield, D.E. (1988) *Sulfate reduction and the diagenesis of iron in anoxic marine sediments*. PhD dissertation, Yale University, New Haven, Connecticut, USA, 248 pp.

Canfield, D.E. (1989a) Sulfate reduction and oxic respiration in marine sediments: Implications for organic carbon preservation in euxinic environments. *Deep-Sea Research*, **36**, 121–138.

Canfield, D.E. (1989b) Reactive iron in marine sediments. *Geochimica et Cosmochimica Acta*, **53**, 619–632.

Canfield, D.E. & Raiswell, R. (1991a) Pyrite formation and fossil preservation. In: *Taphonomy: Releasing the Data Locked in the Fossil Record* (P.A. Allison & D.E.G. Briggs, editors). *Geobiology*, **9**, Plenum Press, New York, pp. 337–387.

Canfield, D.E. & Raiswell, R. (1991b) Carbonate precipitation and dissolution. Its relevance to fossil preservation. In: *Taphonomy: Releasing the Data Locked in the Fossil Record* (P.A. Allison & D.E.G. Briggs, editors). *Geobiology*, **9**, Plenum Press, New York, pp. 411–453.

Canfield, D.E., Bottrell, S.B. & Raiswell, R. (1992) The reactivity of sedimentary iron minerals toward sulfide. *American Journal of Science*, **292**, 659–683.

Canfield, D.E., Jørgensen, B.B., Fossing, H., Glud, R., Gundersen, J., Ramsing, N.B., Thamdrup, B., Hansen, J.W., Nielsen, L.P. & Hall, P.O.J. (1993a) Pathways of organic carbon oxidation in three continental margin sediments. *Marine Geology*, **113**, 27–40.

Canfield, D.E., Thamdrup, B. & Hansen, J.W. (1993b) The anaerobic degradation of organic matter in Danish coastal sediments: iron reduction, manganese reduction and sulfate reduction. *Geochimica et Cosmochimica Acta*, **57**, 3867–3883.

Chamley, H. (1989) *Clay Sedimentology*. Springer-Verlag, Berlin 623 pp.

Chester, R. (1990) *Marine Geochemistry*. Unwin Hyman, London, 698 pp.

Chester, R., Elderfield, H. & Griffin, J.J. (1971) Dust transported in the North-East and South-East trade winds in the Atlantic Ocean. *Nature*, **233**, 474–476.

Chester, R. and Johnson, L.R. (1971) Atmospheric dusts collected off the Atlantic coasts of North Africa and the Iberian Peninsula. *Marine Geology*, **11**, 251–260.

Chiapello, I., Bergametti, G., Chatenet, B., Bousquet, P., Dulac, F. & Santos Soares, E. (1997) Origins of African dust transported over the North-Eastern Tropical Atlantic. *Journal of Geophysical Research*, **102**, 13701–13709.

Claypool, G.E. & Kaplan, I.R. (1974) The origin and distribution of methane in marine sediments. In: *Natural Gas in Marine Sediments* (I.R. Kaplan, editor). Plenum Press, New York, pp. 97–138.

Cole, T.G. (1983) Oxygen isotope geothermometry and origin of smectites in the Atlantis II Deep, Red Sea. *Earth and Planetary Science Letters*, **66**, 166–176.

Cole, T.G. (1985) Composition, oxygen isotope geochemistry, and origin of smectite in the metalliferous sediments of the Bauer Deep, southeast Pacific. *Geochimica et Cosmochimica Acta*, **49**, 221–235.

Cole, T.G. (1988) The nature and origin of smectite in the Atlantis II Deep, Red Sea. *The Canadian Mineralogist*, **26**, 755–763.

Cole, T.G. & Shaw, H.F. (1983) The nature and origin of authigenic smectites in some recent marine sediments. *Clay Minerals*, **18**, 239–252.

Coleman, M.L. & Raiswell, R. (1993) Microbial mineralisation of organic matter – mechanisms of self organisation and inferred rates of precipitation of diagenetic minerals. *Philosophical Transactions of the Royal Society London*, **A344**, 69–87.

Compton, J.S. (1988) Degree of supersaturation and precipitation of organogenic dolomite. *Geology*, **16**, 318–321.

Corliss, J.B., Lyle, M., Dymond, J. & Crane, K. (1978) The chemistry of hydrothermal mounds near the Galapagos Rift. *Earth and Planetary Science Letters*, **40**, 12–24.

Cronan, D.S. (1980) *Underwater Minerals*. Academic Press, London.

Curtis, C.D., Coleman, M.L. & Love, L.G. (1986) Pore water evolution during sediment burial from isotopic and mineral chemistry of calcite, dolomite and siderite concretions. *Geochimica et Cosmochimica Acta*, **50**, 2321–2334.

Davies, T.A. & Gorsline, D.S. (1976) Oceanic sediments and sedimentary processes. In: *Chemical Oceanography*, **5** (J.P. Riley & R. Chester, editors). Academic Press, London, pp. 1–80.

Davison, W., Lishman, J.P. & Hilton, J. (1985) Formation of pyrite in freshwater sediments: Implications for C/S ratios. *Geochimica et Cosmochimica Acta*, **49**, 1615–1620.

Davison, W., Grime, G.W., Morgan, J.A.W. & Clarke, K. (1991) Distribution of dissolved iron in sediment pore waters at submillimetre resolution. *Nature*, **352**, 323–325.

Degens, E.T. & Ross, D.A. (editors) (1969) *Hot Brines and Recent Heavy Metal Deposits in the Red Sea*. Springer-Verlag, New York, 599 pp.

Desprairies, A. (1982) Authigenic minerals in volcanogenic sediments cored during Deep Sea Drilling Project leg 60. In: *Initial reports of the Deep Sea Drilling Project* (D.M. Hussong & S. Uyeda, editors). US Government printing office, Washington D.C., pp. 455–466.

Dymond, J. (1981) Geochemistry of Nazca plate surface sediments: An evaluation of hydrothermal, biogenic, detrital, and hydrogenous sources. In: *Nazca Plate: Crustal Formation and Andean Convergence* (L.D. Kulm, J. Dymond, E.J. Dasch & D.M. Hussong, editors). GSA Memoir, **154**, Geological Society of America, Boulder, Colorado, pp. 133–173.

Dymond, J., Lyle, M., Finney, B., Piper, D.Z., Murphy, K., Conrad, R. & Pisias, N. (1984) Ferromanganese nodules from MANOP Sites H, S and R – control of mineralogical and chemical composition by multiple accretion processes. *Geochimica et Cosmochimica Acta*, **48**, 931–949.

Edmond, J.M. (1974) On the dissolution of carbonate and silicate in the deep ocean. *Deep-Sea Research*, **21**, 455–480.

Edmond, J.M., Von Damm, K.L., McDuff, R.E. & Measures, C.I. (1982) Chemistry of hot springs on the East Pacific Rise and their effluent dispersal. *Nature*, **297**, 187–191.

Elderfield, H. (1976) Hydrogenous material in marine sediments, excluding manganese nodules. In: *Chemical Oceanography*, **5** (J.P. Riley & R. Chester, editors). Academic Press, London, pp. 137–215.

Emeis, K. (1985) Particulate suspended matter in major world rivers – II: Results on the rivers Indus, Waikato, Nile, St. Lawrence, Yangtse, Parana, Orinoco, Caroni and Mackenzie. *Mitt. Geol.-Paläontol. Inst. Univ. Hamb.*, **58** (SCOPE/UNEP Sonderbd.), 593–617.

Filippelli, G.M. & Delaney, M.L. (1996) Phosphorus geochemistry of equatorial Pacific sediments. *Geochimica et Cosmochimica Acta*, **60**, 1479–1495.

Foscolos, A.E. & Powell, T.G. (1980) Mineralogical and geochemical transformation of clays during catagenesis and their relation to oil generation. In: *Facts and Principles of World Petroleum Occurrence* (A.D. Miall, editor). *Canadian Society of Petroleum Geology Memoir*, **6**, 153–172.

Froelich, P.N., Klinkhammer, G.P., Bender, M.L., Luedtke, N.A., Heath, G.R., Cullen, D., Dauphin, P., Hammond, D., Hartman, B. & Maynard, V. (1979) Early oxidation of organic matter in pelagic sediments of the eastern equatorial Atlantic: suboxic diagenesis. *Geochimica et Cosmochimica Acta*, **43**, 1075–1090.

Froelich, P.N., Kim, K.-H., Jahnke, R.A., Burnett, W.C., Soutar, A. & Deakin, M. (1983) Pore water fluoride in Peru continental margin sediments: Uptake from seawater. *Geochimica et Cosmochimica Acta*, **47**, 1605–1612.

Froelich, P.N., Arthur, M.A., Burnett, W.C., Deakin, M., Hensley, V., Jahnke, R., Kaul, L., Kim, K.-H., Roe, K., Soutar, A. & Vathakanon, C. (1988) Early diagenesis of organic matter in Peru continental margin sediments: phosphorite precipitation. *Marine Geology*, **80**, 309–343.

Gac, J.Y., Droubi, A., Fritz, B. & Tardy, Y. (1977) Geochemical behaviour of silica and magnesium during the evaporation of waters in Lake Chad. *Chemical Geology*, **19**, 215–228.

Gardner, L.R. (1973) Chemical models for sulfate reduction in closed anaerobic marine environments. *Geochimica et Cosmochimica Acta*, **37**, 53–68.

Garrels, R.M. (1986) Sediment cycling and diagenesis. *U.S. Geological Survey Bulletin*, **1578**, 1–11.

Garrison, R.E., Kastner, M. & Zenger, D.H. (editors) (1984) *Dolomites of the Monterey Formation and other organic-rich units.* Pacific Section of Society of Economic Paleontologists and Mineralogists Special Publication, **41**, 215 pp.

Geets, J., Borremans, B., Diels, L., Springael, D., Vangronsveld, J., van der Lelie, D. & Vanbroekhoven, K. (2006) DsrB gene-based DGGE for community and diversity surveys of sulfate reducing bacteria. *Journal of Microbiology Methods*, **66**, 194–205.

Gibbs, R.J. (1977) Clay mineral segregation in the marine environment. *Journal of Sedimentary Petrology*, **47**, 237–243.

Glasby, G.P. (edior) (1977) *Marine Manganese Deposits.* Elsevier, Amsterdam.

Goldhaber, M.B. & Kaplan, I.R. (1974) The sulfur cycle. In: *The Sea,* vol. **5** (E.D. Goldberg, editor). Wiley, New York, pp. 569–655.

Goldhammer T., Brüchert, V., Ferdelman, T.G. & Zabel, M. (2010) Microbial sequestration of phosphorus in anoxic upwelling sediments. *Nature Geoscience*, **3**, 557–561.

Goudie, A.S. & Middleton, N.J. (2001) Saharan Dust Storms: nature and consequences. *Earth-Science Reviews*, **56**, 179–204.

Griffin, J.J., Windom, H. & Goldberg, E.D. (1968) The distribution of clay minerals in the World Ocean. *Deep-Sea Research*, **15**, 433–459.

Gulbrandsen, R.A., Roberson, C.E. & Neil, S.T. (1984) Time and the crystallization of apatite in seawater. *Geochimica et Cosmochimica Acta*, **48**, 213–218.

Gurnell, A.M. (1987) Suspended sediment. In: *Glacio-Fluvial Sediment Transfer* (A.M. Gurnell & M.J. Clarke, editors). Wiley, Chichester, UK, pp. 305–354.

Hamdan, L.J., Gillivet, P.M., Pohlman, J.W., Sikaroodi, M., Greinert, J. & Coffin, R.B. (2011) Diversity and biogeochemical structuring of bacterial communities across the Porangahau ridge accretionary prism, New Zealand. *FEMS Microbiology Ecology*, **77**, 518–532.

Hathaway, J.C. & Sachs, P.L. (1965) Sepiolite and clinoptilolite from the mid-Atlantic ridge. *American Mineralogist*, **50**, 852–867.

Hein, J.R., Yeh, H.W. & Alexander, E. (1979) Origin of Fe-rich montmorillonite from the Mn nodule belt of the North Equatorial Pacific. *Clays and Clay Minerals*, **27**, 185–194.

Henrichs, S.M. & Reeburgh, W.S. (1987) Anaerobic mineralization of marine organic matter: Rates and the role of anaerobic processes in the ocean carbon economy. *Geomicrobiology Journal*, **5**, 191–237.

Hoffert, M., Perseil, A., Hekinian, R., Choukroune, P. & Needham, H.D. (1978) Hydrothermal deposits sampled by diving saucer in transform fault A near 37 degrees N on the mid-Atlantic Ridge, Famous area. *Oceanologica Acta*, **1**, 73–86.

Homoky, W.B., Severmann, S., Mills, R.A., Statham, P.J. & Fones, G.R. (2009) Pore fluid Fe isotopes reflect the extent of benthic Fe redox recycling: Evidence from continental shelf and deep-sea sediments. *Geology*, **35**, 751–754.

Hover, V., Walter, L.M., Peacor, D.R. & Martini, A.M. (1999) Mg-smectite authigenesis in a marine evaporative environment, Salina Ometepac, Baja California. *Clays and Clay Minerals*, **47**, 252–268.

Hurd, D.C. (1973) Interactions of biogenic opal, sediment and seawater in the central equatorial Pacific. *Geochimica et Cosmochimica Acta*, **37**, 2257–2282.

Irwin, H., Coleman, M.L. & Curtis, C.D. (1977) Isotopic evidence for source of diagenetic carbonates formed during burial of organic-rich sediments. *Nature*, **269**, 209–213.

Iversen, N. & Jørgensen, B.B. (1985) Anaerobic methane oxidation rates at the sulfate-methane transition in marine sediments from Kattegat and Skaggerak (Denmark). *Limnology Oceanography*, **30**, 944–955.

Jahnke, R.A. & Christiansen, M.B. (1989) A free-vehicle benthic chamber instrument for sea-floor studies. *Deep-Sea Research*, **36**, 625–637.

Jahnke, R.A., Emerson, S.R., Roe, K.K. & Burnett, W.C. (1983) The present day formation of apatite in Mexico continental margin sediments. *Geochimica et Cosmochimica Acta*, **47**, 259–266.

Jarvis, I., Burnett, W.C., Nathan, Y., Almbaydin, F.S.M., Attia, A.K.M., Castro, L.N., Flicoteaux, R., Hilmy, M.E., Husain, V., Qutawnah, A.A., Serjani, A. & Zanin, Y.N. (1994) Phosphorite geochemistry – state of- the-art and environmental concerns. *Eclogae Geologica Helvetica*, **87**, 643–700.

Jørgensen, B.B. (1982) Mineralization of organic matter in the sea bed – the role of sulfate reduction. *Nature*, **296**, 643–645.

Jickells, T., An, Z.S. Andersen, K.K., Baker, A.R., Bergametti, G., Brooks, N., Cao, J.J., Boyd, P., Duce, R., Hunter, K.K. Kubilay, N., LaRoche, J., Liss, P., Prospero, J.M., Ridgwell, A.J., Tegen, I. & Torres, R. (2005) Global iron connections between desert dust, ocean biogeochemistry, and climate. *Science*, **308**, 67–71

Kastner, M. (1981) Authigenic silicates in deep-sea sediments: formation and diagenesis. In: *The Sea*, vol. 7 (C. Emiliani, editor). Wiley, New York, pp. 915–980.

Kastner, M. (1984) Controls of dolomite formation. *Nature*, **311**, 410–411.

Kato, Y., Fujinaga, K., Nakamura, K., Takaya, Y., Kitamura, K., Ohta, J., Toda, R., Nakashima, T. and Iwamori, H. (2011) Deep sea mud in the Pacific Ocean as a potential resource for rare-earth elements. *Nature Geoscience*, **4**, 535–539.

Kim, D., Schuffert, J.D. & Kastner, M. (1999) Francolite authigenesis in California continental slope sediments and its implications for the marine P cycle. *Geochimica et Cosmochimica Acta*, **63**, 3477–3485.

Kolla, V.R., Be, A.W.H. & Biscaye, P.E. (1976) Calcium carbonate distribution in surface sediments of the Indian Ocean. *Journal of Geophysical Research*, **81**, 2605–2616.

Konta, J. (1985) Mineralogy and chemical maturity of suspended matter in major rivers sampled under the SCOPE/UNEP project. *Mitt. Geol.-Paläontol. Inst. Univ. Hamb.*, **58** (SCOPE/UNEP Sonderbd.) 569–592.

Krom, M.D., Davison, P., Zhang, H. & Davison, W. (1994) High resolution pore-water sampling with a gel sampler. *Limnology Oceanography*, **39**, 1967–1972.

Ku, T.C.W., Walter, L.M., Coleman, M.L., Blake, R.E. & Martini, A.M. (1999) Coupling between sulfur recycling and syndepositional carbonate dissolution: Evidence from oxygen and sulfur isotope composition of pore water sulfate, South Florida Platform, U.S.A. *Geochimica et Cosmochimica Acta*, **63**, 2529–2546.

Ku, T.C.W. & Walter, L.M. (2003) Syndepositional formation of iron rich clays in tropical shelf sediments, San Blas Archipelago, Panama. *Chemical Geology*, **197**, 197–213.

Land, L.S. (1998) Failure to precipitate dolomite at 25 degrees C from dilute solution despite 1000-fold over-saturation for 32 years. *Aquatic Geochemistry*, **4**, 361–368.

Lennie A.R. & Vaughan D.J. (1996) Spectroscopic studies of iron sulfide formation and phase relations at low temperatures. In: *Mineral Spectroscopy: A Tribute to Roger G. Burns* (M.D. Dyar, C. McCammon & M.W. Schaefer, editors). Geochemical Society Special Publication No. **5**, Geochemical Society, Houston, Texas, USA, pp. 117–131.

Leppard, G.G. (1992) Evaluation of electron microscopic techniques for the description of aquatic colloids. In: *Environmental Particles, vol. 1* (J. Buffle & H.P. van Leeuwen, editors). Lewis Publishers, Boca Raton, Florida, USA, pp. 231–289.

Li, Y.-H., Teraoka, H., Yang, T.-S. & Chen, J.-S. (1984) The elemental composition of suspended particles from the Yellow and Yangtse rivers. *Geochimica et Cosmochimica Acta*, **48**, 1561–1564.

Lindsay, W.L. (1979) *Chemical Equilibria in Soils*. Wiley, New York, 449 pp.

Liu, C., Kota, S., Zachara, J.M., Fredrickson, J.K. & Brinkman, C.K. (2001) Kinetic analysis of the bacterial reduction of goethite. *Environmental Science & Technology*, **35**, 2482–2490.

Lovley, D.E. & Phillips, E.J.P. (1987) Competitive mechanisms for inhibition of sulfate reduction and methane production in the zone of ferric iron reduction in sediments. *Applied Environmental Microbiology*, **53**, 2636–2641.

Lovley, D.E. & Phillips, E.J.P. (1988) Novel mode of microbial energy metabolism – organic carbon oxidation coupled to dissimilatory reduction of iron or manganese. *Applied Environmental Microbiology*, **54**, 1472–1480.

Luther, G.W., III (1991) Pyrite synthesis via polysulfide compounds. *Geochimica et Cosmochimica Acta*, **55**, 2839–2849.

Lyle, M. (1983) The brown–green color transition in marine sediments: a marker of the Fe(II)–Fe(III) redox boundary. *Limnology Oceanography*, **28**, 106–1033.

Mackenzie F. T. & Garrels R. M. (1966) Chemical mass balance between rivers and oceans. *American Journal of Science*, **264**, 507–525.

Mackin, J.E. & Aller, R.C. (1984a) Dissolved Al in sediments and waters from the East China Sea: Implications for authigenic mineral formation. *Geochimica et Cosmochimica Acta*, **48**, 281–297.

Mackin, J.E. & Aller, R.C. (1984b) Diagenesis of dissolved aluminium in organic-rich estuarine sediments. *Geochimica et Cosmochimica Acta*, **48**, 298–313.

Macquaker, J.H.S., Curtis, C.D. & Coleman, M.L. (1997) The role of iron in mudstone diagenesis: Comparison of Kimmeridge clay formation mudstones from onshore and offshore (UKCS) localities. *Journal of Sedimentary Research*, **67**, 871–878.

Maher, B.A., Prospero, K.M., Mackie, D., Gaiero, D., Hesse, P.P. & Balkanski, Y. (2010) Global connections between aeolian dust, climate and ocean biogeochemistry at the present day and at the last glacial maximum. *Earth-Science Reviews*, **99**, 61–97

Martin, J.H. (1990) Glacial-interglacial CO_2 change: the iron hypothesis. *Palaeoceanography*, **5**, 1–13.

Martin, J.M. & Meybeck, M. (1979) Elemental mass balance of material carried by major world rivers. *Marine Chemistry*, **7**, 173–206.

Martin, J.-M. & Whitfield, M. (1983) The significance of the river input of chemical elements to the ocean. In: *Trace Metals in Seawater* (C.S. Wong, E. Boyle, K.W. Bruland, J.D. Burton & E.D. Goldberg (editors). Plenum Press, New York, pp. 265–296.

McArthur, J.M. (1985) Francolite geochemistry – compositional controls during formation, diagenesis, metamorphism and weathering. *Geochimica et Cosmochimica Acta*, **49**, 23–35.

McMurtry, G.M., Wang, C.H. & Yeh, H.W. (1983) Chemical and isotopic investigations into the origin of clay minerals from the Galapagos hydrothermal mounds field. *Geochimica et Cosmochimica Acta*, **47**, 475–489.

Michalopoulos, P. & Aller, R.C. (2004) Early diagenesis of biogenic silica in the Amazon delta: Alteration, authigenic clay formation, and storage. *Geochimica et Cosmochimica Acta*, **68**, 1061–1085.

Michard, G., Albarede, F., Michard, A., Minster, J.-F., Charlou, J.-L. & Tan, N. (1984) Chemistry of solutions from the 13_N East Pacific Rise hydrothermal site. *Earth and Planetary Science Letters*, **67**, 297–307.

Middelburg, J.J., de Lange, G.J. & Kreulen, R. (1990) Dolomite formation in anoxic sediments of Kau bay, Indonesia. *Geology*, **18**, 399–402.

Moorby, S.A., Cronan, D.S. & Glasby, G.P. (1984) Geochemistry of hydrothermal Mn-oxide deposits from the S.W. Pacific island arc. *Geochimica et Cosmochimica Acta*, **48**, 433–441.

Moore, W.S. & Vogt, P.R. (1976) Hydrothermal manganese crust from two sites near the Galapagos spreading axis. *Earth and Planetary Science Letters*, **29**, 349–356.

Moreno, T., Querol, X., Castillo, S., Alastuey, A., Cuevas, E., Herrmann, L., Mounkaila, M., Elvira, J., & Gibbons, W. (2006) Geochemical variations in aeolian mineral particles from the Sahara–Sahel Dust Corridor. *Chemosphere*, **65**, 261–270.

Mossmann, J.-R., Aplin, A.C., Curtis, C.D. & Coleman, M.L. (1991) Geochemistry of inorganic and organic sulphur in organic-rich sediments from the Peru Margin. *Geochimica et Cosmochimica Acta*, **55**, 3581–3595.

Myers, C.R. & Nealson, K.H. (1988) Microbial reduction of manganese oxides – interactions with iron and sulfur. *Geochimica et Cosmochimica Acta*, **52**, 2727–2732.

Newman, A.C.D. (editor) (1987) *Chemistry of Clays and Clay Minerals.* Monograph, **6**, Mineralogical Society of Great Britain and Ireland, 480 pp.

Odin, G.S. (1985) La verdine, facies granulaire vert, marin et cotier, distinct da la glauconie: distribution actuelle et composition. *Comptes Rendus de l'Académie des Sciences*, **301**, 105–108.

Odin, G.S. (editor) (1988a) *Green Marine Clays.* Developments in Sedimentology, **45**. Elsevier, Amsterdam, 445 pp.

Odin, G.S. (1988b) Glaucony from the Gulf of Guinea. In: *Green Marine Clays* (G.S. Odin, editor). Developments in Sedimentology, **45**, Elsevier, Amsterdam, pp. 225–247.

Odin, G.S. (1990) Clay mineral formation at the continent-ocean boundary: the verdine facies. *Clay Minerals*, **25**, 477–483.

Odin, G.S. & Lamboy, M. (1988) Glaucony from the margin off northwestern Spain. In: *Green Marine Clays* (G.S. Odin, editor). Developments in Sedimentology, **45**, Elsevier, Amsterdam, pp. 249–275.

Odin, G.S. & Masse, J.P. (1988) The verdine facies from the Senegalese continental shelf. In: *Green Marine Clays* (G.S. Odin, editor). Developments in Sedimentology, **45**, Elsevier, Amsterdam, pp. 83–104.

Odin, G.S. & Matter, A. (1981) De glauconiarum origine. *Sedimentology*, **28**, 611–641.

Orians, K.J. & Bruland, K.W. (1986) The biogeochemistry of aluminum in the Pacific Ocean. *Earth and Planetary Science Letters*, **78**, 397–410.

Pédro, G., Carmouze, J.P. & Velde, B. (1978) Peloidal nontronite formation in recent sediments of Lake Chad. *Chemical Geology*, **23**, 139–149.

Perdue, M., Beck, K.C. & Reuter, J.H. (1976) Organic complexes of iron and aluminium in natural waters. *Nature*, **260**, 418–420.

Perry, C.T. & Taylor, K.G. (2006) Inhibition of dissolution within shallow water carbonate sediments: impacts of terrigenous sediment input on syn-depositional carbonate diagenesis. *Sedimentology*, **53**, 495–513.

Perry, C.T., Salter, M.A., Harbourne, A.R., Crowley, S.F., Jelks, H.L. & Wilson, R.W. (2011) Fish as major carbonate mud producers and missing components of the tropical carbonate factory. *Proceedings of the National Academy of Science*, **108**, 3865–3869.

Persons, B.S. (1970) *Laterite: Genesis, Location, Use.* Plenum Press, New York, 103 pp.

Porrenga, D.H. (1966) Clay minerals in recent sediments of the Niger delta. In: *Clays and Clay Minerals. Proceedings of the Fourteenth National Conference, Berkeley, California, USA* (S.W. Bailey, editor). Pergamon Press, Oxford, UK, pp. 221–233.

Porrenga, D.H. (1967) Glauconite and chamosite as depth indicators in the marine environment. *Marine Geology*, **5**, 495–501.

Postma, D. (1981) Formation of siderite and vivianite and the pore water composition of a recent bog sediment in Denmark. *Chemical Geology*, **31**, 225–244.

Postma, D. (1982) Pyrite and siderite formation in brackish and freshwater swamp sediments. *American Journal of Science*, **282**, 1151–1183.

Postma, D. (1985) Concentration of Mn and separation from Fe in sediments. 1. Stoichiometry of the reaction between birnessite and dissolved Fe(II) at 10°C. *Geochimica et Cosmochimica Acta*, **49**, 1023–1033.

Poulton, S.W, Krom, M.D. & Raiswell, R. (2004) A revised scheme for the reactivity of iron (oxyhydr)oxide minerals towards dissolved sulfide. *Geochimica et Cosmochimica Acta*, **68**, 3703–3715.

Poulton, S.W. & Canfield, D.E. (2011) Ferruginous conditions: a dominant features of the Ocean through Earth's history. *Elements*, **7**, 107–112.

Pyzik, A.J. & Sommer, S.E. (1981) Sedimentary iron monosulphides: Kinetics and mechanism of formation. *Geochimica et Cosmochimica Acta*, **45**, 687–698.

Raiswell, R. (1988) Evidence for surface-reaction controlled growth of carbonate concretions in shales. *Sedimentology*, **35**, 571–575.

Raiswell, R. (2011) Iron transport from the continents to the open ocean: the ageing-rejuvination cycle. *Elements*, **7**,101–106.

Raiswell, R. & Canfield, D.E. (1996) Rates of reaction between silicate iron and dissolved sulfide in Peru Margin sediments. *Geochimica et Cosmochimica Acta*, **60**, 2777–2787.

Raiswell, R., Buckley, F., Berner, R.A. & Anderson, T.F. (1988) Degree of pyritisation as a palaeoenvironmental indicator of bottom water oxygenation. *Journal of Sedimentary Petrology*, **58**, 812–819.

Raiswell, R., Whaler, K., Dean, S., Coleman, M.L. & Briggs, D.E.G. (1993) A simple 3-dimensional model of diffusion-with-precipitation applied to localized pyrite formation in framboids, fossils and detrital iron minerals. Marine Geology, 113, 89–100.

Raiswell, R., Tranter, M., Benning, L.G., Siegert, M., De'ath, R., Huybrechts, P., Payne, T. (2006) Contributions from glacially derived sediment to the global iron (oxyhydr)oxide cycle: Implications for iron delivery to the oceans. *Geochimica et Cosmochimica Acta*, **70**, 2765–2780.

Reimers, C.E. (1987) An in situ microprofiling instrument for measuring interfacial pore water gradients: methods and oxygen profiles from the North Pacific Ocean. *Deep-Sea Research*, **34**, 2019–2035.

Reimers, C.E., Ruttenberg, K.C., Canfield, D.E., Christiansen, M.B. & Martin, J.B. (1996) Porewater pH and authigenic phases formed in the uppermost sediments of the Santa Barbara Basin. *Geochimica et Cosmochimica Acta*, **60**, 4037–4057.

Revsbech, N.P. & Jørgensen, B.B. (1986) Microelectrodes – their use in microbial ecology. *Advances in Microbial Ecology*, **9**, 293–352

Rickard, D.T. (1975) Kinetics and mechanisms of pyrite formation at low temperatures. *American Journal of Science*, **275**, 636–652.

Rickard, D.T. (1997) Kinetics of pyrite formation by the H_2S oxidation of iron(II) monosulfide in aqueous solutions between 25°C and 125°C: the rate equation. *Geochimica et Cosmochimica Acta*, **61**, 115–134.

Rickard, D.T. & Luther, G.W., III (1997) Kinetics of pyrite formation by the H_2S oxidation of iron(II) monosulfide in aqueous solutions between 25°C and 125°C: the mechanism. *Geochimica et Cosmochimica Acta*, **61**,135–147.

Rickard, D.T. & Luther, G.W., III (2007) Chemistry of iron sulfides. *Chemical Reviews*, **107**, 514–562.

Roden, E.E. (2003) Fe(III) oxide reactivity towards biological versus chemical reduction. *Environmental Science & Technology*, **37**, 1319–1324.

Roden, E.E. (2004) Analysis of long-term bacterial vs chemical Fe(III) oxide reduction kinetics. *Geochimica et Cosmochimica Acta*, **68**, 3205–3216.

Rosenthal, Y., Lam, P., Boyle, E.A. & Thomson, J. (1995) Authigenic cadmium enrichments in suboxic sediments: Precipitation and postdepositional mobility. *Earth and Planetary Science Letters*, **132**, 99–111.

Ruttenberg, K.C. & Berner, R.A. (1993) Authigenic apatite formation and burial in sediments from non-upwelling, continental-margin environments. *Geochimica et Cosmochimica Acta*, **57**, 991–1007.

Schieber, J. (2007) Oxidation of detrital pyrite as a cause for marcasite formation in marine lag deposits from the Devonian of the eastern US. *Deep-Sea Research II*, **54**, 1312–1326.

Schuffert, J.D., Jahnke, R.A., Kastner, M., Leather, J., Sturz, A. & Wing, M.R. (1994) Rates of formation of modern phosphorite off western Mexico. *Geochimica et Cosmochimica Acta*, **58**, 5001–5010.

Schuffert, J.D., Kastner, M. & Jahnke, R.A. (1998) Carbon and phosphorus burial associated with modern phosphorite formation. *Marine Geology*, **146**, 21–31.

Schulz, H.N. & Schulz, H.D. (2005) Large sulfur bacteria and the formation of phosphorite, *Science*, **307**, 416–418.

Scott, M.R., Scott, R.B., Rona, P.R., Butler, L.W. & Nalwak, A.J. (1974) Rapidly accumulating manganese deposit from the median valley of the Mid-Atlantic Ridge. *Geophysical Research Letters*, **1**, 355–358.

Severmann, S., Lyons, T.W., Anbar, A., McManus, J. & Gordon, G. (2008) Modern iron isotope perspective on the benthic iron shuttle and the redox evolution of ancient oceans. *Geology*, **36**, 487–490.

Sherman, D.M. & Peacock, C.L. (2010) Surface complexation of Cu on birnessite (δ-MnO_2) controls on Cu in the deep ocean. *Geochimica at Cosmochimica Acta*, **74**, 6721–6730.

Shi, Z, Krom, M.D., Bonneville, S., Baker, A.R., Jickells, T.D. & Benning, L.G. (2009) Formation of iron nanoparticles and increase in iron reactivity in mineral dust during simulated cloud processing. *Environmental Science & Technology*, **43**, 6592–6596.

Sholkovitz, E.R. (1976) Flocculation of dissolved organic and inorganic matter during the mixing of river water and seawater. *Geochimica et Cosmochimica Acta*, **40**, 831–845.

Sibley, D.F., Dedoes, R.E. & Bartlett, R. (1987) Kinetics of dolomitization. *Geology*, **15**, 1112–1114.

Sigleo, A.C. & Helz, G.R. (1981) Composition of estuarine colloidal material: major and trace elements. *Geochimica et Cosmochimica Acta*, **45**, 2501–2509.

Slomp, C.P., Epping, E.H.G., Helder, W. & Van Raaphorst, W. (1996) A key role for iron-bound phosphorus in authigenic apatite formation in North Atlantic continental platform sediments. *Journal of Marine Research*, **54**, 1179–1205.

Sørensen, J. & Jørgensen, B.B. (1987) Early diagenesis in sediments from Danish coastal waters – microbial activity and Mn-Fe-S geochemistry. *Geochimica et Cosmochimica Acta*, **51**, 1583–1590.

Stumm, W. & Morgan, J.J. (1981) *Aquatic Chemistry*. Wiley-Interscience, New York.

Sweeney, R.E. & Kaplan, I.R. (1973) Pyrite framboid formation: Laboratory synthesis and marine sediments. *Economic Geology*, **68**, 618–634.

Tardy, Y., Cheverry, C. & Fritz, B. (1974) Neoformation d'une argile magnesienne dans les depressions interdunaires du lac Tchad. Application aux domaines de stabilite des phyllosilicates alumineux, magnesiens et ferrifères. *Comptes Rendus de l'Académie des Sciences, Paris*, **27**, 1999–2002.

Taylor, K.G. & Curtis, C.D. (1995) The stability and facies association of early diagenetic mineral assemblages: an example from a Jurassic ironstone -mudstone succession, U.K. *Journal of Sedimentary Research*, **A65**, 358–368.

Taylor, K.G. & Macquaker, J.H.S. (2000) Early diagenetic pyrite morphology in a mudstone-dominated succession: the Lower Jurassic Cleveland Ironstone Formation, eastern England, *Sedimentary Geology*, **131**, 77–86

Taylor, K.G., Perry, C.T., Greenaway, A.M. & Machent, P.G. (2007) Bacterial iron oxide reduction in a terrigenous-sediment impacted tropical shallow marine carbonate system, north Jamaica. *Marine Chemistry*, **107**, 449–463.

Taylor, K.G. & Macquaker, J.H.S. (2011) Iron in marine sediments: minerals as records of chemical environments. *Elements*, **7**, 83–88.

Thamdrup, B., Fossing, H. & Jørgensen, B.B. (1994) Manganese, iron, and sulfur cycling in a coastal marine sediment, Aarhus Bay, Denmark. *Geochimica et Cosmochimica Acta*, **58**, 5115–5129.

Todorov, J.R., Chistoserdov, A.Y. & Aller, J.Y. (2000) Molecular analysis of microbial communities in mobile deltaic muds of southeastern Papua New Guinea. *FEMS Microbiology Ecology*, **33**, 147–155.

Toth, D.J. & Lerman, A. (1977) Organic matter reactivity and sedimentation rates in the ocean. *American Journal of Science*, **277**, 265–285.

Toth, J.R. (1980) Deposition of submarine crusts rich in manganese and iron. *Geological Society of America Bulletin*, **91**, 44–54.

Van Cappellen, P. & Wang, Y. (1996) Cycling of iron and manganese in surface sediments: A general theory for the coupled transport and reaction of carbon, oxygen, nitrogen, sulfur, iron and manganese. *American Journal of Science*, **296**,197–243.

Van Capellen, P. & Ingall, E.D. (1996) Redox stabilization of the atmosphere and oceans by phosphorus-limited marine productivity. *Science*, **271**, 493–496.

Vaughan, D.J. & Ridout, M.S. (1971) Mossbauer studies of some sulfide minerals. *Journal of Inorganic and Nuclear Chemistry*, **33**, 741–746.

Viers, J., Dupre, B. & Gaillardet, J. (2009) Chemical composition of suspended sediments in World Rivers: New insights from a new database. *Science of the Total Environment*, **407**, 853–868.

Vorhies, J.S. & Gaines, R.R. (2009) Microbial dissolution of clay minerals as a source of iron and silica in marine sediments. *Nature Geoscience*, **2**, 221–226.

Wallace, H.E., Thomson, J., Wilson, T.R.S., Weaver, P.P.E., Higgs, N.C. & Hydes, J.D. (1988) Active diagenetic formation of metal-rich layers in N.E. Atlantic sediments. *Geochimica et Cosmochimica Acta*, **52**, 1557–1569.

Walling, D.E. (2006) Human impact on land–ocean sediment transfer by the world's rivers. *Geomorphology*, **79**,192–216.

Walling, D.E. & Fang, D. (2003) Recent trends in the suspended sediment loads of the world' s rivers. *Global and Planetary Change*, **39**, 111–126.

Wallmann, K., Aloisi, G., Haeckel, M., Tishchenko, P., Pavlova, G., Greinert, J., Kutterolf, S. & Eisenhauer, A. (2008) Silicate weathering in anoxic marine sediments. *Geochimica et Cosmochimica Acta*, **72**, 2895–2918.

Walter, L.M. & Burton, E.A. (1990) Dissolution of recent platform carbonate sediments in marine pore fluids. I, *American Journal of Science*, **290**, 601–643.

Wang, Y. & Van Cappellen, P. (1996) A multicomponent reactive transport model of early diagenesis: Application to redox cycling in coastal marine sediments. *Geochimica et Cosmochimica Acta*, **60**, 2993–3014.

Wellmann, K., Aloisi, G., Haeckel, M., Tischenko, P., Pavlova, G., Greinert, J. & Eisenhauer, A. (2008) Silicate weathering in anoxic marine sediments. *Geochimica et Cosmochimica Acta*, **72**, 3067–3090.

Whitten, D.G.A., & Brooks, J.R.V. (1972) *The Penguin Dictionary of Geology*. Penguin Books, London, 514 pp.

Wilkin, R.T., Barnes, H.L. & Brantley, S.L. (1996) The size distribution of framboidal pyrite in modern sediments: an indicator of redox conditions. *Geochimica et Cosmochimica Acta*, **60**, 3897–3912.

Wilkinson, K.J., Balnois, E., Leppard, G.G. & Buffle, J. (1999) Characteristic features of the major components of freshwater colloidal organic matter revealed by transmission electron and atomic force microscopy. *Colloids and Surfaces*, **A155**, 287–310.

Wilson, T.R.S., Thomson, J., Colley, S., Hydes, D.J. & Higgs, N.C. (1985) Early organic diagenesis: the significance of progressive subsurface oxidation fronts in pelagic sediments. *Geochimica et Cosmochimica Acta*, **49**, 811–822.

Wilson, T.R.S., Thomson, J., Hydes, D.J., Colley, S., Culkin, F. & Sørensen, J. (1986) Oxidation fronts in pelagic sediments: diagenetic formation of metal-rich layers. *Science*, **232**, 972–975.

Windom, H.L. (1976) Lithogenous material in marine sediments. In: *Chemical Oceanography,* **5** (J.P. Riley & R. Chester, editors). Academic Press, London, pp. 103–135.

Wogelius, R.A. & Vaughan, D.J. (2013) Analytical, experimental and computational methods in environmental mineralogy. In: *Environmental Mineralogy II* (D.J. Vaughan & R.A. Wogelius, editors). EMU Notes in Mineralogy, **13**, European Mineralogical Union and The Mineralogical Society of Great Britain & Ireland, London, pp. 5–102.

Zhang, H., Davison, W., Miller, S. & Tych, W. (1995) In-situ high resolution measurements of fluxes of Ni, Cu, Fe, and Mn and concentrations of Zn and Cd in porewaters by DGT. *Geochimica et Cosmochimica Acta*, **59**, 4181–4192.

Zierenberg, R.A. & Shanks, W.C. III. (1983) Mineralogy and geochemistry of epigenetic features in metalliferous sediment, Atlantis II Deep, Red Sea. *Economic Geology*, **78**, 57–72.

Microbial controls on the mineralogy of the environment

Susan A. WELCH[1] and Jillian F. BANFIELD[2]

[1] *School of Earth Sciences and Byrd Polar Research Center,
The Ohio State University, Columbus OH 43210, USA,
e-mail: welch.318@osu.edu*
[2] *Department of Earth and Planetary Science, University
of California Berkeley, Berkeley CA 94720, USA,
e-mail: jbanfield@berkeley.edu*

Microbes have a dramatic impact on the mineralogy of the environment as a result of their metabolic processes. Microorganisms can control both acidity and redox chemistry as a result of respiration, photosynthesis, chemoautotrophy or anaerobic respiration. The net impact of these biogeochemical reactions can result in formation of minerals that have distinctive 'bulk' and surface chemical characteristics, morphology and trace element or isotopic composition that is distinct from those produced abiotically. This chapter describes several examples of both direct and indirect biomineralization including the role of neutrophilic and acidophilic iron oxidizers on the biogeochemistry of Fe minerals in two mine sites, the role of microbial metabolic processes on mineral weathering and nutrient cycling, and the enhanced mobility and enrichment of economically important trace elements such as zinc, gold and the rare earth elements.

1. Introduction

The purpose of this chapter is to provide some insight into how microorganisms affect the chemistry, mineralogy and physical characteristics of their surroundings. This chapter is not intended to be an exhaustive review, though some basic concepts are summarized and specific examples to illustrate in more detail the intimate connections between biological and mineralogical processes are given.

It is critical that studies of environmental mineralogy and geochemistry include analysis of the types of microorganism present and the ways in which they influence reaction kinetics in natural systems. Consequently, this paper concludes with a brief summary of microbiological methods that find special relevance to environmental mineralogy and a discussion of the importance of recent breaks through in genome analysis for Earth Sciences.

2. Microorganisms: what, when, where, how many?

Microorganisms are single-celled life forms that are found in great abundance in a wide variety of environments near the Earth's surface. They are subdivided into three groups,

© Copyright 2013 the European Mineralogical Union and the Mineralogical Society of Great Britain & Ireland
DOI: 10.1180/EMU-notes.11.5

or domains: Eukaryotes, Bacteria, and Archaea (Woese *et al.*, 1990; Winkler & Woese, 1991). Of these, the prokaryotes (Bacteria and Archaea) have the most ancient origins, probably extending back over at least 3.48 Ga (Schopf, 1993; Schopf & Packer, 1987; Allwood *et al.*, 2009). Prokaryotes have extremely diverse metabolic capabilities. They can utilize a large range of inorganic and organic compounds as energy sources and terminal electron acceptors, and can occupy niches that are not colonized by higher life forms.

Microbial metabolisms can be subdivided into several groups. Heterotrophs (many eukaryotes and prokaryotes) utilize fixed organic carbon as energy sources. Photoautotrophs (eukaryotes and prokaryotes) use light energy (photosynthesis) to fix CO_2 and produce organic carbon, and chemoautotrophs (prokaryotes) utilize chemical reactions (such as the oxidation of ferrous iron or ammonia) to generate energy and fix CO_2. Chemotrophic metabolism involves catalysis of energetically possible, but kinetically hindered, redox reactions. Because there are many chemical redox reactions in nature that meet these criteria under some conditions, microorganisms have considerable potential to dramatically alter the geochemistry and mineralogy of their surroundings (Lovely & Chapelle, 1995).

Microorganisms are nearly ubiquitous in the Earth's surface and subsurface environments. Microbes have been detected several kilometres below the Earth's surface (Stevens & McKinley 1995; Stevens *et al.*, 1993; Colwell *et al.*, 1997; Ghiorse, 1997; Onstott *et al.*, 1998; Fyfe, 1996; Fredrickson & Onstott, 1996; Pedersen, 1993, 1997) and microbial alteration of sediments and basalt occurs many hundreds of metres below the seafloor (Parkes *et al.*, 1994; Thorseth *et al.*, 1995; Fisk *et al.*, 1998; Santelli *et al.*, 2008; Schrenk *et al.*, 2010). Some microorganisms are capable of surviving in boiling water (associated with undersea hydrothermal vent systems above 100°C (Deming & Baross, 1993; Summit & Baross, 1998; Edwards *et al.*, 2011; Orcutt *et al.*, 2011) and in terrestrial hot springs, *e.g.* Barns *et al.*, 1994; Gihring & Banfield, 2001). Microbes can also withstand extreme cold (*e.g.* in Antarctic ice, Priscu *et al.*, 1999; Sun & Friedmann 1999; Vorobyova *et al.*, 1997), extreme salinity (*e.g.* in solutions saturated with NaCl; Eisenberg, 1995; Onstott *et al.*, 1998; Grote & O'Malley, 2011; Oren, 2008), very high radiation doses (1.0–1.5 Mrad gamma-irradiation; Minton & Daly 1995) extraordinary acidity (to pH values of ~0; Schleper *et al.*, 1996; Edwards *et al.*, 1999, 2000, high metal concentrations (*e.g.* thousands of ppm of arsenic: Gihring *et al.*, 2003; Gihring & Banfield, 2001), and intense desiccation (*e.g.* active endolithic populations in the soils and rocks of the dry valleys of Antarctica; see Johnston & Vestal, 1993; Sun & Friedmann, 1999).

It was once thought that the abundance, diversity and activity of organisms in subsurface regions below the soil zone are all low. However, surveys of subsurface environments show that microbial numbers are relatively high, $\sim 10^5 - 10^8$ cells/cm^3 (Onstott *et al.*, 1998; Sinclair & Ghiorse, 1989; Bone & Balkwill, 1988; Balkwill, 1989; Fredrickson *et al.*, 1989). Microbial abundance, activity, and diversity in sediments and aquifers are all variable, but do not decrease systematically with depth. Between $\sim 10^5$ and 10^6 cells/ml are typically reported from deep igneous rock aquifers (Haveman *et al.*, 1999; Stevens and McKinley, 1999; Santelli *et al.*, 2008; Schrenk *et al.*, 2010).

Surface waters and oceans contain cell numbers ranging between 10^4 and 10^7 cells/ml (*e.g* Moon-van der Staay *et al.*, 2001; Frias-Lopez *et al.*, 2008; Heidelberg *et al.*, 2010). Microorganism densities can be as high as 10^{12} cells/g in biofilms (*e.g* Bond *et al.*, 2000). Populations in soils typically range from $\sim 10^6$ to 10^9 microbial cells per cm^3 (in the rhizosphere, the region surrounding the root, cell abundances can be much higher (Hinsinger *et al.*, 2009 and references therein). Mineral surfaces may be populated by cells at densities that approach, and even exceed, one cell/μm^2 (Fig. 1). With the advent of molecular biology, we are not only able to determine total microbial cell numbers, but

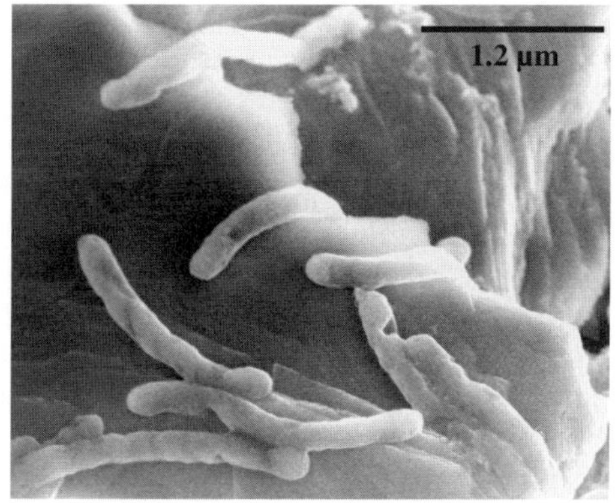

Fig. 1. SEM image illustrating very high cell densities ($>1/\mu$m^2) on an arsenopyrite surface. Colonization occurred following exposure of a prepared FeAsS surface to natural solutions for a period of a few months (samples of T. Gihring).

we also have the unprecedented opportunity to determine microbial cell metabolic diversity and activity. (*e.g.* Tyson *et al.*, 2004; Huber *et al.*, 2007; Heidleberg *et al.*, 2010; Wilmes & Bond, 2009) This is important for understanding microbial control of element mobility and biomineralization because these reactions can be impacted directly and indirectly by microorganisms, and understanding microbial community structure is important for understanding microbial control of geochemical reactions.

3. Microbial biomineralization

Biomineralization refers to a broad spectrum of biogeochemical processes whereby cells can exert direct or indirect control of geochemical processes in their environments, leading to precipitation or redistribution of elements (Kirschvink & Hagadorn, 2000; Skinner & Jahren, 2003; Frankel & Bazylinski, 2003). Examples of biomineralization can range from metal sorption to ligands on cell surfaces, to microbial control of redox conditions from energy generation or respiration leading to dissolution or precipitation of minerals, to the complex weathering reactions that result from increased acidity or organic exudates, to the formation of complex organic-mineral precipitates (*e.g.* shells) or complex combinations of these processes. These reactions may occur as a consequence of cell metabolism or they may proceed even if the cell is inactive. Biologically precipitated minerals may have distinctive 'bulk' and surface chemical characteristics, morphology, and a chemical or isotopic composition that is distinct from those produced abiotically. Many biominerals have particle sizes in the few to few tens of nanometers size range, resulting in unusual reactivity compared to abiotically formed phases. Phase stability and reaction kinetics are particle-size dependent, so that biomineralization products behave in ways that differ from macroscopic materials. These issues are dealt with below.

Because of their small size and large abundance, cells can be a significant source of reactive surface area in the environment. This is especially important in near-neutral solutions where abundant cell-surface carboxyl groups are deprotonated and readily able to adsorb metals from solution (Fowle & Fein, 1999; Fortin *et al.*, 1997; Reith *et al.*, 2007; Templeton & Knowles, 2009; Ginn *et al.*, 2010). Experimental studies of metal sorption on microbial cells show that metal sorption is both pH and species dependent, reflecting the concentrations of carboxyl, hydroxyl, phosphate, amine, and sulfur functional groups on the cell surfaces, and the affinity of metals for these sites. In metal-rich environments, metal sorption can lead to complete encrustation and fossilization of microbial cells (*e.g.* Reith *et al.*, 2006).

4. Microbial control of redox chemistry and elemental distribution patterns

An important category of biomineralization reactions are those that occur due to enzymatic redox reactions, a subset of which are involved in energy generation. Chemolithotropic microorganisms are able to obtain metabolic energy from inorganic compounds including reduced metals, predominately Fe and Mn, sulfur compounds, nitrogen compounds (ammonia, nitrite), or from oxidation of hydrogen or methane formation. Iron biomineralization and utilization of oxidized iron minerals as terminal electron acceptors for respiration are also important (Konhauser, 1998). A brief case study dealing with iron is given below (see Section 7.1). Considerable additional information about biological interactions with specific elements can be found in the excellent geomicrobiology text of Ehrlich (1996).

A variety of mineral dissolution and mineral precipitation reactions accompany biologically-mediated diagenesis in sediments. Organic carbon is respired by layered microbial communities that utilize a well-studied sequence of electron acceptors with increasing distance below the sediment–water interface (see Nealson & Stahl, 1997). Following depletion of oxygen by aerobic heterotrophs, dissimilatory nitrate reduction occurs, resulting in the formation of biomass and $N_{2(g)}$. Microbial oxidation of organic compounds using nitrate as a terminal electron acceptor yields almost as much energy as using oxygen. Below the nitrate reduction zone, utilization of manganese and iron as terminal electron acceptors leads to reductive dissolution of Mn(IV) and Fe(III) minerals. Below this zone, sulfate-reducing microorganisms predominate (Figure 2) and their activity leads to the precipitation of metal sulfides such as pyrite. Under extremely reducing conditions, fermentation and methanogenesis occur. These metabolic processes impact the sediment mineralogy and lead to an increase in Fe^{2+}, Mn^{2+} and S^{2-} in pore water. Magnetite production can occur external to iron-reducing bacterial cells due to reactions between ferrous iron released by the cells and ferric iron minerals. In the sulfate-reduction zone, where concentrations

Fig. 2. Examples of sulfide biomineralization. SEM image of microbially produced (**a**) ZnS (spheres) and (**b**) pyrite (pyritohedra and framboids) formed on wood surfaces from an abandoned mine near Tennyson, Wisconsin, USA; (**c**) framboidal pyrite, pyritohedra and marcasite from marine sediments, and (**d**) iron sulfide (bright phase)-encrusted microbial cells from Lake Fryxell sediments.

of dissolved sulfide and ferrous iron are elevated, metal sulfides such as greigite and mackinawite precipitate.

Microbial sulfate reduction in sediments and biofilms can result in saturation with respect to a variety of mineral phases that contain other elements in major and trace quantities. For example, transition metal sulfides may be precipitated from metal-rich fluids in proximity to sulfate-reducing organisms. If solutions contain high Zn^{2+} for example, sphalerite may result. This may be significant in the generation of economically important mineral deposits (*e.g.* ZnS accumulations, as reported by Labrenz *et al.*, 2000; Fig. 3). Other examples include deposition of copper sulfide minerals (such as digenite and djurleite) as corrosion products within biofilms on metal surfaces (see Little

Fig. 3. SEM image of ZnS (shown by electron diffraction to be nanocrystalline sphalerite) developed in association with sulfate-reducing bacteria from a Mississippi Valley-type, carbonate hosted Pb-Zn deposit (Labrenz *et al.*, 2001).

et al., 1997). These are remarkable in that their formation is unexpected on thermodynamic grounds, given the Eh and pH of the bulk fluid (through, they are expected based on the microenvironment within the biofilm).

In some cases, redox reactions may result in environmentally significant reduction in toxic metal abundance in solution. For example, symbiotic associations of fungi and algae and/or bacteria (lichens) can immobilize large quantities of uranium *via* intra- and extracellular uranium mineral precipitation (Barker *et al.*, 1998b; Haas *et al.*, 1998; Suzuki & Banfield, 1999). Experimental evidence suggests microbial sulfate and arsenic reduction may be accompanied by precipitation of arsenic sulfides (Newman *et al.*, 1997) and that sulfur-depositing microbes can sequester very significant concentrations of Se (Nelson *et al.*, 1996).

5. Microbes and mineral weathering

Microorganisms can dramatically modify the rates and products of mineral weathering reactions by controlling mineral solubility and surface reactivity (*e.g.* dissolution rates). This has great environmental importance because it leads to new patterns of mineral and, therefore, element distribution in soils and sediments (*e.g.* Neaman *et al.*, 2005a, b; Little *et al.*, 2005a, b, c; Goyne *et al.*, 2010; Rosling *et al.*, 2007, 2009; Welch *et al.*, 2007; Finlay *et al.*, 2009). Of course, by modifying mineral weathering rates, microorganisms also exert fundamental controls on soil formation and fertility, as well as on groundwater chemistry.

Specific attention has been paid to the effect of acidity, small organic molecules (organic acids) and microbial polymers in mineral dissolution and precipitation kinetics and metal mobility in weathering reactions. Most minerals exhibit a pH dependence of both solubility and dissolution rate with a minimum reactivity around the pH of the zero point of charge, and then increasing with increasing pH. The most basic mechanisms by which microbes affect mineral weathering rates are by generating CO_2 or carbonic acid, from respiration although there is considerable debate over just how important this phenomenon is in enhancing mineral reactivity (*e.g.* see Drever, 1994), even small changes in acidity and pCO_2 can affect mineral solubility, metal speciation, metal sorption or metal reactivity.

Microbes that control N, S, or Fe redox chemistry (either oxidation or reduction) can produce large changes in pH in the environment. Chemolithotrophic nitrifying bacteria generate nitric acid from reduced nitrogen compounds. For example, Lebedeva *et al.* (1979) noted substantial alteration of basalt associated with nitrate production by nitrifying bacteria. Conversely, both nitrate uptake and subsequent reduction and denitrification (anaerobic respiration) consume protons and will therefore increase solution pH. Sulfur and iron oxidizing prokaryotes can accelerate production of sulfuric acid (see Sections 7.4 and 7.5) by promoting oxidation of metal sulfide minerals, and this can dramatically accelerate weathering of silicate minerals and impact trace metal mobility (*e.g.* Welch *et al.*, 2009). Organisms also have other, more indirect, mechanisms for producing or consuming acidity. For example, the iron oxidation reaction (catalysed by iron oxidizing bacterial genera such as *Gallionella* and *Leptothrix*) consumes protons. However, subsequent precipitation of iron oxyhydroxides generates acidity (see Section 7.1 below, for more details).

Consideration of the impact of organic molecules on mineral-surface reactions includes analysis of the ways in which organic functional groups, often carboxylic acids, complex ions in solution and bind ions on surfaces. This can lead to dramatic changes in the rates of retreat of surface steps, leading to novel surface morphologies (*e.g.* Teng & Dove, 1997). In Section 7.3 below, the roles of organic complexation, microbial uptake, and organic ligand-promoted dissolution in solubilizing silicates and extremely insoluble secondary phosphates are considered. These processes are especially relevant to understanding lichen–mineral interactions and microbial impact on geochemical processes in the rhizosphere.

Microbial dissolution may be largely responsible for other important geochemical processes, some of which may lead to environmental damage. Section 7.4 illustrates how iron and sulfur oxidizing prokaryotes can affect the rates of metal sulfide dissolution, and thereby impact the terrestrial iron and sulfur geochemical cycles. The phenomenon of formation of sulfuric acid-rich and metal-rich solutions is often associated with mining because ore extraction increases access of oxygen and water to high concentrations of sulfide minerals. The acid mine drainage example also illustrates that through control of mineral dissolution kinetics, microorganisms can ensure optimal conditions for their survival.

6. Direct biomineralization

Metabolism-induced mineralization reactions can be both intra- and extracellular. In some cases, mineral precipitation occurs under strict genetic control. One example is the formation of $CaCO_3$ coccoliths within intracellular compartments (golgi) by Coccolithophoridae (de Vrind-de Jong & de Vrind, 1997). A second example is the internal precipitation of magnetite or greigite magnetosomes by bacteria, apparently for navigation (Fig. 4, as described by Devouard *et al.*, 1998). Magnetic oxide precipitation has special significance because the few tens of nanometre crystal diameter confers single domain magnetic properties. Thus, magnetite derived from magnetosomes exerts an important control on the magnetic properties of sediments (see Bazylinski & Moscowitz, 1997). A third example is the deposition of silica tests by diatoms (Fig. 5). These are remarkable in that they form in water that is under-saturated with respect to amorphous SiO_2.

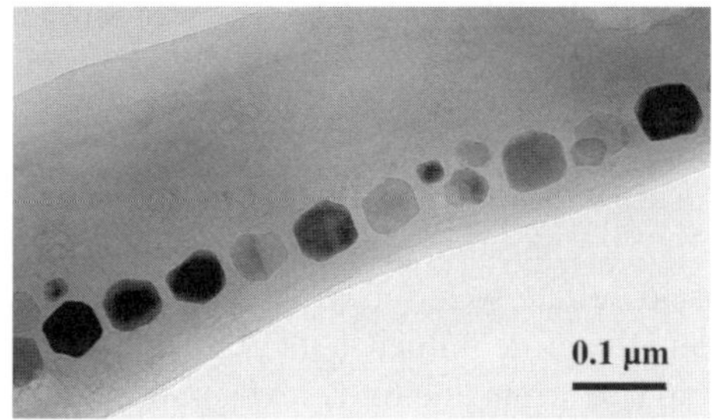

Fig. 4. TEM image of bacterial magnetite crystals (dark, euhedral particles) inside the magnetosome of a magnetite-producing bacterium. (Image provided by B. Devouard *et al.*; details as reported by Devouard *et al.*, 1998.)

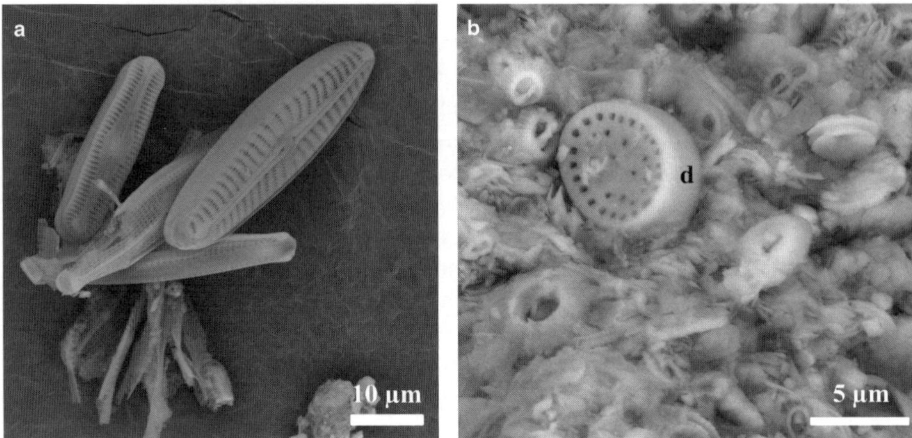

Fig. 5. SEM image illustrating the very elaborate forms of silica precipitated by diatoms: (**a**) pennate diatoms from surface sediments of Lake Fryxell, Antarctica, and (**b**) centric diatom (d) in a coccolith-rich calcareous ooze.

7. Examples

7.1. Biomineralization due to enzymatic iron oxidation

Iron biogeochemistry is one of the currencies of life. The capacity for Fe-based metabolism, either as an electron donor in energy generation or as a terminal electron acceptor in respiration is broadly distributed among bacteria and archaea. It is thought to be one of the earliest evolved metabolic pathways, resulting in the earliest biominerals (Kirschvink & Hagadorn, 2000). Microbially mediated reactions can result in the formation of Fe-bearing phases with different mineralogy, morphology, geochemistry or an isotope composition that is characteristic of their biogenic origin and distinct from those formed through abiotic processes (*e.g.* Chan *et al.*, 2004, 2009, 2011; Welch & Banfield, 2002) (Fig. 6). The activities of Fe bacteria exert critical influence on many major elemental cycles, including the carbon cycle, and presumably have done so over geological time. For example, the formation of the widespread Archaean Banded Iron Formations has been attributed to microbially mediated Fe oxidation (Kappler & Straub, 2005). The same process is a major factor contributing to acid mine drainage and the acid sulfate soils today. All organisms need traces of Fe for their metabolic processes and therefore influence its geochemistry. Fe is often a limiting trace nutrient in oxidized environments due to its low solubility at neutral pH. Many microbes have the ability to produce Fe-binding compounds, siderophores, to sequester trace amounts of Fe from the environment (*e.g.* Kraemer *et al.*, 2005). Other microorganisms, such as the magnetotactic bacteria, produce single-domain magnetite crystals within their magnetosomes (Kirschvink & Hagadorn, 2000; Fortin & Langley, 2005) that are characteristic of their biogenic origin.

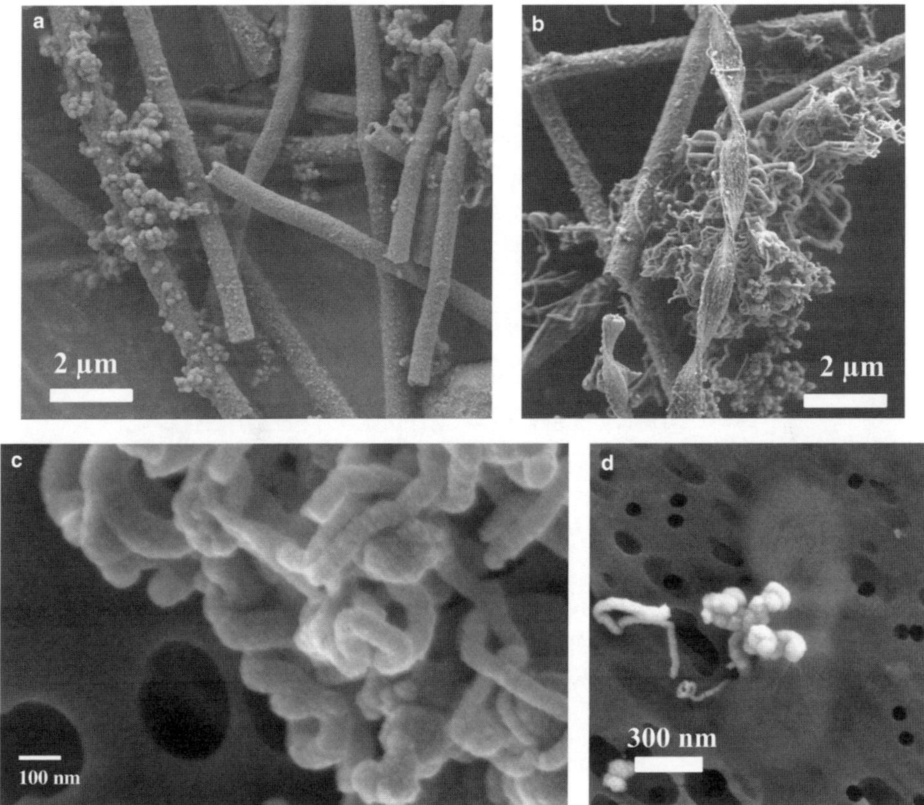

Fig. 6. SEM images of biogenic iron precipitates from neutrophilic iron oxidizers *Gallionella* sp and *Leptothrix* sp: (**a**) hollow tubes of *Leptothrix*; (**b**) twisted iron filament from *Gallionella*; (**c**) intricately twisted filaments from iron oxidizing biofilm, and (**d**) iron precipitates with associated microbial cell.

Iron is the most abundant redox-active element in the Earth's crust. Because of the very rapid rate of ferrous iron oxidation in oxygenated, near-neutral solutions, reduced iron is not bioavailable for metabolic energy generation. However, there are two commonly occurring habitats for iron oxidizing microbes where inorganic oxidation rates are suppressed. Firstly, iron oxidizers can thrive at near-neutral pH in micro-aerophilic (oxygen poor) solutions. Secondly, iron oxidizers are found in very acidic systems such as acid mine drainage sites (as discussed in Section 7.4, below), where inorganic iron oxidation rates are suppressed. Biologically induced precipitation of iron phases can be important in both of these environments. In the example below we focus on neutrophilic iron oxidation.

Neutrophilic iron-oxidizing organisms are difficult to grow, and this has limited their biological characterization. It is probable that the pH gradient between the cell interior and the external solution is much reduced compared to that for acidophiles. This makes metabolism based on iron oxidation more challenging as it reduces the proton motive force (pmf) used generate adenosine triphosphate(ATP) (Fig. 7). However, iron oxidation at near-neutral pH yields more energy than that at low pH. Iron oxidizing bacterial genera such as *Gallionella* appear to derive energy from iron oxidation, but can also grow heterotrophically. In the case of *Gallionella*, CO_2 fixation linked to iron oxidation has been demonstrated (Hallbeck & Pedersen, 1991). Recent detection of new iron oxidizing neutrophilic bacteria by several authors suggests that this metabolism may be far more prevalent than previously appreciated (*e.g.* Emerson & Moyer, 1997; Emerson *et al.*, 1999).

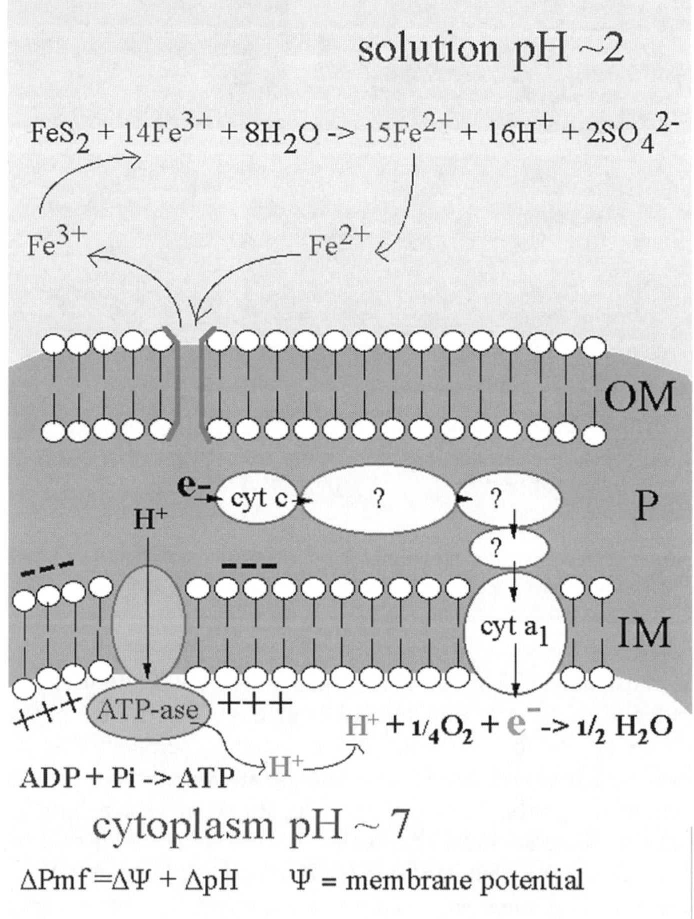

Fig. 7. Diagram illustrating the coupling between sulfide dissolution, ferrous iron oxidation, and energy generation in *Thiobacillus ferrooxidans*. The figure is a compilation from figures and information reported in Ehrlich (1996) and other sources. The electron generated by iron oxidation passes down the electron transfer chain to a terminal oxidase, where it is combined with a proton and oxygen to make water. Formation of ATP, an important energy molecule, occurs when the proton passes through the protein complex in the membrane.

The example described here is from the Tennyson mine, an underground Pb-Zn mine hosted in Ca-Mg carbonate rocks. Reduced, iron-rich ground water that has interacted with metal sulfides in the ore deposit enter the flooded mine tunnels through fractures. These fluids mix with slightly more oxygenated pH 7.2–8.6 water in the tunnels. Iron oxidizing microorganisms colonize the redox interface. Species such as *Gallionella* and *Leptothrix*, easily identified based on their morphology and association with finely crystalline iron oxyhydroxide products, live in the gradient region. Other (as yet unidentified) microorganisms are also present. In the 30 years since the mine was flooded, a thick layer (many tens of

Fig. 8. Photograph of an ~ 20 cm wide iron accumulation on the wall of a tunnel in the flooded Piquette Mine, Tennyson, Wisconsin, USA. This image was captured from a digital video recorded by Tom O'Connor with the assistance of Tamara Thomsen-Ebert and her colleagues.

cm) of polymer-loaded colloidal ferrihydrite, feroxyhyte and goethite has accumulated on the tunnel floors (Banfield *et al.*, 2000). Rust-coloured stalactites and stalagmites have also developed on the roof and tunnel walls at the points of entry of iron-bearing anoxic groundwater (Fig. 8).

Because it is difficult to generate pmf under neutral pH conditions, it is important to consider the possible role of mineral precipitation in energy generation. At $\sim$pH 8, the relevant aqueous iron species is $Fe^{2+} \cdot (H_2O)_{6 \, (aq)}$. Half reaction 1 occurs on enzymes located in the periplasm or at the cell surface. The electrons generated in reaction 1 are passed down an electron chain to a terminal oxidase, where half reaction 2 occurs. The solubility of ferric iron at near-neutral pH is perishingly small, so reaction 1 is followed rapidly by reaction 3.

$$2Fe^{2+}_{(aq)} \longrightarrow 2Fe^{3+}_{(aq)} + 2e^- \tag{1}$$

$$2H^+_{(aq)} + 1/2O_{2(g)} + 2e^- \longrightarrow H_2O_{(aq)} \tag{2}$$

$$2Fe^{3+}_{(aq)} + 4H_2O_{(aq)} \longrightarrow 2FeOOH_{(s)} + 6H^+_{(aq)} \tag{3}$$

The overall reaction is:

$$2Fe^{2+}_{(aq)} + 1/2O_{2(g)} + 3H_2O_{(aq)} \longrightarrow 2FeOOH_{(s)} + 4H^+_{(aq)} \tag{4}$$

A key feature of reaction 4 is that protons are generated on the cell surface simultaneously with precipitation of iron oxyhydroxide. This enhances the pH gradient between the cytoplasm and cell exterior, and therefore may play a key role in generation of pmf. Note that the oxygen in the goethite (or ferrihydrite, which contains extra water

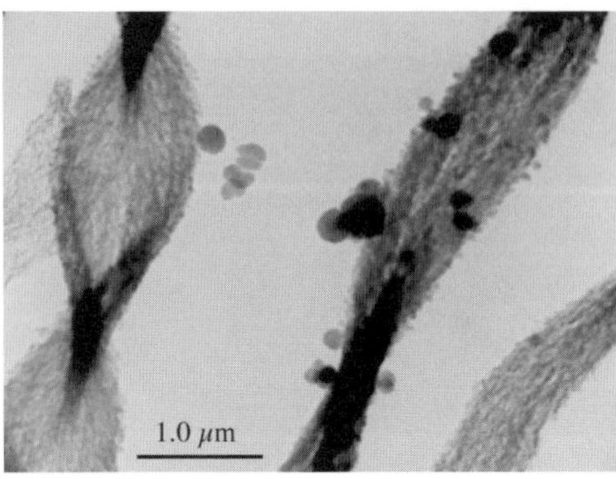

Fig. 9. TEM image showing iron oxyhydroxide particles coating stalk products of bacteria belonging to the *Gallionella* genera. Colloidal accumulations of iron oxyhydroxides are also common. These materials comprise the bulk of reddish materials such as shown in Figure 6.

in the formula) necessarily must come from water and not oxygen, because the half reactions are spatially separated.

Homogeneously sized 2–3 nm diameter nanoparticles nucleate in solution from dissolved ferric iron, which has minimal solubility at pH ∼8. Most electron diffraction patterns are typical of 2-line ferrihydrite, but some indicate 6-line ferrihydrite (2-line ferrihydrite and 6-line ferrihydrite probably share essentially the same structure; Drits *et al.*, 1993a,b; Manceau & Drits, 1993). A few diffraction patterns are consistent with feroxyhyte (Drits *et al.*, 1993a). Ferrihydrite and feroxyhyte, plus minor hematite and goethite formed by transformation from ferrihydrite and feroxyhyte after coarsening, are typical biomineralization products of iron oxidizing organisms.

In near-neutral or acidic solutions, ferrihydrite and goethite surfaces are positively charged. Consequently, some nanoparticles bind to the negatively charged organic cell surface polymers. In low-resolution TEM images of a *Gallionella* stalks (Fig. 9) the dark contrast is due to a few nm-thick coatings of these particles. However, the majority of small particles flocculate to form sub-micron diameter colloidal aggregates (Fig. 9). Flocculation is promoted by the similarity between the pH of the solution and the pH values at which ferrihydrite surfaces have zero net charge. Flocculation is important for subsequent crystal growth and microstructure development, as described in section 7.2.

Positively charged oxyhydroxide surfaces are important sites for adsorption of negatively charged chemical species such as arsenate, phosphate, carbonate and sulfate. Thus, Fe oxyhydroxide phases can control the distribution of many components, including toxic elements and biologically essential nutrients in the environment.

Ferric iron oxyhydroxides plus organic matter produced in the micro-aerophilic zone provide the basis for a second metabolism of geochemical and mineralogical importance. Briefly, in the anoxic zone, iron-reducing organisms (such as those discussed above in the context of sedimentary environments) couple respiration of organic debris to iron reduction. In the flooded Tennyson mine system, the ferrous iron diffuses up to more oxidized regions, where it is again utilized by iron-oxidizing prokaryotes or it combines with sulfide released by sulfate-reducing organisms to generate nanophase iron sulfide minerals.

7.2. Consequences of small particle size for biomineral behaviour

A fundamental property of many products of low-temperature reactions, including biomineralization reactions, is very small particle size. In the case described above, small particle size results from rapid nucleation due to supersaturation following iron oxidation. Small nucleus size is preserved due to low solubility of ions, which limits diffusion-based crystal growth.

Materials scientists have long recognized that small particle size confers novel materials properties, including unusual optical and electronic behaviour and modified surface structure and reactivity compared to the bulk phase. These phenomena are also extremely relevant to environmental sciences because nanophase materials are abundant in solutions, sediments, and soils. We do not provide a comprehensive overview of work on nanocrystalline materials here; rather, we draw upon a few case studies to illustrate some important consequences of small particle size.

Working with TiO_2 as a model system, Gribb & Banfield (1997), Zhang & Banfield (1998, 1999, 2000), Zhang *et al.* (1999) and Penn & Banfield (1998a, 1998b, 1999a, 1999b) explored the interrelationships between crystal size, thermodynamic stability, phase-transformation kinetics, and crystal-growth kinetics. The TiO_2 system is characterized by a variety of polymorphic structures, including anatase, brookite, and rutile. Although phenomena involving these specific minerals are of only minor geological importance, analysis of this system provides clues as to how other complex polymorphic compounds (*e.g.* oxides and oxyhydroxides of Fe, Al and Mn) may behave.

Surface energy is surface-structure dependent. Consequently, specific crystallographic surfaces on a single crystal have different surface free energies, probably scaling approximately with the number of unsatisfied bonds. Furthermore, because different polymorphs have different bulk and surface structures, they must have different average surface energies. Surface energy becomes a significant contributor to total energy when surface area is large (*i.e.* particle size is small). Because many members of polymorphic groups have bulk structure energies that differ by only a few kJ, differences in surface energy can lead to stabilization of different phases than expected for macrocrystalline materials. To our knowledge, Langmuir (1971) first considered reversal of phase stability due to small particle size. The phenomenon in titania was analysed in detail, both experimentally and theoretically, by Gribb & Banfield (1997) and Zhang & Banfield (1998, 1999). Phase-stability reversal was demonstrated using calorimetric surface energy measurements by McHale *et al.* (1997). It is probable that surface energies are critically important in determination of structures of products of biomineralization reactions, although other kinetic factors are certainly important too.

Typically, crystal growth is considered to occur by dissolution of small particles and atom by atom growth of larger, more stable particles. However, it is now clear that an additional coarsening mechanism can also operate when particle size is small. This involves rotation of adjacent particles so that they share the same orientation, followed by elimination of surface-bound water to create a coherent interface. The driving force for both atom-by-atom growth and growth *via* oriented aggregation of small particles (self-assembly), is surface-energy reduction. Because many small nuclei result from high degrees of supersaturation encountered in the vicinity of metabolizing cells, oriented aggregation can be an important path for crystal growth in biomineralization products (Banfield *et al.*, 2000). Growth *via* oriented aggregation should be most important for phases such as iron oxyhydroxides and iron oxides, where atom-by-atom growth is inhibited by the extremely low solubility of ferric iron.

There is a variety of geochemical and microstructural consequences if biomineralization products grow by oriented aggregation. Firstly, impurities adsorbed to the surfaces that are eliminated by oriented aggregation may be metastably incorporated into the bulk (Banfield *et al.*, 2000). Secondly, if adjacent surfaces involved in attachment are atomically rough, dislocations form (Penn & Banfield, 1998a; Banfield *et al.*, 2000). Finally, twins and polytypic intergrowths can result if coherence is achieved only in the two dimensions of the interface (Penn & Banfield, 1998b). Interfaces may be structurally distinct from the bulk phase, thus can play important roles in nucleating subsequent phase transformations, *e.g.* the formation of rutile on anatase crystal surfaces (see Penn & Banfield, 1998b, 1999b). All microstructures that arise from nanocrystal aggregation have the potential to subsequently modify the reactivity, and thus the form and fate of biomineralization products. Factors that restrict particle mobility, such as attachment to a polymer substrate, will inhibit growth and microstructure development. This may lead to preservation of cell-related morphology, which is of significance to fossilization and biosignature preservation (Banfield *et al.*, 2000).

7.3. Biological dissolution in soils and sediments

Microorganisms can dramatically affect the rates, mechanisms, and products of mineral weathering reactions. This can occur as a by-product of microbial metabolism or as the direct result of release of a specific molecular product designed to increase bioavailability of a desired nutrient. Dissolution rates can be enhanced due to change in the solution pH, especially in microenvironments surrounding attached cells. Rates can also be enhanced through production of organic ligands that complex with aqueous ions or ions attached to mineral surfaces, or by changing redox conditions. In addition to these direct chemical effects, organisms can have indirect effects on mineral weathering reactions. For example, rates may be modified due to uptake of ions released from minerals, thus modifying the solution saturation state, or by retaining water at the mineral surface. We illustrate these effects using two case studies. Firstly, we discuss the impact of microbial communities on silicate mineral weathering using the lichen–mineral interface as an example. Secondly, we discuss how microbes can impact dissolution of primary and secondary phosphate minerals to illustrate how multiple factors contribute to unexpected mineral and elemental distribution patterns in microbially impacted environments. Processes documented in both cases are directly relevant to soil formation and plant nutrition.

7.3.1. The lichen–mineral interface and silicate-mineral weathering

Electron microscopic observations of assemblages of minerals in rocks colonized by lichens (symbiotic associations of fungi and photosynthetic bacteria or algae) showed that lichen-associated minerals are extensively weathered compared to rock surfaces adjacent to the lichens (*e.g.* see Barker & Banfield, 1996, 1998). At the base of the lichen thallus, minerals are in intimate contact with microbial cells (primarily fungal hyphae) and their extracellular polymers. Increased weathering intensity is, in part, due to increase in reactive surface area due to physical disaggregation of the rock surface by the fungal hyphae. Clays and oxyhydroxides, with structures and compositions different to those formed at distance from the lichen thallus, precipitate at corroded surfaces and on organic polymers. Extensive etching of mineral surfaces coated by the extracellular polymers, as well as enhanced alteration of mineral surfaces at distance from the lichen-mineral interface, have been observed (see Barker & Banfield, 1996, 1998 for details). These data suggest that both direct microbial impact *via* polymeric compounds and indirect biological impact through dissolved, low-molecular-weight organic by-products are important.

Accelerated dissolution in proximity to lichen communities may be due largely to production of acidic compounds. Mineral dissolution rates for most silicates are highest under acidic conditions and decrease as pH increases. The rate of mineral dissolution, R, as a function of pH can be described as:

$$R = k_1 a_{H+}{}^n + k_2 + k_3 a_{OH-}{}^m. \tag{5}$$

(from Blum & Lasaga, 1988; Drever, 1994; Chou & Wollast, 1985).

The pH of the rate minimum corresponds approximately to the average pH of the point of zero net charge of the mineral surface, reflecting the degree of protonation/deprotonation of surface oxygens. For silicate minerals, the value is typically 4–8, and thus in the pH range of typical weathering solutions. Additional discussion of the rate constants k_1, k_2 and k_3 and the pH dependence of rates in acid and basic solutions (n and m) can be found elsewhere (*e.g.* see White & Brantley, 1995; Drever, 1994; Brady & Walther, 1989).

Lichens and their associated bacteria can affect mineral weathering reactions by production of carbonic acid due to respiration. Degradation of organic carbon in soils and groundwater can lead to pCO_2 levels that are several orders of magnitude greater than atmospheric CO_2 levels (Chapelle *et al.*, 1987; Keller & Wood 1993). However, this increase in acidity should only cause a small increase in mineral weathering/dissolution rates because carbonic acid is a weak acid and water in equilibrium with relatively elevated pCO_2 levels will only have pH decreased to $\sim$4.5 (Drever, 1994).

Decrease of pH in microenvironments within cleavage spaces in biotite has been demonstrated experimentally. Barker *et al.* (1998a) measured a pH of $\sim$3–4 around bacterial cells that had colonized microenvironments within biotite cleavage planes, whereas the pH of the bulk solution remained near-neutral. Assuming that bacteria in these microenvironments covered only $\sim$1% of the mineral surface area, they accounted for $\sim$50–90% of the all material released to solution. pH values of <3 also have been measured in thick microbial biofilms (Parasuraman, 1995).

It is probable that all organisms in lichens and analogous microbial communities associated with roots (rhizosphere communities, primarily bacteria and fungi) produce organic ligands. These affect silicate weathering reactions, both due to their acidity and complexing ability. Oxalic acid is often cited as the main contributor to biogeochemical weathering. However, microbes produce and excrete other low-molecular-weight organic acids *e.g.* fermentation products (such as lactate, formate) and Krebs cycle compounds (such as citrate, succinate, α-ketoglutarate, oxalacetate, pyruvate). Low-molecular-weight organic acids increase soil and mineral dissolution rates by a factor of 2 to 100 depending on experimental conditions (Welch & Ullman, 1993, 1996; Ullman *et al.*, 1996; Stillings *et al.*, 1996; Little *et al.*, 2005a, b, c). Organic acids catalyse the reaction by forming complexes with Al and Fe, either at the mineral surface (thereby weakening metal–oxygen bonds), or in solution (thus, lowering the solution saturation state). Di-functional organic ligands, those that can form bidentate complexes with metals, had a much larger effect than mono-functional ligands such as acetate or formate. Many other laboratory studies have demonstrated correlations between microbial production of low-molecular-weight organic acids and increased mineral dissolution rates (see Barker *et al.*, 1997 for a review).

Microorganisms within lichens and other microbial communities produce abundant high-molecular-weight organic molecules. These can directly or indirectly affect mineral weathering reactions. Polysaccharides and cell-wall components contain functional groups such as carboxyl, phosphate, and teichoic or teichuronic acids. Because these can complex with dissolved ions, they lower the solution saturation state and promote dissolution.

Welch *et al.* (1999) conducted abiotic mineral dissolution experiments with polysaccharides to evaluate the magnitude of the effect inferred from characterization of the lichen–mineral interface. Their results showed that polysaccharides enhance or inhibit mineral dissolution rates, depending upon the experimental conditions. The magnitude of the effect depended on pH, polymer composition and polymer size. At near-neutral pH, acid polysaccharides (alginates, pectin, gum xanthan) inhibited the net release of ions from the mineral to solution by irreversibly binding to the mineral surfaces. However, neutral polysaccharides (starch, cellulose) had no effect. Under mildly acidic conditions (pH 3 to 4.5) acid polysaccharides were able to accelerate mineral dissolution by a factor of 2 to 100 times compared to the control experiments. The maximum enhancement occurred near the pH of the pK_a of the acid functional groups, where there were sufficient protons to react with the mineral surface, and available free ligands to complex with metal ions released to solution. Although the reaction was not stoichiometric and solutions were greatly supersaturated with respect to many possible aluminosilicate phases, crystalline secondary products were not detected.

Phases developed at lichen–mineral interfaces are compositionally distinct compared with those formed by mineral weathering in the rock below (Barker & Banfield 1996, 1998). In part, this may be due to the production of acidity by microorganisms. For example, when exposed to oxygenated, near-neutral pH solutions at distance from the soil zone, biotite in granite is converted to vermiculite and interlayered biotite-vermiculite, primarily *via* leaching of interlayer K and oxidation of ferrous iron, with minimal restructuring of the silicate 2:1 layers (Banfield & Eggleton, 1988). Similar results have been reported from experimental studies conducted at near neutral pH (*e.g.* Malstrom & Banwart, 1997; Kalinowski & Schweda, 1996). Under more extreme weathering conditions, in highly microbially colonized regions in proximity to the soil zone, biotite and vermiculite convert to halloysite (Welch *et al.*, unpublished data) in a reaction that requires disassembly of the silicate sheets (halloysite forms by reprecipitation).

Experimental studies have also shown that at near-neutral pH in inorganic solutions, Si is released from feldspar to solution, leaving Al-rich secondary minerals. Aluminium-enriched secondary minerals are typical products of natural weathering of feldspars (*e.g.* see Banfield & Eggleton, 1990). In contrast, at near-neutral pH, experimental studies suggest that organic acids preferentially mobilize Al from the mineral surfaces, leaving products relatively enriched in silica. Thus, it is clear that biologically (or organically)-mediated mineral weathering can lead to the formation of different secondary minerals than expected if the mineral weathered abiotically.

7.3.2. Microbial weathering of phosphate minerals

Phosphate is a vital nutrient for all organisms. However, in many environments, phosphate is a limiting nutrient. The concentration of dissolved phosphate in natural waters is commonly controlled by the solubility of phosphate minerals such as apatite, $Ca_5(PO_4)_3OH$, or insoluble Al-, Fe- or lanthanide-phosphates such as primary monazite and secondary vivianite, wavellite, florencite, or rhabdophane (*e.g.* see Banfield & Eggleton, 1989 and references therein).

Mineral phosphate-solubilizing microorganisms are common in soils and sediments, especially in rhizosphere soil (Hinsinger & Gilkes, 1997; Lapeyrie *et al.*, 1991; Anderson *et al.*, 1985). The abundance of microbial cells in phosphate-limited environments may be correlated with the distribution of phosphate minerals (*e.g.* Rogers *et al.*, 1998; Taunton *et al.*, 2000a, 2000b; Rosling *et al.*, 2007). This association may impact the stability of associated minerals. For example, in an *in situ* study of feldspar weathering, Rogers *et al.* (1998) noted an increase in microbial colonization and weathering of feldspars that contained apatite inclusions.

The dissolution rates and solubilities of phosphate minerals are complex functions of pH. Dissolution rates can be increased by complexation of ions by organic acids. Organic molecules form metal−ligand bonds at the mineral surface or in solution, thereby weakening surface metal−oxygen bonds. Alternatively, they can decrease the solution saturation state and, thus, increase the rate of PO_4^{3-} release to solution.

Many experimental studies have demonstrated that microbially produced inorganic and organic acids accelerate phosphate mineral (primarily apatite) dissolution rates (*e.g.* Welch *et al.*, 2002; Illmer & Schinner, 1995; Illmer *et al.*, 1995; Hinsinger & Gilkes, 1997; Anderson *et al.*, 1985; Lapeyrie *et al.*, 1991; Azcon *et al.*, 1976). Apatite dissolution experiments with assemblages of organisms cultured from a weathered granite (Welch *et al.*, 2002) have shown that microbes can increase phosphate release from apatite by ten to one hundred times compared to abiotic controls, by producing complexing organic ligands and generating acidity. However, these organisms are also able to solubilize phosphate without a significant change in pH or production of organic acids. The explanation for this phenomenon is unclear, but it is possible that the organisms produce enzymes that catalyse dissolution (these enzymes may be analogous to phosphatases that extract phosphorus from organic polymers). Microbes also have other mechanisms for increasing the weathering rates of phosphate minerals.

Phosphate uptake by the cells will decrease the solution saturation state, and thereby promote mineral phosphate dissolution (*e.g.* Hinsinger & Gilkes 1997).

Because apatite dissolution rates are so sensitive to acidity (Welch *et al.*, 2002), acid production readily explains the disappearance of apatite as rocks weather. However, a subset of the phosphate released by apatite dissolution may be sequestered into alteration products. For example, Banfield and Eggleton (1989) showed that apatite weathering in granite is accompanied by dissolution of primary lanthanide silicates (allanite). Lanthanides released to solution were sequestered into secondary phosphate minerals such as rhabdophane [hydrous (La, Ce, Nd)-phosphate; $K_{sp} \sim 10^{-24}$] and florencite. In regions close to the soil, organic carbon is available to sustain significant populations of bacteria and fungi. Taunton *et al.* (2000a,b) noted that although secondary phosphates persist in very highly weathered samples at distance from the soil zone, they are rapidly dissolved in weathered rock below the soil zone. Their disappearance correlates with microbial colonization of the surfaces of the secondary lanthanide-phosphate minerals. Extremely insoluble phosphates are difficult to dissolve by acid production alone. Lanthanide phosphate solubilization is explained as the result of the interacting effects of microbial phosphate uptake, extremely efficient organic complexation of dissolved trivalent lanthanide ions (dramatically increasing the solubility of the lanthanide phosphates), and accelerated dissolution kinetics *via* acid or ligand-promoted pathways (Taunton *et al.,* in prep.).

7.4. Microorganisms and acid mine drainage mineralogy

When rocks containing abundant metal sulfides (often mostly pyrite, FeS_2) dissolve, sulfuric acid-rich solutions result. This effect is most pronounced when the rocks contain abundant metal sulfides (*e.g.* ore bodies and coal deposits) and can lead to solutions with pH values <0 (*e.g.* see Nordstrom *et al.*, 2000). When acid production is associated with ore deposits, the problem is typically referred to as acid mine drainage.

The overall pyrite dissolution reaction is often written to show oxygen or ferric iron as the oxidants and ferrous iron, protons, and sulfate (sulfuric acid) as products. However, the transformation involves many steps because seven electrons must be lost to convert sulfide (in pyrite) to sulfate (Rimstidt & Vaughan, 2003). There are many potential intermediate sulfur and sulfoxy compounds generated in the process, including elemental sulfur, thiosulfate, tetrathionate, and sulfite. Some of the redox reactions in the $S^0 \rightarrow SO_4^{2-}$ pathway are kinetically inhibited. Thus, certain microorganisms can utilize oxidation of some sulfur-bearing compounds (by oxygen) to generate metabolic energy. Aqueous Fe^{2+} is also an important source of metabolic energy, as noted above. Iron- and sulfur-oxidizing organisms fix CO_2, thus providing organic compounds that can be consumed by bacterial heterotrophs, heterotrophic fungi, and more complex eukaryotes (protists).

Most work suggests that the critical coupling between sulfide dissolution rates and microbial populations involve catalysis of iron-oxidation by certain bacterial and archaeal species. These organisms can affect reaction kinetics because ferric iron is the primary surface oxidant at low pH (see Moses *et al.*, 1997 and the review by Nordstrom & Southam, 1997) and its inorganic reoxidation is kinetically inhibited.

One of the most studied systems is that involving the bacterial iron oxidizing species *Acidothiobacillus ferrooxidans*. However, our studies at the Iron Mountain acid mine drainage site (Edwards *et al.*, 1999) suggest that this organism's main role may be iron oxidation downstream from acid-generating sites, and that their primary effect may be enhanced precipitation of iron oxyhydroxides (Schrenk *et al.*, 1998; Edwards *et al.*, 1999). More recent work has documented the abundance and distribution of other iron-oxidizing bacterial (*e.g. Leptospirillum ferrooxidans*) and archaeal (*e.g. Thermoplasmales*) species in proximity to the pyrite ore (Edwards *et al.*, 2000; Bond *et al.*, 2000) where they probably dramatically impact pyrite dissolution.

The basis for energy generation *via* iron oxidation in the other more relevant prokaryotes is unclear, but it is probably similar to that utilized by *A. ferrooxidans* (as summarized in reactions 1 and 2, above, and Fig. 6). Protons are consumed according to reaction 2, thus avoiding acidification of the cytoplasm.

Biological sulfur oxidation can be especially important in acid generation under conditions where significant elemental S accumulation occurs on pyrite surfaces (McGuire *et al.*, 2001). The initial reactions probably primarily utilize ferric irons as the oxidant (as argued in an alternative pathway by Moses *et al.*, 1997) and do not generate protons. They can be summarized as follows:

$$FeS_{2(s)} + 2Fe^{3+}_{(aq)} \longrightarrow 3Fe^{2+}_{(aq)} + 2S^0_{(s)} \tag{6}$$

Organisms using intermediate sulfur compounds and oxygen as the terminal electron acceptor generate acidity *via* a reaction summarized as:

$$S^0_{(s)} + H_2O_{(aq)} + 1.5O_{2(g)} \longrightarrow H_2SO_{4(aq)} \tag{7}$$

The combined reaction

$$FeS_{2(s)} + 2Fe^{3+}_{(aq)} + 2H_2O_{(aq)} + 3O_{2(g)} \longrightarrow 3Fe^{2+}_{(aq)} + 2H_2SO_{4(aq)} \tag{8}$$

differs in important ways from the more typically cited (*e.g.* Nordstrom and Southam 1997) summary reaction:

$$FeS_{2(s)} + 8H_2O_{(aq)} + 14Fe^{3+}_{(aq)} \longrightarrow 15Fe^{2+}_{(aq)} + 2SO_4^{2-}_{(aq)} + 16H^+_{(aq)} \tag{9}$$

Elemental sulfur accumulation can also be important on arsenopyrite, galena and sphalerite (McGuire *et al.*, 2001).

Faster dissolution rates have been reported in cultures of mixed iron- and sulfur-oxidizing organisms compared to iron-oxidizing organisms alone (see review of Nordstrom & Southam, 1997). Although some authors have argued that sulfur-oxidizing species increase sulfide dissolution rates *via* removal of diffusion-limiting sulfur-rich deposits on metal sulfide surfaces, recent results of Edwards *et al.* (2001) and McGuire *et al.* (2001) indicate that this is not the basis of the effect.

Acidophiles are optimized for low-pH conditions, and cannot operate effectively (or survive) at pH values significantly above their growth optima. However, the periodic nature of rainfall ensures that acidic systems are dynamic and fluctuations in pH are inevitable. The coupling between metabolism and acid generation ensures that the environment is regulated within a range optimal for growth of resident microbial populations.

In addition to the Fe in pyrite, metal sulfide deposits contain significant concentrations of other metals (*e.g.* As, Cu, Cd, Zn) that are released upon sulfide mineral dissolution. Microorganisms have genetically coded pathways (both chromosomal and plasmid-borne) that confer tolerance to metals. In fact, many can grow directly upon surfaces rich in toxic elements (*e.g.* Fig. 1 shows organisms on arsenopyrite, FeAsS). Tolerance mechanisms typically use energy-consuming pump systems and can involve redox reactions, and thus influence metal speciation in the environment (*e.g.* arsenic tolerance involves intracellular reduction of As^{5+} to As^{3+} before it is pumped out of the cell; see Silver & Phung, 1996).

7.5. Biogeochemistry of trace metals

Microorganisms are important drivers of trace-metal mobility and mineral formation under Earth surface conditions. Metal–microbe interactions include bio-sorption, biologically mediated leaching, uptake and assimilation, enzyme catalyzed redox reactions and biomineralization.

Microorgansims have been shown to impact the geochemistry of economically important trace elements including Ag, Au, Cd, Cu, Hg, Ni, Pt, U, Zn, Mo and the *REE* (see Suzuki & Banfield, 1999; 2004; Reith & Rogers, 2008 and references therein; Templeton & Knowles, 2009) and toxic elements such as Cr, As and Pb (*e.g.* Gihring *et al.*, 2003; Girhring & Banfield, 2001). Essentially all of these elements are capable of chemically bonding to carboxylic, amino, hydroxyl phosphate or sulfhydryl (R-S-H) groups on microbial cell surfaces, and, therefore, microbes can potentially concentrate these elements in the environment (*e.g.* Andres *et al.*, 2003; Fein *et al.*, 2005; Takahashi *et al.*, 2005). Some of these elements, such as Mo and Fe, are vital trace nutrients that are taken up by microbial cells. Others are toxic, and microbes have developed defense mechanisms to immobilize, detoxify, or precipitate these elements, decreasing adverse impacts (Lloyd & Lovely, 2001). Some are used in microbial redox cycles; *e.g.* Fe and Mn are well known electron donors and acceptors utilized for energy generation and respiration, respectively (described above). However, microbes can utilize, or at least regulate, redox reactions of other elements such as Ce, Cr, As, and U (Suzuki & Banfield, 2004; Gihring & Banfield, 2001; Gihring *et al.*, 2003). Although often it is not clear if there is a direct microbial control of redox state, *i.e.* metal binding to microbe surfaces and transfer of electrons across the cell membrane, or if the element redox chemistry occurs indirectly, *e.g.* electron transfer from a microbially reduced metal to a more oxidized species. Whereas other trace metals are cycled along with these elements, such as by sorption or co-precipitation of mineral phases.

7.5.1. The biogeochemistry of gold

Recently there has been considerable interest in the role of microorganisms in the formation and accumulation of secondary gold grains in Earth surface conditions. Although gold is generally considered to be a relatively inert, insoluble and therefore fairly immobile element, it has three stable valence states (3, 1, 0) and forms strong complexes with halides, sulfides, thiosulfate and amino complexes. Therefore its mobility can be impacted directly or indirectly by biological processes.

Even before the advent of molecular methods, there was at least tenuous evidence of the role microbes in the formation of these near surface gold deposits. The gold associated with secondary (weathered) environments is commonly coarser grained than those associated with the source rocks; although the mechanism for this coarsening is unclear, dissolution and recrystallization must occur. More compelling evidence for direct biomineralization comes from SEM imaging of gold surfaces. These tiny nuggets are often covered with organic compounds and colloidal gold, or small rounded forms that are similar in size and shape to bacteria (bacterioform gold) suggestive of a biogenic origin (Watterson, 1992; Reith *et al.*, 2006, 2007, 2009; Southam *et al.*, 2009; Lengke & Southam, 2005, 2006, 2007).

Several researchers have conducted biogeochemical and genomic studies to better understand how microbes can solubilize, concentrate and precipitate gold from the environment, and how the presence of elevated concentrations of gold can impact community structure and function. Results of these experiments show that microbes that exhibit a broad range of metabolic strategies are able to influence gold geochemistry by ligand- and acid-promoted solubilzation, sorption, and intra- or extracellular gold precipitation. Microbially mediated gold solubilization depends on the ability of microbes to promote gold oxidation and produce ligands that form stable gold complexes. For example, Southam & Saunders (2005) showed that chemolithotrophic microbes are able to solubilize gold from gold-bearing pyrite as thiosulfate complexes. However, in another study, a known chemolithotrophic bacterium, *Acidothibacillus thiooxidans*, was able to precipitate gold intracellularly from solutions containing Au-thiosulfate complexes, whereas sulfate-reducing bacteria were capable of utilizing Au(I)thiosulfate complexes and precipitating nano-colloidal gold within and surrounding the cells (Lengke & Southam, 2005, 2006). This indicates that microbes active in sulfur redox geochemistry can directly, or indirectly, mediate gold geochemistry resulting in a redistribution of gold from a trace element in metal sulfides to a nearly pure metal.

Iron-reducing bacteria may also play a role in gold geochemistry. Kasefi *et al.* (2001) demonstrated that several species of Fe(III) reducing bacteria and archea are able to precipitate Au(III) complexes from solution, and accumulate colloidal gold on their surfaces, whereas other known Fe-reducers have no discernable affect on gold reduction and precipitation. This is important because it demonstrates that microbes that can reduce Au(III) complexes may have a specific mechanism for Au reduction that is distinct from the mechanism for Fe reduction.

Recent work by Reith and coworkers (Reith *et al.*, 2006, 2009; Reith & McPhail 2006, 2007) has focused on the role of metal-tolerant heterotrophic microbes in controlling gold geochemistry in the regolith. In their work they used confocal and scanning electron microscopy to characterize the evolution of microbial biolfilms on gold nuggets and the formation of "bacterioform gold". They used molecular tools (16s rDNA extraction and analysis to identify species, and DGGE fingerprinting to determine population diversity), to show the microbial communities associated with the bacterioform gold nuggets are distinctly different from those in the surrounding soils. The B-proteobacterium *C. metallidurans* was found on all the biofilms, and experiments with the organism shows that it is capable of reduction of Au(III) complexes from solution and formation of either discrete nano-colloids of gold at the cell surface or finely disseminated or adsorbed gold on microbial cell surfaces. Further experiments with this microorganism (Reith *et al.*, 2009) showed that it has a specific genetic response to gold with up-regulation of genes related to oxidative stress, metal reduction and efflux.

Reith & McPhail (2006, 2007) used microcosm experiments to demonstrate that heterotrophic microbial populations in gold-rich soils were able to solubilize up to 80% of the gold over a period of several weeks. Although the results of the individual treatments varied substantially, the trends were similar for all soils; gold concentrations in solution initially increased and then decreased, which was coincident with a change in amino acid concentrations in solution, and a change in carbon utilization patterns over time. In another study of gold-bearing calcrete, Reith *et al.* (2009) demonstrated that microbial utilization of amino-complexes can result in co-precipitation of calcium carbonate and nanocrystalline gold. This finding is significant because genetic studies and carbon-utilization surveys of both the microcosms and the environmental samples were dominated by alkaphylic microbes capable of utilizing urea and amino acids. The activity of these organisms would destabilize gold-amino complexes, and increase pH, showing a clear mechanistic link between the microbially mediated gold-carbonate precipitates in laboratory experiments and gold-enriched calcrete in the field.

7.5.2. *Biogeochemical control of trace metals in acid sulfate soils*

Acid Sulfate Soils (ASS), are a widespread environmental problem throughout coastal areas in Australia and worldwide. Most coastal ASS are formed from sulfidic sediments originating in the Holocene period as a result of the postglacial marine transgression. In Australia, $\sim$70% of the low-lying coastal areas have either potential (containing sufficient pyrite that the sediments would become acidic if oxidized) or actual acid sulfate soil. However, prolonged drought and increased salinization of surface and groundwaters has resulted in extensive acid sulfate soil conditions adjacent to inland waterways (Fitzpatrick & Shand, 2008). Anthropogenic alteration of natural drainage regimes in these areas has occurred as a result of pressures to increase the productivity of land and mitigate flood impact. Hydrological alteration has exposed previously reduced sulfidic sediments to oxidizing environments. This, in turn, causes the oxidation of pyrite to sulfuric acid, iron oxyhydroxides and jarosite. The generation of acidity from pyrite oxidation or jarosite dissolution is only one of the environmental hazards in these systems. Acidity in the groundwater can react with the sediments and solubilize major and trace elements, either by dissolution of the minerals within the sediments or leaching ions from sorption sites, greatly increasing the release into solution of previously immobile metal cations and salts occurring naturally in these sediments (Preda & Cox, 2000). An investigation of a coastal acid-sulfate-soil site undergoing remediation by induced flooding showed that biological control of iron and sulfur redox chemistry had a profound impact on major and trace element composition of sediments and pore waters (Welch *et al.*, 2009).

The field site was a low-lying coastal back-swamp located near Kemspey, NSW, Australia. The area had been drained in the late 1970s to mitigate flooding, and then acidic conditions and extensive scalding soon developed. During the time of the surveys (2004–2007), the pH of the surface water varied dramatically, from values <3 in scalded areas to near-neutral pH after extensive flooding, development of anoxic conditions, active sufate and iron reduction and revegetation.

Analysis of several cores that were ~ 1 m long from the area all yielded similar results. The sediment profile consists of an organic rich sulfidic A1 horizon, an acidic alluvial A2 horizon, an oxidized jarositic mottled clay zone that transitions to an iron-oxyhydroxide mottled clay zone The trends in pH in the pore water were similar for all the sediment cores. The maximum acidity occurred in the middle of the profile associated with the jarosite and goethite mottled zones, B_1 and B_2 horizons, where pH measurements were as low as ~ 3 to 3.5. The acidity here is due to the biologically mediated oxidation of pyrite that occurred when the sediments were drained and dried, and the transformation of jarosite to goethite when the sediments were inundated. The pH increased towards the top, approximately upper 10 cm or A_1 and A_2 horizon, of the profile due to biologically mediated iron and sulfate reduction reactions which consume acidity and produce pyrite, and dilution from the overlying neutral pH surface water. Before the area was inundated, acidic conditions were prevalent at the surface. The pH increases down profile in the B_3 and C_1 horizon due to the buffering capacity of the fossil bivalve shells in these layers, and because the pyrite in these horizons has not been oxidized (Figure 10). Microbial populations from these cores were characterized by cell counts with epiflourescence microscopy, enrichment culturing targeted at growing iron and sulfur bacteria (Kehoe *et al.*, 2004) and DNA extraction (Welch unpublished). The results of the enrichment cultures were somewhat unexpected in that ferrous iron and sulfur oxidizers and ferric iron and sulfate reducers were culturable throughout the sediment profiles, though microbial cell numbers in cultures for dissimilatory sulfate and iron reducers were initially greater from the more organic- and sulfide-rich horizon. Cell numbers in enrichment cultures for iron and sulfur oxidizers generally increased with depth and were greatest in the acidic samples where the initial pyrite had not completely oxidized.

The concentration of major and trace ions in porewaters in the cores is directly related to microbial control of iron and sulfur redox chemistry which controls acidity and the abundance and distribution of Fe- and S- bearing mineral phases (Welch *et al.*, 2008; 2009; Isaacson *et al.*, 2006) (Fig. 10). The concentration of dissolved Fe is controlled by acidity and redox. The elevated Fe concentrations in the surface samples are greatest in the surface sediments where anoxic conditions exist, decrease slightly with depth, and

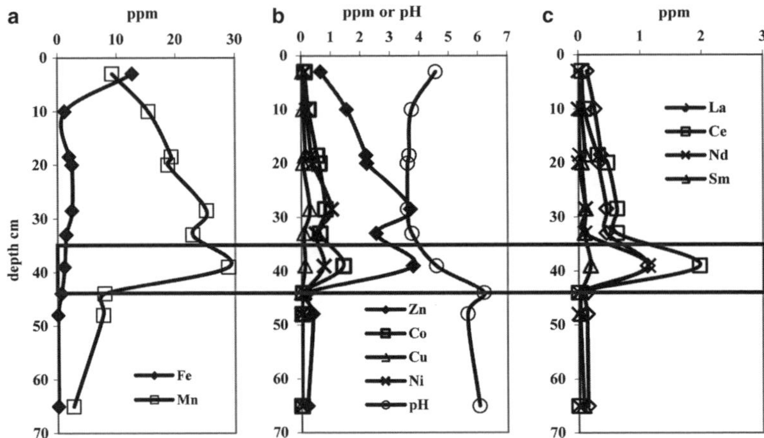

Fig. 10. Depth profiles of metal concentrations in porewaters from an acid sulfate soils site in coastal New South Wales, Australia: (**a**) Fe and Mn; (**b**) Zn, Co, Cu, Ni and pH, and (**c**) REE La, Ce, Nd, Sm.

then increase with depth where microbially mediated pyrite oxidation and jarosite precipitation buffers acidity at pH $\sim$3.5. Trace-metal (Mn, Zn, Co, Cu, Ni) concentrations in pore waters in the oxidized zones are high, on the order of hundreds of ppb to tens of ppm reflecting pyrite oxidation and acid leaching from the surrounding sediments. The concentrations of the *REE* in the pore waters show a complex relationship with acidity and iron mineralogy (Fig. 10c). In the upper part of the core, the concentrations of the *LREE* increase with increasing acidity from $\sim$10s of ppb in the near surface to $\sim$100s of ppb in the more acidic zone where jarosite is the dominant pyrite oxidation product. The concentrations then increase significantly over a few cm depth, to $\sim$ppm level, as pH increases and the modal iron mineralogy shifts from jarosite-rich to goethite-rich. This change in *REE* geochemistry in the porewater reflects the biologically mediated formation of secondary iron phases. The *REE* substitute into the K site of jarosite, so, the lower concentrations under acidic conditions reflect *REE* incorporation into the structure (Welch *et al.*, 2007, 2009). Jarosite forms only under extremely acidic conditions (pH $<$3.5) where concentrations of the iron (FeIII) sulfate and cations (Na$^+$, K$^+$, or H$_3$O$^+$) reach saturation with respect to jarosite. Although jarosite can, and does, form abiotically from pyrite oxidation when the solution is acidic enough for jarosite to precipitate; the abiotic iron oxidation from Fe(II) to Fe(III) is kinetically inhibited, limiting jarosite precipitation. Chemolithotrophic iron oxidizers can catalyze this reaction under acidic conditions.

8. New frontiers for environmental mineralogy and geochemistry

An important goal for environmental mineralogy and geochemistry is to understand how physical-chemical and biological processes interact to shape the Earth's near-surface environments. The ancient origin of life implies a coupling over most of Earth's history. Thus, microbial metabolisms must have been fundamental forces in Earth's evolution. Understanding the types of organisms and the mechanisms by which they have impacted the geochemistry of the lithosphere, hydrosphere and atmosphere requires understanding of metabolic evolution, calibrated to the geological time scale. For this analysis, a molecular level of understanding of biochemical pathways is critical.

Most people accept that only a tiny subset of the microbial species present in natural environments has been cultivated to date. DNA and RNA-based approaches (*e.g.* Woese *et al.*, 1990, Giovannoni *et al.*, 1988; Delong *et al.*, 1989; Weisburg *et al.*, 1991; Hugenholz & Pace, 1996; Amann *et al.*, 1996; also see review by Barns & Nierzwicki-Bauer, 1997) have provided methods that break through the barrier presented by non-culturability. Over the past 10 years or so, these microbiological methods have become straightforward enough to allow their deployment in interdisciplinary studies by non-specialist groups. This has allowed numerous advances in understanding of microbial diversity in natural environments, as well as the ways microbial communities are shaped by environmental chemistry.

Amplification and sequencing of genes (especially highly conserved genes such as those that code for 16S rRNA) from mixed populations in environmental samples provides a starting point for microbial and metabolic analyses of populations. Based on these data, individual cells can be labelled by species-specific oligonucleotide (RNA) probes to evaluate the distribution of both cultured and uncultured species in natural samples. However, the usefulness of such studies is limited when there is insufficient biochemical information about the microorganisms detected. For this reason, isolation and metabolic characterization of microbial species remains important. Geochemical information about the growth environment can provide important insights to guide isolation efforts, or enrichment strategies that may also be informative in the case where growth in pure culture is very difficult or impossible. Biochemical information is of enormous importance in understanding how organisms interact with each other and their environments. A few examples of enzymes that may be present in individual species and that directly affect the geochemistry of their surroundings include those involved in metal oxidation, sulfur oxidation, nitrogen or CO_2 fixation, reductases for sulfite or arsenate. Other cellular products formed under genetic control include molecules such as siderophores (for scavenging of iron; *e.g.* Hersman *et al.*, 1996) and proteinaceous cell surface layers that play a role in nucleation of minerals (*e.g.* Schultze-Lam *et al.*, 1996).

The task of complete characterization of the biochemistry of all environmental species *via* traditional methods is virtually impossible. However, the extraordinary growth in speed of sequence analysis has led to new methods for investigation of functional molecules involved in numerous metabolic pathways. An important approach involves shotgun sequencing of chromosomal fragments recovered directly from the environment. This is most effective when near-complete genomes are recovered (*e.g.* Tyson *et al.* 2004), a task that requires both significant assembly of the genomic sequences and accurate binning to classify the information to correct taxonomy (see Dick *et al.*, 2009). Even for organisms distant from cultivated taxa, this approach can be informative, as much of core metabolism appears to be reasonably conservative.

A difficulty, however, is that many genes have low similarity to genes from previously studied organisms, so much additional biochemical work is required. An important subset of these are mobile-element associated, but others probably encode proteins for key metabolic tasks, some probably involved in environmental adaptation or inter-species interactions. Data showing protein expression and correlating protein abundance with geochemical or community composition information can highlight a subset of these for targeted analysis.

Genomic data will certainly assist in the task of characterizing new proteins and metabolic pathways. Sequence data for target molecules can be employed in many ways. For example, it can be used to design primers, allowing amplification of a gene of interest from DNA from the environmental sample, culture or isolate. The gene can then be engineered into a host organism (*e.g. E. coli*). If the gene can then be expressed by *E. coli* to generate large quantities of protein, the molecule itself is then available for complete characterization (see Banfield & Marshall, 2000, for additional discussion).

Genetic data can enrich the Earth sciences in many additional ways. For example, sequences of functional genes provide ways to study microbial evolution and to place metabolic developments in the context of Earth history. Many microbial metabolic innovations are likely to have had their origins during the period of time when the atmosphere became oxygenated. The fossil record offers few insights into these developments. Biochemical-phylogenetic approaches provide methods to explore the sequence of appearance and modification of functional genes, and the appropriation of old genes into new pathways.

Microbiological investigations rely upon a tree of life to provide a context within which functional genes can be interpreted. Genetic transfer is increasingly recognized, and this perturbs the framework for analysis. Primarily, it necessitates a more flexible way of thinking about how we define organism evolution. Yet it does not displace the basic truth that once there were no biochemical pathways carrying out geochemical processes, and now there are many. With information about the structure and content of microbial genomes, it should be possible to piece together the evolution of pathways for nitrogen fixation, sulfate reduction, sulfide oxidation, iron oxidation, iron reduction, methanogenesis, photosynthesis, respiration, *etc.*, and to index these developments in terms of geological time.

A future goal for geobiological studies is to be able to directly assay the extent to which a reaction in the environment is microbially *vs.* inorganically mediated. Measurement of the activity of a specific pathway (*e.g.* iron oxidation) is complex, especially when multiple species may carry out the reaction, possibly using different enzyme systems. Sequence data for functional genes from large numbers of organisms will provide the information needed to understand the diversity of pathways employed, and ultimately to develop assays for their expression, *in situ*, in modern environments.

To date, hundreds of microbial genomes have been analysed (Binnewies *et al.*, 2006). Simultaneous analysis of genomes of multiple organisms in cultures and environmental samples is underway (*e.g.* Tyson *et al.*, 2004; Venter *et al.*, 2004; Wilmes *et al.*, 2008; Dick *et al.*, 2009; Denef *et al.*, 2009; Mueller *et al.*, 2010). With the continuously accelerating rate of accumulation of sequence data, the promise of genomic data from the majority of organisms in natural environments may become a reality in the not too distant future.

9. Concluding statement

We have moved from a time when most geochemists proposed essentially inorganic models to explain the distribution and speciation of elements at or near Earth's surface to an era where the role of microbial metabolisms is widely advocated. Biological data must be combined with modern geochemical approaches that measure the physical and chemical characteristics of the environment at the micron scale, and with experimental studies that contribute quantitative kinetic measurements. Ultimately, tools to probe directly and *in situ*, the capabilities of organisms in microenvironments in rocks, soils and sediments, and the level of activity of these organisms are needed. We hope to resolve the true magnitudes of the microbial *vs.* inorganic contributions to processes shaping many environments, and their dynamic nature. Realization of the full potential of combined biochemical, mineralogical and geochemical analyses lies in the future.

Acknowledgements

The authors thank B. Devouard for permission to reproduce the modified Figure 4, and Bob Hamers, Molly McGuire, Tom Gihring, and Katrina Edwards for their role in cited work on acid mine drainage. Philip Bond is thanked for helpful discussions. Tamara Thomsen-Ebert, Greg Druschel, and Matthias Labrenz are acknowledged for their contributions to the Tennyson mine study, and Tom O'Connor for permission to reproduce the image used in Figure 8. The authors also thank editor, David Vaughan, for his patience with us in the production of this paper and for other assistance. Partial funding support was provided by a USA National Science Foundation grant (Geochemistry Program) from the Basic Energy Sciences Program, United States Department of Energy. Figures 2c, d and 5a, b were obtained by SAW using the SEMCAL facilities at the OSU.

References

Allwood, A.C., Grotzinger, J.P., Knoll, A.H., Burch, I.W., Anderson, M.S., Coleman, M.L. & Kanik, I. (2009) Controls on development and diversity of early Archean stromatolites. *Proceedings of the National Academy of Sciences*, **106**, 9548–9555.

Amann, R., Snaidr, J., Wagner, M., Ludwig, W. & Schleifer, K.H. (1996) In situ visualization of high genetic diversity in a natural microbial community. *Journal of Bacteriology*, **178**, 3496–3500.

Anderson, D., Krussow, W.R. & Corey, R. (1985) Phosphate rock dissolution in soil: Implications from plant growth studies. *Soil Science Society of America Journal*, **49**, 918–925.

Andres, Y., Texier, A.C. & Le Cloirec, P. (2003) Rare earth elements removal by microbial biosorption: A review. *Environmental Technology*, **24**, 1367–1375.

Azcon, R., Barea, J.M. & Hayman, D.S. (1976) Utilization of rock phosphate in alkaline soils by plants inoculated with mycorrhizal fungi and phosphate-solubilizing bacteria. *Soil Biology and Biochemistry*, **8**, 135–138.

Balkwill, D.J. (1989) Numbers, diversity, and morphological characteristics of aerobic, chemoheterotrophic bacteria in deep subsurface sediments from a site in South Carolina. *Geomicrobiology Journal*, **7**, 33–52.

Banfield, J.F. & Eggleton, R.A. (1988) A transmission electron microscope study of biotite weathering. *Clays and Clay Minerals*, **36**, 47–60.

Banfield, J.F. & Eggleton, R.A. (1989) Apatite replacement and rare earth mobilization, fractionation and fixation during weathering. *Clays and Clay Minerals*, **37**, 113–127.

Banfield, J.F. & Eggleton, R.A. (1990) Analytical transmission electron microscope studies of plagioclase, muscovite and K-feldspar weathering. *Clays and Clay Minerals*, **38**, 77–89.

Banfield, J.F. & Marshall, C.R. (2000) Genomics and the geosciences. Invited perspective. *Science*, **287**, 605–606.

Banfield, J.F., Welch, S.A., Zhang, H., Ebert, T.T. & Penn, R.L. (2000) Crystal growth and microstructural evolution of FeOOH biomineralization products. *Science*, **289**, 751–754.

Barker, W.W. & Banfield, J.F. (1996) Biologically versus inorganically-mediated weathering reactions: Relationships between minerals and extracellular microbial polymers in lithobiontic communities. *Chemical Geology*, **132**, 55–69.

Barker, W.W. & Banfield, J.F. (1998) Zones of chemical and physical interaction at interfaces between microbial communities and minerals. *Geomicrobiology Journal*, **15**, 223–244.

Barker, W.W., Welch, S.A. & Banfield, J.F. (1997) Geomicrobiology of silicate mineral weathering. In: *Geomicrobiology: Interactions between Microbes and Minerals* (J.F. Banfield and K.H Nealson, editors). Reviews in Mineralogy, **35**. Mineralogical Society of America, Washington, D.C., pp. 391–428.

Barker, W.W., Welch, S.A., Chu, S. & Banfield, J.F. (1998a) Experimental observations of the effects of bacteria on aluminosilicate weathering. *American Mineralogist*, **83**, 1551–1563.

Barker, W.W., Haas, J.R., Suzuki, Y. & Banfield, J.F. (1998b) U-phosphate biomineralization as a mechanism of U fixation by lichen. *Abstracts with Programs – Geological Society of America*, **30**, p. 205.

Barns, S.M. & Nierzwicki-Bauer, D.H. (1997) Microbial diversity in ocean, surface and subsurface environments. In: *Geomicrobiology: Interactions between Microbes and Minerals* (J.F. Banfield and K.H Nealson, editors). Reviews in Mineralogy, **35**. Mineralogical Society of America, Washington, D.C., pp. 35–79.

Barns, S.M., Fundyga, R.E., Jeffries, M.W. & Pace, N.R. (1994) Remarkable archaeal diversity detected in a Yellowstone National Park hot spring environment. *Proceedings of the National Academy of Sciences USA*, **91**, 1609–1613.

Bazylinski, D.A. & Moscowitz, B.M. (1997) Microbial biomineralization of magnetic iron minerals: Microbiology, magnetism, and environmental significance. In: *Geomicrobiology: Interactions between Microbes and Minerals* (J.F. Banfield and K.H Nealson, editors). Reviews in Mineralogy, **35**. Mineralogical Society of America, Washington, D.C., pp. 181–223.

Binnewies, T.T., Motro, Y., Hallin, P.F., Lund, O., Dunn, D., La, T., Hampson, D.J., Bellgard, M., Wassenaar, T.M. & Ussery, D.W. (2006) Ten years of bacterial genome sequencing: comparative-genomics-based discoveries. *Functional & Integrative Genomics*, **6**, 165–185.

Blum, A. & Lasaga, A.C. (1988) Role of surface speciation in the low temperature dissolution of minerals. *Nature*, **331**, 431–433.

Bond, P.L., Smriga, S.P. & Banfield, J.F. (2000) Phylogeny of microorganisms populating a thick, subaerial lithotrophic biofilm at an extreme acid mine drainage site. *Applied and Environmental Microbiology*, **66**, 3842–3849.

Bone, T.L. & Balkwill, D.L. (1988) Morphological and cultural comparison of microorganisms in surface soil and subsurface sediments at a pristine study site in Oklahoma. *Microbial Ecology*, **16**, 49–64.

Brady, P.V. & Walther, J.V. (1989) Controls on silicate dissolution rates in neutral and basic pH solutions at 25°C. *Geochimica et Cosmochimica Acta*, **53**, 2823–2830.

Chan, C.S., De Stasio, G., Welch, S.A., Girasole, M., Frazer, B.H., Nesterova, M.V., Fakra. S. & Banfield, J.F. (2004) Microbial Polysaccharide Templation of Nanocrystal Fibers. *Science*, **303**, 1656–1658.

Chan, C.S., Fakra, S.C., Edwards, D.C., Emerson, D. & Banfield, J.F. (2009) Iron oxyhydroxide mineralization on microbial extracellular polysaccharides. *Geochimica et Cosmochimica Acta*, **73**, 3807–3818.

Chan, C.S., Fakra, S.R., Emerson, D., Fleming, E.J. & Edwards, K.F. (2011) Lithotrophic iron-oxidizing bacteria produce organic stalks to control mineral growth: implications for biosignature formation. *International Society for Microbial Ecology Journal*, **5**, 717–727.

Chapelle, F.H., Zelibor, J.L., Jr, Grimes, D.Y. & Knobel, L.L. (1987) Bacteria in deep coastal plain sediments of Maryland: A possible source of CO_2 to groundwater. *Water Resources Research*, **23**, 1625–1632.

Chou, L. & Wollast, R. (1985) Steady-state kinetics and dissolution mechanisms of albite. *American Journal of Science*, **285**, 963–993.

Colwell, F.S., Onstott, T.C., Delwiche, M.E., Chandler, D., Fredrickson, J.K, Yao, Q.J., McKinley, J.P., Boone, D.R., Griffiths, R., Phelps, T.J., Ringelberg, D., White, D.C., LaFreniere, L., Balkwill, D., Lehrman, R.M., Konisky, J. & Long, P.E. (1997) Microorganisms from deep, high temperature sandstones: Constraints on microbial colonization. *Federation of European Microbiological Societies Microbiology Reviews*, **20**, 425–435.

Delong, E.F., Wickham, G.S. & Pace, N.R. (1989) Phylogenetic stains – ribosomal RNA-based probes for the identification of single cells. *Science*, **243**, 1360–1363.

Deming, J.W. & Baross, J.A. (1993) Deep-sea smokers – Windows to a subsurface biosphere. *Geochimica et Cosmochimica Acta*, **57**, 3219–3230.

Denef, V.J., Kalnejais, L.H., Mueller, R.S., Wilmes, P., Baker, B.J., Thomas, B.C., VerBerkmoes, N.C., Hettich, R.L. & Banfield, J.F. (2009) Proteogenomic basis for ecological divergence of closely related bacteria in natural acidophilic microbial communities. *Proceedings of the National Academy of Sciences*, **107**, 2383–2390.

Devouard, B., Pósfai, M., Hua, X., Bazylinski, D.A., Frankel, R.B. & Buseck, P.R. (1998) Magnetite from magnetotactic bacteria: Size distributions and twinning. *American Mineralogist*, **83**, 1387–1398.

de Vrind-de Jong, E.W. & de Vrind, J.P.M. (1997) Algal deposition of carbonates and silicates. In: *Geomicrobiology: Interactions between Microbes and Minerals* (J.F. Banfield and K.H Nealson, editors). Reviews in Mineralogy, **35**. Mineralogical Society of America, Washington, D.C., pp. 267–307.

Dick, G.J., Andersson, A., Baker, B.J., Simmons, S.S., Thomas, B.C., Yelton, A.P. & Banfield, J.F. (2009) A community-wide analysis of microbial genome sequence signatures. *Genome Biology*, **10**, R85.

Drever, J.I. (1994) The effect of land plants on the weathering rate of silicate minerals. *Geochimica et Cosmochimica Acta*, **58**, 2325–2332.

Drits, V.A., Sakharov, B.A., Salyn, A.L. & Manceau, A. (1993a) Structural model for ferrihydrite. *Clay Minerals*, **28**, 185–207.

Drits, V.A., Sakharov, B.A. & Manceau, A. (1993b) Structure of feroxyhyte as determined by simulation of X-ray diffraction curves. *Clay Minerals*, **28**, 209–222.

Edwards, K.J., Gihring, T.M. & Banfield, J.F. (1999) Seasonal variations in microbial populations and environmental conditions in an extreme acid mine drainage environment. *Applied and Environmental Microbiology*, **65**, 3627–3632.

Edwards, K.J., Bond, P.L., Gihring, T.M. & Banfield, J.F. (2000) An archaeal iron-oxidizing extreme acidophile important in acid mine drainage. *Science*, **287**, 1796–1799.

Edwards, K.J., McGuire, M.M. Hamers, R.J. & Banfield, J.F. (2001) Kinetics and surface microstructural evolution of microbially mediated sulfide dissolution: Implications for modeling acid mine drainage generation. *Geochimica et Cosmochimica Acta*, **65**, 1243–1258.

Edwards, K.J., Glazer, B.T., Rouxel, O.J., Bach, W., Emerson, D., Davis, R.E., Toner, B.M., Chan, C.S., Tebo, B.M., Staudigel, H. & Moyer, C.L. (2011) Ultra-diffuse hydrothermal venting supports Fe-oxidizing bacteria and massive umber deposition at 5000 m off Hawaii. *The International Society of Microbial Ecology Journal*, **5**, 1748–1758

Ehrlich, H.L. (1996) *Geomicrobiology*. 3rd edition, Marcel Dekker, New York, 719 pp.

Eisenberg, H. (1995) Life in extreme environments – progress in understanding the structure and function of enzymes from extreme halophilic bacteria. *Archives of Biochemistry and Biophysics*, **318**, 1–5.

Emerson, D. & Moyer, C. (1997) Isolation and characterization of novel iron-oxidizing bacteria that grow at circumneutral pH. *Applied Environmental Microbiology*, **63**, 4784–4792.

Emerson, D., Weiss, J.V. & Megonigal, J.P. (1999) Iron-oxidizing bacteria are associated with ferric hydroxide precipitates (Fe-plaque) on the roots of wetland plants. *Applied Environmental Microbiology,* **65**, 2758–2761.

Fein, J.B., Boily, J.F., Yee, N., Gorman-Lewis, D. & Turner, B.F. (2005) Potentiometric titrations of *Bacillus subtilis* cells to low pH and a comparison of modeling approaches. *Geochimica et Cosmochimica Acta,* **69**, 1123–1132.

Finlay, R., Wallander, H., Smits, M., Holmstrom, S., VanHees, P., Lian, B. & Rosling, A. (2009) The role of fungi in biogenic weathering in boreal forest soils. *Fungal Biology Reviews,* **23**, 101–106.

Fisk, M.R., Giovannoni, S.J. & Thorseth, I.H. (1998) Alteration of oceanic volcanic glass: Textural evidence of microbial activity. *Science,* **281**, 978–980.

Fitzpatrick, R. & Shand, P. (2008) *Inland Acid Sulfate Soil Systems Across Australia.* CRC LEME Open File Report No. 249. (Thematic Volume) CRC LEME, Perth, Australia, 304 pp.

Fortin, D., Ferris, F.G. & Beveridge, T.J. (1997) Surface-mediated mineral development by bacteria. In: *Geomicrobiology: Interactions between Microbes and Minerals* (J.F. Banfield and K.H Nealson, editors). Reviews in Mineralogy, **35**. Mineralogical Society of America, Washington, D.C., pp. 162–180.

Fortin, D.L. & Langley, S. (2005) Formation and occurrence of biogenic iron-rich minerals. *Earth-Science Reviews,* **72**, 1–19.

Fowle, D.A. & Fein, J.B. (1999) Competitive adsorption of metal cations onto two gram positive bacteria: Testing the chemical equilibrium model. *Geochimica Cosmochimica Acta,* **63**, 3059–3067.

Frankel, R.B. and Bazylinski, D.A. (2003) Biologically Induced Mineralization by Bacteria. In: *Geomicrobiology: Interactions between Microbes and Minerals* (J.F. Banfield and K.H. Nealson, editors). Reviews in Mineralogy, **35**. Mineralogical Society of America, Washington, D.C., pp. 95–114.

Fredrickson, J.K. & Onstott, T.C. (1996) Microbes deep inside the Earth. *Scientific American,* **275**, 68–73.

Fredrickson, J.K., Garland, T.R., Hicks, R.J., Thomas, J.M., Li, S.W. & McFadden, K.M. (1989) Lithotrophic and heterotrophic bacteria in deep subsurface sediments and their relation to sediment properties. *Geomicrobiology Journal,* **7**, 53–66.

Frias-Lopez, J., Shi, Y., Tyson, G.W., Coleman, M.L., Schuster, S.T., Chisholm, S.W. & DeLong, E.F. (2008) Microbial community gene expression in ocean surface waters. *Proceedings of the National Academy of Sciences,* **105**, 3805–3810.

Fyfe, W.S. (1996) The biosphere is going deep. *Science,* **273**, 448.

Ghiorse, W.C. (1997) Subterranean life. *Science,* **275**, 789–790.

Ginn, B.R., Szymanowski, J.E.S & Fein, J.B. (2010) Calibration of a linear free energy estimation approach for estimating stability constants for metal-bacterial surface complexes. *Geomicrobiology Journal,* **27**, 321–328.

Girhring, T.M. & Banfield, J.F. (2001) Arsenite oxidation and arsenate respiration by a new Thermus isolate. *Federation of the European Microbiological Societies Microbiology Letters,* **204**, 335–340.

Gihring, T.M., Bond, P.L., Peters, S.C. & Banfield, J.F. (2003) Arsenic resistance in the archaeon "Ferroplasma acidarmanus": new insights into the structure and evolution of the ars genes. *Extremeophiles,* **7**, 123–130.

Giovannoni, S.J., Delong, E.F., Olsen, G.J. & Pace, N.R. (1988) Phylogenetic group-specific oligodeoxynucleotide probes for identification of single microbial cells. *Journal of Bacteriology,* **170**, 720–726.

Goyne, K.W., Brantley, S.L. & Chorover, J. (2010) Rare earth element release from phosphate minerals in the presence of organic acids. *Chemical Geology,* **278**, 1–14.

Gribb, A.A. & Banfield, J.F. (1997) Particle size effects on transformation kinetics and phase stability in nanocrystalline TiO_2. *American Mineralogist,* **82**, 717–728.

Grote, M. & O'Malley, M.A. (2011) Enlightening the life sciences: the history of halobacterial and microbial rhodopsin research. *Federation of European Microbiological Societies Microbiology Reviews,* **35**, 1082–1099.

Haas, J.R., Bailey, E.H. & Purvis, O.W. (1998) Bioaccumulation of metals by lichens; uptake of aqueous uranium by *Peltigara membranacea* as a function of time and pH. *American Mineralogist,* **83**, 1494–1502.

Hallbeck, L. & Pedersen, K. (1991) Autotrophic and mixotrophic growth of *Gallionella ferruginea*. *Journal of General Microbiology,* **137**, 2657–2661.

Haveman, S.A., Pedersen, K. & Ruotsalainen, P. (1999) Distribution and metabolic diversity of microorganisms in deep igneous rock aquifers of Finland. *Geomicrobiology Journal,* **16**, 277–294.

Heidelberg, K.B., Gilbert, J.A. & Joint, I. (2010) Marine genomics: at the interface of marine microbial ecology and biodiscovery. *Microbial Biotechnology,* **3**, 531–543

Hersman, L., Maurice, P. & Sposito, G. (1996) Iron acquisition from hydrous Fe(III) oxides by an aerobic *Pseudomonas* sp. *Chemical Geology,* **132**, 25–31.

Hinsinger, P. & Gilkes, R.J. (1997) Dissolution of phosphate rock in the rhizosphere of five plant species grown in an acid, P fixing mineral substrate. *Geoderma,* **75**, 231–249.

Hinsinger, P.A., Bengough, A.G., Vetterlein, D. & Young, I.M. (2009) Rhizosphere: biophysics, biogeochemistry and ecological relevance. *Plant and Soil,* **321**, 117–152.

Huber, J.A., Welch, D., Morrison, H.G., Huse, S.M., Neal, P.R., Butterfield, D.A. & Sogin, M.L. (2007) Microbial population structures in the deep marine biosphere. *Science,* **318**, 97–100.

Hugenholtz, P. & Pace, N.R. (1996) Identifying microbial diversity in the natural environment: A molecular phylogenetic approach. *Trends in Biotechnology,* **14**, 190–197.

Illmer, P. & Schinner, F. (1995) Solubilization of inorganic calcium phosphates-solubilization mechanisms. *Soil Biology and Biochemistry* **27**, 257–263.

Illmer, P., Barbato, A. & Schinner, F. (1995) Solubilization of hardly-soluble $AlPO_4$ with P-solubilizing microorganisms. *Soil Biology and Biochemistry,* **27**, 265–270.

Isaacson, L., Kirste, D., Beavis, S. & Welch, S. (2006) Controls of Acid, Salt and Metal Distribution at a Coastal Acid Sulfate Soils Site. in Regolith 2006: Consolidation and Dispersion of Ideas. *Proceedings of the CRC LEME Regolith Symposium,* November 2006 (R.W. Fitzpatrick and P. Shand, editors).

Johnston, C.G. & Vestal, J.R. (1993) Biogeochemistry of oxalate in Antarctic cryptoendolithic lichen-dominated community. *Microbial Ecology,* **25**, 305–319.

Kalinowski, B.E. & Schweda, P. (1996) Kinetics of muscovite, phlogopite, and biotite dissolution and alteration at pH 1–4, room temperature. *Geochimica et Cosmochimica Acta,* **60**, 367–385.

Kappler, A. & Straub, K.L. (2005) Geomicrobiological cycling of iron. In: *Molecular Geomicrobiology* (J.F. Banfield, J. Cervini-Silva & K.M. Nealson, editors). Reviews in Mineralogy & Geochemistry, **59**. Mineralogical Society of America, Chantilly, Virginia, USA, pp. 85–108.

Kashefi, K., Tor, J.M., Nevin, K.P. & Lovley, D.R. (2001) Reductive precipitation of gold by dissimilatory Fe(III)-reducing bacteria and archaea. *Applied and Environmental Microbiology,* **67**, 3275–3279.

Keller, J.D. & Wood, W. (1993) Possibility of chemical weathering before the advent of vascular plants. *Nature,* **364**, 223–225.

Kehoe, M., Beavis, S. & Welch, S. (2004) Investigating the role of biotic versus abiotic processes in the generation of acid sulfate soils in coastal NSW. In *Regolith 2004, Proceedings of the CRC LEME Conference,* November 2004. (I. Roach editor).

Kirschvink, J.L. & Hagadorn, J.W. (2000) A grand unified theory of biomineralization. In: *The Biomineralisation of Nano- and Micro-Structures* (E. Bäuerlein, editor). Wiley-VCH Verlag GmbH, Weinheim, Germany, pp. 139–150.

Konhauser, K.O. (1998) Diversity of bacterial iron biomineralization. *Earth-Science Reviews,* **43**, 91–121.

Kraemer, S.M., Butler, A., Borer, P. & Cervini-Silva, J. (2005) Siderophores and the dissolution of iron-bearing minerals in marine systems. In: *Molecular Geomicrobiology* (J.F. Banfield, J. Cervini-Silva & K.M. Nealson, editors). Reviews in Mineralogy & Geochemistry, **59**. Mineralogical Society of America, Chantilly, Virginia, USA, pp. 53–84.

Labrenz, M., Druschel, G.K., Thomsen-Ebert, T., Gilbert, B., Welch, S.A., Kemner, K.M., Logan, G.A., Summons, R.E., DeStasio, G., Bond, P.L., Lai, B., Kelley, S.D. & Banfield, J.F. (2000) Formation of sphalerite (ZnS) deposits in natural biofilms of sulfate-reducing bacteria. *Science,* **290**, 1744–1747.

Langmuir, D. (1971) Particle size effect on the reaction goethite = hematite + water. *American Journal of Science,* **271**, 147–156.

Lapeyrie, F., Ranger, J. & Vairelles, D. (1991) Phosphate-solubilizing activity of ectomycorrhizal fungi in vitro. *Canadian Journal of Botany,* **69**, 342–346.

Lebedeva, E.V., Lyalikova, N.N. & Bugel'skii, Y.Y. (1979) Participation of nitrifying bacteria in the weathering of serpentized ultrabasic rock. *Mikrobiologiya (Mosk.)*, **47**, 898–904.

Lengke, M.F. & Southam, G. (2005) The effect of thiosulfate-oxidizing bacteria on the stability of the gold-thiosulfate complex. *Geochimica et Cosmochimica Acta*, **69**, 3759–3772.

Lengke, M.F. & Southam, G. (2006) Bioaccumulation of gold by sulfate-reducing bacteria cultured in the presence of gold(I)-thiosulfate complex. *Geochimica et Cosmochimica Acta*, **70**, 3646–3661

Lengke, M.F. & Southam, G. (2007) The deposition of elemental gold from gold(I)-thiosulfate complexes mediated by sulfate-reducing bacterial conditions. *Economic Geology*, **102**, 109–126.

Little, D.A., Field, J.B. & Welch, S.A. (2005a) Metal dissolution from rhizosphere and non-rhizosphere soils using low molecular weight organic acids. In *Regolith 2005: Ten years of the CRC LEME. Proceedings of the CRC LEME Regional Symposia*, November 2005 (I.C. Roach, editor).

Little, D.A., Welch, S.A. & Field, J. (2005b) Trace elements in the rhizosphere of mature Acacia falciformis. In *Regolith 2005: Ten years of the CRC LEME. Proceedings of the CRC LEME Regional Symposia*, November 2005 (I.C. Roach, editor).

Little, D.A., Welch, S.A., MacDonald, L.M. and Rogers, S.L. (2005c) Microbial community structural and functional diversity in the rhizosphere of co-occurring forest trees. In: *Regolith 2005: Ten years of the CRC LEME. Proceedings of the CRC LEME Regional Symposia*, November 2005 (I.C. Roach, editor).

Little, B.J., Wagner, P.A. & Lewandowski, Z. (1997) Spatial relationships between bacteria and mineral surfaces. In: *Geomicrobiology: Interactions between Microbes and Minerals* (J.F. Banfield and K.H Nealson, editors). Reviews in Mineralogy, **35**. Mineralogical Society of America, Washington, D.C., pp. 123–159.

Lloyd, J.R. & Lovley, D.R. (2001) Microbial detoxification of metals and radionuclides. *Current Opinion in Biotechnology* **12**, 248–253.

Lovely, D.R. & Chapelle, F.H. (1995) Deep subsurface microbial processes. *Reviews in Geophysics*, **33**, 365–381.

Malstrom, M. & Banwart, S. (1997) Biotite dissolution at 25°C: The pH dependence of dissolution rate and stoichiometry. *Geochimica et Cosmochimica Acta*, **61**, 2779–2799.

Manceau, A. & Drits, V.A. (1993) Local structure of ferrihydrite and feroxyhyte by EXAFS spectroscopy. *Clay Minerals*, **28**, 165–184.

McGuire, M.M., Edwards, K.J., Banfield, J.F. & Hamers, R.J. (2001) Surface chemistry of sulfide minerals during microbial-mediated oxidative dissolution. *Geochimica et Cosmochimica Acta*, **65**, 1243–1258.

McHale, J.M., Auroux, A., Perrotta, A.J. & Navrotsky, A. (1997) Surface energies and thermodynamic phase stability in nanocrystalline aluminas. *Science*, **227**, 788–791.

Minton, K.W. & Daly, M.J. (1995) A model for repair of radiation-induced DNA double-strand breaks in the extreme radiophile *Deinococcus radiodurans*. *Bioessays*, **17**, 457–464.

Moon-van der Staay, S.Y., De Wachter, R. & Vaulot, D. (2001) Oceanic 18S rDNA sequences from picoplankton reveal unsuspected eukaryotic diversity. *Nature*, **409**, 607–610.

Moses, C.O., Nordstrom, D.K., Hersman, J.S. & Mills, A.L. (1997) Aqueous pyrite oxidation by dissolved oxygen and by ferric iron. *Geochimica et Cosmochimica Acta*, **51**, 1561–1571.

Mueller, R., Denef, V.J., Kalnejais, L., Suttle, B., Thomas, B.C., Wilmes, P., Smith, R., Nordstrom, D.K., McCleskey, B., Shah, M., VerBerkmoes, N. & Banfield, J.F. (2010) Ecological distribution and population physiology defined by proteomics in a natural microbial community. *Molecular Systems Biology*, **6**, 4819–4828.

Nealson, K.H. & Stahl, D.H. (1997) Microorganisms and biogeochemical cycles: What can we learn from stratified communities. In: *Geomicrobiology: Interactions between Microbes and Minerals* (J.F. Banfield and K.H. Nealson, editors). Reviews in Mineralogy, **35**. Mineralogical Society of America, Washington, D.C., pp. 5–34.

Neaman, A., Chorover, J. & Brantley, S.L. (2005a) Element mobility patterns record organic ligands in soils on early Earth. *Geology*, **33**, 117–120.

Neaman, A., Chorover, J. & Brantley, S.L. (2005b) Implications of the evolution of organic moieties for basalt weathering over geological time. *American Journal of Science*, **305**, 147–185.

Nelson, D.C., Casey, W.H., Sison, J.D., Mack, E.E., Ahmad, A. & Pollack, J.S. (1996) Selenium uptake by sulfur-accumulating bacteria. *Geochimica et Cosmochimica Acta*, **60**, 3531–3539.

Newman, D.K., Beveridge, T.J. & Morel, F.M.M. (1997) Precipitation of arsenic trisulfide by *Desulfotomaculum auripigmentum*. *Applied and Environmental Microbiology*, **63**, 2022–2028.

Nordstrom, D.K. & Southam, G. (1997) Geomicrobiology of sulfide mineral oxidation. In: *Geomicrobiology: Interactions between Microbes and Minerals* (J.F. Banfield and K.H Nealson, editors). Reviews in Mineralogy, **35**. Mineralogical Society of America, Washington, D.C., pp. 361–390.

Nordstrom, D.K., Alpers, C.N., Ptacek, C.J. & Blowes, D.W. (2000) Negative pH and extremely acidic mine waters from Iron Mountain, California. *Environmental Science & Technology*, **34**, 254–258.

Onstott, T.C., Phelps, T.J., Colwell, F.S., Ringelberg, D., White, D.C., Boone, D.R., McKinley, J.P., Stevens, T.O., Long, P.E., Balkwill, D.L., Griffin, W.T. & Kieft, T. (1998) Observations pertaining to the origin and ecology of microorganisms recovered from the deep subsurface of Taylorsville Basin, Virginia. *Geomicrobiology Journal*, **15**, 353–385.

Orcutt, B.N., Bach, W., Becker, K., Fisher, A.T., Hentscher, M., Toner, B.M., Wheat, C.G. & Edwards, K.J. (2011) Colonization of subsurface microbial observatories deployed in young ocean crust. *International Society of Microbial Ecology Journal*, **5**, 692–703.

Oren, A. (2008) Microbial life at high salt concentrations: phylogenetic and metabolic diversity. *Saline Systems*, **4**, 1–13.

Parasuraman, C.S. (1995) *Mechanism of potential enoblement on passive metals due to biofilms in seawater*. Doctoral dissertation, Univ. of Delaware, Newark, USA.

Parkes, R.J., Cragg, B.A., Bale, S.J., Getliff, J.M., Goodman, K., Rochelle, P.A., Fry, J.C., Weightman, A.J. & Harvey, S.M. (1994) Deep bacterial biosphere in Pacific Ocean sediments. *Nature*, **371**, 410–413.

Pedersen, K. (1993) The deep subterranean biosphere. *Earth-Science Reviews*, **34**, 243–260.

Pedersen, K. (1997) Microbial life in deep granitic rock. *Federation of European Microbiological Societies Microbiology Reviews*, **20**, 399–414.

Penn, R.L. & Banfield, J.F. (1998a) Imperfect oriented attachment: a mechanism for dislocation generation in defect-free nanocrystals. *Science*, **281**, 969–971.

Penn, R.L. & Banfield, J.F. (1998b) Oriented attachment and growth, twinning, polytypism, and formation of metastable phases: insights from nanocrystalline TiO_2. *American Mineralogist*, **83**, 1077–1082.

Penn, R.L. & Banfield, J.F. (1999a) Morphology development and crystal growth in nanocrystalline aggregates under hydrothermal conditions: insights from titania. *Geochimica et Cosmochimica Acta*, **63**, 1549–1557.

Penn, R.L. & Banfield, J.F. (1999b) Formation of rutile nuclei at anatase {112} twin interfaces and the phase transformation mechanism in nanocrystalline titania. *American Mineralogist*, **84**, 871– 876.

Preda, M. and Cox, M.E. (2000) Sediment-water interaction, acidity and other water quality parameters in a subtropical setting, Pimpama River, southeast Queensland. *Environmental Geology*, **34**, 319–329

Priscu, J.C., Adams, E.E., Lyons, W.B., Voytek, M.A., Mogk, D.W., Brown, R.L., McKay, C.P., Takacs, C.D., Welch, K.A., Wolf, C.F., Kirshtein, J.D. & Avci, R. (1999) Geomicrobiology of subglacial ice above Lake Vostok, Antarctica. *Science*, **286**, 2141–2144.

Reith, F. & McPhail, D.C. (2006) Effect of resident microbiota on the solubilization of gold in soils from the Tomakin Park Gold Mine, New South Wales, Australia. *Geochimica et Cosmochimica Acta*, **70**, 1421–1438.

Reith, F. & McPhail, D.C. (2007) Mobility and microbially mediated mobilization of gold and arsenic in soils from two gold mines in semi-arid and tropical Australia. *Geochimica et Cosmochimica Acta*, **71**, 1183–1196.

Reith, F. & Rogers, S. (2008) Assessment of bacterial communities in auriferous and non-auriferous soils using genetic and functional fingerprinting. *Geomicrobiology Journal*, **25**, 203–215.

Reith, F., Rogers, S.L., McPhail, D.C. & Webb, D. (2006) Biomineralization of gold: Biofilms on bacterioform gold. *Science*, **313**, 233–236.

Reith, F., Lengke, M.F., Falconer, D., Craw, D. & Southam, G. (2007) The geomicrobiology of gold. *International Society of Microbial Ecology Journal*, **1**, 567–584.

Reith, F., Etschmann, B., Grosse, C., Moors, H., Benotmane, M.A., Monsieurs, P., Grass, G., Doonan, C., Vogt, S., Lai, B., Martinez-Criado, G., George, G.N., Nies, D.H., Mergeay, M., Pring, A., Southam,

G. & Brugger, J. (2009) Mechanisms of gold biomineralization in the bacterium *Cupriavidus metallidurans*. *Proceedings of the National Academy of Sciences*, **106**, 17757–17762.

Rimstidt, J.D. & Vaughan, D.J. (2003) Pyrite oxidation: A state-of-the-art assessment of the reaction mechanism. *Geochimica et Cosmochimica Acta*, **67**, 873–880.

Rogers, J.R., Bennett, P.C. & Choi, W.J. (1988) Feldspars as a source of nutrients for microorganisms. *American Mineralogist*, **83**, 1532–1540.

Rosling, A., Suttle, K.B., Johansson, E., Van Hees, P.A.W. & Banfield, J.F. (2007) Phosphorus availability influences the dissolution of apatite by soil fungi. *Geobiology*, **5**, 265–280.

Rosling, A., Roose, T., Herrmann, A.M., Davidson, F.A., Finlay, R.D. & Gadd, G.M. (2009) Approaches to modeling mineral weathering by fungi. *Fungal Biology Reviews*, **23**, 138–144.

Santelli, C.M., Orcutt, B.N., Banning, E., Bach, W., Moyer, C.L., Staudigel, H. & Edwards, K.J. (2008) Abundance and diversity of microbial life in the ocean crust. *Nature*, **453**, 653–657.

Schleper, C., Pühler, G., Klenk, H.P. & Zillig, W. (1996) *Picrophilus oshimae* and *Picrophilus torridus* fam. nov., gen. nov., sp. nov., two species of hyperacidophilic, thermophilic, heterotrophic, aerobic archaea. *International Journal of Systematic Bacteriology*, **46**, 814–816.

Schopf, J.W. (1993) Microfossils of the early Archean Apex chert: New evidence of the antiquity of life. *Science*, **260**, 640–646.

Schopf, J.W. & Packer, B.M. (1987) Early Archean microfossils (2.2 billion to 3.5 billion-year-old) microfossils from Warrawoona Group, Australia. *Science*, **237**, 70–73.

Schrenk, M.O., Edwards, K.J., Goodman, R.M., Hamers, R.J. & Banfield, J.F. (1998) Distribution of *Thiobacillus ferrooxidans* and *Leptospirillum ferrooxidans*: Implications for generation of acid mine drainage. *Science*, **279**, 1519–1522.

Schrenk, M.O., Huber, J.A., Edwards, K.J. (2010) Microbial Provinces in the Subseafloor. *Annual Reviews of Marine Science*, **2**, 279–304.

Schultze-Lam, S., Fortin, D., Davis, B.S. & Beveridge, T.J. (1996) Mineralization of bacterial surfaces. *Chemical Geology*, **132**, 171–181.

Silver, S. & Phung, L.T. (1996) Bacterial heavy metal resistance: New surprises. *Annual Review of Microbiology*, **50**, 753–789.

Sinclair, J.L. & Ghiorse, W.C. (1989) Distribution of aerobic bacteria, protozoa, algae, and fungi in deep subsurface sediments. *Geomicrobiology Journal*, **7**, 15–31.

Skinner, H.C.W. & Jahren, A.H. (2003) Biomineralization. *Treatise on Geochemistry*, **8**, 117–184.

Southam, G. & Saunders, J.A. (2005) The geomicrobiology of ore deposits. *Economic Geology*, **100**, 1067–1084.

Southam, G., Lengke, M.F., Fairbrother, L. & Reith, F. (2009) The Biogeochemistry of Gold. *Elements*, **5**, 303–307.

Stevens, T.O. & McKinley, J.P. (1995) Lithoautotrophic microbial ecosystems in deep basalt aquifers. *Science*, **270**, 450–454.

Stevens, T.O., McKinley, J.P. & Fredrickson, J.K. (1993) Bacteria associated with deep alkaline anaerobic groundwaters in southeast Washington. *Microbial Ecology*, **25**, 35–50.

Stillings, L.L., Drever, J.I., Brantley, S.L., Sun, Y. & Oxburgh, R. (1996) Rates of feldspar dissolution at pH 3–7 with 0–8 mM oxalic acid. *Chemical Geology*, **132**, 79–90.

Summit, M. & Baross, J.A. (1998) Thermophilic subseafloor microorganisms from the 1996 north Gorda Ridge eruption. *Deep Sea Research II*, **45**, 2751–2766.

Sun, H.J. & Friedmann, E.I. (1999) Growth on geological time scales in the Antarctic cryptoendolithic microbial community. *Geomicrobiology Journal*, **16**, 193–202.

Suzuki, Y. & Banfield, J.F. (1999) Geomicrobiology of uranium. In: *Uranium: Mineralogy, Geochemistry, and the Environment* (P.C. Burns & R. Finch, editors). Reviews in Mineralogy, **38**. Mineralogical Society of America, Washington, D.C., pp. 393–432.

Suzuki, Y. and Banfield, J.F. (2004) Resistance to, and accumulation of, uranium by bacteria from a uranium contaminated site. *Geomicrobiology Journal*, **21**, 113–121

Takahashi, Y., Chatellier, X.T., Hattori, K.H., Kato, K. & Fortin, D. (2005) Adsorption of rare earth elements onto bacterial cell walls and its implication for REE sorption onto natural microbial mats. *Chemical Geology*, **219**, 53–67.

Taunton, A.E., Welch, S.A. & Banfield, J.F. (2000a) Geomicrobiological controls on lanthanide distributions during granite weathering and soil formation. *Journal of Alloys and Compounds*, **303–304**, 30–36.

Taunton, A.E., Welch, S.A. & Banfield, J.F. (2000b) Microbial controls on phosphate weathering and lanthanide distributions during granite weathering and soil formation. *Chemical Geology*, **169**, 371–382.

Templeton, A. & Knowles, E. (2009) Microbial transformations of minerals and metals: Recent advances in geomicrobiology derived from synchrotron-based X-ray spectroscopy and X-ray microscopy. *Annual Reviews in Earth and Planetary Science*, **37**, 367–91.

Teng, H.H. & Dove, P.M. (1997) Surface site-specific interactions of aspartate with calcite during dissolution: Implications for biomineralization. *American Mineralogist*, **82**, 878–887.

Thorseth, I.H., Torsvik, T., Furnes, H. & Muehlenbachs, K. (1995) Microbes play and important role in the alteration of oceanic crust. *Chemical Geology*, **126**, 137–146.

Tyson, G.W., Chapman, J., Hugenholtz, P., Allen, E.E., Ram, R.J., Richardson, P.M., Solovyvev, V.V., Rubin, E.M., Rokhsar, D.S. & Banfield, J.F. (2004) Community structure and metabolism through reconstruction of microbial genomes from the environment. *Nature*, **428**, 37–43.

Ullman, W.J., Kirchman, D.L., Welch, S.A. & Vandevivere, P. (1996) Laboratory evidence for microbially mediated silicate mineral dissolution in nature. *Chemical Geology*, **132**, 11–17.

Venter, J.C., Remington, K., Heidelberg, J.F., Halpern, A.L., Rusch, D., Eisen, J.A., Wu, D., Paulsen, I., Nelson, K.E., Nelson, W., Fouts, D.E., Levy, S., Knap, A.H., Lomas, M.W., Nealson, K., White, O., Peterson, J., Hoffman, J., Parsons, R., Baden-Tillson, H., Pfannkoch, C., Rogers, Y.H. & Smith, H.O. (2004) Environmental genome shotgun sequencing of the Sargasso Sea. *Science*, **304**, 66–77.

Vorobyova, E., Soina, V., Gorlenko, M., Minkovskaya, N., Zalinova, N., Mamukelashvili, A., Gilichinsky, D., Rivkina, E. & Vishnivetskaya, T. (1997) The deep cold biosphere: facts and hypothesis. *Federation of European Microbiological Societies Microbiology Reviews*, **20**, 277–290.

Watterson, J.R. (1992) Preliminary evidence for the involvement of budding bacteria in the origin of Alaskan placer gold. *Geology*, **20**, 315–318.

Weisburg, W.G., Barns, S.M., Pelletier, D.A. & Lane, D.J. (1991) 16S ribosomal DNA amplification for phylogenetic study. *Journal of Bacteriology*, **173**, 697–703.

Welch, S.A. & Banfield, J.F. (2002) Modification of olivine surface morphology and reactivity during natural and experimental chemical weathering. *Geochimica et Cosmochimica Acta* **66**, 213-221.

Welch, S.A., Barker, W.W. & Banfield, J.F. (1999) Microbial extracellular polysaccharides and plagioclase dissolution. *Geochimica et Cosmochimica Acta*, **63**, 1405–1419.

Welch, S.A. & Ullman, W.J. (1993) The effect of organic acids on plagioclase dissolution rates and stoichiometry. *Geochimica et Cosmochimica Acta*, **57**, 2725–2736.

Welch, S.A. & Ullman W.J. (1996) Feldspar dissolution in acidic and organic solutions: Compositional and pH dependence of dissolution rate. *Geochimica et Cosmochimica Acta*, **60**, 2939–2948.

Welch, S.A., Taunton, A.E. & Banfield, J.F. (2002) Effect of microorganisms and microbial metabolites on apatite dissolution. *Geomicrobiology Journal*, **19**, 343–367.

Welch, S.A., Christy, A., Kirste, D., Beavis, F. & Beavis, S. (2007) Jarosite dissolution. I. Trace metal geochemistry. *Chemical Geology*, **245**, 183–197.

Welch, S.A., Christy, A., Isaacson, L. & Kirste, D. (2009) Mineralogical control of rare earth elements in acid sulfate soils. *Geochimica et Cosmochimica Acta*, **73**, 44–64.

White, A.F. & Brantley, S.L. (editors) (1995) *Chemical Weathering Rates of Silicate Minerals*. Reviews in Mineralogy, **31**, Mineralogical Society of America, Washington, D.C.

Wilmes, P. & Bond, P.L. (2009) Microbial community proteomics: elucidating the catalysts and metabolic mechanisms that drive the Earth's biogeochemical cycles. *Current Opinion in Microbiology*, **12**, 310–317

Wilmes, P., Andersson, A.F., Lefsrud, M.G., Wexler, M., Shah, M., Zhang, B., Hettich, R.L., VerBerkmoes, N.C. & Banfield, J.F. (2008) Community proteogenomics highlights microbial strain-variant protein expression within activated sludge performing enhanced biological phosphorus removal. *The International Society of Microbial Ecology Journal*, **2**, 853–864.

Winkler, S. & Woese, C.R. (1991) A definition of the domains *Archaea, Bacteria,* and *Eucarya* in terms of small subunit ribosomal RNA characteristics. *Systematics in Applied Microbiology,* **14**, 305–310.

Woese, C.R., Kandler, O. & Wheelis, M.L. (1990) Towards a natural system of organisms: Proposal for the domains Archaea, Bacteria, and Eucarya. *Proceedings of the National Academy of Sciences USA,* **87**, 4576–4579.

Zhang, H. & Banfield, J.F. (1998) Phase stability in the nanocrystalline TiO_2 system. In: *Phase Transformations and Systems Driven far from Equilibrium* (E. Ma, P. Bellon, M. Atzmon & R. Trivedi, editors). Materials Research Society Symposia Proceedings, **481**. Materials Research Society, Warrendale, Pennsylvania, USA, pp. 619–624.

Zhang, H. & Banfield, J.F. (1999) A new kinetic model for the anatase-to-rutile phase transformation in nanocrystalline material revealing a second order dependence on the number of particles. *American Mineralogist*, **84**, 528–535.

Zhang, H. & Banfield, J.F. (2000) Understanding polymorphic phase transformation behavior during growth of nanocrystalline aggregates: insights from TiO_2. *Journal of Physical Chemistry B*, **104**, 3481–3487.

Zhang, H., Penn, R.L., Hamers, R.J. & Banfield, J.F. (1999) Enhanced adsorption on surfaces of nanocrystalline materials. *Journal of Physical Chemistry B*, **103**, 4656–4662.

Atmospheric aerosol particles: a mineralogical introduction

MIHÁLY PÓSFAI[1,*] and ÁGNES MOLNÁR[2]

[1]*Department of Earth and Environmental Sciences,
University of Pannonia, Veszprém, POB 158, H-8201 Hungary*
[2]*Department of Earth and Environmental Sciences, MTA-PE Air Chemistry
Research Group, University of Pannonia, Veszprém, POB 158, H-8201 Hungary*
**e-mail: mihaly.posfai@gmail.com*

Aerosol particles in the atmosphere interact with sunlight and initiate cloud formation, thereby affecting radiation transfer and modifying our climate. Aerosol particles also influence air quality and play important roles in various environmental processes. As the tropospheric aerosol is a heterogeneous mixture of various particle types, its climate and environmental effects can only be fully understood through detailed knowledge of the physical and chemical properties of the particles. Here, we review the formation and removal mechanisms of aerosol particles, the major approaches to study their physical and chemical properties, and discuss the most important categories of particle types. The focus of this review is on the 'mineralogical' identification and characterization of individual particles. We review the sources, transport and transformation mechanisms of the various particle types, their interactions with radiation and clouds, focusing on the results of the last 15 years.

1. Introduction

The Earth's atmosphere is a colloidal system that contains liquid and solid aerosol particles as well as gas-phase components. Aerosol particles are ubiquitous and play an important role in the physics and chemistry of the atmosphere, especially in the lower 10–15-km layer, the troposphere.

There is great scientific interest in atmospheric aerosols, stemming from a recognition of their significance in affecting our weather and climate. Aerosol particles control Earth's radiative properties ('heat balance') both directly and indirectly. They scatter and/or absorb solar radiation, which is generally referred to as the aerosol 'direct effect'. Aerosol particles also act as 'cloud condensation nuclei' (CCN), thereby modifying the physical and radiative properties of clouds; this is known as the 'indirect effect' (Fig. 1). The most important consequence of these two effects is the modification of the planetary albedo. The term 'radiative forcing' refers to changes in the planetary radiation budget, caused by anthropogenic or external influences; it is measured in watts per square meter (Wm^{-2}), and a positive value means net warming, whereas a negative value indicates net cooling of an air column above the Earth's surface (Forster *et al.*, 2007). The importance of aerosol radiative effects is clearly indicated by the trend of

© Copyright 2013 the European Mineralogical Union and the Mineralogical Society of Great Britain & Ireland
DOI: 10.1180/EMU-notes.13.3

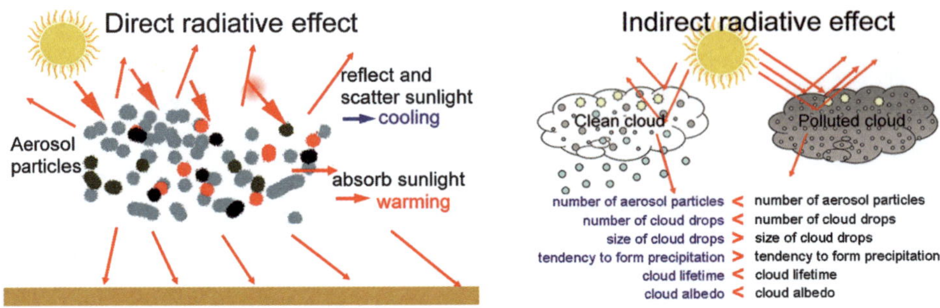

Fig. 1. Atmospheric aerosol particles scatter and absorb the incoming solar radiation, changing the heat balance of Earth (direct climate effect), and provide cloud condensation nuclei and thereby modify the properties of clouds (indirect climate effect). (Increased aerosol particle concentrations result in greater cloud droplet concentrations, thereby enhancing cloud reflectivity and inhibiting precipitation development, which cause clouds to persist longer and reflect still more sunlight.)

solar radiation at the Earth's surface. In the second half of the 20th century, a decline in incident solar radiation and an increase in planetary albedo were observed worldwide until about 1990. This phenomenon is referred to as 'global dimming', and is thought to have resulted from increasing aerosol forcing. After 1990, the trends changed signs and a brightening (a decrease in planetary albedo) occurred (Wild *et al.*, 2005), probably resulting from a reduction of the global aerosol burden.

In addition to being agents of climate change, aerosol particles affect our environment in various ways. For example, high concentrations of particles can cause serious visibility degradation (DeBell *et al.*, 2006; Huang *et al.*, 2009); some particle types are notable for their contribution to atmospheric acidity ('acid rain' or 'acid deposition') (Mészáros, 1992), whereas other types are important because of their adverse health effects when inhaled (Ibald-Mulli *et al.*, 2002; Huang *et al.*, 2009). Aerosols play important roles in the formation of deep-sea sediments and are intermediaries in global nutrient cycles. The atmospheric fluxes of essential nutrients (*e.g.* iron and nitrogen compounds) and other trace elements (Zn, Cu, Mn, Mo, B, Ni, Se, *etc.*) can be very important both in marine and terrestrial ecosystems (Duce *et al.*, 1991; Adriano, 2001; Jickells *et al.*, 2005).

Some aerosol particles form by the erosion of Earth's crust; these are true minerals. Other particle types typically do not conform to the definition of a mineral, as some are in a liquid state most of the time, and others are of anthropogenic origin. However, when aerosol particles are collected onto solid substrates, many of their components crystallize and then they are indistinguishable from minerals; more importantly, they can be studied using 'mineralogical' methods such as electron microscopy or X-ray diffraction. As the atmospheric effects of aerosols depend on the properties of individual particles, a mineralogical perspective is often useful in studies of atmospheric aerosols because it helps in assessing the properties of single particles, as opposed to those of bulk samples.

In some respects atmospheric aerosols can be compared to rocks; *e.g.* we refer to typical 'marine' and 'continental' aerosols, like we distinguish basalt and granite as typical rocks of the oceanic and continental crust, respectively. Both marine and continental aerosols consist of several types of particles that are comparable to the minerals in rocks. Whereas the major rock types have been known for a long time and detailed information on their minerals is readily available, our knowledge of the main aerosol particle types still has significant gaps. Even though some basic information is still missing, climate system models are used to predict how the climate of Earth will change when one or another parameter of atmospheric aerosols changes, because of either anthropogenic or natural perturbations. Clearly, more research is needed on the nature and properties of aerosol particles.

In this chapter we briefly describe the processes that form, modify, and remove aerosol particles from the atmosphere. Then the main methods of particle collection and analysis are discussed, with an emphasis on individual-particle techniques. Four sections are devoted to the most important groups of tropospheric aerosol types, including mineral dust, sea salt, sulfates and carbonaceous particles, in order of diminishing mineralogical character. Our goal is to show which particle types are globally or regionally important, their makeup, how they interact with other aerosol species and gases in the atmosphere, and their principal effects on climate and the environment. We highlight current issues and problems in atmospheric science, especially those that can be addressed using mineralogical methods.

As this is intended to be a 'mineralogical introduction' to tropospheric aerosols, many important aspects of atmospheric aerosol science are not treated here. For a more complete coverage of atmospheric aerosol particles the reader is referred to any of several excellent treatises on atmospheric chemistry, including those by Warneck (1988), Hobbs (1993), Charlson & Heintzenberg (1995), Seinfeld & Pandis (2006), Mészáros (1999), Jacobson (2002a), Gelencsér (2004) and Möller (2010). In the last decade, atmospheric science has found its place in mainstream mineralogy, as illustrated by various reviews of the field that were published in mineralogical textbooks and magazines (Anastasio & Martin, 2001; Buseck & Adachi, 2008; Gieré & Querol, 2010; Pósfai & Buseck, 2010).

2. The formation, transport, and removal of atmospheric aerosol particles

New aerosol particles form continuously in the atmosphere and are also emitted by surface sources. The particles are transported in a turbulent atmosphere by advection. As the relatively large surface area of aerosol particles provides excellent reaction sites, the particles take part in chemical transformations during their transport. Finally, particles are removed from the air either by dry or wet deposition. Removal processes constitute the sink of atmospheric aerosol particles; at the same time, they provide an important material input for the lithosphere, pedosphere, hydrosphere and biosphere.

The physical and chemical behaviour, as well as the fate of aerosol particles in the air are governed by many factors, among which the concentration and size distribution are very important. The size of aerosol particles ranges from several nanometres to some tens of micrometres, varying over five orders of magnitude. On the basis of detailed size-distribution measurements, Husar & Whitby (1973) and Whitby (1978) proposed that the sizes of particles from different sources (and formed by different processes) follow distinct log-normal distributions. According to the classical terminology introduced by Whitby (1978), the three prominent modes of the size distribution of the atmospheric aerosol are termed 'nucleation' ($d < 0.1$ μm), 'accumulation' ($d = 0.1-1$ μm) and 'coarse' ($d > 0.1$ μm) 'modes' (Fig. 2). Particles with $d < 1$ μm constitute the

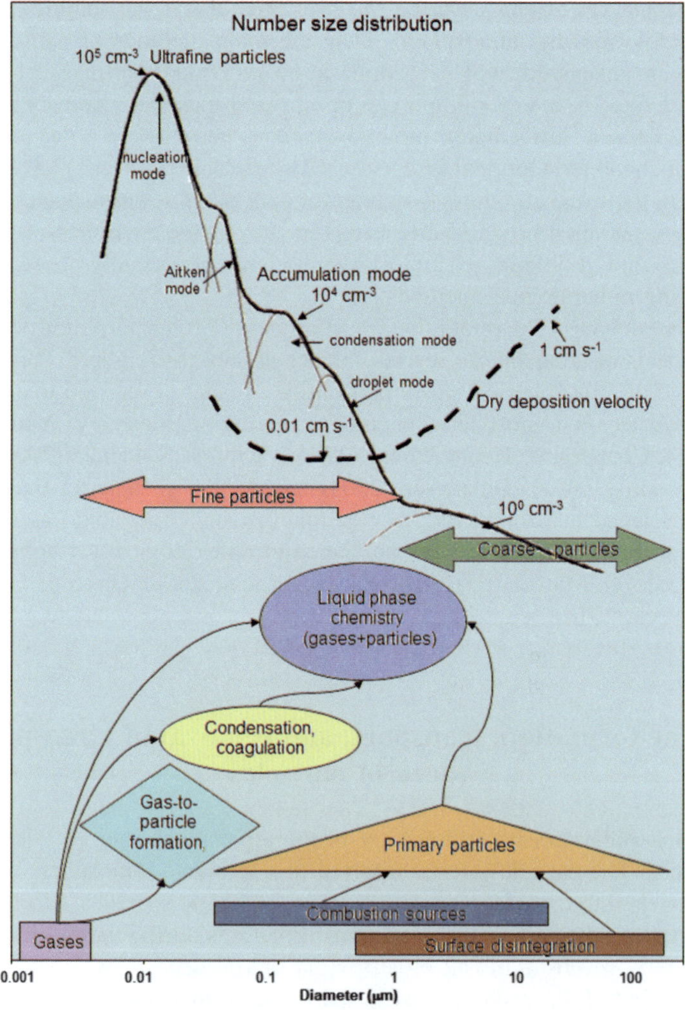

Fig. 2. Schematic representation of the size distribution of aerosol particles (solid curve). The numbers above the modes give typical total number concentrations. Dry deposition velocities (with characteristic deposition values) and the main formation and dynamic processes are also shown.

class of 'fine' particles. Subsequent studies, however, showed that both the nucleation and accumulation modes are frequently bimodal (Covert *et al.*, 1996; Hering *et al.*, 1997; Dal Maso, 2006). According to current classification, particles with $d<90$ nm belong to the ultrafine (UF) mode that is generally composed of nucleation (<25 nm) and Aitken (>25 nm) size ranges. Within the accumulation mode, condensation (up to ~200 nm) and droplet modes ($0.2-1$ μm) are distinguished.

Ultrafine particles form by nucleation from gaseous precursors. Small particles (which are generally liquid) in the nucleation mode coagulate to create particles in the Aitken mode and in the $0.09-1.0$ μm size range. Nucleation-mode particles can also grow into the accumulation mode by direct vapour condensation onto their surfaces. The larger particles in the droplet mode form in cloud processes (see below) (Hering *et al.*, 1997; Hoppel *et al.*, 1994; John, 2011). Owing to these processes, the chemical composition of particles in the fine mode is, in general, similar. The coarse mode (Fig. 2) consists of particles formed by the disintegration of Earth's surface and vegetation. Particles from combustion sources also contribute to this size range. It follows from the different formation processes that the compositions of fine and coarse particles are different.

2.1 Production of primary particles

As about two thirds of the surface of our planet is covered by oceans, sea-salt particles constitute the major fraction of primary coarse particles under marine conditions (O'Dowd *et al.*, 1997). Although water droplets can form directly from sea spray, most sea-salt particles result from the bursting of gas bubbles that reach the water surface. The larger droplets fall back into the water, while the smaller ones remain airborne and can be transported over large distances and to high altitudes.

Large amounts of natural primary aerosol particles are emitted on the continents, particularly in arid and semiarid areas (Prospero, 1999; Denman *et al.*, 2007). Just as the composition of sea-salt particles in the air reflects the composition of ocean water, the chemical nature of mineral dust particles is determined by the composition of Earth's crust. As human activities such as overgrazing and changes in land use can increase the extent of arid regions, the extent and intensity of dust production is modified by mankind. The disintegration and dispersion of bulk plant material also produce primary aerosol particles that consist of leaf debris, viruses, bacteria, spores and pollen, protozoa, algae and humic substances. The data in Table 1 give an indication of the mass of particles produced by surface disintegration.

Combustion processes produce primary particles that consist mostly of carbonaceous materials such as soot (also called elemental or black carbon) and organic compounds

Table 1. Estimates of the annual mass flux of aerosol particles from surface disintegration.

Ocean surface	$10,000-30,000$ Tg y^{-1}	Erickson & Duce (1988); Gong *et al.* (2002)
	$16,300$ Tg y^{-1} $\pm\ 200\%$	Denman *et al.* (2007)
Crustal surface	$1,000-3,000$ Tg y^{-1}	Denman *et al.* (2007)

(Gelencsér, 2004), and that typically belong in the fine fraction of the aerosol. Fossil fuels contain residual mineral materials, including clays, sulfides, carbonates, chlorides and various trace metals that are also emitted in particulate form within fly-ash particles in the coarse size range (Ondov & Biermann, 1980). On the other hand, the condensation of volatile metallic and non-metallic oxides (including SiO_2) produce fine particles, resulting in a bimodal size distribution of fly-ash particles. Soot particles are also emitted by mobile sources, particularly by diesel engines, producing either a unimodal (accumulation mode) or a bimodal (accumulation and coarse mode) aerosol population. Biomass burning (mainly in the tropics) is an important source of primary, typically fine-mode aerosol particles (Crutzen & Andreae, 1990; Gelencsér, 2004).

2.2. Production of secondary particles

In the multiphase system of the atmospheric aerosol there is continuous material exchange between condensed and gas-phase materials; vapours condense to form liquid particles, and particles evaporate as a function of ambient conditions. The condensation or nucleation can proceed either homogeneously (when only condensing vapours take part in the process; see nucleation mode particles) or heterogeneously on existing aerosol particles (see Aitken and condensation mode particles). In most cases, vapours nucleate simultaneously with water molecules, resulting in minute solution droplets.

In general, nucleation can occur when a gas with low saturation vapour pressure is present in the air as a result of direct emissions or atmospheric chemical reactions; such gases can condense to form particles under ambient conditions. Another aerosol nucleation mechanism is active when hot air cools after combustion, leading to a high supersaturation of certain vapours. Sulfate and organic particles are produced by the first nucleation process, whereas the cooling of combustion vapours is responsible for the nucleation of carbonaceous materials and metal oxides.

2.2.1. Homogeneous nucleation

Consider a thermodynamic system that consists of a single condensable substance. The free energy (G) of this initial gas-phase system can be described as:

$$G = G_1 = \mu_1 N_1$$

where N_1 and μ_1 represent the number and the chemical potential of the gas molecules, respectively. Vapour molecules are in continuous motion as a function of temperature; they collide to form, at least temporarily, aggregates ('droplets') of different sizes. Kinetic forces act to destroy the aggregates, while attractive forces among the molecules tend to stabilize them. The free energy of the system after nucleation is the sum of the free energies of the gas-phase and condensed components:

$$G = G_1 + G_2 = \mu_1(N_1 - N_2) + \mu_2 N_2 + \pi d_p^2 \sigma$$

where N_2 and μ_2 are the numbers and chemical potentials of the liquid phase molecules, respectively, d_p is the particle diameter, and σ is the surface tension. The change of free energy (ΔG) due to the nucleation of a particle is given by

$$\Delta G = \mu_1(N_1 - N_2)$$
$$+ \mu_2 N_2 + \pi d_p^2 \sigma$$
$$- \mu_1 N_1$$

which yields:

$$\Delta G = (\mu_2 - \mu_1)N_2$$
$$+ \pi d_p^2 \sigma$$

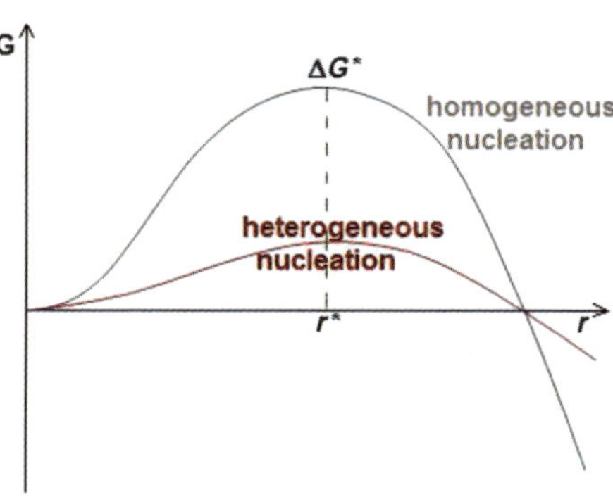

Fig. 3. Schematic diagram illustrating the difference in activation energy (ΔG) for homogeneous and heterogeneous nucleation (r^*: critical droplet radius).

Numerical calculations using the above equation show that ΔG increases continuously with d_p, as long as the air is undersaturated with respect to the vapour. On the other hand, in the case of supersaturation, ΔG has a maximum (ΔG^*) at a critical droplet radius r^*, where the droplet is in a metastable equilibrium with its surroundings (Fig. 3). If such a droplet (germ of condensation) gains vapour molecules, it increases its size, whereas if it loses molecules, it evaporates completely.

Typically, two or several vapours condense together during aerosol formation in the atmosphere. In such cases particle formation rates can be relatively high even if the air is undersaturated with respect to both vapours, because the saturation vapour pressure of the homogeneous mixture is lower than those of the individual substances. A well known example of this phenomenon is the condensation of H_2SO_4 and water (Mirabel & Katz, 1974). The basic aims of aerosol nucleation research are to determine the free energy necessary for nucleus formation and to establish the relationships between particle size and supersaturation values under atmospheric conditions (Mészáros, 1999; Warneck, 1988). When conditions favor the formation of new particles, a nucleation event occurs: nm-sized, stable particles (droplets) form from gaseous compounds, as indicated by a sudden increase of the nucleation-mode ($d < 25$ nm) particle concentration. Recently, great advances have been made in constraining the effects of several parameters on nucleation, by using state-of-the-art experimental facilities (Kirkby *et al.*, 2011). For example, atmospherically relevant mixing ratios of ammonia increase 100–1,000-fold the nucleation rate of sulfuric acid particles. The freshly formed particles then grow by coagulation and by the condensation of gases, which is called heterogeneous nucleation (Dal Maso, 2006).

2.2.2. Heterogeneous nucleation

The surfaces of freshly formed particles (droplets) are continuously bombarded by the surrounding vapour molecules, and this gas flux (condensation) is the most significant mechanism that increases the size of the aerosol particles (Dal Maso, 2006; Seinfeld & Pandis, 2006). Heterogeneous nucleation is particularly efficient if the nucleus (particle) is soluble in the liquid of the condensing vapour(s). As water is the most important condensable material in the air, water-soluble atmospheric particles play a dominant role in atmospheric nucleation. Some nuclei are entirely water soluble, whereas others are originally hydrophobic but can acquire a hygroscopic coating and become efficient nuclei for heterogeneous aerosol formation. The presence of non-volatile solutes such as $(NH_4)_2SO_4$ and NaCl in an aqueous solution considerably reduces the saturation vapour pressure; this effect is responsible for the existence of solution droplets that are in equilibrium with air well below a relative humidity of 100% (Wayne, 1991). Such stable solution droplets, called haze particles, control the optical properties of the air when the air is undersaturated with respect to liquid water.

Heterogeneous processes are responsible for the formation of a significant fraction of atmospheric sulfate and nitrate on sea-salt particles (Sievering *et al.*, 1992), and on mineral dust. Organic compounds probably condense onto sulfate particles. Even though these heterogeneous processes are widespread in the atmosphere, their details are not always known satisfactorily; this will be discussed in more detail below in the sections on sulfate and carbonaceous particles.

2.2.3. *Cloud processes*

Liquid-phase transformation in cloud and fog droplets is another important pathway of heterogeneous secondary aerosol formation. Fog and cloud droplets, formed on cloud condensation nuclei, absorb trace gases from the atmosphere from the moment of their formation. Some gases dissolve in the water by forming ions that can react in the liquid phase with other absorbed species, creating new compounds. If the cloud evaporates without forming precipitation, the contents of the droplets remain airborne in cloud-processed aerosol particles. Typically, aerosol particles go through several cloud cycles before they are finally removed from the air (Hoppel *et al.*, 1996; Rosenfeld *et al.*, 2008). The processes in cloud and fog droplets are in many respects similar to the reactions that take place in the liquid water associated with aerosol particles that are outside clouds; there are significant differences, however, as far as sizes (the diameters of cloud droplets range from about 4 to 150 μm) and the environments (*e.g.* water saturation) of cloud droplets and particles are concerned.

Concerning aerosol formation, the most important chemical process in fog and cloud water is the conversion of absorbed SO_2 into SO_4^{2-}. According to laboratory experiments and theoretical studies, the oxidation of S(IV) to S(VI) can proceed through four processes (Salmon, 1994; Möller, 2010): oxidation by O_2 in the presence of catalysts such as Mn and Fe; oxidation by absorbed O_3; this can be the dominant process in marine clouds, because the droplets form on alkaline sea-salt particles, resulting in a relatively high pH of such droplets (Hegg *et al.*, 1992); oxidation by dissolved H_2O_2 (most effective under pH 5); and OH-initiated oxidation by O_2.

In general, the pH of cloud water is <5; under these conditions, H_2O_2 is the main oxidising agent (Snider & Vali, 1994). According to atmospheric sulfur budget calculations by Langner & Rodhe (1991), the majority of SO_2 (84% of global emissions) is oxidized in cloud water. Wojcik & Chang (1997) estimate that over North America $\sim$70% of S(VI) production results from aqueous processes, and 30% is due to gas-phase reactions.

This conclusion is in good agreement with data of Jaenicke (1993), who found that gas-to-particle conversion of precursor natural gases produces a global aerosol mass of 1300 $Tg \cdot y^{-1}$, whereas reactions in clouds result in a mass of 3000 $Tg \cdot y^{-1}$. Locally and regionally, the relative importance of the two processes is certainly a function of weather conditions.

2.3. Dynamic processes

Besides homogeneous and heterogeneous nucleation and cloud processes, coagulation also plays an important role in the control of aerosol properties. Coagulation is the process of collision and coalescence of particles as a result of their movement. The rate of coagulation depends on the particle concentration and on the coagulation efficiency, a parameter that is a function of the relative velocities of the colliding particles and the interception cross area. The differences in particle velocities result from gravitational settling, turbulence and Brownian motion (Dal Maso, 2006; Seinfeld & Pandis, 2006). The coagulation efficiency can be modified by electrical forces that enhance or retard coagulation as a function of the charge of the particles. Besides, electrical forces are thought to be important in aerosol filtration, in the enhancement of the natural lifting of mineral dust aerosols (Kok & Renno, 2008), as well as in the scavenging of aerosol particles by cloud droplets (*e.g.* Tinsley *et al.*, 2000).

Brownian coagulation is one of the main sinks of ultrafine aerosol particles. The particles in the air are agitated by gas molecules that are in permanent motion as a function of temperature. The velocity of Brownian motion is in inverse relationship with particle sizes; thus, this thermal effect is important only in the ultrafine particle size range, mostly for particles with diameters of 0.001–0.01 μm (Fig. 2). Thermal coagulation results in a decrease of the number and an increase of the sizes of particles in the system.

Besides controlling the particle-size distribution, thermal coagulation is of crucial importance in regulating the chemical composition. If particles of different compositions coagulate, the aerosol becomes 'internally mixed', meaning that different substances occur in a single particle. If particles of different compositions occur in the aerosol, but each individual particle contains only one homogeneous substance, the aerosol is said to be 'externally mixed'. Mixing plays an essential part in controlling the hygroscopic and optical properties of the particles (Fassi-Fihri *et al.*, 1997; Fuller *et al.*, 1999; Freney *et al.*, 2010).

The atmospheric effects of aerosol particles also depend on their 'residence time', the average time they spend in the air. Owing to gravitation, particles move towards the Earth's surface. The settling velocity is determined by particle mass; consequently, particles $>\sim$10 μm cannot remain in the air for a long time (see below). Whereas the number of coarse particles is reduced by sedimentation, particles with diameters $<$0.1 μm are removed rather rapidly from the air by thermal coagulation. As a result of these two processes, particles in the accumulation size range (Fig. 2) remain in the air the longest and constitute the most stable fraction of aerosol particles. Particles in this size interval, mostly those composed of water-soluble materials, provide efficient condensation nuclei that make cloud formation possible. As such particles are removed from the air mainly in precipitation, their residence time is about the same as that of water, which is $\sim$10 days (Mészáros, 1999).

The residence time of aerosol particles also depends on their elevation; the higher they are injected into the atmosphere, the longer the time they spend in the air. For example, mineral dust can reach higher altitudes above regions where intensive vertical convection occurs; this is typical along the thermal equator, the Intertropical Convergence Zone (ITCZ). In general, wind speed in the troposphere also increases with altitude, being higher in the free troposphere (FT) than in the boundary layer (BL)[1]. The combination of longer residence time and greater wind speed at high altitudes results in the long-range transport of dust particles. In spite of these cases, accumulation-mode particles are the most likely objects of long-range transport because of their longer residence times.

2.4. Removal of particles from the atmosphere

2.4.1. Dry deposition

Because of their sizes and masses, gravitational settling is an important sink for coarse particles. The velocity of sedimentation is given by the Stokes formula, in which the sedimentation velocity (v_s) is proportional to particle size (d_p) and density (ρ_p), and inversely proportional to the dynamic viscosity (η) of the air:

$$v_s = \frac{d_p^2 \rho_p g}{18\eta}$$

The values of v_s for spherical particles with a unit density indicate that the sedimentation velocity becomes significant ($> \sim 0.003$ ms^{-1}) for particles with diameters >10 μm (Mészáros, 1999).

Smaller, fine particles are transported vertically by irregular 'turbulent diffusion' that is proportional to the concentration gradient. If the concentration in the air increases with height, the particle flux is directed downwards and *vice versa*. Turbulent diffusion itself, however, does not remove particles from the atmosphere because soil and other surfaces are bordered by a thin laminar layer ($\sim$1 mm thick), across which the particles must be transported by other processes, such as sedimentation or Brownian diffusion (Mészáros, 1999). As the characteristics of this layer are different for various surfaces, particle removal also depends on the physical and chemical properties of the underlying surface.

2.4.2. Wet deposition

Atmospheric aerosol particles and water clouds interact continuously with one another. Cloud droplets form on aerosol particles known as 'cloud condensation nuclei' (CCN), because the supersaturation in clouds is too low for initiating homogeneous nucleation of water vapour (Twomey, 1977; Rosenfeld *et al.*, 2008). The structure of clouds is thus

---

[1]The BL is a 1–2 km thick, turbulent layer next to the Earth's surface; most particles emitted by surface sources can be found in this layer. The BL and the FT are usually separated by a thermal inversion that restricts the exchange of aerosol particles between these two regimes.

controlled by the physical and chemical nature of CCN aerosol particles, as well as by dynamic factors such as updraft velocity. On the other hand, cloud droplet formation removes a major proportion of aerosol particles by 'in-cloud scavenging'. A smaller fraction of particles is removed by thermal coagulation between the nucleation-mode particles and cloud elements, while particles in the accumulation size range can enter the cloud droplets by means of phoretic forces (Grover & Pruppacher, 1977). When clouds precipitate, additional particles are removed from the air layer below the clouds by falling precipitation elements such as drops or snow crystals; this process is termed 'below-cloud scavenging'. Particles $<\sim 0.5-1.0$ μm collide with precipitation elements due to their Brownian motion, whereas coarse particles are scavenged owing to their lower sedimentation velocity that results in their inertial deposition onto the surfaces of drops or snow crystals.

In conclusion, aerosol residence time in the atmosphere is primarily controlled by the removal of particles by clouds and precipitation. The geochemical cycle of particulate matter is mainly determined by wet deposition, which cleans the atmosphere and provides nutrients for terrestrial and aquatic ecosystems.

3. Methods for studying atmospheric aerosols

3.1. Properties of aerosol particles

The roles of aerosol particles in atmospheric processes depend on their physical and chemical properties. To understand the behaviour of aerosol particles, proper measurements of the various aerosol characteristics are required. In this section we review the most important properties of aerosol particles, then describe sampling and analytical methods that are typically used in the study of atmospheric aerosols.

3.1.1. Concentration and size distribution

The simplest way to characterize an aerosol is to give the total number or total mass of particles by unit gas volume ('number' or 'mass concentration', respectively). The dimension of the number concentration is generally cm^{-3} (more exactly particles/ cm^3), whereas the mass concentration is expressed in $\mu g \cdot m^{-3}$, $mol \cdot m^{-3}$, or mol/ mol. As the size of atmospheric aerosol particles varies by several orders of magnitude, the total number or mass concentrations are not sufficient to characterize the particles (Fig. 4). Therefore, we use differential size-distribution functions to describe the number and mass concentrations (denoted N and M, respectively) as a function of particle diameter (d_p) (see Mészáros, 1999):

$$\frac{dN}{d(d_p)} = f(d_p) \quad \text{or} \quad \frac{dM}{d(d_p)} = F(d_p),$$

where dN and dM are the number and mass of aerosol particles in the size interval between d_p and $d_p + d(d_p)$, respectively. It follows from this definition that the integral

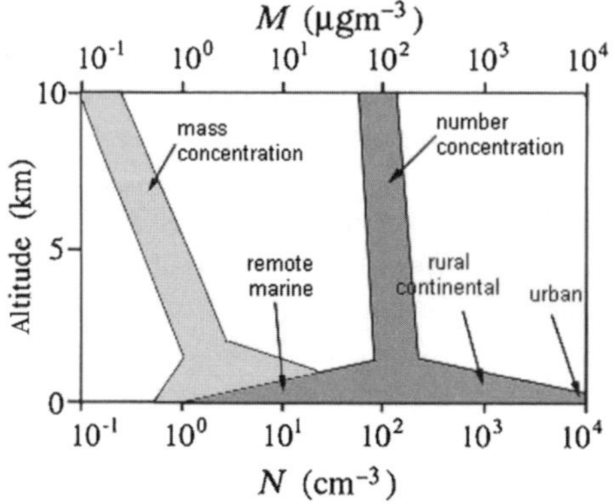

Fig. 4. Estimated number, *N*, (dark area) and mass (lightly shaded area) concentration ranges, *M*, for tropospheric aerosols as a function of altitude (after Heintzenberg, 1994).

of the size distribution function is equal to the total number/mass concentration:

$$N = \int_0^\infty f(d_\mathrm{p})d(d_\mathrm{p}) \quad \text{or}$$

$$M = \int_0^\infty F(d_\mathrm{p})d(d_\mathrm{p}).$$

If we know the number-size distribution and assume that the particles are spherical with known density (ρ_p), then the number-size distribution can be converted into mass-size distribution:

$$F(d_\mathrm{p}) = f(d_\mathrm{p})\frac{\pi}{6}d_\mathrm{p}^3\rho_\mathrm{p}.$$

A similar approach can be used to define surface and volume size distributions. By means of the size-distribution functions, the average size of particles can be determined. The number (d_pn) or mass (d_pm) average diameters are calculated by weighting particle diameters according to the corresponding size-distribution function:

$$\overline{d}_\mathrm{pn} = \frac{\int_0^\infty f(d_\mathrm{p})d_\mathrm{p}d(d_\mathrm{p})}{\int_0^\infty f(d_\mathrm{p})d(d_\mathrm{p})} \quad \text{or} \quad \overline{d}_\mathrm{pm} = \frac{\int_0^\infty F(d_\mathrm{p})d_\mathrm{p}d(d_\mathrm{p})}{\int_0^\infty F(d_\mathrm{p})d(d_\mathrm{p})}.$$

We note again that the schematic multimodal size distribution in Figure 2 is the result of different-size distribution functions.

3.1.2. Composition, external/internal mixing, shape, liquid/solid state

Owing to their different formation mechanisms, the chemical compositions of fine and coarse particles, and continental and marine aerosols are substantially different (Fig. 5). Also, as the consequence of dynamic and chemical transformation processes, the aerosol generally consists of internally mixed particles. Concerning their optical and hygroscopic properties, as well as their environmental effects, internally and externally mixed aerosols behave differently (Haywood & Shine, 1995). The compositions and internal/external mixing of arosol particles are discussed in detail below, in the sections devoted to the major particle groups.

Depending on their origins, particles also differ in shape and physical state. Whereas condensation aerosols are generally liquid and have spherical shapes, particles of

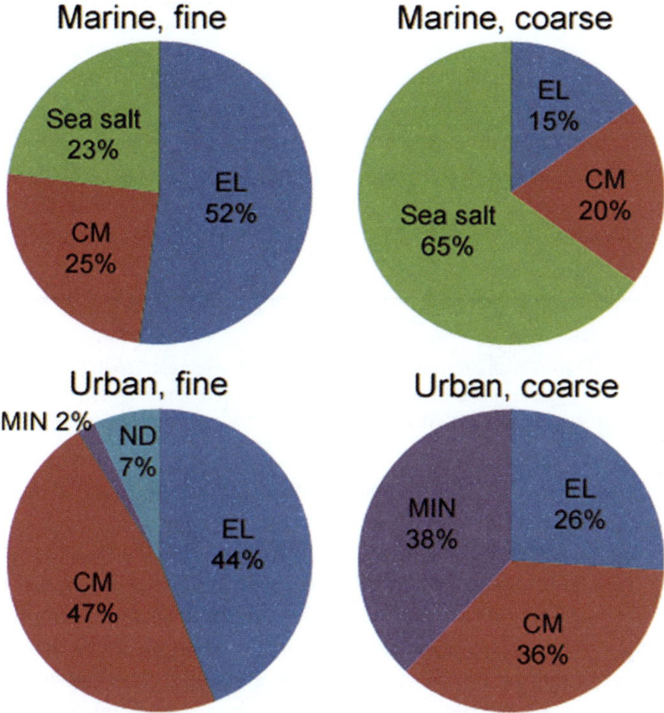

Fig. 5. Selected examples of fine *vs.* coarse and urban *vs.* marine aerosol compositions. The data for marine and urban aerosol samples are from the North Atlantic, $\varphi = 40°$ (Virkkula *et al.*, 2006), and from Vienna, Austria (Puxbaum *et al.*, 2004), respectively. MIN: mineral matter; CM: carbonaceous matter, ND: not determined; EL: electrolytes (including SO_4^{2-}, NO_3^-, NH_4^+, Na^+ and K^+ in the marine samples, and SO_4^{2-}, NO_3^-, NH_4^+ and non-sea-salt Na^+ in the urban samples). The size cut between fine and coarse particles is at 2 μm diameter.

surface origin are solid and have irregular shapes (Buseck *et al.*, 2000; Pósfai & Buseck, 2010). Combustion processes produce both solid and liquid particles that can have variable morphologies. In many cases, irregular particles in the fine size range result from coagulation processes. Knowledge of the average shapes of particles is essential for assessing their optical properties (Bohren & Singham, 1991).

3.1.3. Hygroscopic behaviour

The hygroscopic behaviour of aerosol particles is related to the hydrophilic or hydrophobic character of their compounds. As the relative humidity (RH) increases, a water-soluble, solid particle grows very little only by the adsorption of water molecules onto its surface. When the RH reaches a critical value, the deliquescence point, the particle becomes a solution droplet because at that RH the saturated solution of the compound is in equilibrium with its surroundings. This critical RH is as low as 62% for

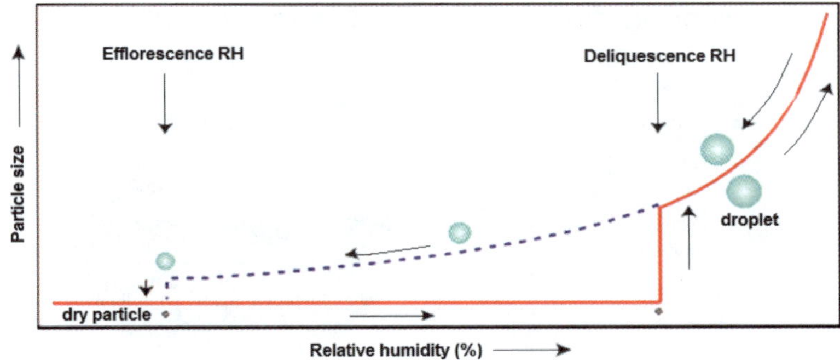

Fig. 6. Schematic hydration–dehydration curves for a water-soluble inorganic salt; the arrows below the solid line mark the direction of hygroscopic growth with increasing relative humidity (RH), and the arrows above the dashed line mark the dehydration path that the particle follows with decreasing RH.

NH_4NO_3 particles (at 25°C), and 81 and 75% for $(NH_4)_2SO_4$ and NaCl, respectively (Tang, 1996; Wise *et al.*, 2005). Further increase of the RH results in the growth of the particle (droplet). On the other hand, when the RH decreases, the particle remains in a metastable liquid state down to well below its deliquescence point (Fig. 6). There-fore, water-soluble aerosol particles are liquid throughout most of their atmospheric life-times under tropospheric conditions (Rood *et al.*, 1989). We note that in the case of particles of mixed composition, particularly when organic material is involved in the mixture, this deliquescence/efflorescence behaviour may not be observed. Instead, a continuous growth/reduction of particle volume takes place with changing RH. As the size distribution of particles varies with RH, their optical properties are also affected. This is well illustrated by the everyday experience that visibility is inversely pro-portional to RH.

The most important consequence of aerosol hygroscopicity is cloud and precipita-tion formation. Clouds are produced in rising, cooling air that becomes slightly super-saturated with respect to water, leading to the condensation of water vapour. The condensation of atmospheric water under environmental conditions is exclusively het-erogeneous; it takes place on aerosol particles that are active as CCN (Rosenfeld *et al.*, 2008). Each aerosol particle is activated at a critical supersaturation value that depends on composition and particle size. Cloud droplets can form only on those par-ticles that have lower (or equal) critical supersaturation than the supersaturation in the forming cloud. Using thermodynamic considerations, an inverse relationship is found between critical supersaturation and critical particle (droplet) size (Fig. 7); the larger the particles, the more active they are as CCN (Dusek *et al.*, 2006). Note that water-insoluble particles can also act as condensation nuclei if they are wettable; however, at a fixed supersaturation their critical sizes are much larger than those of water-soluble nuclei (for further details see Seinfeld & Pandis, 2006).

The hygroscopic growth curves for many inorganic aerosol compounds are known (Tang & Munkelwitz, 1993); however, various internally mixed inorganic/organic par-ticles are common in the atmosphere, and the hygroscopic properties of such particles

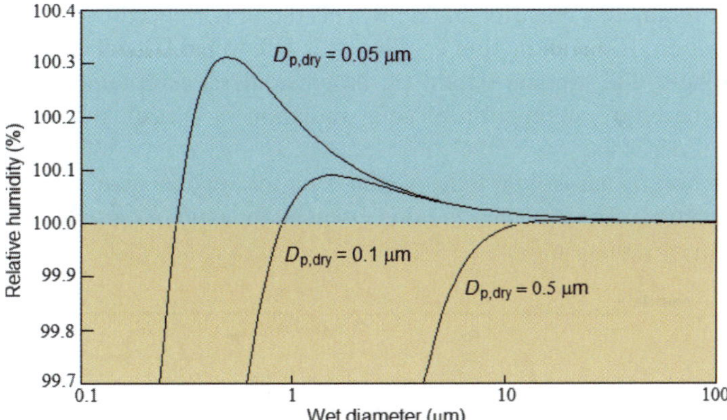

Fig. 7. Activation of aerosol particles as cloud condensation nuclei as a function of particle diameter (Köhler-curves). Larger particles activate at lower supersaturations.

can profoundly differ from those of homogeneous, pure substances (Andrews & Larson, 1993; Saxena *et al.*, 1995; Semeniuk *et al.*, 2007; Freney *et al.*, 2010). A better understanding of the hygroscopic behaviour of internally mixed particles is crucial for evaluating their roles in the atmosphere.

3.1.4. Optical properties

Direct aerosol climate forcing relates to the interaction of aerosol particles with the incoming solar radiation (Fig. 1), an effect that is determined by the optical properties of the aerosol. Particles interact with light in two different ways (Horvath, 1992; Mészáros, 1999; Seinfeld & Pandis, 2006): 'scattering' means that the energy received by a particle is re-radiated without a change in the wavelength, whereas if the energy of the received radiation is changed into another form (*e.g.* heat) or is re-radiated at a different wavelength, the process is termed 'absorption'. The efficiency of light scattering and absorption depend primarily on the size and chemical composition of the particle, respectively. Water or solution droplets are purely scattering, whereas soot particles (and, to a lesser extent, mineral dust) constitute the most important group of absorbing materials in the atmosphere.

The sum of scattering and absorption is called 'extinction' or 'attenuation'. Scattering and absorption efficiencies are the ratios of the amounts of scattered and absorbed energies to the energy received by the particles, respectively. Thus, the 'extinction efficiency' is defined as the sum of these two efficiencies. Extinction can be characterized by the extinction (or attenuation) coefficient that gives the amount of extinction over unit pathlength, as the sum of the scattering and absorption coefficients:

$$\sigma_{ext} = \sigma_{scat} + \sigma_{abs}.$$

These important parameters have the dimension of inverse length. If the value of the scattering coefficient is divided by the extinction coefficient, the 'single-scatter albedo' (ω) is obtained, a quantity that represents the importance of scattering processes in light attenuation. If the extinction coefficient is integrated from the surface to the top of the atmosphere, the 'optical depth' (τ, dimensionless) is obtained, a quantity that characterizes the entire attenuation of solar radiation in a vertical air column due to the presence of aerosols.

Light extinction by aerosol particles depends on the 'optical size' and the 'refractive index' of the particle. The optical size (x) is given by the ratio of particle diameter (d_p) to the wavelength of radiation (λ):

$$x = \pi \cdot d_p / \lambda$$

If x is small ($d_p \ll \lambda$), the spatial distribution of the scattered light is approximately spherical (Rayleigh scattering). If the value of d_p is comparable to or larger than λ, then the angular distribution of scattered radiation is a complicated function of particle size, as described by Mie theory (see, *e.g.* Haywood & Ramaswamy, 1998).

For the refractive index (RI), a complex number is used, the real part of which represents scattering (this is the customary RI, the ratio of the velocity of light in vacuum to that in the material), and the imaginary part describes the absorption properties of the particle. The real parts of RIs of the main aerosol particle types typically range from 1.4 to 1.6 (for pure water it is 1.33) (Table 2).

The RI of the total aerosol can be approximated by taking the sum of individual RIs, multiplied by the corresponding volume fractions. The problem with this approach is that it inadequately describes the optical behaviour of an aerosol that may consist of internally mixed particles (Bond & Bergstrom, 2006). Detailed calculations show that the RI of an internally mixed particle can differ significantly from what could be expected from a weighted summation of individual RIs, and depends on subtle parameters such as the position of one species within the other (Haywood & Shine, 1995; Fuller *et al.*, 1999). For a proper characterization of aerosol optical properties, the RIs of internally mixed particles need to be better understood and more precisely measured.

Table 2. Complex refractive indices ($m = n - ki$) of some major components of atmospheric aerosol at 589 nm wavelength; $i = (-1)^{1/2}$ (HULIS: humic-like substances; OM: organic matter).

Material	n	ki	Reference
Ammonium sulfate	1.521	0	Weast (1987)
Ammonium bisulfate	1.473	0	Weast (1987)
Sulfuric acid	1.426	0	Weast (1987)
Ammonium nitrate	1.413	0	Weast (1987)
Sodium chloride	1.544	0	Weast (1987)
Quartz (at 550 nm)	1.550	0	Seinfeld & Pandis (2006)
Carbonaceous matter			
elemental carbon	1.50	0.47	Horvath (1998)
soot (at 500–550 nm)	1.95	0.79	Bond & Bergstrom (2006)
HULIS (at 500–550 nm)	1.67	0.017	Hoffer *et al.* (2006)
tar ball (at 500–550 nm)	1.67	0.27	Alexander *et al.* (2008)
non-absorbing OM	1.40	0	Turpin & Lim (2001)

The indirect climatic effect of the aerosol is related to the scattering of sunlight by cloud droplets. This light scattering is strongly dependent on the number and size of the droplets. In polluted air masses a larger number of cloud condensation nuclei are present than in a clean atmosphere. As a consequence of the higher CCN concentration, more numerous and smaller cloud droplets are formed from the available water vapour (often called the 'Twomey-effect'), resulting in enhanced light scattering (Fig. 1).

3.2. Sampling

Aerosol samples are typically collected in sampling campaigns, although at some locations aerosol properties are measured regularly. For example, daily aerosol measurements have been performed at the Mauna Loa Observatory, Hawaii, since the 1950s, and at Barbados and Miami since 1965 and 1974, respectively (Prospero, 1999). Aerosol concentrations and optical properties have been monitored by the Global Atmosphere Watch of the World Meteorological Organization. Other environmental (including air pollution) monitoring programs are also active both in Europe (EMEP: European Monitoring and Evaluation Programme) and in the US (NOAA-ESRL: National Oceanic and Atmospheric Administration - Earth System Research Laboratory, EPA: United States Environmental Protection Agency). Satellite observations can continuously provide data on aerosol properties such as optical depth over the oceans and single-scatter albedo (Kaufman *et al.*, 2002).

Because of aerosol climate effects, the nature of the global background aerosol and how it is changed by anthropogenic influences is of considerable interest; therefore, large sampling campaigns are organized for studying atmospheric aerosols in remote regions, where the natural and anthropogenic effects can be better distinguished than in densely inhabited, polluted areas. In such international campaigns, a combination of land, ship, and aircraft-based sampling platforms are usually used. As the methods of aerosol sampling are devised to provide data on the physical and, in some cases, chemical properties of the particles, commonly used sampling devices are described in the section below.

3.3. Probes for studying the physical properties of the aerosol

The total number concentration of aerosol particles is generally measured with a 'condensation particle counter' (CPC). In this device, the air is saturated and suddenly expanded to produce a significant vapour (water or alcohol) supersaturation. In the supersaturated environment the vapour condenses onto the aerosol particles. The number of droplets formed in this way equals the number of aerosol particles with a critical supersaturation less than that generated by the instrument; such particles are typically >3–10 nm. The droplets are counted, after calibration, by the extinction of a light beam through the chamber (Gras *et al.*, 2002).

The size distribution of particles <0.1 μm can be determined with 'electrical mobility analysers' (see, *e.g.* http://www.tsi.com/differential-mobility-analyzers/). In these instruments (DMA: differential mobility analyser, DMPS: differential mobility particle sizer, SMPS: scanning mobility particle sizer) the particles are electrically charged, then passed through an electric field and then counted; as size and electrical mobility are related, the size distribution of particles can be deduced. The smaller aerosol particles can also be collected from the air by means of 'thermal precipitators'. In these instruments, metal wires are heated to produce a temperature gradient. Because of the impact of more energetic gas molecules from the heated side, aerosol particles move away from the wire in the direction of a cold surface.

Several methods can be used to detect larger particles. They can be studied *in situ* in a gaseous medium using 'single-particle optical counters' in which the particles are illuminated and the light scattered individually by each particle is measured photoelectrically at a certain angle. The number of scattering signals is related to the particle number, whereas the amplitude of each signal gives the particle size. Aerodynamic particle sizer spectrometers provide high-resolution, real-time aerodynamic measurements of coarse particles (from 0.5 to 20 μm) (http://www.tsi.com).

In general, particles >0.1 μm can be collected by inertial deposition. Inertial sampling is carried out by placing an obstacle or collector into the air stream. As the particles have larger inertia than the air molecules, they strike the collector while the air flows around the obstacle. With increasing air-stream velocity and decreasing collector size, the size of collected particles decreases. As wind speed is generally not large enough for the direct collection of particles, a suitable particle velocity can be achieved by pumping air through a tube and then passing it through narrow slits or jets. Size-segregated particles can be collected by means of a 'cascade impactor', a device that contains slits of different sizes. 'Low-pressure impactors' can be used to collect particles with sizes down to $\sim$30 nm. The particles collected onto impactor stages can be analysed chemically or with a microscope.

For chemical analysis, particles are often sampled on filters. Air is pumped through a filter substrate placed in a suitable filter holder (which can be a cascade impactor, for example). Filters for aerosol sampling have two main categories: fibrous filters consist of mats of fibres made of glass, quartz or cellulose, whereas membrane filters are composed of thin films of polymeric materials that contain small pores of controlled sizes. When using filters, in the absence of electric forces the larger and smaller particles are captured from the air by impaction and diffusion, respectively.

Aerosol mass concentration can be measured by gravimetry or by means of 'PM monitors' (beta-gauge monitor; Chueinta & Hopke, 2001) or using a 'tapered element oscillating microbalance' (TEOM; Patashnick & Rupprecht, 1991). The 'aerosol mass spectrometer' (AMS) is capable of providing real-time size and chemical information on non-refractory components (Canagaratna *et al.*, 2007).

Several methods can be used for studying the hygroscopic growth of particles. The size distribution of aerosol particles as a function of relative humidity (RH) can be deduced by charging individual dry particles and measuring their electric mobility spectrum with a DMA. Then, after humidifying the air sample, the size distribution is measured with another DMA (the system is called the tandem differential mobility analyser, TDMA). Another way of estimating hygroscopic growth is to measure the mass of loaded filters or impactor stages under different RHs. The hygroscopic growth of particles can be calculated from the mass change relative to dry particles.

For the measurement of the optical properties of aerosol particles, 'nephelometry' and 'absorption photometry' are used. The scattering coefficient of the aerosol is determined by using an integrating nephelometer, in which the light scattered by a cloud of particles is measured at a fixed solid angle. The absorption coefficient is measured by the change in optical transmission caused by particle deposition on a suitable collecting surface such as a filter. By using nephelometers at different RHs, the RH-dependence of optical properties can be measured (McInnes *et al.*, 1998). The absorption of an ensemble of airborne particles as opposed to those on a filter can be measured using a 'photoacoustic spectrometer' (PAS) (Moosmüller *et al.*, 1998) in which absorbed light energy is converted to pressure. Absorption by airborne, single particles can be determined using a 'single-particle soot photometer' (SP^2) (Schwarz *et al.*, 2008) that measures incandescent light emanating from particles that are exposed to a laser beam and absorb. Remote sensing methods measure the 'aerosol optical depth', a measure of extinction of light through a column of the atmosphere. Both ground-based and satellite-born systems are used for studies of aerosols and clouds. 'Light detection and ranging' (LIDAR) is a technique that measures backscattered light and is used for obtaining the extinction coefficient of the scattering object (Holben *et al.*, 2001).

Records of both past climate changes and of urban air pollution can be analysed by measuring the magnetic susceptibility of dust. As the strongest signal is produced by iron-bearing particles that are typically emitted by vehicles and industrial combustion sources, magnetic measurements of leaves can robustly trace pollution in a city environment (Maher, 2009).

3.4. Analyses of the composition and structure of aerosol particles

Because of their small mass and size, the chemical analysis of atmospheric aerosol particles has always been a challenge for analytical chemists. Analytical methods applied in aerosol research can be grouped into three main categories on the basis of whether elements, inorganic ions, or organic compounds are determined (Lodge, 1986; Adams, 1994). Another way of classifying aerosol analytical methods is whether a large number of particles are collectively analysed (bulk analyses), or the composition of particles is determined individually (single-particle or individual-particle methods). The analysis of single particles is of great importance because the atmospheric significance of the particles is determined by their individual chemical and physical properties (Buseck & Pósfai, 1999; Pósfai & Buseck, 2010).

For the elemental analysis of bulk aerosol samples, the components of the particles are usually extracted using a suitable solvent, and the solution is analysed. In order to induce the emission of characteristic radiation, elements have to be excited by either spraying the solution into a flame (flame photometry) or into an argon plasma (ICP: inductively coupled plasma; the name refers to the plasma generation). The characteristic absorption of atomized elements is measured using atomic absorption spectroscopy (AAS). Neutron-activation analysis (NAA) has also been used widely in aerosol studies since the 1960s (Duce *et al.*, 1966; Rahn *et al.*, 1971). Among the analytical methods that use X-rays (Markowitz & Van Grieken, 1991), it is mostly X-ray emission spectroscopy (XRS) which is employed in aerosol research. The elements in the sample can be excited with photons, protons (or other heavy charged particles), or electrons; accordingly, we distinguish X-ray fluorescence (XRF), particle-induced X-ray emission (PIXE), and electron probe microanalysis (EPMA; see Wogelius & Vaughan, 2013, this volume).

For the bulk analysis of inorganic ions, the aerosol samples are leached and the solutions typically analysed using ion chromatography (Mulik & Sawicki, 1979), although standard wet chemistry can also be used. An alternative and complementary version of ion chromatography is the capillary electrophoresis technique in which ions are separated according to their electric mobility (Dabek-Zlotorzynska *et al.*, 1995). The presence of H_2SO_4, $(NH_4)_2SO_4$, sea salt and soot particles in aerosol samples can be inferred from their physical properties, either by measuring their hygroscopic growth or thermal decomposition temperatures in TDMAs (McInnes *et al.*, 1996; Clarke *et al.*, 1996; see the previous section on physical probes).

Various groups of organic compounds have been identified in atmospheric aerosol particles (Rogge *et al.*, 1993). Owing to the large number of individual compounds and their very low concentrations, the chemical analysis of these species is a difficult task. In the majority of cases various chromatographic techniques are used; gas chromatography can be used for the analysis of organic compounds with relatively low molecular weight, whereas larger organic molecules can be separated by any of the various techniques of liquid chromatography. The water-soluble fraction of organic compounds (*e.g.* carboxylic acids) can be analysed using ion chromatography or capillary zone electrophoresis (Mészáros *et al.*, 1997). Beside gas and liquid chromatography, infrared spectroscopy is used for obtaining information on the functional groups in organic molecules. Total organic carbon (TOC) and organic carbon (OC) are determined by evolved gas analysis, which measures the concentration of CO_2 that forms from aerosol carbon as the sample is heated over a preset temperature range (Andreae & Gelencsér, 2006). Although this method is useful for obtaining global estimates the organic compounds present in the sample, it does not reveal the nature of the species.

Both inorganic and organic ions of individual aerosol particles can be analysed using mass spectrometry. Various versions of the technique have been developed over the past 15 years, resulting in the use of several names (and acronyms), including 'aerosol mass spectrometry' (AMS), 'particle analysis by laser mass spectrometry' (PALMS), 'aerosol time-of-flight mass spectrometry' (ATOFMS), and 'matrix-assisted laser desorption/ionization mass spectrometry' (MALDI) (Noble & Prather, 1996; Murphy *et al.*, 1998a, 1998b; Hughes *et al.*, 1999; Sullivan & Prather, 2005; Canagaratna *et al.*, 2007). The particles are sized and then ionized either with a laser beam or by electron impact, then the resulting fragments are immediately inserted into and analysed in a time-of-flight mass spectrometer. Both chemical and size information are obtained rapidly and on-line; however, as the parent species are inferred from their ionized fragments, the interpretation of complex spectra and determination of original species can be difficult. The mass spectrometry technique can also be used for studying the surface composition of aerosol particles, by separating the desorption and ionisation steps (Lazar *et al.*, 1999).

None of the methods discussed above gives information about the crystal structures of the solids in aerosols. The identification of the structures of crystalline aerosol particles (such as minerals) is useful for obtaining information on their physical properties including refractive indices, and also on their chemical compositions. Originally non-crystalline (liquid) particles may crystallize during or after sampling on the collection surface; the structures of such particles provide indirect information on their chemical compositions. X-ray powder diffraction (XRD) is a convenient method for identifying mineral phases in high-volume aerosol samples (Caquineau *et al.*, 1997; Leinen *et al.*, 1994; Kandler *et al.*, 2009). As the volume of the material available is usually very small, special specimen preparation procedures have been developed for the analysis of mineral aerosols (Caquineau *et al.*, 1997). Attempts to use XRD for the analysis of non-mineral particles resulted in the identification of a large number of hydrous and anhydrous sulfates and nitrates (Harrison & Sturges, 1984; Sturges *et al.*, 1989). The diffractograms are complex and the unambiguous identification of phases difficult; moreover, it is questionable whether these compounds were originally present in the air as crystals or formed on the filter substrate from liquid particles by fractional crystallization. Therefore, the use of XRD for analysing non-mineral aerosol particles has not gained much popularity in aerosol science.

Electron microscopes are the most commonly used instruments in individual-particle studies. Scanning electron microscopy (SEM) provides morphological and compositional information; SEMs can be automated and thus produce large datasets (Anderson *et al.*, 1992; Osán *et al.*, 2000; Krueger *et al.*, 2003; Laskin *et al.*, 2006). Automated SEMs are generally used for studying particles $> \sim 0.1$ μm; the results do not give structural information, and the resolution does not permit the identification of distinct constituents in internally mixed particles. Analytical transmission electron microscopy (ATEM) combines morphological and compositional information at high spatial resolution (Buseck & Pósfai, 1999; Sheridan *et al.*, 1993; Pósfai & Buseck, 2010). The elemental compositions of particles can be studied with the ATEM using either energy-dispersive X-ray spectrometry (EDS) or electron energy-loss spectroscopy (EELS) (Xhoffer *et al.*, 1989; Katrinak *et al.*, 1992; Maynard, 1995). In addition, electron diffraction in the TEM provides structural data on single particles or even on the individual constituents of internally mixed particles (Pósfai *et al.*, 1995). The limitations of ATEM are that the number of analysed particles is relatively small, owing to the manual, labour-intensive operation of the instrument, and that the specimen is inserted into vacuum that causes the loss of adsorbed water and other volatile substances. Ideally, individual particles are studied first using an SEM, then the specific details of typical particle types can be analysed using an ATEM. Three-dimensional morphological data can be obtained from atmospheric particles using electron tomography (ET) (van Poppel *et al.* 2005; Adachi *et al.*, 2007). This method involves the acquisition of a series of images taken at different specimen tilt angles. If the tilt range is large enough (at least $\sim \pm 70°$), and images are obtained at $1°$ or $2°$ intervals, the shape of the particle can be reconstructed from the series of images.

In conventional SEM and TEM, the sample is in a vacuum and, thus, dehydrates and its morphology may change, potentially making it impossible to recognize the particles. These problems have been partly overcome by the recent development of electron microscopes in which the sample can be studied in low-vacuum conditions. The environmental SEM (ESEM) is now an established tool in the study of atmospheric particles, and has been used for characterizing the hygroscopic behavior of a variety of particle types and for studying heterogeneous surface reactions (Ebert *et al.*, 2002; Krueger *et al.*, 2003; Kaegi & Holzer, 2003). Environmental TEM (ETEM) is an emerging technique that has only been used in a handful of atmospheric studies but, on account of the superior resolution of TEM, appears to hold great promise for the analysis of the hygroscopic properties of particles (Wise *et al.*, 2005; Semeniuk *et al.*, 2007; Freney *et al.*, 2010).

Atomic force microscopy (AFM; see Wogelius & Vaughan, 2013, this volume) has been used in a few studies on atmospheric aerosols (Friedbacher *et al.*, 1995; Köllensperger *et al.*, 1997; Pósfai *et al.*, 1998). This microscope method is useful for measuring particle sizes and morphologies under ambient conditions (although the particles are not in the air but on a collection surface). However, the lack of direct compositional information and artifacts arising from the interactions between the cantilever tip and the specimen have hindered the widespread use of AFM in atmospheric science. AFM can be put to use in the study of the hygroscopic behaviour of individual particles, by measuring size changes in an environmental cell, under precisely controlled relative humidity conditions (Köllensperger *et al.*, 1999).

4. Major particle types in the troposphere

4.1. Mineral dust

In terms of mass, mineral grains ('dust particles') constitute the largest fraction of atmospheric aerosols, with an estimated annual production of 1500–2600 Tg (Cakmur *et al.*, 2006). Dust plumes are often the most prominent features in satellite imagery (Kaufman *et al.*, 2005). As the main source areas of dust particles are geographically distinct, their atmospheric distribution and climate effects also show strong regional variation. The climate effects of mineral aerosols are complex. Depending on their mineralogical composition, dust particles can both scatter and absorb solar radiation, and may absorb some of the terrestrial infrared radiation (Liao & Seinfeld, 1998; Yoshioka *et al.*, 2005). Dust particles can be efficient cloud condensation nuclei (CCN) (Twohy *et al.*, 2009; Koehler *et al.*, 2009), as well as ice nuclei (DeMott *et al.*, 2003; Möhler *et al.*, 2008), thereby contributing to the indirect climate effects of the aerosol. The deposition of airborne dust far from its sources is a function of the prevailing wind direction, and can control the primary productivity in remote oceanic regions (Mahowald *et al.*, 2005) and affect local environment and health (Griffin *et al.*, 2001; Prospero *et al.*, 2008). The emission and atmospheric transport of mineral dust was reviewed by Prospero (1999) and Prospero *et al.* (2002), and summaries of dust climate effects were given by Liao & Seinfeld (1998) and Mahowald *et al.* (2006).

The health effects of mineral dust were reviewed by Guthrie & Mossman (1993) and are the subject of another chapter in this volume (Skinner, 2013).

4.1.1. Emission mechanism and main source regions

Dust-particle formation over the continents is caused by the action of wind on the soil surface. According to Alfaro *et al.* (1998), the size distribution of freshly produced mineral dust is remarkably constant. Independent of the soil type, three lognormally distributed populations can be distinguished, centred around median diameters of 1.5, 6.7 and 14.2 μm. The relative abundance of particles within the three distinct populations is a function of wind speed. At low wind speeds mostly large aerosol particles are released that are aggregates of soil particles; as the friction velocity increases, particles in the two finer populations are emitted. The emission mechanism for fine particles is called sand-blasting; this means that larger, saltating grains impart enough momentum to soil aggregates to overcome the binding forces between particles within aggregates (Lu & Shao, 1999). In global circulation models a large uncertainty is associated with estimates of the fraction of clay-sized (<2 μm) particles that are produced by this brittle fragmentation process (Kok, 2010). In addition to wind speed, the production of dust particles depends on the state of the surface (*e.g.* soil particle and aggregate size, soil moisture and vegetation cover) (Marticorena *et al.*, 1997). The occurrence of dust storms in a given region varies seasonally (Duce, 1995).

Atmospheric dust is generated primarily in arid or semiarid regions. The most active sources of dust plumes are distributed within a band that includes North Africa, the Middle East, and Central Asia. Interestingly, some other arid regions such as Australia, the Kalahari and the Atacama deserts do not produce prominent dust plumes (Prospero, 1999). Some regions such as the Chad basin and the south side of the Sahara Atlas mountains in Africa, the Aral Sea region and the Tarim Basin in Asia, appear to be persistent sources of large-scale dust events. A common feature of these regions is that they lie in topographical lows, where seasonal runoffs deposit fine-grained sediments. During the dry season, these deposits become exposed. Dust 'hot spots' are, thus, created by the coincidence of factors that result in strong surface winds and a large amount of easily deflatable material (Washington & Todd, 2005).

Mineral dust has been considered a natural, 'background' component of tropospheric aerosols. However, human activity changes the flux of soil particles into the atmosphere. Changing land use, such as deforestation and over-cultivation, particularly in the Sahel region of Africa, were estimated to result in a twofold increase in the global flux of the atmospheric mineral dust burden (Tegen & Fung, 1995). However, it is difficult to deconvolute the effects of changing land use and natural, interannual climatic variability in African dust transport. New observations and updated models suggest much smaller anthropogenic effects in the production of mineral dust (Tegen *et al.*, 2004; Yoshioka *et al.*, 2005), prompting the IPCC to estimate the human-induced atmospheric dust input in the 0 to 20% range (Forster *et al.*, 2007).

A distinct source of mineral dust aerosol is volcanic activity. Explosive volcanic eruptions emit $\sim$200 Tg y^{-1} fine ash into the troposphere that consists mostly of mineral and rock fragments (Durant *et al.*, 2010). Although major eruptions are sporadic events, the particles of volcanic origin affect atmospheric composition and climate, as well as human activities, as illustrated by the major disruption in European aviation in April 2010, caused by ash plumes emanating from the Eyjafjallajökull volcano in Iceland (Johnson *et al.*, 2012).

4.1.2. Long-range transport patterns

Mineral dust can be carried in the atmosphere over thousands of kilometres. There are prominent pathways of dust transport over the subtropical North Atlantic, the western North Pacific, and the Arabian Sea. The atmospheric concentration of mineral dust at Barbados and Miami shows a strong seasonal pattern with a maximum in the summer months, in agreement with the periodic nature of Saharan dust outbreaks (Prospero, 1999). The dust plumes are carried from Africa to the Americas by the north-east trade winds (Fig. 8); dust episodes extend over several days or more. Depending on

Fig. 8. A satellite image of a Saharan dust plume over the eastern North Atlantic, obtained in the NASA SeaWIFS (Sea-viewing WIde Field-of-view Sensor) Project. The white features over the ocean are clouds.

the presence of cyclones, Saharan dust is also transported over the Mediterranean Sea to Europe (Papayannis *et al.*, 2008).

Another major route of atmospheric dust transport extends over the North Pacific, from China to North America. Dust outbreaks occur in the Asian deserts during spring, resulting in seasonally elevated aerosol concentrations as far downwind as Hawaii and the north-western United States (Uematsu *et al.*, 1983; Fischer *et al.*, 2009). About 80% of the annual aerosol mass deposition on Mauna Loa is due to mineral dust from Asian sources (Parrington *et al.*, 1983), and Asian dust seriously impacts air quality in the western United States (Zhao *et al.*, 2008).

The vertical distribution of dust is never uniform in the atmosphere; a summary of observations is given by Duce (1995). Most of the long-range transport over the oceans occurs above the marine boundary layer (MBL), and dust plumes can consist of one or more layers that may be at altitudes as high as 8 km. The characteristic layering of dust was observed for both African (Prospero & Carlson, 1972; Ben-Ami *et al.*, 2009) and Asian (Iwasaka *et al.*, 1983) dust plumes at Barbados and Hawaii, respectively. The heterogeneous, layered distribution of dust is typical for emissions from explosive volcanic eruptions (Schumann *et al.*, 2011).

4.1.3. The mineralogy of dust plumes

Although the amount of continental dust, inferred either from the Al content of the aerosol (Duce, 1995) or indirectly from remote sensing measurements (Holben *et al.*, 2001) is continuously monitored at several locations, the mineralogy is not. Therefore, we do not know much about the seasonal variability of the dust mineralogy. XRD data are available from several regions; however, these studies can be regarded as one-time experiments that provide snapshots of the mineralogy of plumes originating in major dusty areas of the Northern Hemisphere. Not surprisingly, the major constituents of dust plumes are common rock-forming minerals, including quartz, feldspars, clay minerals, mica, carbonates, oxides, and evaporite minerals (Engelbrecht & Derbyshire, 2010). In general, clay minerals are enriched in plumes that are transported over long distances; this is because the smaller the particles, the longer they stay in the air.

The mineralogical composition of Saharan dust plumes is certainly influenced by the variable mineralogies of distinct source areas (Krueger *et al.*, 2004). Most studies on Saharan mineral dust found that illite and kaolinite are the main clay minerals (Adedokun *et al.*, 1989; Caquineau *et al.*, 1998; Chou *et al.*, 2008; Kandler *et al.*, 2009). Caquineau *et al.* (1998) found that the illite/kaolinite ratio of the dust is a good tracer for distinguishing dust plumes from three distinct Saharan source regions (the Sahel, south and central Sahara, and north and west Sahara). The illite/kaolinite ratio increases from south to north, and it does not change during long-range transport, as indicated by the similar ratios obtained from Sal Island (Cape Verde) and Barbados. The latitudinal change in the illite/kaolinite ratio is consistent with the zonal variations of clay mineral ratios in marine sediments (Biscaye, 1965; Chamley, 1989). Calcite- and palygorskite- (Schütz & Sebert, 1987), as well as smectite-rich plumes (Avila *et al.*, 1997; Coz *et al.*, 2009) seem to originate from the northern Sahara.

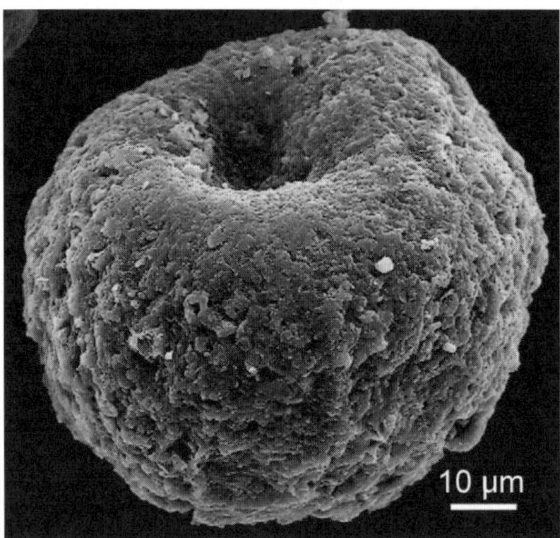

Fig. 9. SEM image of an iberulite particle, a large aggregate of various minerals with a typical funnel-shaped depression, collected in Spain, from Saharan dust (reprinted from a paper by Diáz-Hernandez & Párraga, 2008: *Geochimica et Cosmochimica Acta*, **72**, The nature and tropospheric formation of iberulites: pinkish mineral microspherulites, pp. 3883–3896, © 2009 with permission from Elsevier).

A distinct particle type, termed iberulite, was found in Saharan dust that reached southern Europe (Díaz-Hernández & Párraga, 2008). Iberulites are large (several tens of μm) aggegates that consist primarily of clay and other silicate minerals, with occasional biogenic debris such as plant and diatom fragments. A distinctive feature of iberulites is their spherical shape that has a vortex-shaped depression (Fig. 9). Díaz-Hernández & Párraga (2008) speculated that such particles can form in clouds only, and that their shape results from their movement within falling cloud drops.

Comparison of the mineralogies of Asian dust plumes and their potential source soils can offer insight into the controls over dust entrainment. For example, based on the analyses of individual mineral grains in the aerosol and in typical desert soils, the source of a dust plume was traced back to the silty soils of the Loess Plateau (Jeong, 2008), whereas dust in Beijing was observed to have originated in the saline soils of northern China (Yuan *et al.*, 2006). Two different types of calcite were observed by SEM in Asian dust; whereas micron-sized calcite was produced by bedrock erosion, calcite nanofibers originated from loess (Jeong & Chun, 2006). Modern loess deposition was observed in Beijing, when a dust plume arrived containing silt-size grains of quartz, feldspar and calcite; quartz and feldspar grains were typically coated by carbonate or iron oxide (Sheng *et al.*, 1981). This dust plume originated in the gravel desert region to the west of the Chinese Loess Plateau, so it was not redeposited but freshly formed loess. In dust plumes above the North Pacific Ocean, the most abundant minerals were clays, quartz and plagioclase (Leinen *et al.*, 1994; Merrill *et al.*, 1994), with illite being the dominant clay mineral in both the western and eastern North Pacific.

Dust aerosol from the deserts in the Middle East has quite variable mineralogical composition, with abundant halite and sulfate particles from evaporated seawater and playas (Engelbrecht & Derbyshire, 2010). Central Asian deserts seem to produce dust plumes rich in quartz, feldspar, and carbonates (Andronova *et al.*, 1993). In addition to these common minerals, relatively high concentrations of kyanite, akermanite, and spinel were observed; these probably indicate a specific metamorphic source.

Most solid particles emitted by explosive volcanic eruptions consist of volcanic glass. Such particles typically have angular, irregular surfaces because they form by the collapse and fragmentation of bubble walls in rapidly rising magma (Durant *et al.*, 2010). In addition, common rock-forming minerals such as feldspars, pyroxenes, and forms of silica such as quartz and tridymite also occur in volcanic plumes (Gíslason *et al.*, 2011).

4.1.4. Changes during transport, chemical reactions

When dust is transported great distances, the peak of the size-distribution curve shifts towards smaller particle diameters. Larger particles settle closer to the source (see the section on dry deposition), and the plume will consist mostly of clay-sized particles. At Barbados and Miami, $\sim 1/3 - 1/2$ of the African dust mass had an aerodynamic diameter of $<2.0 - 2.5$ μm (Prospero, 1999), whereas most of the aerosol mass was found in the particle population with diameters >10 μm above the Sahara (Duce, 1995). As distinct minerals typically have different grain-size distributions, the change in the total size distribution affects the mineralogical composition of the aerosol; for example, the concentration of quartz in Saharan dust decreased during the transit of the Atlantic, while that of clay minerals increased (Glaccum & Prospero, 1980). Kaolinite and smectite were found to be the main mineral particles above the Equatorial Pacific (Pósfai *et al.*, 1994) and the North Atlantic (Pósfai *et al.*, 1995; Buseck *et al.*, 2000), respectively, probably due to the relative enrichment of the aerosol in clay particles during long-range transport (Fig. 10). Despite the general trend of smaller particles being transported over larger distances, a significant coarse mass mode (at ~ 25 μm diameter) can occur even far from the source of dust (McConnell *et al.*, 2008).

Particle reactions and internal mixing during the transport of mineral dust can substantially change the composition of the original aerosol (Dentener *et al.*, 1996). Heterogeneous reactions of SO_2 on Ca-rich mineral aerosols result in a significant amount of atmospheric sulfate being associated with larger dust particles. Even relatively close to the source, calcite and halite partially reacted to sulfate in a dust storm in the Taklamakan desert (Okada & Kai, 2004). In a TEM study of North Atlantic aerosol particles (Pósfai *et al.*, 1995), many internally mixed sea-salt/mineral particles contained anhydrite, but none contained calcite; during transport over the ocean, calcite reacted with SO_2 or sulfate and formed anhydrite. Dust types that contain a significant fraction of carbonate minerals also act as effective sinks for nitric acid (Krueger *et al.*, 2004). As global climate models are based on the assumption that sulfates occur in the accumulation mode, a change of the size distribution of sulfate particles as a result of their association with minerals means that the models overestimate the sulfate climate-cooling effect. An even larger fraction of gas-phase nitric acid may be associated with, and neutralized by, mineral aerosol.

As dust travels from China to Japan, a large fraction of the mineral particles become internally mixed with sea salt (Zhang *et al.*, 2003). Pollution by fossil-fuel combustion and urban or agricultural activities was also shown to contribute to the complexity of mixed mineral/sea-salt particles above the Pacific (Gao *et al.*, 2007) and in Israel (Falkovich *et al.*, 2001). The significance of these studies is that they identify mineral dust as a potential cleansing agent for organic pollutants.

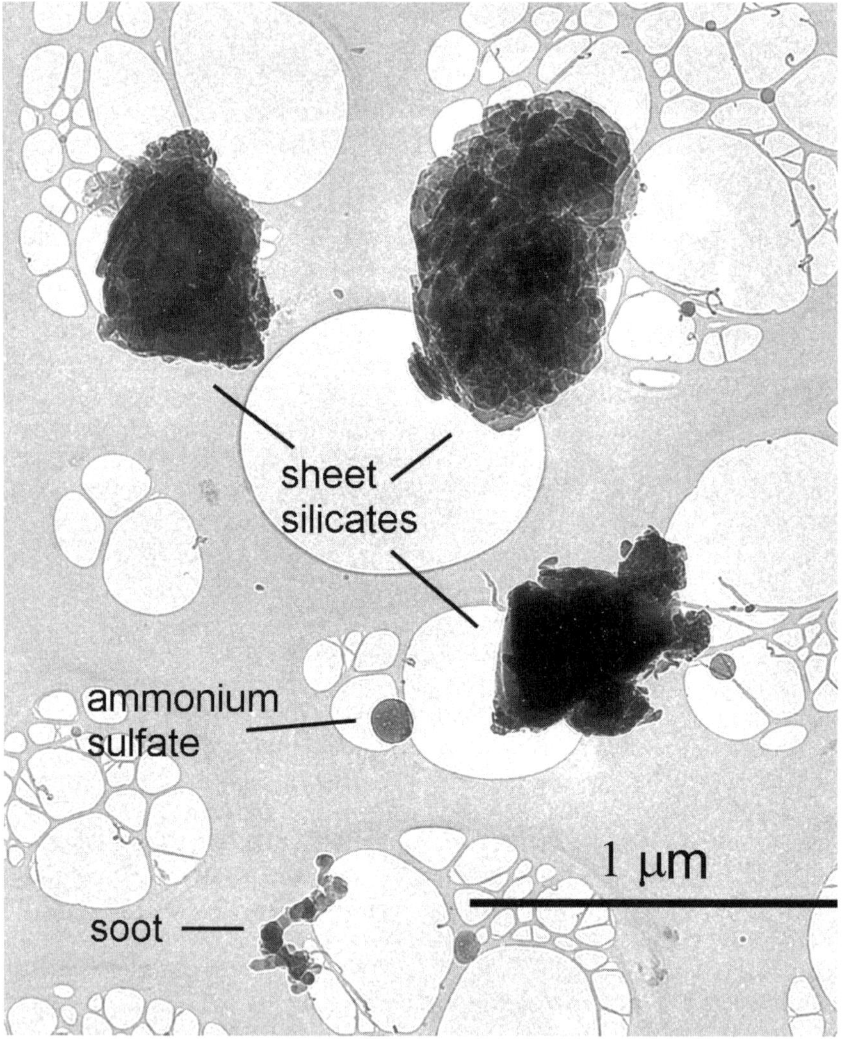

Fig. 10. Electron micrograph of typical mineral dust particles (smectites) of African origin, collected at an altitude of 2.1 km on a mountaintop on Izana, Canary Islands, during the 'Second Aerosol Characterization Experiment'. The supporting substrate is a lacey carbon film onto which the particles were deposited.

4.1.5. *Atmospheric and environmental effects of metal-bearing particles*

Reactions on the surfaces of Fe-bearing mineral particles can affect the concentrations of reactive gas-phase species in the atmosphere. Zhuang *et al.* (1992) pointed out that Fe^{3+} can be photochemically reduced to Fe^{2+} in the water associated with aerosol particles, and this reaction may provide a source of OH, the key atmospheric oxidant. An alternative mechanism for the formation of Fe^{2+} in the aerosol is photoreduction of Fe^{3+}

organo-complexes, resulting in the formation of H_2O_2, another important atmospheric oxidant (Jacob, 2000). Because of the significance of these reactions, it is useful to know which minerals in a dust plume contain Fe, and in what oxidation state.

Fe-bearing mineral particles have another important role in global biogeochemical cycles: in large regions of the oceans, Fe is the limiting nutrient for phytoplankton (Martin & Gordon, 1988), and aerosol particles supply the Fe that is necessary for the growth of organisms at the base of the food chain (Duce, 1986). When mineral particles are acidified in the atmosphere, their iron content may dissolve and become available for marine organisms once the particles deposit in the ocean (Cwiertny *et al.*, 2008). According to computer-controlled SEM studies of dust source materials of various mineralogical compositions, neither the total iron content, nor the amount of iron oxides and hydroxides can be used to predict how much iron will be solubilized. Contrary to the commonly held view of iron oxides being the dominant contributors, in most samples sheet silicates were the major sources of dissolved iron (Journet *et al.*, 2008). In order to understand the stimulation of phytoplankton growth by atmospheric dust, not only is a detailed understanding of the mineralogy required, but also a knowledge of particle reactions and histories of air masses (Cwiertny *et al.*, 2008).

Trace metals (other than Fe) can also influence chemical reactions in aerosol water and cloud drops, and some represent a health hazard (Nriagu & Davidson, 1986). Al, Ti and Mn are generally considered natural mineral components of the aerosol and mostly occur in the coarse fraction (Milford & Davidson, 1985), whereas Pb, Zn, Cu, Cr and Cd are typically of anthropogenic origin, and occur in particles with diameters < 1 μm (Molnár *et al.*, 1993; Hlavay *et al.*, 1996; Mészáros *et al.*, 1997). The environmental effects of metal-bearing aerosols range from local to regional, and show strong variations over historical times (Patterson & Settle, 1987). For example, the phasing-out of leaded gasoline resulted in a drastic reduction of the concentration of Pb-bearing halides in the urban aerosol. Whether of mineral or anthropogenic origin, trace metal-bearing aerosol particles can be transported long distances, affecting remote ecosystems. According to Duce *et al.* (1991), atmospheric and riverine input into the oceans is of equal importance for Fe, Cu, Ni and As, whereas Zn, Cd and Pb enter the ocean mostly from the atmosphere. A notable historical example of metal pollution in a remote region was provided by the analysis of Greenland ice cores that showed clear maxima of Pb concentrations during the times of ancient Greek and Roman cultures (Hong *et al.*, 1994).

The urban atmosphere contains particles from a multitude of industrial and traffic sources (Grobéty *et al.*, 2010), some of which can be confused with soil-derived minerals. Cement factories and construction/demolition activities emit primarily aluminosilicate and carbonate particles, whereas iron oxide particles are typical contaminants near railway lines and in underground metro stations (Fig. 11a,b) (Lorenzo *et al.*, 2006; Salma *et al.*, 2009). Power plants emit fly-ash particles that consist of amorphous silica or various metal oxides (Ramsden & Shibaoka, 1982; Silva *et al.*, 2009), and smelter and other industrial emissions can produce metal-rich and carbonaceous particles (Bradley & Buseck, 1983). Submicron rutile particles observed above the Equatorial Pacific (Pósfai *et al.*, 1994) may have been abrasion products of aircraft paint

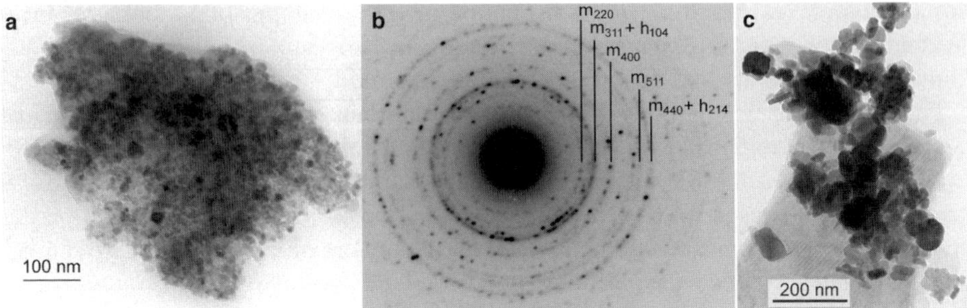

Fig. 11. Iron oxide particles from pollution. (a) TEM image of an aggregate of hematite and magnetite nanocrystals from the atmosphere of an underground metro station in Budapest (Salma *et al.*, 2009), with (b) its associated electron diffraction pattern (m: magnetite, h: hematite reflections; unmarked rings correspond to hematite). (c) A particle consisting mainly of hematite nanocrystals from fugitive dust from a bauxite residue (red mud) spill near Ajka, Hungary (Gelencsér *et al.*, 2011).

finish. Al oxide particles form in the exhaust plumes of rockets and the space shuttle and can be observed in the stratosphere (Ross *et al.*, 1999). A special case of air pollution is presented by reservoirs of bauxite residue. When dehydrated, these red mud deposits can emit copious quantities of dust that consists mostly of aggregates of hematite nanoparticles (Fig. 11c) and aluminosilicates that form as by-products of the Bayer process (Gelencsér *et al.*, 2011). As the composition of bauxite varies, the mineralogy of fugitive dust from distinct red mud deposits over the world may also differ significantly.

4.1.6. Climate effects of mineral dust

Mineral dust has a complex direct climate effect because dust particles scatter and absorb both incoming (visible) and outgoing (infrared) radiation (Andreae, 1996). The single scattering albedo (ω) of mineral dust is between 0.90 and 0.99 (Forster *et al.*, 2007), suggesting that absorption by dust in the visible part of the spectrum is globally less significant than previously thought. However, in the infrared spectral region, minerals act like the greenhouse gases. This complex effect may result in heating of the atmosphere and cooling of the surface, thereby changing atmospheric circulation patterns (Tegen *et al.*, 1996). In large regions of the Earth, mineral dust can be the dominant cooling agent in the troposphere. Such regions include the North Atlantic west of Africa (Li *et al.*, 1996), the Arabian Sea, and the western North Pacific (Tegen *et al.*, 1996). Uncertainties in estimates of direct climate forcing by mineral dust result partly from the variability of the mineralogical compositions of dust plumes, and partly from the gaps in our knowledge of the chemical and physical properties of dust particles (Tegen & Fung, 1994).

For an assessment of the mineral aerosols' radiative impact on climate, one needs to know the complex refractive indices at all wavelengths of interest for major types of particles that are present in a dust plume (Sokolik *et al.*, 1993; Claquin *et al.*, 1998). The absorption properties of a mineral dust plume are especially strongly dependent on the particular species in the aerosol (Volz, 1983). Both scattering and absorption are affected by particle shapes (Mishchenko, 2008). While most radiative calculations use Mie theory that is applicable for spherical particles, mineral particles are rarely spheres or ellipsoids but have quite irregular shapes. The presence of sharp-edged, angular-type particles results in various differences in optical depth and single scattering albedo compared to those of the volume-equivalent spheres (Kalashnikova & Shokolik, 2002). The sensitivity of direct radiative forcing by dust to some of the above parameters was studied by Liao & Seinfeld (1998) and Pilinis & Li (1998).

In the last ten years, several large-scale measurement campaigns were conducted to study the properties and climate effects of mineral dust aerosol, including the Aeolian Dust Experiment on Climate (ADEC) (Mikami *et al.*, 2006), the Dust Outflow and Deposition to the Ocean (DODO) (McConnell *et al.*, 2008), the African Monsoon Multi-disciplinary Analysis (AMMA) (Chou *et al.*, 2008), and the Saharan Mineral Dust (SAMUM) (Heintzenberg, 2009) experiments. A burst of new results from these and other campaigns has refined our understanding of mineral dust properties, providing useful data for the interpretation of optical measurements and the validation of models that calculate radiative transfer. For example, shortwave absorption by desert dust appears to be mainly controlled by the concentrations or iron oxides (hematite and goethite) (Arimoto *et al.*, 2002; Kandler *et al.*, 2009). Particle shapes are indeed irregular and complex (Coz *et al.*, 2009; Iwasaka *et al.*, 2003) and affect both short- and longwave absorption (Hudson *et al.*, 2008). Aspect ratios and complex refractive indices were determined for each major mineral species in Saharan aerosol (Kandler *et al.*, 2009); thus, size, shape and composition-dependent optical properties of fresh and transported dust particles could be compared (Ansmann *et al.*, 2011). The last IPCC report estimated a global direct radiative forcing by mineral dust in the range from -0.56 to $+0.1$ W m^{-2} (Forster *et al.*, 2007). Based on the wealth of new data assembled in the last few years, estimates of the direct radiative forcing by mineral aerosol will probably be better constrained in future reports.

Concerning indirect climate effects, mineral particles influence the radiative properties and lifetimes of clouds by serving as both cloud condensation nuclei and ice nuclei (CCN and IN, respectively) (Andreae & Rosenfeld, 2008). Whether a dust particle can act as a CCN or IN depends primarily on its composition (water solubility) and size. Recent results suggest that even insoluble mineral dust particles can be efficient CCN (Twohy *et al.*, 2009; Koehler *et al.*, 2009), particularly if they contain some hygroscopic material. As discussed above, internal mixing of dust particles with sulfates and nitrates during atmospheric transport is a common phenomenon that enhances the chances of the particles to act as CCN. Insoluble minerals that survive transport within clouds without being activated as CCN can reach the middle and upper troposphere and become IN (DeMott *et al.*, 2003; Möhler *et al.*, 2008). As pointed out by Andreae & Rosenfeld (2008), measurements and model predictions disagree about the size distributions of dust particles. An important property of mineral aerosol is that large dust particles can act as giant CCN (Laskin *et al.*, 2005), *i.e.* they are among the first particles to activate on cloud formation, thereby enhancing the tendency of the cloud to produce rain (Levin *et al.*, 2005). Although in recent climate models the CCN-forming potential of mineral dust is taken into account (Forster *et al.*, 2007), the complicated relationships between the properties of dust particles and the microphysics of clouds are not yet being considered.

The relationships between mineral aerosol and climate have existed over geological times. Detailed stratigraphic records are available for the past half a million years and indicate that the concentration of atmospheric dust has varied in association with natural climate change. Loess deposits (Ding *et al.*, 1994), marine sediments (Rea, 1994), and polar ice cores (Legrand, 1995) all show that dust deposition rates were much higher during glacials than in warmer periods. According to Petit *et al.* (1999), concentrations of dust particles in the Antarctic Vostok ice core rise from $\sim$50 ng $\cdot$ g^{-1} during interglacials to 1000–2000 ng $\cdot$ g^{-1} during glacial stages. The study of past and future climate changes and associated dust deposition patterns is a scientific discipline in its own right (Mahowald & Luo, 2003), and even a cursory discussion is beyond the scope of this chapter.

4.2. Sea salt

In terms of mass, sea-salt particles are the most abundant aerosol species in the marine atmosphere; they are also transported to great distances above the continents. Because of the significant amount of organic matter associated with sea-salt particles, the term 'sea-spray aerosol' is also used for this particle type. As sea-salt particles are efficient scatterers of the solar radiation (Winter & Chylek, 1997) and, because of their hygroscopic behaviour, act as CCN (O'Dowd *et al.*, 1999), they exert a cooling effect on Earth's atmosphere. Even though the formation mechanism and the basic physical and chemical properties of sea-salt aerosol particles have been known for more than half a century (Blanchard & Woodcock, 1957), the details of their physical properties including size distributions, their atmospheric reactions and climate effects are still receiving much attention (Mårtensson *et al.*, 2003; Lewis & Schwarz, 2004; Clarke *et al.*, 2006; Keene *et al.*, 2007; Nilsson *et al.*, 2007). Below we discuss the formation, chemistry and climatic significance of sea-spray aerosol, with a focus on its inorganic constituents.

4.2.1. Generation of sea-salt aerosols and their initial compositions

Sea-salt aerosol particles are produced by the bursting of air bubbles at the surface of the ocean. The number of sea-salt particles in the marine boundary layer (MBL) is approximately exponentially related to wind speed. The stronger the winds, the more whitecaps are produced, resulting in a larger number of bubbles rising to the surface. A bursting bubble produces as many as 10 larger jet drops and up to several hundred, submicrometre-sized film drops (Woodcock, 1972). In addition, the tearing of droplets from wave crests also produces particles (Lewis & Schwarz, 2004). Both laboratory and field studies indicate that the size distribution of sea-salt particles has two modes, one at 2.5 μm corresponding to the larger particles produced by jet drops and another at $\sim$0.1 μm that consists of particles produced by film drops (Murphy *et al.*, 1998a; Mårtensson *et al.*, 2003; Clarke *et al.*, 2006).

The initial composition of sea-salt aerosol particles is a direct consequence of their production mechanism: they are drops of seawater. They contain dissolved ions in the

Table 3. Atomic ratios of major seawater elements relative to Na (Millero & Sohn, 1992).

Cl/Na	S/Na	Mg/Na	K/Na	Ca/Na
1.16	0.06	0.11	0.022	0.022

same relative amounts as seawater (Table 3). As the surface of the ocean is enriched in organic compounds and microorganisms relative to the bulk of seawater (Sieburth, 1982), and rising bubbles scavenge some of these organic substances (Blanchard, 1978), sea-salt aerosols typically contain some organic compounds in addition to dissolved inorganic ions (Middlebrook *et al.*, 1998; Aller *et al.*, 2005). Based on hygroscopic behavior, Clarke *et al.* (2006) suggested that the lower end of the size distribution of marine aerosol (particles with a number maximum at 30–40 nm) consists mostly of sea salt. On the other hand, other observations in remote regions were interpreted to indicate that the mass fraction of organics is significant in this size range (O'Dowd *et al.*, 2004; Bigg & Leck, 2008). The organic components of submicron sea-spray particles are almost completely water insoluble and consist of lipid-like material exuded by phytoplankton (Keene *et al.*, 2007; Facchini *et al.*, 2008; Decesari *et al.*, 2011). The composition of the organic matter varies with the dominant type of phytoplankton, and the biogenic exudate affects both the size distribution and hygroscopicity of sea-spray aerosol (Fuentes *et al.*, 2010; 2011).

Once in the atmosphere, the seawater droplets can be lofted into higher altitudes, where the relative humidity is lower, causing water to evaporate from the particles (Wise *et al.*, 2007). When the particles reach the upper troposphere, some constituents of the original sea-salt particle can crystallise, forming halite and hydrous sulfates. The transition from liquid to solid state affects the optical properties of sea-salt aerosol, as the scattering and backscattering coefficients of the crystalline, cube-shaped particles differ significantly from those of spherical ones (Chamaillard *et al.*, 2003).

Recrystallized sea-salt particles can be observed in the vacuum of a TEM (Fig. 12). The bulk of an unaltered, fresh sea-salt particle is NaCl (the large black crystals in Figure 12); in addition, typically rod-shaped or veil-like, thin sulfate crystals occur in every particle. The sulfates contain variable ratios of Na, Mg, Ca, K, the four main metals of seawater; rod-shaped crystals typically have a high Ca content. The mixed-cation sulfates are easily damaged in the electron beam. Selected-area electron diffraction (SAED) patterns obtained from such particles show the presence of several distinct crystalline phases even within the same particle (Pósfai *et al.*, 1995). Even though the four metals are fractionated into distinct sulfate crystals, the compositions of entire, unaltered particles are the same, reflecting their seawater origin. In addition to the inorganic salts, carbon-bearing, amorphous fibers occur in the particle, probably of biogenic origin.

4.2.2. Interactions of sea salt with other aerosol and gas species

Sea-salt particles react with gases in the atmosphere, resulting in the release of Cl and the formation of additional sulfate and nitrate on the particles (Horváth *et al.*, 1981;

Fig. 12. TEM images of sea-salt particles from the Southern Ocean atmosphere, collected at Cape Grim, Tasmania, during the First Aerosol Characterization Experiment. 'Mixed-cation sulfates' contain Na, Mg, K and Ca in various relative amounts.

Kutsuna & Ibusuki, 1994; Kerminen *et al.*, 1997). Unlike the sulfate that originates from seawater, the excess sulfate is called 'non-sea-salt (nss) sulfate' in the aerosol literature. The mechanism for nss sulfate formation on sea-salt particles includes the condensation, dissolution, and subsequent oxidation of SO_2 in the water associated with sea-salt aerosol particles. Outside clouds, the main oxidizing agent is O_3 (Sievering *et al.*, 1992; Chameides & Stelson, 1992). Much additional sulfate forms on sea salt in cloud droplets (Clegg & Toumi, 1998). Thus, sea-salt particles provide a significant sink for SO_2 in the marine atmosphere. As less SO_2 is available for forming pure sulfate aerosols, the number of sulfate CCN is diminished. N-bearing species such as NO_x during the day and HNO_3 both day and night can also react with sea-salt particles. The result of these reactions is that $NaNO_3$ forms from NaCl, and Cl is liberated. Another mechanism that releases Cl (and Br) from sea-salt particles is the reaction of gas-phase, reactive iodine species with sea-salt. The ocean is a source of alkyl iodides that photolyse rapidly in the marine boundary layer, leading to the formation of inorganic iodine species (Vogt *et al.*, 1999). The uptake of hypoiodous acid (HOI) by sea-salt can lead to a characteristic pulse of Cl production after sunrise (McFiggans *et al.,*. 2002). Therefore, sea-salt particles play an important role in the atmospheric cycles of halogens, S and N (Allan *et al.*, 2009).

Fully reacted sea-salt particles have lost all of their Cl contents; such particles form characteristic crystals in a TEM specimen. On the basis of SAED patterns, the crystal structures of sulfates that form from sea salt differ from those of 'mixed-cation sulfates' in unreacted sea-salt particles, and are identical to that of a high-temperature polymorph of Na_2SO_4. This structure is probably stabilized by the presence of the other seawater cations Mg, K and Ca (Pósfai *et al.*, 1995). Sea-salt particles that have completely reacted with NO_x or HNO_3 contain $NaNO_3$ (Li *et al.*, 2003a); indeed, SAED patterns obtained from such particles show them to consist mainly of nitratite, $NaNO_3$ (Pósfai *et al.*, 1995). Even though the presence of these crystalline phases in a TEM may be irrelevant to atmospheric conditions, the identification of particle constituents by electron diffraction is useful for obtaining information on crystals that contain light elements and are thus difficult to analyse using EDS.

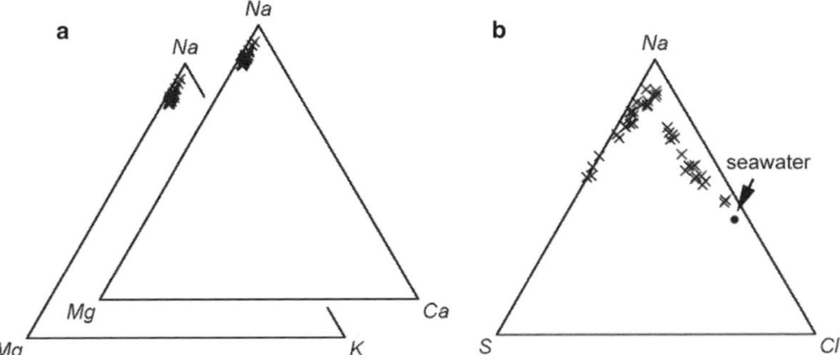

Fig. 13. Ternary plots showing sea-salt particle compositions (in atomic ratios), obtained from EDS spectra. Each datapoint represents an individual particle collected from the North Atlantic troposphere; the three triangles in (**a**) and (**b**) show the same particles. Uniform ratios of K, Mg, Ca and Na in (a) indicate that these elements remain unchanged during atmospheric reactions, whereas the wide range of compositions in (b) shows Cl loss and S excess relative to seawater compositions.

The reactions of sea-salt particles and the temporal and spatial changes of their compositions in the troposphere can be studied by analysing individual particles (Li *et al.*, 2003a). In such studies care must be taken to analyse entire particles instead of their distinct crystalline components which may have compositions that result from fractionation on the collection surface. A sea-salt particle can be identified on the basis of its typical cation ratios; as atmospheric reactions do not remove or deposit Na, Mg, K, and Ca ions from or onto the particles, the relative amounts of these elements remain unchanged (Fig. 13a). On the other hand, variations in the Cl/Na and S/Na ratios can be used to study the reactions and atmospheric histories of sea-salt particles (Fig. 13b).

The temporal changes of sea-salt particle compositions can be illustrated by the example of a sample set that was collected in the North Atlantic, during the ASTEX/MAGE campaign (Pósfai *et al.*, 1995). Aerosol samples were taken from a ship (18 m above sea level), near the Azores Islands. Data reported in the original paper are replotted in ternary diagrams in Figure 14. As the position of the ship was almost stationary, the changes in compositions reflect the results of local sea salt production and ageing, as well as the effects of long-range transport of particles; air masses that arrived at the sampling site may have brought strongly altered sea-salt particles. Figure 14a contains data from ∼30 particles, all of which have compositions characteristic of freshly produced, unreacted sea salt. A few days later a polluted continental air mass arrived, resulting in high SO_2 concentrations and the formation of excess sulfate on sea-salt particles (Fig. 14b). Most compositions in Figure 14b plot near a line that ties unreacted sea salt with Na_2SO_4, indicating that many particles with intermediate compositions are in the process of transforming from unreacted sea salt into sulfate. Completely reacted particles have compositions close to Na_2SO_4. Over the same day, new sea-salt particles were produced, as indicated by the unreacted particles in Figure 14c that have compositions close to NaCl; in the

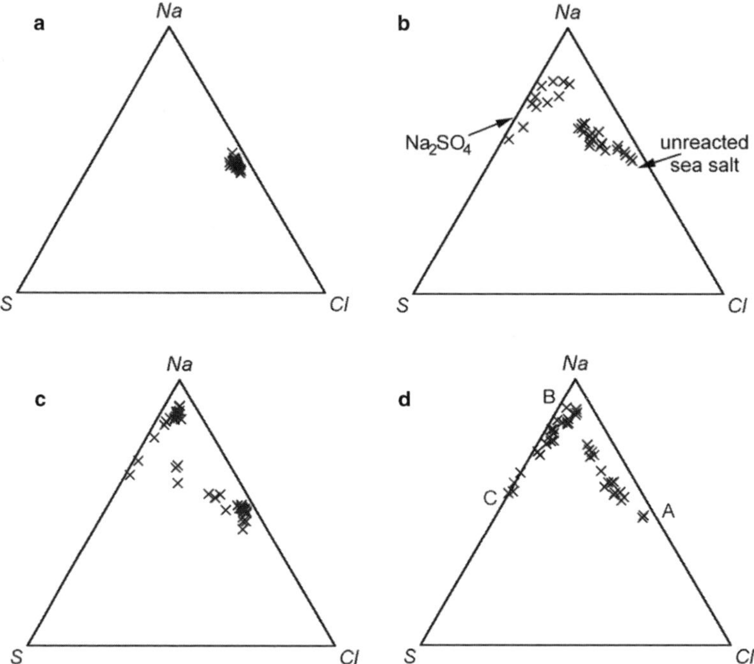

Fig. 14. Temporal changes of sea-salt particle compositions in a series of samples collected in the North Atlantic, near the Azores Islands (see text for details).

same sample, a group of particles appears near the apex of the triangle, indicating the presence of sea-salt particles that completely reacted to form $NaNO_3$. The relatively distinct groups of unreacted and completely reacted particles in the same sample probably indicate the mixing of locally produced, fresh sea-salt aerosol with reacted particles that were transported from longer distances. The compositions along line AB in Figure 14d may indicate the reaction of fresh sea salt with NO_x, whereas the particles that plot along line BC are in the process of transforming from nitrate-rich, completely reacted sea salt into Na_2SO_4.

Similar ternary plots were used to study the changes of particle compositions with altitude above the Southern Ocean, during the ACE-1 (First Aerosol Characterization Experiment) project, south of Tasmania (Buseck & Pósfai, 1999). Sea-salt particles were unreacted near the ocean surface, whereas partially and fully reacted, sulfate-rich particles occurred at higher altitudes (at 1200 and 2100 m above sea level). The particles at higher altitudes had more time to react with SO_2 than the particles near the ocean surface. In the remote and clean troposphere of the Southern Ocean, N-bearing species were insignificant; therefore, no nitrate-rich sea-salt particles occurred. The amount of Cl loss and the formation of sulfate and nitrate on individual sea-salt particles were used to trace the pollution history of air masses over the North Atlantic (Li *et al.*, 2003a) and in Finland (Niemi *et al.*, 2006).

Sea-salt particles can interact in the troposphere not only with gases but with other aerosol species as well. In continentally affected marine air masses, many sea-salt particles are aggregated with minerals, mainly clays (Levin *et al.*, 2005). Such aggregates can form in clouds, where both sea salt and hygroscopic mineral particles serve as nuclei for cloud droplets. When cloud drops coalesce and then the cloud evaporates without forming precipitation, the residues of the droplets will be aggregated (internally mixed) sea-salt/ mineral particles (Andreae *et al.*, 1986; Pósfai *et al.*, 1995; Li *et al.*, 2003a).

4.2.3. *The climatic significance of sea-salt aerosols*

Under clean marine conditions and moderate-to-high wind speeds, sea-salt particles are a major fraction of the accumulation mode aerosol above the oceans (O'Dowd *et al.*, 1997) and can dominate both direct and indirect aerosol climate effects. In order to identify the influence of tropospheric aerosols on the global radiation budget, Haywood *et al.* (1999) compared satellite observations of clear-sky, top-of-the-atmosphere solar irradiances with those computed using a general circulation model. They found that over the oceans, sea salt is the leading aerosol contributor to the global mean clear-sky radiation balance. Above the Southern Ocean, the concentration of sea-salt particles is much larger than those of sulfates, and sea-salt particles are responsible for ~80% of the light scattering (Murphy *et al.*, 1998a).

Sea-salt particles are excellent CCN; however, most of the sea-salt aerosol mass is in the coarse mode, which has relatively minor contribution to the fraction of cloud-active particles due to the low number concentration of these particles (Andreae & Rosenfeld, 2008). Sea-salt particles can be the dominant CCN fraction over the ocean under special conditions, such as very high wind speeds and/or weak sources of other aerosol types. For example, ~60% of the CCN population consisted of sea-salt particles above the windy Southern Ocean (Murphy *et al.*, 1998a). The role of sea-salt particles in the formation of clouds and precipitation can be amplified by the presence of large particles that can be giant CCN and enhance precipitation. However, this effect has not yet been quantified. The CCN potential of sea-salt particles prompted suggestions that the effects of global warming could be mitigated by seeding clouds artificially with sea salt, thereby enhancing the albedo and longevity of maritime clouds (Latham *et al.*, 2008). The development of wind-driven spray vessels was proposed (Salter *et al.*, 2008) for the pumping of sea spray into the air. Cloud seeding with sea salt appears to be the most benign concept among the many geoengineering ideas that have emerged in the last ten years; however, its efficiency remains to be demonstrated.

4.3. Sulfates

Water-soluble, inorganic compounds (other than sea salt) have long been recognized as major constituents of tropospheric aerosols in both continental (Junge, 1963; Mészáros, 1968; Charlson *et al.*, 1978) and marine atmospheres (Mészáros & Vissy, 1974). Although sulfates, nitrates and other species occur among these substances, here we discuss only the most common particles, which are the sulfates.

Sulfates are secondary particles: they form when volatile precursors are oxidized to gaseous H_2SO_4 that forms particles by nucleation and condensation. When pure, sulfate particles scatter but do not absorb visible light. As sulfates are water-soluble, they are important agents of cloud droplet nucleation; thus, sulfates were recognized in the 1990s as the main global cooling agents among the aerosol species (Charlson *et al.*, 1992; Kiehl & Briegleb, 1993; Chuang *et al.*, 1997).

In this section, we discuss particles that have variable compositions between H_2SO_4 and $(NH_4)_2SO_4$, depending on the degree of neutralization of H_2SO_4 by NH_3. These particles form in the atmosphere from gas-phase precursors; their nucleation, subsequent growth and mixing with other species (Zhang *et al.*, 200; Liu *et al.*, 2005), as well as their climate effects (King *et al.*, 2007; Wang *et al.*, 2008) are the subject of intense research.

4.3.1. Precursor gases

The precursor gases of sulfates are emitted by three types of sources, biogenic, volcanic and anthropogenic (Möller, 1995; Dentener *et al.*, 2006). While human activities and volcanic eruptions release SO_2, biogenic emissions result in reduced sulfur gases such as dimethyl sulfide (DMS) and carbonyl sulfide (COS) (Andreae & Crutzen, 1997). The origins of precursors and the pathways of sulfate formation can be traced on the basis of the S isotopic signature of the particles (Sinha *et al.*, 2008).

S-containing gases are emitted naturally by marine phytoplankton (Andreae, 1986); dimethyl sulfide (DMS) is thought to be the major precursor to sulfate aerosol in remote oceanic regions. DMS is oxidized through intermediaries (such as methane sulfonate) into SO_2. As both DMS and SO_2 oxidation are related to the presence of photochemical processes, the rate of sulfate formation varies as a direct function of solar radiation intensity. The relationship between aerosol formation and photochemical activity is supported by numerous atmospheric observations (Heintzenberg, 1985, Mészáros & Vissy, 1974, Mészáros, 1973).

It was suggested that a feedback mechanism operates in the marine atmosphere between planktonic DMS emissions and climate (Charlson *et al.*, 1987); enhanced DMS emissions produce more sulfate aerosol particles with a concomitant increase in CCN concentrations and, thus, cloud albedo above the oceans. This enhancement would reduce global temperatures and reduce marine productivity and DMS emission, closing a negative feedback loop. However, despite significant research efforts, the magnitude and even the sign of this feedback remains disputed (Andreae & Rosenfeld, 2008). Nevertheless, in remote marine regions the majority of CCN are known to be sulfates that form by the oxidation of DMS (Davison *et al.*, 1996; Andreae *et al.*, 2003); thus, biogenic emissions must play an important role in regulating cloud properties over the oceans. Another, temporally highly variable, natural source of SO_2 are volcanoes that eject large amounts of this gas into high altitudes. During periods of high volcanic activity, the sulfate aerosol formed from SO_2 emitted by volcanoes can have a significant effect on the global albedo and, thus, temperature. Well known examples are the 1983 eruption of El Chichón in Mexico that caused a hemispheric cooling of up to 0.5°C between 1983 and 1985 (Michalsky *et al.*, 1990), and the 1991 eruption of Mt. Pinatubo in the Philippines that also caused a surface cooling of about 0.5°C in 1992 (Lacis & Mishchenko, 1995). Extremely S-rich emissions by volcanic eruptions are caused by the fractionation of S-containing volatiles on the top of the magma chamber just before eruptions (Keppler, 1999).

At the end of the 20th century, anthropogenic emissions were responsible for ~60–80% of the S in the atmosphere (Chuang *et al.*, 1997), most of which was emitted in the Northern Hemisphere. Power plants that burn S-containing coal or oil are major emitters of SO_2 and are well known for their local and regional effects on acidity of aerosols and rain. As a result of emission controls, in Europe and North America, the release of SO_2 into the atmosphere from such industrial sources has declined significantly in the last three decades, as indicated by the decrease of S concentrations in the continental troposphere (Mészáros, 1999), and a spectacular recovery from acidification of aquatic ecosystems (Stoddard *et al.*, 1999). At the same time, rapid economic growth resulted in increasing rates of S emissions in Asia. Globally, the emission of anthropogenic SO_2 decreased by about a quarter from 1980 to 2000 (Stern, 2005), and is expected to decrease further as more stringent emission controls are introduced in China. The reduction of the amount of sulfate aerosol was suggested to be partly responsible for the observed increased transparency ("global brightening") of the atmosphere (Wild, 2009). However, in many regions, most spectacularly in China, increasing emissions of NO_x and NH_3 may result in enhanced nitrate aerosol production (Bauer *et al.*, 2007), partly compensating for the radiative changes caused by the reduction of sulfate in the atmosphere.

4.3.2. *Homogeneous and heterogeneous nucleation of sulfate particles*

Nucleation processes of sulfate particles are of great climatic significance, particularly above the oceans, where the concentration of CCN is much smaller than above the continents (Charlson *et al.*, 1987). In the last 20 years our understanding of sulfate particle nucleation has changed considerably. According to classical nucleation theory, water molecules and sulfuric acid vapour, formed in the air by homogeneous gaseous reactions between SO_2 and OH, condense to produce ultrafine sulfuric acid solution droplets. Although there was observational evidence for the binary homogeneous sulfate nucleation within the marine boundary layer (MBL) (Covert *et al.*, 1992; Clarke *et al.*, 1998), several studies have shown that other processes need to be invoked to account for the number of observed sulfate particles in the MBL (Hoppel *et al.*, 1994; Bigg, 1997).

Ammonia was shown to have a stabilizing effect on newly formed sulfate nuclei (Coffman & Hegg, 1995; Marti *et al.*, 1997); *e.g.* large numbers of newly formed particles were observed downwind of NH_3-emitting penguin colonies on the Macquarie Islands, in the Southern Ocean (Weber *et al.*, 1998). In addition to the presence of NH_3, the speciation of ammonium salts also affects nuclation rates. Models produce a better match with experimental observations when the formation of ammonium bisulfate in the H_2SO_4-NH_3-H_2O ternary system is taken into account (Merikanto *et al.*, 2007). Recently, the effects of neutral and/or ionic clusters (Kulmala & Kerminen, 2008), as well as cosmic rays (Carslaw *et al.*, 2002) were invoked to explain nanoparticle nucleation rates. Experiments at the synchrotron facility of CERN modeled the interactions of cosmic rays with atmospheric gases to study aerosol nucleation (Kirkby *et al.*, 2011). Atmospherically relevant NH_3 concentrations were found to increase sulfate nucleation rates 100- to 1000-fold over the rates observed in a binary system, while ions produced at ground-level galactic-cosmic-ray intensity enhanced nucleation by a factor of 2 to 10. Nevertheless, it is still debated whether cosmic rays can have significant effects on aerosol formation at the Earth's surface.

In the vicinity of clouds, conditions can be favourable for bursts of new particle formation; clean, aerosol-free air, relatively low temperature, and high relative humidity is necessary for such events (Perry & Hobbs, 1994; Hoppel *et al.*, 1994). In the free troposphere, conditions for sulfate nucleation are easier met than in the MBL (Clarke, 1993); newly formed particles can find their ways into the lower atmosphere. Modelling studies by Raes (1995) have shown that the upper troposphere is a major source of sulfate nuclei in the MBL.

Results from the past 10 years clearly show that new particle formation in the continental troposphere is enhanced by the presence of volatile organics. In the polluted urban environment, the photodegradation of aromatic hydrocarbons results in the formation of stable aromatic acid–sulfuric acid complexes that appear to reduce the nucleation barrier of sulfuric acid (Zhang *et al.*, 2004). Volatile organic compounds (VOCs) also occur in pristine continental environments and were found to participate in particle nucleation processes. Oxidation products of sesquiterpenes emitted by boreal forests interact with sulfuric acid and form new particles (Kulmala *et al.*, 2007; Bonn *et al.*, 2008; Smith *et al.*, 2010; Zhang *et al.*, 2012a).

Biological emission may produce the fluctuations in nuclei counts in coastal regions, where bursts of new particle formation were observed at low tide (Grenfell *et al.*, 1999). The concentrations of sulfuric acid and organic vapours were too low to explain the production of new particles. As the concentration of iodine can be 2 to 3 orders of magnitude larger in nucleation-mode marine aerosol than in sea salt (Mäkelä *et al.*, 2002), the focus of research narrowed on iodine-bearing, condensable species (O'Dowd *et al.*, 2002; Jimenez *et al.*, 2003) that are emitted by macroalgae when they are exposed to the atmosphere. The precursor vapours were first thought to be mainly organic species (such as CH_2I_2) but later, molecular iodine (I_2) was identified as the major precursor to iodine oxide particle formation (McFiggans *et al.*, 2004; Saiz-Lopez *et al.*, 2006). Because of its complexities and relevance to the nucleation of marine aerosol particles, iodine chemistry in the marine boundary layer remains an actively studied topic (McFiggans *et al.*, 2010).

Heterogeneous nucleation is also important in sulfate formation. When other particles are present in the air, sulfate formation on their surfaces is thermodynamically favoured over homogeneous nucleation. As discussed above, sulfate nucleation and condensation on both mineral dust and sea-salt particles is a widespread process in the troposhere. Because of the different oxidation states of S in methanesulfonate (a product of biogenic DMS oxidation) and in sulfate (a product of the oxidation of anthropogenic SO_2), the ratios of these two species could be determined in individual sea-salt particles by fitting empirical curves to the S L-edge in NEXAFS spectra, obtained using STXM (Hopkins *et al.* 2008). Methanesulfonate salts were the dominant form of non-sea-salt sulfur in large particles, and sulfate appeared to be more common in smaller particles. Soot from combustion sources may also provide a surface for the nucleation of sulfate, resulting in aggregated

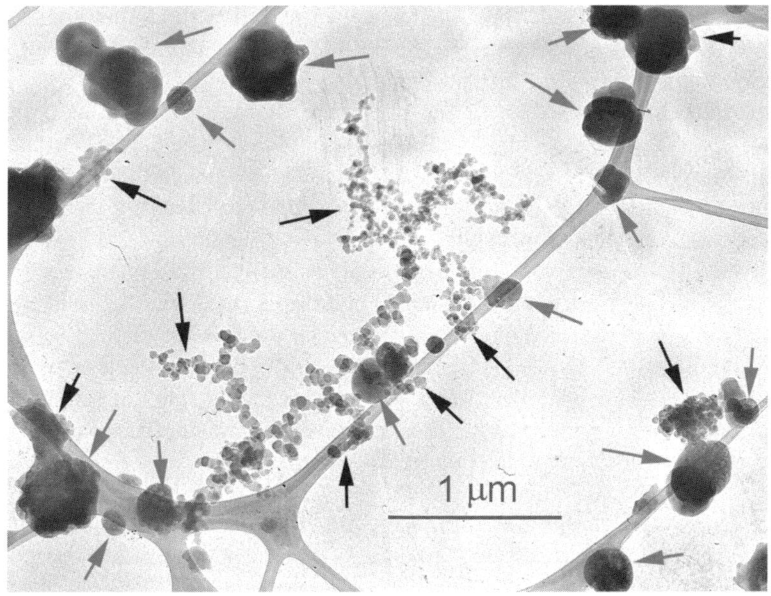

Fig. 15. TEM image of internally mixed soot/sulfate particles from the continental atmosphere (K-puszta, Hungary). The black and grey arrows point to soot and sulfate particles, respectively.

soot/sulfate particles (Fig. 15). The widespread occurrence and potential physical properties of such heterogeneous particles are discussed in more detail below.

4.3.3. The 'mineralogical' identities of sulfate aerosol particles

Sulfate particles are typically composed of $(NH_4)_2SO_4$ in the continental aerosol because NH_3 is abundant above the continents. In the marine aerosol, the composition of sulfates varies between H_2SO_4 and $(NH_4)_2SO_4$ (Clarke *et al.*, 1987), as known from chemical analyses and particle decomposition temperatures obtained from *in situ* sequential heating of ambient particles. H_2SO_4 volatilizes between 100 and 150°C, whereas $(NH_4)_2SO_4$ completely decomposes when the sample is heated to 300°C (Kreidenweis *et al.*, 1998). Sulfate particles can be crystalline or liquid while in the atmosphere, depending on temperature, relative humidity (RH) and their degree of neutralization by NH_3 (Martin, 2000). As discussed above in the section on aerosol physical properties, pure and dry $(NH_4)_2SO_4$ particles grow nearly linearly up to 81% RH (Fig. 6); such particles may be considered solid phases that are coated by water. At 81% RH $(NH_4)_2SO_4$ particles deliquesce, *i.e.* they become solution droplets. While the particles dehydrate, they remain in a metastable liquid state until they effloresce at ~30% RH. As most $(NH_4)_2SO_4$ particles in the lower troposphere experience RHs that exceed 81%, they are deliquesced droplets most of the time, owing to the hysteresis loop between hydration and dehydration curves. The solid or liquid phase of sulfate particles affects their optical properties. As aqueous particles have a larger mass-extinction efficiency but produce a smaller backscattered fraction of solar radiation than their solid counterparts, hysteresis effects result in a $\sim$20% uncertainty of calculated sulfate direct radiative forcing (Wang *et al.*, 2008).

When particles are collected on a filter or any other substrate, the originally liquid particles may crystallize, producing hydrous sulfate phases which may not have been present in the atmosphere. Further changes occur when the particles are exposed to the vacuum of a TEM; some volatile species may be lost, changing the compositions of hydrated sulfates. These problems illustrate that the presence of certain sulfate species in a TEM specimen does not necessarily mean that the same species occurred in the atmosphere. Indirect evidence, such as the presence of rings of satellite "drops" around a central particle have been used in several studies for distinguishing sulfate species (Mamane & de Pena, 1978; Qian & Ishizaka, 1993). On collection, hydrated particles spread on the hydrophobic carbon surface, producing rings of satellite particles around a larger central particle (Fig. 16); the development of such rings is related to the acidity of the original particles (Buseck & Pósfai, 1999).

4.3.4. Aggregates with other aerosol species affect the physical properties of sulfates

Among the main types of aerosol particles, sulfates were the first to be included in global climate models, owing to their simple optical and chemical properties and their large concentrations in the troposphere (Kiehl & Briegleb, 1993; Kasibhatla *et al.*, 1997; Chuang *et al.*, 1997). These early models were based on the assumption that sulfate particles are composed purely of sulfate. Mineralogical (individual-particle) studies have been most useful in showing that a large fraction of sulfate particles is, however, internally mixed with other species (Mészáros, 1984; Okada, 1985; Pósfai *et al.*, 1999). These heterogeneous aerosol particles may have different optical and hygroscopic properties and, consequently, climate effects compared to pure sulfates. The dynamics of interactions between sulfate and non-sulfate aerosol components are now included in global models (Liu *et al.*, 2005).

Many sulfate particles are aggregated with soot, even in the remote marine troposphere (Pósfai *et al.*, 1999). Depending on the altitude and the particular sample, 12–50% of sulfate particles contained soot inclusions above the Southern Ocean. In a pollution plume above the Atlantic Ocean near the Azores Islands, 90% of the larger ($>$ 0.5 µm diameter) ammonium sulfate particles were aggregated with soot and associated fly-ash particles that were probably emitted by a coal-burning power plant. In continental samples that were collected at a background site in Hungary the ratio of soot-containing *vs.* soot-free sulfates varied in about the same range as above the Southern Ocean. However, a clear difference between remote marine and continental samples was that in the polluted continental aerosol, large branching soot aggregates occurred (as in Fig. 15), whereas the soot particles above the oceans were generally smaller.

The presence of soot inclusions in sulfates raises several problems that should be addressed in climate models. Soot is the most efficient absorber of solar radiation in the atmosphere, and if included in sulfate, the optical properties of the mixed particle will differ from that of the pure sulfate particle. The absorption of light by the soot particle can be even more effective when enclosed by a sulfate droplet than if floating by itself

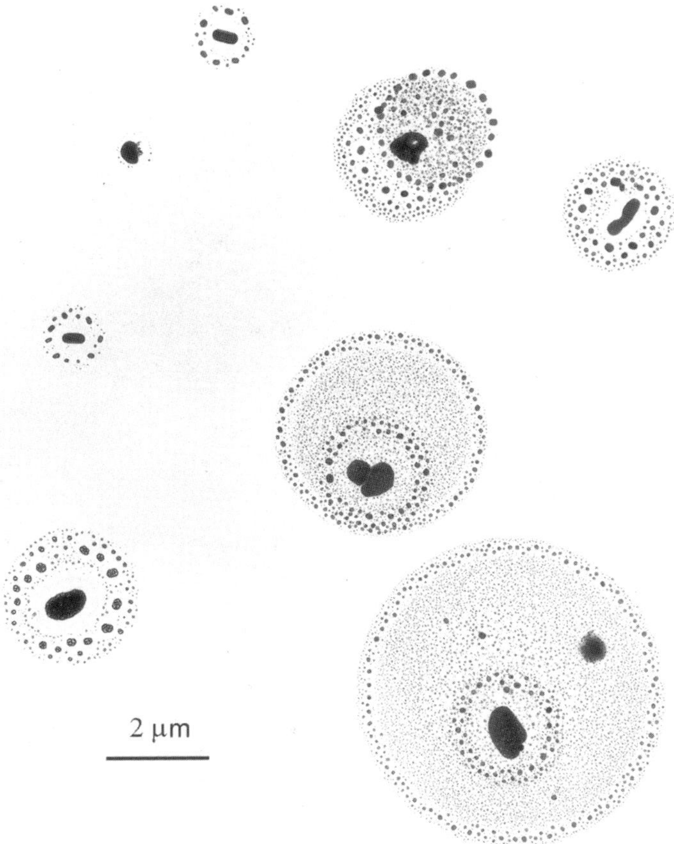

Fig. 16. TEM image of rings of satellite particles around larger cores of sulfate particles collected from an aircraft, 2.1 km above the Southern Ocean. The development of halos on the collection substrate is related to the acidity of the original hydrated sulfate particles.

in the air, because the sulfate-containing droplet can act as a lens that focuses more sunlight onto the absorbing particle. The larger the soot/sulfate mass ratio within the same particle, the more the optical properties of the original sulfate will change. If the soot particle is encased in sulfate, its specific absorption can increase by a factor of 4 (Fuller *et al.*, 1999), and the global direct radiative forcing (warming) by soot can be doubled (Jacobson, 2001). Aggregates of sulfate with soot and other aerosol species are now considered in global climate models. By averaging the results of these models, the last IPCC report estimates a global direct radiative forcing of -0.41 Wm^{-2} by sulfate (Forster *et al.*, 2007). More recent calculations for 2010 show a slightly stronger negative forcing of -0.62 Wm^{-2} (Skeie *et al.*, 2011).

Sulfate particles are typically coated by a film that is composed of organic compounds. Such coatings on sulfates were revealed by TEM studies (Pósfai *et al.*, 1998;

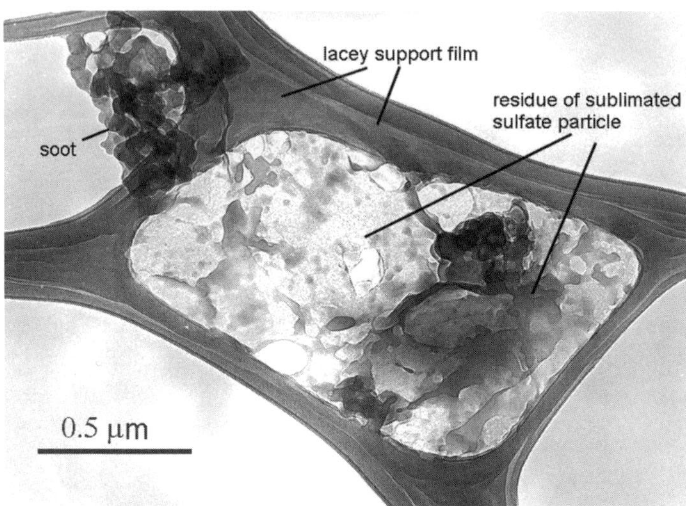

Fig. 17. TEM image of a thick residue that remained after a sulfate particle was sublimated with the electron beam (K-puszta, Hungary).

Buseck & Pósfai, 1999) that showed the presence of residues that were left behind after the volatile ammonium sulfate particles were intentionally sublimated with the electron beam (Fig. 17). The widespread use of various versions of aerosol mass spectrometry since the late 1990s resulted in the recognition that sulfates and organics are invariably mixed within individual particles in the troposphere (Murphy *et al.*, 2006).

The association of sulfate with organic compounds within the same particle affects the hygroscopic behaviour of the original sulfate. Organics can either enhance or delay the hygroscopic growth of sulfates, depending on their chemical character (Saxena *et al.*, 1995). They can also retard the evaporation of water from sulfate particles by forming a protective coating on the surface (Shulman *et al.*, 1996), as was also observed using a combination of AFM and TEM methods (Pósfai *et al.*, 1998). The development of environmental electron microscopy (both ESEM and ETEM, see the section on methods above) resulted in a breakthrough in the analysis of the hygroscopic behaviour of complex particles (Fig. 18).

Sulfates play an important role in cloud nucleation, as their water-soluble composition makes them excellent CCN. Sulfates are the most important CCN above the oceans and are also significant contributors to cloud formation above the continents (Andreae & Rosenfeld, 2008). Even a modest amount of sulfate in an otherwise insoluble particle can activate the particle as CCN (Freney *et al.*, 2010; Fig. 18). On the other hand, when sulfates are internally mixed with organic compounds, their cloud nucleation ability can change; if the organic compounds have lower solubility than sulfate, they cause a significant delay in the cloud activation of droplets (Shulman *et al.*, 1996). In addition, measurements on real cloud samples have shown that organics can lower the surface tension of cloud droplets that form on mixed sulfate/organic

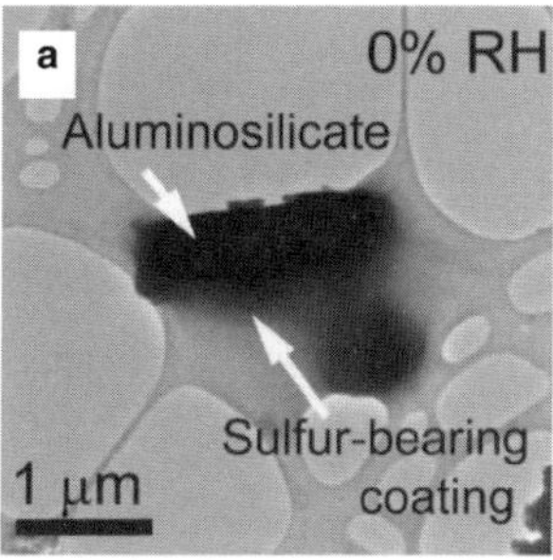
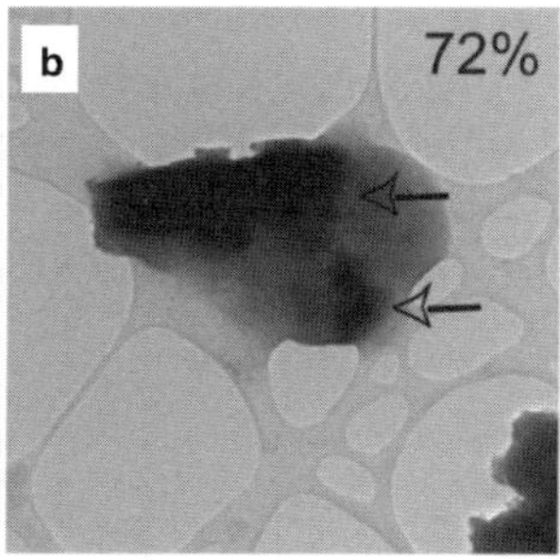
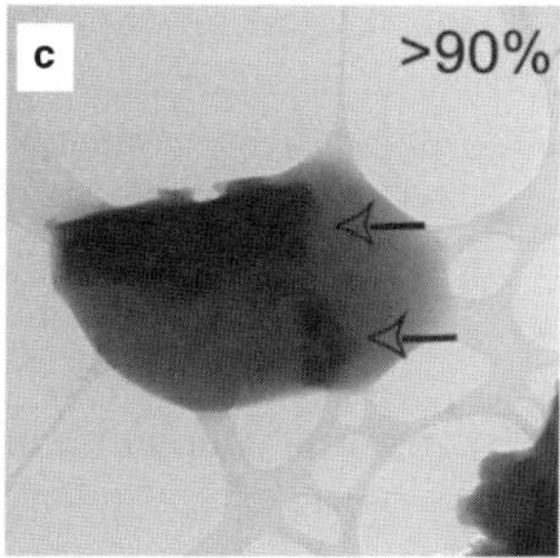

Fig. 18. Hygroscopic growth of a mixed particle, as observed under controlled, changing relative humidity (RH) in an environmental TEM. (**a**) The original particle at 0% RH; (**b**) and (**c**) the sulfur-bearing, probably organic coating collects water and the particle grows but its insoluble components retain their shapes even at RH > 90%, as indicated by the arrows (from Freney *et al.*, 2010; reproduced with the kind permission of the American Geophysical Union from the *Journal of Geophysical Research*).

aerosol particles, resulting in a large decrease of equilibrium supersaturation (Facchini *et al.*, 1999). At a lower critical supersaturation, more and smaller droplets form from the same available liquid water content than at a higher critical value. As the albedo of clouds increases with droplet number concentration (Twomey, 1977), the surface-active organic compounds cause a globally important increase in cloud albedo.

4.3.5. A note on nitrates

Nitrates are similar to sulfates in terms of formation mechanisms and climate effects. Nitrogen oxide gases (NO_x) are emitted by traffic and industry and are oxidized by OH radicals or O_3 to form nitric acid or peroxyacetyl nitrate (PAN) (Jacobson, 2002a). These low-volatility compounds exist either in the gas or aqueous phase and can be adsorbed onto existing particles or nucleate ammonium nitrate particles when excess ammonia is available after ammonium sulfate formation (Grobéty *et al.*, 2010).

With the focus on sulfates, in the past century relatively few studies have addressed the properties and radiative effects of nitrates. The relative significance of nitrates is increasing with diminishing SO_2 emissions but essentially constant or even increasing emissions of NO_x, especially in urban atmospheres (Dall'Osto *et al.*, 2009) and in regions with high NH_3 concentrations (Zhang *et al.*, 2012b). Nitrate aerosol loads in the future will depend most strongly on changes in ammonia sources (Bauer *et al.*, 2007). Even today, over North-Western Europe ammonium nitrate was found to be the dominant pollution aerosol beside organic matter (Morgan *et al.*, 2010). Interestingly, the concentration of ammonium nitrate increased with altitude, as a result of the partitioning of this semi-volatile material to the particle phase at cold temperature and enhanced relative humidity.

Ammonium nitrate particles are non-absorbing in the visible spectrum and are highly hygroscopic, making them excellent CCN. As the radiative properties of nitrates also depend on their physical state, whether they are present in aqueous or solid phases (Martin *et al.*, 2004), estimates of their climate effects vary strongly. Models have used the optical properties of sulfates for nitrates; however, this practice may have resulted in an underestimation of the direct radiative forcing by nitrates, as the wavelength-dependent scattering of ammonium nitrate differs from that of sulfate (Zhang *et al.*, 2012b). The direct radiative forcing of nitrate aerosol was tentatively estimated at -0.1 Wm^{-2} by the fourth IPCC report (Forster *et al.*, 2007), and is calculated to increase to -0.14 Wm^{-2} by 2030 (Bauer *et al.*, 2007).

4.4. Carbonaceous particles

The aerosol particles discussed in this section are composed mainly of carbon. Apart from this basic feature, the distinct particle types within this group have little in common; they are emitted by various sources, both anthropogenic and natural, and their atmospheric roles and climate effects are highly variable (Novakov & Penner, 1993; Penner, 1995; Gelencsér, 2004; Forster *et al.*, 2007). In terms of their concentrations and climate effects, they probably include the most important particle types above the continents. Even though most carbonaceous particles or their precursor gases are emitted on land, the particles can be transported above the ocean and distributed in the upper troposphere all over the globe, exerting an influence on global climate (Haywood & Ramaswamy, 1998; Schult *et al.*, 1997; Ramanathan & Carmichael, 2008). Among the major aerosol types, the physical and chemical properties of carbonaceous (particularly organic) particles are the least known and the most intensely studied.

Below we will distinguish the major carbonaceous particle types on the basis of their formation mechanisms and properties. Combustion produces soot, the major absorber of shortwave radiation in the atmosphere. The burning of biofuels and biomass emits a variety of carbonaceous particle types, some of which can be regarded primary, others secondary particles. Both natural and anthropogenic emissions of volatile organics can result in the formation of secondary organic aerosol particles. Finally, primary biogenic particles such as viruses, bacteria and pollen are also important constituents of the atmosphere.

4.4.1. Soot or black carbon (BC)

The incomplete combustion of fossil fuels or vegetation produces carbonaceous particles that are commonly called soot. Soot can be regarded a primary particle type, as it forms from aromatic precursors in the flame. Vast quantities of soot are produced by biomass burning practices in the tropics (Crutzen & Andreae, 1990; Forster *et al.*, 2007), as well as by industry, automobile, aircraft and ship traffic (Penner, 1995; Bond *et al.*, 2004). As soot particles are the strongest absorbers of solar radiation among all atmospheric aerosol types (Jacobson, 2002b), often their optical properties are emphasized and they are referred to as black carbon (BC) particles. Even though soot consists of a mixture of elemental (or 'graphitic') and organic carbon, the terms 'BC', 'elemental carbon', and 'soot' are often used interchangeably, creating much confusion in the literature, as pointed out by Bond & Bergstrom (2006) and Andreae & Gelencsér (2006). Here we use 'soot' when the morphologies, microstructures and compositions of the particles are discussed, and use 'BC' when citing the results of studies that referred to BC particles on the basis of their strong absorption of visible light.

The estimated annual emission of BC is $\sim$8 Tg y^{-1} (Bond *et al.*, 2004), with $\sim$20%, 40% and 40% from biofuels, fossil fuels and open biomass burning, respectively. The global pattern of BC particle concentrations in the troposphere results from the geographical distribution of their major emission sources. Both observations (Cachier *et al.*, 1995; Liousse *et al.*, 1993) and models (Cooke & Wilson, 1996; Liousse *et al.*, 1996) show high concentrations of BC in tropical regions during the biomass burning season, which peaks in January in Sub-Saharan Africa, and August and September in South America and South Africa (Levine, 1991). Other major sources of BC include densely populated, industrialized areas. Not surprisingly, BC concentrations are much higher in urban atmospheres and under polluted continental conditions than in remote marine regions (Nunes & Pio, 1993; Penner *et al.*, 1993). Commercial aircraft traffic contributes to high-altitude BC that shrouds the whole globe at an altitude between $\sim$10 and 11 km (Blake & Kato, 1995; Pueschel *et al.*, 1992). This high-altitude source of BC particles can cause globally significant changes in the optical properties of aerosols over the oceans.

Soot particles have typical morphologies and microstructures. Combustion produces hydrocarbons that condense immediately at the source and form 20 to 50 nm sized, solid spherules (Lahaye, 1992). These carbonaceous spherules then coagulate and form branching aggregates that are easily recognizable in electron micrographs (Fig. 19a). The number of spherules within an aggregate depends on several factors; for example, the air/fuel ratio and the speed of the engine affect the sizes of soot aggregates emitted by a diesel engine (Roessler *et al.*, 1981). Individual spherules within a soot particle have characteristic microstructures of turbostratic graphitic layers (Fig. 19b). The spacings between the onion-like, wrapped layers are typically larger than the 3.34 Å value of d_{001} in ordered graphite. According to quantitative electron diffraction analysis,

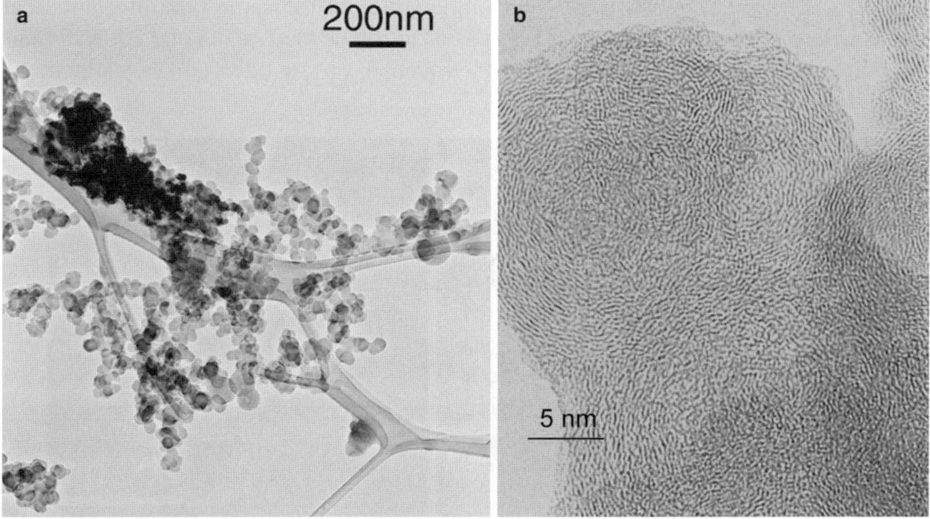

Fig. 19. (a) A soot aggregate from savanna smoke. (b) High-resolution TEM image obtained from a soot particle. The wavy graphitic layers form onion-like spherules.

first-neighbour atomic distances in soot particles are shorter than the C—C distances in graphite (Kis *et al.*, 2006), suggesting the presence of H-bearing, aromatic rings. The structures of the individual layers in soot spherules probably consist of islands of a few contiguous aromatic rings. In some soot aggregates, the individual spherules share a few common outer layers that envelope several spherules. There is a noticeable variation in the degree of order of the turbostratic layers within individual spherules, as well as in the extent to which the spherules are attached to one another by shared graphitic layers (Pósfai *et al.*, 1999). Such microstructural variations may be related to the conditions of soot formation. TEM studies have shown that nanostructural features of soot particles, particularly their graphitic content and lattice fringe spacings, reflect their distinct sources (including oil boilers, jet aircraft, diesel engines) (Vander Wal *et al.*, 2010).

The fractal properties of soot aggregates are also influenced by the type of fuel and the combustion conditions (Katrinak *et al.*, 1992; Wentzel *et al.*, 2003; van Poppel *et al.*, 2005). The morphology of soot strongly affects its optical properties. Based on 3D reconstructions from electron tomography, Adachi *et al.* (2007, 2010) found that the mass-normalized scattering cross section of soot was 20 to 28 times larger than that of unaggregated individual spherules of the same mass. The absorption cross-section was enhanced by 15% compared to unaggregated particles.

The possible sources of soot particles can be inferred from their compositions. However, most studies use analytical techniques that cannot distinguish the compositions of the carbonaceous spherules from those of additional constituents, such as coatings or non-soot nanoparticles within the same aggregate. Therefore, reports of atmospheric soot compositions should be treated with caution, as in most cases, it is unknown whether the identified composition refers to soot or to a mixture of several aerosol constituents, including soot. In general, the controlled technological burning of fossil fuels emits almost 'pure' soot, although even the same type of engine can emit a variety of particles ranging from 'naked' to 'embedded' soot (Fig. 20) as a function of air/fuel ratio and engine thrust. Much greater variety of carbonaceous particles

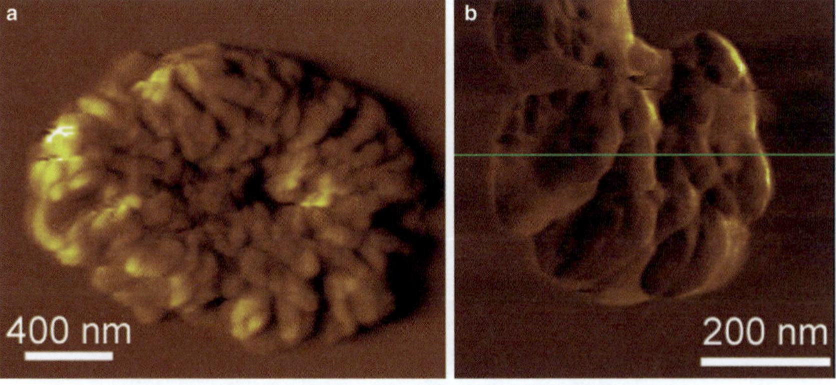

Fig. 20. AFM images of soot particles emitted by the same petrol-powered car. (**a**) An aggregate of "naked" soot spherules; (**b**) soot spherules are embedded in some material, probably organics.

results from biofuel, agricultural, and uncontrolled biomass fires (Li *et al.*, 2003b; Reid *et al.*, 2005). Soot from biomass burning invariably contains K (Andreae, 1983; Turn *et al.*, 1997; Li *et al.*, 2003b). Soot from coal burning is associated with metal-bearing fly-ash particles (Ramsden & Shibaoka, 1982; Sheridan, 1989), whereas oil combustion typically produces particles with significant V content in associated fly-ash. In a study of soot/sulfate aggregates in the marine troposphere, it was found that most soot particles collected above the clean Southern Ocean do not contain any elements other than C and O, whereas soot from the polluted North Atlantic is associated with Fe, Mn, or Zn-bearing fly-ash particles (Pósfai *et al.*, 1999). Presumably most of the Southern Ocean soot originated from aircraft emissions, whereas the North Atlantic soot was emitted by a coal-burning power plant. Soot particles collected at a continental location (K-puszta, Hungary) may also have different compositions, probably reflecting their different sources (Fig. 21).

Soot particles are originally hydrophobic, as shown by an environmental TEM study (Semeniuk *et al.*, 2007). However, there is ample evidence that atmospheric ageing results in

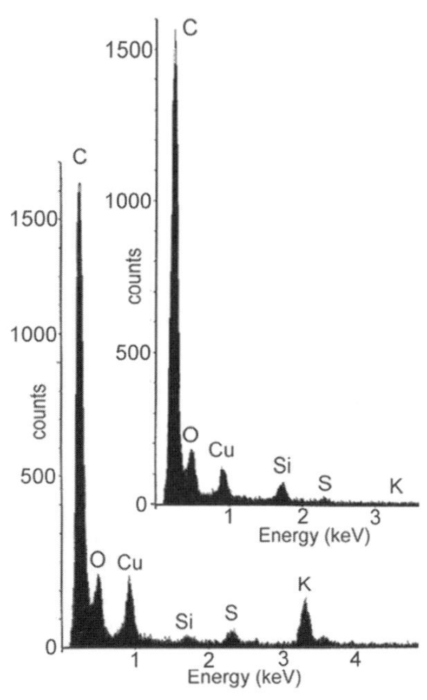

Fig. 21. EDS spectra of two soot particles (K-puszta, Hungary). One of the particles contains significant K, whereas the other does not, probably reflecting their different origins. (The Cu peaks are artefacts generated by the TEM grid.)

aggregates of soot particles with other, hygroscopic species that make the mixed particles grow by water-vapour condensation (Chugtai *et al.*, 1996). This growth results in bimodal size distributions for soot (or BC) particles, with a peak in the ultrafine mode representing newly formed particles, and another peak in the accumulation mode, representing soot particles that have grown by water condensation (Berner *et al.*, 1996). Metals associated with soot may act as catalysts for surface reactions (Chugtai *et al.*, 1996; Grgic *et al.*, 1993), resulting in further atmospheric reactions.

Soot particles are major constituents of the 'classical' or 'London' smog, a combination of smoke and fog that is produced by pollution from coal combustion. In such smog a high concentration of SO_2 is present; SO_2 itself is a respiratory irritant, but when associated with particles (*e.g.* soot), a three- to four-fold enhancement of the irritant response results (Wayne, 1991). Although emission controls alleviated pollution problems in London after the 1952 smog disaster, when more than 4000 deaths from respiratory illness were attributed to a smog lasting several days, classical smog still poses health problems in many rapidly developing urban areas, particularly in Asia.

The climate effects of BC particles are dominated by their strong absorption of visible light. The absorbed energy is released into the atmosphere as thermal radiation, heating the atmosphere at the level of the absorbing particles and reducing solar irradiance at ground level. In addition to exerting a net warming effect on Earth, BC particles can, thus, change the vertical temperature profile in the troposphere, resulting in more stable meteorological conditions (Crutzen & Andreae, 1990; Ramanathan *et al.*, 2005). The presence of BC in the atmosphere above highly reflective surfaces such as snow, ice, or clouds, may cause a significant positive radiative forcing (Forster *et al.*, 2007). Soot can be involved in cloud formation, resulting in brown clouds with enhanced absorption of solar radiation, and also the heating and drying out of cloud droplets, producing a 'semi-direct' radiative forcing (Ramanathan & Carmichael, 2008). Thus, the climate effects of soot have many aspects that need to be considered in climate models.

The globally averaged direct radiative forcing due to BC aerosol from fossil fuel combustion (and excluding BC from biomass burning) was estimated as $\sim +0.2 \, \mathrm{Wm}^{-2}$, not including the semi-direct effect or the impact of BC on snow and cloud albedo (Forster *et al.*, 2007). More recent calculations resulted in an even larger positive forcing of $+0.43 \, \mathrm{Wm}^{-2}$ (Skeie *et al.*, 2011). This warming effect partially counteracts the cooling caused by sulfates. When internal mixtures of soot and sulfate particles are considered, the net cooling caused by the total aerosol is further significantly diminished because of the more efficient absorption by internally mixed particles (Haywood *et al.*, 1997) (for a discussion on the optical properties of soot/sulfate particles, see the previous section on sulfates, and Fuller *et al.*, 1999). It is also important to note that the direct radiative forcing by BC particles near the surface is generally weaker than at higher altitudes because of the shielding effects of clouds (Haywood & Ramaswamy, 1998). Therefore, aircraft-emitted soot may be climatically more significant than its relatively small mass would suggest, as these particles are injected into the atmosphere at high altitude where they have a greater probability of being above clouds than particles close to the Earth's surface. As soot emissions from the controlled combustion sources, particularly vehicles, could be better constrained than greenhouse gas emissions, Jacobson (2002b) has suggested that the technologically most feasible way to slow the warming of the Earth is to reduce soot emission from traffic sources.

4.4.2. *Organic particles from biomass burning*

Biomass burning produces atmospheric particles in such amounts that they affect both regional atmospheric processes and global climate (Forster *et al.*, 2007). In addition to soot, biomass smoke contains other carbonaceous particles. Individual-particle TEM studies identified a new particle type, called 'tar ball' (Pósfai *et al.*, 2003a, 2004). Although tar balls are spherical, they are larger than individual soot spherules (ranging from $\sim$30 to 800 nm) and are completely amorphous, *i.e.* they lack the nanostructural order of soot (Fig. 22a). Compared to soot, another distinct feature of tar balls is that they do not form grape-like aggregates. Tar balls consist mostly of C, and can contain O, N and traces of S, K, Si and some other elements (Hand *et al.*, 2005; Adachi & Buseck, 2011). Tivanski *et al.* (2007) noted that the compositions of tar balls resemble those of atmospheric humic-like substances. In contrast to the primary formation of soot, tar balls can be considered secondary particles. They form within

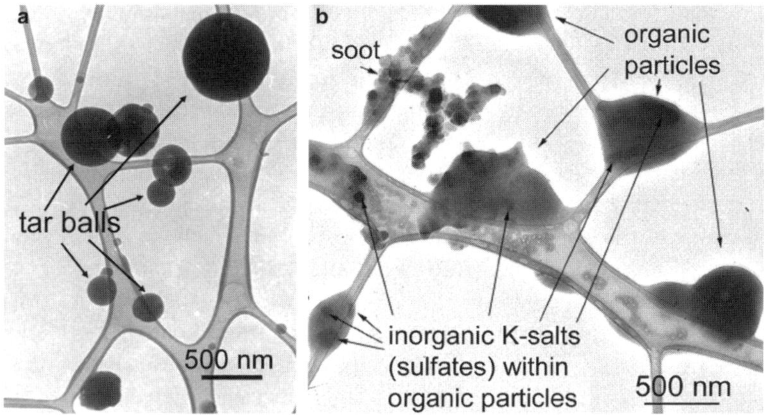

Fig. 22. TEM images of typical carbonaceous particles from biomass burning, from the smoke of a savanna fire in Southern Africa: (**a**) tar balls; (**b**) organic particles with inorganic salt inclusions (from Pósfai *et al.*, 2003a, reproduced with the kind permission of the American Geophysical Union from the *Journal of Geophysical Research*).

the smoke of biomass or biofuel fires, as indicated by the increase of their relative abundance as the smoke ages. Their relative concentration peaked in 30-min-old savanna smoke in southern Africa (Pósfai *et al.*, 2003a), and increased from ~1% in a >4-min-old smoke to 14% in a <10-min-old plume from Mexican shrub fires (Adachi & Buseck, 2011).

Given the abundance of tar balls in biomass smoke, their potential climate effects need to be considered. Tar balls appear to be hydrophobic, according to environmental TEM observations (Semeniuk *et al.*, 2007; Adachi & Buseck, 2011); thus, they are probably not effective as CCN. On the other hand, their direct climate effects can be significant, as their sizes are in the range where scattering of solar radiation is most efficient. The complex refractive index of tar balls was determined for the entire visible spectrum, using information derived from electron energy-loss spectra obtained from individual particles (Alexander *et al.*, 2008). According to this study, tar balls are brown; thus, their absorption of solar radiation is non-negligible. Adachi & Buseck (2011) found that the absorption of solar radiation by tar balls was almost as large as that of soot in a 10-min-old smoke from a shrub fire.

The most abundant particle type in young biomass smoke consists of carbonaceous, presumably organic compounds and does not have the characteristic shapes of soot or tar ball particles in TEM images. Instead, such particles have irregular shapes, suggesting that they were hydrated when collected, and then spread on the collection surface as they dried (Fig. 22b). Typically, these particles contain inorganic salt inclusions (Martins *et al.*, 1998; Li *et al.*, 2003b; Pósfai *et al.*, 2003a; Hand *et al.*, 2005). While the exact compositions of these particles vary with the type of vegetation burnt and the age of the smoke, the major components of the particles are C, O and crystalline K-salts (KCl, K_2SO_4 or KNO_3 (Li *et al.*, 2003b). The presence of K in carbonaceous particles is usually taken as evidence for the biomass-burning origin of the aerosol. In contrast to soot and tar ball, organic particles with inorganic inclusions show significant hygro-scopic growth at RH values $>\sim 55\%$ (Semeniuk *et al.*, 2007). Hand *et al.* (2010) found that inorganic species were responsible for most of the hygroscopic growth of biomass smoke particles. The hygroscopic nature of these particles makes them excellent CCN.

The distinct particle types such as soot, tar ball and other carbonaceous particles in biomass smoke can have distinct formation mechanisms, atmospheric histories, or both. Knowing the properties of each particle type can help to assess their climate effects; on the other hand, the presence of certain particle types in the air may be diagnostic of the sources of the aerosol. For example, satellite images and measurements clearly show that biomass smoke plumes from Africa and South America travel long distances over the ocean. In addition, individual-particle studies revealed that biomass burning (agricultural fire) severely affects atmospheric composition and air quality even in developed regions in Europe (Niemi *et al.*, 2005, 2006) and North America (Hudson *et al.*, 2004).

The optical properties of aerosol from biomass burning depend on several factors, including the type of vegetation, type of fire (flaming or smoldering), and the age of the smoke. For example, $<20\%$ of smoke particles emitted by savanna burning contained soot (Pósfai *et al.*, 2003a), the rest consisting mostly of organics. Thus, the dominant optical effect of such smoke is scattering, which was long thought to cause a net cooling effect of aerosol plumes produced by biomass burning. The global-mean direct radiative forcing due to smoke from biomass burning was estimated to be ~ -0.3 by Hobbs *et al.* (1997) and -0.2 Wm^{-2} by the third report by the IPCC (1996). However, extensive aerosol measurement and remote sensing campaigns in the past ten years resulted in a revised estimate of a slight positive direct radiative forcing of $+0.03 \pm 0.12$ Wm^{-2} (Forster *et al.*, 2007), then Skeie *et al.* (2011) again estimated a slight negative forcing (-0.07 Wm^{-2}) for the year 2010. Because biomass smoke can produce a strong positive forcing when above clouds, and the vertical structure of the smoke is highly variable, a large error is associated with these estimates. Bulk measurements show that smoke particles produced by biomass fires can serve as CCN (Crutzen & Andreae, 1990; Martin *et al.*, 2010). In light of the individual-particle TEM studies mentioned above, this effect probably arises from the dominant presence of organic particles that contain inorganic inclusions. As a result, biomass smoke has an indirect climate forcing effect that is probably significant but difficult to quantify. Biomass smoke particles can have contrasting effects on clouds: they can brighten clouds in the early morning by increasing the number of CCN, whereas they 'burn off' cloud droplets in the afternoon as a result of their light absorption (Ten Hoeve *et al.*, 2012). Such complexities are not treated by current climate models.

4.4.3. Secondary organic aerosol (SOA)

The significance of secondary organic aerosol (SOA) was recognized less than 20 years ago (Novakov & Penner, 1993), and the formation and effects of SOA particles are among the most studied aspects of current aerosol science (Gelencsér, 2004). SOA particles form from volatile organic compounds (VOCs) that are emitted by both biogenic and anthropogenic sources, with the biogenic contribution generally believed to be globally dominant (Tsigaridis & Kanakidou, 2007). However, the highest organic mass loadings are in regions of significant anthropogenic emissions (Jimenez *et al.*, 2009), suggesting that the interactions of biogenic VOCs with anthropogenic gases are a prerequisite for high loadings of SOA. Estimates of the global strengths of various SOA sources vary tremendously; the global atmospheric budget of SOA is estimated between 12 and 1820 Tg a^{-1} (Spracklen *et al.*, 2011), indicating the need for more detailed studies on this class of particles.

A major source of anthropogenic organic particles is fossil fuel combustion. The best studied among such particles are those that form the 'photochemical smog'. Urban air pollution in Los Angeles initiated many studies on the chemical compositions and atmospheric reactions of organic aerosol particles that form from non-methane hydrocarbons emitted primarily by automobiles (Rogge *et al.*, 1996; Odum *et al.*, 1997). Such particles cause degraded visibility and undesirable health effects, by carrying mutagenic and carcinogenic compounds (Finlayson-Pitts & Pitts, 1997).

Organic aerosol is also produced by natural, biogenic sources. The mass of SOA formed annually from biogenic precursors was estimated to be 18.5 Tg (Griffin *et al.*, 1999), making SOA particles one of the principal constituents of the continental aerosol. Isoprene and terpene emitted by terrestrial plants are considered the most important precursors for SOA formation (Yu *et al.*, 1999; Claeys *et al.*, 2004). Photo-oxidation of VOCs reduces their vapor pressure and organic gases condense onto existing aerosol particles; photo-oxidation and polymerization processes continue in the condensed phase and result in semi- or non-volatile organic species (Gelencsér, 2004).

Bursts of new particle formation have been observed in a variety of environments, including boreal forests, rural and urban areas (Kulmala *et al.*, 2004; Salma *et al.*, 2011). A significant fraction of such nm-sized particles are probably SOA/sulfate mixtures. Laboratory studies showed that sulfate nucleation is enhanced by the presence of aromatic acids (Zhang *et al.*, 2004); thus, organics are already present in the sulfate particles at nucleation. As the freshly nucleated particles can grow by the further condensation of organics, accumulation-mode SOA/sulfate mixed particles result (Smith *et al.*, 2008).

In addition to vegetation, the ocean is also a source of organic aerosol. Some oceanic organic compounds occur in primary particles (Middlebrook *et al.*, 1998; Bigg & Leck, 2008) that are mainly associated with sea salt that dominates the optical and chemical properties of such mixed particles. However, secondary organic particles also occur in the marine environment, particularly during phytoplankton blooms (O'Dowd *et al.*, 2004). A new SOA particle-formation mechanism was suggested to occur from isoprene emitted by algae during phytoplankton blooms (Meskhidze & Nenes, 2006). However, the conclusions of this study have been disputed.

As SOA particles are amorphous and composed of light elements, they are the least amenable major particle type for 'mineralogical' methods such as TEM and SEM. Various combinations of spectroscopic and microscope methods seem the most promising for studying the structure and composition of the carbonaceous components in

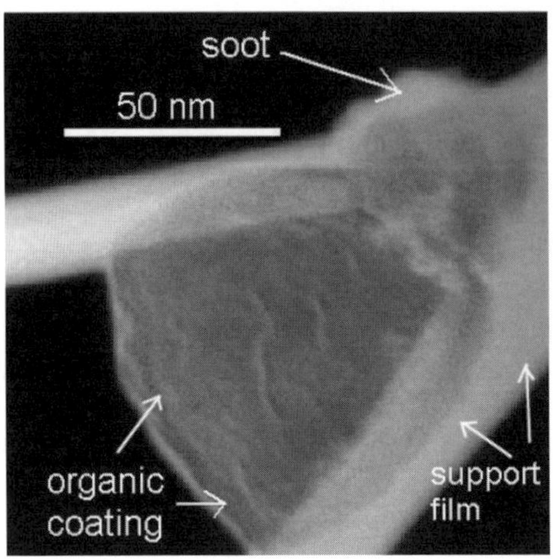

Fig. 23. Electron energy-loss map showing the distribution of C within a residue that remained after a sulfate particle was sublimated (K-puszta, Hungary). The coating on the sulfate consists mainly of C, confirming that organics are internally mixed with sulfates in this aerosol.

internally-mixed, individual sulfate/ carbonaceous particles. Using a transmission electron microscope, energy-loss spectra and energy-filtered images can be used to analyse C in the particles (Katrinak *et al.*, 1992; Pósfai *et al.*, 2003a, 2004). Elemental maps such as in Figure 23 prove unambiguously that the coatings on sulfates are composed mainly of C; hence, these coatings represent SOA that condensed onto the sulfate particles.

One of the major questions of the last decade about the radiative effects of atmospheric particles has been whether carbonaceous species other than soot contribute significantly to light absorption. As reviewed by Andreae & Gelencsér (2006), mounting evidence suggests that "brown carbon" is a significant component of the aerosol. Some of the brown carbon species form in combustion, but others may be SOA particles Gelencsér (2004). Attempts have been made to measure or infer refractive indices for distinct, reasonably well defined organic particle types that may qualify as brown carbon. For example, humic-like substances (HULIS) appear to be abundant constituents of SOA particles (Gelencsér, 2004), and HULIS absorb light at $\sim$300 nm wavelength (Hoffer *et al.*, 2006). A common property of brown carbon species is a strong wavelength dependence of their absorption (Kirchstetter *et al.*, 2004; Bond & Bergstrom, 2006; Hoffer *et al.*, 2006; Schnaiter *et al.*, 2006). The absorption of downward UV radiation (Jacobson, 1999) by atmospheric brown carbon may turn out to be a globally important component of direct radiative forcing.

The hygroscopic behavior of organic/inorganic mixed particles has been examined in a large number of studies, resulting in a general conclusion that the deliquescence and efflorescence behaviors of such mixed particles are difficult to predict from the properties of their individual components. Organic acids reduced the water absorption of NaCl but enhanced that of $(NH_4)_2SO_4$ (Choi & Chan, 2002; Cruz & Pandis, 2000). Malonic and citric acid caused the inorganic salts to absorb a significant amount of water before deliquescence, whereas glutaric acid caused them to deliquesce gradually (Choi & Chan, 2002). If the organic fraction was significant in mixed organic/ $(NH_4)_2SO_4$ particles, the efflorescence relative humidity was lower than of pure sulfate, and the particles were more likely to remain in the liquid state (Pósfai *et al.*, 1998; Pant *et al.*, 2004). To date, only a handful of individual-particle ETEM studies have addressed the problems related to the hygroscopic behaviour of organic/inorganic mixtures (Semeniuk *et al.*, 2007; Freney *et al.*, 2010), highlighting complexities such as the gradual deliquescence of particles and the common presence of mixed solid/liquid particles. According to recent model calculations, the hygroscopic behaviour of SOA-bearing particles can be reasonably predicted form the properties of pure components if the particle components have not re-equilibrated on drying (Topping *et al.*, 2011).

Organic constituents in mixed particles can have several specific effects on CCN activity. Surface-active organic material can reduce the surface tension or form hydrophobic surface films, both of which can change the CCN activity of the particle (Kiss *et al.*, 2005; Topping *et al.*, 2007; Möhler *et al.*, 2007). In addition, even a small amount of soluble inorganic material can make a slightly soluble organic particle CCN (Bilde & Svenningsson, 2004). At the moment, the CCN activity of mixed particles cannot be reliably predicted from their sub-saturated hygroscopicity (Good *et al.*, 2010).

Observational evidence suggests that a large fraction of organic particles are able to become CCN, both under clean and polluted, marine and continental conditions (Novakov & Penner, 1993; Andrews *et al.*, 1997; Novakov *et al.*, 1997; Mazurek *et al.*, 1997). Water-soluble organic compounds were found to form a significant fraction of the fine aerosol at various locations in Europe (Zappoli *et al.*, 1999; Krivácsy *et al.*, 2001). In the Amazonian rainforest, the majority of CCN were submicrometer SOA particles of biogenic origin during normal (non-biomass-burning) periods (Gunthe *et al.*, 2009; Pöschl *et al.*, 2010). Even though the particular organic species in the aerosol may be different at these sites and some remain unknown, the observations show that whether anthropogenic or natural, a large fraction of SOA particles are effective as CCN.

Based on aerosol mass spectrometry measurements of the submicron aerosol from a large number of northern Hemisphere locations, Jimenez *et al.* (2009) proposed a model framework for the atmospheric evolution of organic aerosol. Despite the variability of the sources of precursors, the large number of possible aerosol-forming processes, and the resulting complexity of components in SOA, an aerosol with relatively uniform properties results form the atmospheric ageing of particles. In general, ageing causes the particles to become more oxidized, less volatile, increasingly hygroscopic, and their light absorption increases. Fortunately, the remarkable convergence of physical and chemical properties of aged SOA particles reduces the complexity that needs to be represented in models that study aerosol climate effects (Andreae, 2009).

4.4.4. Bioaerosol

Primary biogenic particles in the atmosphere include viruses, bacterial cells, spores, pollen, and biogenic debris such as marine colloids and plant fragments. Biogenic particles can be released into the atmosphere from water surfaces (Blanchard & Syzdek, 1982), vegetation and soil (Artaxo *et al.*, 1994), and by anthropogenic activities (Gregory, 1973). Biological constituents of atmospheric aerosols have long been recognized, *e.g.* by Darwin, and large numbers of spores, pollen and bacteria were found even at high altitudes and in remote oceanic and polar regions (as reviewed by Gregory, 1973). Even giant biological particles, such as 30−55 μm long pine and spruce pollen can be transported over thousands of kilometres (Campbell *et al.*, 1999). According to anecdotal evidence, dead African locusts arriving with the trade winds constitute a non-negligible fraction of the aerosol at Barbados. Observations of the emissions of bacteria from various ecosystems, and the global distribution of microorganisms in the atmosphere were reviewed by Burrows *et al.* (2009b). The total number of annually emitted particles that contain bacteria was estimated to be about a 'mole' (*i.e.* 7.6×10^{23} to 3.5×10^{24}), and strong latitudinal and seasonal variations were found in the concentration of bacterial cells (Burrows *et al.*, 2009a).

Tropical vegetation is a major source of primary biological particles that may include plant debris, leaf abrasion products (waxes), fungi, pollen, bacteria and viruses. These biological particles occur both in the coarse and fine aerosol fractions, and with $\sim$35% of the total number concentration, dominate the Amazonian aerosol during the wet season (Elbert *et al.*, 2007); they provide the main source of atmospheric P, and are important in the cycles of K, Zn and some other elements (Artaxo *et al.*, 1994; Echalar *et al.*, 1998). Primary biogenic particles occur in highly variable concentrations in continental settings (Burrows *et al.*, 2009a,b; Winiwarter *et al.*, 2009), and are abundant in temperate urban atmospheres, where they can constitute up to 30% of the aerosol number and volume concentration (Matthias-Maser & Jaenicke, 1994). Even in the traffic emission-dominated atmosphere of Los Angeles, 1–3% of the total fine aerosol burden originates from urban vegetative detritus (Hildemann *et al.*, 1996).

The ocean is also a source of microorganisms to the atmosphere; individual cells of bacteria were observed in the aerosol above the Southern Ocean, a region of high biological productivity (Pósfai *et al.*, 2003b). The morphologies of the bacteria suggest an aquatic origin. The cells were presumably ejected into the air together with sea-salt particles; however, they were not attached to sea salt on the TEM grids, indicating their likely hydrophobic character. In addition to bacteria, diatom and coccolith fragments (Pósfai *et al.*, 1994; Sievering *et al.*, 1999) also occur in the marine aerosol (Fig. 24).

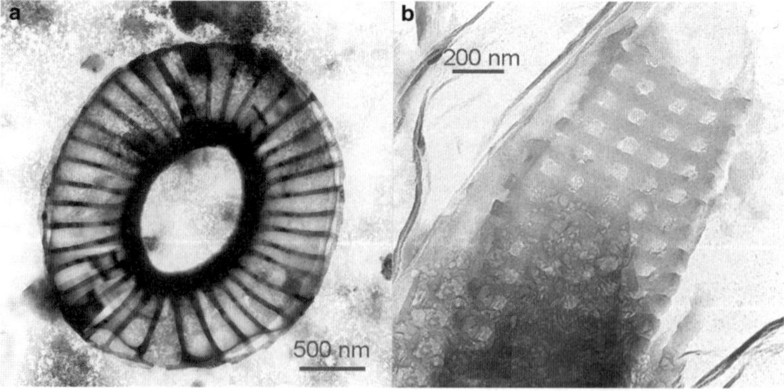

Fig. 24. Biogenic aerosol particles collected from sea spray on a beach in Oregon; **(a)** a coccolith and **(b)** a diatom fragment.

Biological particles in the atmosphere are of interest for their roles in the spread of diseases, ice nucleation in clouds and on plants, and for their potential importance in cloud droplet nucleation (Morris *et al.*, 2008). Certain species of bacteria are known to be effective ice nuclei at temperatures as high as $-1.5°C$ (Schnell, 1977). 'Ice nuclei' make metastable supercooled water drops freeze at higher temperatures than they would without the presence of such nuclei. Ice-nucleating bacteria were found living on plant leaves (Lindemann & Upper, 1985) and in the surface layer of the ocean (Schnell & Vali, 1976), places from which they can be easily transported into the atmosphere. Although generally mineral dust particles are believed to be the major ice nucleating agents in supercooled clouds, it was suggested that biological particles, primarily bacteria, are also important in atmospheric ice nucleation processes (Vali, 1985). Recent results strongly suggest that bioaerosol particles are indeed important in ice formation in clouds. In the Amazonian rainforest during the wet season, ice nuclei in the supermicrometer size range were found to be predominantly primary biogenic particles (Pöschl *et al.*, 2010). Biological ice nucleators were ubiquitous in snow (Christner *et al.*, 2008), and even the ice nucleation ability of soil dust was attributed to biological particles attached to the mineral particles (Conen *et al.*, 2011).

Concerning the role of primary biogenic particles in nucleating cloud droplets, similar statements can be made as about the CCN-forming ability of other organic particles. One of the most intriguing questions is whether microorganisms are passively and intermittently transported in the air, or whether at least some of them actually live in the atmosphere. There are indications that some bacteria show metabolic activity in cloud droplets (Sattler *et al.*, 2001). If the activity of microorganisms in the atmosphere is confirmed by further studies, new pathways of aerosol and cloud-droplet chemistry will have to be considered (Husárová *et al.*, 2011). In addition to bacteria, fungal spores occur in large concentrations in the atmosphere and probably influence the hydrological cycle (Fröhlich-Nowoisky *et al.*, 2009). Our knowledge of bioaerosol particles is rapidly expanding as a result of exciting methodological developments in the last ten years (Georgakopoulos *et al.*, 2009). Instead of using the traditional method of counting colony-forming units on cultivating plates, the number of viable microorganisms could be estimated by measuring the fluorescence of biological particles (Huffman *et al.*, 2010). Useful results on the types and concentrations of microorganisms have come from studies that combine microscopy with cultivation (*i.e.* Bauer *et al.*, 2002). Methods adopted from molecular environmental biology provide unique data on airborne organisms (Maron *et al.*, 2005). DNA fragments from aerosol samples were amplified using polymerase chain reaction, and analyzed by sequencing or terminal restriction fragment length polymorphism analysis (T-RFLP) (Després *et al.*, 2007), with the result that the diversity of all biological material could be assessed. Although such methods offer insight into the ensemble makeup of the aerosol, genetic information on individual particles may be available in the future if organism-specific fluorescent probes are constructed and combined with microscopy. The use of modern biological methods has triggered a revolutionary change in the solid earth sciences and led to the development of a new field, geomicrobiology. A similar explosion of research involving the study of individual bioaerosol particles appears imminent in atmospheric science.

5. Conclusions

The radiative climate forcing by both aerosol particles and anthropogenic greenhouse gases (such as CO_2, CH_4, N_2O and halocarbons) is regularly estimated by the Intergovernmental Panel on Climate Change and published in influential reports. According to the last report (Forster *et al.*, 2007), the total forcing from 1750 to 2005 by the long-lived anthropogenic greenhouse gases is $+2.9 \pm 0.3$ Wm^{-2}. Anthropogenic aerosols are estimated to produce a combined direct and indirect forcing of -2.2 to -0.5 Wm^{-2} in the 90% confidence range, with a median value of -1.3 Wm^{-2}. Although more recent calculations of aerosol radiative forcing slightly differ form the values in the IPCC report (Myhre *et al.*, 2009; Skeie *et al.*, 2011), all studies agree that anthropogenic aerosol particles have had a major role in climate modification since the industrial revolution. Aerosol particles may counteract little or most of the global warming effect of greenhouse gases, depending on where in the uncertainty range is the true magnitude of aerosol forcing. The observed warming over the past century has been ~0.6 K, a value that probably results from the combined and opposite effects of greenhouse gases and aerosols. By examining Earth's energy balance in the last 60 years, Murphy *et al.* (2009) found that $\sim50\%$ and 20% of the warming effect of greenhouse gases has been balanced by tropospheric and volcanic aerosols, respectively. Only $\sim10\%$ of the positive forcing has gone into heating the Earth, almost all into the oceans. With powerful policies being implemented in the developed world to reduce atmospheric pollution by particles, the aerosol burden in the troposphere is expected to diminish, resulting in 'global brightening' (Wild, 2009). This trend could lead to the end of significant 'climate protection' by aerosols (Andreae, 2007).

Our knowledge of any estimates of climate change are complicated by the fact that the effects of greenhouse gases are global, whereas those of aerosol particles range from local through regional to global. In addition, the typical residence times of aerosol particles are around a week, while those of greenhouse gases range from decades to centuries. Increasingly detailed data are available about the tropospheric aerosol, and sophisticated climate system models are being developed in order to be able to understand and predict human-induced climate change (Forster *et al.*, 2007). However, as seen from the above IPCC estimates, large uncertainties still exist in such models, resulting from the complex and partially known effects of tropospheric aerosol particles. In particular, the spatial and temporal variability, size-dependent chemical and physical properties, optical, cloud- and ice-nucleating effects of atmospheric aerosol particles need to be better understood.

In several aspects of the above problems, mineralogical studies can provide much needed information. Standard mineralogical tools such as XRD and automated light and electron optical methods can be used to great effect for monitoring composition and physical properties of some aerosol types (Kandler *et al.*, 2009; Grobéty *et al.*,

2010). As pointed out by Buseck (2010) and Pósfai & Buseck (2010), speciation and typical shapes of complex particles can be obtained by mineralogical studies that involve advanced electron microscope methods, and from these the optical properties of individual particles can be derived. Reactions on particle surfaces and the hygroscopic properties of particles can be studied using environmental SEM and TEM. Knowledge of the physical relationships between distinct aerosol species (their mixing state) is very important for a number of problems in atmospheric aerosol science. This is an area where the use of individual-particle methods (especially analytical TEM) can significantly increase our knowledge of both aerosol properties and atmospheric processes that are active in particle formation.

Acknowledgements

The authors thank Drs. Peter Buseck, Ernő Mészáros and András Gelencsér for many years of joint work on atmospheric aerosol particles, and invaluable discussions during the preparation of this chapter. They also thank Renáta Simonics and Éva Tompa for their contributions to the sections that discuss carbonaceous particles. They thank Prof. David Vaughan and an anonymous referee for their comments that improved the manuscript significantly. This research was supported by a joint EU-Hungarian grant (TÁMOP 4.2.1/B-09/1/KONV-2010-0003).

References

Adachi, K. & Buseck, P.R. (2011) Atmospheric tar balls from biomass burning in Mexico. *Journal of Geophysical Research*, **116**, D05204.

Adachi, K., Chung, S.H., Friedrich, H. & Buseck, P.R. (2007) Fractal parameters of individual soot particles determined using electron tomography: Implications for optical properties. *Journal of Geophysical Research*, **112**, D14202.

Adachi, K., Chung, S.H. & Buseck, P.R. (2010) Shapes of soot aerosol particles and implications for their effects on climate. *Journal of Geophysical Research*, **115**, D15206.

Adams, F. (1994) *Chemical characterization of atmospheric particles.* In: *Topics in atmospheric and interstellar physics and chemistry*, (C.F. Boutron editor). Les Editions de Physique, Les Ulis, France, pp. 271–289.

Adedokun, J.A., Emofurieta, W.O. & Adedeji, O.A. (1989) Physical, mineralogical and chemical properties of harmattan dust at Ile-Ife, Nigeria. *Theoretical and Applied Climatology*, **40**, 161–169.

Adriano, D.C. (2001) *Trace Elements in Terrestrial Environments. Biogeochemistry, Bioavailability, and Risk of Metals.* Springer-Verlag, Berlin, Heidelberg, New York.

Alexander, D.T.L., Crozier, P.A. & Anderson, J.R. (2008) Brown carbon spheres in East Asian outflow and their optical properties. *Science*, **321**, 833–836.

Alfaro, S.C., Gaudichet, A., Gomes, L. & Maillé, M. (1998) Mineral aerosol production by wind erosion: Aerosol particle sizes and binding energies. *Geophysical Research Letters*, **25**, 991–994.

Allan, J.D., Topping, D.O., Good, N., Irwin, M., Flynn, M., Williams, P.I., Coe, H., Baker, A.R., Martino, M., Niedermeier, N., Wiedensohler, A., Lehmann, S., Müller, K., Herrmann, H. & McFiggans, G. (2009) Composition and properties of atmospheric particles in the eastern Atlantic and impacts on gas phase uptake rates. *Atmospheric Chemistry and Physics*, **9**, 9299–9314.

Aller, J.Y., Kuznetsova, M.R., Jahns, C.J. & Kemp, P.F. (2005) The sea surface microlayer as a source of viral and bacterial enrichment in marine aerosols. *Journal of Aerosol Science*, **36**, 801–812.

Anastasio, C. & Martin, S.T. (2001) Atmospheric nanoparticles. In: *Nanoparticles in the Environment* (J.F. Banfield and A. Navrotsky, editors). Reviews in Mineralogy and Geochemistry, **44**. Mineralogical Society of America, Washington, D.C., pp. 293–349.

Anderson, J.R., Buseck, P.R., Saucy, D.A. & Pacyna, J.M. (1992) Characterization of individual fine-fraction particles from the Arctic aerosol at Spitzbergen, May–June 1987. *Atmospheric Environment*, **26A**, 1747–1762.

Andreae, M.O. (1983) Soot carbon and excess fine potassium: Long-range transport of combustion-derived aerosols. *Science*, **220**, 1148–1151.

Andreae, M.O. (1986) The ocean as a source of atmospheric sulfur compounds. In: *The Role of Air-Sea Exchange in Geochemical Cycling* (P. Buat-Menard, editor). D. Reidel, Dordrecht, The Netherlands, pp. 331–362.

Andreae, M.O. (1996) Raising dust in the greenhouse. *Nature*, **380**, 389–390.

Andreae, M.O. (2007) Atmospheric aerosols versus greenhouse gases in the 21st century. *Philosophical Transactions of the Royal Society London*, **A365**, 1915–1923.

Andreae, M.O. (2009) A new look at aging aerosols. *Science*, **326**, 1493–1494.

Andreae, M.O. & Crutzen, P.J. (1997) Atmospheric aerosols: Biogeochemical sources and role in atmospheric chemistry. *Science*, **276**, 1052–1058.

Andreae, M.O. & Gelencsér, A. (2006) Black carbon or brown carbon? The nature of light-absorbing carbonaceous aerosols. *Atmospheric Chemistry and Physics*, **6**, 3131–3148.

Andreae, M.O. & Rosenfeld, D. (2008) Aerosol-cloud-precipitation interactions. Part 1. The nature and sources of cloud-active aerosols. *Earth-Science Reviews*, **89**, 13–41.

Andreae, M.O., Charlson, R.J., Bruynseels, F., Storms, H., Van Grieken, R. & Maenhaut, W. (1986) Internal mixture of sea salt, silicates, and excess sulphate in marine aerosols. *Science*, **232**, 1620–1623.

Andreae, M.O., Andreae, T.W., Meyerdierks, D. & Thiel, C. (2003) Marine sulfur cycling and the atmospheric aerosol over the springtime North Atlantic. *Chemosphere*, **52**, 1321–1343.

Andrews, E. & Larson, S.M. (1993) Effect of surfactant layers on the size changes of aerosol particles as a function of relative humidity. *Environmental Science & Technology*, **27**, 857–865.

Andrews, E., Kreidenweis, S.M., Penner, J.E. & Larson, S.M. (1997) Potential origin of organic cloud condensation nuclei observed at marine site. *Journal of Geophysical Research*, **102**, 21,997–22,012.

Andronova, A.V., Gomes, L., Smirnov, V.V., Ivanov, A.V. & Shukurova, L.M. (1993) Physico-chemical characteristics of dust aerosols deposited during the Soviet-American experiment (Tadzhikistan, 1989). *Atmospheric Environment*, **27A**, 2487–2493.

Ansmann, A., Petzold, A., Kandler, K., Tegen, I., Wendisch, M., Müller, D., Weinzierl, B., Müller, T. & Heintzenberg, J. (2011) Saharan mineral dust experiments SAMUM-1 and SAMUM-2: What have we learned? *Tellus B*, **63**, 403–429.

Arimoto, R., Balsam, W. & Schloesslin, C. (2002) Visible spectroscopy of aerosol particles collected on filters: iron-oxide minerals. *Atmospheric Environment*, **36**, 89–96.

Artaxo, P., Gerab, F., Yamasoe, M.A. & Martins, J.V. (1994) Fine mode aerosol composition at three long-term atmospheric monitoring sites in the Amazon Basin. *Journal of Geophysical Research*, **99**, 22,857–822,868.

Avila, A., Queralt-Mitjans, I. & Alarcón, M. (1997) Mineralogical composition of African dust delivered by red rains over northeastern Spain. *Journal of Geophysical Research*, **102**, 21,977–21,996.

Bauer, H., Kasper-Giebl, A., Loflund, M., Giebl, H., Hitzenberger, R., Zibuschka, F. & Puxbaum, H. (2002) The contribution of bacteria and fungal spores to the organic carbon content of cloud water, precipitation and aerosols. *Atmospheric Research*, **64**, 109–119.

Bauer, S.E., Koch, D., Unger, N., Metzger, S.M., Shindell, D.T. & Streets, D.G. (2007) Nitrate aerosols today and in 2030: A global simulation including aerosols and tropospheric ozone. *Atmospheric Chemistry and Physics*, **7**, 5043–5059.

Ben-Ami, Y., Koren, I. & Altaratz, O. (2009) Patterns of North African dust transport over the Atlantic: winter vs. summer, based on CALIPSO first year data. *Atmospheric Chemistry and Physics*, **9**, 7967–7874.

Berner, A., Sidla, S., Galambos, Z., Kruisz, C., Hitzenberger, R., ten Brink, H.M. & Kos, G.P.A. (1996) Modal character of atmospheric black carbon size distributions. *Journal of Geophysical Research*, **101**, 19,559–19,565.

Bigg, E.K. (1997) A mechanism for the formation of new particles in the atmosphere. *Atmospheric Research*, **43**, 129–137.

Bigg, E.K. & Leck, C. (2008) The composition of fragments of bubbles bursting at the ocean surface. *Journal of Geophysical Research*, **113**, D11209.

Bilde, M. & Svenningsson, B. (2004) CCN activation of slightly soluble organics: The importance of small amounts of inorganic salt and particle phase. *Tellus, Series B: Chemical and Physical Meteorology*, **56**, 128–134.

Biscaye, P.E. (1965) Mineralogy and sedimentation of recent deep-sea clay in the Atlantic Ocean and adjacent seas and oceans. *Geological Society of America Bulletin*, **76**, 803–832.

Blake, D.F. & Kato, K. (1995) Latitudinal distribution of black carbon soot in the upper troposphere and lower stratosphere. *Journal of Geophysical Research*, **100**, 7195–7202.

Blanchard, D.C. (1978) Jet drop enrichment of bacteria, virus, and dissolved organic material. *Pure and Applied Geophysics*, **116**, 302–308.

Blanchard, D.C. & Syzdek, L.D. (1982) Water-to-air transfer and enrichment of bacteria in drops from bursting bubbles. *Applied and Environmental Microbiology*, **43**, 1001–1005.

Blanchard, D.C. & Woodcock, A.H. (1957) Bubble formation and modification in the sea and its meteorological significance. *Tellus, Series B: Chemical and Physical Meteorology*, **9**, 145–158.

Bohren, C.F. & Singham, S.B. (1991) Backscattering by nonspherical particles: A review of methods and suggested new approaches. *Journal of Geophysical Research*, **96**, 5269–5277.

Bond, T.C. & Bergstrom, R.W. (2006) Light absorption by carbonaceous particles: An investigative review. *Aerosol Science and Technology*, **40**, 1–41.

Bond, T.C., Streets, D.G., Yarber, K.F., Nelson, S.M., Woo, J.-H. & Klimont, Z. (2004) A technology-based global inventory of black and organic carbon emissions from combustion. *Journal of Geophysical Research*, **109**, D14203.

Bonn, B., Kulmala, M., Riipinen, I., Sihto, S.-L. & Ruuskanen, T.M. (2008) How biogenic terpenes govern the correlation between sulfufric acid concentrations and new particle formation. *Journal of Geophysical Research*, **113**, D12209.

Bradley, J.P. & Buseck, P.R. (1983) Catalytically grown carbon filaments from a smelter aerosol. *Nature*, **306**, 770–772.

Burrows, S.M., Butler, T., Jöckel, P., Tost, H., Kerkweg, A., Pöschl, U. & Lawrence, M.G. (2009a) Bacteria in the global atmosphere – Part 2: Modeling of emissions and transport between different ecosystems. *Atmospheric Chemistry and Physics*, **9**, 9281–9297.

Burrows, S.M., Elbert, W., Lawrence, M.G. & Pöschl, U. (2009b) Bacteria in the global atmosphere – Part 1: Review and synthesis of literature data for different ecosystems. *Atmospheric Chemistry and Physics*, **9**, 9263–9280.

Buseck, P.R. (2010) Atmospheric-particle research: past, present, and future. *Elements*, **6**, 208–209.

Buseck, P.R. & Adachi, K. (2008) Nanoparticles in the atmosphere. *Elements*, **4**, 389–394.

Buseck, P.R. & Pósfai, M. (1999) Airborne minerals and related aerosol particles: Effects on climate and the environment. *Proceedings of the National Academy of Sciences of the USA*, **96**, 3372–3379.

Buseck, P.R., Jacob, D.J., Pósfai, M., Li, J. & Anderson, J.R. (2000) Minerals in the air: An environmental perspective. *International Geology Review*, **42**, 577–593.

Cachier, H., Liousse, C., Buat-Ménard, P. & Gaudichet, A. (1995) Particulate content of savanna fire emissions. *Journal of Atmospheric Chemistry*, **22**, 123–148.

Cakmur, R.V., Miller, R.L., Perlwitz, J., Geogdzhayev, I.V., Ginoux, P., Koch, D., Kohfeld, K.E., Tegen, I. & Zender, C.S. (2006) Constraining the magnitude of the global dust cycle by minimizing the difference between a model and observations. *Journal of Geophysical Research*, **111**, D06207.

Campbell, I.D., McDonald, K., Flannigan, M.D. & Kringayark, J. (1999) Long-distance transport of pollen into the Arctic. *Nature*, **399**, 29–30.

Canagaratna, M.R., Jayne, J.T., Jimenez, J.L., Allan, J.D., Alfarra, M.R., Zhang, Q., Onasch, T.B., Drewnick, F., Coe, H., Middlebrook, A., Delia, A., Williams, L.R., Trimborn, A.M., Northway, M.J., DeCarlo, P.F., Kolb, C.E., Davidovits, P. & Worsnop, D.R. (2007) Chemical and microphysical characterization of ambient aerosols with the aerodyne aerosol mass spectrometer. *Mass Spectrometry Reviews*, **26**, 185–222.

Caquineau, S., Gaudichet, A., Gomes, L., Magonthier, M.-C. & Chatenet, B. (1998) Saharan dust: Clay ratio as a relevant tracer to assess the origin of soil-derived aerosols. *Geophysical Research Letters*, **25**, 983–986.

Caquineau, S., Magonthier, M.-C., Gaudichet, A. & Gomes, L. (1997) An improved procedure for the X-ray diffraction analysis of low-mass atmospheric dust samples. *European Journal of Mineralogy*, **9**, 157–166.

Carslaw, K.S., Harrison, R.G. & Kirkby, J. (2002) Atmospheric science: Cosmic rays, clouds, and climate. *Science*, **298**, 1732–1737.

Chamaillard, K., Jennings, S.G., Kleefeld, C., Ceburnis, D. & Yoon, Y.J. (2003) Light backscattering and scattering by nonspherical sea-salt aerosols. *Journal of Quantitative Spectroscopy & Radiative Transfer*, **79–80**, 577–597.

Chameides, W.L. & Stelson, A.W. (1992) Aqueous-phase chemical processes in deliquescent sea-salt aerosols: a mechanism that couples the atmospheric cycles of S and sea salt. *Journal of Geophysical Research*, **97D**, 20,565–520,580.

Chamley, H. (1989) *Clay Sedimentology*. Springer, New York.

Charlson, R.J. & Heintzenberg, J. (1995) *Aerosol Forcing of Climate*. John Wiley & Sons, New York.

Charlson, R.J., Covert, D.S., Larson, T.V. & Waggoner, A.P. (1978) Chemical properties of tropospheric sulfur aerosols. *Atmospheric Environment*, **12**, 39–53.

Charlson, R.J., Lovelock, J.E., Andreae, M.O. & Warren, S.G. (1987) Oceanic phytoplankton, atmospheric sulfur, cloud albedo and climate. *Nature*, **326**, 655–661.

Charlson, R.J., Schwartz, S.E., Hales, J.M., Cess, R.D., Coakley, J.A., Jr., Hansen, J.E. & Hofmann, D.J. (1992) Climate forcing by anthropogenic aerosols. *Science*, **255**, 423–430.

Choi, M.Y. & Chan, C.K. (2002) The effects of organic species on the hygroscopic behaviors of inorganic aerosols. *Environmental Science & Technology*, **36**, 2422–2428.

Chou, C., Formenti, P., Maille, M., Ausset, P., Helas, G., Harrison, M. & Osborne, S. (2008) Size distribution, shape, and composition of mineral dust aerosols collected during the African Monsoon Multidisciplinary Analysis Special Observation Period 0: Dust and Biomass-Burning Experiment field campaign in Niger, January 2006. *Journal of Geophysical Research*, **113**, D00C10.

Christner, B.C., Morris, C.E., Foreman, C.M., Cai, R.M. & Sands, D.C. (2008) Ubiquity of biological ice nucleators in snowfall. *Science*, **319**, 1214.

Chuang, C.C., Penner, J.E., Taylor, K.E., Grossman, A.S. & Walton, J.J. (1997) An assessment of the radiative effects of anthropogenic sulfate. *Journal of Geophysical Research*, **102**, 3761–3778.

Chueinta, W. & Hopke, P.K. (2001) Beta gauge for aerosol mass measurement. *Aerosol Science & Technology*, **35**, 840–843.

Chugtai, A.R., Brooks, M.E. & Smith, D.M. (1996) Hydration of black carbon. *Journal of Geophysical Research*, **101**, 19,505–519,514.

Claeys, M., Grahem, B., Vas, G., Wang, W., Vermeylen, R., Pashynska, V., Cafmeyer, J., Guyon, P., Andreae, M.O., Artaxo, P. & Maenhaut, W. (2004) Formation of secondary organic aerosols through photooxidation of isoprene. *Science*, **303**, 1173–1176.

Claquin, T., Schulz, M., Balkanski, Y. & Boucher, O. (1998) Uncertainties in assessing radiative forcing by mineral dust. *Tellus, Series B: Chemical and Physical Meteorology*, **50B**, 491–505.

Clarke, A.D. (1993) Atmospheric nuclei in the Pacific midtroposphere: Their nature, concentration, and evolution. *Journal of Geophysical Research*, **98**, 20,633–620,647.

Clarke, A.D., Ahlquist, N.C. & Covert, D.S. (1987) The Pacific marine aerosol: evidence for natural acid sulphates. *Journal of Geophysical Research*, **92**, 4179–4190.

Clarke, A.D., Porter, J.N., Valero, F.P.J. & Pilewskie, P. (1996) Vertical profiles, aerosol microphysics, and optical closure during the Atlantic Stratocumulus Transition Experiment: Measured and modeled column optical properties. *Journal of Geophysical Research*, **101**, 4443–4453.

Clarke, A.D., Davis, D., Kapustin, V.N., Eisele, F., Chen, G., Paluch, I., Lenschow, D., Bandy, A.R., Thornton, D., Moore, K., Mauldin, L., Tanner, D., Litchy, M., Carroll, M.A., Collins, J. & Albercook, G. (1998) Particle nucleation in the tropical boundary layer and its coupling to marine sulfur sources. *Science*, **282**, 89–92.

Clarke, A.D., Owens, S.R. & Zhou, J.C. (2006) An ultrafine sea-salt flux from breaking waves: implications for cloud condensation nuclei in the remote marine atmosphere. *Journal of Geophysical Research*, **111**, D06202.

Clegg, N.A. & Toumi, R. (1998) Non-sea-salt-sulphate formation in sea-salt aerosol. *Journal of Geophysical Research*, **103**, 31,095.

Coffman, D.J. & Hegg, D.A. (1995) A preliminary study of the effect of ammonia on particle nucleation in the marine boundary layer. *Journal of Geophysical Research*, **100**, 7147–7160.

Conen, F., Morris, C.E., Leifeld, J., Yakutin, M.V. & Alewell, C. (2011) Biological residues define the ice nucleation properties of soil dust. *Atmospheric Chemistry and Physics*, **11**, 9643–9648.

Cooke, W.F. & Wilson, J.J.N. (1996) A global black carbon aerosol model. *Journal of Geophysical Research*, **101**, 19,395–319,409.

Covert, D.S., Kapustin, V.N., Quinn, P.K. & Bates, T.S. (1992) New particle formation in the marine boundary layer. *Journal of Geophysical Research*, **97**, 20,581–520,589.

Covert, D.S., Wiedensohler, A., Aalto, P., Heintzenberg, J., McMurry, P.H. & Leck, C. (1996) Aerosol number size distributions from 3 to 500 nm diameter in the Arctic marine boundary layer during summer and autumn. *Tellus, Series B: Chemical and Physical Meteorology*, **48B**, 197–212.

Coz, E., Gomez-Moreno, F.J., Pujadas, M., Casuccio, G.S., Lersch, T.L. & Artano, B. (2009) Individual particle characteristics of North African dust under different long-range transport scenarios. *Atmospheric Environment*, **43**, 1850–1863.

Crutzen, P.J. & Andreae, M.O. (1990) Biomass burning in the tropics: Impact on atmospheric chemistry and biogeochemical cycles. *Science*, **250**, 1669–1678.

Cruz, C.N. & Pandis, S.N. (2000) Deliquescence and hygroscopic growth of mixed inorganic – Organic atmospheric aerosol. *Environmental Science and Technology*, **34**, 4313–4319.

Cwiertny, D.M., Baltrusaitis, J., Hunter, G.J., Laskin, A., Scherer, M.M. & Grassian, V.H. (2008) Characterization and acid-mobilization study of iron-containing mineral dust source materials. *Journal of Geophysical Research*, **113**, D05202.

Dabek-Zlotorzynska, E., Dlouhy, J.F., Houle, N., Piechowski, M. & Ritchie, S. (1995) Comparison of capillary zone electrophoresis with ion cromatography and standard photometric methods for the determination of inorganic ions in atmospheric aerosols. *Journal of Chromatography A*, **706**, 469–478.

Dal Maso, M. (2006) *Analysis of atmospheric particle formation events*. Academic Dissertation, University of Helsinki, Helsinki.

Dall'Osto, M., Harrison, R.M., Coe, H., Williams, P.I. & Allan, J.D. (2009) Real time chemical characterization of local and regional nitrate aerosols. *Atmospheric Chemistry and Physics*, **9**, 3709–3720.

Davison, B., Hewitt, C.N., O'Dowd, C.D., Lowe, J.A., Smith, M.H., Schwikowski, M., Baltensperger, U. & Harrison, R.M. (1996) Dimethyl sulfide, methane sulfonic acid and physicochemical aerosol properties in Atlantic air from the United Kingdom to Halley Bay. *Journal of Geophysical Research*, **101**, 22,855–22,867.

DeBell, L.J., Gebhart, K.A., Hand, J.L., Malm, W.C., Pitchford, M.L., Schichtel, B.A. & White, W.H. (2006) IMPROVE (Interagency Monitoring of Protected Visual Environments) Spatial and seasonal patterns and temporal variability of haze and its constituents in the United States. Report IV CIRA Report ISSN: 0737-5352-74, Colorado State University, Fort Collins.

Decesari, S., Finessi, E., Rinaldi, M., Paglione, M., Fuzzi, S., Stephanou, E.G., Tziaras, T., Spyros, A., Ceburnis, D., O'Dowd, C., Dall'Osto, M., Harrison, R.M., Allan, J., Coe, H. & Facchini, M.C. (2011) Primary and secondary marine organic aerosols over the North Atlantic Ocean during the MAP experiment. *Journal of Geophysical Research*, **116**, D22210.

DeMott, P.J., Cziczo, D.J., Prenni, A.J., Murphy, D.M., Kreidenweis, S.M., Thomson, D.S., Borys, R. & Rogers, D.C. (2003) Measurements of the concentration and composition of nuclei for cirrus formation. *Proceedings of the National Academy of Sciences of the United States of America*, **100**, 14655–14660.

Denman, K.L., Brasseur, G., Chidthaisong, A., Ciais, P., Cox, P.M., Dickinson, R.E., Hauglustaine, D., Heinze, C., Holland, E., Jacob, D., Lohmann, U., Ramachandran, S., da Silva Dias, P.L., C., W.S. & Zhang, X. (2007) *Couplings between changes in the climate system and biogeochemistry.* In: *Climate Change 2007: The Physical Science Basis. Contribution of Working Group I to the Fourth Assessment Report of the Intergovernmental Panel on Climate Change*, (S. Solomon, D. Qin, M. Manning, Z. Chen, M. Marquis, K.B. Averyt, M. Tignor and H.L. Miller editors). Cambridge University Press, Cambridge, UK, pp. 499–587. .

Dentener, F.J., Carmichael, G.R., Zhang, Y., Lelieveld, J. & Crutzen, P.J. (1996) Role of mineral aerosol as a reactive surface in the global troposphere. *Journal of Geophysical Research*, **101**, 22,869–22,889.

Dentener, F., Kinne, S., Bond, T., Bucher, O., Cofala, J., Generoso, S., Ginoux, P., Gong, S., Hoelzemann, J.J., Ito, A., Marelli, L., Penner, J.E., Putaud, J.-P., Textor, C., Schulz, M., van der Werf, G.R. & Wilson, J. (2006) Emissions of primary aerosol and precursor gases in the years 2000 and 1750 prescribed data-sets for AeroCom. *Atmospheric Chemistry and Physics*, **6**, 4321–4344.

Després, V., Nowoisky, J., Klose, M., Conrad, R., Andreae, M.O. & Pöschl, U. (2007) Characterization of primary biogenic aerosol particles in urban, rural, and high-alpine air by DNA sequence and restriction fragment analysis of ribosomal RNA genes. *Biogeosciences*, **4**, 1127–1141.

Díaz-Hernández, J.L. & Párraga, J. (2008) The nature and tropospheric formation of iberulites: Pinkish mineral microspherulites. *Geochimica et Cosmochimica Acta*, **72**, 3883–3906.

Ding, Z., Yu, Z., Rutter, N.W. & Liu, T. (1994) Towards an orbital time scale for Chinese loess deposits. *Quaternary Science Reviews*, **13**, 39–70.

Duce, R.A. (1986) *The impact of atmospheric nitrogen, phosphorus, and iron species on marine biological productivity.* In: *The role of air-sea exchange in geochemical cycling*, (P. Buat-Ménard editor). D. Reidel, Dordrecht, Holland, pp. 497–529.

Duce, R.A. (1995) *Sources, distributions, and fluxes of mineral aerosols and their relationship to climate.* In: *Aerosol forcing of climate*, (R.J. Charlson & J. Heintzenberg editors). John Wiley & Sons, New York, pp. 43–72.

Duce, R.A., Winchester, J.W. & Nahl, T.W. (1966) Iodine, bromine, and chlorine in winter aerosols and snow from Barrow, Alaska. *Tellus, Series B: Chemical and Physical Meteorology*, **18**, 238–248.

Duce, R.A., Liss, P.S., Merrill, J.T., Atlas, E.L., Buat-Ménard, P., Hicks, B.B., Miller, J.M., Prospero, J.M., Arimoto, R., Church, T.M., Ellis, W., Galloway, J.N., Hansen, L., Jickells, T.D., Knapp, A.H., Reinhardt, K.H., Schneider, B., Soudine, A., Tokos, J.J., Tsunogai, S., Wollast, R. & Zhou, M. (1991) The atmospheric input of trace species to the world ocean. *Global Biogeochemical Cycles*, **5**, 193–254.

Durant, A.J., Bonadonna, C. & Horwell, C.J. (2010) Atmospheric and environmental impacts of volcanic particulates. *Elements*, **6**, 235–240.

Dusek, U., Frank, G.P., Hildebrandt, L., Curtius, J., Schneider, J., Walter, S., Chand, D., Drewnick, F., Hings, S., Jung, D., Borrmann, S. & Andreae, M.O. (2006) Size matters more than chemistry for cloud-nucleating ability of aerosol particles. *Science*, **312**, 1375–1378.

Ebert, M., Inerle-Hof, M. & Weinbruch, S. (2002) Environmental scanning electron microscopy as a new technique to determine the hygroscopic behaviour of individual aerosol particles. *Atmospheric Environment*, **36**, 5909–5916.

Echalar, F., Artaxo, P., Martins, J.V., Yamasoe, M., Gerab, F., Maenhaut, W. & Holben, B. (1998) Long-term monitoring of atmospheric aerosols in teh Amazon Basin: Source identification and apportionment. *Journal of Geophysical Research*, **103**, 31,849.

Elbert, W., Taylor, P.E., Andreae, M.O. & Pöschl, U. (2007) Contribution of fungi to primary biogenic aerosols in the atmosphere: Wet and dry discharged spores, carbohydrates, and inorganic ions. *Atmospheric Chemistry and Physics*, **7**, 4569–4588.

Engelbrecht, J.P. & Derbyshire, E. (2010) Airborne mineral dust. *Elements*, **6**, 241–246.

Erickson, D.J. & Duce, R.A. (1988) On the global flux of atmospheric sea salt. *Journal of Geophysical Research*, **93**, 14,079–14,088.

Facchini, M.C., Mircea, M., Fuzzi, S. & Charlson, R.J. (1999) Cloud albedo enhancement by surface-active organic solutes in growing droplets. *Nature*, **401**, 257–259.

Facchini, M.C., Rinaldi, M., Decesari, S., Carbone, C., Finessi, E., Mircea, M., Fuzzi, S., Ceburnis, D., Flanagan, R., Nilsson, E.D., de Leeuw, G., Martino, M., Woeltjen, J. & O'Dowd, C.D. (2008) Primary submicron marine aerosol dominated by insoluble organic colloids and aggregates. *Geophysical Research Letters*, **35**, L17814.

Falkovich, A.H., Ganor, E., Levin, Z., Formenti, P. & Rudich, Y. (2001) Chemical and mineralogical analysis of individual mineral dust particles. *Journal of Geophysical Research*, **106**, 18029–18036.

Fassi-Fihri, A., Suhre, K. & Rosset, R. (1997) Internal and external mixing in atmospheric aerosols by coagulation: Impact on the optical and hygroscopic properties of the sulphate-soot system. *Atmospheric Environment*, **31**, 1393–1402.

Finlayson-Pitts, B.J. & Pitts, J.N., Jr. (1997) Tropospheric air pollution: Ozone, airborne toxics, polycyclic aromatic hydrocarbons, and particles. *Science*, **276**, 1045–1052.

Fischer, E.V., Hsu, N.C., Jaffe, D.A., Jeong, M.-J. & Gong, S.L. (2009) A decade of dust: Asian dust and springtime aerosol load in the U.S. Pacific Northwest. *Geophysical Research Letters*, **36**, L03821.

Forster, P., Ramaswamy, V., Artaxo, P., Berntsen, T., Betts, R., Fahey, D.W., Haywood, J., Lean, J., Lowe, D.C., Myhre, G., Nganga, J., Prinn, R., Raga, G., Schulz, M. & Van Dorland, R. (2007) *Changes in atmospheric constituents and in radiative forcing*. In: *Climate Change 2007: The Physical Science Basis. Contribution of Working Group I to the Fourth Assessment Report of the Intergovernmental Panel on Climate Change*, (S. Solomon, D. Qin, M. Manning, Z. Chen, M. Marquis, K.B. Averyt, M. Tignor and H.L. Miller, editors). Cambridge University Press, Cambridge, UK, New York, USA, pp. 129–234.

Freney, E.J., Adachi, K. & Buseck, P.R. (2010) Internally mixed atmospheric aerosol particles: Hygroscopic growth and light scattering. *Journal of Geophysical Research*, **115**, D19210.

Friedbacher, G., Grasserbauer, M., Meslmani, Y., Klaus, N. & Higatsberger, M.J. (1995) Investigation of environmental aerosol by atomic force microscopy. *Analytical Chemistry*, **67**, 1749–1754.

Fröhlich-Nowoisky, J., Pickersgill, D.A., Després, V.R. & Pöschl, U. (2009) High diversity of fungi in air particulate matter. *Proceedings of the National Academy of Sciences, USA*, **106**, 12814–12819.

Fuentes, E., Coe, H., Green, D., De Leeuw, G. & McFiggans, G. (2010) On the impacts of phytoplankton-derived organic matter on the properties of the primary marine aerosol – Part 1: Source fluxes. *Atmospheric Chemistry and Physics*, **10**, 9295–9317.

Fuentes, E., Coe, H., Green, D. & McFiggans, G. (2011) On the impacts of phytoplankton-derived organic matter on the properties of the primary marine aerosol – Part 2: Composition, hygroscopicity and cloud condensation activity. *Atmospheric Chemistry and Physics*, **11**, 2585–2602.

Fuller, K.A., Malm, W.C. & Kreidenweis, S.M. (1999) Effects of mixing on extinction by carbonaceous particles. *Journal of Geophysical Research*, **104**, 15,941–15,954.

Gao, Y., Anderson, J.R. & Hua, X. (2007) Dust characteristics over the North Pacific observed through shipboard measurements during the ACE-Asia experiment. *Atmospheric Environment*, **41**, 7907–7922.

Gelencsér, A. (2004) *Carbonaceous Aerosol*. Springer, Berlin, Heidelberg, New York.

Gelencsér, A., Kováts, N., Turóczi, B., Rostási, Á., Hoffer, A., Imre, K., Nyirő-Kósa, I., Csákberényi-Malasics, D., Tóth, Á., Czitrovszky, A., Nagy, A., Nagy, S., Ács, A., Kovács, A., Ferincz, Á., Hartyáni, Z. & Pósfai, M. (2011) The red mud accident in Ajka (Hungary) characterization and potential health effects of fugitive dust. *Environmental Science & Technology*, **45**, 1608–1615.

Georgakopoulos, D.G., Després, V., Fröhlich-Nowoisky, J., Psenner, R., Ariya, P.A., Pósfai, M., Ahern, H.E., Moffett, B.F. & Hill, T.C.J. (2009) Microbiology and atmospheric processes: Biological, physical and chemical characterization of aerosol particles. *Biogeosciences*, **6**, 721–737.

Gieré, R. & Querol, X. (2010) Solid particulate matter in the atmosphere. *Elements*, **6**, 215–222.

Gíslason, S.R., Hassenkam, T., Nedel, S., Bovet, N., Eiriksdottir, E.S., Alfredsson, H.A., Hem, C.P., Balogh, Z.I., Dideriksen, K., Oskarsson, N., Sigfusson, B., Larsen, G. & Stipp, S.L.S. (2011) Characterization of Eyjafjallajökull volcanic ash particles and a protocol for rapid risk assessment. *Proceedings of the National Academy of Sciences, USA*, **108**, 7307–7312.

Glaccum, R.A. & Prospero, J.M. (1980) Saharan aerosols over the tropical North Atlantic – mineralogy. *Marine Geology*, **37**, 295–321.

Gong, S.L., Barrie, L.A. & Lazare, M. (2002) Canadian Aerosol Module (CAM) a size-segregated simulation of atmospheric aerosol processes for climate and air quality models — 2. Global sea-salt aerosol and its budgets. *Journal of Geophysical Research*, **107**, 4779.

Good, N., Topping, D.O., Duplissy, J., Gysel, M., Meyer, N.K., Metzger, A., Turner, S.F., Baltensperger, U., Ristovski, Z., Weingartner, E., Coe, H. & McFiggans, G. (2010) Widening the gap between measurement and modelling of secondary organic aerosol properties? *Atmospheric Chemistry and Physics*, **10**, 2577–2593.

Gras, J.L., Podzimek, J., O'Connor, T.C. & Enderle, K.-H. (2002) Nolan – Pollak type CN counters in the Vienna aerosol workshop. *Atmospheric Research*, **62**, 239–254.

Gregory, P.H. (1973) *The Microbiology of the Atmosphere*. John Wiley & Sons, New York.

Grenfell, J.L., Harrison, R.M., Allen, A.G., Shi, J.P., Penkett, S.A., O'Dowd, C.D., Smith, M.H., Hill, M.K., Robertson, L., Hewitt, C.N., Davison, B., Lewis, A.C., Creasey, D.J., Heard, D.E., Hebestreit, K., Alicke, B. & James, J. (1999) An analysis of rapid increases in condensation nuclei concentrations at a remote coastal site in western Ireland. *Journal of Geophysical Research*, **104**, 13,771–13,780.

Griffin, R.J., Dabdub, D., Cocker III, D.R. & Seinfeld, J.H. (1999) Estimate of global atmospheric organic aerosol from oxidation of biogenic hydrocarbons. *Geophysical Research Letters*, **26**, 2721–2724.

Griffin, D.W., Garrison, V.H., Herman, J.R. & Shinn, E.A. (2001) African desert dust in the Caribbean atmosphere: Microbiology and public health. *Aerobiologia*, **17**, 203–213.

Grobéty, B., Gieré, R., Dietze, V. & Stille, P. (2010) Airborne particles in the urban environment. *Elements*, **6**, 229–234.

Grover, S.N. & Pruppacher, H.R. (1977) A numerical determination of efficiency with which spherical aerosol particles collide with spherical water drops due to inertial impaction and phoretic and electrical forces. *Journal of Atmospheric Science*, **34**, 1655–1663.

Gunthe, S.S., King, S.M., Rose, D., Chen, Q., Roldin, P., Farmer, D.K., Jimenez, J.L., Artaxo, P., Andreae, M.O., Martin, S.T. & Pöschl, U. (2009) Cloud condensation nuclei in pristine tropical rainforest air of Amazonia: size-resolved measurements and modeling of atmospheric aerosol composition and CCN activity. *Atmospheric Chemistry and Physics*, **9**, 7551–7575.

Guthrie, G.D. & Mossman, B.T., editors (1993) *Health Effects of Mineral Dusts*. Reviews in Mineralogy, **28**. Mineralogical Society of America, Washington, D.C.

Hand, J.L., Day, D.E., McMeeking, G.M., Levin, E.J.T., Carrico, C.M., Kreidenweis, S.M., Malm, W.C., Laskin, A. & Desyaterik, Y. (2010) Measured and modeled humidification factors of fresh smoke particles from biomass burning: Role of inorganic constituents. *Atmospheric Chemistry and Physics*, **10**, 6179–6194.

Hand, J.L., Malm, W.C., Laskin, A., Day, D., Lee, T., Wang, C., Carrico, C., Carrillo, J., Cowin, J.P., Collett, J., Jr. & Iedema, M.J. (2005) Optical, physical, and chemical properties of tar balls observed during the Yosemite Aerosol Characterization Study. *Journal of Geophysical Research*, **110**, D21210.

Harrison, R.M. & Sturges, W.T. (1984) Physico-chemical speciation and transformation reactions of particulate atmospheric nitrogen and sulfur compounds. *Atmospheric Environment*, **18**, 1829–1833.

Haywood, J.M. & Ramaswamy, V. (1998) Global sensitivity studies of the direct radiative forcing effect due to anthropogenic sulfate and black carbon aerosols. *Journal of Geophysical Research*, **103**, 6043–6058.

Haywood, J.M. & Shine, K.P. (1995) The effect of anthropogenic sulfate and soot aerosol on the clear sky planetary radiation budget. *Geophysical Research Letters*, **22**, 603–606.

Haywood, J.M., Ramaswamy, V. & Soden, B.J. (1999) Tropospheric aerosol climate forcing in clear-sky satellite observations over the oceans. *Science*, **283**, 1299–1303.

Haywood, J.M., Roberts, D.L., Slingo, A., Edwards, J.M. & Shine, K.P. (1997) General circulation model calculations of the direct radiative forcing by anthropogenic sulfate and fossil-fuel soot aerosol. *Journal of Climate*, **10**, 1562–1577.

Hegg, D.A., Yuen, P.-F. & Larson, T.V. (1992) Modeling the effects of heterogeneous cloud chemistry on the marine particle size distribution. *Journal of Geophysical Research*, **97**, 12,927–12,933.

Heintzenberg, J. (1985) What can we learn from aerosol measurements at baseline stations? *Journal of Atmospheric Chemistry*, **3**, 153–169.

Heintzenberg, J. (1994) The life cycle of the atmospheric aerosol. In: *Topics in Atmospheric and Interstellar Physics and Chemistry* (C.F. Boutron, editor). Les Editions de Physique, Les Ulis, France, pp. 251–270.

Heintzenberg, J. (2009) The SAMUM-1 experiment over Southern Morocco: Overview and introduction. *Tellus, Series B: Chemical and Physical Meteorology*, **61**, 2–11.

Hering, S., Eldering, A. & Seinfeld, J.H. (1997) Bimodal character of accumulation mode aerosol mass distribution in southern California. *Atmospheric Environment*, **31**, 1–11.

Hildemann, L.M., Rogge, W.F., Cass, G.R., Mazurek, M.A. & Simoneit, B.R.T. (1996) Contribution of primary aerosol emissions from vegetation-derived sources to fine particle concentrations in Los Angeles. *Journal of Geophysical Research*, **101**, 19,541–19,549.

Hlavay, J., Polyák, K., Bódog, I., Molnár, A. & Mészáros, E. (1996) Distribution of trace metals in filter-collected aerosol samples. *Fresenius Journal of Analytical Chemistry*, **354**, 227–232.

Hobbs, P.V. (1993) *Aerosol–cloud–climate interactions*. International Geophysics Series, **54**, Academic Press, San Diego, California, USA.

Hobbs, P.V., Reid, J.S., Kotchenruther, R.A., Ferek, R.J. & Weiss, R. (1997) Direct radiative forcing by smoke from biomass burning. *Science*, **275**, 1776–1778.

Hoffer, A., Gelencsér, A., Guyon, P., Kiss, G., Schmid, O., Frank, G.P., Artaxo, P. & Andreae, M.O. (2006) Optical properties of humic-like substances (HULIS) in biomass-burning aerosols. *Atmospheric Chemistry and Physics*, **6**, 3563–3570.

Holben, B.N., Tanré, D., Smirnov, A., Eck, T.F., Slutsker, I., Abuhassan, N., Newcomb, W.W., Schafer, J.S., Chatenet, B., Lavenu, F., Kaufman, Y.J., Vande Castle, J., Setzer, A., Markham, B., Clark, D., Frouin, R., Halthore, R., Karneli, A., O'Neill, N.T., Pietras, C., Pinker, R.T., Voss, K. & Zibordi, G. (2001) An emerging ground-based aerosol climatology: aerosol optical depth from AERONET. *Journal of Geophysical Research*, **106**, 12,067–12,097.

Hong, S., Candelone, J.P., Patterson, C.C. & Boutron, C.F. (1994) Greenland ice evidence of hemispheric lead pollution two millenia ago by Greek and Roman civilization. *Science*, **265**, 1841–1843.

Hopkins, R.J., Desyaterik, Y., Tivanski, A.V., Zaveri, R.A., Berkowitz, C.M., Tyliszczak, T., Gilles, M.K. & Laskin, A. (2008) Chemical speciation of sulfur in marine cloud droplets and particles: Analysis of individual particles from the marine boundary layer over the California current. *Journal of Geophysical Research*, **113**, D04209.

Hoppel, W.A., Frick, G.M., Fitzgerald, J.W. & Larson, R.E. (1994) Marine boundary layer measurements of new particle formation and the effects nonprecipitating clouds have on aerosol size distribution. *Journal of Geophysical Research*, **99**, 14,443–14,459.

Hoppel, W.A., Frick, G.M. & Fitzgerald, J.W. (1996) Deducing droplet concentration and supersaturation in marine boundary layer clouds from surface aerosol measurements. *Journal of Geophysical Research*, **101**, 26,553–26,565.

Horvath, H. (1992) Effects on visibility, weather and climate. In: *Atmospheric Acidity. Sources, Consequences and Abatement* (M. Radojevic and R.M. Harrison, editors). Elsevier, London, pp. 435–466.

Horvath, H. (1998) Influence of atmospheric aerosols upon the global radiation balance. In: *Atmospheric Particles IUPAC Series on Analytical and Physical Chemistry of Environmental Systems* (R.M. Harrison & R. van Grieken, editors). John Wiley, New York, pp. 543–596.

Horváth, L., Mészáros, E., Antal, E. & Simon, A. (1981) On the sulfate, chloride and sodium concentration in maritime air around the Asian continent. *Tellus, Series B: Chemical and Physical Meteorology*, **33**, 382–386.

Huang, W., Tan, J., Kan, H., Zhao, N., Song, W., Song, G., Chen, G., Jiang, L., Jiang, C., Chen, R. & Chen, B. (2009) Visibility, air quality and daily mortality in Shanghai, China. *Science of the Total Environment*, **407**, 3295–3300.

Hudson, P.K., Murphy, D.M., Cziczo, D.J., Thomson, D.S., de Gouw, J.A., Warneke, C., Holloway, J., Jost, H.-J. & Hübler, G. (2004) Biomass-burning particle measurements: Characteristics composition and chemical processing. *Journal of Geophysical Research*, **109**, 1–11.

Hudson, P.K., Young, M.A., Kleiber, P.D. & Grassian, V.H. (2008) Coupled infrared extinction spectra and size distribution measurements for several non-clay components of mineral dust aerosol (quartz, calcite, and dolomite). *Atmospheric Environment*, **42**, 5991–5999.

Huffman, J.A., Treutlein, B. & Pöschl, U. (2010) Fluorescent biological aerosol particle concentrations and size distributions measured with an Ultraviolet Aerodynamic Particle Sizer (UV-APS) in Central Europe. *Atmospheric Chemistry and Physics*, **10**, 3215–3233.

Hughes, L.S., Allen, J.O., Kleeman, M.J., Johnson, R.J., Cass, G.R., Gross, D.S., Gard, E.E., Galli, M.E., Morrical, B.D., Fergenson, D.P., Dienes, T., Noble, C.A., Liu, D.-Y., Silva, P.J. & Prather, K.A. (1999) Size and composition distribution of atmospheric particles in Southern California. *Environmental Science & Technology*, **33**, 3506–3515.

Husar, R.B. & Whitby, K.T. (1973) Growth mechanism and size spectra of photochemical aerosols. *Environmental Science & Technology*, **7**, 241–247.

Husárová, S., Vaitilingom, M., Deguillaume, L., Traikia, M., Vinatier, V., Sancelme, M., Amato, P., Matulová, M. & Delort, A.-M. (2011) Biotransformation of methanol and formaldehyde by bacteria isolated from clouds. Comparison with radical chemistry. *Atmospheric Environment*, **45**, 6093–6102.

Ibald-Mulli, A., Wichmann, H.E., Kreyling, W. & Peters, A. (2002) Epidemiological evidence on health effects of ultrafine particles. *Journal of Aerosol Medicine: Deposition, Clearance, and Effects in the Lung*, **15**, 189–201.

IPCC (1996) *Climate Change 1995: The Science of Climate Change*. Cambridge University Press, Cambridge, UK.

Iwasaka, Y., Minoura, H. & Nagaya, K. (1983) The transport and special scale of Asian dust-storm cloud: A case study of the dust-storm event of April 1979. *Tellus, Series B: Chemical and Physical Meteorology*, **35B**, 189–196.

Iwasaka, Y., Shibata, T., Nagatani, T., Shi, G., Kim, Y.S., Matsuki, A., Trochkine, D., Zhang, D., Yamada, M., Nagatani, M., Nakata, H., Shen, Z., Li, G., Chen, B. & Kawahira, K. (2003) Large depolarization ratio of free tropospheric aerosols over the Taklamakan Desert revealed by lidar measurements: Possible diffusion and transport of dust particles. *Journal of Geophysical Research*, **108**, 8652.

Jacob, D.J. (2000) Heterogeneous chemistry and tropospheric ozone. *Atmospheric Environment*, **34**, 2131–2159.

Jacobson, M.Z. (1999) Isolating nitrated and aromatic aerosols and nitrated aromatic gases as sources of unltraviolet light absorption. *Journal of Geophysical Research*, **104**, 3527–3542.

Jacobson, M.Z. (2001) Strong radiative heating due to the mixing state of black carbon in atmospheric aerosols. *Nature*, **409**, 695–697.

Jacobson, M.Z. (2002a) *Atmospheric Pollution. History, Science, and Regulation*. Cambridge University Press, Cambridge, UK.

Jacobson, M.Z. (2002b) Control of fossil-fuel particulate black carbon and organic matter, possibly the most effective method of slowing global warming. *Journal of Geophysical Research*, **107**, 4410.

Jaenicke, R. (1993) *Tropospheric aerosols*. In: *Aerosol–cloud–climate interactions*, (P.V. Hobbs editor). Academic Press, San Diego, California, USA, pp. 1–31.

Jeong, G.Y. (2008) Bulk and single-particle mineralogy of Asian dust and a comparison with its source soils. *Journal of Geophysical Research*, **113**, D02208.

Jeong, G.Y. & Chun, Y.S. (2006) Nanofiber calcite in Asian dust and its atmospheric roles. *Geophysical Research Letters*, **33**, L24802.

Jickells, T.D., An, Z.S., Andersen, K.K., Baker, A.R., Bergametti, C., Brooks, N., Cao, J.J., Boyd, P.W., Duce, R.A., Hunter, K.A., Kawahata, H., Kubilay, N., LaRoche, J., Liss, P.S., Mahowald, N., Prospero, J.M., Ridgwell, A.J., Tegen, I. & Torres, R. (2005) Global iron connections between desert dust, ocean biogeochemistry, and climate. *Science*, **308**, 67–71.

Jimenez, J.L., Bahreini, R., Cocker III, D.R., Zhuang, H., Varutbangkul, V., Flagan, R.C., Seinfeld, J.H., O'Dowd, C.D. & Hoffman, T. (2003) New particle formation from photooxidation of diiodomethane (CH 2I 2). *Journal of Geophysical Research*, **108**, AAC 5-1–5-25.

Jimenez, J.L., Canagaratna, M.R., Donahue, N.M., Prevot, A.S.H., Zhang, Q., Kroll, J.H., DeCarlo, P.F., Allan, J.D., Coe, H., Ng, N.L., Aiken, A.C., Docherty, K.S., Ulbrich, I.M., Grieshop, A.P., Robinson, A.L.,

Duplissy, J., Smith, J.D., Wilson, K.R., Lanz, V.A., Hueglin, C., Sun, Y.L., Tian, J., Laaksonen, A., Raatikainen, T., Rautiainen, J., Vaattovaara, P., Ehn, M., Kulmala, M., Tomlinson, J.M., Collins, D.R., Cubison, M.J., Dunlea, E.J., Huffman, J.A., Onasch, T.B., Alfarra, M.R., Williams, P.I., Bower, K., Kondo, Y., Schneider, J., Drewnick, F., Borrmann, S., Weimer, S., Demerjian, K., Salcedo, D., Cottrell, L., Griffin, R., Takami, A., Miyoshi, T., Hatakeyama, S., Shimono, A., Sun, J.Y., Zhang, Y.M., Dzepina, K., Kimmel, J.R., Sueper, D., Jayne, J.T., Herndon, S.C., Trimborn, A.M., Williams, L.R., Wood, E.C., Middlebrook, A.M., Kolb, C.E., Baltensperger, U. & Worsnop, D.R. (2009) Evolution of organic aerosols in the atmosphere. *Science*, **326**, 1525–1529.

John, W. (2001) Size Distribution Characteristics of Aerosols. In: *Aerosol Measurement: Principles, Techniques, and Applications*, (P. Kulkarni, P.A. Baron & K. Willeke, editors). Wiley and Sons Inc., Hoboken, New Jersey, USA, DOI: 10.1002/9781118001684.ch4.

Johnson, B., Turnbull, K., Brown, P., Burgess, R., Dorsey, J., Baran, A.J., Webster, H., Haywood, J., Cotton, R., Ulanowski, Z., Hesse, E., Woolley, A. & Rosenberg, P. (2012) In situ observations of volcanic ash clouds from the FAAM aircraft during the eruption of Eyjafjallajokull in 2010. *Journal of Geophysical Research*, **117**, D00U24.

Journet, E., Desboeufs, K.V., Caquineau, S. & Colin, J.-L. (2008) Mineralogy as a critical factor of dust iron solubility. *Geophysical Research Letters*, **35**, L07805.

Junge, C.E. (1963) *Air Chemistry and Radioactivity*. Academic Press, New York.

Kaegi, R. & Holzer, L. (2003) Transfer of a single particle for combined ESEM and TEM analyses. *Atmospheric Environment*, **37**, 4353–4359.

Kalashnikova, O.V. & Sokolik, I.N. (2002) Importance of shapes and compositions of wind-blown dust particles for remote sensing at solar wavelengths. *Geophysical Research Letters*, **29**, 1398.

Kandler, K., Schütz, L., Deutscher, C., Ebert, M., Hofmann, H., Jäckel, S., Jaenicke, R., Knippertz, P., Lieke, K., Massling, A., Petzold, A., Schladitz, A., Weinzierl, B., Wiedensohler, A., Zorn, S. & Weinbruch, S. (2009) Size distribution, mass concentration, chemical and mineralogical composition and derived optical parameters of the boundary layer aerosol at Tinfou, Morocco, during SAMUM 2006. *Tellus, Series B: Chemical and Physical Meteorology*, **61**, 32–50.

Kasibhatla, P., Chameides, W.L. & John, J.S. (1997) A three-dimensional global model investigation of seasonal variations in the atmospheric burden of anthropogenic sulfate aerosols. *Journal of Geophysical Research*, **102**, 3737–3759.

Katrinak, K.A., Rez, P. & Buseck, P.R. (1992) Structural variations in individual carbonaceous particles from an urban aerosol. *Environmental Science & Technology*, **26**, 1967–1976.

Kaufman, Y.J., Tanré, D. & Boucher, O. (2002) A satellite view of aerosols in the climate. *Nature*, **419**, 215–223.

Kaufman, Y.J., Boucher, O., Tanré, D., Chin, M., Remer, L.A. & Takemura, T. (2005) Aerosol anthropogenic component estimated from satellite data. *Geophysical Research Letters*, **32**, L17804.

Keene, W.C., Maring, H., Maben, J.R., Kieber, D.J., Pszenny, A.A.P., Dahl, E.E., Izaguirre, M.A., Davis, A.J., Long, M.S., Zhou, X., Smoydzin, L. & Sander, R. (2007) Chemical and physical characteristics of nascent aerosols produced by bursting bubbles at a model air–sea interface. *Journal of Geophysical Research*, **112**, D21202.

Keppler, H. (1999) Experimental evidence for the source of excess sulfur in explosive volcanic eruptions. *Science*, **284**, 1652–1654.

Kerminen, V.-M., Wexler, A.S. & Potukuchi, S. (1997) Growth of freshly nucleated particles in the troposphere: Roles of NH_3, H_2SO_4, HNO_3, and HCl. *Journal of Geophysical Research*, **102**, 3715–3724.

Kiehl, J.T. & Briegleb, B.P. (1993) The relative role of sulfate aerosols and greenhouse gases in climate forcing. *Science*, **260**, 311–314.

King, M.D., Thompson, K.C., Ward, A.D., Pfrang, C. & Hughes, B.R. (2007) Oxidation of biogenic and water-soluble compounds in aqueous and organic aerosol droplets by ozone: A kinetic and product analysis approach using laser Raman tweezers. *Faraday Discussions*, **137**, 173–192.

Kirchstetter, T.W., Novakov, T. & Hobbs, P.V. (2004) Evidence that the spectral dependence of light absorption by aerosols is affected by organic carbon. *Journal of Geophysical Research*, **109**, D21208.

Kirkby, J., Curtius, J., Almeida, J., Dunne, E., Duplissy, J., Ehrhart, S., Franchin, A., Gagné, S., Ickes, L., Kürten, A., Kupc, A., Metzger, A., Riccobono, F., Rondo, L., Schobesberger, S., Tsagkogeorgas, G., Wimmer, D., Amorim, A., Bianchi, F., Breitenlechner, M., David, A., Dommen, J., Downard, A., Ehn, M., Flagan, R.C., Haider, S., Hansel, A., Hauser, D., Jud, W., Junninen, H., Kreissl, F., Kvashin, A., Laaksonen, A., Lehtipalo, K., Lima, J., Lovejoy, E.R., Makhmutov, V., Mathot, S., Mikkilä, J., Minginette, P., Mogo, S., Nieminen, T., Onnela, A., Pereira, P., Petäjä, T., Schnitzhofer, R., Seinfeld, J.H., Sipilä, M., Stozhkov, Y., Stratmann, F., Tomé, A., Vanhanen, J., Viisanen, Y., Vrtala, A., Wagner, P.E., Walther, H., Weingartner, E., Wex, H., Winkler, P.M., Carslaw, K.S., Worsnop, D.R., Baltensperger, U. & Kulmala, M. (2011) Role of sulphuric acid, ammonia and galactic cosmic rays in atmospheric aerosol nucleation. *Nature*, **476**, 429–435.

Kis, V.K., Pósfai, M. & Lábár, J.L. (2006) Nanostructure of atmospheric soot particles. *Atmospheric Environment*, **40**, 5533–5542.

Kiss, G., Tombácz, E. & Hansson, H.-C. (2005) Surface tension effects of humic-like substances in the aqueous extract of tropospheric fine aerosol. *Journal of Atmospheric Chemistry*, **50**, 279–294.

Koehler, K.A., Kreidenweis, S.M., DeMott, P.J., Petters, M.D., Prenni, A.J. & Carrico, C.M. (2009) Hygroscopicity and cloud droplet activation of mineral dust aerosol. *Geophysical Research Letters*, **36**, L08805.

Kok, J.F. (2010) A scaling theory for the size distribution of emitted dust aerosols suggests climate models underestimate the size of the global dust cycle. *Proceedings of the National Academy of Sciences, USA*, **108**, 1016–1021.

Kok, J.F. & Renno, N.O. (2008) Electrostatics in wind-blown sand. *Physical Review Letters*, **100**, 014501.

Köllensperger, G., Friedbacher, G., Grasserbauer, M. & Dorffner, L. (1997) Investigation of aerosol particles by atomic force microscopy. *Fresenius Journal of Analytical Chemistry*, **358**, 268–273.

Köllensperger, G., Friedbacher, G., Kotzick, R., Niessner, R. & Grasserbauer, M. (1999) In-situ atomic force microscopy investigation of aerosols exposed to different humidities. *Fresenius Journal of Analytical Chemistry*, **364**, 296–304.

Kreidenweis, S.M., McInnes, L.M. & Brechtel, F.J. (1998) Observations of aerosol volatility and elemental composition at Macquarie Island during the First Aerosol Characterization Experiment (ACE 1). *Journal of Geophysical Research*, **103**, 16,511–16,524.

Krivácsy, Z., Gelencsér, A., Kiss, G., Mészáros, E., Molnár, A., Hoffer, A., Mészáros, T., Sárvári, Z., Temesi, D., Varga, B., Baltensperger, U., Nyeki, S. & Weingartner, E. (2001) Study on the chemical character of water soluble organic compounds in fine atmospheric aerosol at the Jungfraujoch. *Journal of Atmospheric Chemistry*, **39**, 235–259.

Krueger, B.J., Grassian, V.H., Iedema, M.J., Cowin, J.P. & Laskin, A. (2003) Probing heterogeneous chemistry of individual atmospheric particles using scanning electron microscopy and energy-dispersive X-ray analysis. *Analytical Chemistry*, **75**, 5170–5179.

Krueger, B.J., Grassian, V.H., Cowin, J.P. & Laskin, A. (2004) Heterogeneous chemistry of individual mineral dust particles from different dust source regions: The importance of particle mineralogy. *Atmospheric Environment*, **38**, 6253–6261.

Kulmala, M. & Kerminen, V.M. (2008) On the formation and growth of atmospheric nanoparticles. *Atmospheric Research*, **90**, 132–150.

Kulmala, M., Vehkamäki, H., Petäjä, T., Dal Maso, M., Lauri, A., Kerminen, V.-M., Birmili, W. & McMurry, P.H. (2004) Formation and growth rates of ultrafine atmospheric particles: A review of observations. *Journal of Aerosol Science*, **35**, 143–176.

Kulmala, M., Riipinen, I., Sipilä, M., Manninen, H.E., Petäjä, T., Junninen, H., Dal Maso, M., Mordas, G., Mirme, A., Vana, M., Hirsikko, A., Laakso, L., Harrison, R.M., Hanson, I., Leung, C., Lehtinen, K.E.J. & Kerminen, V.M. (2007) Toward direct measurement of atmospheric nucleation. *Science*, **318**, 89–92.

Kutsuna, S. & Ibusuki, T. (1994) Fourier transform infrared measurement of the formation of nitrogen compounds on sodium chloride particles exposed to the ambient air in the Arctic. *Journal of Geophysical Research*, **99**, 25,479–25,488.

Lacis, A.A. & Mishchenko, M.I. (1995) Climate forcing, climate sensitivity, and climate response: A radiative modeling perspective on atmospheric aerosols. In: *Aerosol Forcing of Climate* (R.J. Charlson and J. Heintzenberg editor) John Wiley & Sons, New York, pp. 11–42.

Lahaye, J. (1992) Particulate carbon from the gas phase. *Carbon*, **30**, 309–314.

Langner, J. & Rodhe, H. (1991) A global three-dimensional model of the tropospheric sulfur cycle. *Journal of Atmospheric Chemistry*, **13**, 225–264.

Laskin, A., Cowin, J.P. & Iedema, M.J. (2006) Analysis of individual environmental particles using modern methods of electron microscopy and X-ray microanalysis. *Journal of Electron Spectroscopy and Related Phenomena*, **150**, 260–274.

Laskin, A., Wietsma, T.W., Krueger, B.J. & Grassian, V.H. (2005) Heterogeneous chemistry of individual mineral dust particles with nitric acid: A combined CCSEM/EDX, ESEM, and ICP-MS study. *Journal of Geophysical Research*, **110**, 1–15.

Latham, J., Rasch, P., Chen, C.-C., Kettles, L., Gadian, A., Gettelman, A., Morrison, H., Bower, K. & Choularton, T. (2008) Global temperature stabilization via controlled albedo enhancement of low-level maritime clouds. *Philosophical Transactions of the Royal Society London*, **A366**, 3969–3987.

Lazar, A.C., Reilly, P.T.A., Whitten, W.B. & Ramsey, J.M. (1999) Real-time surface analysis of individual airborne environmental particles. *Environmental Science & Technology*, **33**, 3393–4001.

Legrand, M. (1995) Atmospheric chemistry changes versus past climate inferred from polar ice cores. In: *Aerosol Forcing of Climate* (R.J. Charlson and J. Heintzenberg, editors). John Wiley & Sons, New York, pp. 123–151.

Leinen, M., Prospero, J.M., Arnold, E. & Blank, M. (1994) Mineralogy of aeolian dust reaching the North Pacific Ocean 1. Sampling and analysis. *Journal of Geophysical Research*, **99**, 21,017–21,023.

Levin, Z., Teller, A., Ganor, E. & Yin, Y. (2005) On the interactions of mineral dust, sea-salt particles, and clouds: A measurement and modeling study from the Mediterranean Israeli Dust Experiment campaign. *Journal of Geophysical Research*, **110**, D20202.

Levine, J.S. (1991) *Global Biomass Burning*. MIT Press, Cambridge, Massachusetts, USA.

Lewis, E.R. & Schwartz, S.E. (2004) *Sea Salt Aerosol Production: Mechanisms, Methods, Measurements and Models — a Critical Review*. Geophysical Monograph, **152**, American Geophysical Union, Washington D.C.

Li, J., Anderson, J.R. & Buseck, P.R. (2003a) TEM study of aerosol particles from clean and polluted marine boundary layers over the North Atlantic. *Journal of Geophysical Research*, **108**, 4189.

Li, J., Pósfai, M., Hobbs, P.V. & Buseck, P.R. (2003b) Individual aerosol particles from biomass burning in southern Africa: 2. Compositions and aging of inorganic particles. *Journal of Geophysical Research*, **108**, 8484.

Li, X., Maring, H., Savoie, D., Voss, K. & Prospero, J.M. (1996) Dominance of mineral dust in aerosol light-scattering in the North Atlantic trade winds. *Nature*, **380**, 416–419.

Liao, H. & Seinfeld, J.H. (1998) Radiative forcing by mineral dust aerosols: Sensitivity to key variables. *Journal of Geophysical Research*, **103**, 31,637–31,645.

Lindemann, J. & Upper, C.D. (1985) Aerial dispersal of epiphytic bacteria over bean plants. *Applied and Environmental Microbiology*, **50**, 1229–1232.

Liousse, C., Cachier, C. & Jennings, S.G. (1993) Optical and thermal measurements of black carbon aerosol content in different environments: Variation of the specific attenuation cross-section, sigma (σ). *Atmospheric Environment*, **27**, 1203–1211.

Liousse, C., Penner, J.E., Chuang, C., Walton, J.J., Eddleman, H. & Cachier, H. (1996) A global three-dimensional model study of carbonaceous aerosols. *Journal of Geophysical Research*, **101**, 19,411–19,432.

Liu, X., Penner, J.E. & Herzog, M. (2005) Global modeling of aerosol dynamics: Model description, evaluation, and interactions between sulfate and nonsulfate aerosols. *Journal of Geophysical Research*, **110**, 1–37.

Lodge, J.P. (1986) Chemical identification of individual particles. In: *Physical and Chemical Characterization of Individual Airborne Particles* (K.R. Spurny, editor). Wiley & Sons, New York, pp. 116–126.

Lorenzo, R., Kaegi, R., Gehrig, R. & Grobéty, B. (2006) Particle emissions of a railway line determined by detailed single particle analysis. *Atmospheric Environment*, **40**, 7831–7841.

Lu, H. & Shao, Y. (1999) A new model for dust emission by saltation bombardment. *Journal of Geophysical Research*, **104**, 16,827.

Maher, B.A. (2009) Rain and dust magnetic records of climate and pollution. *Elements*, **5**, 229–234.

Mahowald, N.M. & Luo, C. (2003) A less dusty future? *Geophysical Research Letters*, **30**, 1903.

Mahowald, N.M., Baker, A.R., Bergametti, G., Brooks, N., Duce, R.A., Jickells, T.D., Kubilay, N., Prospero, J.M. & Tegen, I. (2005) Atmospheric global dust cycle and iron inputs to the ocean. *Global Biogeochemical Cycles*, **19**, GB4025.

Mahowald, N.M., Muhs, D.R., Levis, S., Rasch, P.J., Yoshioka, M., Zender, C.S. & Luo, C. (2006) Change in atmospheric mineral aerosols in response to climate: Last glacial period, preindustrial, modern, and doubled carbon dioxide climates,. *Journal of Geophysical Research*, **111**, D10202.

Mäkelä, J.M., Hoffmann, T., Holzke, C., Vakeva, M., Suni, T., Mattila, T., Aalto, P.P., Tapper, U., Kauppinen, E.I. & O'Dowd, C.D. (2002) Biogenic iodine emissions and identification of end-products in coastal ultrafine particles during nucleation bursts. *Journal of Geophysical Research – Atmospheres*, **107**, 8110.

Mamane, Y. & de Pena, R.G. (1978) A quantitative method for the detection of individual submicrometer size sulfate particles. *Atmospheric Environment*, **12**, 69–82.

Markowitz, A. & Van Grieken, R. (1991) X-ray methods. In: *Instrumental Analysis of Pollutants* (C.N. Hewitt, editor). Elsevier, London, pp. 173–242.

Maron, P.-A., Lejon, D.P.H., Carvalho, E., Bizet, K., Lemanceau, P., Ranjard, L. & Mougel, C. (2005) Assessing genetic structure and diversity of airborne bacterial communities by DNA fingerprinting and 16S rDNA clone library. *Atmospheric Environment*, **39**, 3687–3695.

Mårtensson, E.M., Nilsson, E.D., de Leeuw, G., Cohen, L.H. & Hansson, H.-C. (2003) Laboratory simulations and parameterization of the primary marine aerosol production. *Journal of Geophysical Research*, **108**, AAC 15–11 AAC 15–12.

Marti, J.J., Jefferson, A., Cai, X.P., Richert, C., McMurry, P.H. & Eisele, F. (1997) H_2SO_4 vapor pressure of sulfuric acid and ammonium sulfate solutions. *Journal of Geophysical Research*, **102**, 3725–3735.

Marticorena, B., Bergametti, G. & Aumont, B. (1997) Modeling the atmospheric dust cycle 2. Simulation of Saharan dust sources. *Journal of Geophysical Research*, **102**, 4387–4404.

Martin, J.H. & Gordon, R.M. (1988) Northeast Pacific iron distributions in relation to phytoplankton productivity. *Deep-Sea Research*, **35**, 177–196.

Martin, S.T. (2000) Phase transitions of aqueous atmospheric particles. *Chemical Reviews*, **100**, 3403–3453.

Martin, S.T., Hung, H.M., Park, R.J., Jacob, D.J., Spurr, R.J.D., Chance, K.V. & Chin, M. (2004) Effects of the physical state of tropospheric ammonium-sulfate-nitrate particles on global aerosol direct radiative forcing. *Atmospheric Chemistry and Physics*, **4**, 183–214.

Martin, S.T., Andreae, M.O., Artaxo, P., Baumgartner, D., Chen, Q., Goldstein, A.H., Guenther, A., Heald, C.L., Mayol- Bracero, O.L., McMurry, P.H., Pauliquevis, T., Pöschl, U., Prather, K.A., Roberts, G.C., Saleska, S.R., Dias, M.A.S., Spracklen, D.V., Swietlicki, E. & Trebs, I. (2010) Sources and properties of Amazonian aerosol particles. *Reviews in Geophysics*, **48**, RG2002.

Martins, J.V., Artaxo, P., Liousse, C., Reid, J.S., Hobbs, P.V. & Kaufman, Y.J. (1998) Effects of black carbon content, particle size, and mixing on light absorption by aerosols from biomass burning in Brazil. *Journal of Geophysical Research – Atmospheres*, **103**, 32,041–032,050.

Matthias-Maser, S. & Jaenicke, R. (1994) Examination of atmospheric bioaerosol particles with radii >0.2 μm. *Journal of Aerosol Science*, **25**, 1605–1613.

Maynard, A.D. (1995) The application of electron-energy-loss spectroscopy to the analysis of ultrafine aerosol-particles. *Journal of Aerosol Science*, **26**, 757–777.

Mazurek, M., Masonjones, M.C., Masonjones, H.D., Salmon, L.G., Cass, G.R., Hallock, K.A. & Leach, M. (1997) Visibility-reducing organic aerosols in the vicinity of Grand Canyon National Park: Properties observed by high resolution gas chromatography. *Journal of Geophysical Research*, **102**, 3779–3793.

McConnell, C.L., Highwood, E.J., Coe, H., Formenti, P., Anderson, B., Osborne, S., Nava, S., Desboeufs, K., Chen, G. & Harrison, M.A.J. (2008) Seasonal variations of the physical and optical characteristics of

Saharan dust: Results from the Dust Outflow and Deposition to the Ocean (DODO) experiment. *Journal of Geophysical Research*, **113**, D14S05.

McFiggans, G., Cox, R.A., Mössinger, J.C., Allan, B.J. & Plane, J.M.C. (2002) Active chlorine release from marine aerosols: Roles for reactive iodine and nitrogen species. *Journal of Geophysical Research*, **107**, 10-11–10-13.

McFiggans, G., Coe, H., Burgess, R., Allan, J., Cubison, M., Alfarra, M.R., Saunders, R., Saiz-Lopez, A., Plane, J.M.C., Wevill, D.J., Carpenter, L.J., Rickard, A.R. & Monks, P.S. (2004) Direct evidence for coastal iodine particles from Laminaria macroalgae – Linkage to emissions of molecular iodine. *Atmospheric Chemistry and Physics*, **4**, 701–713.

McFiggans, G., Bale, C.S.E., Ball, S.M., Beames, J.M., Bloss, W.J., Carpenter, L.J., Dorsey, J., Dunk, R., Flynn, M.J., Furneaux, K.L., Gallagher, M.W., Heard, D.E., Hollingsworth, A.M., Hornsby, K., Ingham, T., Jones, C.E., Jones, R.L., Kramer, L.J., Langridge, J.M., Leblanc, C., LeCrane, J.P., Lee, J.D., Leigh, R.J., Longley, I., Mahajan, A.S., Monks, P.S., Oetjen, H., Orr-Ewing, A.J., Plane, J.M.C., Potin, P., Shillings, A.J.L., Thomas, F., Von Glasow, R., Wada, R., Whalley, L.K. & Whitehead, J.D. (2010) Iodine-mediated coastal particle formation: An overview of the Reactive Halogens in the Marine boundary layer (RHaMBLe) Roscoff coastal study. *Atmospheric Chemistry and Physics*, **10**, 2975–2999.

McInnes, L., Bergin, M., Ogren, J. & Schwartz, S. (1998) Apportionment of light scattering and hygriscopic growth to aerosol composition. *Geophysical Research Letters*, **25**, 513–516.

McInnes, L.M., Quinn, P.K., Covert, D.S. & Anderson, T.L. (1996) Gravimetric analysis, ionic composition, and associated water mass of the marine aerosol. *Atmospheric Environment*, **30**, 869–884.

Merikanto, J., Napari, I., Vehkamäki, H., Anttila, T. & Kulmala, M. (2007) New parameterization of sulfuric acid-ammonia-water ternary nucleation rates at tropospheric conditions. *Journal of Geophysical Research*, **112**, D15207.

Merrill, J., Arnold, E., Leinen, M. & Weaver, C. (1994) Mineralogy of aeolian dust reaching the North Pacific Ocean 2. Relationship of mineral assemblages to atmospheric transport patterns. *Journal of Geophysical Research*, **99**, 21,025–21,032.

Meskhidze, N. & Nenes, A. (2006) Phytoplankton and cloudiness in the southern ocean. *Science*, **314**, 1419–1423.

Mészáros, A. (1984) The number concentration and size distribution of the soot particles in the 0.02–0.5 μm radius range at sites of different pollution levels. *Science of the Total Environment*, **36**, 283–288.

Mészáros, A. & Vissy, K. (1974) Concentration, size distribution and chemical nature of atmospheric aerosol particles in remote oceanic areas. *Journal of Aerosol Science*, **5**, 101–109.

Mészáros, E. (1968) On the size distribution of water soluble particles in the atmosphere. *Tellus, Series B: Chemical and Physical Meteorology*, **20**, 443–448.

Mészáros, E. (1973) Evidence of the role of indirect photochemical processes in the formation of atmospheric sulphate particulate. *Journal of Aerosol Science*, **4**, 429–434.

Mészáros, E. (1992) Occurrence of atmospheric acidity. In: *Atmospheric Acidity, Sources, Consequences and Abatement* (M. Radojevic and R.M. Harrison, editors). Elsevier Applied Science, London, New York, pp. 1–37.

Mészáros, E. (1999) *Fundamentals of Atmospheric Aerosol Chemistry*. Akadémiai Kiadó, Budapest.

Mészáros, E., Barcza, T., Gelencsér, A., Hlavay, J., Kiss, G., Krivácsy, Z., Molnár, A. & Polyák, K. (1997) Size distributions of inorganic and organic species in the atmospheric aerosol in Hungary. *Journal of Aerosol Science*, **28**, 1163–1175.

Michalsky, J.J., Pearson, E.W. & LeBaron, B.A. (1990) An assessment of the impact of volcanic eruptions on the Northern Hemisphere's aerosol burden during the last decade. *Journal of Geophysical Research*, **95**, 5677–5688.

Middlebrook, A.M., Murphy, D.M. & Thomson, D.S. (1998) Observations of organic material in individual marine particles at Cape Grim during the First Aerosol Characterization Experiment (ACE 1). *Journal of Geophysical Research*, **103**, 16,475–16,483.

Mikami, M., Shi, G.Y., Uno, I., Yabuki, S., Iwasaka, Y., Yasui, M., Aoki, T., Tanaka, T.Y., Kurosaki, Y., Masuda, K., Uchiyama, A., Matsuki, A., Sakai, T., Takemi, T., Nakawo, M., Seino, N., Ishizuka, M.,

Satake, S., Fujita, K., Hara, Y., Kai, K., Kanayama, S., Hayashi, M., Du, M., Kanai, Y., Yamada, Y., Zhang, X.Y., Shen, Z., Zhou, H., Abe, O., Nagai, T., Tsutsumi, Y., Chiba, M. & Suzuki, J. (2006) Aeolian dust experiment on climate impact: An overview of Japan–China joint project ADEC. *Global and Planetary Change*, **52**, 142–172.

Milford, J.B. & Davidson, C.I. (1985) The size of particulate trace elements in the atmosphere – a review. *Journal of the Air Pollution Control Association*, **37**, 1249–1260.

Millero, F.J. & Sohn, M.L. (1992) *Chemical Oceanography*. CRC Press, Ann Arbor, Michigan, USA.

Mirabel, P. & Katz, J.L. (1974) Binary homogeneous nucleation as a mechanism for the formation of areosols. *Journal of Chemical Physics*, **60**, 1138–1144.

Mishchenko, M.I. (2008) Multiple scattering by particles embedded in an absorbing medium. 2. Radiative transfer equation. *Journal of Quantitative Spectroscopy & Radiative Transfer*, **109**, 2386–2390.

Molnár, A., Mészáros, E., Bozó, L., Borbély-Kiss, I., Koltay, E. & Szabó, G. (1993) Elemental composition of atmospheric aerosol particles under different conditions in Hungary. *Atmospheric Environment*, **27A**, 2457–2461.

Moosmüller, H., Arnott, W.P., Rogers, C.F., Chow, J.C., Frazier, C.A., Sherman, L.E. & Dietrich, D.L. (1998) Photoacoustic and filter measurements related to aerosol light absorption during the Northern Front Range Air Quality Study (Colorado 1996/1997). *Journal of Geophysical Research*, **103**, 28,149–28,157.

Morgan, W.T., Allan, J.D., Bower, K.N., Esselborn, M., Harris, B., Henzing, J.S., Highwood, E.J., Kiendler-Scharr, A., McMeeking, G.R., Mensah, A.A., Northway, M.J., Osborne, S., Williams, P.I., Krejci, R. & Coe, H. (2010) Enhancement of the aerosol direct radiative effect by semi-volatile aerosol components: Airborne measurements in North-Western Europe. *Atmospheric Chemistry and Physics*, **10**, 8151–8171.

Morris, C.E., Sands, D.C., Bardin, M., Jaenicke, R., Vogel, B., Leyronas, C., Ariya, P.A. & Psenner, R. (2008) Microbiology and atmospheric processes: an upcoming era of research on bio-meteorology. *Biogeosciences Discussions*, **5**, 191–212.

Möhler, O., Benz, S., Saathoff, H., Schnaiter, M., Wagner, R., Schneider, J., Walter, S., Ebert, V. & Wagner, S. (2008) The effect of organic coating on the heterogeneous ice nucleation efficiency of mineral dust aerosols. *Environmental Research Letters*, **3**, 025007.

Möhler, O., DeMott, P.J., Vali, G. & Levin, Z. (2007) Microbiology and atmospheric processes: The role of biological particles in cloud physics. *Biogeosciences*, **4**, 1059-1071.

Möller, D. (1995) *Sulfate aerosols and their atmospheric precursors*. In: *Aerosol Forcing of Climate*, (R.J. Charlson and J. Heintzenberg, editors). John Wiley & Sons, New York, pp. 73–90.

Möller, D. (2010) *Chemistry of the Climate System*. De Gruyer, Berlin.

Mulik, J.D. & Sawicki, E. (1979) Ion chromatography. *Environmental Science & Technology*, **13**, 804–809.

Murphy, D.M., Anderson, J.R., Quinn, P.K., McInnes, L.M., Brechtel, F.J., Kreidenwels, S.M., Middlebrook, A.M., Pósfal, M., Thomson, D.S. & Buseck, P.R. (1998a) Influence of sea-salt on aerosol radiative properties in the Southern Ocean marine boundary layer. *Nature*, **392**, 62–65.

Murphy, D.M., Thomson, D.S. & Mahoney, M.J. (1998b) In situ measurements of organics, meteoritic material, mercury, and other elements in aerosols at 5 to 19 kilometers. *Science*, **282**, 1664–1669.

Murphy, D.M., Cziczo, D.J., Froyd, K.D., Hudson, P.K., Matthew, B.M., Middlebrook, A.M., Peltier, R.E., Sullivan, A., Thomson, D.S. & Weber, R.J. (2006) Single-particle mass spectrometry of tropospheric aerosol particles. *Journal of Geophysical Research*, **111**, D23S32.

Murphy, D.M., Solomon, S., Portmann, R.W., Rosenlof, K.H., Forster, P.M. & Wong, T. (2009) An observationall based energy balance for the Earth since 1950. *Journal of Geophysical Research*, **114**, D17107.

Myhre, G., Berglen, T.F., Johnsrud, M., Hoyle, C.R., Berntsen, T.K., Christopher, S.A., Fahey, D.W., Isaksen, I.S.A., Jones, T.A., Kahn, R.A., Loeb, N., Quinn, P., Remer, L., Schwarz, J.P. & Yttri, K.E. (2009) Modelled radiative forcing of the direct aerosol effect with multi-observation evaluation. *Atmospheric Chemistry and Physics*, **9**, 1365–1392.

Niemi, J.V., Tervahattu, H., Vehkamäki, H., Martikainen, J., Laakso, L., Kulmala, M., Aarnio, P., Koskentalo, T., Sillanpää, M. & Makkonen, U. (2005) Characterization of aerosol particle episodes in Finland caused by wildfires in Eastern Europe. *Atmospheric Chemistry and Physics*, **5**, 2299–2310.

Niemi, J.V., Saarikoski, S., Tervahattu, H., Mäkelä, T., Hillamo, R., Vehkamäki, H., Sogacheva, L. & Kulmala, M. (2006) Changes in background aerosol composition in Finland during polluted and clean periods studied by TEM/EDX individual particle analysis. *Atmospheric Chemistry and Physics*, **6**, 5049–5066.

Nilsson, E.D., Martensson, E.M., Van Ekeren, J.S., de Leeuw, G., Moerman, M. & O'Dowd, C. (2007) Primary marine aerosol emissions: size resolved eddy covariance measurements with estimates of the sea salt and organic carbon fractions. *Atmospheric Chemistry and Physics Discussions*, **7**, 13,345–313,400.

Noble, C.A. & Prather, K.A. (1996) Real-time measurement of correlated size and composition profiles of individual atmospheric aerosol particles. *Environmental Science & Technology*, **30**, 2667–2680.

Novakov, T. & Penner, J.E. (1993) Large contribution of organic aerosols to cloud-condensation-nuclei concentrations. *Nature*, **365**, 823–826.

Novakov, T., Hegg, D.A. & Hobbs, P.V. (1997) Airborne measurements of carbonaceous aerosols on the East Coast of the United States. *Journal of Geophysical Research*, **102**, 30,023–30,030.

Nriagu, J.O. & Davidson, C.I. (1986) *Toxic Metals in the Atmosphere*. Wiley & Sons, New York.

Nunes, T.V. & Pio, C.A. (1993) Carbonaceous aerosols in industrial and coastal atmospheres. *Atmospheric Environment*, **27**, 1339–1346.

O'Dowd, C.D., Smith, M.H., Consterdine, I.E. & Lowe, J.A. (1997) Marine aerosol, sea-salt, and the marine sulfur cycle: A short review. *Atmospheric Environment*, **31**, 73–80.

O'Dowd, C.D., McFiggans, G., Creasey, D., Pirjola, L., Hoell, C., Smith, M., Allen, B., Plane, J., Heard, D., Lee, J., Pilling, M. & Kulmala, M. (1999) On the photochemical production of new particles in the coastal boundary layer. *Geophysical Research Letters*, **26**, 1707.

O'Dowd, C.D., Jimenez, J.L., Bahreini, R., Flagan, R.C., Seinfeld, J.H., Hämerl, K., Pirjola, L., Kulmala, M., Jennings, S.G. & Hoffmann, T. (2002) Marine aerosol formation from biogenic iodine emissions. *Nature*, **417**, 632–636.

O'Dowd, C.D., Facchini, M.C., Cavalli, F., Ceburnis, D., Mircea, M., Decesari, S., Fuzzi, S., Yoon, Y.J. & Putaud, J.-P. (2004) Biogenically driven organic contribution to marine aerosol. *Nature*, **431**, 676–680.

Odum, J.R., Jungkamp, T.P.W., Griffin, R.J., Flagan, R.C. & Seinfeld, J.H. (1997) The atmospheric aerosol-forming potential of whole gasoline vapor. *Science*, **276**, 96–99.

Okada, K. (1985) Number-size distribution and formation process of submicrometer sulfate-containing particles in the urban atmosphere of Nagoya. *Atmospheric Environment*, **19**, 743–757.

Okada, K. & Kai, K. (2004) Atmospheric mineral particles collected at Qira in the Taklamakan Desert, China. *Atmospheric Environment*, **38**, 6927–6935.

Ondov, J.M. & Biermann, A.-H. (1980) Physical and Chemical Characterization of Aerosol Emissions from Coal-fired Power Plants. In: *Environmental and Climate Impact of Coal Utilization. Aerosol Emissions from Coal Plants* (J.J. Singh and A. Deepak, editors). Academic Press, New York, pp. 1–17.

Osán, J., Szalóki, I., Ro, C.-U. & Van Grieken, R. (2000) Light element analysis of individual microparticles using thin-window EPMA. *Mikrochimica Acta*, **132**, 349–355.

Pant, A., Fok, A., Parsons, M.T., Mak, J. & Bertram, A.K. (2004) Deliquescence and crystallization of ammonium sulfate-glutaric acid and sodium chloride glutaric acid particles *Geophysical Research Letters*, **31**, L12111.

Papayannis, A., Amiridis, V., Mona, L., Tsaknakis, G., Balis, D., Bösenberg, J., Chaikovski, A., De Tomasi, F., Grigorov, I., Mattis, I., Mitev, V., Müller, D., Nickovic, S., Pérez, C., Pietruczuk, A., Pisani, G., Ravetta, F., Rizi, V., Sicard, M., Trickl, T., Wiegner, M., Gerding, M., Mamouri, R.E., D'Amico, G. & Pappalardo, G. (2008) Systematic lidar observations of Saharan dust over Europe in the frame of EARLINET (2000–2002). *Journal of Geophysical Research*, **113**, D10204.

Parrington, J.R., Zoller, W.H. & Aras, N.K. (1983) Asian dust: seasonal transport to the Hawaiian Islands. *Science*, **220**, 195–197.

Patashnick, H. & Rupprecht, E.G. (1991) Continuous PM-10 measurement susing the tapered element oscillating microbalance. *Journal of the Air and Waste Management Association*, **41**, 1079–1083.

Patterson, C.C. & Settle, D.M. (1987) Review of data on eolian fluxes of industrial and natural lead to the lands and seas in remote regions ona global scale. *Marine Chemistry*, **22**, 137–162.

Penner, J.E. (1995) *Carbonaceous aerosols influencing atmospheric radiation: Black and organic carbon*. In: *Aerosol forcing of climate*, (R.J. Charlson and J. Heintzenberg editors). John Wiley & Sons, New York, pp. 91–109.

Penner, J.E., Eddleman, H. & Novakov, T. (1993) Towards the development of a global inventory for black carbon emissions. *Atmospheric Environment*, **27**, 1277–1295.

Perry, K.D. & Hobbs, P.V. (1994) Further evidence for particle nucleation in clear air adjacent to marine cumulus clouds. *Journal of Geophysical Research*, **99**, 22,803–22,818.

Petit, J.R., Jouzel, J., Raynaud, D., Barkov, N.I., Barnola, J.-M., Basile, I., Bender, M., Chapellaz, J., Davis, M., Delaygue, G., Delmotte, M., Kotlyakov, V.M., Legrand, M., Lipenkov, V.Y., Lorius, C., Pépin, L., Ritz, C., Saltzman, E. & Stievenard, M. (1999) Climate and atmospheric history of the past 420,000 years from the Vostok ice core, Antarctica. *Nature*, **399**, 429–435.

Pilinis, C. & Li, X. (1998) Particle shape and internal inhomogeneity effects on the optical properties of tropospheric aerosols of relevance to climate forcing. *Journal of Geophysical Research*, **103**, 3789–3800.

Pósfai, M. & Buseck, P.R. (2010) Nature and climate effects of individual tropospheric aerosol particles. *Annual Reviews of Earth and Planetary Sciences*, **38**, 17–43.

Pósfai, M., Anderson, J.R., Buseck, P.R., Shattuck, T.W. & Tindale, N.W. (1994) Constituents of a remote Pacific marine aerosol: a TEM study. *Atmospheric Environment*, **28**, 1747–1756.

Pósfai, M., Anderson, J.R., Buseck, P.R. & Sievering, H. (1995) Compositional variations of sea-salt-mode aerosol particles from the North Atlantic. *Journal of Geophysical Research*, **100**, 23,063–23,074.

Pósfai, M., Xu, H., Anderson, J.R. & Buseck, P.R. (1998) Wet and dry sizes of atmospheric aerosol particles: A combined AFM-TEM study. *Geophysical Research Letters*, **25**, 1907–1910.

Pósfai, M., Anderson, J.R., Buseck, P.R. & Sievering, H. (1999) Soot and sulfate aerosol particles in the remote marine troposphere. *Journal of Geophysical Research*, **104**, 21,685–21,693.

Pósfai, M., Simonics, R., Li, J., Hobbs, P.V. & Buseck, P.R. (2003a) Individual aerosol particles from biomass burning in southern Africa: 1. Compositions and size distributions of carbonaceous particles. *Journal of Geophysical Research*, **108**, 8483.

Pósfai, M., Li, J., Anderson, J.R. & Buseck, P.R. (2003b) Aerosol bacteria over the Southern Ocean during ACE-1. *Atmospheric Research*, **66**, 231–240.

Pósfai, M., Gelencsér, A., Simonics, R., Arató, K., Li, J., Hobbs, P.V. & Buseck, P.R. (2004) Atmospheric tar balls: Particles from biomass and biofuel burning. *Journal of Geophysical Research*, **109**, D06213.

Pöschl, U., Martin, S.T., Sinha, B., Chen, Q., Gunthe, S.S., Huffman, J.A., Borrmann, S., Farmer, D.K., Garland, R.M., Helas, G., Jimenez, J.L., King, S.M., Manzi, A., Mikhailov, E., Pauliquevis, T., Petters, M.D., Prenni, A.J., Roldin, P., Rose, D., Schneider, J., Su, H., Zorn, S.R., Artaxo, P. & Andreae, M.O. (2010) Rainforest aerosols as biogenic nuclei of clouds and precipitation in the Amazon. *Science*, **329**, 1513–1516.

Prospero, J.M. (1999) Long-range transport of mineral dust in the global atmosphere: Impact of African dust on the environment of the southeastern United States. *Proceedings of the National Academy of Sciences of the USA*, **96**, 3396–3403.

Prospero, J.M. & Carlson, T.N. (1972) Vertical and areal distribution of Saharan dust over the western equatorial North Atlantic Ocean. *Journal of Geophysical Research*, **77**, 5255–5265.

Prospero, J.M., Blades, E., Naidu, R., Mathison, G., Thani, H. & Lavoie, M.C. (2008) Relationship between African dust carried in the Atlantic trade winds and surges in pediatric asthma attendances in the Caribbean. *International Journal of Biometeorology*, **52**, 823–832.

Prospero, J.M., Ginoux, P., Torres, O., Nicholson, S.E. & Gill, T.E. (2002) Environmental characterization of global sources of atmospheric soil dust identified with the Nimbus 7 Total Ozone Mapping Spectrometer (TOMS) absorbing aerosol product. *Reviews in Geophysics*, **40**, 1002.

Pueschel, R.F., Blake, D.F., Snetsinger, K.G., Hansen, A.D.A., Verma, S. & Kato, K. (1992) Black carbon (soot) aerosol in the lower stratosphere and upper troposphere. *Geophysical Research Letters*, **19**, 1659–1662.

Puxbaum, H., Gomiscek, B., Kalina, M., Bauer, H., Salam, A., Stopper, S., Preining, O. & Hauck, H. (2004) A dual site study of PM2.5 and PM10 aerosol chemistry in the larger region of Vienna, Austria. *Atmospheric Environment*, **38**, 3949–3958.

Qian, G.-W. & Ishizaka, Y. (1993) Electron microscope studies of methane sulfonic acid in individual aerosol particles. *Journal of Geophysical Research*, **98**, 8459–8470.

Raes, F. (1995) Entrainment of free tropospheric aerosols as a regulating mechanism for cloud condensation nuclei in the remote marine boundary layer. *Journal of Geophysical Research*, **100**, 2893–2903.

Rahn, K.A., Dams, R., Robbin, J.A. & Winchester, J.W. (1971) Diurnal variation of aerosol trace element concentrations as determined by non-desctructive neutron activation analysis. *Atmospheric Environment*, **5**, 413–422.

Ramanathan, V. & Carmichael, G. (2008) Global and regional climate changes due to black carbon. *Nature Geoscience*, **1**, 221–227.

Ramanathan, V., Chung, C., Kim, D., Bettge, T., Buja, L., Kiehl, J.T., Washington, W.M., Fu, Q., Sikka, D.R. & Wild, M. (2005) Atmospheric brown clouds: Impacts on South Asian climate and hydrological cycle. *Proceedings of the National Academy of Sciences of the USA*, **102**, 5326–5333.

Ramsden, A.R. & Shibaoka, M. (1982) Characterization and analysis of individual fly-ash particles from coal-fired power stations by a combination of optical microscopy, electron microscopy and quantitative electron microprobe analysis. *Atmospheric Environment*, **16**, 2191–2206.

Rea, D. (1994) The paleoclimatic record provided by eolian deposition in the deep sea: The geologic history of wind. *Reviews of Geophysics*, **32**, 159–195.

Reid, J.S., Koppmann, R., Eck, T.F. & Eleuterio, D.P. (2005) A review of biomass burning emissions part II: intensive physical properties of biomass burning particles. *Atmospheric Chemistry and Physics*, **5**, 799–825.

Roessler, D.M., Faxvog, F.R., Stevenson, R. & Smith, G.W. (1981) *Optical properties and morphology of particulate carbon: Variation with air/fuel ratio*. In: *Particulate Carbon: Formation During Combustion*, (D.C. Siegla & G.W. Smith editors). Plenum Press, New York, pp. 57–84.

Rogge, W.F., Mazurek, M.A., Hildemann, L.M. & Cass, G.R. (1993) Quantification of urban organic aerosols at a molecular level: Identification, abundance and seasonal variation. *Atmospheric Environment*, **27**, 1309–1330.

Rogge, W.F., Hildemann, L.M., Mazurek, M.A. & Cass, G.R. (1996) Mathematical modeling of atmospheric fine particle-associated primary organic compound concentrations. *Journal of Geophysical Research*, **101**, 19,379–19,394.

Rood, M.J., Shaw, M.A., Larson, T.V. & Covert, D.S. (1989) Ubiquitous nature of ambient metastable aerosol. *Nature*, **337**, 537–539.

Rosenfeld, D., Lohmann, U., Raga, G.B., O'Dowd, C.D., Kulmala, M., Fuzzi, S., Reissell, A. & Andreae, M.O. (2008) Flood or drought: how do aerosols affect precipitation? *Science*, **321**, 1309–1313.

Saiz-Lopez, A., Plane, J.M.C., McFiggans, G., Williams, P.I., Ball, S.M., Bitter, M., Jones, R.L., Hongwei, C. & Hoffmann, T. (2006) Modelling molecular iodine emissions in a coastal marine environment: The link to new particle formation. *Atmospheric Chemistry and Physics*, **6**, 883–895.

Salma, I., Pósfai, M., Kovács, K., Kuzmann, E., Homonnay, Z. & Posta, J. (2009) Properties and sources of individual particles and some chemical species in the aerosol of a metropolitan underground railway station. *Atmospheric Environment*, **43**, 3460–3466.

Salma, I., Borsós, T., Weidinger, T., Aalto, P., Hussein, T., Dal Maso, M. & Kulmala, M. (2011) Production, growth and properties of ultrafine atmospheric aerosol particles in an urban environment. *Atmospheric Chemistry and Physics*, **11**, 1339–1353.

Salmon, G.A. (1994) Chemical reactions in cloud droplets. In: *Physico-chemical Behaviour of Atmospheric Pollutants* (G. Angeletti & G. Restelli, editors). European Commission Report (EUR 15609/2 EN), Brussels, pp. 854–863.

Salter, S., Sortino, G. & Latham, J. (2008) Sea-going hardware for the cloud albedo method of reversing global warming. *Philosophical Transactions of the Royal Society London*, **13**, 3989–4006.

Sattler, B., Puxbaum, H. & Psenner, R. (2001) Bacterial growth in supercooled cloud droplets. *Geophysical Research Letters*, **28**, 243–246.

Saxena, P., Hildemann, L.M., McMurry, P.H. & Seinfeld, J.H. (1995) Organics alter hygroscopic behavior of atmospheric particles. *Journal of Geophysical Research*, **100**, 18,755–18,770.

Schnaiter, M., Gimmler, M., Llamas, I., Linke, C., Jäger, C. & Mutschke, H. (2006) Strong spectral dependence of light absorption by organic carbon particles formed by propane combustion. *Atmospheric Chemistry and Physics*, **6**, 2981–2990.

Schnell, R.C. (1977) Ice nuclei in seawater, fog water and marine air off the coast of Nova Scotia: Summer 1975. *Journal of the Atmospheric Sciences*, **34**, 1299–1305.

Schnell, R.C. & Vali, G. (1976) Biogenic ice nuclei: Part I. Terrestrial and marine sources. *Journal of the Atmospheric Sciences*, **33**, 1554–1564.

Schult, I., Feichter, J. & Cooke, W.F. (1997) Effect of black carbon and sulfate aerosols on the global radiation budget. *Journal of Geophysical Research*, **102**, 30,107–30,117.

Schumann, U., Weinzierl, B., Reitebuch, O., Schlager, H., Minikin, A., Forster, C., Baumann, R., Sailer, T., Graf, K., Mannstein, H., Voigt, C., Rahm, S., Simmet, R., Scheibe, M., Lichtenstern, M., Stock, P., Rüba, H., Schäuble, D., Tafferner, A., Rautenhaus, M., Gerz, T., Ziereis, H., Krautstrunk, M., Mallaun, C., Gayet, J.-F., Lieke, K., Kandler, K., Ebert, M., Weinbruch, S., Stohl, A., Gasteiger, J., Gross, S., Freudenthaler, V., Wiegner, M., Ansmann, A., Tesche, M., Olafsson, H. & Sturm, K. (2011) Airborne observations of the Eyjafjalla volcano ash cloud over Europe during air space closure in April and May 2010. *Atmospheric Chemistry and Physics*, **11**, 2245–2279.

Schütz, L. & Sebert, M. (1987) Mineral aerosols and source identification. *Journal of Aerosol Science*, **18**, 1–10.

Schwarz, J.P., Spackman, J.R., Fahey, D.W., Gao, R.S., Lohmann, U., Stier, P., Watts, L.A., Thomson, D.S., Lack, D.A., Pfister, L., Mahoney, M.J., Baumgardner, D., Wilson, J.C. & Reeves, J.M. (2008) Coatings and their enhancement of black carbon light absorption in the tropical atmosphere. *Journal of Geophysical Research*, **113**, D03203.

Seinfeld, J.H. & Pandis, S.N. (2006) *Atmospheric Chemistry and Physics: From Air Pollution to Climate Change*. John Wiley, New York.

Semeniuk, T.A., Wise, M.E., Martin, S.T., Russell, L.M. & Buseck, P.R. (2007) Hygroscopic behavior of aerosol particles from biomass fires using environmental transmission electron microscopy. *Journal of Atmospheric Chemistry*, **56**, 259–273.

Sheng, L.T., Fei, G.X., Sheng, A.Z. & Xiang, F.Y. (1981) The dust fall in Beijing, China on April 18, 1980. *Geological Society of America Special Paper*, **186**, 149–157.

Sheridan, P.J. (1989) Characterization of size segregated particles collected over Alaska and the Canadian high Arctic, AGASP-II flights 204–206. *Atmospheric Environment*, **23**, 2371–2386.

Sheridan, P.J., Schnell, R.C., Kahl, J.D., Boatman, J.F. & Garvey, D.M. (1993) Microanalysis of the aerosol collected over south-central New Mexico during the ALIVE field experiment, May–December, 1989. *Atmospheric Environment*, **27A**, 1169–1183.

Shulman, M.L., Jacobson, M.C., Carlson, R.J., Synovec, R.E. & Young, T.E. (1996) Dissolution behavior and surface tension effects of organic compounds in nucleating cloud droplets. *Geophysical Research Letters*, **23**, 277–280.

Sieburth, J.M. (1982) Microbiological and organo-chemical processes in the surface and mixed layers. In: *Air-sea Exchange of Gases and Particles* (P.S. Liss & W.G.N. Slinn, editors). D. Reidel, Dordrecht, The Netherlands, pp. 121–172.

Sievering, H., Boatman, J., Gorman, E., Kim, Y., Anderson, L., Ennis, G., Luria, M. & Pandis, S. (1992) Removal of sulfur from the marine boundary layer by ozone oxidation in sea-salt aerosol. *Nature*, **360**, 571–573.

Sievering, H., Lerner, B., Slavich, J., Anderson, J., Pósfai, M. & Cainey, J. (1999) O_3 oxidation of SO_2 in sea-salt aerosol water: Size distribution of non-sea-salt sulfate during the First Aerosol Characterization Experiment (ACE 1). *Journal of Geophysical Research*, **103**, 21,707–21,717.

Silva, L.F.O., Moreno, T. & Querol, X. (2009) An introductory TEM study of Fe-nanominerals within coal fly ash. *Science of the Total Environment*, **407**, 4972–4974.

Sinha, B.W., Hoppe, P., Huth, J., Foley, S. & Andreae, M.O. (2008) Sulfur isotope analyses of individual aerosol particles in the urban aerosol at a central European site (Mainz, Germany). *Atmospheric Chemistry and Physics*, **8**, 7217–7238.

Skeie, R.B., Berntsen, T.K., Myhre, G., Tanaka, K., Kvaleväg, M.M. & Hoyle, C.R. (2011) Anthropogenic radiative forcing time series from pre-industrial times until 2010. *Atmospheric Chemistry and Physics*, **11**, 11827–11857.

Skinner, C.W. (2013) *Minerals and Human Health*. In: *Environmental Mineralogy II* (D.J. Vaughan & R.A. Wogelius, editors). EMU Notes in Mineralogy, **13**, European Mineralogical Union and the Mineralogical Society of Great Britian & Ireland, pp. 441–484.

Smith, J.N., Dunn, M.J., VanReken, T.M., Iida, K., Stolzenburg, M.R., McMurry, P.H. & Huey, L.G. (2008) Chemical composition of atmospheric nanoparticles formed from nucleation in Tecamac, Mexico: Evidence for an important role for organic species in nanoparticle growth. *Geophysical Research Letters*, **35**, L04808.

Smith, J.N., Barsantia, K.C., Friedlia, H.R., Ehnd, M., Kulmala, M., Collins, D.R., Scheckman, J.H., Williams, B.J. & McMurry, P.H. (2010) Observations of aminium salts in atmospheric nanoparticles and possible climatic implications. *Proceedings of the National Academy of Sciences of the USA*, **107**, 6634–6639.

Snider, J.R. & Vali, G. (1994) Sulfur dioxide oxidation in winter orographic clouds. *Journal of Geophysical Research*, **99**, 18,713–18,733.

Sokolik, I., Andronova, A. & Johnson, T.C. (1993) Complex refractive index of atmospheric dust aerosols. *Atmospheric Environment*, **27A**, 2495–2502.

Spracklen, D.V., Jimenez, J.L., Carslaw, K.S., Worsnop, D.R., Evans, M.J., Mann, G.W., Zhang, Q., Canagaratna, M.R., Allan, J., Coe, H., McFiggans, G., Rap, A. & Forster, P. (2011) Aerosol mass spectrometer constraint on the global secondary organic aerosol budget. *Atmospheric Chemistry and Physics*, **11**, 12109–12136.

Stern, D.I. (2005) Global sulfur emissions from 1850 to 2000. *Chemosphere*, **58**, 163–175.

Stoddard, J.L., Jeffries, D.S., Lükewille, A., Clair, T.A., Dillon, P.J., Driscoll, C.T., Forsius, M., Johannessen, M., Kahl, J.S., Kellogg, J.H., Kemp, A., Mannio, J., Monteith, D.T., Murdoch, P.S., Patrick, S., Rebsdorf, A., Skjelkvlle, B.L., Stainton, M.P., Traaen, T., Van Dam, H., Webster, K.E., Wieting, J. & Wilander, A. (1999) Regional trends in aquatic recovery from acidification in North America and Europe. *Nature*, **401**, 575–578.

Sturges, W.T., Harrison, R.M. & Barrie, L.A. (1989) Semi-quantitative X-ray diffraction analysis of size fractionated atmospheric particles. *Atmospheric Environment*, **23**, 1083–1098.

Sullivan, R.C. & Prather, K.A. (2005) Recent advances in our understanding of atmospheric chemistry and climate made possible by on-line aerosol analysis instrumentation. *Analytical Chemistry*, **77**, 3861–3885.

Tang, I.N. (1996) Chemical and size effects of hygroscopic aerosols on light scattering coefficients. *Journal of Geophysical Research*, **101**, 19,245–19,250.

Tang, I.N. & Munkelwitz, H.R. (1993) Composition and temperature dependence of the deliquescence properties of hygroscopic aerosols. *Atmospheric Environment*, **27A**, 467–473.

Tegen, I. & Fung, I. (1994) Modeling of mineral dust in the atmosphere: Sources, transport, and optical thickness. *Journal of Geophysical Research*, **99**, 22,89–22,914.

Tegen, I. & Fung, I. (1995) Contribution to the atmospheric mineral aerosol load from land surface modification. *Journal of Geophysical Research*, **100**, 18,707–18,726.

Tegen, I., Lacis, A.A. & Fung, I. (1996) The influence on climate forcing of mineral aerosols from disturbed soils. *Nature*, **380**, 419–422.

Tegen, I., Werner, M., Harrison, S.P. & Kohfeld, K.E. (2004) Relative importance of climate and land use in determining present and future global soil dust emission. *Geophysical Research Letters*, **31**, L05105.

Ten Hoeve, J.E., Jacobson, M.Z. & Remer, L.A. (2012) Comparing results from a physical model with satellite and in situ observations to determine whether biomass burning aerosols over the Amazon brighten or burn off clouds. *Journal of Geophysical Research*, **117**, D08203.

Tinsley, B.A., Rohrbaugh, R.P., Hei, M. & Beard, K.V. (2000) Effects of image charges on the scavenging of aerosol particles by cloud droplets and on droplet charging and possible ice nucleation processes. *Journal of the Atmospheric Sciences*, **57**, 2118–2134.

Tivanski, A.V., Hopkins, R.J., Tyliszczak, T. & Gilles, M.K. (2007) Oxygenated interface on biomass burn tar balls determined by single particle scanning transmission X-ray microscopy. *Journal of Physical Chemistry A*, **111**, 5448–5458.

Topping, D.O., Barley, M.H. & McFiggans, G. (2011) The sensitivity of Secondary Organic Aerosol component partitioning to the predictions of component properties-Part 2: Determination of particle hygroscopicity and its dependence on "apparent" volatility. *Atmospheric Chemistry and Physics*, **11**, 7767–7779.

Topping, D.O., McFiggans, G.B., Kiss, G., Varga, Z., Facchini, M.C., Decesari, S. & Mircea, M. (2007) Surface tensions of multi-component mixed inorganic/organic aqueous systems of atmospheric significance: Measurements, model predictions and importance for cloud activation predictions. *Atmospheric Chemistry and Physics*, **7**, 2371–2398.

Tsigaridis, K. & Kanakidou, M. (2007) Secondary organic aerosol importance in the future atmosphere. *Atmospheric Environment*, **41**, 4682–4692.

Turn, S.Q., Jenkins, B.M., Chow, J.C., Pritchett, L.C., Campbell, D., Cahill, T. & Whalen, S.A. (1997) Elemental characterization of particulate matter emitted from biomass burning: Wind tunnel derived source profiles for herbaceous and wood fuels. *Journal of Geophysical Research*, **102**, 3683–3699.

Turpin, B.J. & Lim, H.J. (2001) Species contributions to PM2.5 mass concentrations: Revisiting common assumptions for estimating organic mass. *Aerosol Science and Technology*, **35**, 602–610.

Twohy, C.H., Kreidenweis, S.M., Eidhammer, T., Browell, E.V., Heymsfield, A.J., Bansemer, A.R., Anderson, B.E., Chen, G., Ismail, S., DeMott, P.J. & Van Den Heever, S.C. (2009) Saharan dust particles nucleate droplets in eastern Atlantic clouds. *Geophysical Research Letters*, **36**, L01807.

Twomey, S. (1977) *Atmospheric aerosols*. Elsevier, New York.

Uematsu, M., Duce, R.A., Prospero, J.M., Chen, L., Merrill, J.T. & McDonald, R.L. (1983) Transport of mineral aerosol from Asia over the North Pacific Ocean. *Journal of Geophysical Research*, **88**, 5343–5352.

Vali, G. (1985) Atmospheric ice nucleation – a review. *Journal de Recherches Atmospheriques*, **19**, 105–115.

van Poppel, L.H., Friedrich, H., Spinsby, J., Chung, S.H., Seinfeld, J.H. & Buseck, P.R. (2005) Electron tomography of nanoparticle clusters: Implications for atmospheric lifetimes and radiative forcing of soot. *Geophysical Research Letters*, **32**, 1–4.

Vander Wal, R.L., Bryg, V.M. & Hays, M.D. (2010) Fingerprinting soot (towards source identification). Physical structure and chemical composition. *Aerosol Science*, **41**, 108–117.

Virkkula, A., Teinilä, K., Hillamo, R., Kerminen, V.-M., Saarikoski, S., Aurela, M., Viidanoja, J., Paatero, J., Koponen, I.K. & Kulmala, M. (2006) Chemical composition of boundary layer aerosol over the Atlantic Ocean and at an Antarctic site. *Atmospheric Chemistry and Physics*, **6**, 3407–3421.

Vogt, R., Sander, R., Von Glasow, R. & Crutzen, P.J. (1999) Iodine chemistry and its role in halogen activation and ozone loss in the marine boundary layer: A model study. *Journal of Atmospheric Chemistry*, **32**, 375–395.

Volz, F.E. (1983) Infrared optical constants of aerosols at some locations. *Applied Optics*, **22**, 3690–3700.

Wang, J., Jacob, D.J. & Martin, S.T. (2008) Sensitivity of sulfate direct climate forcing to the hysteresis of particle phase transitions. *Journal of Geophysical Research*, **113**, D11206.

Washington, R. & Todd, M.C. (2005) Atmospheric controls on mineral dust emission from the Bodele Depression, Chad: the role of the low level jet. *Geophysical Research Letters*, **32**, L17701.

Wayne, R.P. (1991) *Chemistry of atmospheres*. Clarendon Press, Oxford.

Weast, R.C. (1987) *Physical constants of organic compounds*. CRC Press, Boca Raton, Florida.

Weber, R.J., McMurry, P.H., Mauldin, L., Tanner, D.J., Eisele, F.L., Brechtel, F.J., Kreidenweis, S.M., Kok, G.L., Schillawski, R.D. & Baumgardner, D. (1998) A study of new particle formation and growth involving biogenic and trace gas species measured during ACE 1. *Journal of Geophysical Research*, **103**, 16,385–16,396.

Wentzel, M., Gorzawski, H., Naumann, K.H., Saathoff, H. & Weinbruch, S. (2003) Transmission electron microscopical and aerosol dynamical characterization of soot aerosols. *Journal of Aerosol Science*, **34**, 1347–1370.

Whitby, K.T. (1978) The physical characteristics of sulfur aerosols. *Atmospheric Environment*, **12**, 135–159.

Wild, M. (2009) Global dimming and brightening: a review. *Journal of Geophysical Research*, **114**, D00D16.

Wild, M., Gilgen, H., Roesch, A., Ohmura, A., Long, C.N., Dutton, E.C., Forgan, B., Kallis, A., Russak, V. & Tsvetkov, A. (2005) From dimming to brightening: Decadal changes in solar radiation at earth's surface. *Science*, **308**, 847–850.

Winiwarter, W., Bauer, H., Caseiro, A. & Puxbaum, H. (2009) Quantifying emissions of primary biological aerosol particle mass in Europe. *Atmospheric Environment*, **43**, 1403–1409.

Winter, B. & Chylek, P. (1997) Contribution of sea salt aerosol to the planetary clear-sky albedo. *Tellus, Series B: Chemical and Physical Meteorology*, **49B**, 72–79.

Wise, M.E., Biskos, G., Martin, S.T., Russell, L.M. & Buseck, P.R. (2005) Phase transitions of single salt particles studied using a transmission electron microscope with an environmental cell. *Aerosol Science and Technology*, **39**, 849–856.

Wise, M.E., Semeniuk, T.A., Bruintjes, R., Martin, S.T., Russell, L.M. & Buseck, P.R. (2007) Hygroscopic behavior of NaCl-bearing natural aerosol particles using environmental transmission electron microscopy. *Journal of Geophysical Research*, **112**, D10224.

Wogelius, R.A. & Vaughan, D.J. (2013) Analytical, experimental, and computational methods in environmental mineralogy. In: *Environmental Mineralogy II* (D.J. Vaughan & R.A. Wogelius, editors). EMU Notes in Mineralogy, **13**, European Mineralogical Union and the MineralogicalSociety of Great Britian & Ireland, pp. 5–102.

Wojcik, G.S. & Chang, J.S. (1997) A re-evaluation of sulfur budgets, lifetimes, and scavenging ratios for eastern North America. *Journal of Atmospheric Chemistry*, **26**, 109–145.

Woodcock (1972) Smaller salt particles in oceanic air and bubble behavior in the sea. *Journal of Geophysical Research*, **77**, 362–371.

Xhoffer, C., Jacob, W. & Van Grieken, R. (1989) Application of electron energy loss spectroscopy to aerosols. *Journal of Aerosol Science*, **20**, 1617–1619.

Yoshioka, M., Mahowald, N., Dufresne, J.-L. & Lou, C. (2005) Simulation of absorbing aerosol indices for African dust. *Journal of Geophysical Research*, **110**, 1–22.

Yu, J., Griffin, R.J., Cocker, D.R., Flagan, R.C., Seinfeld, J.H. & Blanchard, P. (1999) Observation of gaseous and particulate products of monoterpene oxidation in forest atmospheres. *Geophysical Research Letters*, **26**, 1145.

Yuan, H., Zhuang, G., Rahn, K.A., Zhang, X. & Li, Y. (2006) Composition and mixing of individual particles in dust and nondust conditions of North China, spring 2002. *Journal of Geophysical Research*, **111**, D20208.

Zappoli, S., Andracchio, A., Fuzzi, S., Facchini, M.C., Gelencsér, A., Kiss, G., Krivácsy, Z., Molnár, A., Mészáros, E., Hansson, H.-C., Rosman, K. & Zebühr, Y. (1999) Inorganic, organic and macromolecular components of fine aerosol in different areas of Europe in relation to their water solubility. *Atmospheric Environment*, **33**, 2733–2743.

Zhang, D., Iwasaka, Y., Shi, G., Zang, J., Matsuki, A. & Trochkine, D. (2003) Mixture state and size of Asian dust particles collected at southwestern Japan in spring 2000. *Journal of Geophysical Research*, **108**, ACH 9-1–ACH 9-12.

Zhang, R., Khalizov, A., Wang, L., Hu, M. & Xu, W. (2012a) Nucleation and growth of nanoparticles in the atmosphere. *Chemical Reviews*, **112**, 1957–2011.

Zhang, H., Shen, Z., Wei, X., Zhang, M. & Li, Z. (2012b) Comparison of optical properties of nitrate and sulfate aerosol and the direct radiative forcing due to nitrate in China. *Atmospheric Research*, **113**, 113–125.

Zhang, R., Suh, I., Zhao, J., Zhang, D., Fortner, E.C., Tie, X., Molina, L.T. & Molina, M.J. (2004) Atmospheric new particle formation enhanced by organic acids. *Science*, **304**, 1487–1490.

Zhao, T.L., Gong, S.L., Zhang, X.Y. & Jaffe, D.A. (2008) Asian dust storm influence on North American ambient PM levels: observational evidence and controlling factors. *Atmospheric Chemistry and Physics*, **8**, 2717–2728.

Zhuang, G., Yi, Z., Duce, R.A. & Brown, P.R. (1992) Link between iron and sulfur cycles suggested by detection of Fe(II) in remote marine aerosols. *Nature*, **355**, 537–539.

Mineralogy of mine wastes and strategies for remediation

DAVID W. BLOWES[1,*] CAROL J. PTACEK[1,‡] and JOHN L. JAMBOR[1,†]

[1]*Department of Earth and Environmental Sciences, University of Waterloo, Waterloo, Ontario N2 L 3G1, Canada*
**e-mail: blowes@uwaterloo.ca*
‡e-mail: ptacek@uwaterloo.ca

The release of acid mine drainage (AMD) from mines and mine wastes is an environmental problem of international scale. The AMD results from the oxidation of sulfide minerals and the release of acidity at rates that exceed the capacity of carbonate and aluminosilicate gangue minerals to neutralize the pH. Static testing protocols used to assess the potential for AMD generation are based on chemical characteristics, and generally do not consider the mineralogy of the waste materials. Static testing of pure minerals and samples of typical host rocks indicate that aluminosilicate minerals are unlikely to provide neutralization at rates that are sufficient to prevent the release of AMD in wastes containing more than a very modest sulfide content. The products of sulfide oxidation include a broad range of (oxy)hydroxide, sulfate and hydroxysulfate minerals. The most common of these secondary minerals are goethite, gypsum and jarosite. Covellite is common in wastes derived from copper deposits, and marcasite and elemental sulfur are common in waste deposits containing pyrrhotite. The mechanisms of abiotic and microbially mediated sulfide oxidation have been studied extensively. Microbially-mediated oxidation of acid-soluble sulfide minerals is initiated by the release of H_2S, and follows the polysulfide pathway, whereas the oxidation of acid-insoluble sulfide minerals is initiated by oxidation by Fe(III) and follows the thiosulfate pathway. The paragenetic sequence for the oxidation of sulfide-bearing mine wastes has been defined. This sequence includes more extensive accumulation of elemental sulfur associated with wastes containing acid-soluble sulfide minerals. A variety of active and passive approaches exists for the remediation of AMD-generating waste-disposal sites. Although many of these techniques have been implemented at mine sites, other novel approaches are currently being evaluated. Passive treatment systems are desirable because of the lower costs associated with these systems and the long duration of AMD generation.

1. Introduction

Acid rock drainage is a world-wide phenomenon that may occur naturally, as in the formation of gossans that are associated with oxidized mineral deposits, or in less complex situations in which streams traverse sulfide-bearing, typically pyritiferous, rocks whose minerals are incapable of neutralizing the acidity that is generated by the oxidation of the host-rock sulfides. At many mine sites, however, anthropogenic activity has been directly responsible for causing or exacerbating acidic drainage by exposing or increasing the exposure of sulfide-bearing materials, so that they are susceptible to atmospheric oxidation and the attendant acid generation.

[†]Deceased

© Copyright 2013 the European Mineralogical Union and the Mineralogical Society of Great Britain & Ireland
DOI: 10.1180/EMU-notes.13.7

The release of acidic effluents can be directly harmful to aquatic life, and there may also be damage to subaqueous and subaerial habitat. A second effect, which probably has a greater environmental impact than the acidity itself, is that the low-pH solutions increase the solubilization of potentially toxic metals and semi-metals such as Pb, Zn, Cu, Cd, Cr, Ni, Co, Al, Sb and As. Treatment of acidic effluents is commonly feasible, albeit expensive, through simple procedures such as the addition of lime to raise the pH of the collected waters. Control of dissolved metals, however, is generally more complex and costly, as is discussed in the second part of this paper.

Figure 1 provides a perspective on AMD in relation to acid rain and some familiar acidic substances. Pure rain has a pH of 5.6, which represents equilibrium between pure water and atmospheric CO_2. AMD is generally regarded to be present if effluent affected by sulfide mineral oxidation has a pH lower than between 5 and 5.5 because in this pH range there is a negative impact on biota. Thus, there is but little difference between the pH of rain and the upper pH limit for AMD, but this small difference emphasizes the delicate balance that must be achieved if natural environments are not to be harmed by anthropogenic activity.

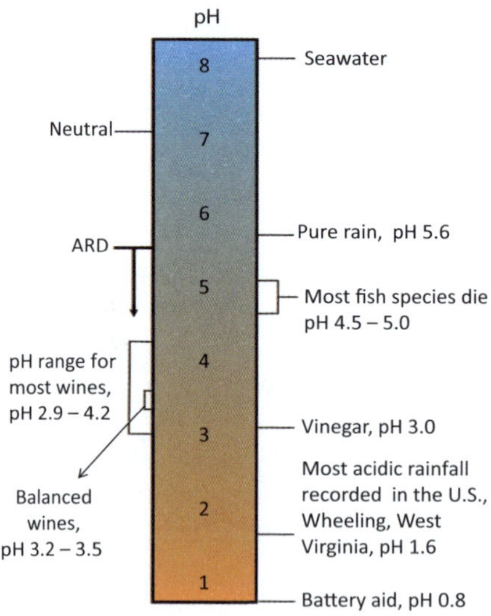

Fig. 1. Comparison of the pH of AMD, of acid rain, and of some familiar acids. The Adirondack lakes and Wheeling West Virginia, referred to on the right side of the diagram, are in the northeastern United States, and the pH of rain on the right side of the diagram refers to that region. By 1975, the pH of the Adirondack lakes had decreased to 4.8. Data are from Rodel & Navidi (1990) and Robinson (1994).

2. Prediction of AMD

In most localities, it is necessary to submit to regulatory agencies detailed plans as to how operations will be conducted at a potential mining site and to have the plans approved by these agencies. Among the many facets that are examined by these agencies are those that deal with waste storage and disposal, both for solids and liquids. The principal concerns from the environmental mineralogy perspective are the solids, namely tailings and waste rock. In mine planning, it is necessary to predict whether the solids will have the potential to be acid-generating and the potential to release deleterious constituents, or will be environmentally benign. If the wastes are determined to be acid-generating or have the potential to release contaminants, mine plans must be adjusted to ensure that any effluents to be released have a pH, and metal and metalloid concentrations that are within environmentally acceptable limits as dictated by the particular jurisdiction.

Numerous static test methods have been proposed, or are in use, to predict whether wastes will be acid-generating (*e.g.* Sobek *et al.,* 1978; MEND, 1989, 1991a, 1991b; Miller *et al.,* 1994). The various methods, which are described here only briefly, have a common purpose: the determination, by chemical means, of how the minerals in a sample might behave upon weathering. No determinative mineralogy is involved. The assumptions are that the sulfide-sulfur in a sample is present as pyrite (FeS_2) that will oxidize in accordance with the reaction:

$$FeS_2 + 15/4O_2 + 7/2H_2O \rightarrow Fe(OH)_3 + 2SO_4^{2-} + 4H^+ \tag{1}$$

and that some of the non-sulfide minerals in the accompanying mineral assemblage will react to attenuate or neutralize the acidity. Hence, the two terms in common use are acid-producing potential (AP) and neutralization potential (NP). If the AP of a sample exceeds its NP, then the sample will be acid-generating. The determination of AP and NP values thus provides a numerical accounting with regard to prediction, and the numerical data are generally referred to as an acid-base accounting (ABA). The AP and NP values are calculated in kilograms of $CaCO_3$ equivalent per tonne of material, but to avoid this somewhat cumbersome unit and to give a better perspective of the net potential of a sample, increasingly the ABA results are expressed simply as the ratio of NP:AP. Values of NP:AP ratio of greater than 2 are required to maintain pH > 6.3 (Price, 2009). The 'safe' value for this ratio, *i.e.* deemed to be not acid-producing, varies with jurisdiction but is commonly set at 2:1 to 3:1 (Lawrence & Wang, 1996; Jambor & Blowes, 1998; Price, 2009).

One of the fundamental limitations of ABA measurements is that they are derived from a one-time, short-duration test; hence, these tests are widely referred to as static tests. Static tests do not incorporate kinetic aspects, *i.e.* there is no accommodation of the fact that different minerals react at different rates when exposed to weathering or low-pH conditions. Thus, in the evaluation of potential AMD at a site, almost invariably, the static tests are followed by kinetic tests, from which an attempt is made to decipher mineral-weathering behaviour over time. The kinetic tests generally involve numerous cycles of leaching of selected samples in an apparatus such as a humidity cell or column, followed by chemical analysis of the leachate after each, or many, of the cycles (*e.g.* Lapakko, 2003; Ardau *et al.,* 2008; Price, 2009; Matthies *et al.,* 2011).

Because static tests do not yield information about reaction rates, this deficiency has led to recently increased usage of terms such as "available NP" or "effective NP" (Price *et al.,* 1997; Jambor, 2003; Price, 2009). Such usage is a tacit acceptance that NP values, as currently measured, are not satisfactorily fulfilling their role as a predictive tool for AMD. The usage bespeaks a recognition that some of the measured NP will, in reality, provide negligible neutralization until some time after the low-pH conditions have come into existence, *i.e.* after an AMD problem already exists. Examination of laboratory experimental data on the dissolution rates of minerals (Table 1) shows that most of the non-carbonate minerals which are typically associated with metalliferous deposits, other than those in skarns and ultramafic rocks, are extremely slow to react. One could predict, therefore, that the non-carbonate minerals associated with most

Table 1. Relative dissolution rates of non-sulfide minerals*

	$\dfrac{\text{Rate}}{\text{Rate for calcite}} \times 10^5$			$\dfrac{\text{Rate}}{\text{Rate for calcite}} \times 10^5$
Calcite	100,000	Plagioclase	An_{76}	0.25
Dolomite	6,000		An_{46}	0.12
Forsterite	4		An_{13}	0.02
Diopside	1.4		An_0	0.02
Enstatite	0.93	Sanidine		0.03
Talc	0.06	microcline		0.01
Chrysotile	0.06			
Biotite	0.01–0.03			
Phlogopite	0.02			
Chlorite	0.02			
Kaolinite	0.006–0.02			
Muscovite	0.006			
Montmorillonite	0.002			
Quartz	0.0005			

*In laboratory experiments at pH 5, far from equilibrium. Rates relative to calcite were converted from data in Drever & Clow (1995) and from Nagy (1995), using in the latter the rates relative to muscovite.

metalliferous ore deposits, including the feldspars, will provide a negligible amount of 'available NP'. On this basis, it was suggested by Jambor (2000) that, in static tests, those based exclusively on carbonate NP (excluding siderite) would be the most realistic indicator of neutralization capacity in terms of potential AMD. In measuring $NP_{carbonate}$ it is necessary to exclude siderite, or more specifically the solid-solution Fe component of the carbonate minerals, because the net neutralizing effect of that component is zero (Skousen *et al.*, 1997).

The most widely used static test for determining NP is that of Sobek *et al.* (1978). The procedure involves initial testing by adding a drop of 25% HCl to the sample to check its degree of effervescence, or 'fizz rating' (nil, slight, moderate, or strong). The resultant rating governs the amount and concentration of HCl that is added to react with the sample, which is then heated to almost boiling, diluted, boiled briefly, and is subsequently titrated with NaOH to determine the amount of acid that has been consumed. The vigorous reaction in the acidification stage typically consumes some silicate minerals, thus resulting in NP values that are high relative to those obtained at low temperatures. Which minerals react to provide the high-temperature NP, and how much NP each of these selectively reacted minerals contributes to the total NP of a sample, are not determined by the procedure. Thus, in the concept of 'available NP', one is again left to speculate as to what proportion to discount from the total NP.

The NP measurements for specific minerals indicate the NP contribution of each mineral to a Sobek test (Jambor *et al.*, 2002). Table 2 provides an indication of what might be expected in terms of the magnitude of the NP results for common rock-forming minerals. To put these results into perspective, a common regulatory guideline prior to the adoption of NP/AP values or ratios was that a rock containing 0 wt.% $S_{sulfide}$ would require a NP of at least 20 kg $CaCO_3$ equivalent per tonne of material to be safely considered as non acid-generating. For example, quartz does not contribute to NP, and

Table 2. Results of Sobek NP tests of individual minerals[a].

Mineral	Sobek results	BET (m^2/g)	Normalized NP	Mineral	Sobek results	BET (m^2/g)	Normalized NP
Anorthite An_{93}	10.7	0.75	14.3	Muscovite	0.3	2.12	0.1
Anorthite An_{50}	2.6	0.55	4.7	Phlogopite	2.7	1.72	1.6
Albite Ab_{80}	0.6	0.57	1.1	Phlogopite	8.5	1.37	6.2
Albite Ab_{98}	0.5	0.27	1.9				
K-feldspar	1.4	0.77	1.8	Clinochlore	10.3	7.04	1.5
K-feldspar	1.0	0.65	1.5	Clinochlore	0.8	1.50	0.5
K-feldspar	1.3	0.67	1.9				
Microcline	0.5	0.54	0.9	Talc	1.7	3.04	0.6
				Antigorite	15.1	39.3	0.4
Enstatite	3.2	1.25	2.6	Clinocrysotile[d]	87.6	40.43	2.2
Diopside	4.5	0.50	9.0	Lizardite	16.1	2.05	7.9
Hendenbergite	6.6	18.97	0.3				
Augite	4.6	3.91	1.2	Kaolinite	0.0	17.1	0.0
				Pyrophyllite	0.3	7.57	0.0
Tremolite	5.2	1.23	4.2	Montmorillonite	13.8	31.4	0.4
Pargasite	4.4	0.34	12.9				
				Forsterite	23.9	0.12	1.99
Wollastonite[b]	440	0.199	2211	Siderite	34.4	2.37	14.5
Wollastonite[b,c]	567	17.09	33.2				
				Epidote	1.0	0.069	14.5
				Zoisite	3.0	0.058	51.7
Goethite	1.5	5.353	0.3	Jarosite	−3.9	1.785	−2.2
Hematite	2.0	3.013	0.7	Jarosite	2.9	1.985	1.5

[a]Values for Sobek tests are tonnes of $CaCO_3$ equivalent per tonne of mineral or rock. BET values are measured surface areas, and normalized NP is the value obtained after standardization to account for the different surface areas. All samples were crushed to 80% minus 200 mesh Tyler. Data from Jambor *et al.* (2000b) and Jambor *et al.* (2002).
[b]Tested with 40 mL of 0.1 N HCl.
[c]Leached with acetic acid solution to remove calcite.
[d]Tested with 80 mL of 0.5 N HCl.

if the samples in Table 2 are, or approximately, 100% of the sample, it is evident that most of the minerals do not contribute appreciable amounts of NP in terms of the regulatory guideline. The NP contribution of 55 monomineralic samples was determined by Jambor *et al.* (2002). The NP contribution of most rock-forming minerals, including pyroxenes, amphiboles, feldspars, micas, chlorites and clays were insufficient to surpass a threshold value of 20 kg $CaCO_3$ equivalent per tonne of material. However, olivine, epidote, zoisite and wollastonite were exceptions, with wollastonite giving NP values that were similar to some carbonate minerals. In addition, the NP associated with plagioclase is variable and dependent on the degree of Ca substitution, with NP < 1 for the Na end-member and approaching 14 for compositions near the Ca-end member.

The NP measurements presented in Table 2 were obtained using the Sobek test method (Sobek *et al.*, 1978). In addition, Jambor *et al.* (2000b, 2002) measured the surface area of the test charges using the BET method (Brunauer *et al.*, 1938), and normalized the NP values to the measured surface area. A comparison of the normalized NP values to the initial BET values (Table 2) illustrates the impact of surface area on NP values and demonstrates that the apparently large NP values associated with chlorite and some clay minerals may be due to the large surface area of these minerals.

Carbonate minerals represent the only significant source of NP for most rock-forming mineral assemblages associated with ore bodies (Jambor *et al.*, 2002). For example, barren granitoid rocks, if devoid of sulfides and carbonate minerals, would fall within the 'uncertain' category rather than being rated as not acid-generating. Jambor *et al.* (2006, 2007) conducted NP measurements on common rock types and compared the measured NP values with those estimated on the basis of quantitative mineralogy combined with NP values assigned to the constituent minerals. Comparison of the measured NP values with those computed on the basis of quantitative mineralogy (Table 3) shows reasonable consistency between the two approaches, with the NP measurements typically indicating higher NP values. The greatest discrepancy is observed for rocks that contain olivine, serpentine or calcic plagioclase. Because the rates of dissolution of these minerals are slow relative to the rate of sulfide mineral oxidation, the contributions of these minerals to NP are not likely to be available in settings where the sulfide mass is sufficient to result in rapid acidification. Jambor *et al.* (2006) noted that the dissolution rates of silicate and aluminosilicate minerals are likely to be sufficient to prevent acidic drainage only under unusual conditions; for example, at sites where the sulfide content is very low (*e.g.* <0.2 wt.% S) and sulfide minerals are encapsulated in the aluminosilicate assemblage.

In terms of regulatory guidelines, it would be convenient to establish a minimum sulfide content below which acid generation would be unlikely. However, inspection of Table 2 shows that a rock such as granite, quartz monzanite or quartz diorite, which consists largely of K-feldspar and albite, with most of the remainder as quartz and up to 10% mica or amphibole, will have a very modest NP. The feldspars will contribute <3 units of NP, and amphibole or biotite will add <2 units. Thus, the total NP will be <5 units and NP:AP will be $<1:1$ at a sulfide content of 0.3 wt.% S. Even at 0.2 wt.% S such a rock will be acid generating. Recent guidance documents have recommended that there be no lower limit assigned to the mass of sulfide minerals considered to be

Table 3. Computed NP values from mineralogy, and measured whole-rock NP values[a].

No.	Rock	Whole rock		Total carbon		Mineral	Main NP source
		NP_{meas}	NP_{comp}	C_{ppm}	NP^c	NP^a_{carb}	
1	Granite	10.1	8.9	850	7.2	9	Calcite
2	Granodiorite	3.7	8.1	100	<1	2	Calcite
3	Syenite	2.4	13.2	100	<1		Nepheline
4	Monzanite porph.	4.1	7.1	<100	<1	3	Calcite
5	Hornblende diorite	10.5	16.1	245	2.1	7	Calcite
6	Diabase	5.3	8.1	<100	<1		Plagioclase
7	Hornblende gabbro	9.5	17.5	350	3.0	5	Plagioclase
8	Gabbro	8.8	12.7	<100	<1		Plagioclase
9	Anorthosite[b]	23.4	20.2	1500	12.6		Calcite, plagioclase
10	Pyroxene	30.3	26.3	400	3.4		Olivine
11	Peridotite	19.1	28.1	500	4.2	5	Calcite, olivine, serp.
12	'Andesine'	9.5	8.5	400	3.4	4	Plagioclase, calcite
13	Rhyolite	2.9	0.7	70	<1	–	K-feldspar
14	Dacite	15	5.7	240	2.1	–	Plagioclase, An 57
15	Hornblende andesite	6.7	5.2	130	1.1	–	Plagioclase, An 53
16	Olivine andesite	11	8.0	125	1.1	–	Plagioclase, An 60
17	Basalt	15	3.2	165	1.4	–	Pyroxene
18	Olivine basalt	19	6.0	60	<1	–	Olivine
19	Argillaceous shale	78[d]	85	–	–	84	Calcite
20	Carbonaceous shale	1.3	1.0	–	–	–	Chlorite
21	Siltstone	13[e]	10	3710	31	9.2, [7.9]	Dolomite
22	Greywacke	5.3	1.2	370	3.1	–	Chlorite
23	Sandstone	16	18	2715	23	18, [4.3]	Dolomite
24	Gray slate	59[f]	65	10950	92	64	Calcite, dolomite
25	Green slate	5.1[g]	4.8	570	4.8	3.3	Dolomite
26	Garnet mica schist	3.9	1.9	1310	11	–	Biotite
27	Hornblende schist	4.7	2.4	95	<1	–	Amphibole
28	Amphibolite	34	26	95	<1	–	Olivine, serpentine
29	Greenstone	8.0	3.5	180	1.5	–	Pyroxene, chlorite
30	Biotite gneiss	12[h]	5.1	1205	10	4	Calcite
31	Cordierite hornfels	7.3	1.9	135	1.1	–	Biotite, cordierite

NP_{meas} – whole-rock Sobek value; NP_{comp} – computed value from mineralogy, uncorrected for the Fe component of carbonate minerals; C_{ppm} – measured total carbon in ppm; NP_c – calculated NP assuming all C is present as $CaCO_3$.

[a]Mineral NP_{carb} = NP values from mineralogical carbonate content as would appear in the Sobek results assuming no hydrolysis to $Fe(OH)_3$. The value in [ ] indicates the true NP after correction for the Fe content, if present, of the carbonate mineral(s).

[b]Calcite content not determinable by mineralogy, and NP_{calc} includes the value from C_{Total}. The NP value is for a 'non'-fizz rating; for a test at 'slight' fizz, the NP is 29.3.

[c]NP for a 'moderate' fizz rating; for a 'strong' fizz, NP = 125.

[d]NP for a 'non' fizz rating; for a 'slight' fizz rating, NP = 14.

[e]NP for a 'non' fizz rating.

[f]NP for a 'slight' fizz rating; for a 'strong' fizz rating, NP = 78.

[g]NP for a 'non' fizz rating: for a 'slight' fizz rating, NP = 14.

[h]Matrix plagioclase strongly chloritized, resulting in anonymously low (An 5) compositions for analysable grains.

Values from Jambor *et al.* (2006, reproduced with permission; 2007, reproduced with kind permission from Springer Science + Business Media).

non-acid generating and that assessments of acid-generating potential be established on the basis of an evaluation of the available AP and the effective NP (Price, 2009).

Three significant points can be deduced from the combination of the mineral-dissolution rates (Table 1) and the NP results (Tables 2, 3). First, unless they contain carbonate minerals, most igneous, metamorphic and sedimentary rocks will have insufficient NP to qualify them for the non-acid-producing category, even if very low sulfide concentrations of minerals are present. Exceptions in the igneous suite are rocks such as dunites and those which are feldspathoidal; it is noteworthy that orthopyroxene, which is rated as highly susceptible to natural weathering (Goldrich, 1938), does not contribute much Sobek NP (Table 2). A second point is that grain size has an obvious impact on NP values (Jambor *et al.*, 2000b). Thus, argillaceous rocks or fine-grained, clay-rich arenites may furnish appreciable NP if they contain abundant smectite rather than kaolinite. The third point is that silicate and aluminosilicate minerals are slow to react (Table 1), and that their NP potential will not be 'available' until after AMD has commenced.

A final comment concerning NP values is that protocols concerning sample preparation need to be more specific because of the apparently large effect that grain size may have on measured NP. The Sobek test stipulates only that the sample for analysis be minus 60 mesh ($<$ 250 µm) (Sobek *et al.*, 1978), and the British Columbia Modified Technique (MEND, 1991a) specifies a grain-size distribution with 80% finer than 200 mesh (74 µm), but with some flexibility for tailings samples (Lawrence and Wang, 1996, 1997). It is known well that finer grain sizes promote the rate of chemical reactivity, and it is evident from Table 2 that some of the NP values were substantially affected by the grain size of the mineral tested.

3. Oxidation products

Table 4, which is an updated version of that in Jambor & Blowes (1998) and Jambor *et al.* (2000a), shows the secondary minerals that have been identified in sulfide-rich tailings. Additions are from Bertorino *et al.* (1995), Benvenuti *et al.* (1997) and Agnew (1998), the last of whom has considerably expanded the previous list by examination of numerous tailings impoundments in Australia. The purpose of Table 4 is not merely to furnish a compilation of minerals. What is of more importance is that the nature of the minerals listed provides an insight into the products and the reactions that occur between the solids and pore waters, before the waters escape as effluents from a tailings impoundment. It should be emphasized that Table 4 does not include minerals formed in waste-rock piles, or minerals that have formed as down-stream precipitates, and the list does not include minerals whose identification has not been verified by powder X-ray diffraction (XRD) methods.

Despite the numerous minerals listed in Table 4, only six are indicated as commonly present. Of these, goethite occurs universally. Gypsum is also common, in part because many mills process calcareous ores or add lime to increase the pH of the tailings slurry prior to its discharge to a containment facility. At mines that have processed cupriferous ores, the tailings almost always contain traces of covellite at or near the interface between the saturated and unsaturated zones, which is also the zone of maximum Fe-oxyhydroxide accumulation. Marcasite and native sulfur are common only in pyrrhotite-bearing deposits, wherein both minerals form as an early-stage alteration of pyrrhotite [$Fe_{(1-x)}S$], and are subsequently replaced by goethite (Figs 2, 3, 4).

Table 4. Secondary minerals identified in sulfide-rich tailings.

Oxides, oxyhydroxides		*Sulfates (cont'd)*	
*Goethite	α-FeO(OH)	Anglesite	$PbSO_4$
Lepidocrocite	γ-FeO(OH)O	Thenardite	Na_2SO_4
Akaganéite	β-FeO(OH,Cl)	*Jarosite	$KFe_3(SO_4)_2(OH)_6$
Maghemite	γ-Fe_2O_3	Natrojarosite	$NaFe_3(SO_4)_2(OH)_6$
Hematite	Fe_2O_3	Hydronium jarosite	$(H_3O)Fe_3(SO_4)_2(OH)_6$
Ferrihydrite	$FeO[O_5(OH)_{1-x}\gamma\square_x$	Alunite	$KAl_3(SO_4)_2(OH)_6$
Diaspore	AlO(OH)	Copiapite	$Fe^{2+}Fe_4^{3+}(SO_4)_6(OH)_2 \cdot 20H_2O$
		Metahohmannite	$Fe_2^{3+}(SO_4)_2(OH)_2 \cdot 3H_2O$
Carbonates		Fibroferrite	$Fe^{3+}(SO_4)(OH) \cdot 5H_2O$
Cerussite	$PbCO_3$	Alunogen	$Al_2(SO_4)_3 \cdot 17H_2O$
Azurite	$Cu_3(CO_3)_2(OH)_2$	Hydrobasaluminite	$Al_4(SO_4)(OH)_{10} \cdot nH_2O$
Hydrozincite	$Zn_5(CO_3)_2(OH)_6$	Halotrichite	$Fe^{2+}Al_2(SO_4)_4 \cdot 22H_2O$
		Pickeringite	$MgAl_2(SO_4)_4 \cdot 22H_2O$
Sulfates		Blödite	$Na_2Mg(SO_4)_2 \cdot 4H_2O$
Bassanite	$2CaSO_4 \cdot H_2O$	Kaolinite	$KAl(SO_4)_2 \cdot 11H_2O$
*Gypsum	$CaSO_4 \cdot 2H_2O$		
Goslarite	$ZnSO_4 \cdot 7H_2O$	*Other minerals*	
Melanterite	$FeSO_4 \cdot 7H_2O$	*Covellite	CuS
Ferrohexahydrite	$FeSO_4 \cdot 6H_2O$	*Marcasite	FeS_2
Siderotil	$FeSO_4 \cdot 5H_2O$	*Sulfur	S
Rozenite	$FeSO_4 \cdot 4H_2O$	Chrysocolla	$(Cu^{2+},Al)_2H_2Si_2O_5(OH)_4 \cdot nH_2O$
Epsomite	$MgSO_4 \cdot 7H_2O$	Hemimorphite	$Zn_4Si_2O_7(OH)_2 \cdot H_2O$
Hexahydrite	$MgSO_4 \cdot 6H_2O$	Cristobalite	SiO_2
Pentahydrite	$MgSO_4 \cdot 5H_2O$	Opal	$SiO_2 \cdot nH_2O$
Starkeyite	$MgSO_4 \cdot 4H_2O$	*Hydrobiotite-vermiculite*	
		Smectite	$X_{0.3}Y_{2-3}(Si,Al)_4O_{10}(OH)_2 \cdot nH_2O$
		Kaolinite	$Al_2Si_2O_5(OH)_4$
		Erythrite	$Co_3(AsO_4)_2 \cdot 8H_2O$

*common, widespread.

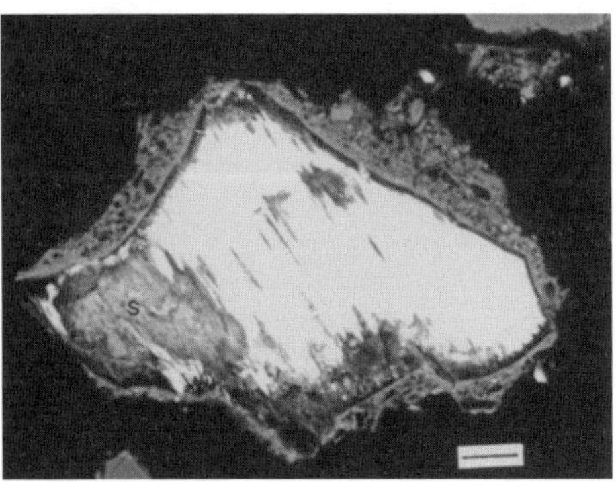

Fig. 2. Backscattered-electron (BSE) image of secondary-mineral formation in the Nickel Rim tailings, Sudbury, Canada. Pyrrhotite (white) is in the initial stage of alteration, partly replaced by native sulfur (S) and largely surrounded by a rim of Fe oxyhydroxide cementing tailings fines. Scale bar = 20 μm.

A succinct but wide-ranging review of the reactions that occur on a micro-scale at the surfaces of oxidizing sulfide minerals was given by Nordstrom & Southam (1997). It has been shown, mainly by X-ray photoelectron spectroscopy (XPS) and Auger electron spectroscopy (Richardson & Vaughan, 1989a,b; Pratt *et al.*, 1994a,b; Nesbitt *et al.*, 1995; Sasaki *et al.*, 1995; Yin *et al.*, 1995; Legrand *et al.*, 1997; Pratt & Nesbitt, 1997; Mikhlin *et al.*, 1998; Thomas *et al.*, 1998) that iron is readily leached from the surface of minerals such as pyrite, pyrrhotite, chalcopyrite [$CuFeS_2$], pentlandite [$(Fe,Ni)_9S_8$], and arsenopyrite [$FeAsS$], thus commonly leaving behind an interior S-enriched layer and, in some cases, forming an oxidized-Fe layer at the surface.

For pyrite that had been reacted with water vapour and air, XPS has indicated that formation of Fe^{3+} oxyhydroxides precedes the formation of Fe sulfates (Nesbitt & Muir, 1994). In pyrrhotite, the newly formed S-rich layer was concluded to have a composition and structure similar to those of marcasite (Mycroft *et al.*, 1995). Examples of the alteration of pyrrhotite to form marcasite, native sulfur, and Fe oxyhydroxides in naturally weathered tailings, are shown in Figures 2–8.

Although native sulfur is commonly associated with pyrrhotite in its alteration rims and pseudomorphs in oxidized tailings, sulfur has only rarely been observed in association

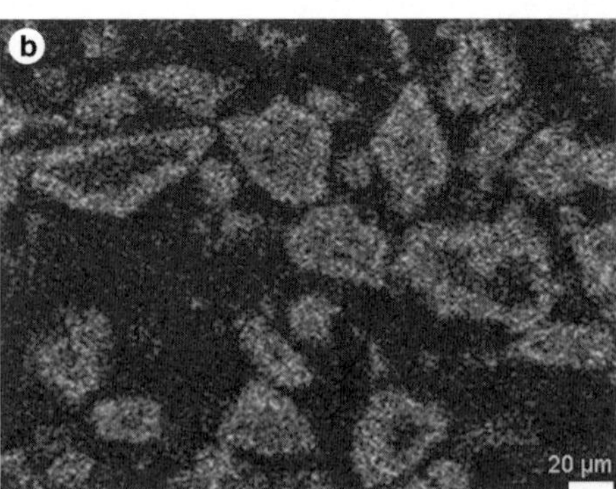

Fig. 3. BSE image (**a**) of pyrrhotite-rich tailings from the Waite Amulet site, Quebec, Canada (from Blowes & Jambor, 1990; figure reprinted from *Applied Geochemistry*, The pore water geochemistry and the mineralogy of the vadose zone of sulfide tailings, Waite Amulet, Quebec, Canada, pp. 327–346, Copyright (1990), with permission from Elsevier), showing relict pyrrhotite (po) rimmed by slightly darker marcasite; (**b**) X-ray map of S for the same area. The texture represents an early stage in the alteration sequence.

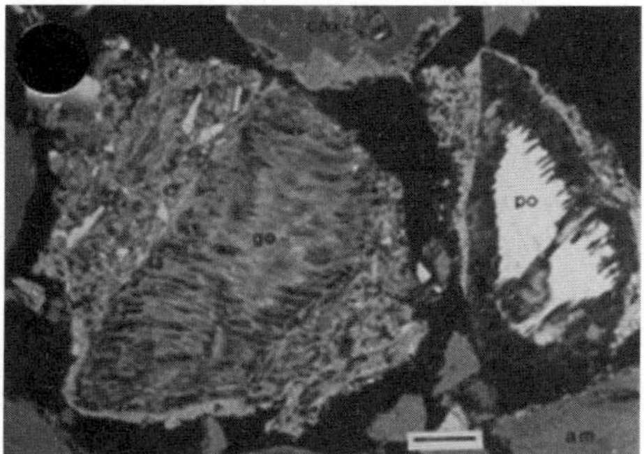
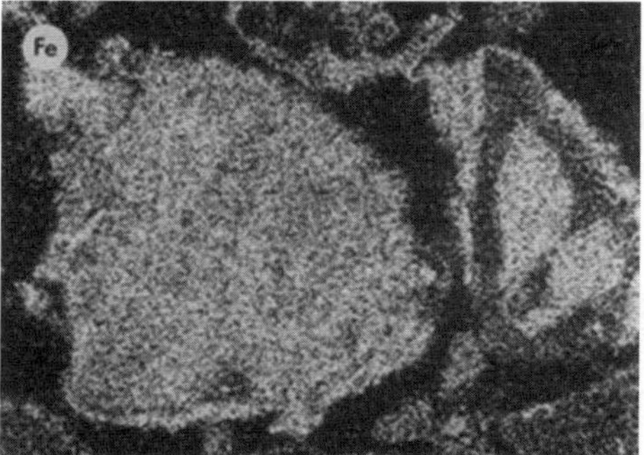

Fig. 4. BSE image and corresponding X-ray maps for Fe and S. Pyrrhotite (po) remnant on the right is rimmed by native sulfur (black), which is in turn partly rimmed by goethite. The grain to the left is a pseudomorph of goethite (go) after pyrrhotite. Associated grains are intergrown albite, augite (cpx) and amphibole (am). Nickel Rim, Sudbury, Canada. Scale bar = 25 μm.

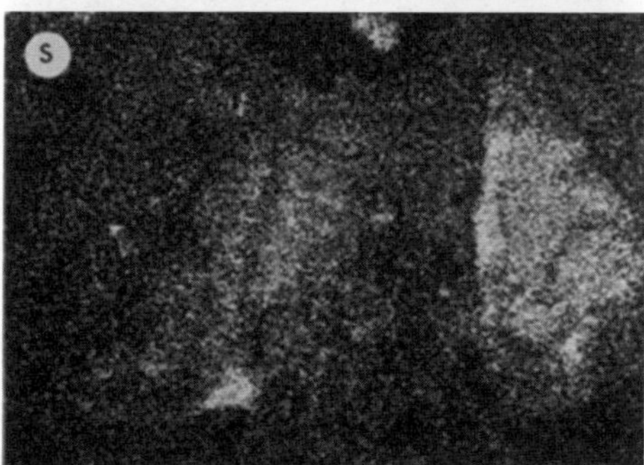

with chalcopyrite and pentlandite, and has not been observed with pyrite and arseno-pyrite. However, sulfur has been observed as a reaction product during bioleaching of copper sulfide ores (Watling, 2006). According to Nordstrom & Southam (1997), the metal monosulfide minerals such as sphalerite [(Zn,Fe)S], millerite [NiS], galena [PbS], and pyrrhotite, react with acid to form H_2S, whereas the metal disulfide minerals such as pyrite), under similar conditions, do not produce H_2S. Contact with air rapidly oxidizes the H_2S to sulfur, thiosulfate and sulfite, and these are then more slowly converted to sulfate. Thus, although H_2S is not produced in all conditions, its generation by acidic attack of pyrrhotite can lead to the formation of native sulfur, whereas the same does not occur with pyrite. The rarity of observed occurrences of native sulfur in association with chalcopyrite [$CuFeS_2$] is appropriate in that, as noted by Nordstrom & Southam (1997), the mineral is not a true disulfide because the sulfur ions are bonded to the metals rather than to other sulfur ions.

Mine wastes, and surface water impacted by mine wastes can be colonized by a wide diversity of microorganisms, including acidophilic bacteria and archea, as well as acidophilic and acid-tolerant micro-algae, fungi and yeasts, and protozoa (Johnson, 2007). Acidophilic microorganisms can be divided into three groups, depending on their pH for optimal growth. The optimum condition for extreme acidophiles is pH < 3, the optimum condition for moderate acidophiles is pH 3–5, while acid-tolerant species have pH optima at pH > 5, but are able to grow at pH 3 (Johnson, 2007). In addition to pH, temperature is an important environmental variable. Most acidophiles are mesophilic and grow between 15 and 40°C. Moderately thermophilic acidophiles grow best from 40 to 60°C and extreme acidophiles grow optimally in the range between 60 and 85°C (Gould and Kapoor, 2003). Thermophilic bacteria may play an important role in the oxidation of sulfide minerals in waste rock piles, where convective gas flow may enhance oxygen transport, increasing the rate of the exothermic sulfide oxidation reactions (Ritchie, 2003). Psychrotolerant species of acidophiles have been identified in northern environments (Gould and Kapoor, 2003), but have been studied less extensively.

The mechanisms of sulfide oxidation by microorganisms have been examined extensively over several decades. Recently, differing mechanisms have been proposed to describe the microbially mediated dissolution and oxidation of acid-soluble sulfides, *e.g.* sphalerite [(Fe,Zn)S], galena [PbS], arsenopyrite, and chalcocite [Cu_2S], *vs.* acid-insoluble sulfide minerals such as pyrite and tungstenite (WS_2) (Schippers and Sand, 1999; Rohwerder *et al.*, 2003).

The dissolution of acid-soluble sulfide minerals, proceeds through the 'polysulfide process' which is initiated by the dissolution of the sulfide mineral in acidic conditions releasing dissolved metals and H_2S and subsequent oxidation of H_2S by Fe(III):

$$MS + 2H^+ \rightarrow M^{2+} + H_2S \tag{2}$$

$$MS + Fe^{3+} + 2H^+ \rightarrow M^2 + + H_2S^{\bullet+} + Fe^{2+} \tag{3}$$

Dissociation of the $H_2S^{\bullet+}$ initiates the formation of polysulfides through the formation of disulfide:

$$H_2S^{\bullet+} \rightarrow H^+ + HS^{\bullet} \tag{4}$$

$$2HS^{\bullet} \rightarrow H_2S_2 \tag{5}$$

The disulfide may react further with Fe(III) or with other $HS^{\bullet}$ radicals. In acidic solutions the polysulfides decompose, ultimately forming elemental sulfur typically in S_8 rings (Schippers and Sand, 1999). The formation of polythionates and sulfate subsequently occurs through side reactions.

Microbially mediated oxidation of acid-insoluble sulfides, *e.g.* pyrite, proceeds through the 'thiosulfate pathway', which is initiated by oxidation by Fe(III) (Schippers and Sand, 1999). A series of six single electron transfers are required prior to separation of the sulfur and metal moiety (Rohweder *et al.*, 2003), resulting in the release of the dissolved metal and thiosulfate:

$$FeS_2 + 6Fe^{3+} + 3H_2O \rightarrow S_2O_3^{2-} + 7Fe^{2+} + 6H^+ \tag{6}$$

Thiosulfate is subsequently oxidized to sulfate *via* tetrathionate and other polythionate intermediaries:

$$S_2O_3^{2-} + 8Fe^{3+} + 5H_2O \rightarrow 2SO_4^{2-} + 8Fe^{2+} + 10H^+ \tag{7}$$

Significant amounts of elemental sulfur may be produced in the absence of sulfur oxidizing bacteria. In the thiosulfate process, the presence of Fe(II) oxidizing bacteria is required because the rate of abiotic Fe(III) oxidation slows under acidic conditions (Singer and Stumm, 1970; Rohwender *et al.*, 2003).

4. Paragenesis

In the current context, the term paragenesis refers to the sequence of deposition of secondary minerals. The behaviour of sulfide minerals under oxidizing conditions is governed by the local macro- and micro-environments. For example, in most methods of kinetic testing, the sulfides and the host mineral assemblage are in an apparatus in which the initial dissolved oxidation products are repeatedly flushed away. Thus, textures such as those shown in Figure 2 and in the following diagrams have little opportunity to develop. With each washing cycle, 'fresh' sulfide surfaces are exposed, and the general result is to have a diminution of the sulfide minerals without the accumulation of solid-phase replacement minerals.

In the more natural setting of a tailings impoundment that is exposed to existing climatic conditions, and in which a water table is allowed to form, the early stage of oxidation of sulfides leads to the crystallization of Fe oxyhydroxides and simple soluble sulfates ($\pm$ sulfur) as replacements of the primary sulfides (*e.g.* Blowes & Jambor, 1990; Blowes *et al.*, 1991, 1992; Moncur *et al.*, 2005; Gunsinger *et al.*, 2006). The Fe oxyhydroxides may include minute amounts of lepidocrocite [γ-FeOOH], but by far the most abundant secondary products are goethite [α-FeOOH] and smaller amounts of ferrihydrite (Table 4). The proportion of ferrihydrite relative to that of goethite is difficult to estimate because identification of ferrihydrite in bulk tailings samples is hampered by the poor XRD properties of the mineral; therefore, recourse to differential dissolution using ammonium oxalate (Schwertmann, 1964, 1973; Schwertmann *et al.*, 1982) is commonly made. The solutions, however, dissolve the water-soluble sulfates, the tertiary minerals (Jambor, 1994), and magnetite and siderite (Golden *et al.*, 1994; Rhoton *et al.*, 1981), thus potentially overestimating the amount of ferrihydrite that is present.

Following the initial development of goethite and the soluble sulfates, and the establishment of an acidic domain in the vadose zone, the oxidation front moves downwards, in concert with the low-pH front. In the zone nearer the surface, jarosite may form because the acidic solutions have been in place for sufficient time to begin attacking the aluminosilicates (Fig. 5). Extensive removal of aluminosilicates, including chlorite, biotite and smectite, are also observed in the zone nearer the surface (Moncur *et al.*, 2009). The feldspars are relatively inert, but trioctahedral micas are susceptible to loss of K^+ to form hydrobiotite. The Fe(III) oxyhydroxides, which in the initial stages seem to be the main 'sinks' for the deleterious elements *via* adsorption (Figs 7, 8), apparently undergo a recycling or recrystallization in which the adsorbed species are released, together with the products from soluble sulfates, and move downward. With maturation, the near-surface sulfides are depleted, and pH rises gradually so that jarosite is unstable and susceptible to dissolution. Although no tailings impoundment of sufficiently advanced age is known, presumably a quartz-enriched ferruginous capping analogous to that of a mature gossan, would develop over the long term.

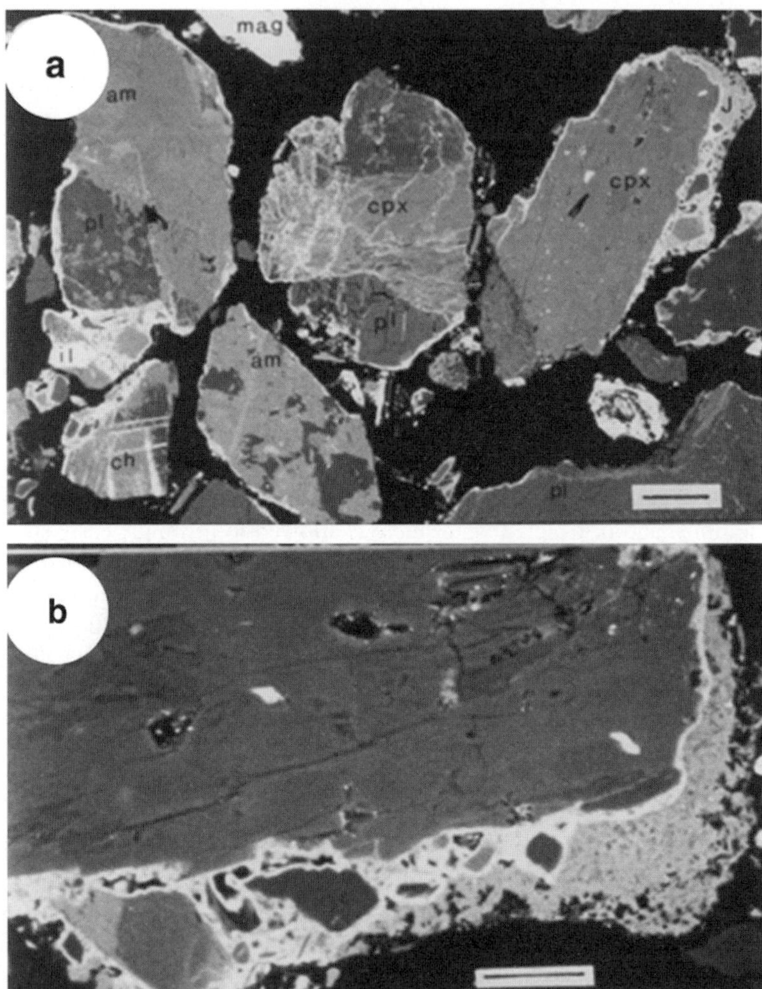

Fig. 5. BSE images of Nickel Rim tailings, Sudbury, Canada. (**a**) Magnetite (mag), amphibole (am), plagioclase (pl), ilmenite (il), chlorite (cl), and augite (cpx), with some grains partly rimmed by a thin rind of Fe oxyhydroxide, and near the top right, by jarosite (J). Scale bar = 50 μm. (**b**) Enlargement of grain of cpx at the right of part a, showing a thin rind of Fe oxyhydroxide (white) overlain by greyish jarosite, with the presence of jarosite indicating that alteration of aluminosilicates has commenced. Scale bar = 20 μm.

4.1. Simple sulfate salts

The list of minerals in Table 4 is dominated by sulfates of the type $M^{2+}SO_4 \cdot nH_2O$. These salts are unstable, and their hydration states are dependent on the local humidity or moisture conditions. The heptahydrates precipitate in moist conditions, readily losing some of their water of crystallization in drier situations, but the process is reversible. Hence,

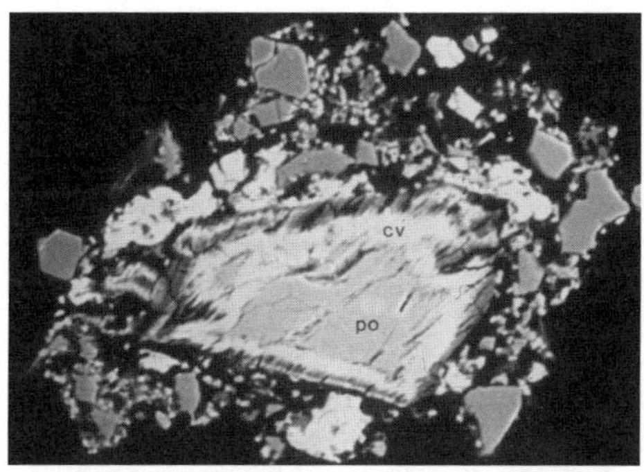

Fig. 6. BSE image of pyrrhotite (po) surrounded by a rim of secondary covellite (cv) in tailings from the Heath Steele mine, New Brunswick, Canada (after Blowes *et al.*, 1991; Blowes *et al.*, 1992). The deposition of covellite serves to control pore-water concentrations of Cu. Scale bar = 10 μm.

there is no physical impediment to restrict the formation of the lower hydrates of Zn, such as bianchite [$ZnSO_4 \cdot 6H_2O$] and boyleite [$ZnSO_4 \cdot 4H_2O$], and it is probably only a matter of time until their occurrence is reported. The monohydrates, especially szomolnokite [$FeSO_4 \cdot H_2O$], occur on the surfaces of coal spoils (Bayless & Olyphant, 1993; Kopsick & Kopsick, 1982; Gruner & Hood, 1971) and should also be present as efflorescences on oxidized tailings impoundments.

The $M^{2+}SO_4 \cdot nH_2O$ salts in Table 4 belong to the Zn, Mg and Fe series, but equivalent members also exist for Ni^{2+}, Co^{2+} and Mn^{2+}. The absence to date of

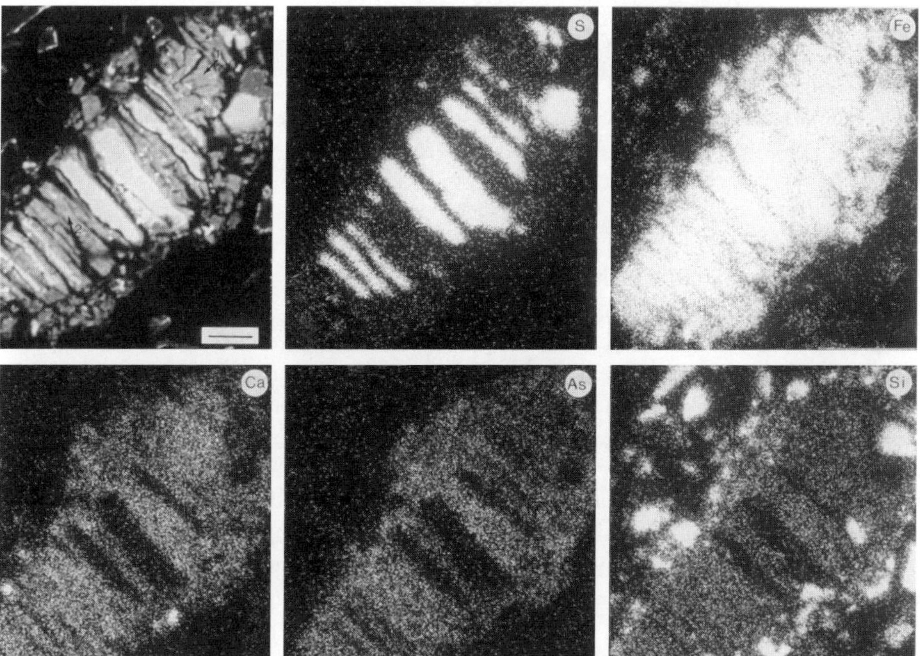

Fig. 7. BSE image of pyrrhotite (white) surrounded and penetrated by an Fe oxyhydroxide, probably goethite; numbers and arrows refer to microprobe-analysed spots. X-ray maps indicate that the alteration rim has been a sink for Ca, As and Si; analyses for points 1 and 2 gave CaO 3.2, 2.5, SiO_2 1.1, 1.3, As_2O_5 8.0, 8.7, SO_3 3.5, 3.0 wt.%, respectively. Tailings are from the former Delnite gold mine tailings near Timmins, Ontario, Canada (from Jambor & Blowes, 1991; reproduced with permission). Scale bar = 10 μm.

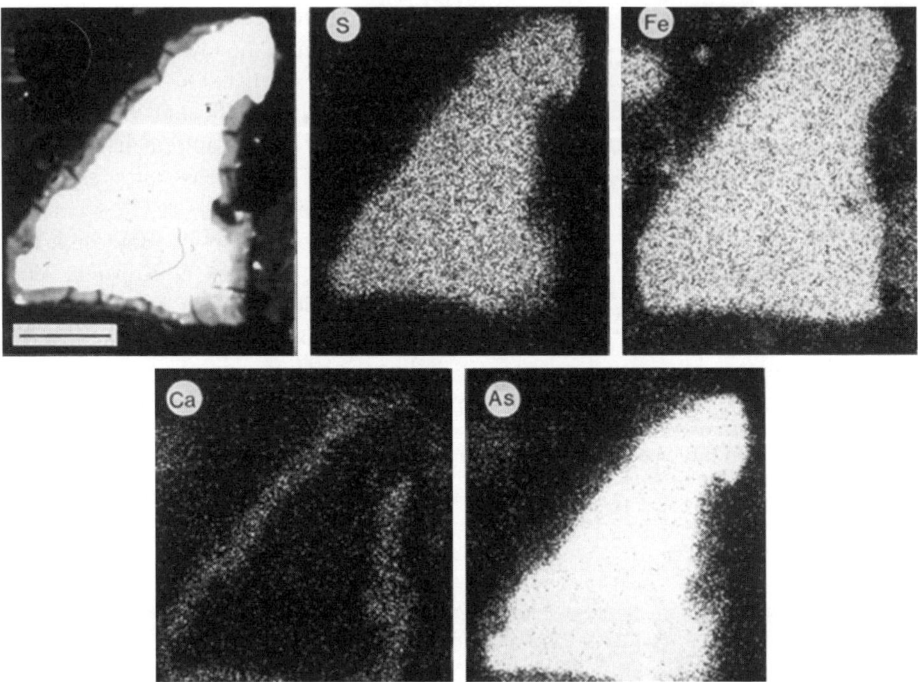

Fig. 8. BSE image and corresponding X-ray maps for arsenopyrite rimmed by Fe oxyhydroxide. Compare with Figure 8 and note that the sharp demarcation for As in the arsenopyrite grain shows that little of its As has been retained in the alteration rim. Delnite tailings. Scale bar = 10 μm.

the latter group in tailings merely reflects the importance of the role played by particular elements in the composition of the primary minerals. In waste-rock piles, the primary ore-mineral assemblage has not been extracted, hence the opportunities to form a diverse assemblage of secondary minerals are greater. The strict restriction of the Table 4 minerals to those that occur on and within tailings is physically constrained, but is largely arbitrary from a geochemical viewpoint. Nevertheless, the restriction is useful in that it is easy to pick out from Table 4 many of the minerals that represent a more advanced stage in the paragenetic sequence, because these minerals contain cations that are derived by dissolution of siliceous gangue minerals. Moreover, whether on a micro or macro scale, the general trend for the simple, non-hydroxylated Fe salts is to form the minerals of Fe^{2+} initially, followed by mixed Fe^{2+}–Fe^{3+} minerals, and thence to the Fe^{3+} minerals (Jambor *et al.*, 2000b). This general trend of $Fe^{2+} \rightarrow Fe^{3+}$ has been observed in unperturbed supergene zones (*e.g.* Bolshakov & Ptushko, 1969; Jacobsen, 1989), in arid-climate sulfate deposits in Chile (Bandy, 1938), in underground precipitates at Iron Mountain, California (Nordstrom & Alpers, 1999), and in laboratory studies (Buurman, 1975).

5. Mineralogical methodology

The approaches used to investigate the mineralogy of mine wastes have been discussed by Jambor (1994) and Jambor & Blowes (1998), and methods of study in environmental mineralogy are described by Wogelius & Vaughan (2000, 2013, this volume).

It is evident from the preceding text that the hydrous sulfate salts are extremely sensitive to variations in temperature and humidity. After cores of tailings are removed from an impoundment, the samples should be frozen immediately to minimize potential reactions between the pore water and solids. The cores can subsequently be thawed and dried at room temperature. Although typically higher than that within the impoundment, the ambient temperature is the closest approximation that is practical for handling the bulk samples. Samples which are anticipated to contain Fe^{2+} can be prepared by drying under an inert atmosphere to prevent oxidation of reduced metals, including Fe^{2+} and Mn^{2+}.

To hasten the drying process, some investigators have heated their core samples to 60°C, or even higher. This procedure will change the hydration states of some of the sulfate salts, and heating also promotes reactions between the pore water and the solids.

A final comment on determinative mineralogy pertains to the necessity of verifying identifications by alternative methods. As pointed out by Jambor & Blowes (1998) "A quantitative microprobe analysis of an iron-oxide alteration rim will not unequivocally identify the rim as goethite, lepidocrocite, ferrihydrite, schwertmannite, or mixtures of these and possibly other minerals. As no optical microscopic discrimination is possible, in the absence of corroborative data there is no justification for a specific identification." Microprobe compositions, like powder XRD patterns, do not by themselves necessarily give unequivocal identification of minerals that can have a high adsorptive capacity, or minerals in which extensive or non-binary solid solutions are common. It would be inadvisable, for example, to identify by XRD pattern, a mineral in the alunite-jarosite family, unless additional data to support the identification were obtained. These caveats are not restricted to, but are especially applicable to, the mineralogy of oxidation products in mine wastes.

6. Rates of AMD generation

The rapidity at which sulfide minerals react depends on the rate at which oxidants are transported to the mineral grain and the rate of reaction at the sulfide surface. The acid generated through sulfide mineral oxidation can be neutralized through dissolution of carbonate, (oxy)hydroxide and silicate minerals. The rate at which these neutralizing minerals dissolve varies considerably and can be gauged from Table 5. The extent of reaction will depend on local conditions, including the quantities of sulfide minerals present, but rates of sulfide oxidation can, from a practical viewpoint, be considered as almost instantaneous, provided the oxidant can be transported to the unoxidized mineral surface. For example, pyrrhotite may oxidize so rapidly in ordinary atmospheric conditions that concentrates to be used for experimental purposes must be kept in a desiccator to minimize oxidation. In field conditions, the rate of pyrrhotite oxidation

Table 5. The reactivity of various minerals expressed as the lifetime of a 1 mm crystal in aqueous solution with the pH poised at 5. This table includes the reactivity of calcite for a comparison.

Mineral	Lifetime (y)	Mineral	Lifetime (y)
Calcite	0.43	Albite	575,000
Wollastonite	79	Microcline	921,000
Forsterite	2300	Epidote	923,000
Diopside	6800	Muscovite	2,600,000
Enstatite	10,100	Kaolinite	6,000,000
Sanidine	291,000	Quartz	34,000,000

From Lasaga & Berner (1998), reprinted from *Chemical Geology*, Fundamental aspects of quantitative models for geochemical cycles, 161–175, Copyright 1998, with permission from Elsevier.

is rapid, and the time scale for potential AMD can be thought of in terms of weeks or months rather than years or decades.

Pyrite is slower to react than pyrrhotite. Nevertheless, in wastes that contain abundant pyrite, as is typical for the rejects from massive sulfide ores, the time scale for potential AMD generation is not much different from that of pyrrhotite-bearing wastes. The initial oxidation of pyritiferous wastes begins almost immediately after exposure to atmospheric conditions, and the initial acidity that is generated has the effect of catalyzing further reaction. Once the acidity is generated, however, the rate of dissolution of carbonates also increases rapidly. If carbonates are not present, attenuation of the acidity will lag because of the slow dissolution rates of most silicates and aluminosilicates (Table 5). In laboratory column transport experiments, in which tailings were exposed to sulfuric acid at flow rates representative of field conditions, the dissolution of carbonate and ferric and aluminum oxyhydroxide minerals in the tailings was sufficiently rapid to be treated as equilibrium reactions, whereas dissolution of chlorite was kinetically limited (Jurjovec *et al.*, 2002, 2004). In the prevention or control of potential AMD, therefore, there is no hiatus for the implementation of mitigation measures. Remediation, as the name indicates, applies to situations in which an environmentally damaging scenario already exists. Such situations are common at abandoned mines and, at some sites the mine wastes have been generating AMD for centuries.

7. Strategies for remediation

Since recognition of the problems of low-quality drainage associated with tailings impoundments, the mining industry has attempted to develop reliable techniques for the remediation or prevention of these problems. The development of these techniques requires an understanding of the complex mineralogical characteristics of tailings impoundments. Understanding the mechanisms controlling tailings pore-water chemistry has led to the development and testing of innovative approaches for the disposal and treatment of mine tailings. As our understanding improves, additional techniques undoubtedly will be put forward.

Johnson & Hallberg (2005) reviewed the abatement techniques that have been developed to date, and prepared a cost-benefit comparison for the various approaches, which can be grouped into three broad categories: (1) collection and treatment of tailings-derived discharge; (2) infiltration controls; and (3) sulfide-oxidation controls. An alternative abatement approach not included in the above are *in situ* mixing of reactive materials within tailings impoundments.

The effluent from many tailings impoundments in North America is collected and treated prior to discharge from the mine-site. Acceptable effluent quality is achieved, but the capital and consumable costs associated with perpetual maintenance of treatment facilities are sufficiently large that the development of alternative prevention and abatement techniques is highly desirable.

7.1. Remediation of existing sources of acidic drainage

At many locations, mining has been active for several decades. Although mine wastes at these locations presumably were deposited in accordance with the accepted practices available at the time of disposal, many of these wastes are now the source of low-quality drainage waters. Although, at some of these sites, the period of most intense sulfide oxidation and associated release of dissolved metals has passed, the transport of the products of sulfide oxidation will continue through the tailings or underlying aquifers for several decades (*e.g.* Coggans *et al.*, 1999; Bain *et al.*, 2000). Remedial programs for these sites focus on the collection and treatment of tailings-derived discharge, or on reducing the infiltration and movement of recharge water through the tailings. At tailings impoundments in which sulfide oxidation is in its early stages, it is beneficial to inhibit these reactions through the use of sulfide-oxidation controls, such as saturated soil covers or organic carbon covers, by which oxygen ingress is inhibited. For older tailings areas in which the peak period of sulfide oxidation has passed, a potential alternative approach to remediation is to install porous, permeable sulfate-reducing reactive barriers in the path of plumes of tailings-derived water.

7.2. Collection and treatment

Collection and treatment facilities are in place at many tailings sites throughout the world. Conventional plants for water treatment vary from relatively simple, to automated, computer-controlled facilities. Even relatively primitive facilities in some instances can provide an effective control on the release of acidic drainage and dissolved metals. The cost of maintenance of these facilities, however, is large and there is a strong desire, if not a necessity, to develop and implement alternative treatment systems. Moreover, where lime is used, the disposal of the resulting sludge residues has become an environmental concern, both in terms of disposal and in the potential release of dissolved metals through subsequent leaching.

Alternative passive-treatment systems which have been proposed include downstream polishing or metal scavenging by using wetlands, wood-waste, or peat. These systems of passive treatment may be enhanced through the use of anoxic limestone drains to precondition water prior to discharge to the wetland systems (Hedin & Watzlaf, 1994; Robins *et al.*, 1999). In some situations, anoxic limestone drains can be used as a stand-alone treatment system. Other proposals focus on concentrating the beneficial biological activity observed in a wetland within a constructed facility. For many years, wetlands and peat bogs have been recognized as areas of metal accumulation (Kalin & van Everdingen, 1987; Ritcey, 1989; Brown, 1991; Machemer & Wildeman, 1992). Passive treatment of acidic coal-mine drainage by constructed wetlands has been demonstrated to be efficient and effective. In the Appalachian region of the United States and elsewhere, hundreds of wetland systems have been treating low-quality drainage waters derived from coal mine sites (Kleinmann *et al.*, 1991; Walton-Day, 2003; Hedin *et al.*, 2010; Matthies *et al.*, 2010). Constructed wetlands for treatment of drainage from base- and precious-metal tailings impoundments have also been proposed (Kleinmann *et al.*, 1991; Gould *et al.*, 1994). Passive polishing by downstream wetlands requires the establishment and maintenance of wetlands that not only will remain stable over long periods, but will survive the metal loadings associated with base- and precious-metal tailings. As metal loadings from many tailings impoundments are expected to increase over future decades (Blowes & Jambor, 1990; Blowes *et al.*, 1992; Coggans *et al.*, 1999; Moncur *et al.*, 2005), the wetland must withstand both current and future loadings to be suitable as a long-term treatment system. One of the major concerns with this concept is the gradual infilling of the wetland basin and the potential for the remobilization of contaminants associated with the tailings impoundment.

The mechanisms resulting in the immobilization of dissolved metals in wetlands have been the focus of much research in recent years. Kleinmann *et al.* (1991) noted that a significant attenuation of dissolved metals results from the bacterially catalyzed reduction of sulfate to sulfide and the accompanying re-precipitation of metal sulfides through reactions of the form:

$$2CH_2O + SO_4^{2-} + 2H^+ \rightarrow H_2S + H_2CO_3 \tag{8}$$
$$Me^{2+} + HS^- \rightarrow MeS_{(s)} + H^+ \tag{9}$$

where CH_2O represents a labile source of organic carbon, and *Me*S represents an amorphous or poorly crystalline sulfide precipitate. These reactions result in an increase in alkalinity of the tailings discharge-water, and a decrease in the concentrations of dissolved metals. In wetlands, organic carbon is supplied and replenished through the annual growth of wetland plants and their degradation in the wetland. Sulfate, Fe(II) and other dissolved metals are supplied by the captured acidic mine-drainage waters (Matthies *et al.*, 2010).

Understanding of the controlling reactions in wetland systems has resulted in the proposal and development of treatment systems in which acidic drainage is directed through continuous-flow reactors. Dvorak *et al.* (1991) proposed that such reactors should contain solid-phase organic carbon to induce sulfate reduction and precipitation of metal sulfides. The results of pilot-scale experiments demonstrate significant decreases in the concentrations of SO_4 and the dissolved metals Fe, Zn, Mn, Ni, and Cd; these decreases are accompanied by increases in effluent pH and alkalinity. Dvorak *et al.* (1991) indicated that scaling up of the reactors to treat acidic drainage from minesites is economically feasible. A further enhancement to this system was proposed by

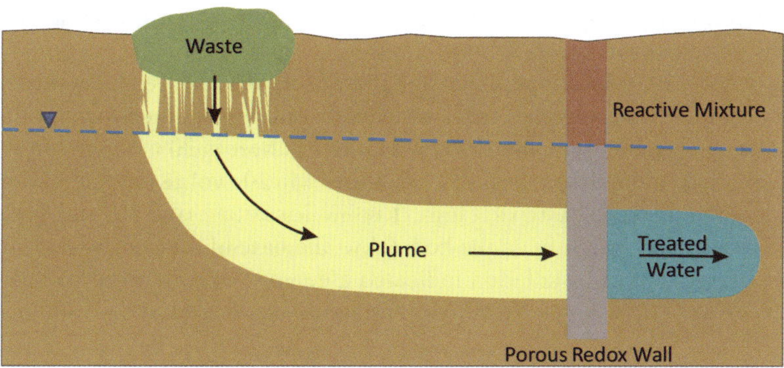

Fig. 9. Schematic diagram of a porous reactive wall for *in situ* treatment of contaminated groundwater.

Kepler & McCleary (1994), in which treatment using sulfate reduction is combined with treatment using anoxic limestone drains.

Plumes of tailings-derived water moving in aquifers affected by mine-tailings impoundments are anaerobic, and typically the pH is near-neutral and the temperature is relatively constant throughout the year (Morin, 1983; Bain *et al.*, 2000). These conditions are well suited for the *in situ* treatment of mine drainage waters. Blowes & Ptacek (1992a, 1994) proposed intercepting Fe(II)- and SO_4-laden groundwater prior to discharge to the surface-water flow system through the use of permeable reactive barriers containing labile organic carbon (Figs 9, 10). Treatment using permeable reactive barriers requires that the contaminant be sufficiently amenable to reaction as the plume of contaminated water passes through the reactive barrier. The barrier material must also be sufficiently reactive during the limited residence time of the contaminant plume, yet sufficiently stable so that the lifespan of the barrier makes it economically competitive with other treatment systems. The results of laboratory bench-scale experiments using various potential barrier components have suggested that the sulfate-reduction and metal-sulfide precipitation reactions proceed rapidly enough that treatment of

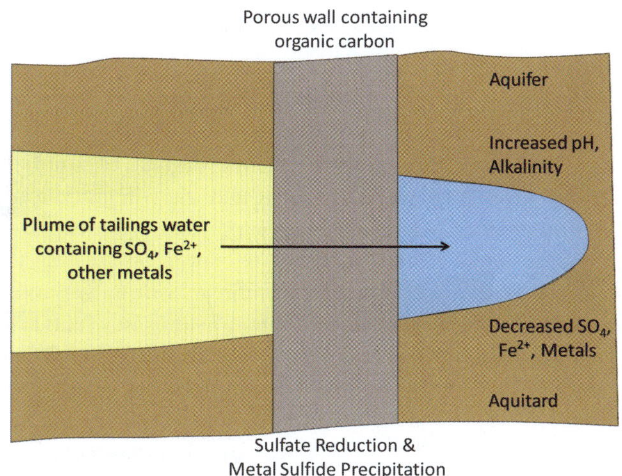

Fig. 10. Schematic diagram showing relevant reactions for treatment of tailings water containing SO_4, Fe(II) and other metals.

tailings-derived water by sulfate-reducing reactive barriers is feasible (Waybrant *et al.*, 1997, 1998, 2002; Blowes *et al.*, 1998).

Benner *et al.* (1997) described the installation of a full-scale reactive barrier for the treatment of groundwater derived from an inactive tailings impoundment at the Nickel Rim mine site near Sudbury, Ontario (Fig. 11). The barrier is 20 m long, 4 m thick, and 3.5 m deep, and contains a reactive mixture of 20 vol.% municipal compost, 20 vol.% leaf mulch, 9 vol.% wood chips, 1 vol.% limestone, and 50 vol.% pea gravel. The barrier was placed directly on the bedrock at the base of the aquifer. Monitoring of the barrier subsequent to installation indicated a decrease of dissolved sulfate concentrations by 2000–3000 mg/l, iron concentrations decreased 270–1300 mg/l, and

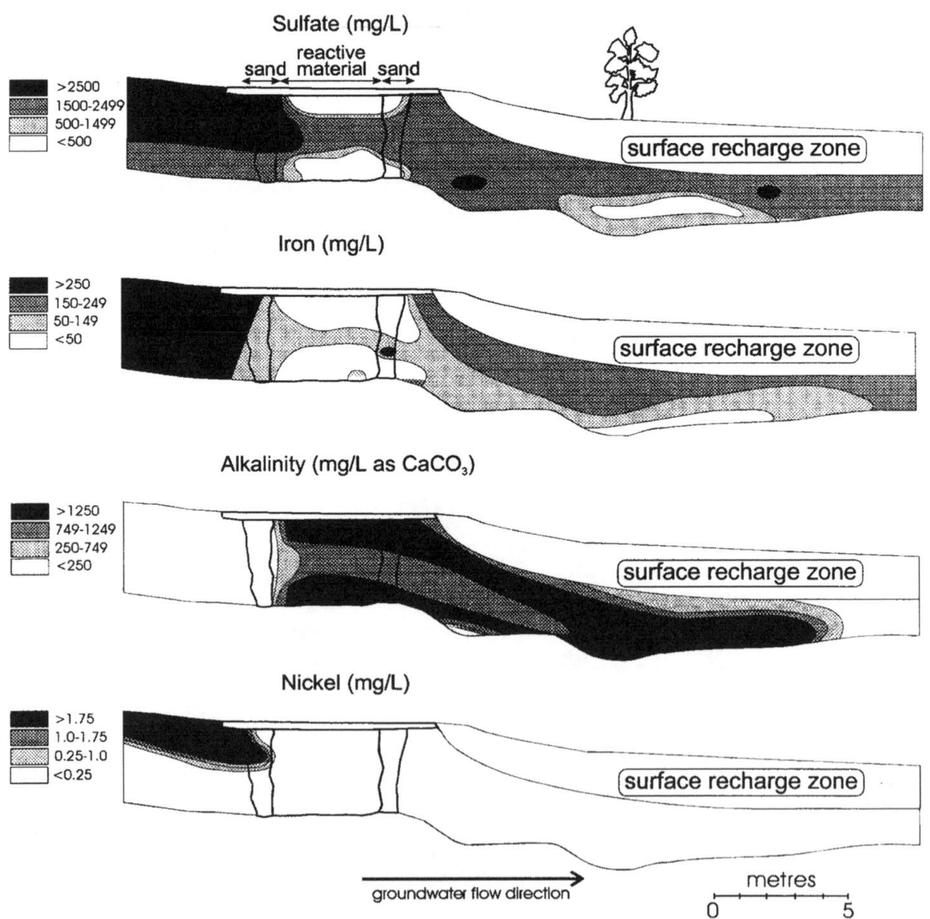

Fig. 11. Removal of SO₄, Fe and Ni in a permeable reactive wall installed at the Nickel Rim mine site, Ontario (from Benner 2000, reproduced with permission).

alkalinity increased 800–2700 mg/l. These changes resulted in significant improvements in the water chemistry down-gradient of the reactive barrier (Benner *et al.*, 1999). These improvements in water chemistry were attributed to microbially mediated sulfate reduction and metal-sulfide mineral precipitation. Sulfate reduction was indicated by increased numbers of sulfate-reducing bacteria, increased concentrations of dissolved sulfate, and enrichment in ^{34}S relative to ^{32}S (Benner *et al.*, 1999). Mineralogical study of precipitates formed on the surfaces of sampling tubes installed into the barrier indicated the accumulation of ferrous sulfide precipitates with a structure similar to mackinawite [$Fe_{(1+x)}S$] (Herbert *et al.*, 1998).

Monitoring of the barrier performance over a period of ~5 years indicated a gradual decline in the rate of sulfate reduction within the barrier, with the rate being ~50% of the initial rate 4.5 years after installation (Benner, 2000). Benner (2000) also noted that the groundwater temperature and the residence time within the barrier affected the rate of sulfate reduction. Core samples of the barrier material were collected ~6 years after installation of the barrier. A mineralogical study of the barrier material (Jambor *et al.*, 2005) showed that iron within the barrier was present in a variety of forms, the most common being iron sulfides and iron (oxy)hydroxides. Framboidal pyrite was observed but was relatively rare. Other iron sulfides had the appearance of greigite [Fe_3S_4] and pyrrhotite. Although Jambor *et al.* (2005) reported an overlap in the precipitation of various minerals, the general progression was Fe (oxy)hydroxide, followed by greigite (or pyrrhotite), followed by pyrite. The principal sink for nickel was observed to be precipitation as a native element (Fig. 12). Secondary carbonate minerals were uncommon.

The performance of the Nickel Rim reactive barrier was simulated using the reactive transport model MIN3P (Mayer *et al.*, 2006). These simulations which included the reaction mechanisms proposed by Benner *et al.* (1999), the secondary minerals identified by Jambor *et al.* (2005), and the flow-system parameters and temperature conditions measured at the Nickel Rim site, are consistent with the field observations. The mineralogical analyses provided an important constraint on these calculations.

A pilot-scale permeable reactive barrier was installed at an industrial site in Vancouver, British Columbia, Canada (McGregor *et al.*, 1999). At this site, the target contaminants, principally dissolved copper and zinc, decreased to very low concentrations during the residence time in the barrier (McGregor *et al.*, 1999). Analysis of the results from the pilot-scale installation indicated that sulfate-reducing bacteria were present within the barrier and that dissolved metals were removed through precipitation of metal sulfide minerals (Ludwig *et al.*, 2002). Samples of the reactive materials retrieved from the pilot-scale barrier ~4 years after installation were examined by Jambor *et al.* (2005). Sulfide minerals observed within the reactive material were dominated by framboidal pyrite. Copper sulfides present were covellite, small amounts of chalcopyrite, and bornite. The secondary chalcopyrite occurred as oscillating rims on other phases (Fig. 13). Following 4 years of monitoring of the pilot-scale barrier, a full-scale permeable reactive barrier was installed at the site in Vancouver, British Columbia (Mountjoy and Blowes, 2002). This barrier was built in a series of sections that totalled 400 m

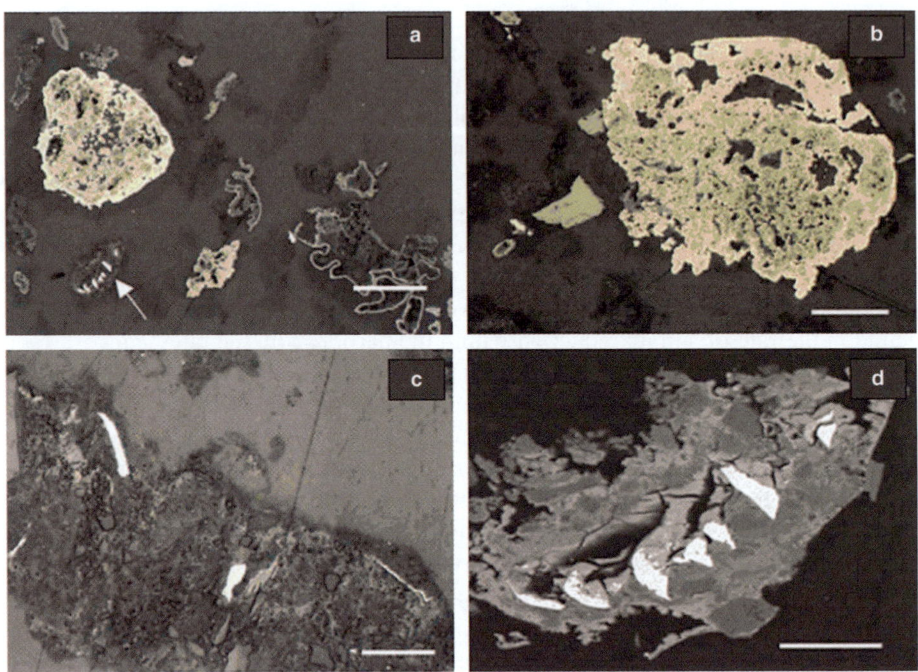

Fig. 12. Photomicrographs of solid samples collected from the Nickel Rim permeable reactive barrier. (**a**) The large particle on the left is greigite rimmed by pyrite; the thin rinds on other particles are Fe oxyhydroxide with an exterior film of Fe sulfide that is probably greigite (composition obtained in analysis is $Fe_{3.2}S_4$). The arrow points to a group of white grains of native nickel. Reflected light, scale bar = 100 μm. (**b**) The main mass is greigite, as confirmed by Powder XRD, and the outer, yellow rim is pyrite. The adjacent grey grains on the left are primarily illmenite. Reflected light, scale bar = 50 μm. Figures (*a*) and (*b*) represent some of the largest grains of secondary Fe sulfide observed in the barrier samples. (**c**) The white grains, including the thread-like grain at the lower right, are native nickel which is present in organic matter. Reflected light, scale bar = 100 μm. (**d**) Back-scattered electron image of the native nickel shown in part a; scale bar = 10 μm. The nickel (white) is enclosed within a medium-grey, cracked Ni > Fe sulfate or similar compound. The slightly darker inclusions are quartz and Al-bearing silicates (from Jambor *et al.*, 2005, reproduced with the permission of the Mineralogical Association of Canada).

long and 17 m deep. The reactive material in the full-scale barrier consisted of a mixture of gravel, municipal compost and zero-valent iron.

A permeable reactive barrier was installed in Charleston, South Carolina, USA to treat As and Pb derived from a fertilizer production plant which utilized sulfide minerals in the fertilizer production (Ludwig *et al.*, 2002). Full-scale reactive barriers have been installed at a variety of mine sites. A permeable reactive barrier was used to treat coal-spoil-heap drainage in Northumberland, UK (Jarvis *et al.*, 2006). This barrier typically removed 50% Fe and 40% sulfate from the drainage water. In Aznalcóllar, Spain, the removal of metals in a permeable reactive barrier was attributed to an increase in pH rather than formation of metal sulfides (Gibert *et al.*, 2011). Because of relatively

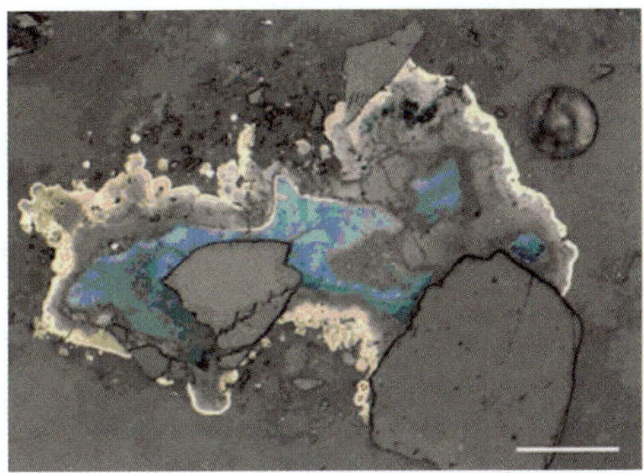

Fig. 13. Reflected-light photomicrograph of a grain from a pilot-scale, industrial-site permeable reactive barrier. Scale bar = 100 μm. Secondary covellite (blue, confirmed by XRD) with peripheral secondary chalcopyrite and secondary brownish bornite which show oscillatory zoning (from Jambor *et al.*, 2005, reproduced with the permission of the Mineralogical Association of Canada).

high flow rates at the site, sulfate-reducing conditions were not established throughout the barrier, leading to less effective treatment.

8. Infiltration controls

Although numerous design strategies to restrict the movement of tailings pore-water from the impoundment to adjacent groundwater and surface-water flow systems have been proposed, the most effective strategy to control the movement of tailings pore-water from the impoundment is to restrict the entry of meteoric water, surface water and groundwater into the impoundment. This, in turn, will reduce the quantity of tailings-derived seepage leaving the impoundment.

Entry of groundwater and surface water into the tailings can be avoided by appropriate site selection. Where suitable impoundment locations are not available, or where tailings are already placed in an undesirable location, the use of synthetic liners, cut-off walls, and diversion trenching may be considered. Infiltration of meteoric water can also be restricted by surface contouring and by emplacement of low-permeability covers, either of natural geological or synthetic materials, over the tailings. Optimal cover design would provide a barrier both to infiltration of meteoric water and to atmospheric O_2. 'Store and release' covers are an alternative cover design that is well suited to sites located in arid climates, or sites with prolonged wet and dry seasons (Williams *et al.*, 2006). These cover designs are intended to intercept and collect infiltrating precipitation during the wet season and promote evapotranspiration of the stored water during the dry season. A store-release-cover was installed at the Kidston Gold Mine, Queensland, Australia and monitored. The cover design includes a 0.5 m thick compacted layer overlain by a vegetated rocky soil layer, with a dozed mounded surface and a minimum thickness of 1.5 m. The compacted layer limits percolation and the vegetated layer enhances evaporative losses from the cover. Monitoring indicated percolation of water through the cover was 1.1% of incident rainfall over a 9 year period (Williams *et al.*, 2006).

Although it is desirable to limit the entry of meteoric water, groundwater and surface water into tailings areas, this may not be economically feasible given the large areas (>200 ha) of some tailings impoundments. In these cases, it may still be desirable to restrict the movement of tailings water into underlying aquifers to prevent degradation of the quality of the aquifer water, and to direct tailings-derived water to collection points prior to treatment. Cutoff walls, diversion trenching, and impermeable barriers can be used to direct flowing groundwater towards *in situ* treatment zones (Starr & Cherry, 1994) or toward treatment systems on the tailings surface.

9. Sulfide-oxidation controls

Abatement techniques that limit sulfide oxidation are bactericidal controls, precipitates and biopolymers that coat sulfide surfaces, and O_2-diffusion barriers. The first involves inhibition of the naturally occurring sulfide-oxidizing bacteria through the use of bactericides applied either directly to the tailings surface, or as an intimate mixture with the tailings in the impoundment (Erickson & Ladwig, 1985). In the absence of these bacteria, the rate of sulfide-mineral oxidation decreases as the pH decreases, and the concentrations of dissolved metals remain relatively low. A variety of bactericides has been tested, and show short-term inhibition of bacterial activity (Sobek, 1987; Stichbury *et al.*, 1995; Lortie *et al.*, 1999). The requirement for continual reapplication, however, suggests that bactericides may be better suited to short-term disposal sites, such as locations where tailings are temporarily exposed during impoundment construction. Lortie *et al.* (1999) suggested using bactericides to prevent acid generation in waste rock stored in exposed locations prior to disposal in open pits.

The second type of oxidation control armours the sulfide-mineral surface with a coating of an insoluble, non-reactive precipitate, thereby isolating the sulfide mineral from oxidants (*e.g.* O_2 and Fe^{3+}). Among the various armouring phases that have been proposed are ferric phosphate (Huang & Evangelou, 1994) and ferric oxyhydroxides (Ahmed, 1991). Huang & Evangelou (1994) observed decreased rates of sulfide oxidation in samples amended with phosphate compared to control samples. In field settings, ferric oxyhydroxide rims are observed on altered sulfide minerals (Blowes & Jambor, 1990; Jambor & Blowes, 1991). Studies which model sulfide oxidation in mine wastes using 'shrinking core' models (Davis & Ritchie, 1986; Davis *et al.*, 1986) assume that similar coatings affect the rate of sulfide oxidation. These field and modelling studies suggest that the observed oxidation rates in the presence of these altered rims remain sufficiently high to represent an environmental concern. To be an effective remedial system the armouring coating must maintain rates below those observed for naturally occurring alteration. Zhang & Evangelou (1998) and Fytas *et al.* (1999) evaluated coatings composed of ferric silicate. Ferric silicate was considered to be a preferable coating material because it binds ferric iron to the surface of the sulfide mineral in a low-solubility form, because it is resistant to acidic surroundings, and because the by-products from the reaction are more environmentally benign than phosphate.

Bioshrouding is an alternative to coating sulfides with low-solubility mineral coatings (Johnson *et al.*, 2008). By promoting the growth of heterotrophic iron-reducing bacteria on sulfide mineral surfaces, the rate of sulfide mineral oxidation can be reduced by as much as 75%. The heterotrophic bacteria produce extrapolymer substances (EPS) which form a layer on the sulfide surface which inhibits attachment of iron-oxidizing bacteria. The iron-reducing bacteria also remove ferric iron, a primary oxidant of sulfide minerals, further limiting the potential for sulfide mineral oxidation.

The third type of sulfide-oxidation control restricts the entrance of gas-phase O_2 into the impoundment by placing a diffusion barrier between the atmosphere and the reactive sulfide tailings. Among the several approaches to the barrier technique that have been proposed are dry covers composed of fine-grained materials, which may be layered to maintain high water contents and low O_2 diffusion coefficients (Collin & Rasmuson, 1986; Sodermark & Lundgren, 1988; Rasmuson & Collin, 1988; Nicholson *et al.*, 1989; Yanful *et al.*, 1994); covers composed of synthetic materials of low O_2 diffusivity (Sodermark & Lundgren, 1988; Malhotra, 1991); or covers containing O_2-consuming materials (Reardon & Poscente, 1984; Reardon & Moddle, 1985; Broman *et al.*, 1991; Tassé *et al.*, 1994).

Covers constructed of fine-grained material rely on the moisture-retaining characteristics of these materials to maintain high moisture contents several metres above the water table (Collin & Rasmuson, 1986; Nicholson *et al.*, 1989). Because of the moisture-retaining capabilities of fine-grained materials, it is possible to layer these materials (silt- and clay-size sediments) over coarser-grained tailings and to maintain saturated or near-saturated condition several metres above the water table (Fig. 14).

Covers incorporating synthetic layers have been proposed (Sodermark & Lundgren, 1988; Lundgren and Lindahl, 1991). For example, a composite cover including, from the base upwards: a filter layer composed of sand or geotextile, a sealing layer constructed

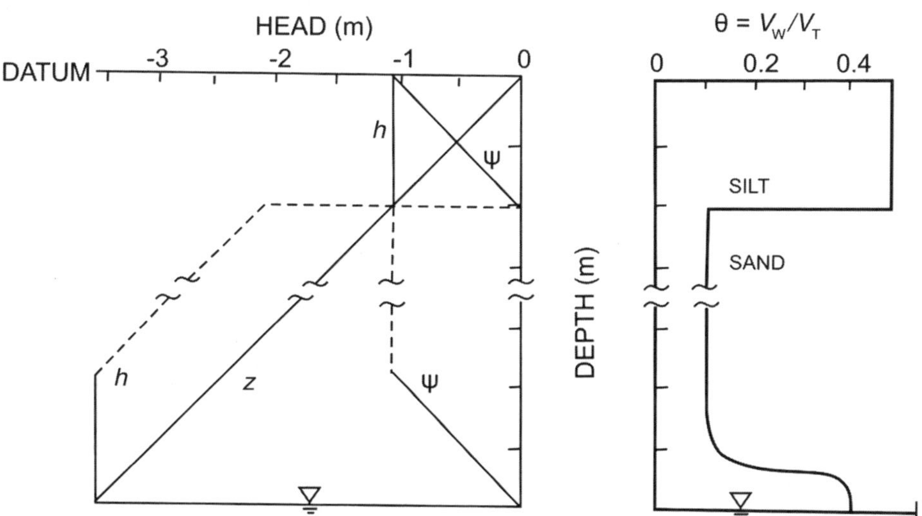

Fig. 14. Hypothetical head distribution (hydraulic head, *h*; elevation head, *z*, and pressure head, ψ) and expected moisture contents through a silt and an underlying coarse sand layer (from Nicholson *et al.*, 1989; © 2008 Canadian Science Publishing, reproduced with permission).

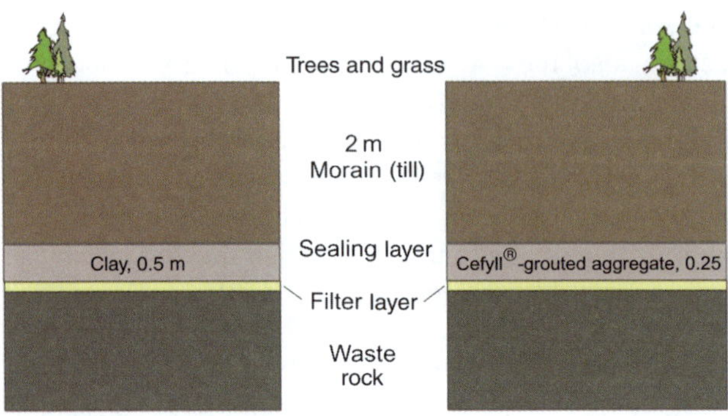

Fig. 15. Composition of cover materials placed over the Bersbo waste rock pile. Figure on left shows the addition of a clay sealing layer, and on the right, the addition of a Cefyll® sealing layer to limit sulfide oxidation.

of 25 cm of Cefyll®, which is a flyash-stabilized concrete, or 50 cm of clay, and a protective layer composed of 2 m of moraine (glacial till) material was installed over a waste-rock pile at the Bersbo site in Sweden (Fig. 15; Lundgren & Lindahl, 1991). Such covers are effective at limiting O_2 ingress into the wastes, but are expensive to construct (Sodermark & Lundgren, 1988; Lundgren, 2001) and may be suscep-tible to cracking after installation because of desiccation or subsidence. A sensitivity analysis conducted by Collin (1987) indicated that cracking of soil covers would lead to the oxidation of the underlying sulfide minerals; thus, it was recommended that soil covers be protected from root penetration, frost action, and desiccation. The effec-tiveness of covers for tailings areas relies on long-term integrity, but because the covers are located on the tailings surface, they are at locations highly susceptible to erosion. Lewis & Gallinger (1999) described the installation of a geomembrane layer covered by 1 m of protective soil cover on a tailings impoundment in Quebec, Canada. The objec-tive of the cover system is to reduce infiltration and limit O_2 migration into the tailings impoundment.

In addition to covers with low O_2-diffusion characteristics, several authors have pro-posed covers that consume O_2 (Reardon & Poscente, 1984; Blenkinsopp, 1991; Broman *et al.* 1991; Tassé *et al.*, 1994). Covers rich in organic carbon in the form of wood-waste, sewage sludge and industrial by-products have been proposed. The intent is to consume O_2 by reaction with organic carbon, thus preventing O_2 contact with the underlying sulfide minerals. Several studies (*e.g.* Tassé *et al.*, 1994) have demonstrated that these covers are able to consume O_2, thereby illustrating their potential effectiveness. Reardon & Poscente (1984) calculated the mass of organic carbon required for covers com-posed of wood waste to serve for long-term prevention of acidic drainage. It was concluded that the mass of organic carbon required was prohibitive, and alternative techniques were suggested. Pierce *et al.* (1994) proposed the use of composted sewage sludge as a potential tailings cover. Advantages include the fine-grained nature of the sludge, its high content of organic carbon, and its high pH.

As with all tailings covers, those involving organic carbon must be applied shortly after tailings deposition has ended. Unlike other dry covers, however, those of organic carbon have the potential to release high concentrations of organic acids to the tailings surface. Interaction between these low-molecular-weight organic com-pounds and ferric oxyhydroxide minerals precipitated during previous periods of

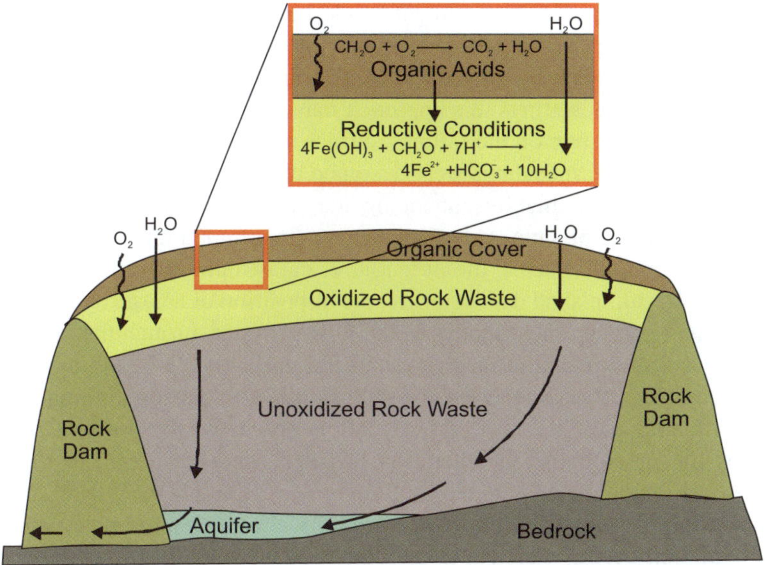

Fig. 16. Schematic diagram of a tailings impoundment covered with organic matter.

oxidation has the potential to result in reductive dissolution of the ferric oxyhydroxides through reactions of the form:

$$4Fe(OH)_3 + CH_2O + 7H^+ \rightarrow 4Fe^{2+} + HCO_3^- + 10H_2O \qquad (10)$$

Such reactions will also release trace and heavy metals adsorbed on, or coprecipitated with, these ferric oxyhydroxides. Reductive leaching experiments by Ribet *et al.* (1995) demonstrated that significant masses of metals were present in the shallow, altered zone of the Nickel Rim tailings impoundment (Sudbury, Canada) in a form that was susceptible to attack by reductive dissolution (Fig. 16).

An alternative to the use of dry covers is the use of wet covers. Development of a wet cover can be achieved by disposal of tailings into deep lakes (*e.g.* Pedersen *et al.*, 1993), or by the construction of wet covers on existing tailings impoundments (*e.g.* Robertson, 1992). In the absence of a site suitable for the development of a subaqueous tailings disposal system, wet covers can be established either as ponded water maintained behind a water-retention dam, or as a bog on the tailings surface (Kalin & van Everdingen, 1987). In such cases, O_2 ingress into the tailings is limited by the slow diffusion of O_2 through the cover. The establishment of a bog on the tailings surface may also lead to the development of reducing conditions in the cover overlying the tailings, similar to the conditions encountered in natural bogs, which would further lessen the movement of O_2 into the tailings. An attempt to establish bog deposits on an existing tailings surface required that water continually be added to the tailings surface to maintain water-saturated conditions; without this addition, the tailings drain because of their coarse-grained nature and the high permeability of the sediments underlying the tailings (Michelutti, 1988). Although the bog cover decreases the aqueous concentrations of dissolved metals, the total metal loadings increase because of the large volume of water added to the test plot. In other hydrogeological settings this approach may be more effective. In all cases, the emplacement of a wet cover on the tailings surface will increase the hydraulic gradient across the dam. Many tailings dams were designed to be permeable to allow the tailings to drain and consolidate, thereby enhancing the structural stability of the impoundment. The maintenance of saturated conditions in such impoundments may reduce the stability of the dam and increase the potential of catastrophic dam failure. In addition to the potential for diminished dam stability, the imposition of a large hydraulic gradient through a tailings impoundment increases the flow of water through the tailings.

The rate of sulfide oxidation is greatest in the early stages of tailings weathering because, at this time, the surfaces of the sulfide minerals are pristine and the length of the O_2-diffusion path is short. As oxidation proceeds, the sulfide minerals are armoured by the precipitation of alteration products, and the length of the diffusion path increases. Simulations conducted by Bain *et al.* (2000) for the Nickel Rim tailings at Sudbury, using the *MINTOX* model of Wunderly *et al.* (1996), showed that the peak period of oxidation occurs shortly after tailings deposition ends. The rate then declines quickly to a relatively low value, and continues to decline further for several years. The depletion of sulfides in the shallow tailings has the effect of developing an expanding silt-size cover on the surface of the tailings. Field determination of Fe and SO_4 concentrations from the Nickel Rim site have shown that maximum Fe and SO_4 concentrations occur 5–8 m below the tailings surface (Johnson *et al.*, 2000). These higher concentrations represent Fe and SO_4 that were derived from oxidation shortly after tailings disposal ended, and which have been displaced into the deeper tailings by infiltrating rain and snowmelt. Concentrations of Fe and SO_4 from the vicinity of the tailings surface are lower, indicating that the rate of oxidation has declined.

10. Prevention of sulfide oxidation

Recognition that existing tailings management programs have the potential to generate low-quality drainage for long periods has lead to alternative proposals for tailings management. These proposals include, but are not limited to, disposal of tailings in deep lakes, separation of sulfide minerals for separate disposal, disposal of tailings as a thickened slurry with improved moisture-retaining capabilities, and enhanced sulfate reduction within tailings impoundments through the addition of solid-phase organic carbon.

Deep-water disposal of tailings has several advantages. The tailings are located at the base of the flow system and are thereby isolated from the effects of erosion and catastrophic dam failure. The sulfide content is isolated from atmospheric O_2 by a thick water cover, limiting oxidation to the mass of O_2 dissolved in the overlying water cover ($\sim$9 mg/l at 25°C and atmospheric O_2 concentrations and pressures). Studies of tailings deposited in lakes in northern Canada suggest that sulfide minerals are stable in deep-lake environments, and that the concentrations of dissolved metals associated with these wastes are low (Pedersen *et al.*, 1993, 1994). In these field studies, Pedersen *et al.* (1999) observed that the tailings sequestered metals through the precipitation of diagenetic sulfide minerals and that metals were released from resubmerged tailings following exposure to air as a result of periodic wet and dry cycles, thus emphasizing the need to maintain a permanent water cover over sulfide-bearing mine wastes.

Concentration of sulfide minerals can be achieved by more completely extracting the sulfide minerals in the final stage of flotation in a mill. The separation process would result in two tailings streams, one a large-volume, low-sulfur tailings product, and the other a smaller volume but sulfide-rich product (Stuparyk *et al.*, 1995). The sulfur-rich portion of the tailings must be handled in a manner that will prevent contact with atmospheric O_2. The larger volume of low-sulfur tailings may be disposed of by conventional techniques, or may be used for backfill, dam construction, or other surface-construction applications. Monitoring of a controlled experiment at a tailings disposal area at Sudbury, Canada, indicated that the decreased sulfide content of low-sulfur tailings (0.35 wt.% S) resulted in decreased concentrations of dissolved metals and a significant delay in the onset of acidification relative to tailings with 0.98 wt.% S, and tailings with 2.3 wt.% S (Hanton-Fong *et al.*, 1997).

Separation of the sulfide portion of the tailings represents 'geochemically engineering' the tailings properties to beneficial conditions for long-term disposal. Amendments made to the tailings represent an alternative form of geochemical engineering. Addition of organic carbon to tailings as they are deposited in the impoundment was proposed by Blowes & Ptacek (1992b). The addition of organic carbon enhances sulfate-reduction reactions and reprecipitation of metal sulfides. If the zone rich in organic carbon is located below the equilibrium water-table position (Fig. 17), the reprecipitated sulfide

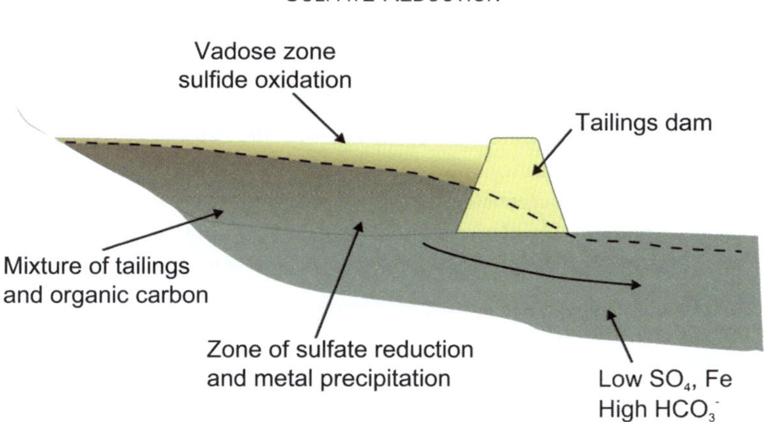

Fig. 17. Addition of organic carbon dispersed throughout the tailings to enhance sulfate reduction and precipitation of sulfide minerals.

minerals will be isolated from atmospheric O_2 by the overlying tailings solids and water column. Mass-balance calculations conducted by Blowes & Ptacek (1992b) suggest that it is feasible to add sufficient organic carbon to stabilize much of the Fe and SO_4 contained in a tailings impoundment which has low to moderately high sulfide contents (5–20 wt.% S). Addition of organic carbon to sulfide-rich tailings was evaluated in pilot-test experiments conducted at the Kidd Creek Zn-Cu metallurgical site, Timmins, Ontario (Hulshof *et al.*, 2003, 2006) and at the Greens Creek Zn-Pb mine, near Anchorage, Alaska (Lindsay *et al.*, 2009, 2011a,b,c). At the Kidd Creek metallurgical site, organic carbon was added to the sulfide-rich tailings in two *in situ* test cells. The first test cell contained organic carbon in the form of pulp waste, derived from a nearby pulp and paper plant, the second test cell contained woodwaste, principally residual tree branches and bark. The test cells were constructed by excavating the tailings material to a depth of 1 m, mixing the organic carbon thoroughly with the tailings and placing the mixture in the excavation. Flow in the test cells was constrained through the use of sheet piling walls, with the base of the test cells remaining open and connected to the underlying tailings. The changes in water chemistry were monitored over a two year period. The addition of organic carbon promoted the growth and activity of sulfate-reducing bacteria, resulting in the precipitation of secondary sulfide minerals, and a decline in the concentrations of dissolved metals, principally Fe and Zn. The rate and extent of metal removal was greater in the test cell containing pulp waste than in the test cell containing wood waste (Hulshof *et al.*, 2006).

At the Greens Creek mine, the effectiveness of the addition of organic carbon in the form of peat, spent brewing grain and municipal sewage sludge was evaluated through the construction and monitoring of seven test cells (Lindsay *et al.*, 2009, 2011a,b,c). The test cells were installed by excavating tailings materials, mixing with organic carbon and placing the mixture within the excavation. Flow in the test cells was constrained through the addition of a low permeability liner on the sides of the test cells, whereas the base of the test cells remained open and connected to the underlying tailings impoundment. Monitoring over four years indicated that populations of iron and sulfate reducing bacteria were established in all of the test cells within the first year of installation, but the largest and most active populations were observed in the test cells containing spent brewing grain (Lindsay *et al.*, 2011b). The precipitation of metal sulfides resulted in substantial decreases in the concentrations of sulfate and dissolved metals, principally Zn. The decreases in Zn concentrations were accompanied by an increase in dissolved As, which was attributed to microbially mediated reduction of iron oxyhydroxides and release of co-precipitated As. Mineralogical study of samples collected from the test cells indicated the accumulation of poorly crystalline Zn and Fe sulfides (Lindsay *et al.*, 2011b,c).

The ongoing addition of organic carbon to tailings impoundments following deposition is potentially costly and undesirable. Integrating carbon amendments with the addition of low-permeability barriers or oxygen-diffusion barriers would potentially extend the longevity of the mitigation system. In addition, Ñancucheo & Johnson (2011) proposed manipulating microbial communities through the addition of phototrophic acidophilic microalgae to provide an ongoing source of organic carbon, combined with inoculation with acidophilic iron and sulfur-reducing bacteria. Microcosm experiments indicated higher pH values and lower concentrations of Cu and Zn in microcosms inoculated with iron-reducing and sulfate-reducing bacteria and micro-algae versus microcosms inoculated with only pyrite oxidizing chemolithotrophs.

In contrast to geochemically engineering a tailings-management system, it has been proposed that the physical hydrogeological characteristics of the tailings be modified as they are deposited in the impoundment. Two of the proposed alternatives are thickened tailings deposition (Robinsky, 1975, 1979; Robinsky *et al.*, 1991), and the intentional formation of cemented 'hardpan' layers (Blowes *et al.*, 1991; Ahmed, 1991). Deposition of thickened tailings prevents the segregation of grain sizes that is observed in conventional tailings areas (Robinsky *et al.*, 1991); thus, the tailings are poorly sorted and have unique moisture-retaining characteristics that result in an extensive tension-saturated zone above the water table. Tension-saturated zones measured at the Kidd Creek tailings impoundment (Timmins, Ontario Canada), at which the thickened-tailings disposal technique is employed, extend >4 m above the water table.

The oxygen diffusion coefficient of unconsolidated materials is dependent on the moisture content. Empirical equations have been developed to describe the relationship between gas-diffusion coefficients and the gas-filled porosity of unconsolidated sediments. Reardon & Moddle (1985) developed the equation

$$D(O_2) = 3.98 \cdot 10^{-5}[(\varepsilon - 0.05)/0.95]^{1.7} \; T^{3/2} \tag{11}$$

where $D(O_2)$ is the diffusion coefficient for O_2 gas, ε is the air-filled porosity, and T is the temperature in Kelvin. This relationship indicates that the rate of the diffusion of

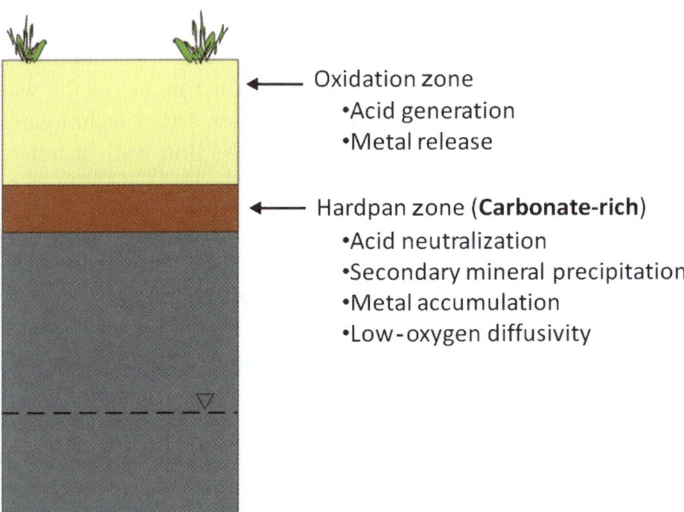

Fig. 18. Geochemical blending of carbonate with sulfide-rich tailings to promote chemical precipitation of a low O_2-diffusivity barrier within an impoundment.

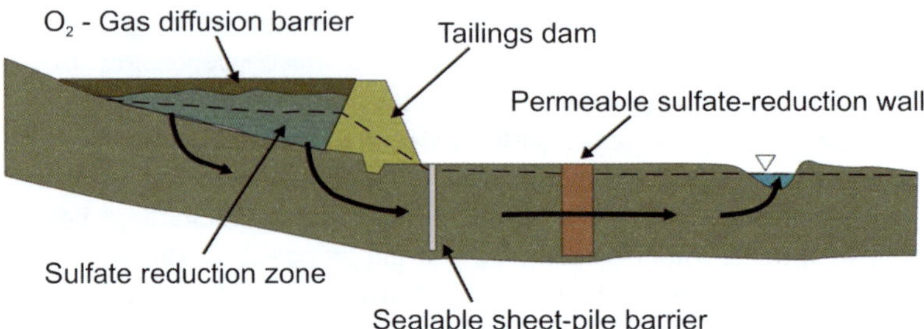

Fig. 19. Schematic diagram showing the use of a combination of treatment techniques including a low-O_2 gas diffusion barrier, a sulfate reduction treatment zone in the tailings, an impermeable sheet pile barrier and a permeable sulfate reduction wall installed along the groundwater flowpath.

gas-phase O_2 through saturated tailings is much less than that through unsaturated tailings. The large tension-saturated zone observed in thickened tailings deposits (Woyshner & St. Arnaud, 1994; Al *et al.*, 1994; Al & Blowes, 1999) limits the zone of rapid oxidation to near the tailings surface.

The occurrence of cemented or 'hardpan' layers has been documented at many mine-waste sites (Boorman & Watson, 1976; McSweeny & Madison, 1988; Blowes *et al.*, 1991). The formation of those layers inhibits the diffusion of pore gases into and out of tailings impoundments. Blowes *et al.* (1991) proposed enhancing the formation of these layers through the selective layering of tailings during the late stages of deposition (Fig. 18). Ahmed (1991) proposed the addition of $FeSO_4$ solutions to enhance the formation of cemented layers. Although the formation of cemented layers has been proposed by several authors, it has yet to be demonstrated on a field scale.

For sites where extremely low release rates are required, a combination of prevention and treatment techniques can be used (Fig. 19). For example, geochemical blending of the tailings with organic carbon to promote sulfate reduction below the water table can be used in conjunction with an O_2-gas diffusion barrier. Other techniques, such as the use of a sealable sheet-pile barrier or a permeable reaction wall installed in the path of the plume, can be used in the event that one of these techniques fails.

11. Conclusions

Conventional approaches to tailings deposition ensure the structural stability of inactive tailings impoundments. Although revegetation programs stabilize the tailings with respect to aeolian and water erosion, these programs do little to prevent sulfide-mineral oxidation or the transport of dissolved constituents through inactive tailings. Among the programs intended to remediate the environmental effects of existing tailings areas are the collection and treatment of drainage waters using conventional water-treatment facilities, and passive downstream treatment using constructed wetlands or permeable reactive barriers.

Sulfide-oxidation controls include the use of bactericides, surface armouring processes, and O_2-diffusion barriers. Bactericides have been demonstrated to be useful over short time periods, but their ability to prevent sulfide oxidation over long periods has yet to be demonstrated. Surface-armour systems result in the precipitation of inert coatings on sulfide minerals. The stability of these coatings over the long term, and their effectiveness relative to natural surface coatings, is not yet known.

Oxygen-diffusion barriers, including water covers, layered soil covers and synthetic covers have been demonstrated to limit diffusion of O_2 gas over short time periods (months to years) in field settings. The ability of these barriers to prevent sulfide oxidation over the longer term (decades to centuries) remains unknown. Among the various techniques to prevent the oxidation of sulfide minerals currently under study, deep-water disposal seems to be most thoroughly evaluated. Separation of sulfide-rich and sulfide-poor tailings is technically feasible. The benefits of thickened-tailings disposal and the addition of organic carbon during tailings deposition are currently being examined at field and laboratory scales.

References

Agnew, M. (1988) *The formation of hardpans within tailings as possible inhibitors of acid mine drainage, contaminant release and dusting.* PhD thesis, University of Adelaide, Australia.

Ahmed, S.M. (1991) Electrochemical and surface chemical methods for the prevention of the atmospheric oxidation of sulphide tailings and acid generation. In: *Proceedings Second International Conference on the Abatement of Acidic Drainage,* Montreal: MEND, Natural Resources Canada, pp. 305–319.

Al, T.A. & Blowes, D.W. (1999) The hydrogeology of tailings impoundment formed by central discharge of thickened tailings: implications for tailings management. *Journal of Contaminant Hydrology,* **38**, 489–505.

Al, T.A., Blowes, D.W. & Jambor, J.L. (1994) A geochemical study of the main tailings at the Falconbridge Limited, Kidd Creek Division Metallurgical Site, Timmins, Ontario. In: *The Environmental Geochemistry of Sulfide Mine-wastes* (J.L. Jambor & D.W. Blowes, editors). Short Course Volume, **22**, Mineralogical Association of Canada, Ottawa, Ontario, pp. 333–364.

Ardau, C., Blowes, D.W. & Ptacek, C.J. (2008) Comparison of laboratory testing protocols to field observations of the weathering of sulfide-bearing mine tailings. *Journal of Geochemical Exploration,* **100**, 182–191.

Bain, J.G., Blowes, D.W., Robertson, W.D. & Frind, E.O. (2000) Modelling of sulfide oxidation with reactive transport at a mine drainage site. *Journal of Contaminant Hydrology,* **41**, 23–47.

Bandy, M.C. (1938) Mineralogy of three sulphate deposits of northern Chile. *American Mineralogist,* **23**, 669–760.

Bayless, E. & Olyphant, G.A. (1993) Acid-generating salts and their relationship to the chemistry of groundwater and storm runoff at an abandoned mine site in southwestern Indiana, U.S.A. *Journal of Contaminant Hydrology,* **12**, 313–328.

Benner, S.G. (2000) *Hydrogeology, geochemistry and microbiology of a reactive barrier for acid mine drainage.* PhD thesis, University of Waterloo, Ontario, Canada.

Benner, S.G., Blowes, D.W. & Ptacek, C.J. (1997) A full-scale porous reactive wall for prevention of acid mine drainage. *Ground Water Monitoring and Remediation,* **XVII**, 99–107.

Benner, S.G., Blowes, D.W., Gould, W.D., Herbert, R., Jr. & Ptacek, C.J. (1999) Geochemistry of a permeable reactive barrier for metals and acid mine drainage. *Environmental Science & Technology,* **33**, 2793–2799.

Benvenuti, M., Mascaro, I., Corsini, F., Lattanzi, P., Parrini, P. & Tanelli, G. (1997) Mine waste dumps and heavy metal pollution in abandoned mining district of Boccheggiano (southern Tuscany, Italy). *Environmental Geology*, **30**, 238–243.

Bertorino, G., Caredda, A.M., Ibba, A. & Zuddas, P. (1995) Weathering of Pb-Zn mine tailings in pH buffered environment. In: *Proceedings of the 8ᵗʰ International Symposium on Water-Rock Interaction* (Y.K. Kharaka. & O.V. Chudaev, editors). Balkema, Rotterdam, pp. 859–862.

Bigham, J.M., Schwertmann, U., Carlson, L. & Murad, E. (1990) A poorly crystallized oxyhydroxysulfate of iron formed by bacterial oxidation of Fe(II) in acid mine waters. *Geochimica et Cosmochimica Acta*, **54**, 2743–2758.

Bigham, J.M., Carlson, L. & Murad, E. (1994) Schwertmannite, a new iron oxyhydroxysulphate from Pyhäsalmi, Finland, and other localities. *Mineralogical Magazine*, **58**, 641–648.

Blenkinsopp, S.A. (1991) The use of biofilm bacteria to exclude oxygen from acidogenic mine tailings. In: *Proceedings Second International Conference on the Abatement of Acidic Drainage,* Montreal: MEND, Natural Resources Canada, pp. 369–377.

Blowes, D.W. & Jambor, J.L. (1990) The pore-water geochemistry and the mineralogy of the vadose zone of sulfide tailings, Waite Amulet, Quebec, Canada. *Applied Geochemistry*, **5**, 327–346.

Blowes D.W. & Ptacek, C.J. (1992a) Geochemical remediation of groundwater by permeable reactive walls: Removal of chromate by reaction with iron-bearing solids. In: *Proceedings Subsurface Restoration Conference, 3rd International Conference on Ground Water Quality Research,* June 21–24, 1992, Dallas, Texas, pp. 214–216.

Blowes, D.W. & Ptacek, C.J. (1992b) Treatment of mine tailings. US Patent 4990031, Canadian Patent 1327027, UK Patent GB 2219617 B.

Blowes, D.W. & Ptacek, C.J. (1994) System for treating contaminated groundwater. US Patent 5362394, Canadian Patent 2062204, European Patent 92103559.8.

Blowes, D.W., Reardon, E.J., Jambor, J.L. & Cherry, J.A. (1991) The formation and potential importance of cemented layers in inactive sulfide mine tailings. *Geochimica et Cosmochimica Acta*, **55**, 965–978.

Blowes, D.W., Jambor, J.L., Appleyard, E.C., Reardon, E.J. & Cherry, J.A. (1992) Temporal observations of the geochemistry and mineralogy of a sulfide-rich mine-tailings impoundment, Heath Steele Mines, New Brunswick. *Exploration Mining and Geology*, **1**, 251–264.

Blowes, D.W., Ptacek, C.J., Benner, S.G., Waybrant, K.R. & Bain, J.G. (1998) Porous reactive walls for the prevention of acid mine drainage: a review. *Mineral Processing and Extractive Metallurgy Reviews,* **19**, 25–37.

Bolshakov, A.P. & Ptushko, L.I. (1969) Alteration products of melanterite from the Nikitov mercury ore deposits. *International Geology Reviews*, **13**, 849–854.

Boorman, R.S. & Watson, D.M. (1976) Chemical processes in abandoned sulfide tailings dumps and environmental implications for Northeastern New Brunswick. *Canadian Institute of Mineralogy Bulletin,* **69**, 86–96.

Broman, P.G., Haglund P., & Mattsson, E., (1991) Use of sludge for sealing purposes in dry covers– Development and field experiences. In: *Proceedings Second International Conference on the Abatement of Acidic Drainage,* Montreal: MEND, Natural Resources Canada, pp. 515–528.

Brown, A. (1991) Proposal for the mitigation of acid leaching from tailings using a cover of muskeg peat. In: *Proceedings Second International Conference on the Abatement of Acidic Drainage,* Vol. 2, Montreal: MEND, Natural Resources Canada, pp. 517–527.

Brunauer, S., Emmett P.H. & Teller, E. (1938) Adsorption of gases in multimolecular layers. *Journal of the American Chemical Society*, **60**, 309–319.

Buurman, P. (1975) In vitro weathering products of pyrite. *Geologie en Mijnbouw*, **54**, 101–105.

Coggans, C.J., Blowes, D.W., Robertson, W.D. & Jambor, J.L. (1999) The hydrogeochemical evolution of a nickel mine tailings impoundment, Copper Cliff, Ontario. *Economic Geology Special Volume, Case Studies in Acid Mine Drainage*, 447–465.

Collin, M. (1987) *Mathematical Modelling of Water and Oxygen Transport in Layered Soil Covers for Deposits of Pyritic Mine Tailings*. Licentiate Treatise, Royal Institute of Technology, Stockholm.

Collin, M. & Rasmuson, A. (1986) *Distribution and flow of water in unsaturated layered cover materials for waste rock*. National Swedish Environmental Protection Board, Report **3088**.

Davis, G.B. & Ritchie, A.I.M. (1986) A model of oxidation in pyritic mine wastes. I. Equations and approximate solution. *Applied Mathematical Modelling*, **10**, 314–322.

Davis, G.B., Doherty, G. & Ritchie, A.I.M. (1986) A model of pyrite oxidation in mine wastes. II. Comparison of numerical and approximate solutions. *Applied Mathematical Modelling*, **10**, 323–329.

Drever, J.I. & Clow, D.W. (1995) Weathering rates in catchments. In: *Chemical Weathering Rates of Silicate Minerals* (A.F. White & S.L. Brantley, editors), Reviews in Mineralogy, **31**, Mineralogical Society of America, Washington, D.C., pp. 463–483.

Dvorak, D.H., Hedin, R.S., Edenborn, H.M. & Gustafson, S.L. (1991) Treatment of metal-contaminated water using bacterial sulfate-reduction: Results from pilot-scale reactors. In: *Proceedings of the Second International Conference on the Abatement of Acidic Drainage*, Vol. 1, Montreal: MEND, Natural Resources Canada, pp. 301–314.

Erickson, P.M. & Ladwig, J. (1985) Control of acid formation by inhibition of bacteria and by coating pyrite surfaces. Final Report to the West Virginia Dept. of Natural Resources, 68 pp.

Fytas, K., Bousquet, P. & Evangelou, B. (1999) Application of silicate coatings on pyrite to prevent acid mine drainage. In: *Proceedings Sudbury'99 Mining and the Environment II*, Vol. 3, Sudbury, Ontario, Canada, pp. 1199–1207.

Gibert, O., Rötting, T., Cortina, J.L., de Pablo, J., Ayora, C., Carrera, J. & Bolzicco, J. (2011) In-situ remediation of acid mine drainage using a permeable reactive barrier in Aznalcóllar (SW Spain). *Journal of Hazardous Materials*, **191**, 287–295.

Golden, D.C., Ming, D.W., Bowen, L.H., Morris, R.V. & Lauer, H.V. Jr (1994) Acidified oxalate and dithionite solubility and color of synthetic, partially oxidized Al-magnetites and their thermal oxidation products. *Clays and Clay Minerals*, **42**, 53–62.

Goldrich, S.S. (1938) A study in rock weathering. *Journal of Geology*, **46**, 17–58.

Gould, W.D. & Kapoor, A. (2003) The microbiology of acid mine drainage. In: *Environmental Aspects of Mine Wastes* (J.L. Jambor, D.W. Blowes, A.I.M. Ritchie, editors). Mineralogical Association of Canada, *Short Course Vol.* **31**, pp. 203–226.

Gould, W.D., Béchard, G. & Lortie, L. (1994) The nature and role of microorganisms in the tailings environment. In: *The Environmental Geochemistry of Sulfide Mine-wastes* (J.L. Jambor & D.W. Blowes, editors).Short Course, Vol. **22**, Mineralogical Association of Canada, Ottawa, pp. 185–199.

Gruner, D. & Hood, W.C. (1971) Three iron sulfate minerals from coal refuse dumps in Perry County, Illinois. *Transactions of the Illinois State Academy of Science*, **64**, 156–158.

Gunsinger, M.R., Ptacek, C.J., Blowes, D.W., Jambor, J.L. & Moncur, M.C. (2006) Mechanisms controlling acid neutralization and metal mobility within a Ni-rich tailings impoundment. *Applied Geochemistry*, **21**, 1301–1321.

Hanton-Fong, C.J., Blowes, D.W. & Stuparyk, R.A. (1997) Evaluation of low-sulfur tailings in the prevention of acid mine drainage. In: *Proceedings of the Fourth International Conference on Acid Rock Drainage*, Vancouver, British Columbia, pp. 835–846.

Hedin, R.S. & Watzlaf, G.R. (1994) The effects of anoxic limestone drains on mine water chemistry. In: *International Land Reclamation Mine Drainage Conference & Third International Conference on Abatement of Acidic Drainage*, Vol. 1, U.S. Dept. Interior, Bureau of Mines Special Publication **SP 06A-94**, pp. 185–194.

Hedin, R., Weaver, T., Wolfe, N. & Weaver, K. (2010) Passive treatment of acidic gold mine drainage: The Anna S Mine Passive Treatment Complex. *Mine Water and the Environment*, **29**, 165–175.

Herbert, R., Benner, S.G., Pratt, A.R. & Blowes, D.W. (1998) Surface chemistry and morphology of poorly crystalline iron sulfides precipitated in media containing sulfate-reducing bacteria. *Chemical Geology*, **144**, 87–97.

Huang, X. & Evangelou, V.P. (1994) Suppression of pyrite oxidation rate by phosphate addition. In: *Environmental Geochemistry of Sulfide Oxidation* (C.N. Alpers & D.W. Blowes, editors). American Chemical Society Symposium Series, Vol. **550**, Washington, D.C., pp. 562–573.

Hulshof, A., Blowes, D.W., Ptacek, C.J. & Gould, W.D. (2003) Microbial and nutrient investigations into the use of in situ layers for treatment of tailings effluent. *Environmental Science & Technology*, **37**, 5027–5033.

Hulshof, A.H.M., Blowes, D.W. & Gould, W.D. (2006) Evaluation of in situ layers for treatment of acid mine drainage: A field comparison. *Water Research*, **40**, 1816–1826.

Jacobsen, U.H. (1989) Hydrated iron sulfate occurrences at Navarana Fjord, central North Greenland. *Bulletin Geological Survey of Denmark*, **37**, 175–180.

Jambor, J.L. (1994) Mineralogy of sulfide-rich tailings and their oxidation products. In: *The Environmental Geochemistry of Sulfide Mine-Wastes* (J.L. Jambor & D.W. Blowes, editors). Short Course Vol. **22**, Mineralogical Association of Canada, Ottawa, pp. 59–102.

Jambor, J.L. (2000) The relationship of mineralogy to acid- and neutralization-potential values in ARD. In: *Environmental Mineralogy: Microbial Interactions, Anthropogenic Influences, Contaminated Land and Waste Management* (J.D. Cotter-Howell, L.S. Campbell, E. Valsami-Jones, M. Batchelder, editors). Mineral Society Series, **9**, Mineralogical Society of London, pp. 141–159.

Jambor, J.L. (2003) Mine-waste mineralogy and mineralogical perspectives of acid-base accounting. In: *Environmental Aspects of Mine Wastes* (J.L. Jambor, D.W. Blowes, A.I.M. Ritchie, editors). Mineralogical Association of Canada Short Course Vol. **31**, pp. 117–145.

Jambor, J.L. & Blowes, D.W. (1991) Mineralogical study of low-sulfide, high carbonate, arsenic bearing tailings from the Delnite minesite, Timmins area, Ontario. In: *Proceedings Second International Conference on the Abatement of Acidic Drainage*, Vol. 4, Montreal: MEND, Natural Resources Canada, pp. 175–197.

Jambor, J.L. & Blowes, D.W. (1998) Theory and applications of mineralogy in environmental studies of sulfide-bearing mine wastes. In: *Modern Approaches to Ore and Environmental Mineralogy* (L.J. Cabri & D.J. Vaughan, editors). Short Course Vol. **27**, Mineralogical Association of Canada, Ottawa, pp. 367–401.

Jambor, J.L., Alpers, C.N. & Nordstrom, D.K. (2000a) Metal sulfate salts from sulfide mineral oxidation. In: *Sulfate Minerals: Crystallography, Geochemistry, and Environmental Significance* (C.N. Alpers, J.L. Jambor & D.K. Nordstrom, editors). Reviews in Mineralogy, **40**, Mineralogical Society of America, Chantilly, Virgina, USA, pp. 303–350.

Jambor, J.L. Dutrizac, J.E. & Chen, T.T. (2000b) Contribution of specific minerals to neutralization potential in static tests. In: *Proceedings 5th International Conference on Acid Rock Drainage*. Society of Mineral Exploration, Littleton, Colorado, pp. 551–565.

Jambor, J.L., Dutrizac, J.E., Groat, L.A. & Raudsepp, M. (2002) Static tests of neutralization potentials of silicate and aluminosilicate minerals. *Environmental Geology*, **43**, 1–17.

Jambor, J.L., Raudsepp, M. & Mountjoy, K. (2005) Mineralogy of permeable reactive barriers for the attenuation of subsurface contaminants. *The Canadian Mineralogist*, **43**, 2117–2140.

Jambor, J.L., Dutrizac, J.E. & Raudsepp, M. (2006) Comparison of measured and mineralogically predicted values of the Sobek neutralization potential for intrusive rocks. In: *Proceedings 7th International Conference on Acid Rock Drainage*, St. Louis, Missouri, USA, pp. 820–832.

Jambor, J.L., Dutrizac, J.E. & Raudsepp, M. (2007) Measured and computed neutralization potentials from static tests of diverse rock types. *Environmental Geology*, **52**, 1173–1185.

Jarvis, A.P., Moustafa, M., Orme, P.H.A. & Younger, P.L. (2006) Effective remediation of grossly polluted acidic, and metal-rich, spoil heap drainage using a novel, low-cost, permeable reactive barrier in Northumberland, UK. *Environmental Pollution*, **143**, 261–268.

Johnson, D.B. (2007) Physiology and ecology of acidophilic microorganisms. In: *Physiology and Biochemistry of Extremophiles* (C. Gerday & N. Glansdorff, editors). American Society for Microbiology Press, Washington, D.C., pp. 257–270.

Johnson, D.B. & Hallberg, K.B. (2005) Acid mine drainage remediation options: a review. *Science of the Total Environment*, **338**, 3–14.

Johnson, R.H., Blowes, D.W., Robertson, W.D. & Jambor, J.L. (2000) The hydrogeochemistry of the Nickel Rim mine tailings impoundment, Sudbury, Ontario. *Journal of Contaminant Hydrology*, **41**, 49–80.

Johnson, D.B., Yajie, L. & Okibe, N. (2008) "Bioshrouding" – A novel approach for securing reactive mineral tailings. *Biotechnology Letters*, **30**, 445–449.

Jurjovec, J., Ptacek, C.J. & Blowes, D.W. (2002) Acid neutralization mechanisms and metal release in mine tailings. *Geochimica et Cosmochimica Acta,* **66**, 1511–1523.

Jurjovec, J., Blowes, D.W., Ptacek, C.J. & Mayer, K.U. (2004) Multicomponent reactive transport modelling of acid neutralization reactions in mill tailings. *Water Resources Research,* **40**, W1120201–W1120217.

Kalin, M. & van Everdingen, R.O. (1987) Ecological engineering: Biological and geochemical aspects. Phase I experiments. In: *Seminar/Workshop, 23–26 March, 1987*: Halifax, Nova Scotia, Canada, pp. 565–590.

Kepler, D.A. & McCleary, E.C. (1994) Successive alkalinity-producing systems (SAPS) for the treatment of acidic mine drainage. In: *International Land Reclamation Mine Drainage Conference & Third International Conference on Abatement of Acidic Drainage,* Vol. **1**. U.S. Department of the Interior, Bureau of Mines Special Publication **SP 06A-94**, pp. 195–204.

Kleinmann, R.L.P., Hedin, R.S. & Edenborn, H.M. (1991) Biological treatment on mine-water – an overview. In: *Proceedings Second International Conference on Abatement of Acidic Drainage*, Vol. **4**, Montreal: MEND, Natural Resources Canada, pp. 27–42.

Kopsick, D.A. & Kopsick, P.R. (1982) Occurrence of secondary sulfate minerals at a coal processing waste pile. In: *Proceedings 2nd Symposium on Process Mineralogy* (R.D. Hagni, editor) Warrendale, Pennsylvania, USA, pp. 385–394.

Lapakko, K.A. (2003) Developments in humidity-cell tests and their applications. In: *Environmental Aspects of Mine Wastes* (J.L. Jambor, D.W. Blowes & A.I.M. Ritchie, editors). Short Course, Vol. **31**, Mineralogical Association of Canada, Ottawa, pp. 147–164.

Lasaga, A.C. & Berner, R.A. (1998) Fundamental aspects of quantitative models for geochemical cycles. *Chemical Geology,* **145**, 161–175.

Lawrence, R.W. & Wang, Y. (1996) Determination of neutralization potential for acid rock drainage prediction. MEND project 1.16.3. Ottawa: Natural Resources Canada, 89 pp.

Lawrence, R.W. & Wang, Y. (1997) Determination of neutralization potential in the prediction of acid rock drainage. In: *Proceedings Fourth International Conference on Acid Rock Drainage.* Vol. **1**. MEND, Natural Resources Canada, Ottawa, pp. 451–464.

Legrand, D.L., Bancroft, G.M. & Nesbitt, H.W. (1997) Surface characterization of pentlandite, $(Fe,Ni)_9S_8$, by X-ray photoelectron microscopy. *International Journal of Mineral Processing,* **51**, 217–228.

Lewis, B.A. & Gallinger, R.D. (1999) Poirier site reclamation program. In: *Proceedings Sudbury'99 Mining and the Environment II,* Sudbury, Ontario, Canada, Vol. **2**., pp. 439–448.

Lindsay, M.B.J., Blowes, D.W., Condon, P.D. & Ptacek, C.J. (2009) Managing pore-water quality in mine tailings by inducing microbial sulfate reduction. *Environmental Science & Technology,* **43**, 7086–7091.

Lindsay, M.B.J., Blowes, D.W., Ptacek, C.J.& Condon, P.D. (2011a) Transport and attenuation of metal(loid)s in mine tailings amended with organic carbon: Column experiments. *Journal of Contaminant Hydrology,* **125**, 26–38.

Lindsay, M.B.J., Blowes, D.W., Condon, P.D. & Ptacek, C.J. (2011b) Organic carbon amendments for passive in situ treatment of mine drainage: Field experiments. *Applied Geochemistry,* **26**, 1169–1183.

Lindsay, M.B.J., Wakeman, K.D., Rowe, O.F., Grail, B.M., Ptacek, C.J., Blowes, D.W. & Johnson, D.B. (2011c) Microbiology and geochemistry of mine tailings amended with organic carbon for passive treatment of pore water. *Geomicrobiology Journal,* **28**, 229–241.

Lortie, L., Gould, W.D., Stichbury, M., Blowes, D.W. & Thurel, A. (1999) Inhibitors for the prevention of acid mine drainage (AMD). In: *Proceedings Sudbury'99 Mining and the Environment II,* Sudbury, Ontario, Canada, Vol. **3**, pp. 1191–1198.

Ludwig, R.D., McGregor, R.G., Blowes, D.W., Benner, S.G. & Mountjoy, K. (2002) A permeable reactive barrier for treatment of heavy metals. *Ground Water,* **40**, 59–66.

Lundgren, T. (2001) The dynamics of oxygen transport into soil covered mining waste deposits in Sweden. *Journal of Geochemical Exploration,* **74**, 163–173.

Lundgren, T. & Lindahl, L.-A. (1991) The efficiency of covering the sulphidic waste rock at Bersbo, Sweden. In: *Proceedings Second International Conference on the Abatement of Acidic Drainage*, Vol. **3**, MEND, Natural Resources Canada, pp. 239–255.

Machemer, S.D. & Wildeman, T.R. (1992) Adsorption compared with sulfide precipitation as metal removal processes from acid mine drainage in a constructed wetland. *Journal of Contaminant Hydrology*, **9**, 115–131.

Malhotra, V.M. (1991) Fibre-reinforced high-volume fly ash shotcrete for controlling aggressive leachates from exposed rock surfaces and mine tailings. In: *Proceedings Second International Conference on the Abatement of Acidic Drainage* Vol. **1**, MEND: Natural resources Canada, pp. 505–513.

Matthies, R., Aplin, A.C. & Jarvis, A.P. (2010) Performance of a passive treatment system for net-acidic coal mine drainage over five years of operation. *Science of the Total Environment*, **408**, 4877–4885.

Matthies, R., Bowell, R.J. & Williams, K.P. (2011) Geochemical assessment of gold mine tailings proposed for marine tailings disposal. *Geochemistry Exploration Environmental Analysis*, **11**, 41–50.

Mayer, K.U., Benner, S.G. & Blowes, D.W. (2006) Process-based reactive transport modeling of a permeable reactive barrier for the treatment of mine drainage. *Journal of Contaminant Hydrology*, **85**, 195–211.

McGregor, R., Blowes, D., Ludwig, R., Pringle, E., Choi, M., Booth, R., Duchene, M. & Pomeroy, M. (1999) Remediation of heavy metal-contaminated groundwater using a porous reactive wall within an urban setting. In: *Proceedings Sudbury'99 Mining and the Environment II,* Sudbury, Ontario, Canada, Vol. **2**, pp. 645–653.

McSweeny, K. & Madison, F.W. (1988) Formation of a cemented subsurface horizon in sulfidic mine waste. *Journal of Environmental Quality,* **17**, 256–262.

MEND (1989) Investigation of prediction techniques for acid mine drainage. Project 1.16.1a, report by Coastech Research. Ottawa: MEND, Natural Resources Canada.

MEND (1991a) Acid rock drainage prediction manual. Project 1.16.1b, report by Coastech Research. Ottawa: MEND, Natural Resources Canada.

MEND (1991b) New methods for determination of key mineral species in acid generation prediction by acid-base accounting. Project 1.16.1c, report by Norecol Environmental Consultants. Ottawa: MEND, Natural Resources Canada.

Michelutti, R.E. (1988) Minimizing metal contaminated seepages from high sulphide mine tailings by establishing a marsh cover. In: *Proceedings International Conference on Control of Environmental Problems from Metal Mines*, June, 1988, Roros, Norway, 15 pp.

Mikhlin, Yu.L., Tomaskevich, Ye.V., Pashkov, G.L., Okotrub, A.V., Asanov, I.P. & Mazalov, L.N. (1998) Electronic structure of the non-equilibrium iron-deficient layer of hexagonal pyrrhotite. *Applied Surface Science*, **125**, 73–84.

Miller, S., Robertson, A. & Donohue, T. (1994) Advances in acid drainage prediction using the net acid generation (NAG) test. In: *Proceedings Fourth International Conference on Acid Rock Drainage*. Ottawa, Vol. 2: MEND, Natural Resources Canada, pp. 535–549.

Moncur, M.C., Ptacek, C.J., Blowes, D.W. & Jambor, J.L. (2005) Release, transport and attenuation of metals from an old tailings impoundment. *Applied Geochemistry*, **20**, 639–659.

Moncur, M.C., Jambor, J.L., Ptacek, C.J. & Blowes, D.W. (2009) Mine drainage from the oxidation of sulfide minerals and magnetite in mine wastes. *Applied Geochemistry*, **24**, 2362–2373.

Morin, K.A. (1983) *Prediction of subsurface contaminant transport in acidic seepage from uranium tailings impoundments*. PhD thesis, University of Waterloo, Waterloo, Ontario.

Mountjoy, K.J. & Blowes, D.W. (2002) Installation of a full-scale permeable reactive barrier for the treatment of metal-contaminated groundwater. In: *Proceedings Third International Conference on Remediation of Chlorinated and Recalcitrant Compounds*, pp. 561–568.

Mycroft, J.R., Nesbitt, H.W. & Pratt, A.R. (1995) X-ray photoelectron and Auger electron spectroscopy of air-oxidized pyrrhotite: Distribution of oxidized species with depth. *Geochimica et Cosmochimica Acta*, **59**, 721–733.

Nagy, K.L. (1995) Dissolution and precipitation kinetics of sheet silicates. In: *Chemical Weathering Rates of Silicate Minerals* (A.F. White. & S.L. Brantley, editors). Reviews in Mineralogy, **31**, Mineralogical Society of America, Washington, D.C., USA, pp. 173–233.

Nesbitt, H.W. & Muir, I.J. (1994) X-ray photoelectron spectroscopy of a pristine pyrite surface reacted with water vapour and air. *Geochimica et Cosmochimica Acta*, **58**, 4667–4679.

Nesbitt, H.W., Muir, I.J. & Pratt, A.R. (1995) Oxidation of arsenopyrite by air and air-saturated, distilled water, and implications for mechanism of oxidation. *Geochimica et Cosmochimica Acta*, **59**, 1773–1786.

Nicholson, R.V., Gillham, R.W., Cherry, J.A. & Reardon, E.J. (1989) Reduction of acid generation in mine tailings through the use of moisture-retaining cover layers as oxygen barriers. *Canadian Geotechnical Journal*, **26**, 1–8.

Nordstrom, D.K. & Alpers, C.N. (1999) Negative pH, efflorescent mineralogy, and consequences for environmental restoration at the Iron Mountain Superfund site, California. *Proceedings of the National Academy of Science, USA*, **96**, 3455–3462.

Nordstrom, D.K. & Southam, G. (1997) Geomicrobiology of sulfide mineral oxidation. In: *Geomicrobiology: Interactions between Microbes and Minerals* (J.F. Banfield & K.H. Nealson, editors). Reviews in Mineralogy, **35**, Mineralogical Society of America, Chantilly, Virginia, USA, pp. 361–370.

Ňancucheo, I. & Johnson, D.B. (2011) Significance of microbial communities and interactions in safeguarding reactive mine tailings by ecological engineering. *Applied and Environmental Microbiology*, **77**, 8201–8208.

Pedersen, T.F., Mueller, B., McNee, J.J. & Pelletier, C.A. (1993) The early diagenesis of submerged sulphide-rich mine tailings in Anderson Lake, Manitoba. *Canadian Journal of Earth Sciences*, **30**, 1099–1109.

Pedersen, T.F., McNee, J.J., Mueller, B., Flather, D.H. & Pelletier, C.A. (1994) Geochemistry of submerged tailings in Anderson Lake, Manitoba: Recent results. In: *Proceedings of the International Land Reclamation and Mine Drainage Conference & Third International Conference on the Abatement of Acidic Drainage*, Vol. 1, U.S. Dept. Interior, Bureau of Mines Special Publication **SP 06A-94**, pp. 288–296.

Pedersen, T.F., Martin, A.J. & McNee, J.J. (1999) Contrasting trace metal dynamics in contaminated Lago Junin, Peru and submerged tailings deposits in Canadian lakes: The importance of permanent submergence. In: *Proceedings Sudbury'99 Mining and the Environment II*, Sudbury, Ontario, Canada, Vol. 1, pp. 165–175.

Pierce, W.G., Belzile, N., Wiseman, M.E. & Winterhalder, K. (1994) Composted organic wastes as anaerobic reducing covers for long term abandonment of acid-generating tailings. In: *Proceedings of the International Land Reclamation and Mine Drainage Conference & the Third International Conference on the Abatement of Acidic Drainage*, Vol. **2**, U.S. Dept. Interior, Bureau of Mines Special Publication **SP 06A-94**, pp. 148–157.

Pratt, A.R. & Nesbitt, H.W. (1997) Pyrrhotite leaching in acid mixture of HCl and H_2SO_4. *American Journal of Science*, **297**, 807–828.

Pratt, A.R., Muir, I.J. & Nesbitt, H.W. (1994a) X-ray photoelectron and Auger electron spectroscopic studies of pyrrhotite and mechanism of air oxidation. *Geochimica et Cosmochimica Acta*, **58**, 827–841.

Pratt, A.R., Nesbitt, H.W. & Muir, I.J. (1994b) Generation of acids from mine waste: Oxidative leaching in dilute H_2SO_4 solutions at pH 3.0. *Geochimica et Cosmochimica Acta*, **58**, 5147–5159.

Price, W.A. (2009) *Prediction manual for drainage chemistry from sulphidic geologic materials*. MEND, Natural Resources Canada, **1.20.1**.

Price, W.A., Morin, K. & Hutt, N. (1997) Guidelines for the prediction of acid rock drainage and metal leaching for mines in British Columbia: part II. Recommended procedures for static and kinetic testing. In: *Proceedings Fourth International Conference on Acid Rock Drainage*. Ottawa: MEND, Natural Resources Canada, Vol **1**, pp. 15–30.

Rasmuson, A. & Collin, M. (1988) Mathematical modelling of water and oxygen transport in layered soil covers for deposits of pyritic mine tailings. In: *Proceedings International Conference on Control of Environmental Problems from Metal Mines*, June, 1988, Roros, Norway, 32 pp.

Reardon, E.J. & Moddle, P.M. (1985) Gas diffusion measurements on uranium mill tailings: implications to cover layer design. *Uranium* **2**, 111–131.

Reardon, E.J. & Poscente, P.J. (1984) A study of gas composition in sawmill waste deposits. *Reclamation and Revegetation Research* **3**, 109–128.

Rhoton, F.E., Bigham, J.M., Norton, L.D. & Smeck, N.E. (1981) Contribution of magnetite to oxalate extractable iron in soils and sediments from the Maumee River Basin of Ohio. *Soil Science Society of America Journal*, **45**, 645–649.

Ribet, I., Ptacek, C.J. , Blowes, D.W. & Jambor, J.L. (1995) The potential for metal release by reductive dissolution of weathered mine tailings. *Journal of Contaminant Hydrology*, **17**, 239–273.

Richardson, S. & Vaughan, D.J. (1989a) Arsenopyrite: a spectroscopic investigation of altered surfaces. *Mineralogical Magazine*, **53**, 223–229.

Richardson, S. & Vaughan, D.J. (1989b) Surface alteration of pentlandite and spectroscopic evidence for secondary violarite formation. *Mineralogical Magazine*, **53**, 213–222.

Ritchie, A.I.M. (2003) Oxidation and gas transport in piles of sulfidic material. In: *Environmental Aspects of Mine Wastes* (J.L. Jambor, D.W. Blowes. A.I.M. Ritchie, editors). Mineralogical Association of Canada *Short Course Vol. 31*, pp. 73–94.

Ritcey, G.M. (1989) *Tailings Management*. Elsevier Science Publ. Co., New York, 970 pp.

Robertson, J. (1992) Subaqueous disposal: a promising method for the effective control of reactive waste materials. In: *24th Annual Operator's Confernce of the Canadian Mineral Processors*, Paper **3**, 10 pp. Canadian Institute of Mining, Metallurgy and Petroleum, Montreal, Quebec.

Robins, E.I., Cravotta C.A., III, Savela, C.E. & Nord, G.L., Jr. (1999) Hydrobiogeochemical interactions in 'anoxic' limestone drains for neutralization of acidic mine drainage. *Fuel*, **78**, 259–270.

Robinsky, E. (1975) Thickened discharge: a new approach to tailings disposal. *Canadian Institute of Mining Bulletin*, **68**, 47–53.

Robinsky, E. (1979) Tailings disposal by the thickened discharge method for improved economy and environmental control. In: *Tailings Disposal Today* (G.O. Argall, editor). Miller Freeman Publications, San Francisco, California, USA, pp. 76–92.

Robinsky, E., Barbour, S.L., Wilson, G.W., Bordin, D. & Fredlund, D.G. (1991) Thickened sloped tailings disposal: an evaluation of seepage and abatement of acid drainage. In: *Proceedings Second International Conference on Abatement of Acid Drainage*, Vol. **1**: MEND Natural Resources Canada, pp. 529–550.

Robinson, J. (editor) (1994) *The Oxford Companion to Wine*. Oxford University Press, New York.

Rodel, S.R. & Navidi, M.H. (1990) *Chemistry*, 2nd edition. West Publishing Co., New York.

Rohweder, T., Gehrke, T., Kinzler, K. & Sand, W. (2003) Bioleaching review part A: Progress in bioleaching: fundamentals and mechanisms of bacterial metal sulfide oxidation. *Applied Microbiology and Biotechnology*, **63**, 239–248.

Sasaki, K., Tsunekawa, M., Ohtsuka, T. & Konno, H. (1995) Confirmation of a sulfur-rich layer on pyrite after oxidation by Fe(III) ions around pH 2. *Geochimica et Cosmochimica Acta*, **59**, 3155–3158.

Schippers, A. & Sand, W. (1999) Bacterial leaching of metal sulfides proceeds by two indirect mechanisms via thiosulfate or via polusulfides and sulfur. *Applied and Environmental Microbiology*, **65**, 319–321.

Schwertmann, U. (1964) Differentiation of iron oxides in soil through extraction with ammonium oxalate solution. *Zeitschrift für Pflanzenernährung und Bodenkunde*, **105**, 194–202 (in German).

Schwertmann, U. (1973) Use of oxalate for Fe extraction from soils. *Canadian Journal of Soil Science*, **53**, 244–246.

Schwertmann, U., Schulze, D.G. & Murad, E. (1982) Identification of ferrihydrite in soils by dissolution kinetics, differential X-ray diffraction, and Mößbauer spectroscopy. *Soil Science Society of America Journal*, **46**, 869–875.

Singer, P.C. & Stumm, W. (1970) Acidic mine drainage: the rate-determining step. *Science*, **167**, 1121–1123.

Skousen, J., Renton, J., Brown, H., Evans, P., Leavitt, B., Brady, K., Cohen, L. & Ziemkiewicz, P. (1997) Neutralization potential of overburden samples containing siderite. *Journal of Environmental Quality*, **26**, 673–681.

Sobek, A.A. (1987) The use of surfactants to prevent AMD in coal refuse and base metal tailings. In: *Proceedings Acid Mine Drainage Seminar/Workshop*, Halifax, Nova Scotia, March, 1987, pp. 357–390.

Sobek, A.A., Schuller, W.A., Freeman, J.R. & Smith, R.M. (1978) Field and laboratory methods applicable to overburden and minesoils. Cincinnati, Ohio: U.S. Environmental Protection Agency, Report EP-600/2-78-054.

Sodermark, B. & Lundgren, T. (1988) The Bersbo project – The first full scale attempt to control acid-mine drainage in Sweden. In: *Proceedings International Conference on Control of Environmental Problems from Metal Mines*, June, 1988, Roros, Norway, 17 pp.

Starr, R.C. & Cherry, J.A. (1994) In situ remediation of contaminated groundwater: The funnel-and-gate system. *Ground Water*, **32**, 465–476.

Stichbury, M., Bechard, G., Lortie, L. & Gould, W.D. (1995) Use of inhibitors to prevent acid mine drainage. In: *Proceedings Sudbury'95 Mining and the Environment*, Sudbury, Ontario, Canada, Vol. **2**, pp. 613–622.

Stuparyk, R.A., Kipkie, W.B., Kerr, A.N. & Blowes, D.W. (1995) Production and evaluation of low sulphur tailings at INCO's Clarabelle Mill. In: *Proceedings Sudbury'95 Mining and the Environment*, Sudbury, Ontario, Canada, Vol. **1**, pp. 159–169.

Tassé, N., Germain, M.D. & Bergeron, M. (1994) Composition of interstitial gases in wood chips deposited on reactive mine tailings. In: *Environmental Geochemistry of Sulfide Oxidation* (C.N. Alpers & D.W. Blowes, editors). ACS Symposium Series, Vol. **550**, American Chemical Society, Washington, D.C., pp. 631–634.

Thomas, J.E., Jones, C.F., Skinner, W.M. & Smart, R.St.C. (1998) The role of surface sulfur species in the inhibition of pyrrhotite dissolution in acid conditions. *Geochimica et Cosmochimica Acta*, **62**, 1555–1565.

Walton-Day, K. (2003) Passive and active treatment of mine drainage. In: *Environmental Aspects of Mine Wastes* (J.L. Jambor, D.W. Blowes & A.I.M. Ritchie, editors). Short Course, Vol. **31**, Mineralogical Association of Canada, Ottawa, pp. 335–359.

Watling, H.R. (2006) The bioleaching of sulphide minerals with emphasis on copper sulphides – A review. *Hydrometallurgy*, **84**, 81–108.

Waybrant, K.W., Blowes, D.W., Ptacek, C.J. & Benner, S.G. (1997) Selection of reactive mixtures for the prevention of acid mine drainage using porous reactive walls. In: *Proceedings Sudbury '95 Conference, Mining and the Environment*, Sudbury, Ontario, Vol. **3**, pp. 945–953.

Waybrant, K.W., Blowes, D.W. & Ptacek, C.J. (1998) Prevention of acid mine drainage using *in situ* porous reactive walls. *Environmental Science & Technology*, **32**, 1972–1979.

Waybrant, K.R., Ptacek, C.J. & Blowes, D.W. (2002) Treatment of mine drainage using permeable reactive barriers: Column experiments. *Environmental Science & Technology*, **36**, 1349–1356.

Williams, D.J., Stolberg, D.J., & Currie, N.A. (2006) Long-term performance of a "store and release" cover over potentially acid forming waste rock in a semi-arid climate. In: *Unsaturated Soils 2006: Proceedings of the 4th International Conference on Unsaturated Soils, 2–6 April 2006, Carefree, Arizona*. ASCE, Reston, VA. Geotechnical Special Publication No. 147, pp. 765–776.

Wogelius, R.A & Vaughan, D.J. (2000) Analytical, experimental, and computational methods in environmental mineralogy. In: *Environmental Mineralogy* (D.J. Vaughan & R.A. Wogelius, editors). EMU Notes in Mineralogy, **2**, Eötvös University Press, Budapest, pp. 7–87.

Wogelius, R.A & Vaughan, D.J. (2013) Analytical, experimental, and computational methods in environmental mineralogy. In: *Environmental Mineralogy II* (D.J. Vaughan & R.A. Wogelius, editors). EMU Notes in Mineralogy, **13**, European Mineralogical Union and the Mineralogical Society of Great Britian & Ireland, pp. 5–102.

Woyshner, M.R. & St-Arnaud, L. (1994) Hydrogeological evaluation and water balance of a thickened tailings deposit near Timmins, ON, Canada. In: *Proceedings International Land Reclamation and Mine Drainage Conference & Third International Conference on Abatement of Acidic Drainage*, Vol. **2**, U.S. Department of the Interior, Bureau of Mines Special Publication **SP 06A-94**, pp. 198–207.

Wunderly, M., Blowes, D.W., Frind, E.O. & Ptacek, C.J. (1996) A multicomponent reactive transport model incorporating kinetically controlled pyrite oxidation. *Water Resources Research*, **32**, 3173–3187.

Yanful, E.K., Aube, B.C., Woyshner, M. & ST-Arnaud, L.C. (1994) Field and laboratory performance of engineered covers on the Waite Amulet tailings. In: *Proceedings International Land Reclamation and Mine Drainage Conference & Third International Conference on Abatement of Acidic Drainage*, Vol. **2**, U.S. Department of the Interior, Bureau of Mines Special Publication **SP 06A-94**, pp. 138–147.

Yin, Q., Kelsall, G.H., Vaughan, D.J. & England, K.E.R. (1995) Atmospheric and electrochemical oxidation of the surface of chalcopyrite ($CuFeS_2$). *Geochimica et Cosmochimica Acta*, **59**, 1091–1100.

Zhang, Y.L. & Evangelou, V.P. (1998) Formation of ferric hydroxide-silica coatings on pyrite and its oxidation behaviour. *Soil Science*, **163**, 1–10.

Suitability of minerals for controlled landfills and containment

RITA HERMANNS STENGELE[1] and MICHAEL PLÖTZE[2]

[1]*Friedlipartner AG, Nansenstrasse 5, CH-8050 Zurich, Switzerland,*
e-mail: Rita.Hermanns@friedlipartner.ch
[2]*ETH Zurich, Institute for Geotechnical Engineering, CH-8093 Zurich, Switzerland,*
e-mail: michael.ploetze@igt.baug.ethz.ch

In European countries, municipal waste deposits must be compatible with the environment. The 'multi-barrier system', also referred to as the 'three-barrier system', provides the basis for containment. These three barriers are the disposal site, the technical barrier system and the waste itself. The impact of a waste-disposal facility on groundwater quality will depend on the nature of the site, the climate, the type of waste, the local hydrogeology and the presence of a dominant flow path and, perhaps most importantly, on the nature of the barrier with which is intended to limit and control any migration. Here we show the types of landfill design and different mineral barriers (*e.g.* natural clay, cut-off walls) which are usually used in environment protection. Contaminant transport in porous and adsorbing media is explained. Waste-containment barriers are another focus of this chapter. The encapsulation of old landfills using cut-off walls and the *in situ* cleanup of contaminated groundwater using reactive walls and funnel-and-gate systems are also described.

1. Introduction

According to the Swiss Waste Management Standards (1986), and those of most European countries, municipal waste deposits must be compatible with the environment. The multi-barrier system, also referred to as the three-barrier system, provides a basis for containment and consists of three components (Figure 1; Swiss Federal Office for the Environment, 1986). These components must be coordinated in an optimal manner, so that the release and propagation of harmful substances is prevented or reduced to an environmentally acceptable level. The three barrier components are:

1st barrier: the disposal site (geology, hydrogeology); the properties of the site determine the long-term behaviour of the deposit.

2nd barrier: the technical barrier system (sealing, drainage and gas-venting systems, and operational concept); the facilities and operational concept must comply with the state-of-the-art.

3rd barrier: the waste (type, properties, treatment, disintegration); reduction of the risk of pollutant release by means of waste treatment.

The site itself must function as a natural geological barrier and has to fulfil the following requirements: any possible emission of pollutants from the deposit must be reduced

© Copyright 2013 the European Mineralogical Union and the Mineralogical Society of Great Britain & Ireland
DOI: 10.1180/EMU-notes.13.8

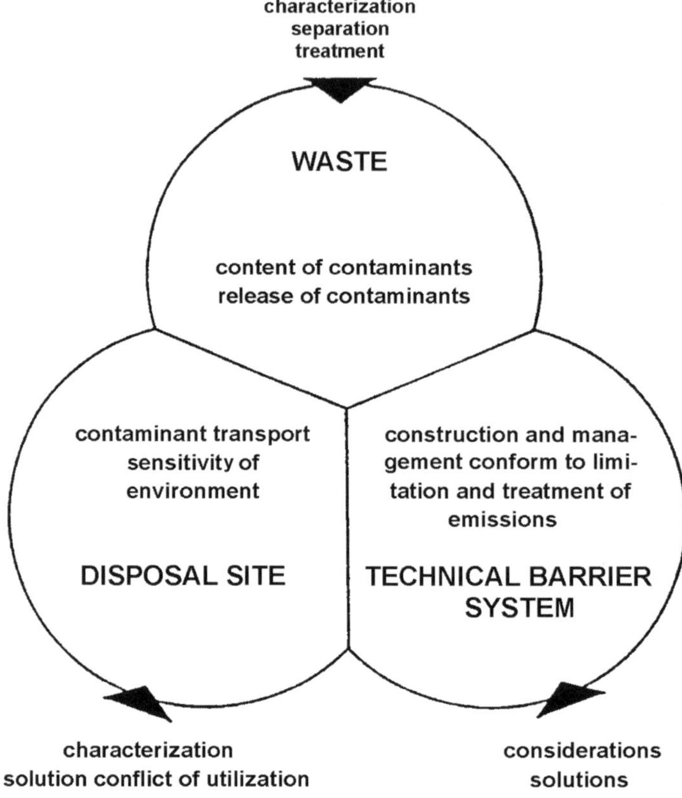

Fig. 1. Philosophy of the multi-barrier system (Swiss Federal Office for the Environment, 1986).

to a minimum; any emission and transport of such pollutants must occur very slowly; propagation of any possible 'pollutant plume' must be limited in distance; and any emissions must be contained quickly and effectively near the site.

The geological barrier (disposal site) should not be considered to be an additional sealing layer of the technical barrier, but a pollutant retention layer in case of failure of the technical barrier. The purpose of the technical barrier system is to enclose and retain the pollutants, *i.e.* to retain, collect and remove leachate and surface water, to guarantee the stability of the deposit and, with an adequate operational concept, control the placement of the waste and the long-term monitoring of the deposit. Technical barriers are a preventative measure to protect the deposit from unsuitable waste and harmful external influences on the one hand and, on the other hand, to protect the geosphere and biosphere from any harmful substances emanating from the deposit. Waste should either be avoided, or reduced and deposited in an environmentally friendly way, in its country of origin. Through treatment with physical, chemical or biological methods the amount of waste should be reduced and possible pollutants immobilized. Waste should be deposited in such a way as to ensure that any landfill is compatible with the environment.

The impact of a waste-disposal facility on groundwater quality will depend on the nature of the site, the climate, the type of waste, the local hydrogeology and presence of a dominant flow path and, perhaps most importantly, on the nature of the barrier with which it is intended to limit and control any migration. Barriers will usually include one or more of the following components: natural clay or clay-rich soils; re-compacted clay-rich liners; cut-off walls; reactive walls; natural bedrock; and geomembranes in composite liner systems.

1.1. Geological barriers

Pollution control and the retardation of pollutants are the most important characteristics required for a geological barrier. The retarding process involves the following three requirements (Török, 1996): (1) the geological formation has to reduce the possibility of pollutant escape from the waste into the geological barrier and, in the worst case, into the environment. (2) If pollutant escape occurs, the geological barrier has to minimize the spreading velocity. (3) The geological barrier has to keep the pollutant within the formation and reduce its mobility and ability to be released into another medium or geological formation.

In most cases water is the medium in which pollutants are transported into the environment. However, contaminants can have different forms and different transport mechanisms; the most important are aqueous-phase transport, non-aqueous-phase liquid transport and solid-particle transport.

The geological and hydrogeological conditions of a landfill location site dictate the risk potential of groundwater contamination in the area downstream of the landfill. The investigation of the site location must be carried out carefully. The aim of this investigation is to obtain sufficient knowledge about the structure of the subsoil and the groundwater conditions. Apart from detailed geological mapping, exploration work such as drilling and geophysical prospecting needs to be done. Both such investigations and subsequent monitoring should include the non-polluted upstream area in the case of contaminated land. Drilling has to be to a sufficient depth and should include both the lithological strata and any key aquifer (with groundwater measuring wells). Following careful recording of the data from drilling, sounding and prospecting (strata table, drill logs providing geological and hydrogeological cross-sections), undisturbed samples should be taken for laboratory tests. During drilling or after installation of a groundwater measuring point, hydraulic tests (pumping tests, borehole tests, slug-tests *etc.*) should be carried out, their nature depending on the subsoil and groundwater conditions. The most important factors to be considered are the geology, hydrogeology, and indirect geological factors and hazards (Török, 1996). The following factors are critical for site-location assessment (after Egloffstein *et al.*, 1996): strata, bedding conditions (thickness, extension, lithological description and stratigraphical classification); tectonic disturbances, tectonic fabrics, joint encrustations, oxidized zones, grades of weathering, thickness of loosened zones; position, thickness and permeability of aquifers and impermeable beds, flow direction(s) of groundwater and apparent flow velocity, any new development of groundwater at the given location; geochemical characteristics regarding contaminant-retention capacity (*e.g.* cation exchange capacity, sorption capacity, retardation factors); and soil and rock physical characteristics necessary to assess the bearing capacity (settlement behaviour) and the stability (base failure, slope failure) of the landfill site subsoil.

From the geological and hydrogeological point of view, the best landfill sites are located on thick formations with low permeability and high contaminant-retention capacity, and which extend beyond the landfill location. These requirements are usually fulfilled by clays or claystones with a sufficient thickness and relatively undeveloped tectonic fabrics.

The main function of the base liner, that of sealing, is defined by the hydraulic conductivity. According to European regulations (Council of the European Union, 1999), a conductivity of $k \leq 1 \times 10^{-7}$ m/s with a minimum thickness of 1 m is required for inert waste, $k \leq 1 \times 10^{-9}$ m/s with a minimum thickness of 1 m for non-hazardous waste, and $k \leq 1 \times 10^{-9}$ m/s with a minimum thickness of 5 m for hazardous waste. These values vary with national regulations. Where the geological barrier does not naturally meet the above conditions, it can be augmented artificially and reinforced by other measures giving equivalent protection. An artificially established geological barrier should not be <0.5 m thick.

1.2. Technical barrier

An engineered landfill should be designed in such a way that no harmful substances reach the biosphere and hydrosphere in unacceptable quantities (German Geotechnical Society of the International Society of Soil Mechanics and Foundation Engineering, 1997). There are different approaches to meeting this requirement in landfill design. One approach is to encapsulate the waste body as perfectly as possible (as in the USA and Germany). Another approach is to use a design leading to limited and environmentally acceptable leaching (as in Sweden) resulting in a long-term neutralization of the waste in the landfill by sorption, degradation and dilution.

The design of capping and basal-lining systems composed of the elements shown in Figure 2 may serve as an example. Such lining systems are built in many countries around the world. They are designed for specific kinds of waste and designs vary with regard to country-specific regulations that consider a number of factors such as geology, hydrogeology, climate and economy (Jessberger, 1997). There are general aspects that have to be considered when designing the landfill (German Geotechnical Society of the International Society of Soil Mechanics and Foundation Engineering, 1997). These include climate, subsoil, disposal-site environment, waste material, lining systems, leachate and gas-management systems, site operation, monitoring, site closure and long-term supervision. Geotechnical engineers are required in the field of contaminated land reclamation and containment. In this context, the following topics must be addressed (Jessberger, 1994): site investigation including sampling and laboratory analyses; evaluation of potential danger; development and evaluation of remediation concepts; performance of the clean-up or containment provided; and long-term monitoring.

1.3. Waste barriers

Although the meaning of the term 'waste' is well understood, the terminology used to characterize the various types of waste materials may vary substantially from one country to another. Waste categories such as municipal, domestic, commercial, industrial, special, hazardous and toxic are all applied to waste, and do not have the same meaning in different countries (Knochenmus *et al.*, 1998). Safety and performance requirements imposed for waste-disposal sites vary between different countries. Furthermore, there may be regional variations in waste-management procedures within a given country. It is difficult to compare in a precise manner the requirements imposed as each country does not have the same legal definition of the term 'waste', or statutory list of the type of waste which can be accepted by a permitted disposal facility, or concept of the role of landfill in the ultimate disposal of waste.

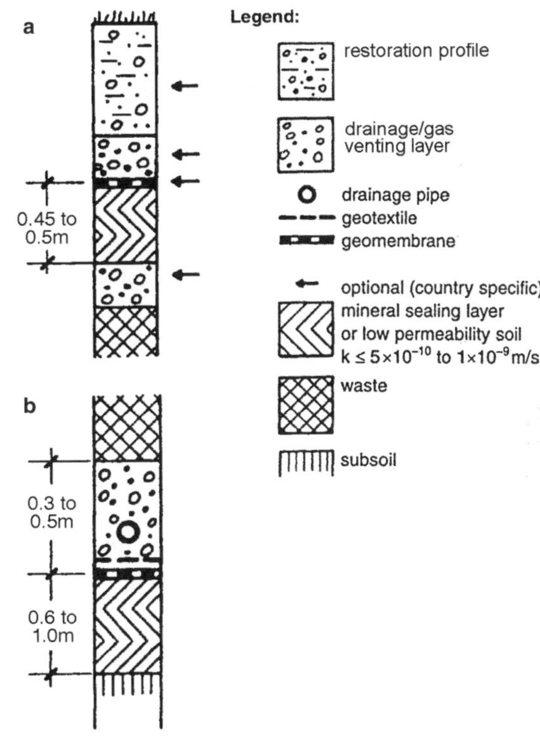

Fig. 2. Elements of (**a**) capping systems and (**b**) basal lining systems (Jessberger, 1997, reproduced with the permission of Balkema, Rotterdam).

Some countries consider that landfilling with untreated waste is an acceptable means of waste disposal, while other countries (such as Switzerland and Germany) only permit landfilling of mechanically or chemically stabilized waste.

It is important not to limit the evaluation of the safety requirements to the geological and hydrogeological conditions of the disposal site and the use of artificial barriers as this can lead to errors. The chemical and mechanical behaviour of the waste should not be overlooked when establishing landfill safety requirements.

Waste which can be accepted by landfills is commonly subdivided into three types, based on its origin: (1) toxic or *hazardous* (industrial) *wastes* – these present the greatest potential hazard; (2) municipal solid waste, or the equivalent industrial wastes, which are similar to urban waste but are produced by industrial activities (*non-hazardous waste*); (3) *inert waste* (such as construction materials).

Landfills are also commonly divided into three types, depending on the waste which they are authorized to accept. The minimum safety requirements for each landfill type increase with the potential hazard associated with the accepted waste.

While the number of landfills in the USA has declined steadily over the years, the average size of landfill site has increased. At the national level, landfill capacity appears to be sufficient, although it is limited in some areas. Since 1990, the total amount of municipal solid waste (MSW) going to landfills has decreased by $\sim$10 million tons, from 143.3 million to 135.5 million tons in 2010. Combustion of MSW for energy recovery has decreased from $\sim$34 million tons in 2000 to 29 million tons in 2010. The percentage of total MSW recovered for recycling (including composting) was $\leq$15% until 1990. The recovery rate by 2010 had increased to $\sim$34% (www.epa.gov, June 2012).

Closing of landfills is an important driver for adopting new waste-treatment options. The number of landfills in the countries and regions in 25 states of the European Union has decreased significantly in the last 10–15 years, mostly through the closure of dumpsites and other low-standard sites. Although this probably implies a reduction in total landfill capacity, data on current waste generation and landfill rates indicate that existing capacity in most countries is sufficient for many years to come. Incineration capacity has increased significantly as governments have tightened emission standards, although the rate of growth has varied widely in the areas studied (*e.g.* in Germany and Belgium, incineration capacity is $\sim$35%, Italy $\sim$15%, but <10% in Finland and Hungary). Mechanical-biological treatment is used as an alternative to incineration to treat mixed municipal waste in some European countries (*e.g.* Estonia, Germany, Italy). Since 1999, capacity at composting and anaerobic digestion plants has increased significantly in Finland, Germany, Hungary and Italy. It rose by 50% in Germany in only 4 years (European Environment Agency, 2009).

1.4. Contaminant transport in porous and adsorbing media

The migration of water and, therefore, the mobility of contaminants (*e.g.* heavy metal ions) through any porous and sorbing media like a geological or mineralogical technical barrier, is affected by permeability, diffusion and sorption properties of the barrier. The transport of contaminants miscible with an aqueous phase in porous media is controlled by a variety of physical and chemical processes. The primary physical processes governing fluid transport are advection, diffusion and dispersion. Advection is mass transport due to bulk fluid flow under a hydraulic gradient. Advection is the main transport process in high flow-rate media. Diffusion is the dominant process in low flow-rate media. The main driving force of diffusion is a concentration gradient, alongside the self-diffusion due to temperature-dependent Brownian motion. Dispersion is a parallel flow process of advection and diffusion and involves spreading due to heterogeneities in the flow field. All these processes involve an equalization process in which fluids in pressure, temperature or concentration gradients flow through a porous medium. Processes which hinder transport through adsorbing media have to be considered in the description of contaminant transport as the retardation factor and the first-order decay. The retardation factor includes the effective or accessible porosity and the sorption. Sorption is a term that sums up chemical processes, *e.g.* adsorption, desorption and dissolution/precipitation at the fluid–solid interface. First-order decay can be the decay of radioactive species or the decay of organic compounds due, for example, to hydrolysis reactions. Oxidation and reduction processes at the solid surface or in the flowing medium influence the retardation behaviour of the contaminants. The relevance of reactive accessory minerals such as carbonates (mainly calcite) and sulfates (mainly gypsum) for dissolution/precipitation processes during transport, *e.g.* through clay liners, has been shown recently by De Soto *et al.* (2012).

There are three transport processes which are important with respect to waste containment by engineered barriers (Manassero *et al.*, 1997): (1) pure diffusion; (2) diffusion with positive advection; and (3) diffusion with negative advection. Pure diffusion may occur when a compacted clay barrier is placed below the water table, or when a slurry wall is placed around an existing contaminated area. The most common situation for a new landfill is diffusion with positive advection, where the compacted clay liner is placed above the water table. Diffusion with negative advection occurs when a clay liner is placed below the water table, or over an artesian aquifer, and when the groundwater table within a contaminated area contained by a slurry wall (*e.g.* by pumping) is lowered to induce inward advective flux (Manassero *et al.*, 1997). In principle, a system involving diffusion with negative advection is the most effective for containment, but is not always used due to practical problems related to the continuous use of dewatering systems.

Although the equations governing transport can be formulated for transport in three dimensions, one-dimensional modelling is simpler and is applicable to many practical problems in environmental geotechnics. Thus, only the one-dimensional formulation of the governing equation is presented here. For the flux in a slab, one can distinguish between unsteady- and steady-state transport. In a semi-infinite slab, when a transport process begins, a time-dependent flow profile develops in the slab. Over a longer time, the system will slip into a steady-state flow. The steady-state mass flux for one-dimensional transport through the aqueous phase of a soil may be written as shown below.

For advection (Darcy's law):

$$q = k \cdot i = -k \cdot \delta h/\delta x = \Theta \cdot v \tag{1}$$

where q is the liquid flux, k is the hydraulic conductivity, i is the hydraulic gradient, h is the hydraulic head, v is the seepage velocity, Θ is the volumetric water content and x is the direction of transport.

For diffusion (Fick's first law):

$$j_s = -D_s \cdot \delta c/\delta x \tag{2}$$

where j_s is the diffusive mass flux, D_s is the steady state diffusion coefficient, c is the contaminant concentration and x is the direction of transport.

For dispersion:

$$j_{ad} = q \cdot c - \Theta \cdot D_s \cdot \delta c/\delta x \tag{3}$$

where j_{ad} is the advective-diffusive mass flux.

These differential equations are subject to two boundary conditions; at $x = 0$ we have a hydraulic head of h_o or c_o, and at the end of a slab with thickness l, a head of h_1 or c_1. The constant flux q or j is then found as follows.

For advection:

$$q = -k \cdot (h_o - h_1)/l = k \cdot i \tag{4}$$

For diffusion:

$$j = -D_s \cdot (c_o - c_1)/l \tag{5}$$

This equation applies for steady-state conditions.

The non-steady state or transient flux (*e.g.* change of pressure or concentration with time) is described by the Terzaghi equation for advection:

$$\delta h/\delta t = k_h \cdot \delta^2 h/M \cdot \delta x^2 \tag{6}$$

where k_h/M is the consolidation factor and t is the time of fluid flow.

For non-steady-state diffusion, Fick's second law is:

$$\delta c/\delta t = D \cdot \delta^2 c/\delta x^2 \tag{7}$$

where t is the diffusion time and D is the non-steady state diffusion coefficient.

Contaminant transport through sandy soil will be mainly *via* the faster process of advection. However, in fine-grained soils, wherein the hydraulic flow is very small (*e.g.* $k < \sim 10^{-9}$ m/s), the transport may be controlled by diffusion. Therefore, diffusive transport is important in clay barriers. In the dispersion regime, one can assume a steady advective fluid flow, v. Then Fick's second law can be modified as follows.

For dispersion:

$$\delta c/\delta t = D \cdot \delta^2 c/\delta \cdot (x - v \cdot t)^2 \tag{8}$$

In the case of diffusion with a source of constant inlet concentration, c_o, the differential equation for Fick's second law can be solved with the inverse error function solution (erfc):

$$c(x, t) = c_o \cdot \text{erfc } x \Big/ \sqrt{4 \cdot D \cdot t} \tag{9}$$

For an initial source of amount Q which relaxes by diffusion, a Gauss-function solution for equation 7 may be found:

$$c(x, t) = Q \Big/ \sqrt{\pi \cdot D \cdot t} \cdot \exp\left(-x^2/4 \cdot D \cdot t\right) \tag{10}$$

In many practical problems, the flux in the slab is of greater interest than the concentration profile itself. Then one can derive the flux across the interface at $x = 0$:

$$j_{(x=0)} = \sqrt{D/\pi \cdot t} \cdot (c_o - c_1) \tag{11}$$

Diffusion through porous soil is slower and more complex than diffusion through a solution, especially when adsorptive clays are present. In a solution, maximum values for the diffusion coefficient are found. The free solution diffusion coefficient, D_0, or aqueous phase diffusion coefficient, D_w, describes the diffusivity of a substance in water at infinite dilution. Numerical values range between 0.5 and 2×10^{-9} m^2/s. The pore diffusion coefficient D_p (or D^*) describes the mobility in pore water under the influence of different geometric restrictions:

$$D_p = \tau_a \cdot D_0 \tag{12}$$

in which τ_a is an apparent tortuosity factor that takes into account the geometric factors termed 'constructivity' (which reduces the diffusion pathway) and 'tortuosity' (which lengthens the diffusion pathway). The geometric factor, τ_a, ranges from ~ 0.1 to 0.4 for most fine-grained soils and increases to values ranging from 0.5 to 0.7 with increasing grain size. For compact clay, τ_a is usually <0.1. The effective or intrinsic diffusion coefficient D_e is related to the pore diffusion coefficient by the accessible porosity, and contains all geometric and physical factors which reduce molecular mobility in the pore space as compared to (free) water.

The apparent diffusion coefficient D_a (or D_{app}) takes into account these adsorption-desorption reactions within the medium. Given the assumption that there is linearity between the amount adsorbed and the equilibrium concentration, it is often written as:

$$D_a = D_p/R \tag{13}$$

The retardation factor R includes the effective or accessible porosity and the sorption and is defined by:

$$R = 1 + \rho_d/\Theta \cdot K_d \tag{14}$$

in which ρ_d is the bulk dry density of the soil, Θ is the volumetric water content or effective porosity and K_d the distribution coefficient.

The distribution coefficient defines the amount of a given constituent that is adsorbed by a soil for a unit increase in the equilibrium concentration in solution. There are some models describing ion binding to oxide mineral surfaces (Solomon & Hawthorne, 1983; Kraepiel *et al.*, 1999). Sorption can be characterized as being reversible or irreversible depending on the adsorption site. A comparison of the performance of the most commonly used surface complexation models (*e.g.* site-binding, two- and triple-layer, Stern model) was given by Venema *et al.* (1996). Distribution coefficients at a certain temperature are usually determined from adsorption isotherms. This shows the functional coherence between the loading of the adsorbent with the adsorbate and the adsorbate concentration of the solution under equilibrium conditions. Sorption isotherms can be generated graphically by recording data over time. They can be analysed using graphical or numerical approximations. To use mathematical expressions for this purpose, an empirical function has to be found which correlates with the measured data. In an ideal case, a 'law' can be derived from a model which outlines the experimental result. This case occurs very rarely. However, one example is the isotherm equation by Langmuir, which is of great practical importance (McBride, 2000; Rowe *et al.*, 2004 chapters 1 and 6). It can be written as follows:

$$q = q_m \cdot (K_L \cdot c)/(1 + K_L \cdot c) \tag{15}$$

where q = concentration of the sorbate on the sorbent (loading) [mol/g], q_m = maximum loading (with mono-molecular occupancy of the adsorbing surface) [mol/ g], K_L = Langmuir constant, c = concentration of the dissolved adsorbate (under equilibrium conditions) [mol/m^3].

This equation can be recast in the following form:

$$1/q = 1/q_m + 1/q_m \cdot K_L/c \tag{16}$$

Recording $1/q$ over $1/c$, every Langmuir isotherm shows a linear coherence. This is a very good way of determining the terms q_m and K_L from the experimental data. The ordinate of the function signifies the reciprocal value of the saturation loading q_m and the function gradient indicates the reciprocal value of the product $q_m \cdot K_L$.

For low concentrations ($K_L \cdot c \ll 1$) and under adsorption equilibrium conditions, adsorbent loading is proportional to the adsorbate residual concentration (Henry's law):

$$q = K_H \cdot c \tag{17}$$

with:

$$K_H = K_L \cdot q_m \tag{18}$$

For this range of concentrations, the isotherm generally shows a linear progression; hence it is called a linear isotherm. For high concentrations ($K_L \cdot c \gg 1$), q is approximately q_m (horizontal isotherm). With both approximations, calculations can be simplified in certain cases (Sonntheimer, 1975).

Another important, entirely empirical, formula is that after Freundlich:

$$q = K_F \cdot c^n \tag{19}$$

with K_F and n = constant.

Very often, this isotherm is valid for adsorption from solutions of intermediate concentrations. For low concentrations, the Freundlich isotherm is unrealistic as it cannot describe the linear increase of adsorption with concentration, according to Henry's law. For high concentrations it is unrealistic as well, as no limit is provided either for concentration or for loading.

For linearization of the Freundlich isotherm, the loading q is recorded over concentration c in a double-logarithmic scale. The straight-line gradient gives the exponent n and the ordinate value gives K_F (with $\log c = 0 \Rightarrow c = 1 \Rightarrow q_c = 1 = K_F$).

In many cases, a simple criterion can be used to decide which of the two limits, steady or non-steady state diffusion, is more closely approached. This criterion hinges on the magnitude of the variable S:

$$S = (x^2/D_a) \cdot t \tag{20}$$

where x is the diffusion length or migration distance, D_a the apparent diffusion coefficient and t the time.

If this variable S is much larger then unity, one can infer a non-steady state. If the variable is less than or approximately equal to unity, one can infer a steady state or equilibrium.

For determination of apparent diffusion coefficients, several experimental techniques are used. Most common are through-diffusion experiments. With these diffusion cells, steady-state diffusion measurements are carried out (Rowe *et al.*, 2004, chapter 6). It should be emphasized that a steady-state diffusion coefficient is usually slightly greater than the non-steady state coefficient. For a steady-state diffusion experiment, a disc of soil with the thickness d, confined by thin filters, is separated from a volume of tracer solution on the one side and a volume of water on the other side. The change in tracer concentration in the water with time is measured on the other side of the membrane. After an initial step of non-steady-state diffusion there follows a step with steady-state transport. The total amount of tracer accumulated in the far side of membrane is given by the asymptotic solution with a slope of $c_o \cdot D_s/d$ and the time lag t_e, *i.e.* the intercept with the axis:

$$t_e = d^2/\, 6 \cdot D_s \tag{21}$$

More sophisticated methods for determination of concentration gradients in the sample, which also allow measurements over short distances and times, are X-ray radiography and Rutherford Backscattering Spectrometry (Cave *et al.*, 2009; Alonso *et al.*, 2009). Only a few such field-based measurements exist. Barbour *et al.* (2012) described long-term measurements over years with deuterium spikes as a tracer showing the transition from advection to diffusion-dominated transport in a clay-rich aquitard.

The formula for estimating contaminant transport due to pure diffusion can also be simplified and expressed as:

$$x = \sqrt[6]{(D_a \cdot t)} \tag{22}$$

where x is the migration distance, t the time and D_a is the apparent diffusion coefficient where all parameters concerning diffusion are grouped together (geometric factors, accessibility and retardation, where the product of the latter two is sometimes called a 'rock capacity factor').

2. Types of landfill design

2.1. Methods of disposal

Waste can be divided into two main groups: municipal (domestic) solid waste (MSW) and industrial waste (without radioactive waste). Depending on the composition of the specific waste material, its mechanical characteristics may differ from those of typical soils. According to GLR Recommendations (German Geotechnical Society of the International Society of Soil Mechanics and Foundation Engineering, 1997), waste can be also classified into two main groups: soil-like waste, defined as particulate material to which soil mechanics principles are applicable; non-soil-like waste, defined as material to which soil mechanics principles are not applicable or only applicable in a limited way. An indication of the waste types belonging to the two groups is given in Table 1.

Industrial waste should also be classified on the basis of the industry which generates it (Kamon, 1997). After treatment, waste materials can be divided into three groups: (1) wastes subjected to incineration or melting (coal ash, iron slag, incinerated ash); (2) wastes subjected to crushing (waste concrete powder, waste rock powder); and (3) wastes left 'as is' without any treatment (waste sludge, waste oil). For residues generated by incineration or melting, their characteristics depend on the raw materials, the incineration temperature and time, and the incinerator, furnace or boiler system used (Bunge, 2007). They may be broadly classified as fly ash collected from fuel gas, bottom ash left at the bottom of a boiler, and slag produced by melting. The second group of wastes is generated in large quantities from construction work (*e.g.* excavation in urban areas for construction of pipelines, subways, tunnels). The wastes in the first and second groups may be stabilized by compaction or by chemical additives to cause solidification, after which they can be used as road materials or in embankments. The last group contains waste sludge, waste plastics and so on, the treatment of which is very difficult for both technical and economic reasons.

Table 1. Geotechnical classification of waste types according to the German Geotechnical Society of the International Society of Soil Mechanics and Foundation Engineering (1997).

Soil-like waste	Non-soil-like waste
Excavated soil	Municipal solid waste (MSW)
Industrial sludge	Bulky waste
Road construction debris	'Green waste'
Incineration residue (slag, ash, dust)	MSW-like industrial waste
Construction debris	Waste from construction sites
Sewage sludge	Solids
	Residues from mechanical-biological treated wastes

For efficient disposal, the properties of the various wastes must be considered, and whether the waste material is inorganic or organic, whether it contains heavy metals, and so on. The aim of the 'mechanical improvement' of waste, particularly municipal solid waste, is to increase its strength and to anticipate settlement, although gas production and leachate migration may be also affected. Applying such improvement techniques to the waste pile (König & Jessberger, 1997) can result in: (1) compaction of the waste, preventing settlement and gaining additional disposal volume; and (2) stabilization of waste – before sealing systems are installed, for extension of an existing landfill, or for re-utilization of the landfill.

Several techniques may be employed: preloading (reduction of void ratio, consolidation, in anticipation of load-induced settlement); sand compaction piles (sand is incorporated into the waste, displacing the waste and filling voids); dynamic compaction (dynamic consolidation); and pressure injection stabilization (grouting, filling of voids using lime, or lime fly-ash composites, or asphalt).

Dynamic compaction is very useful in the improvement of MSW over a depth of 10 m by inducing settlements of 10–30% of the initial thickness. In Switzerland, for example, the incineration of MSW has become a primary waste-reduction procedure. Through incineration, 1 kg of MSW is converted to ~700 g of gas, 270 g of slag and 30 g of filter ash (Nüesch *et al.*, 1996). As filter ash contains leachable heavy metals, this material must be treated before being placed in landfills.

A complete characterization of fly ash and filter residue using X-ray image analysis techniques in an electron microprobe has also been carried out in order to understand the potential for leaching by percolating water (Nüesch *et al.*, 1996). The crystalline materials in the ashes were shown by powder X-ray diffraction (XRD) techniques to be gypsum, anhydrite, quartz, halite, sylvite, calcite and ettringite. Microchemical analysis techniques have shown that the bulk of the Pb and most other heavy metals are associated with silicate glasses of varying compositions.

Traditionally, fly ash has been immobilized with cement. Research has been carried out to investigate the possibilities of immobilization using clay. One type of fly ash and four types of clay were used; relevant data are given in Tables 2 and 3 (Nüesch *et al.*, 1996). Four different tests were carried out to determine the behaviour of the test materials: leaching, compaction, deformation and shearing. Leaching tests with percolation columns were used to study the adsorption capacity of fly ash-clay mixtures. Test series were run using 20, 30 and 40 wt.% of clay mixed with the filter ash for the four clays. It was shown that the immobilization of heavy metals is dependent on both the type and the percentage of the clay added to the fly ash.

2.2. Bottom-lining systems

In the past, hydraulic barriers simply consisting of a layer of compacted clay, or of a single geomembrane, have been used for solid waste containment. In many cases, no specific measures were taken to control pollutant migration in landfills underlain by a low-permeability natural subsoil. Today, the design of a landfill barrier system is generally based on either a prescriptive standard (as in West European Countries) or a

Table 2. Mineralogical composition and physical properties of materials (from Nüesch *et al.*, 1996, reproduced with the permission of Balkema, Rotterdam).

	Unwashed slag (SU)	Washed slag (SW)	Unwashed fly ash (FU)	Washed fly ash (FW)	Opalinus clay (OP)	Montigel clay (MO)
Minerals	Quartz	Quartz	Anhydrite	Quartz	Quartz 30%	Montmorillonite 66%
	Calcite 5%	Calcite 12%	Sylvite	Sylvite	Mixed layer 20%	Quartz 8.3%
	Hydroxylapatite	Hydroxylapatite	Halite	Halite	Illite 15%	Mica 12–15%
			Quartz	Anhydrite	Chlorite 10%	Feldspar 2–4%
	Feldspar	Feldspar	Calcite 9%	Calcite 10%	Kaolinite 10%	Carbonate 3.8%
	Chromite	Chromite	Hematite	Hematite	Carbonate 10%	Kaolinite 2%
	Aluminium	Aluminium	Goethite	Goethite	Accessories 5%	Accessories 3%
	Fe oxide	Fe oxide	Rutile	Rutile		
		Ankerite	Bassanite	Gypsum		
		Mirabilite				
		Melanterite				
		Bassanite				
Surface area (m^2/g)	9.5	10.7	10.7	19.5	21	490
CEC (meq/ 100 g)	–	–	–	–	12	62
γ_s (kN/m^3)	28.52	27.76	26.36	26.71	26.50	28.10
Water content	0.57	0.52	1.45	5.82	2.7	3
(wt.%)	53	52	17	11	2	1.4
>63 μm (wt.%)	38	43	60	59	89	87.9
<20 μm (wt.%)	9	12	1	1	57	77.6
<2 μm (wt.%)	–	–	15	19	–	–
dissolved (wt.%)						

performance standard (as in the USA and Canada). For the design of bottom liners, an important issue concerns the present state of regulations adopted by industrialized countries. In spite of the desirability of common regulations, different approaches are still adopted in different countries. The information in Figure 3 shows that there is a trend towards composite sealing barriers. In the USA, the trend is towards defining

Table 3. Concentration (wt.%) of major elements in washed and unwashed slag and fly ash (from Nüesch *et al.*, 1996, reproduced with the permission of Balkema, Rotterdam).

	Elements	SU*	SW*	FU*	FW*
Major elements	Si	20	20	14.7	15.4
	Ca	13.2	13.3	12.6	14.1
	Fe	9.6	7.3	2.9	3
	Al	5.5	5.6	5.8	7.1
Elements in soluble salts	Na	1.9	1.9	2.5	0.9
	K	0.8	0.8	3.9	1.2
Reactive elements	C	1.3	4	4	7.3
	P	1.76	1.06	0.95	1.02
	S	0.6	0.8	2.7	2.8
Toxic heavy metals	Cd	0	0	0.05	0.05
	Cr	0.05	0.05	0.07	0.08
	Cu	0.26	0.23	0.12	0.15
	Pb	0.16	0.12	0.86	0.99
	Zn	0.22	0.30	3.02	3.47
	Sn	0.02	0.02	0.15	0.18

*SU = unwashed slag; SW = washed slag; FU = unwashed fly ash; FW = washed fly ash.

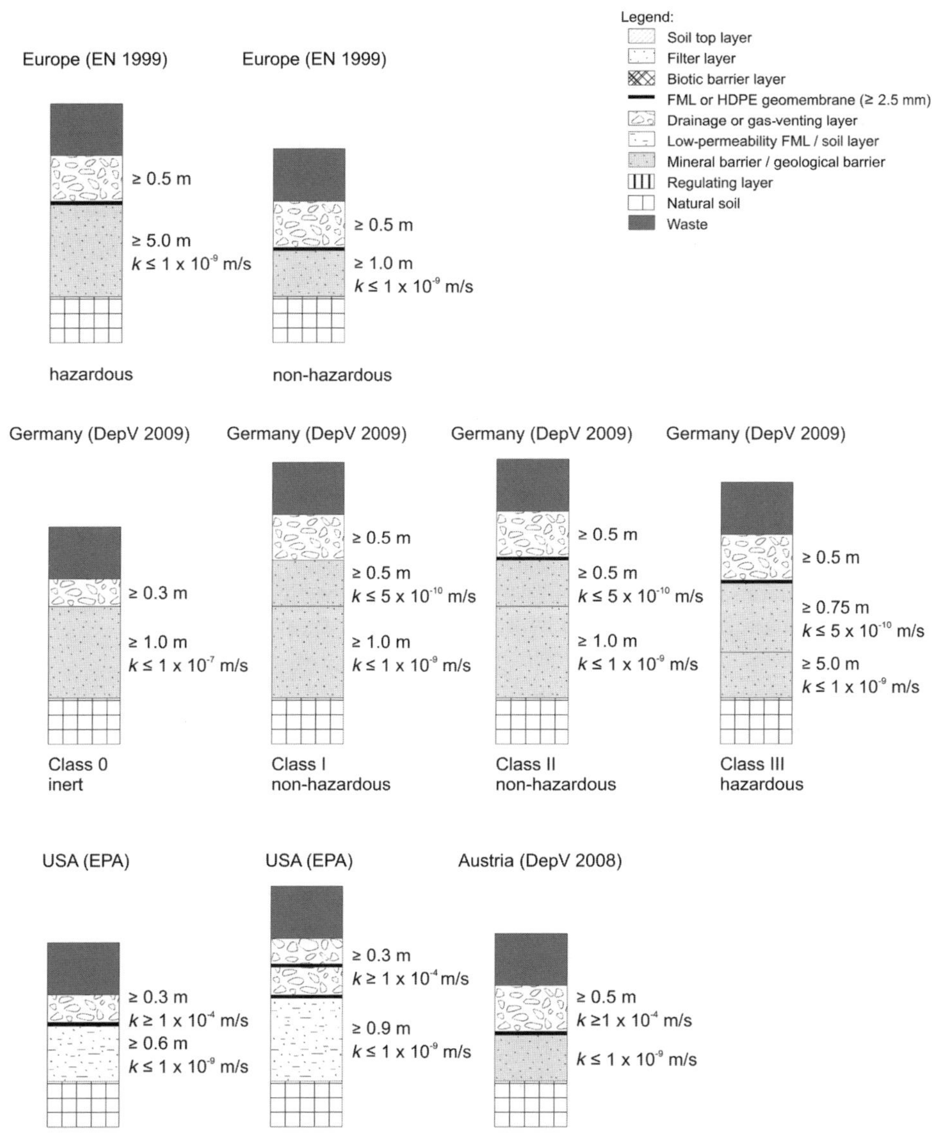

Fig. 3. Bottom-lining systems for municipal-waste landfills based on the regulations and recommendations of various countries.

the sealing lining systems independently from the natural subsoil conditions, whereas some European countries take into account natural subsoil features in defining the lining system.

Aside from regulations and guidelines, the design of modern landfills should be based on the following fundamental principles (after Manassero *et al.*, 1997).

(1)   The mineral barrier should be the basic component of the sealing system for long-term performance ($t > 50$ years).

(2)   The requirements of the mineral sealing layer, in order of importance, are: (a) low hydraulic conductivity at the field scale; (b) long-term compatibility with any chemicals to be contained; (c) large sorption capacity, and (d) low diffusion coefficient.

(3)   Composite lining systems can give important advantages both in the short and long-term because of: (a) reduction of hydraulic conductivity due to the attenuation of local defects of both the geomembrane and compacted clay, as shown in Table 4; (b) enhancement of flow within the drainage layers toward the collection pipes, and (c) the geomembrane on top of the clay-rich barrier delaying direct contact between clay and leachate for long enough for consolidation of the clay portion of the composite system.

(4)   Drainage efficiency is a very important effect in reducing the height of ponding leachate on the barrier and, consequently, on the advective migration.

(5)   Construction details play a fundamental role in the final efficiency of the lining system in terms of full-scale hydraulic conductivity.

2.3. Capping systems

The main purpose of the capping or cover system is (after Manassero *et al.*, 1997): (1) to raise the ground surface and provide appropriate slopes for promoting runoff and controlled drainage of surface water; (2) to separate buried waste from vegetation, animals and humans; (3) to minimize infiltration of water into the waste, and (4) to control the release of gas from the waste. Not all of these functions must be satisfied for all landfills; specific requirements depend on from many factors including: (1) location of the landfill; (2) climate conditions; (3) phase of landfill activity; (4) general strategy of landfill management, and (5) type of waste.

Table 4. Rate of leachate migration through various types of liners.

Rate of leachate (in litres/1000 m^2/day)		
Mineral liner	10,000	for a soil liner having $k \approx 10^{-7}$ m/s
	1,000	for a soil liner having $k \approx 10^{-8}$ m/s
	100	for a soil liner having $k \approx 10^{-9}$ m/s
Geomembrane liner	10	for a geomembrane installed with strict construction quality assurance and containing a typical number of defects
Simple composite liner	1	for a composite liner constructed with a soil having $k \approx 10^{-8}$ to 10^{-7} m/s
	0.01	for a composite liner constructed with a soil having $k \approx 10^{-9}$ m/s
Double composite liner	0	for a double composite liner where each of the two liners is constructed with a soil having $k \approx 10^{-9}$ m/s

Most cover systems involve multiple components which can be grouped into the following five categories (after Daniel & Körner, 1993): (1) *Surface layer* (promotes vegetative growth; promotes evapotranspiration; prevents erosion) [usual materials – topsoil, cobbles, geosynthetic erosion control system]; (2) *Protection layer* (stores water; protects underlying layers from intrusion by plants, animals and humans; protects barrier layer from desiccation and freeze/thaw; maintains stability) [usual materials: mixed soils; cobbles]; (3) *Drainage layer* (drains away infiltrating water to minimize barrier-layer contact and to dissipate seepage forces) [usual materials – sands, gravels, geotextiles, geonets, geocomposites]; (4) *Barrier layer* (minimizes infiltration of water into waste and escape of gas out of water) [usual materials – compacted clay liners; geomembranes; geosynthetic clay liners]; and (5) *Gas collection layer* (transmits gas to collection points for removal and/or recovery) [usual materials – sand; geotextiles; geonets].

The different types of cover systems required by the regulations and recommendations of various countries are shown in Figure 4. Generally, a compacted clay barrier at least 0.45 m thick must be provided in order to minimize water infiltration and gas migration (Manassero *et al.*, 1997).

Single compacted clay liners can be used only in cases where a very thick protection layer is provided to avoid problems due to desiccation and freeze/thaw cycles. Moreover, in the case of reduced overburden pressure, no significant differential settlement can be sustained by compacted clay liners without the appearance of tension cracks and reduction of sealing capacity. For cover systems, the use of geosynthetic clay liners alone is unacceptable for waste that produces gas unless the clay is wetted soon after installation and not then subjected to drying cycles.

A composite liner formed from a compacted clay liner and a geomembrane is a common minimum requirement for certain categories of waste. The primary concern with this kind of liner is the behaviour of the compacted clay if large differential settlements of the underlying waste occur. For such cover systems, acceptable performance is based on three main conditions. These are: (1) the geomembrane lifespan is long enough; (2) neither the geomembrane nor the compacted clay increases greatly in permeability through punching and cracks due to differential settlements, and (3) desiccation problems from below the clay liner do not occur. Composite liners comprising geomembrane and geosynthetic clay liners can be considered an ideal cover system for many reasons, including capability to deal with differential settlement, easy construction and performance control, and cost efficiency. In addition, interest in using geosynthetic and alternative materials, especially recycled products or waste materials, is growing (Genske, 2005; Guyonnet *et al.*, 2009).

Long-term performance of landfill covers consisting of three-layer systems shows that only 1–3% of precipitation percolates through the barrier layer. The long-term effectiveness of the mineral sealing layer depends on the ability of the top layer to protect it from critical loss in soil water/critical increase of suction. The hydraulic conductivity at the time of construction gives an initial (minimum) value. The hydraulic conductivity of the compacted clay layer or of the geosynthetic clay liner may increase substantially if there is no long-lasting protection against desiccation (by a thick soil

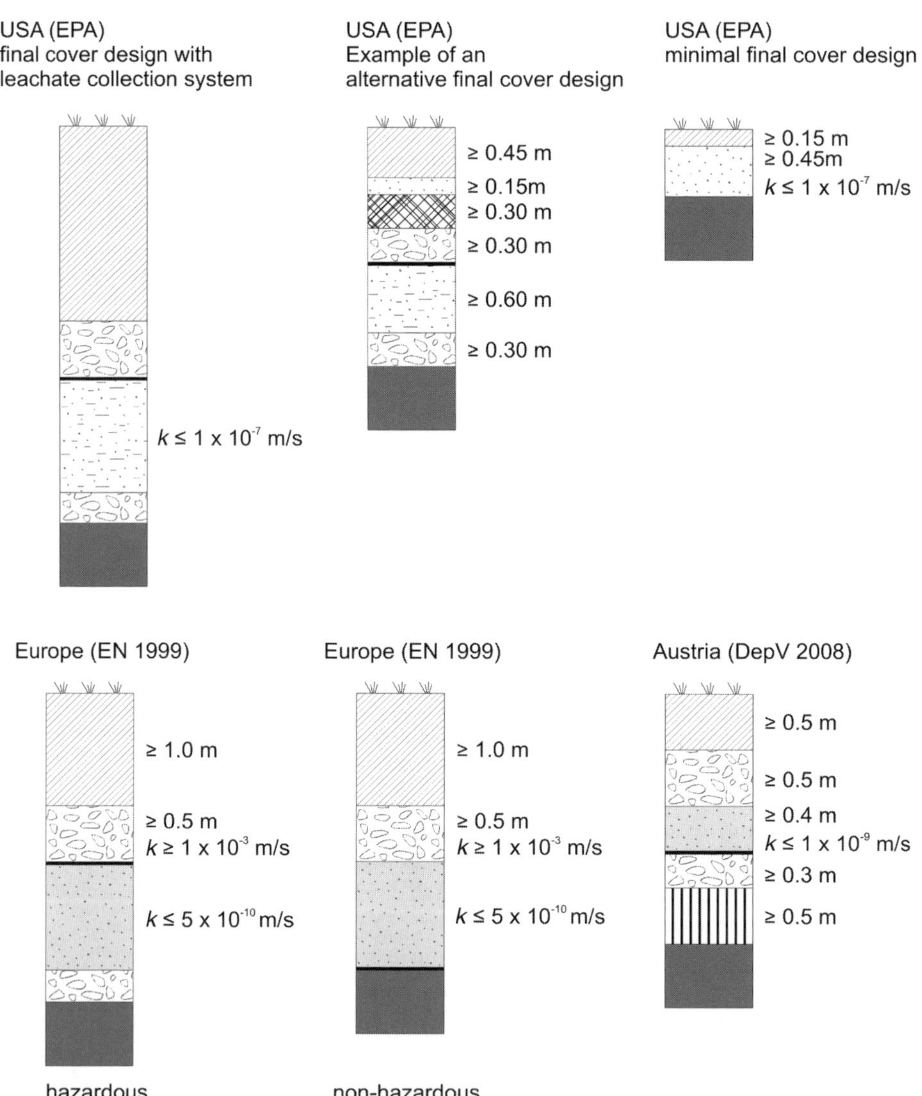

Fig. 4. Cover systems based on the regulations and recommendations of various countries.

cover or by a geomembrane). This has to be taken into account in landfill cover design (Henken-Mellies, 2011).

The advantages of using capillary barriers for cover systems have been pointed out by many authors on the basis of theoretical and experimental studies (*e.g.* Jessberger, 1997). A capillary barrier results when unsaturated flow occurs through a relatively fine layer overlying a relatively coarse layer (*e.g.* clay over sand, silt over gravel, or

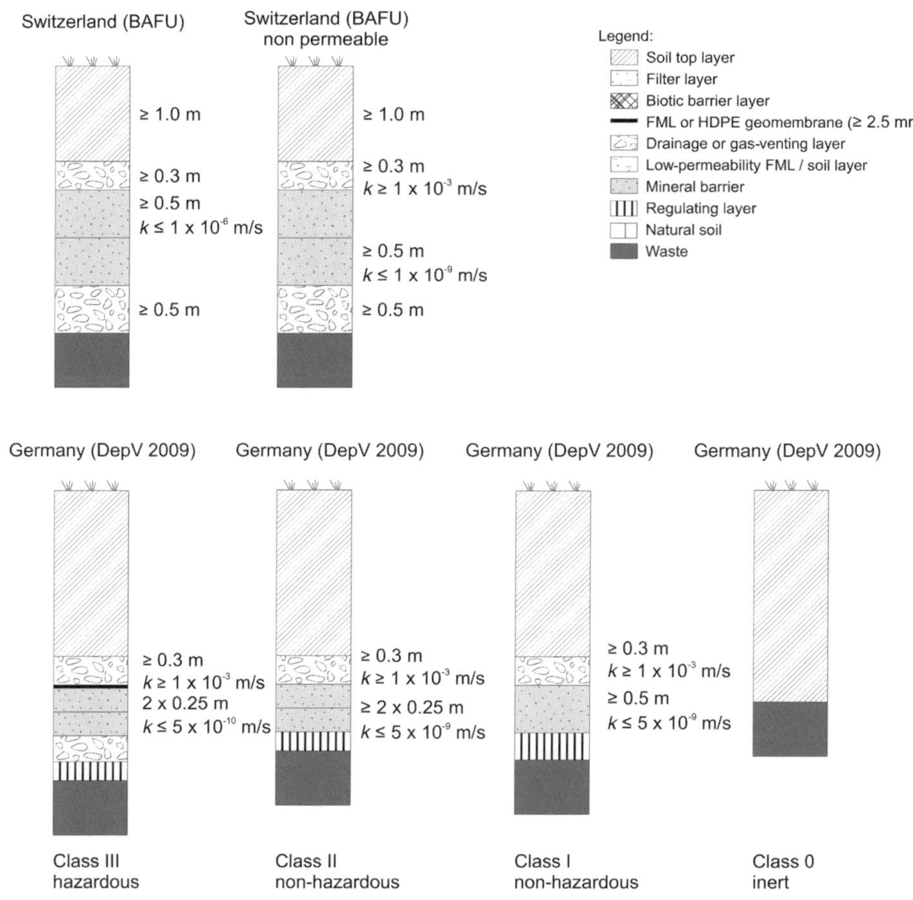

Fig. 4. Continued.

sand over gravel). In a capillary barrier, only a small fraction of the flux associated with the wetting front passing through the finer layer is transmitted into the underlying coarser layer (Shackelford, 1996). A capillary barrier effect occurs when: (1) the unsaturated hydraulic conductivity of the coarser layer falls to a smaller value than the unsaturated hydraulic conductivity of the finer layer as the capillary suction increases, and (2) the residual suction in the finer layer after first passage of the wetting front is larger than the gravitational force on the wetting front. Both of these effects result in an effective 'reflection' of the wetting front at the interface of the finer and coarser layers. Capillary barriers seem to be more effective than the common resistant barrier for reducing both water percolation into the waste and gas migration into the atmosphere.

3. Mineral barriers for landfills

The greatest environmental concern with landfill wastes is the generation of leachate from infiltrating surface water and groundwater, with resultant contamination of groundwater supplies. Therefore, a modern landfill is a designed system which must prevent such contamination. The main function of the base liner is sealing, defined as providing a low hydraulic conductivity and a high toxicant retention potential. There are many practical situations where the geological barrier alone is insufficient to prevent unacceptable long-term environmental impact.

Successful design of a landfill liner requires that consideration be given to the hydraulic performance of the landfill. Within the geological barrier, detailed mineralogical studies are relevant in order to evaluate changes in the properties of the rock in response to the emplacement of waste, irreversible rock-water interactions, the dehydration and transformation of clays, the migration of fluid, and the fractures and fluid pathways. In the multi-barrier concept, the geology of the disposal site plays an important role, and has to be seen as an essential part of an encapsulation system that combines technical with geological barriers. The geological setting may contribute significantly to waste isolation. Factors which should be taken into account are the thickness and extent, and the manner of deposition of sedimentary layers, fabric disturbances, grade of weathering, local hydrology, geochemical characteristics regarding the contaminant retention capacity, and physical characteristics to assess the bearing capacity and stability. From the geological point of view, a landfill site should be located in thick formations with low permeability and high contaminant retention capacity, and which extend beyond the landfill location. Beds or rock bodies which can act as geological barriers include clay horizons, granite bodies, limestones, metamorphic rocks or salt domes. The requirements for both low hydraulic conductivity and high toxicant retention potential are usually fulfilled by clays or claystones with a sufficient undisturbed and continuous thickness and without tectonic fabrics. Clay liners are a key component of barriers (Yong *et al.*, 1992; Rowe *et al.*, 2004).

Although detailed requirements vary from one jurisdiction to another, some criteria generally apply both for geological and engineered barriers, such as low hydraulic conductivity ($k_f \approx 1 \times 10^{-7}$ to 1×10^{-10} m/s), retention potential (*e.g.* the cation exchange capacity, CEC, of clay minerals) and liner thickness. The engineered mineralogical barrier may consist of recompacted clay liners, bentonites, concretes, geomembranes and composite liner systems. Important physico-chemical properties that affect the suitability of a particular clay soil for use in an engineered clay liner are: compaction parameters, swelling and shrinking behaviour, hydraulic conductivity, chemical properties such as CEC and large specific surface area. These physico-chemical properties are largely determined by the mineralogical composition of the liner. Increasing the thickness of the liner isolating the landfill from the groundwater system substantially reduces the maximum contaminant concentration at the base of the liner and increases the time required to reach this maximum. The permeability of clay barriers is significantly influenced by the soil microstructure and macrostructure (Collins & McGown, 1974;

Murray *et al.*, 1997). The total porosity, *n*, of sand is of the same order as that of clay (Lege *et al.*, 1996). However, clays are composed of very fine plate-like particles (<2 μm) with a greater amount of very small interparticle pores which are inaccessible to water. This results in a smaller effective porosity, n_e and, thus, a hydraulic conductivity which is several orders of magnitude smaller. Fissures, laminations and other features making up the macrofabric of clays provide preferential seepage paths. The resulting flow rate may be as great as to totally obscure that through the other pore-space types. Fractured clays or rocks generally involve a series of fractures separated by blocks of intact material. Such discontinuities may be viewed as representing large void spaces with an increased permeability above that of the surrounding intact clay. The primary transport mechanism for dissolved contaminants in such fractured media is usually advective-dispersive transport along the fractures (Rowe *et al.*, 2004). Important factors influencing hydraulic conductivity viewed from the perspective of the fabric of a clayey soil are the water content, the density and compaction of the clay, the layering and dispersion of the particles and the physical and chemical properties of the leachate solution. Wetting and drying of the clay liners can cause liner cracking. On the other hand, so-called 'self-healing' may occur where such shrinkage cracks are closed by clay swelling, which in turn depends on the swelling and shrinkage behaviour of the actual clay minerals and on the physical and chemical properties of the pore solution. The migration of water and, therefore, the mobility of contaminants (such as heavy metal ions) in clay liners is governed by migration and by sorption processes. Sorption processes include adsorption, ion-exchange reactions, complexation reactions, precipitation, and coprecipitation on mineral surfaces. These factors are determined by the mineralogical composition, by the structure and dry density of the clays, and by the water chemistry. The influence of pore-water composition is reflected by lower conductivity with same the porosity for the Na form compared to the Ca form of the clays. The latter form exhibits an aggregated structure, causing a more open microstructure with lower tortuosity (*e.g.* Westsik *et al.*, 1983; Cheung *et al.*, 1987; Hasenpatt *et al.*, 1988; Mitchell, 1993; see section 3.2).

3.1. Suitability of minerals for barrier systems

The most important properties of minerals for use in barrier systems are adsorption capacity and, because of the surface-controlled uptake of contaminants, surface area and chemical activity of the surface due to surface charge. Large surface areas can result from small particle size and layered or open structures. Other characteristics of importance are physical strength and plasticity, resistance to alteration and water uptake, and swelling behaviour. The mineral groups of special interest in waste management are clays, zeolites, carbonates and, in special cases, Fe oxides (Campbell, 2000; Czurda, 2006) (see also Tables 5, 7). Clay minerals combine the necessary properties (Churchman *et al.*, 2006). Due to their structures, surfaces and compositions, clays and clay minerals have many useful properties, such as large adsorption and cation exchange capacities, small diffusion and permeabilities, and large specific surface areas and swelling ability. Therefore, clays liners play a key role in barrier design.

Clay rocks are mainly composed of small particles (<2 μm). The main minerals in clay rocks, alongside others such as quartz, feldspar and carbonates are the clay minerals themselves. Clay minerals are plate-like sheet silicates with a small particle size (commonly $\ll 2$ μm). The properties of clays depend heavily on their precise mineralogical composition. Therefore, determination of the types and relative amounts of the minerals present are important for assessment of the suitability of a particular clay as a barrier material. Measurements of the CEC and specific surface area and, in some cases, of diffusion coefficients and hydraulic conductivity may be required.

Only a brief overview of the mineralogy of clay-mineral species commonly encountered in engineering practice is provided here. More detailed texts on industrial clays and clay mineralogy include Grim (1962, 1953), Brindley & Brown (1980), Bailey (1988), Jasmund & Lagaly (1993), Moore & Reynolds (1997) and Christidis (2011). The fundamental structural elements of clay minerals are the tetrahedral and the octahedral sheets (Fig. 5). The silicate tetrahedron comprises four oxygens (O^{2-}) surrounding the central silicon (Si^{4+}). Substitution in the silicon position is common, predominantly by aluminium (Al^{3+}), and less frequently by Fe^{3+}. In the octahedral sheets, the most common central cations in natural clays are Al, Mg and $Fe^{2+/3+}$, but cations of other elements such as Li, Ti, V, Cr, Mn, Co, Ni, Cu and Zn can also be present. In most clay minerals, two of the surrounding six anion positions are occupied by $(OH)^-$ ions, the rest by O^{2-}. The silicate tetrahedra are linked to each other by sharing three corners in such a way that a network of hexagonal interstices surrounded by oxygens is formed. The unshared corners point in the same direction and form part of an immediately adjacent octahedral sheet in which individual octahedra are linked laterally by sharing octahedra edges.

The structural entity formed by linking one tetrahedral sheet with one octahedral sheet is known as a '1:1 layersilicate', whereas a '2:1 layer silicate' links two tetrahedral

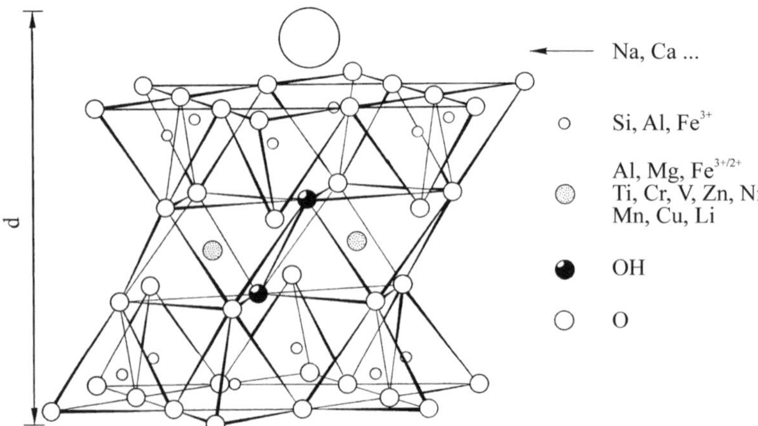

Fig. 5. Diagrammatic sketch of the crystal structure of smectite with the various ion positions in the interlayer and in the tetrahedral and octahedral sheets.

sheets with one octahedral sheet. The unit cell of a single layer phyllosilicate contains four tetrahedra and six octahedra in the case of the 1:1 layer type and eight tetrahedra and six octahedra in the case of the 2:1 layer structures. Structures where all of the octahedral cation positions are occupied are trioctahedral (all three octahedra of the formula unit), whereas in the so-called dioctahedral structures, only 2/3 of the positions are filled. In the octahedral sheet of trioctahedral clay minerals, divalent cations (Mg^{2+}, Fe^{2+} *etc.*) dominate, whereas in the octahedral sheet of dioctahedral clays, trivalent (Al^{3+}, Fe^{3+} *etc.*) cations dominate. On the other hand, natural samples often exhibit a crystal chemistry that is intermediate between these end-members, having between two and three filled octahedral positions. A 1:1 or 2:1 layer plus the interlayer forms the unitstructure.

Any excess layer charge, x, generated by non-equivalent substitution of the central atoms in the octahedral or tetrahedral sites, or by the presence of octahedral vacancies, is neutralized by various interlayer materials including (hydrated) cations, organic molecules, and hydroxide octahedral groups and sheets. The excess layer charge is generated by non-equivalent substitution of the central atoms in the octahedral or tetrahedral sites and by the presence of octahedral vacancies. The deprotonation of OH to O is the reason for the charge of the edge sites of clay minerals, and is dependent on the pH of the solution (negative charge at pH $\geq$ 5.5 and positive at pH $\leq$ 5.5. The permanent layer charge and the pH-dependent modifiable edge charge of clay minerals leads to their facility to adsorb and exchange ions at the surfaces (interlayer surface and outer surface) and at the edge sites of the mineral particles. The numerical value of this cation-exchange capability is described by the CEC which is a measure of the number of positive charges in milliequivalents needed to neutralize 100 g of clay (meq/100 g). Methods to determine the CEC involve the complete exchange of the naturally occurring cations by a cationic species such as ammonium, methylene blue, Co^{3+} hexamine complex, Ag thiourea complex or Cu^{2+} ethylenamine (Bergaya *et al.*, 2006). Exchange with ammonium acetate or with $[Cu(trien)]^{2+}$ complexes are the most versatile methods for CEC determination (MacKenzie, 1951; Meier & Kahr, 1999).

The clay minerals have been classified into several groups, subgroups and species on the basis of layer type (1:1, 1:2), sources and sites of layer charge (x) per formula unit, type of interlayer, and principal interlayer and octahedral cation (Guggenheim *et al.*, 2006). Common clay minerals in natural clays and clayey rocks are kaolinite, smectite, vermiculite, illite, chlorite and interstratified (so-called mixed-layer) clay minerals; these are regularly or randomly interstratified sequences mostly of two different layer types (*e.g.* illite/smectite, various chlorites).

The most obvious property of clays is their change in volume by adsorption of water or other polar solvents. Therefore, in engineering practice, clay minerals have often been classified as swelling (*e.g.* clay minerals of the smectite group with montmorillonite as a common example) and non-swelling types (*e.g.* kaolinite, illite and chlorite) (see Table 5). In swelling clays, two types of swelling are observed. Osmotic swelling results from the difference in ion concentrations close to the clay surfaces and in the pore waters. Inner crystalline swelling is caused by the hydration of the exchangeable interlayer cations of the dry clay. The swellability of clay minerals is primarily

Table 5. Selected properties of different types of clay minerals (compilation of authors' data and data from Yong & Warkentin, 1975; van Olphen, 1979; and Czurda, 2006).

Mineral	Swelling behaviour	CEC (meq/100 g)	N_2-BET specific surface area (m^2/g)
Kaolinite	Non-swelling	5–15	5–25
Montmorillonite	Swelling	65–140	40–100 (–800**)
Vermiculite	(non-)swelling	20–30 (–250*)	120–210
Illite	Non-swelling	20–30	50–100
Chlorite	Non-swelling	20–30	50–80

*depending on the index cation in the CEC measurement; **total water accessible surface area.

determined by their CEC and the layer-charge density and location (Sato *et al.*, 1992). Another factor affecting hydration and swelling is the nature of the interlayer cation (Brindley & Brown, 1980; Plötze & Kahr, 2003; Kaufhold *et al.*, 2010). Due to their small crystal size, clay minerals have a large surface area. The surface area per unit mass or volume, which is measured by the adsorption of gas molecules (after Brunauer *et al.*, 1938, the so-called BET measurements) includes the accessible surface. In swellable clay minerals, however, there is the outer surface and the inner surface of the interlayer.

Clay mineral species normally encountered in engineering practice are kaolinite, smectite, illite and chlorite (see Table 5). The structure and the swelling behaviours of various clay minerals in water and organic liquids allow their determination using XRD methods (Brindley & Brown, 1980; Moore & Reynolds, 1997). The clay mineral species can be distinguished by the basal lattice spacing and its changes after different treatments (intercalation of organic liquids like glycol or collapse by heating). Kaolinite consists of stacked 1:1 layer units which have a c_0 basal lattice spacing d (interlayer to interlayer) of 0.72 nm. There are neutral octahedral sheets and very slightly charged tetrahedral sheets (layer charge $x \sim 0$). As a result, the CEC is very small ($\sim$5 meq/100 g). The strength of the bonding between the layers enables kaolinite to form fairly large crystals of up to 4 μm in diameter in comparison to the other clay minerals, so that kaolinite has a smaller specific surface area of $\sim$15 m^2/g and a greater permeability than other clay minerals. The smaller adsorptive capacity and the larger hydraulic conductivity (k_f rarely $<10^{-8}$ m/s) make kaolinite less well suited as a clay mineral for a liner.

Smectite is a group name for the 2:1 layer swelling clays. Smectites are the main component of bentonites. Chemical and structural features differ between the various smectite species. The di- or trioctahedral nature of the octahedral sheets, the chemical composition and the predominant charge location divides smectite into the related subgroups. The most common smectite is montmorillonite. Smectites have a layer charge x in the range $\sim$0.2–0.6 per formula unit. This charge corresponds to a CEC of 65 to 140 meq/100 g and is too small to fix cations and bind the layers together. The charge-compensating cations adsorbed at the surface and in the interlayer in natural montmorillonite are mainly Ca^{2+} and Na^+. The weaker bonding of these cations makes them exchangeable. Furthermore, the adsorbed cations attract water; therefore, smectites are highly swellable. The interlayer swelling from the hydration of the exchangeable interlayer cations leads to variability in the basal spacing from ~0.1 nm in dry state to >1.7 nm in a water-saturated state (Plötze & Kahr, 2003; Ferrage *et al.*, 2007). Smectites consists of very tiny particles (<0.2 μm), which produce a large specific (BET) surface area (40–100 m^2/g) The total water-accessible surface area can reach up to 800 m^2/g. Thus, smectites can have very low hydraulic conductivities ($k_f < 10^{-11}$ m/s).

Illite is very similar to muscovite, and has a layer structure with two tetrahedral sheets linked to an octahedral sheet (2:1 layer silicate). However, compared to muscovite, the layer charge x is smaller (0.9) but still large enough that the layers are held together by a very strongly bonded potassium. The K$^+$ ions are not exchangeable. A small unbalanced charge leads to a CEC of ~25 meq/100 g. The small hydraulic conductivity, the CEC and the non-swelling character which avoids shrinkage cracks, makes illite, like the smectites, a desirable clay mineral for use in engineered clay liners for municipal solid-waste disposal.

The chlorite structure consists of 2:1 layers bonded together by positively charged octahedral interlayer sheets having Mg, Al or Fe in the central position. Chlorites display a wide range of chemical compositions and a variety of polytypes. The properties of chlorite are similar to illite. The layer bonding is strong enough to prevent c-axis swelling. In soils, chlorite is a common but usually minor component. Iron chlorites are highly susceptible to oxidation on weathering.

3.2. Alteration of clay rock and clay minerals with special reference to their geotechnical properties

The engineering properties of clay rocks depend on several interacting factors. Compositional factors include types and amounts of specific minerals and their characteristics, and pore-water composition. Environmental factors include water content, density, temperature, pressure, fabric and water availability. These all interact. Two kinds of reaction occur with alteration of clays over time. These are short-term and mostly reversible reactions on the one hand, and long-term mostly irreversible reactions on the other hand. Short-term processes involve reactions in the interlayer and at the surface. The adsorption of ions in the interlayer or at the surface is a reversible process on interaction with solutions. The adsorption and exchange behaviour depends on the layer charge and on the charge density and, therefore, on the hydration behaviour of the adsorbed cations. At greater layer charges, the cations are more strongly bound and less exchangeable. The hydration behaviour of cations depends on the charges and sizes of the ions. Smaller or more highly charged cations bind more water molecules. They have greater interlayer hydration energies, are more weakly adsorbed and more easily exchangeable. The adsorption and exchange behaviour of clay minerals is responsible for the potential of clay liners to retain toxins. The sorption capacity of clay rocks for different metals, and the migration behaviour and the anion and cation adsorption behaviour of montmorillonite clays under various chemical conditions have been investigated intensively (*e.g.* Grauer, 1986, 1988;

Higgo, 1987; Czurda & Wagner, 1991; Kruse, 1992; Wagner, 1992; Venema *et al.*, 1996; Kraepiel *et al.*, 1999). Cation transport and retardation processes are influenced by the bonding forms of these cations and by solution composition and by specific rock and mineral parameters.

The adsorption of ions can alter some properties of clays. Changes in hydraulic conductivity and diffusion behaviour of clays in waste-containment barriers have been studied (*e.g.* Madsen & Mitchell, 1989; Madsen, 1998; Madsen & Kahr, 1993; Pusch, 1994). Inorganic contaminants influence the double-layer thickness and, thus, the clay particle association (dispersed or flocculated), clay fabric, and pore space (van Olphen, 1977; Mitchell 1993). In general, greater concentrations and greater valences of cations cause flocculation of clay particles, increase the hydraulic conductivity, and limit the swelling of expandable clay minerals. Acids tend to cause flocculation, whereas bases tend to disperse particles. The clay–organic reactions and the intercalation of organic molecules in clay minerals have been studied intensively (*e.g.* Mortland, 1970; Theng, 1974; Madsen & Mitchell, 1989; Jasmund & Lagaly, 1993). The influence of organic chemicals on clay properties is controlled by their water solubility, polarity and concentration. Pure organic liquids will interact with clays by causing some shrinkage and cracking, with large increases in hydraulic conductivity. Dilute solutions of organics have essentially no effect on the hydraulic conductivities of clays. The exchange of the naturally occurring cations for organic cations in smectites leads to changes in their hydrophylic character. These modified organophilic smectites show interesting properties and are used for adsorption of organic pollutants (Stockmeyer, 1992; Stockmeyer *et al.*, 1995; Meier, 1998).

Another kind of short-term reaction is dehydration of clay minerals at temperatures below 200°C and (less relevant for landfill barriers) the dehydroxylation and decomposition of clay minerals at temperatures >350°C. The loss of water bound in the interlayer and on the surface is a reversible process which leads to the swelling and the shrinkage of clays. Factors controlling hydration behaviour include the layer charge, its density and location, and the nature of interlayer cations. The investigation of dehydroxylation behaviour by various methods of thermal analysis can be used to distinguish between different clay minerals and to characterize the mineralogical and chemical properties of clays (*e.g.* Stucki & Bish, 1990).

Dissolution and phase transformation are long-term reactions of clay minerals under environmental conditions. The effects of chemical attack on clay minerals have been the subject of studies focused on the stabilities and phase transformations of clays (Madsen & Mitchell, 1989; Güven, 1990; Pusch, 1994; Walther, 1996; Echle *et al.*, 1997; Kaufhold & Dohrmann, 2009, 2010, 2011). Clay minerals are relatively stable under Earth's surface conditions but acids and bases may attack the crystal lattice, acids especially the octahedral sheets and bases mainly the tetrahedral sheets. The transformation of smectite to illite through alteration reactions has been the subject of many studies. The primary factors controlling the smectite illitization reaction are temperature and potassium availability. Secondary factors include time, water content, fluid composition and pressure. Illitization at temperatures >80°C leads to a decrease in swellability and adsorption and exchange capacity. The reaction mechanism involved in this transition has been the subject of numerous studies (see Altaner & Ylagan, 1997; Meunier *et al.*, 1998, for reviews). Reaction mechanisms proposed include layer-by-layer transformation in the solid state and dissolution-reprecipitation (neoformation). The conversion requires a significant enhancement of the excess layer charge in smectite, mainly by substitution of Al^{3+} for Si^{4+} in tetrahedral sites, and a supply of K^+ ions which are then bound non-exchangeably in the interlayer. Hydrothermal alteration of bentonite at temperatures $>100°C$ causes clay cementation by precipitation of silica and of aluminium compounds and ferric oxides (Pusch & Güven, 1990; Pusch, 1994; Pusch, 2006). Both illitization and cementation are important for the longevity of buffer performance in radioactive-waste disposal, less so in landfilling of municipal solid waste. The influence of gamma irradiation on clay-mineral properties relevant for barrier performance in radioactive-waste disposal seems not to be significant (Plötze *et al.*, 2003; Allard & Calas, 2007).

4. Waste-containment barriers

4.1. Introduction

4.1.1. Site investigation

For the site investigation of contaminated soils and abandoned landfills, the standard investigation procedure is as follows (Jessberger, 1997; Knödel *et al.*, 2007): (1) Information on the input material, location of demolished factories and other buildings, and on possible accidents, are taken from files or from interviews with former employees or adjacent residents. Aerial photographs, taken over time, are extremely useful in providing detailed information on the historical development of the site. (2) Based on the above, an appropriate sampling strategy is chosen, avoiding, as far as possible, uncertainties with respect to the extent and the type of contamination. Guidelines for the investigation of soil contamination based on both statistical and non-statistical methods were reviewed by Bosman (1993). (3) Groundwater, soil and air sampling has to be conducted in such a way that, for a contaminated site or an abandoned landfill, the initial site conditions can be determined reliably, effects on the environment can be recognized, and remedial methods introduced. Further contamination of soil or groundwater caused by using contaminated sampling equipment must be avoided (German Geotechnical Society of the International Society of Soil Mechanics and Foundation Engineering, 1997). (4) A strategy for the chemical analysis of groundwater, soil and air samples has to be developed, based on the known or expected contamination.

4.1.2. Remediation concept

Different methods for remediation of contaminated sites and abandoned landfills are available. The German Council of Environmental Advisors defines as remedial measures: (1) Restriction in utilization: this can lead to a change in use of the site after remediation in a way that is compatible to the remediation method (*e.g.* use as an industrial park rather than a children's playground); (2) providing containment by cutting-off migration pathways by using capping systems, vertical cut-off walls, basal lining systems, active pneumatic or hydraulic equipment, or by immobilization. (3) Decontamination of contaminated soil by thermal treatment, microbiological treatment, or soil washing. (4) Replacement of contaminated soil or waste.

There are other methods in the area of remediation techniques which are developing rapidly, such as permeable reactive walls, low-permeability barriers formed using bio-polymers, and electrokinetic walls.

This chapter will focus on two methods of waste containment: (1) use of vertical cut-off walls or slurry walls; and (2) permeable reactive walls and funnel-and-gate systems.

4.2. Encapsulation of old landfills using slurry walls ('cut-off' walls)

4.2.1. Introduction

Vertical underground barrier systems may be used to minimize water loss from hydraulic structures, reduce the inflow of water into building excavations, and prevent draw-down of the surrounding groundwater table. It is now also common practice to use vertical barrier systems for environmental protection (Krajewski *et al.*, 2010). The principal uses are in the encapsulation of old waste deposits and of contaminated industrial areas. Typically, encapsulation involves four main elements: (1) reduction in the flow of contaminated water into the surrounding groundwater system using vertical cut-off walls. Such walls have to be keyed into a natural geological barrier. (2) Reduction in the flow of uncontaminated water into the waste deposit using cut-off walls, and/or collection of this water with drainage systems or galleries. (3) Drawdown inside the encapsulated area; the hydraulic gradient is directed from outside to inside, contaminated seepage water is collected, pumped and regenerated. (4) Covering of a contaminated area with sealing materials to reduce the entrance of rainfall and surface water.

Today, a number of different methods exist to construct vertical cut-off wall systems. The three main techniques are (see Figure 6, after Brandl, 1998): (1) permeability reduction of *in situ* soil; (2) soil-displacement methods; (3) excavation methods.

To encapsulate old waste deposits, the barrier system used has to fulfil a number of requirements including low permeability, low diffusion coefficient, long-term stability and significant adsorption of pollution. The cut-off wall constructed also has to deform without cracks during the consolidation or hardening processes.

4.2.2. Constructing vertical barrier systems using the two-phase cut-off wall method

The two-phase method for constructing cut-off walls is similar to the slurry-trench construction system used for building static load-bearing walls (Figure 7). The excavation of the trench is undertaken section by section using special construction tools for slurry trench walls, such as clamp shells. During excavation, the soil walls are supported by a bentonite suspension. Subsequently, the actual sealing material is introduced into the trench during a process of exchange. It displaces the bentonite suspension, which may be pumped out and regenerated, or removed. In order to displace the bentonite suspension with the sealing material, there has to be a sufficiently large difference in density

TECHNOLOGY	CUT-OFF SYSTEM	GROUND PLAN (schematic)	DIMENSIONS d (m)	t_{max} (m)
Permeability reduction of *in situ* soil	compaction wall		$0.4–1.0$ [1]	$10–20$
	grouting wall	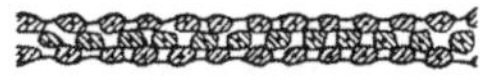	$1.0–2.5$	$20–80$
	soil freezing wall		≥ 0.7	$50–100$
	jet grouting wall		$0.4–2.5$	$30–70$
			$\geq 0.15–0.3$ [2] (lamella)	$20–30$
	soil mix wall	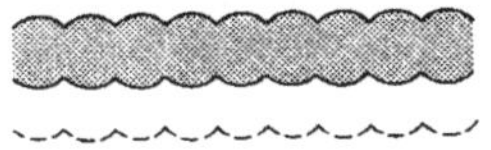	$0.8–1.5$	$30–60$
	cut-mix-grout wall		≥ 0.7	10
Soil-displacement methods	geomembrane wall		≥ 0.005 (0.002) [6]	$20–40$
	sheet pile wall		≈ 0.02	$20–30$
	vibrating beam –slurry wall		$\geq 0.05–0.2$ [3]	$10–35$
	'thin (diaphragm) wall'			
	earth concrete driven- sheet pile wall		≥ 0.4	$15–25$
Excavation methods	secant bored pile wall		$0.4–1.5$	$20–40$
	diaphragm wall (with hydrofraise)		$0.4–1.6$ [4]	$100–170$
	diaphragm wall (with grab)		$0.4–1.0$	$40–70$
	diaphragm wall with incorporated liner(s)		$0.6–1.0$ $(0.4–1.6)$ [5]	$20–50$

1) Vibrocompaction, vibrofloration

2) Total width of the diamond-shaped jet grouting walls ≥ 0.5 m

3) Near the flanges of the vibrating beam significantly wider

4) Up to 3.0 m in special cases

5) In special cases

6) In case of twin walls

Fig. 6. Overview of methods for cut-off wall construction. Approximate values for d (m) and maximum depth t_{max} (m), (Brandl, 1998, reproduced with the permission of Balkema, Rotterdam).

between the two materials. The main advantage of the two-phase method is the extensive range of potential sealing materials which may be used.

In a study of encapsulation of an old hazardous waste deposit in Switzerland, different mineral sealing materials were tested (Hermanns-Stengele, 1997). The amount of

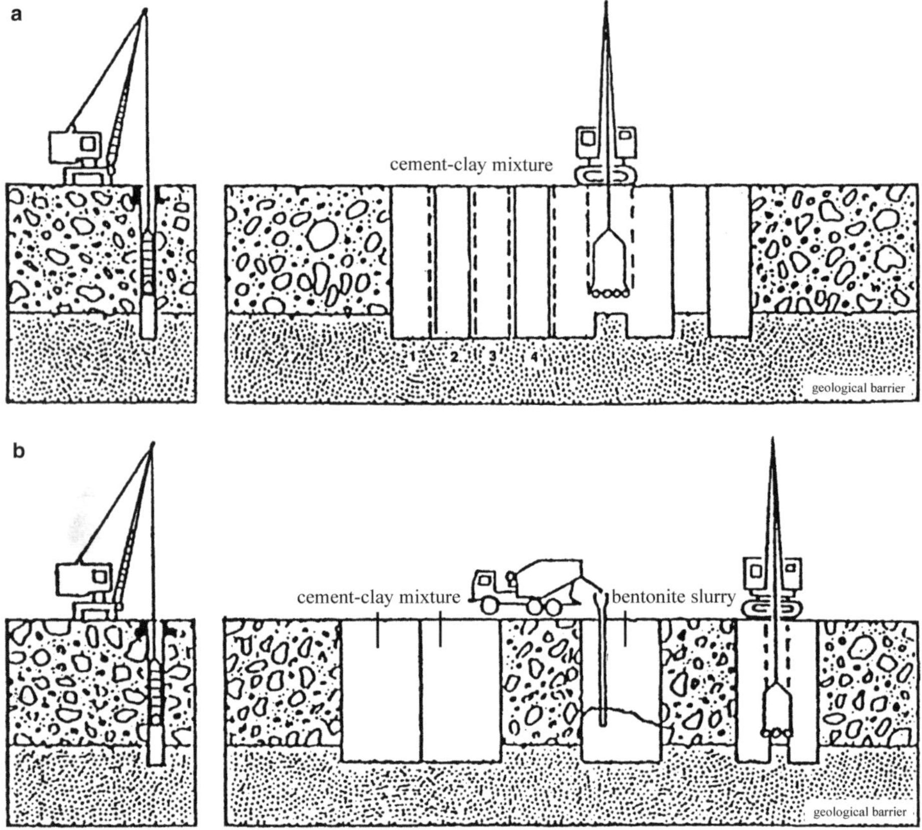

Fig. 7. Construction of slurry walls: (**a**) single phase and (**b**) two-phase cut-off wall.

active material needs to be sufficient to improve the long-term behaviour of the mixture and increase the retention of any polluted liquids. Inert materials such as sand or gravel were not used. Several mixtures were tested using Portland cement, powdered clay and different kinds of filler. The results obtained for two selected mixtures (Mixtures A and B, see Table 6) are presented here. The main difference between the two mixtures was the addition of a filler to A. This filler is described as a fly ash from an anthracite power plant, and is called EFA-Filler. Table 6 also shows different mixtures used in the past for two-phase cut-off walls, especially in Germany (mixtures 1 to 4).

4.2.3. Requirements

The following conditions are required of barrier mixtures: (1) when fresh, the mixtures should be mixable and pumpable; (2) during the process of concreting, the mixture should not become segregated; (3) the barrier material has to replace the bentonite suspension completely; (4) cracks must not appear in the course of the consolidation

Table 6. Two-phase cut-off wall mixtures (from Hermanns-Stengele, 1997, reproduced with the permission of Balkema, Rotterdam).

Mixture	Compounds per m³ of slurry mass	Density ρ (t/m³)	Coefficient of permeability k (m/s)	Reference
1	200 kg of cement 100 kg of clay flour 100 kg of rock flour 810 kg of sand 0/3 mm 490 kg of gravel 2/8 mm 360 kg of water	2.06	1×10^{-10}	Strobl (1989)
2	40 kg of Na-bentonite 280 kg of cement 250 kg of rock flour 1200 kg of sand/gravel 0/8 mm 350 kg of water	2.17	2×10^{-10}	Witt (1987)
3	340 kg of clay flour 350 kg of sand 0/2 mm 990 kg of gravel 2/8 mm 3 kg of Dynagrout DWR A 5 kg of Dynagrout DWR B 35 kg of water glass 180 kg of water	2.20	1×10^{-10}	Hitze (1987)
4	55 kg of Na bentonite 80 kg of cement 210 kg of rock flour 1080 kg of sand/gravel 555 kg of water	1.96	5×10^{-10}	Knappe (1988)
A	368 kg of cement 491 kg of clay flour 123 kg of fly ash 654 kg of water	1.63	$<1 \times 10^{-11}$	Hermanns (1993)
B	279 kg of cement 638 kg of clay flour 678 kg of water	1.60	1×10^{-9}	Hermanns (1993)

(setting) or hardening processes; (5) when hardened, the cut-off wall should be able to deform without cracking; (6) the permeability has to be low, and the resistance to diffusion has to be high, in the presence of contaminated liquids; (7) the long-term behaviour of the wall system has to withstand chemical attack; (8) the barrier material must have significant adsorption properties.

4.2.4. Laboratory test results

Mixtures A and B were prepared in the laboratory, placed in cylindrical containers, and deposited under water (18°C) prior to investigation. After 28 days, permeability tests were carried out using flexible-wall cells; the test liquids were tap water and a simulated contaminated water. The hydraulic gradients used in the test were between $i = 30$ and $i = 100$. The percolated test liquids were collected and the coefficient of permeability k was estimated using Darcy's law.

$$v = k\,i \tag{21}$$

where v is the velocity (m/s), k is the permeability (m/s) and i is the hydraulic gradient.

The collected percolates were analysed chemically. One of the main considerations was the concentrations of calcium ions removed by water movement. A low release of Ca is an important operating condition for barrier materials. A loss due to percolation of 30% of the calcium content is seen as critical with regard to material permeability (Strobl, 1987). The permeability of porous materials depends on the porosity and on the pore radius. The distribution and the sizes of pores were measured using mercury intrusion porosimetry.

The permeability of the vertical barrier can be reduced to extremely small values using appropriate compounds. A reduction of permeability from $k_f = 10^{-9}$ m/s to $k_f < 10^{-11}$ m/s can be obtained, which corresponds to the accumulation of calcium ions during the test period (Mixture A). This accumulation leads to a reduction of pore size. Permeation processes were minimized when the average pore radius was <100 nm. The determination of diffusion coefficients using different test liquids showed that this transport mechanism is relatively unimportant for the two mixtures examined. Because of the test method chosen, the evaluation of the coefficient of diffusion results from Fick's law. Using chlorophenole (1%), a coefficient of diffusion (D_s) of $\sim 5 \times 10^{-11}$ m^2/s was obtained. Heavy metals like zinc were precipitated because of the high pH of the pore water. In the case of Mixture A, diffusion transport processes will not reduce the barrier properties of a cut-off wall over realistic time periods.

4.3. *In situ* cleanup of contaminated groundwater using reactive walls and funnel-and-gate systems

4.3.1. Introduction

Increasingly it has been realized that conventional pump-and-treat systems for remediation of contaminant plumes do not lead to satisfactory results. This is particularly true if remediation periods of several decades are involved, which means a continuous and high input of energy for pumping water from extraction wells and operating water treatment systems, as well as periodic maintenance and monitoring (Köhler *et al.*, 2000).

In the early 1990s, the method known as 'Permeable Reactive Barriers' (PRBs) was developed in Canada and the USA. The PRB is a passive *in situ* technique that runs without the need for permanent and costly operation. Once installed in the subsurface, reactions between the material in the PRB and groundwater contaminants take place without any external interference. In Europe, a study of the clean-up of several contaminated sites with permeable reactive barriers was published by Birke (2006). These studies were continued until 2009, and provide a fundamental basis for future good practice.

4.3.2. Specifications required for permeable reactive barrier (PRB) installation

Permeable reactive barriers may be installed in an aquifer across the flow path of a contaminant plume. As the contaminated groundwater moves through the PRB due to the natural gradient, the contaminants are removed by physical, chemical or biological processes. Depending on what processes take place, the reactive barrier material can remain permanently in the subsurface, or replaceable units can be provided. As the reactions that occur in such systems are affected by many parameters, successful application of this technology requires adequate characterization of contaminants, groundwater-flow regime and subsurface geology. In case of instability of the subsoil or, if great depths are to be reached, techniques of specialized heavy construction are essential for implementation (*i.e.* sheet piling, contiguous bored cased piles).

Important for deciding on the feasibility of a PRB, apart from financial considerations, are the examination of: installation facility and permeability ($k_{f,PRB} > k_{f,aquifer}$); efficiency and clean-up performance; reliability and potentiality for monitoring the performance of the geochemical/geophysical processes; hydraulic and mineralogical long-term stability; and environmental compatibility.

Currently, two basic designs are being used in full-scale implementation of PRBs. A continuous trench, filled with the reactive media, is called a 'reactive wall' or 'treatment wall'. The combination of cut-off walls and permeable *in situ* reactor(s) is known as the 'funnel-and-gate' system (EPA, 1998; Gavaskar *et al.*, 1998; NATO/CCMS, 1998; Starr & Cherry, 1994; Teutsch *et al.*, 1996).

4.3.3. Reactive walls

Groundwater remediation using continuous reactive walls is a particularly environmentally friendly approach. The disturbance of groundwater flow is minimal. The permeability of the filling material must be at least as great as that of the natural aquifer, so that the ground water flow does not change direction and pass around, instead of through, the wall. The horizontal thickness must be designed such that the residence time of the polluted groundwater is sufficient to achieve the required effluent clean-up.

4.3.4. Funnel-and-gate systems

To clean up polluted ground water *in situ*, a continuous reactive wall must be big enough to cover the full range of the contaminant plume. If the plume is wide and reaches great depths, the extent of the reactive wall needs to be very large. To avoid this problem, and to increase the efficiency of the flow-through, one can use funnel-and-gate systems. Here, the contaminated groundwater flow is enclosed and directed towards the permeable reactive zone (the 'gate') by cut-off walls (the 'funnel'). The cut-off walls typically consist of interlocking sheet pilings or slurry walls. It is important to note that, in this method, water upstream of the system generates an increased gradient and hence higher flow velocity. This must be taken into account when considering the contact time of the contaminants with the reactive material and the thickness of the gate. The capture zone of a funnel-and-gate system is related directly to the flow through the gate. Plumes with a mixture of contaminants that require more than one type of *in situ* reactor can be controlled using multiple gates in series. Alternatively, multiple gates in parallel can be used for controlling wide plumes. Generally speaking, the capture zone of a funnel-and-gate system with several (small) gates is larger than that of a system of the same overall extent and only one reactive zone.

An *in situ* PRB should function until either the reactive capacity is consumed or it is clogged by precipitates or microorganisms. A reactor enclosed in a relatively small gate enables easier removal and replacement of the reactive material than does a continuous reactive wall. If the extent of a plume is relatively large, the reactive wall should be designed for a lifetime exceeding the remediation process. Another advantage of funnel-and-gate systems is that they offer a better means of using the reactive filling material in a heterogeneous subsurface, particularly when used in combination with drainage trenches.

4.3.5. Reactive material

Interest in PRBs has increased greatly over recent years, and full-scale projects involving them have been implemented, especially in North America. The predominant filling material used has been zero-valent iron (Fe^0) in the form of chips, jet blasting media, iron-foam or filler material for concrete. These materials can often be found as by-products of the metal-working industry. Some other highly reactive metals and metal alloys are also being developed using more complex procedures. Fe^0 is very suitable for degrading chlorinated organic compounds (Dahmke, 1997; Dahmke *et al.*, 1999; Gillham *et al.*, 1994; Gillham & O'Hannesin, 1994; Johnson & Tratnyek, 1994). Here, an abiotic, heterogeneous surface process occurs which involves adsorption of the contaminant and a progressive, and normally complete reductive dehalogenation of the chlorinated compound. The final degradation products are ethene or ethane, respectively. In small quantities, some problematic low-grade chlorinated compounds, such as cis-dichloroethylene (cDCE) and vinylchloride (VC), may also occur as products. Generally speaking, Fe^0-reactive barriers are an important method for remediation of groundwater contaminated with halogenated compounds. However, there are still questions surrounding their use. These include the nature of the degradation processes, influence of further substances in the ground water, performance of contaminant mixtures with diverse degradation or sorption characteristics, and processes occurring with downgrading of the PRB. Recent research shows that using this method, problems such as clogging or inhibition of chemical reactions may occur in the presence of HCO_3^-, SO_4^{2-}, PO_4^{3-} or NO_3^- or other competing oxidants (Dahmke *et al.*, 1999). Furthermore, the influence of microbiological activities on the degradation process is largely unknown.

Zero-valent iron and other metals are suitable for clean-up of toxic metals as well as metalloids (Cr^{VI}, As^V, As^{III}, Se^{VI} and Tc). Soluble forms of these elements are transformed into insoluble forms by reduction and then precipitated as hydroxides (Blowes *et al.*, 1996). Results of studies (Mak & Lo, 2011) show that compared to using zero-valent iron and a quartz sand mixture, the use of Fe(0) and an iron oxide-coated sand (IOCS) mixture can reduce the environmental impact. IOCS can enhance the removal efficiency of Cr(VI) and As(V). The usage of Fe(0) can be reduced which in turn reduces the impact induced by the reactive media. Other studies using Nanosized Zero Valent Iron (NZVI) show the effect of enhancement of Fenton reactions in organic-load reduction of landfill leachate (Shafieiyoun *et al.*, 2012). This technique has been observed to be a quick and efficient procedure. Less widespread is the use of adsorptive PRBs, *i.e.* those where immobilization of the pollutants is achieved by attaching them chemically to mineral surfaces. With drinking water, treatment by adsorption of organic components on solid surfaces is well established. For this, activated carbon is used predominantly. Activated carbon was also used as filling material in the first adsorptive PRBs (Grathwohl & Peschik, 1997; Teutsch *et al.*, 1996). The tendency of a component to leave the fluid phase and adsorb on a solid surface is quantified by sorption isotherms. This results only in retardation of contaminant transport, and with the reactive material showing a finite capacity for uptake, frequent replacement may be required. Besides these commercial technologies, further methods are being developed. Figure 8 provides an overview of different systems.

4.3.6. Development of adsorptive filling materials

Materials are being generated that can be utilized for adsorption of a wide range of the contaminants which occur at hazardous waste-disposal sites. One goal is to achieve the immobilization of pollutants which have proved difficult to manage with other remediation methods. Ideally, one hopes to find suitable materials among those produced as by-products of industry. Sorptive processes may use specific physical-chemical properties of a substance for separation of certain components from mixtures. Adsorption is greater the larger the surface area per unit mass. It increases rapidly with decreasing

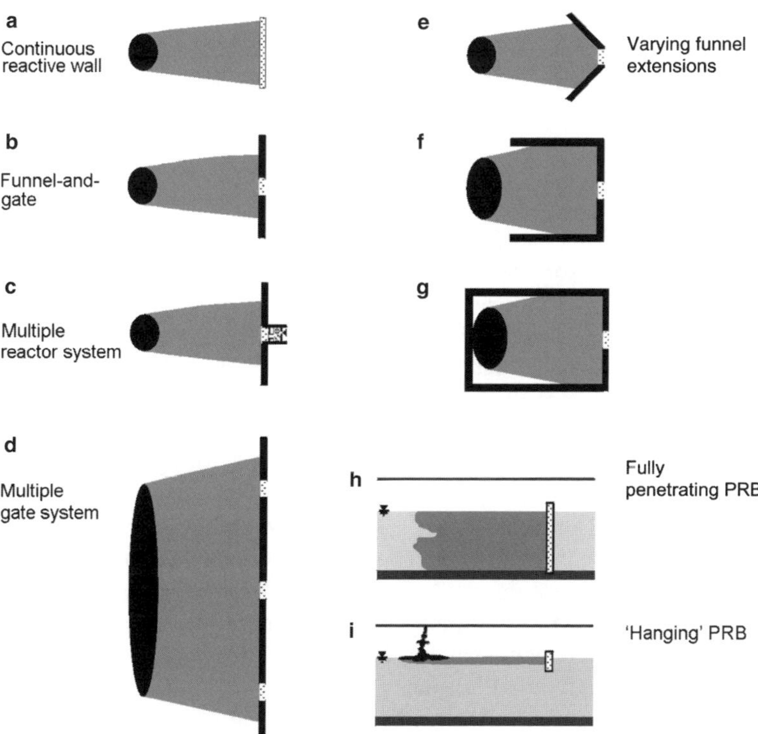

Fig. 8. Various configurations of permeable reactive barriers (after Starr & Cherry, 1994).

grain size. With certain clay minerals, this effect is intensified by uptake involving interlayer exchangeable cations resulting in swelling. However, in PRBs, clays exhibiting swelling characteristics must be excluded as they result in a decrease in the effective porosity and permeability. Effective removal of contaminants from water can only be expected given the necessary affinities between the contaminant and the adsorptive medium. Surface tension of the fluid is also relevant, as it affects the wetting capability of the solid.

The adsorbing solid phase (adsorbent) is commonly deployed in a reactor system (filter) through which the effluent flows. Within the so-called 'adsorption zone' or 'mass-transfer zone' of the filter, the concentration of the component to be removed ('adsorbate') should decrease from the initial value to zero. Over time, the mass-transfer zone moves through the filter until, finally, all of the adsorbent is saturated with the adsorbate which breaks through in its initial concentration. Commonly, the adsorption process is reversible on changing the chemical conditions. Thus, the adsorbent may be recovered by using certain regeneration methods. The following processes can be distinguished (Sonntheimer, 1975): (1) physical adsorption mechanisms due to electrostatic and van der Waals forces interacting between the molecules of adsorbent and adsorbate; with these adsorption processes, the molecules of the adsorbate are not changed; and (2) chemical adsorption mechanisms due to chemical surface reactions between adsorbent and adsorbate, where electron exchange, electron coupling or complexation of chemical species take place.

Chemical-adsorption mechanisms are more specific and less reversible; this leads to mono-molecular occupancy of the adsorbing surface sites by adsorbate molecules. With physical adsorption processes, large degrees of molecular occupancy can occur. Very often, however, physical and chemical adsorption processes cannot be separated clearly. For the so-called active centres, *i.e.* where adsorption processes occur, a competition between the solvent and dissolved material (the adsorbate) takes place. Preferential adsorption of the adsorbate require greater affinity for the adsorbent than for the solvent. Usually, the less soluble a material is in water, the better it can be adsorbed. Polar molecules are more soluble than non-polar ones, so it is more likely for the non-polar materials to become attached to the solid surface. Substances with an intermediate molecular weight are also adsorbed preferentially. Small molecules have too few adsorption sites, whereas big molecules may not fit the intrinsic adsorption pores (intra-crystalline micropores of <2 nm diameter) which form the main part of the inner surface of a clay mineral.

For the design of an adsorption system, *e.g.* a PRB, the following parameters have to be determined (*e.g.* by column testing on laboratory scale): the maximum capacity of the applied adsorbent; the adsorption equilibrium, otherwise the maximum capacity may not be completely exploited during the operation process; the adsorption kinetics, as equilibrium is not achieved immediately; the technical feasibility, *e.g.* with regard to dimensions, exchangeability and life span.

Data from different hazardous sites can be collected simultaneously. As soon as suitable materials are identified, the results gained at the laboratory scale may be verified by implementing a field test, or even a full-scale remediation project.

Table 7. Classification of selected adsorptive materials with regard to contaminant categories (from Köhler *et al.*, 2000, reproduced with the permission of IFSTTAR, Nantes, France).

Material	Sorption characteristics or geochemical function	Contaminant class	Examples of sorbed pollutants
Zeolite	Micropores Surface charge	Cations	Heavy metals Mineral oils
Clays (Bentonite)	Surface charge Ion exchange Micropores	Cations Large organic molecules	Heavy metals Aromatic hydrocarbons Phenols Aniline Pesticides
Silicon oxides (Silicic acid, diatomite)	Micro- and macropores	Large and small organic and inorganic molecules	Mineral oils Aromatic hydrocarbons Halogenides
Alumina clay (Aluminium oxide)	Macropores Complexation	Large and small organic and inorganic molecules	Sulfates Phenols
Hydrotalcite	Surface charge	Anions	Chromate Sulfates Phenols
Activated carbon	Surface charge Ion exchange Macropores	Large and small organic molecules Cations	Heavy metals Aromatic hydrocarbons Phenols Aniline Pesticides
Ferric oxides (siderite)	Adsorption Reduction Oxidation	Halogenated hydrocarbons Cations	Halogenides Phenols Aniline Heavy metals
Calcite, Limestone	pH - Buffer	Acids (neutralization)	Acids, particularly sulfuric acid

4.3.7. Criteria for applicability of adsorptive materials

For utilization in PRBs, relatively coarse-grained materials are required. These guarantee high hydraulic permeability. To be able to use the adsorptive properties of clay minerals, for example, they must have certain soil mechanical and mineralogical characteristics that are unusual, or even undesirable, for many conventional applications. They need to be without any fine-grained material or be available in granulated form (as so-called pellets) and exhibit the following characteristics: hydraulic and mineralogical long-term stability; little or no deformation; erosion stability of pellets; and uniformly graded grain sizes.

A classification of materials that fulfil these characteristics has been proposed (Table 7). Based on the adsorption processes involved, particular pollutant types are assigned to particular adsorptive media. In this table, ferric oxides and calcite (or limestone) are recorded separately, as their actions cannot be classified as adsorption processes. The development of PRBs and research on appropriate filling materials, as well as on-site applications, are being promoted in many countries, especially in the USA (see EPA, 1998, 1999). It would be desirable for this comparatively economical, long-term remediation method to achieve global acceptance.

5. Conclusions

Landfills for municipal and industrial wastes, and the associated potential ground-pollution problems are a challenge for current soil-mechanics concepts and methods of analysing soil behaviour. For this reason, environmental aspects of geotechnology have been further developed to include the study of soil and rock interaction with the atmosphere, biosphere, hydrosphere and lithosphere. This chapter gives a comprehensive overview of the containment practices and technical constructions for controlled landfills. It addresses the special properties of minerals, particularly clay minerals, in transport and containment of contaminants in the geological and different types of technical barriers. Some aspects of suitability and application of minerals in *in situ* cleanup approaches are explained.

Acknowledgements

We thank Günter Kahr for his help in the area of contamination transport and Sven Köhler for his efforts concerning Permeable Reactive Walls (both ETH Zurich, Switzerland). We also thank Lars Mülli for his help concerning tables and figures.

References

Allard, Th. & Calas, G. (2009) Radiation effects on clay mineral properties. *Applied Clay Science*, **43**, 143–149.

Alonso, U., Missana, T., Garcia-Gutierrez, M., Patelli, A., Siitari-Kauppi, M. and Rigato, V. (2009) Diffusion coefficient measurements in consolidated clay by RBS micro-scale profiling. *Applied Clay Science*, **43**, 477–484.

Altaner, S.P. & Ylagan, R.F. (1997) Comparison of structural models of mixed-layer illite/smectite and reaction mechanisms of smectite illitization. *Clays and Clay Minerals*, **45**, 517–533.

Barbour, S.L., Hendry, M.J. and Wassenaar, L.I. (2012) In situ experiment to determine advective-diffusive controls on solute transport in a clay-rich aquitard. *Journal of Contaminant Hydrologyy*, **131**, 79–88.

Bailey, S.W., editor (1988) *Hydrous Phyllosilicates (exclusive of Micas)*. Reviews in Mineralogy, **19**, Mineralogical Society of America, Washington, D.C.

Bergaya, F., Lagaly, G. & Vayer, M. (2006) Cation and anion exchange. In: *Handbook of Clay Science* (F. Bergaya, B.K.G. Theng & G. Lagaly, editors). Developments in Clay Science, **1**, Elsevier, Oxford, pp. 979–1001.

Blowes, D.W., Ptacek, C.J., Brenner, S.G., Waybrant, K.R. & Bain, J.G. (1996) Treatment of mine drainage water using in-situ Permeable Reactive Walls. In: *Passive Systeme zur in-situ Sanierung von Boden und Grundwasser; Workshop – extended abstracts*. Dresden, Germany.

Boochs, P.W., Barovic, G. & Mull, R. (1977) Einfluss der Sorption auf Transportvorgänge im Grundwasser. *DGM 21*, **3**, 60–65.

Bosman, R. (1993) Sampling strategies and the role of geostatistics in the investigation of soil contamination. In: *Contaminated Soil '93* (F. Arendt, G.J. Annokkée, R. Bosman & W.J. van den Brink, editors). Kluwer Academic Publishers, Dordrecht, The Netherlands, pp. 587–597.

Brandl, H. (1998) Risk analyses, quality assurance, and regulations in landfill engineering and environmental protection. In: *Environmental Geotechnics*, **3** (P.S. Sêco e Pinto, editor). Balkema, Rotterdam, pp. 1299–1328.

Brindley, G.W. & Brown, G., editors (1980) *Crystal Structures of Clay Minerals and their Identification.* Monograph, **5**, Mineralogical Society, London.

Brunauer, S., Emmett P.H. & Teller, E. (1938) Adsorption of gases in multimolecular layers. *Journal of the American Chemical Society*, **60**, 309–319.

Bunge, R. (2007) Trockenaustrag von KVA-Schlacken. Hochschule für Technik, Rapperswil HSR, Switzerland.

Burmeier, H. (2006) Anwendung von durchströmten Reinigungswänden zur Sanierung von Altlasten, Teil I und Teil II, Universität Lüneburg, 2006 (also available in English, www.rubin-online.de).

Campbell, L.S. (2000) Minerals and waste management. In: *Environmental Mineralogy: Microbial Interactions, Anthropogenic Influences, Contaminated Land and Waste Management* (J.D. Cotter-Howells, L.S. Campbell, E. Valsami-Jones & M. Batchelder, editors). Mineralogical Society Series, **9**. Mineralogical Society, London, pp. 313–318.

Cave, L., Al, T., Xiang, Y. and Vilks, P. (2009) A technique for estimating one-dimensional diffusion coefficients in low-permeability sedimentary rock using X-ray radiography: Comparison with through-diffusion measurements. *Journal of Contaminant Hydrologyy*, **103**, 1–12.

Cheung, S.C.H., Gray, M.N. & Dixon D.A. (1987) Hydraulic and ionic diffusion properties of bentonite-sand buffer materials. In: *Coupled Processes associated with Nuclear Waste Repositories* (C.F. Tsang, editor). Harcourt Brace Jovanovich, Orlando, Florida, USA, pp. 393–407.

Christidis, G.E. (2011) Industrial clays. In: *Advances in the Characterization of Industrial Minerals* (G.E. Christidis, editor). EMU Notes in Mineralogy, **9**, European Mineralogical Union and the Mineralogical Society of Great Britain and Ireland, London.

Churchman, G.J., Gates, W.P., Theng, B.K.G. & Yuan, G. (2006) Clays and clay minerals for pollution control. In: *Handbook of Clay Science* (F. Bergaya, B.K.G. Theng & G. Lagaly, editors). Developments in Clay Science, **1**. Elsevier, Oxford, UK, pp. 625–675.

Collins, K. & McGown, A. (1974) The form and function of microfabric features in a variety of natural soils. *Géotechnique*, **24**, 223–254.

Czurda, K. (2006) Clay liners and waste disposal. In: *Handbook of Clay Science* (F. Bergaya, B.K.G. Theng & G. Lagaly, editors). Developments in Clay Science, **1**. Elsevier, Oxford, UK, pp. 693–701.

Czurda, K.A. & Wagner, J.F. (1991) Cation-transport and retardation processes in view of the toxic-waste deposition problem in clay rocks and clay liner encapsulation. *Engineering Geology*, **30**, 103–113.

Dahmke, A. (1997) Aktualisierung der Literaturstudie "Reaktive Wände" – pH–Redox–reaktive Wände. In: *Technischer Bericht*. Institut für Wasserbau, Universität Stuttgart, Germany.

Dahmke, A., Ebert, M., Schlicker, O., Körber, R. & Wüst, W. (1999) Fe(0)–Wände zum Abbau organischer Verbindungen. In: *Einsatz passiver reaktiver Wände – Genehmigungsfragen und Sanierungskosten.* ITVA–Seminar ARW/1/99, Magdeburg, Germany.

Daniel, D.E. & Körner, R.M. (1993) Cover systems. In: *Geotechnical Practice for Waste Disposal*. Chapman & Hall, London, pp. 455–495.

De Soto, I.S., Ruiz, A.I., Ayora, C., Garcia, R., Regadio, M. & Cuevas, J. (2012) Diffusion of landfill leachate through compacted natural clay containing small amounts of carbonates and sulphates. *Applied Geochemistry*, **27**, 1202–1213.

Drury, D. (1997) Hydraulic Consideration for choice of landfill liner. In: *Geoenvironmental Engineering, Contaminated ground: Fate of Pollutants and Remediation* (R.N. Yong & H.R. Thomas, editors). Thomas Telford, London, pp. 312–318.

Echle, W., Gorantonaki, A., Dullmann, H. & Obernosterer, I. (1997) Redox-dependent mineralogical and chemical changes in clayey landfill basal liners and their soil mechanical effects. In: *Advanced Landfill Liner Systems* ((H. August, U. Holzlöhner & T. Meggyes, editors). Thomas Telford, London, pp. 155–162.

Egloffstein, Th., Burkhardt, G. & Mainka, A. (1996) Geotechnische Aspekte bei der Standortsuche und Standorterkundung für Abfalldeponien. In: *Das Multibarrierensystem in der Deponiebautechnik, Schriftenreihe Angewandte Geologie Karlsruhe,* **44** (K. Czurda & H. Hötzl, editors). Schriftenreihe Angewandte Geologie Karlsruhe, Germany, pp. 4.1–4.63.

European Environment Agency (2009) *Diverting Waste from Landfill – Effectiveness of Waste-management Policies in the European Union.* EEA Report No 7/2009, Denmark.

EPA United States Environmental Protection Agency (1998) *Permeable Reactive Barrier Technologies for Contaminant Remediation.* In: EPA/600/R-98/125, US EPA, Washington, D.C.

EPA United States Environmental Protection Agency (1999) *Field Applications of in situ Remediation Technologies: Permeable Reactive Barriers.* In: EPA 542-R-99-002, US EPA Washington, D.C.

Ferrage, E., Lanson, B., Sakharov, B.A., Geoffroy, N., Jacquot, E. & Drits, V.A. (2007) Investigation of dioctahedral smectite hydration properties by modeling of X-ray diffraction profiles: Influence of layer charge and charge location. *American Mineralogist,* **92,** 1731–1743.

Gavaskar, A.R., Gupta, N., Sass, B.M., Janosy, R.J. & O'Sullivan, D. (1998) *Permeable Barriers for Groundwater Remediation. Design, Construction and Monitoring.* Batelle Press Columbus, Ohio, USA.

Genske, T. (2005) *Alternative Abdichtungssysteme von Deponien unter Verwendung von Abfällen – Recherche zum Stand der Technik im Deponiebau.* Mitteilung der Ländergemeinschaft Abfall (LAGA) Nr. 20.

German Geotechnical Society of the International Society of Soil Mechanics and Foundation Engineering (1997) and www.gda-online.de: *GLR-Recommendations "Geotechnics of landfill design and remedial Works, Technical Recommendations".* Ernst & Sohn Verlag, Berlin.

Gillham, R.W. & O'Hannesin, S.F. (1994) Enhanced degradation of halogenated aliphatics by zero-valent iron. *Groundwater,* **32,** 958–967.

Gillham, R.W., Blowes, D.W., Ptacek, C.J. & O'Hannesin, S.F. (1994) Use of zero-valent metals in in situ remediation of contaminated ground water. In: *33rd Hanford Symposium on Health and the Environment. In Situ Remediation: Scientific Basis for Current and Future Technologies,* **2,** (G.W. Gee & N.R. Wing, editors). Pasco, Washington, USA, pp. 913–927.

Grathwohl, P. & Peschik, G. (1997) Permeable sorptive walls for treatment of hydrophobic organic contaminant plumes in groundwater. In: *International Conference on Containment Technology,* St. Petersburg, Florida, USA, pp. 711–717.

Grauer, R. (1986) Bentonite as a backfill material in the high-level waste repository: chemical aspects. In: *Technischer Bericht/NAGRA,* NTB 86-12E.

Grauer, P. (1988) *Zum chemischen Verhalten von Montmorillonit in einer Endlagerverfüllung.* NAGRA NTB 88-24, Paul Scherrer Institute, Würenlingen, Switzerland.

Grim, R.E. (1953) *Clay Mineralogy.* McGraw-Hill, New York.

Grim, R.E. (1962) *Applied Clay Mineralogy.* McGraw-Hill, New York.

Guggenheim, S., Adams, J.M., Bain, D.C., Bergaya, F., Brigatti, M.F., Drits, V.A., Formoso, M.L.L., Galán, E., Kogure, T. & Stanjek, H. (2006) Summary of recommendations of nomenclature committees relevant to clay mineralogy: report of the Association Internationale pour l'Etude des Argiles (AIPEA) Nomenclature Committee for 2006. *Clays and Clay Minerals,* **54,** 761–772.

Güven, N. (1990) Longevity of bentonite as buffer material in a nuclear-waste repository. *Engineering Geology,* **28,** 233–247.

Guyonnet, D., Touze-Foltz, N. & Norotte, V. (2009) Performance-based indicators for controlled geosynthetic clay liners in landfill applications. *Geotextiles and Geomembranes,* **27,** 321–331.

Hasenpatt, R., Degen, W. & Kahr, G. (1988) Flow and diffusion in clays. *Applied Clay Science,* **4,** 179–192.

Henken-Mellies, W.-U. (2011) Long-term performance of landfill covers – results of lysimeter test fields in Bavaria (Germany). *Waste Management & Research,* **29,** 59–68.

Hermanns, R. (1993) *Sicherung von Altlasten mit vertikalen mineralischen Barrierensystemen im Zweiphasen–Schlitzwandverfahren.* Veröffentlichungen des Instituts für Geotechnik, **204**, ETH Zürich, Schweiz: vdf–Verlag.

Hermanns-Stengle, R. (1997) Waste deposit encapsulation using vertical barriers. In: Publications Committee of XIV ICSMFE. *Proceedings of the 14th International Conference on Soil Mechanics and Foundation Engineering.* Rotterdam/Brookfield, **3**. Balkema, Rotterdam, The Netherlands, pp. 1911–1914.

Higgo, J.J.W. (1986) Clay as a barrier to radionuclide migration. *Progress in Nuclear Energy*, **19**, 173–207.

Jasmund, K. & Lagaly, G. (1993) *Tone und Tonminerale.* Steinkopff Verlag, Darmstadt, Germany.

Jessberger, H.L. (1991) Geotechnics of Landfills and Contaminated Land. Technical Recommendations "GLC". *Bautechnik 68*, **9**, 294–315.

Jessberger, H.L. (1994) Remedial works in a highly industrialized region. In: *Proceedings of the First International Congress of Environmental Geotechnics*, Edmonton, Alberta, Canada.

Jessberger, H.L. (1997) TC5 activities – Technical Committee on Environmental Geotechnics, ISSMFE. In: *Environmental Geotechnics,* **3** (M. Kamon, editor). Balkema, Rotterdam, The Netherlands, pp. 1227–1254.

Johnson, T.L. & Tratnyek, P.G. (1994) A column study of geochemical factors affecting reductive dechlorination of chlorinated solvents by zero-valent iron. In: *33rd Hanford Symposium on Health and the Environment. In situ Remediation: Scientific Basis for Current and Future Technologies*, **2**. Pasco, Washington, USA, pp. 931–947.

Kamon, M. (1997) Geotechnical utilization of industrial wastes. In: *Environmental Geotechnics,* **3** (M. Kamon, editor). Balkema, Rotterdam, The Netherlands, pp. 1293–1310.

Kaufhold, S. & Dohrmann, R. (2009) Stability of bentonites in salt solutions vertical bar sodium chloride. *Applied Clay Science*, **45**, 171–177.

Kaufhold, S. & Dohrmann, R. (2010) Stability of bentonites in salt solutions II. Potassium chloride solution – Initial step of illitization? *Applied Clay Science*, **49**, 98–107.

Kaufhold, S. & Dohrmann, R. (2011) Stability of bentonites in salt solutions III – Calcium hydroxide. *Applied Clay Science*, **51**, 300–307.

Kaufhold, S., Dohrmann, R. & Klinkenberg, M. (2010) Water uptake capacity of bentonites. *Clays and Clay Minerals*, **58**, 37–43.

Knochenmus, G., Wojnarowicz, M. & Van Impe, W.F. (1998) Stability of municipal solid waste. In: *Environmental Geotechnics*, **3**, (P.S. Sêco e Pinto, editor). Balkema, Rotterdam, The Netherlands, pp. 977–1000.

Knödel, K., Lange, G. & Voigt, H.-J. (2007) *Environmental Geology. Handbook of field methods and case studies.* BGR, Springer Verlag, Berlin, Heidelberg.

Köhler, S., Hermanns Stengele, R. & Laue, J. (2000) Permeable reactive barrier systems for groundwater cleanup. Physical Modelling and Testing in Environmental Geotechnics. *Network of European Centrifuges for Environmental Geotechnic Research*, France, pp. 173–181.

König, D. & Jessberger, H.L. (1997) Waste Mechanics. In: *Environmental Geotechnics* (H.L. Jessberger, editor). Technical Committee TC 5 "Environmental Geotechnics", ISSMFE International Society for Soil Mechanics and Foundation Engineering. Ruhr Universität, Bochum, Germany, pp. 35–76

Kraepiel, A.M.L., Keller, K. & Morel, F.M.M. (1999) A model for metal adsorption on montmorillonite. *Journal of Colloid and Interface Science*, **210**, 43–54.

Krajewski, W., Held, T., Jungbauer, H. & Di Muzio, M. (2010) *Sanierungstechniken und verfahren. Handbuch Altlasten, Band 6 Teil 3.* Hessisches Landesamt für Umwelt und Geologie (editor), Wiesbaden, Germany.

Kruse, K. (1992) *Die Adsorption von Schwermetallen an verschiedenen Tonen.* Thesis, Swiss Federal Institute of Technology, Zurich, Switzerland.

Lege, T., Kolditz, O., Zielke, W. & Kasper, H. (1996) *Handbuch zur Erkundung des Untergrundes von Deponien und Altlasten; Band 2.* Strömungs- und Transportmodellierung, Springer, Berlin.

MacKenzie, R.C. (1951) A micromethod for determination of cation-exchange capacity of clay. *Journal of Colloid Science*, **6**, 219–222.

Madsen, F.T. (1998) Clay mineralogical investigations related to nuclear waste disposal. *Clay Minerals*, **33**, 109–129.

Madsen F.T. & Kahr, G. (1993) Diffusion of ions in compacted bentonite. In: *Proceedings of the International Conference on Nuclear Waste Management and Environmental Remediation*, Prague, pp. 239–246.

Madsen, F.T. & Mitchell, J.K. (1989) Chemical effects on clay hydraulic conductivity and their determination. In:*Mitteilungen des Instituts für Grundbau und Bodenmechanik*, **135**. ETH Zürich, Switzerland pp. 115–121.

Madsen, F.T., Kahr, G. & Hermanns, R. (1996) Mineralogical aspects of diaphragm wall materials for the encapsulation of hazardous waste sites. In: *Clay Minerals Society Meeting*. Gatlinburg, Tennessee, USA.

Mak, M. & Lo, I. (2011) Environmental life cycle assessment of permeable reactive barriers: effects of construction methods, reactive materials and groundwater constituents. In: *Environmental Science & Technology*, **45**, 10148–54.

Manassero, M. (1996) Controlled Landfill Design. In: *Environmental Geotechnics* (H.J. Jessberger, editor). Technical Committee TC 5 "Environmental Geotechnics", ISSMFE International Society for Soil Mechanics and Foundation Engineering. Ruhr Universität, Bochum, Germany, pp. 77–112.

Manassero, M., Van Impe, W.F. & Bouazza, A. (1997) Waste disposal and containment. In: *Environmental Geotechnics*, **3** (M. Kamon, editor). A.A. Balkema, Rotterdam, The Netherlands, pp. 1425–1474.

Meier, L.P. (1998) *Organic cations-related adsorption behaviour of surface modified smectites*. Thesis, Swiss Federal Institute of Technology, Zurich, Switzerland.

Meier, L. & Kahr, G. (1999) Determination of the cation exchange capacity (CEC) of clay minerals using the complexes of copper(II) ion with triethylenetetramine and tetraethylenepentamine. *Clays and Clay Minerals*, **473**, 386–388.

McBride, M.B. (2000) Chemisorption and precipitation reactions. In: *Handbook of Soil Science* (M.E. Sumner, editor). CRC Press, Boca Raton, Florida, USA, B-265–B-302.

Meunier, A., Velde, B. & Griffault, L. (1998) The reactivity of bentonites: a review. An application to clay barrier stability for nuclear waste storage. *Clay Minerals*, **33**, 187–196.

Mitchell, J.K. (1993) *Fundamentals of Soil Behaviour,* 2nd edition. Wiley, New York.

Moore, D.M. & Reynolds R.C. Jr. (1997) *X-ray Diffraction and the Identification and Analysis of Clay Minerals*. Oxford University Press, Oxford, UK, 378 pp.

Mortland, M.M. (1970) Clay-organic complexes and interactions. *Advances in Agronomy*, **22**, 75–117.

Murray, E.J., Jones, R.H. & Rix, D.W. (1997) Relative importance of factors influencing the permeability of clay soils. In: *Geoenvironmental Engineering, Contaminated Ground: Fate of Pollutants and Remediation* (R.N. Yong & H.R. Thomas, editors). Thomas Telford, London, pp. 229–239.

National Academy of Sciences (2011) Assessment of the performance of engineered waste containment barriers (in www.nap.edu).

NATO/CCMS Pilot Study (1998) Evaluation of Demonstrated and Emerging Technologies for the Treatment of Contaminated Land and Groundwater (Phase III). Special Session: *Treatment Walls and Permeable Reactive Barriers*. EPA 542-R-98-00, 108 pp.

Nüesch, R., Amann, P. & Mendoza, A. (1996) Properties and immobilisation from slag municipal waste. In: *Environmental Geotechnics*, **2** (M. Kamon, editor). A.A. Balkema, Rotterdam, The Netherlands, pp. 833–838.

Plötze, M. & Kahr, G. (2003) Swelling pressure and suction of clays. In: *International Workshop on Geotechnics of Soft Soils – Theory and Practice* (P.A. Vermeer, H.F. Schweiger, M. Karstunen & M. Cudny, editors). VGE, Noordwijkerhout, The Netherlands, pp. 479–483.

Plötze, M., Kahr, G. & Hermanns Stengele, R. (2003) Alteration of clay minerals – gamma-irradiation effects on physicochemical properties. *Applied Clay Science*, **23**, 195–202.

Pusch, R. (1994) *Waste Disposal in Rock, Developments in Geotechnical Engineering, 76*. Elsevier, Amsterdam.

Pusch, R. (2006) Clays and nuclear waste management. In: *Handbook of Clay Science* (F. Bergaya, B.K.G. Theng & G. Lagaly, editors). Developments in Clay Science, **1**. Elsevier, Oxford, pp. 703–716.

Pusch, R. & Güven, N. (1990) Electron microscopic examination of hydrothermally treated bentonite clays. *Engineering Geology*, **28**, 303–314.

Rowe, R.K., Quigley, R.M., Brachman, R.W.I. & Booker, J.R. (2004) *Barrier systems for waste disposal*, 2nd edition. Spon Press, London, 591 pp.

Rumer, R.R. & Ryan, M.E. (1995) *Barrier Containment Technologies for Environmental Remediation Application*. John Wiley & Sons, Inc., New York.

Sato, T., Watanabe, T. & Otsuka, R. (1992) Effects of layer charge, charge location and energy change on expansion properties of dioctahedral smectites. *Clays and Clay Minerals,* **40**, 103–113.

Shackelford, C.D. & Chiu, T. (1997) The influence of the capillary barrier effect on the use of sand undrains in waste disposal practice: In: *Geoenvironmental Engineering, Contaminated Ground: Fate of Pollutants and Remediation* (R.N. Yong & H.R. Thomas, editors). Thomas Telford, London, pp. 357–364.

Shafieiyoun, S., Ebaddi, T. & Nikazar, M. (2012) Treatment of landfill leachate by Fenton process with nano sized zero-valent Iton particles. *International Journal of Environmental Research,* **6**, 119–128.

Solomon, D.H. & Hawthorne, D.G. (1983) *Chemistry of Pigments and Fillers.* Wiley, New York.

Sonntheimer, H. (1975) Verfahrenstechnische Grundlagen von Adsorption und Ionenaustausch. Skriptum zur Vorlesung "*Physikalisch-chemische Trennverfahren*", Engler-Bunte-Institut der Technischen Universität Karlsruhe, Germany.

Starr, R.C. & Cherry, J.A. (1994) In situ remediation of contaminated ground water: The Funnel-and-Gate System. *Ground Water,* **2**, 465–475.

Stockmeyer, M.R. (1992) *Organophile bentonite als komponente in deponiebarriere-systemen.* Thesis, Swiss Federal Institute of Technology, Zurich, Switzerland.

Stockmeyer, M.R., Madsen, F.T. & Kahr, G. (1995) Contaminant transport in organophillic waste deposit liners. *Hazardous Waste & Hazardous Materials,* **12**, 149–166.

Strobl, T. (1987) Einsatz von Dichtwänden an der Brombachtalsperre. In: *Mittelungen des Instituts für Grundbau und Bodenmechanik* (H. Meseck, editor), **23**. Technische Universität Braunschweig, Germany, pp. 372–377.

Stucki, J.W. & Bish, D.L. (editors) (1990) *Thermal Analysis in Clay Science.* CMS workshop lectures, **3**, The Clay Minerals Society, Bloomington, Indiana, USA, 192 pp.

Swiss Federal Office for the Environment (1986) Swiss Waste Management Standards.

Teutsch, G. & Grathwohl, P., Schad, H. & Werner, P. (1996) In situ reaktionswände – ein neuer Ansatz zur passiven Sanierung von Boden– und Grundwasserverunreinigungen. *Grundwasser – Zeitschrift der Fachsektion Hydrogeologie,* **1**, 12–20.

The Council of the European Union (1999) Council Directive 1999/31/EC of 26 April 1999 on the landfill of Waste.

Theng, B.K.G. (1974) *The Chemistry of Clay-Organic Reactions.* Adam Hilger, London, 343 pp.

Török, À. (1996) Importance of geological conditions in waste disposal. In: *Das Multibarrierensystem in der Deponiebautechnik,* **44** (K. Czurda & H. Hötzl, editors). Schriftenreihe Angewandte Geologie Karlsruhe, Germany, pp. 5.1–5.15.

Van Olphen, H. (1977) *An Introduction to Clay Colloid Chemistry,* 2nd edition. Wiley, New York.

Van Olphen, H. (editor) (1979) *Data Handbook for Clay Materials and other Non-metallic Minerals.* Pergamon Press, Oxford, UK.

Venema, P., Hiemstra, T. & van Riemsdijk, W.H. (1996) Comparison of different site binding models for cation sorption: Description of pH dependency, salt dependency, and cation-proton exchange. *Journal of Colloid and Interface Science,* **181**, 219–228.

Wagner, J.–F. (1992) *Verlagerung und Festlegung von Schwermetallen in tonigen Deponieabdichtungen. Ein Vergleich von Labor– und Geländestudien,* **22**. Schriftenreihe Angewandte Geologie Karlsruhe, Germany.

Walther, J. (1996) Relation between rates of aluminosilicate mineral dissolution, pH, temperature, and surface charge. *American Journal of Science,* **296**, 693–728.

Westsik, J.H., Hodges, F.N., Kuhn, W.L. & Myers, T.R. (1983) Water migration through compacted bentonite backfills for containment of high-level nuclear waste. *Nuclear and Chemical Waste Management,* **4**, 291–299.

Yong, R.N. & Warkentin, B.P. (1975) *Soil Properties and Behaviour,* 2nd edition. Elsevier, Amsterdam.

Yong, R.N., Mohamed, A.M.O. & Warkentin, B.P. (1992) *Principles of Contaminant Transport in Soils.* Elsevier, New York, 327 pp.

Mineralogy in long-term nuclear waste management

Charles CURTIS[*] and Katherine MORRIS

*Research Centre for Radwaste and Decommissioning, and
Williamson Research Centre for Molecular Environmental Science,
School of Earth, Atmospheric and Environmental Sciences,
The University of Manchester, Oxford Road, Manchester M13 9PL, UK
e-mail address: ccurtis355@aol.com

Globally there is a legacy of radioactive waste associated with more than 50 years of nuclear power generation and weapons production. Typically, this waste is stored at Earth's surface, and nations are beginning to face up to managing these highly radioactive and hazardous materials for the long term. Nuclear-power generation is increasingly considered as a low carbon, politically secure power source and geological disposal is seen as the favoured route for long-term management of the waste materials produced. Thus, timely implementation of geological disposal is a challenge facing nuclear-power generating societies if we are to demonstrate safe management of these waste materials for the future. In this chapter we review the role that environmental mineralogy has in this critical task. We begin with an introduction to the topic, discuss the waste types and strategies for management, and then discuss the environmental mineralogy of spent fuel and high-level vitrified wastes. We then place the discussion in a geological context with coverage of natural analogue sites, of mineral weathering and stability, and of geochemical conditions that will influence mineral stability. Finally, we discuss mineralogy relevant to the engineered disposal facility as well as the natural environment surrounding the facility in terms of retardation of radionuclide mobility and examine the role of geomicrobiology in nuclear-waste management.

1. Introduction

Most radioactive waste derives from nuclear power plants as spent fuel. Apart from a few early reactor types, which utilized uranium metal, most reactors use fuel assemblies based on sintered, microcrystalline UO_2. A small number of reactors are licensed to use a mixed oxide fuel (MOx) based on UO_2 and PuO_2. In one sense the environmental mineralogy of nuclear waste starts here: what are the properties of these oxides and how will they impact on waste management? Globally, some spent fuel is reprocessed to recover U and Pu for further UO_2 fuel and MOx fuel fabrication. The highly radioactive, fission product-bearing acid solutions generated are concentrated, calcined and then incorporated either into glasses as high level waste (HLW) or, perhaps for the future, into various synthetic mineral assemblages (*e.g.* 'SYNROC' and similar materials). For the spent nuclear fuel that is not destined for reprocessing, packaging in high-integrity metal containers is likely. In addition, operation of the nuclear fuel cycle leads to generation of other intermediate level wastes (ILW) which are most likely grouted and stored in steel

© Copyright 2013 the European Mineralogical Union and the Mineralogical Society of Great Britain & Ireland
DOI: 10.1180/EMU-notes.13.9

drums. ILW and HLW together are categorized as higher-activity wastes and in the UK are destined for geological disposal. The objective of conditioning these wastes (vitrification, packaging or grouting) is to create a much more stable waste form, less likely to release radionuclides into the environment. For HLW, expanding clay minerals are being investigated for use in engineered containment systems, to further delay radionuclide release into the accessible environment by limiting groundwater access. Finally, deep geological disposal concepts take account of geochemical reactions between rock minerals and migrating solute radionuclides to retard their escape in groundwater.

Chapters elsewhere in this volume deal with minerals and soil development and minerals in the disposal and containment of industrial and domestic wastes. Here, as there, the processes of mineral-water-gas interactions at low temperature and pressure are central to understanding and crucially important in the design and implementation of management strategies. The public generally believes that radioactive wastes are much more harmful than other kinds: they pose a unique threat to life and the environment and they persist indefinitely. None of these perceptions is really correct but it is hardly surprising that dread of nuclear fission persists: the consequences of the detonation of nuclear devices over Hiroshima and Nagasaki were both dreadful and widely publicized. Nuclear accidents such as those at Three Mile Island, Chernobyl and most recently, during 2011, at Fukushima-Daiitchi, resonate in the media. The heavy metals present in both domestic and industrial wastes, of course, persist indefinitely; only a small proportion of the products of the nuclear power cycle has half-lives to be counted in millions of years. The toxicity of radiation is very real, but so it is equally for many natural and man-made organic compounds. Furthermore, the toxic effects of radiation are very precisely known (contrast many synthetic organic compounds) and monitoring technologies for radionuclides are extremely sensitive, comprehensive and reliable.

So what are the special problems? First, the radiation levels and thus radiotoxicity associated with spent nuclear fuel and HLW are extremely high and remain so for centuries after removal from the reactor. Substantial radiotoxicity and heat generation demand remote handling and cooling in these early decades (non-passive management). Second, the fission process is not symmetric: ^{235}U and ^{239}Pu, the major fissile isotopes in the nuclear reactor, fission to produce a range of fission products with peaks at mass number 97 and 137; this leads to a range of radioactive fission products in the nuclear fuel. For example, typical fission pathways are:

$$^{72}Zn + {}^{162}Sm + 2\,^{1}n$$
$$\nearrow$$
$$^{1}n + {}^{235}U \qquad \rightarrow {}^{80}Sr + {}^{153}Xe + 3\,^{1}n$$
$$\searrow$$
$$^{94}Kr + {}^{139}Ba + 3\,^{1}n$$

Although their presence, identity and radiation characteristics can all be determined readily, the biogeochemistry of radionuclide behaviour in the environment is likely to be very much more complicated than basic chemistry. Once released to the biosphere,

Table 1. Radiation and dose – definitions and background.

Activity	Activity unit is the Becquerel (Bq) (1 Bq = 1 disintegration per second). Takes no account of biological harm of the radiation type.
Dose	Energy absorbed by matter when radioactivity interacts with it (Units = Grey; 1 Grey = 1 Joule of energy per kilo).
Quality factor	Converts dose in Grey to dose in Sieverts (the SI unit of effective dose equivalent). Relates to the biological harm caused by the different types of ionizing radiation. The quality factor is ×20 for alpha radiation and ×1 for beta and gamma radiation.
Effective dose	Dose (Grey) × quality factor – Sieverts, typically mSv is used as the unit of dose for a member of the general public.
Collective dose	Total radiation dose received by a population during a defined period from a defined source.
Dose constraint	Restriction placed on dose received from any single radiation source.
Dose limit	Radiation dose in mSv to the public limit (per year).

UK annual average dose ≈ 2.50 mSv per year, ~88% background, ~12% anthropogenic.

Air	1.30 mSv/y	primarily radon
Ground	0.35 mSv/y	primarily ^{40}K, Th, U
Food/drink	0.30 mSv/y	primarily ^{40}K
Medical	0.30 mSv/y	X-rays, medical isotopes
Cosmic rays	0.25 mSv/y	air travel
Licensed nuclear facilities	0.005 mSv/y	

via accidental or controlled releases, migration of the radionuclides will be complex and heterogeneous.

The risk becomes real when concentrations are such that significant radiation doses could be received by humans or biota. Dose depends on (and takes account of) the type of radiation involved and the susceptibility of the organ irradiated. This dose is then translated into risk – risk of death from cancer or contracting a transmittable genetic malfunction. The Earth is a radioactive planet such that background levels are certainly finite (Table 1). The risk of developing a fatal cancer or transmittable genetic malfunction from a typical western European background dose is about 1 in 10^4; a value that brackets many of the risks we face each day.

It follows that the goal of radioactive waste management must be to ensure that significant doses additional to this background level are not encountered by the general public. Clearly, the nuances of the safety case for geodisposal must take into consideration some large challenges: the intense radioactivity, far in excess of any background radiation at the time of disposal; the operation of the facility over decades to centuries; and the predicted evolution of the facility over geological time, so the required control is maintained such that significant additional dose is not received by humans or biota. This requires mechanistic information that underpins the safety case and the geochemical sciences including mineralogy are at the core of this.

2. General strategies for long-term waste management

The primary requirement for public and worker protection is shielding – shielding the body from radiation. This is a relatively simple matter: less-penetrating radiations (α)

are absorbed by a sheet of paper. More penetrating radiations are effectively absorbed by heavy-metal shielding, or an appropriately thicker barrier of concrete or rock. Burial beneath a few metres of rock provides shielding. Unfortunately, however, this only applies if the source of radiation is fully contained over the long, geological times needed for radioactive decay to occur. Clearly, engineering integrity cannot be guaranteed over geological timescales and, therefore, the safety case needs to consider the scenario where the engineered barrier degrades.

There are two approaches for dealing with potentially mobile radionuclides. The most obvious is containment and this is mostly what radioactive waste management is all about – much like all other toxic waste. Clearly, the longer containment is in place, the longer there will have been for decay to lesser radiation levels and the lower the radiotoxicity of the wastes. However, engineering integrity can only be guaranteed for a finite time (several thousands of years?) and at that point, it becomes necessary to consider whether radionuclides in their disposal environment will dilute to such concentrations that any additional dose so caused is insignificant compared with background dose. The additional risk is likewise then negligible. This 'dilute and disperse' management scenario is sound provided that accumulated concentrations do not yield significant additional doses and also that there are no natural concentrating mechanisms to reverse the dilution process. Thus radioactive-waste management requires an assessment of both the engineered barrier performance and an assessment of the dilution and dispersion behaviour of radionuclides from the engineered geological disposal facility as degradation occurs, and over geological timescales. These two aspects encapsulate the challenges associated with disposal of radioactive wastes: the technical engineering aspects; and the equally challenging safety case development that credibly understands and assesses the risk to biota from geodisposal over millennia and beyond. Finally, it is worth highlighting that 'doing nothing', *i.e.* storing the materials at Earth's surface has its risks, and they are often considered worse than 'doing something', or disposing of the waste in the deep sub-surface. Put another way, a subsurface, engineered vault with an appropriately developed safety case seems inherently safer than a conventional surface store.

In normal operation, present practice in the nuclear power industry comes fairly close to achieving containment. There are authorized gaseous and liquid discharges associated with power plant operation and also with spent-fuel reprocessing; authorization implies minimal risk to society. Much the greater part of the activity ends up as solid wastes and it is these with which we are primarily concerned here. Technology allows the majority of the waste forms generated now to be packaged safely and stored with effective shielding. There is an additional requirement for security for the relatively small volumes but high radiological risk associated with U and Pu recovered from spent fuel *via* reprocessing. Indeed, Pu and U management is highlighted by several nations as a long-term challenge. This assessment is reasonably accurate for current power generation practices but not for those of the early days of the nuclear industry and, particularly, the period of the 'cold war' when the nuclear arms race placed great pressure on reprocessing but less on management of the wastes so generated. There is also a threat posed by nations wishing to join the 'nuclear club'. To do so requires secrecy as international law prohibits such developments. Under these circumstances there will be a lack of scrutiny and heterogeneity in the application of safe management practices. In consequence of this, and of the early days of nuclear development where both knowledge and regulation were limited, the planet has a legacy of contaminated buildings, land and wastes which require urgent attention if risks to Society are to be minimized. Even the current 'safe' containment practices, however, consist of short-term measures only. We can either opt for these active management strategies or seek some more 'permanent' solution. Existing engineered containment involves waste packages that have a design lifetime of several tens to hundreds of years – as have the stores they reside in. Repackaging, transfer and rebuilding of stores will undoubtedly have to be repeated and at relatively frequent intervals (every 100+ years) with industry workers being subjected to finite doses in consequence and with all the risks associated with human management failures. The requirement for active management of surface facilities will remain for at least 10^4 years. It will be costly and that cost is presently being passed on to future generations; this generation having taken the benefit of the electricity produced and, arguably, the security provided. Although providing effective containment, long-term surface stores offer very little protection to acts of warfare or terrorism. One small conventional 'missile' placed on any one of several hundred facilities world-wide would cause serious contamination, and nuclear security has become a major concern for governments since events such as the terrorist attacks on the World Trade Centre and the Pentagon in September 2001. Consequently, many people argue, and governments agree, that deep geological disposal is the safest choice for long-term management of higher-activity radioactive wastes.

3. Common solid radioactive waste forms and the minerals which are present

3.1. General categories

Nuclear power plants are responsible for almost all waste currently being produced. On balance, the 'once through cycle' is currently favoured and this envisages disposal of the spent fuel directly after a period of storage to allow heat generation rates to fall to manageable levels. The alternative strategy is reprocessing to recover fissile materials for further power generation or military manufacture of nuclear devices. Reprocessing is ongoing in several countries and involves dissolution of the fuel in nitric acid, recovery of U and Pu by solvent extraction and then concentration of the fission product-laden high-level liquid waste followed by conversion to a stable solid waste form, typically *via* vitrification to form a 'glass'. Thus during operation of the nuclear fuel cycle, many different waste streams are generated and these require a range of management practices.

Various waste classification schemes are in use which take account of activity (therefore immediate hazard). Some schemes also take account of radioactive half-life and chemical composition. They are more easily linked to disposal routes. At its simplest (practice and classification varies between nations) we have: Spent Fuel (heat generating, contains fission products, uranium and actinides, typically grouped with HLW); High Level Waste (HLW – initially liquid then converted to glass, contains mainly fission products, also heat generating); Transuranic wastes (heavy nuclides generated by neutron capture and which are α-emitters); Intermediate Level Waste (ILW – larger volume, lower-activity wastes containing fission products, uranium and transuranics; some streams contain significant organic matter); Low Level Waste (LLW also larger volumes and usually containing contaminated organic matter: paper, gloves *etc.*), typically disposed of at specialist engineered landfills; and, Very Low Level Waste (VLLW normally disposed of in specialist domestic landfill sites).

U and Pu separated by reprocessing for the purpose of fuel manufacture are not considered wastes by those nations presently engaged in reprocessing, although several nations are considering how to manage these challenging materials if they were to be classified as wastes.

In the UK, VLLW is defined as having an activity (measured in activity units of Becquerels (Bq); see Table 1) of $<4 \times 10^2$ Bq/kg (α, β & γ), LLW activity between 4×10^2 Bq (α, β & γ) and 1.2×10^7 Bq/kg (β & γ) or 4×10^6 Bq/kg (α) and ILW an activity of $>1.2 \times 10^7$ Bq/kg (β & γ); 4×10^6 Bq/kg (α). HLW is a term reserved for reprocessed fission products or spent nuclear fuel and the waste is heat generating. The International Atomic Energy Agency, IAEA, distinguishes between short-lived and long-lived ILW and LLW; the separation being a half-life of >30 years for the long-lived waste category.

3.2. Spent fuel

As mentioned in the introduction, most power plants use fuel assemblies based on sintered, microcrystalline UO_2 (a few use 'MOx' based on UO_2 and PuO_2). These are sealed within relatively inert stainless steel or zircalloy tubes. The spent fuel contains fission product isotopes typically at ~ 3 wt.%, and transuranic radionuclides generated by neutron capture (and their daughters) typically at ~ 1 wt.%. The fuel element cladding also contains radionuclides generated by neutron capture and contribute to the total activity of the spent fuel. In addition, the reactor components will also have neutron capture isotopes associated with them.

The oxide fuel pellets represent primary containment in the case of direct fuel disposal – the 'once through cycle'. The thermodynamic stability of these oxides in water under various environmental conditions has been investigated quite thoroughly, although it should be noted that in spent nuclear fuel, the oxides will be undergoing radiation-induced damage and radiolysis-induced redox change, and stability diagrams will not include these effects. Stability relations were presented by Brookins (1988) as pH-Eh diagrams and summarized in Faure (1991). Both UO_2 and PuO_2 are considered relatively stable and are very sparingly soluble at common environmental pH values (4 to 9) and under reducing conditions. With modestly oxidizing conditions, however, $U(IV)O_2$ (uraninite) is unstable relative to the $U(VI)O_2^{2+}$ (uranyl) species which is significantly soluble under these conditions and forms highly soluble carbonate complexes, important as carbonate is a common environmental anion. PuO_2, by contrast, is thought to be stable and insoluble to high Eh, although this has been questioned with partial oxidation to $PuO_{2+0.27}$ possible and with up to one quarter of the Pu present as the relatively mobile Pu(VI) (Haschke *et al.*, 2000). This has important implications for storage and geological disposal of Pu (Edwards, 2000). Plutonium as Pu^{4+} can also be reduced to the slightly more mobile Pu^{3+} at the low Eh and high pH conditions which may be prevalent in an engineered geological disposal facility; *e.g.* when steel cladding corrodes and generates hydrogen.

Uranium metal with Mg-alloy cladding was used in some early reactors and in the UK this technology was extended to the 'Magnox' series of reactors. From the outset it was appreciated that such materials are fundamentally unstable in the presence of water and reprocessing was always envisaged as the first stage in management of spent Magnox fuel. In addition, the UK, France and Japan have operational reprocessing facilities for UO_2 spent nuclear fuels. The end game for this reprocessed, separated U and Pu is as yet unclear although the UK government has very recently concluded that, for nuclear security reasons, the preferred policy for managing the UK plutonium stockpile is reuse *via* conversion to mixed oxide (MOx) fuels for use in civil nuclear reactors (DECC, 2011).

3.3. Reprocessing waste forms

Following solvent extraction of U and Pu from high-activity waste liquids, the residue represents one of the most dangerous materials ever generated by humans: very high concentrations of numerous fission product radionuclides in an extremely mobile form. It is vital to convert these liquids to a much more stable form, and as quickly as possible. The most straightforward method developed so far and already used on a large scale is 'vitrification'. In Europe, the USA and Japan, borosilicate glass is the chosen matrix. This has a composition based on 45% SiO_2, 14% B_2O_3, 10% Na_2O, 5% Al_2O_3 together with lesser amounts of other oxides and incorporating $\sim$12–25% fission and actinide products (Savage, 1995; Curti *et al.*, 2006). The glass is stored in a dedicated high-security store in steel canisters and cooled by air.

Other glass compositions have been investigated. Based on geological observations of mineral persistence, however, alternative solutions aiming to mimic rock properties have been proposed and researched. It is known that natural glasses (volcanic lavas and ashes) weather more quickly than rocks, and some rocks are particularly resistant. It therefore makes sense to seek to generate a crystalline rock consisting of minerals which are resistant to weathering and have 'flexible' structures capable of accommodating substantial quantities of fission products.

Some work has been done on titanite $(CaTiSiO_3)$-based ceramics, notably by the Canadians, as a long-term management strategy for CANada Deuterium Uranium (CANDU) fuel.

More extensive and systematic studies were undertaken by Ringwood (see Ringwood *et al.*, 1981, for example). 'SYNROC' was the term given to a number of formulations based on several Ti minerals together with some additional phases/glass. The 'core' of the approach is to generate melt compositions from which hollandite $(Ba(Mn^{4+},$

$Mn^{2+})_8O_{16}$), perovskite ($CaTiO_3$) and zirconolite ((Ca, Ce)Zr(Ti, Nb, Fe^{3+})$_2O_7$) will crystallize and incorporate substantial amounts of fission product and actinide wastes. The extent to which these phases are capable of containing fission products and their ultimate stability/resistance to weathering has yet to be tested and the prevalence of vitrification and the associated extensive studies on durability of vitrified wasteforms (Curti *et al.*, 2006; Knauss *et al.*, 2007) means that SYNROC concepts are yet to be widely adopted within the nuclear industry.

3.4. Natural analogues

Much of the science underpinning the design and development of waste-management strategies is based on laboratory or field experiments of short duration; key radionuclides of concern have typically only existed for, at most, 60 years. These are vastly shorter than the periods for which containment must be effected for long-term management. As is suggested above by reference to natural glasses, a different approach to understanding performance is to look for examples of element mobility (or lack of it) over geological time periods and use examples of natural containment as a basis for waste disposal concepts. This approach has been developed substantially and examples are referred to as natural analogue studies. A good review is provided by Savage (1995).

Uranium occurs naturally as uraninite (UO_2) and as a microcrystalline equivalent, pitchblende. Mineral specimens are invariably more oxic than this formula and compositions up to about U_3O_8 have been reported. Under oxidizing conditions the soluble uranyl ion, UO_2^{2+}, is formed and this is often relatively mobile, either as this species or as carbonate or sulfate complexes, depending on the physico-chemical conditions. Secondary enrichment of uranium in ore bodies occurs when uranyl in groundwater encounters a reducing environment: uraninite is then re-precipitated. There are many examples of uraninite deposits which have remained in place for millions of years but, of course, when *in situ* conditions change, *e.g.* from reducing to reoxidizing, there is the potential for further remobilization of the U as UO_2^{2+}. Indeed, within the localized engineered geological disposal environment there are components of the wasteform that have redox activity; *e.g.* zero-valent iron, and the hydrogen that results from iron corrosion, as well as oxidants such as nitrate. More recently it has been recognized that these components, and others such as cellulose, have the potential to provide conditions favourable to microbial activity (West & McKinley, 1984) and that understanding these processes and how they may (positively or negatively) affect or dictate radionuclide fate is of interest to development of a robust safety case (Morris *et al.*, 2011).

At first sight, both uranium oxide and plutonium oxide should be stable as sintered ceramic oxide pellets in the presence of water under reducing conditions and, thus, be highly immobile. Plutonium oxide ($Pu(IV)O_2$) is generally considered to be the more stable under oxic conditions although as discussed above, $PuO_{2+0.27}$ has been reported with the presence of significant plutonyl ($Pu(VI)O_2^{2+}$) in the oxidized phase (Haschke *et al.*, 2000). The plutonyl species is likely to be more mobile under environmental conditions than Pu^{4+} (*e.g.* Nelson & Lovett, 1978). Uranium oxide ($U(VI)O_2$) is prone to oxidation as the much more soluble uranyl ($U(VI)O_2^{2+}$) species (although U(VI) may be strongly sorbed or incorporated into key minerals such as the Fe oxides). The extent to which radiation damage and fission product loadings will destabilize the spent fuel relative to pure ceramic oxides is not known but, as noted above, it must be considerable.

Brookins (1978) also studied the natural reactor at Oklo, Gabon, where a natural ore body went 'critical' and underwent a series of fission events some billion years ago. Numerous scientists have investigated this deposit because of the information potentially available on the stability of minerals and the migration of fission products. This and subsequent studies show that under some crustal conditions, elevated levels of naturally concentrated elements have remained immobile for periods of up to 10^9 years. Such examples are obvious models for the design and location of geological disposal facilities although the inherent uncertainties of a complex, natural system propagated over geological time mean that careful interpretation of these analogue sites is necessary. Furthermore, analogue studies cannot, and do not, replace the extensive research programmes required for any specific geological disposal facility.

3.5. Containment strategies, performance and environmental mineralogy

A geological disposal facility will have failed if radionuclides escape to the surface in concentrations in excess of the appropriate regulatory risk target. Overall, the geological disposal facility will have a series of 'multiple barriers' which act to isolate the waste and retard radionuclide mobility. Each of these multiple barriers has associated 'safety functions' which set out what the barrier component should achieve to ensure safety (NDA-RWMD, 2010). Clearly, environmental mineralogy contributes in several ways to these safety functions. For example, within the engineered vaults of a geological disposal facility, minerals (and glasses) will provide containment by acting as hosts for radionuclides within their structures (retardation) and will also restrict ingress of groundwater to the disposal facility (isolation). Ultimately, groundwater ingress will cause degradation of engineered structures and the evolving heterogeneous environment will provide both new mineral assemblages that can interact with radionuclides (retardation). Outside of the geological disposal facility, evolved and natural surface-active or reactive minerals will also interact with migrating radionuclides by sorption and/or incorporation (retardation).

3.6. Mineral persistence in weathering

The rate at which radionuclides are released from either primary or secondary containment becomes a key performance variable. Since Goldich (1938), there have been numerous studies of the persistence of different natural minerals in surface and near-surface environments likely to be encountered in a geological disposal facility. From these and many experimental studies, a relatively straightforward picture emerges that equates well with calculations of standard state free-energy release in typical weathering reactions (Curtis, 1976, see Table 2). Silicates formed at higher temperatures and pressures are generally less stable in weathering than those formed at lower temperatures and pressures. In surface environments with oxygen present, reduced phases are substantially less stable, with sulfides especially so. The environmental threat posed by acid mine drainage (AMD) is not surprising on energy-release criteria: sulfide minerals are very unstable when exposed to wet, oxic conditions.

Table 2. Weathering reactions calculated for reaction free-energy values (ΔG_r^0), $kJ \cdot gatom^{-1}$.

Common rock-forming minerals of igneous/crustal origin	
1. $Mg_2SiO_4(s) + 4H^+(aq) = 2Mg^{2+}(aq) + 2H_2O(l) + SiO_2(s)$	-4.00
2. $MgSiO_3(s) + 2H^+(aq) = Mg^{2+}(aq) + H_2O(l) + SiO_2(s)$	-2.98
3. $CaMg(SiO_3)_2(s) + 4H^+(aq) = Mg^{2+}(aq) + Ca^{2+}(aq) + 2H_2O(l) + 2SiO_2(s)$	-2.72
4. $Mg_7Si_8O_{22}(OH)_2(s) + 14H^+(aq) = 7Mg^{2+}(aq) + 8H_2O(l) + 8SiO_2(s)$	-2.49
5. $Ca_2Mg_5Si_8O_{22}(OH)_2(s) + 14H^+(aq) = 5Mg^{2+}(aq) + 2Ca^{2+}(aq) + 8H_2O(l) + 8SiO_2(s)$	-2.24
6. $CaAl_2Si_2O_8(s) + 2H^+(aq) + H_2O(l) = Al_2Si_2O_5(OH)_4(s) + Ca^{2+}(aq)$	-1.32
7. $2NaAlSi_3O_8(s) + 2H^+(aq) + H_2O(l) = Al_2Si_2O_5(OH)_4(s) + 4SiO_2(s) + 2Na^+(aq)$	-0.75
8. $2KAlSi_3O_8(s) + 2H^+(aq) + H_2O(l) = Al_2Si_2O_5(OH)_4(s) + 4SiO_2(s) + 2K^+(aq)$	-0.51
9. $2KAl_3Si_3O_{10}(OH)_2(s) + 2H^+(aq) + 3H_2O(l) = 2K^+(aq) + 3Al_2Si_2O_5(OH)_4(s)$	-0.32
Common igneous/crustal minerals – with reduced iron, Fe^{2+}	
10. $Fe_2SiO_4(s) + 0.5O_2(g) = Fe_2O_3(s) + SiO_2(s)$	-6.58
11. $2FeSiO_3(s) + 0.5O_2(g) = Fe_2O_3(s) + 2SiO_2(s)$	-6.64
Highly reduced phases, methane representing organic compounds/hydrocarbons	
12. $2FeS_2(s) + 4H_2O(l) + 7.5O_2(g) = Fe_2O_3(s) + 4SO_4^{2-}(aq) + 8H^+(aq)$	-17.68
13. $CH_4(g) + 2O_2(g) = H_2O(l) + H^+(aq) + HCO_3^-(aq)$	-20.54
Radwaste metals and oxides	
14. $U(m) + 1.5O_2(g) + 2H^+(aq) = UO_2^{2+}(aq) + H_2O(l)$ (oxygen)	-198.45
15. $U(m) + 2H_2O(l) = UO_2(s) + 2H_2(g)$ (anoxic)	-79.64
16. $Mg(m) + 0.5O_2(g) + 2H^+(aq) = Mg^{2+}(aq) + H_2O(l)$ (oxygen)	-172.99
17. $Mg(m) + 2H^+(aq) = Mg^{2+}(aq) + H_2(g)$ (anoxic)	-151.80
18. $UO_2(s) + 0.5O_2(g) + 2H^+(aq) = UO_2^{2+}(aq) + H_2O(l)$ (oxygen)	-26.48
19. $UO_2(s) + 4H^+(aq) = U^{4+}(aq) + 2H_2O(l)$ (anoxic)	$+3.79$
20. $PuO_2(s) + 4H^+(aq) = Pu^{4+}(aq) + 2H_2O(l)$ (anoxic)	$+6.10$

Data details from Curtis (1976), reproduced with permission of Wiley; additional data from Faure (1991); state subscripts: s = solid, g = gas, l = liquid, m = metal, aq = aqueous ion.

Maintaining anoxic conditions for U and Pu disposal as oxides is seen to be very sensible! On the other hand, these elements are very unstable as metals, even under anoxic natural conditions. This would apply to metallic U and Mg cladding used in early Magnox reactors. These calculations are for standard states (temperatures, pressures, unit activity reactants and products). The same calculations can be performed for more realistic conditions (Table 3). Very low-ionic-strength rainwater is very much more aggressive than seawater but the same relative mineral stability sequence is maintained in both fluids. This simple analysis confirms that groundwater compositions likely to be encountered at the depth of a geological disposal facility are much less effective in degrading both primary and secondary minerals (and their containment properties) than groundwaters in near-surface settings.

3.7. The degradation of rocks to soils: (i) climate

The natural land surface is soil: a relatively thin veneer of unconsolidated material covering the solid rocks of the Earth's crust. Physical processes of erosion move and sort

Table 3. Weathering reaction free-energy values (kJ · gatom^{-1}) calculated for igneous mineral reactions with natural waters of different kinds, notably total salt content expressed as ionic strength (from Curtis, 1990, reproduced with the permission of the Geological Society, London).

		Standard State	Rainwater pH = 5 I = 0.0003	River water pH = 6.5 I = 0.002	Seawater pH = 8 I = 0.618
$Mg_2SiO_4(s)$ forsterite	ΔG_r^0	−4.00	−11.94	−7.63	−2.54
$MgSiO_3(s)$ enstatite	ΔG_r^0	−2.98	−8.37	−5.02	−0.98
$CaMg(SiO_3)_2(s)$ diopside	ΔG_r^0	−2.72	−7.52	−3.74	−0.14
$CaAl_2Si_2O_8(s)$ anorthite	ΔG_r^0	−1.32	−4.11	−2.47	−1.23
$NaAlSi_3O_8(s)$ albite	ΔG_r^0	−0.75	−2.59	−1.77	−0.08
$KAlSi_3O_8(s)$ orthoclase	ΔG_r^0	−0.51	−2.22	−1.35	−0.04
$KAl_3Si_3O_{10}(OH)_2(s)$ muscovite	ΔG_r^0	−0.32	−0.96	−0.45	+0.31

these soils to generate sediments. Rainwater breaks down the solid rocks to their mineral-grain constituents and slowly dissolves them. The dissolved salts produced find their way *via* groundwater to streams and eventually to the sea. The residual mineral grains, together with precipitated iron and aluminium hydroxides, constitute the mineral fraction of soil: organic matter and a host of living organisms, mostly micro-organisms, complete the description.

Geomorphologists study the development of land forms and, using a range of methodologies, have quantified rates of denudation: the net annual lowering of the land surface due to a combination of both physical erosion and chemical weathering processes. Both the rate of chemical weathering and the mineral composition of the resulting soils depend markedly on climate. Hot, wet climates maximize the weathering process and lead to the destructive dissolution of virtually all rock-forming minerals. This is why the wet tropics are draped by deep, homogeneous red oxide/hydroxide-rich soils, irrespective of the nature of the underlying rocks from which they are derived. These general relationships were recognized early by Strakhov (1967). On a smaller scale these relationships can be very striking. Sherman (1952) studied the sequence of authigenic minerals developing on Hawaiian basalts. Barshad (1966) documented the sequence of minerals forming under different climatic regimes on different rock-types in California. There is, thus, a strong correlation between climate and both the depth and mineralogy of soil profiles. The effects of altitude are somewhat more complex. Chemical weathering at high altitude is less intense, as also at high latitude. The depth of the profile may be greater, however, due to deeper penetration of meteoric water in response to greater hydrostatic heads.

3.8. The degradation of rocks to soils: (ii) chemical 'aggressiveness' of different depth-related environments within the soil profile

As has been mentioned, the depth of a soil profile may be very great in humid settings where there has been no physical erosion for very long time periods. These periods may extend to geological times (10^6 years) in low-topography tropical and sub-tropical settings where there has been no glaciation. Harder (1952) cites an example of a soil profile developed on basalt in central India where fresh rock is not encountered above ~20 m. Deeper profiles still are not uncommon. Craig & Loughnan (1964), Harder (1952) and Goldich (1938) all describe the sequence of minerals found with depth in residual soils and document the same basic pattern: original assemblages of rock-forming minerals are progressively eliminated (depending on composition and structure) to be replaced by clay minerals and hydrated oxides. The minerals stable at the profile surface are those closest to equilibrium with meteoric water (very low ionic strength – ~0.0003). Fresh rainfall approximates to distilled water, which is chemically extremely aggressive as all substances are soluble to some extent. Such solutes as are present confer acidity due to dissolution of carbon dioxide together with variable amounts of sulfur and nitrogen oxides. As meteoric water percolates down through the unsaturated zone of a soil profile, it reacts with minerals and takes solute products into solution. Its solute content increases progressively and its 'aggressiveness' declines in line with degree of undersaturation. Only the least soluble and least reactive minerals can survive at the surface; higher ionic strength solutions at depth reduce reaction rates and may stabilize certain minerals (if they are thermodynamically stable and the soil water becomes saturated with respect to that phase). Upper soil horizons in hot, wet climates are dominated by hydrated oxides of iron(III) and aluminium – these are more stable than any aluminosilicate. These points are discussed more fully in Curtis (1990).

Goldich (1938) specifically addressed the persistence of different primary bedrock minerals in soil profiles. He first documented a pattern now so well established as to be termed the 'Goldich Stability Series'. Broadly, this shows that minerals persist in soils according to the temperature and pressure of their formation: high-temperature minerals degrade quickly, minerals which crystallized at low temperatures (more closely approaching those of earth-surface environments) persist for longer. This is the reverse of 'Bowen's Reaction Series', which is the observed sequence of crystallization from magma with cooling. Attempts to quantify these relationships in terms of properties of the minerals themselves and with environment are discussed in Curtis (1976).

In summary, original rock-forming minerals do survive for geological times at depths greater than some tens of metres beneath tropical soil profiles and virtually to the surface in cold, arctic or alpine environments (although, of course, physical erosion may then be dramatic). Quantitative studies of the survival of minerals in surface and deeper environments should provide valuable indicators of how man-made materials and structures will behave over time intervals measured in thousands of years, in the absence of any kind of management.

3.9. Thermodynamic calculations of relative mineral stability in weathering

Curtis (1976) reviewed earlier attempts to evaluate mineral persistence in thermochemical terms and suggested that the logical approach was to look at the energetics of weathering reactions – not just the properties of the minerals themselves. Accordingly, he calculated standard-state Gibbs Free Energy values for reactions in which primary minerals decomposed to give only products known to be stable and to occur ubiquitously in the surface environment – solid products, gases and aqueous solutes. He demonstrated that these values did indeed follow persistence observations very closely and, moreover, the methodology allowed comparisons with very different minerals (sulfides, for example) and between different environmental settings (oxidizing or reducing, for example). A key finding was that redox reactions were very much more energetic than simple dissolution-precipitation reactions. Iron(II)-bearing silicates are proportionately unstable relative to Mg-equivalent mafic minerals. The energy liberated by the spontaneous oxidation of pyrite FeS_2 (both Fe and S oxidized) proves to be very great indeed, and of the same order as hydrocarbon combustion. This finding sits well with the observation that it is extremely difficult to constrain acid-mine drainage problems and, also, the spontaneous combustion of old colliery waste which can expose large quantities of pyrite to aerial oxidation.

Reactions 1–13 in Table 2 are reproduced from Curtis (1976) and document the relative stability of common rock-forming minerals. Equations 1–9 tally with Goldich's observations, 10 and 11 illustrate the destabilizing influence of Fe(II). Equations 12 and 13 cover redox reactions where more than one element is involved and complete oxidation releases much energy.

3.10. Degradation of radioactive waste-product phases

The calculation method used above offers an initial approach to predicting the stability of various phases over very long time scales, once primary managed containment has failed. Equations 14 to 17 demonstrate the persistence properties of U-metal and one of the more reactive cladding materials from Magnox fuel. Weathering-reaction free-energy estimates are an order of magnitude greater than pyrite oxidation or hydrocarbon combustion which are themselves an order of magnitude higher than typical rock-forming mineral decomposition reaction energies.

This prediction will be no surprise to the managers of spent metallic fuel storage facilities such as those employed in UK Magnox plants. The argument in favour of conditioning is extremely strong, and it is exceedingly difficult to see any advantage in long-term storage where extremely close atmospheric compositional control would be necessary and failure would lead to very rapid reaction. Indeed, over the next few years the remaining UK Magnox fuel from the now obsolete Magnox fleet will be reprocessed to recover U and Pu with the resultant HLW converted to borosilicate glass.

Lastly, equations 18, 19 and 20 provide an indication, on the same footing, of how spent fuel might behave if disposed directly, as in the 'once-through cycle'. The instability of UO_2 relative to soluble UO_2^{2+} is clear and a strong indicator that storage or disposition under oxidizing and wet conditions is to be avoided; radionuclides would be rapidly released to the environment. This contrasts with reducing environments where UO_2 is seen to be stable and extremely insoluble – more so than any of the natural rock-forming minerals considered here, although recent work suggests that for ILW, under hyperalkaline reducing conditions, U(VI) may dominate and be significantly incorporated into poorly soluble mineral phases, *e.g.* Ca-uranate and uranyl silicates (Tits *et al.*, 2011). Overall, this would seem to argue, as other approaches have, in favour of management under strictly anoxic conditions. Such conditions would require constant management at the Earth's surface but not with deep burial within groundwater-saturated rock formations. Interestingly, an extended operational period (tens to hundreds of years) where the sub-surface facility will presumably be exposed to air and thus "oxic" is planned for most modern geological disposal facilities to accommodate the legacy and future materials destined for geodisposal from national power programmes.

The analysis of Pu data suggests that plutonium dioxide (PuO$_2$; as utilized in MOx reactors) may be much more stable than UO$_2$ in oxidizing conditions: it is perhaps somewhat less stable under reducing conditions, but still very stable compared with other waste forms. Interestingly, Haschke *et al.* (2000) suggest that higher oxidation states of Pu may be more stable than formerly supposed and this implies something less than the stability indicated above. Again, this would suggest that storage or disposal under reducing conditions is the safer approach, and highlights the need for further experimental and theoretical work on these challenging systems.

It must also be that spent fuel, having been subjected to extreme irradiation and radiolysis and having trapped significant quantities of fission products within the lattice, will be less stable than the pure oxides considered here. Indeed, spent nuclear fuel is considered to have an 'instant release fraction' of radionuclides which is that fraction which has migrated to crystal boundaries or even outside the pellets in the fuel during fission. The 'instant release fraction' radionuclides are considered to be mobilized very quickly when the fuel is contacted with water in the repository as the alloy tubes corrode and will move quickly into the engineered barriers surrounding the spent fuel. Thus it is clear again that the conclusions are optimistic.

This simple analysis has perhaps most value in that the conclusions based on very simple, standard-state calculations, are calibrated against the observations of numerous soil scientists over many years. The conclusions reached for radwaste 'proxies' represent an extrapolation. This having been said, the data do suggest several fairly clear strategies: metallic phases (uranium and magnesium here) are so very unstable that it would be irresponsible to consider anything other than conditioning to some silicate or oxide matrix; vitrified waste forms (and synthetic rock materials) are somewhat less stable than natural rock-forming minerals but certainly comparable; oxides of U and Pu are very stable, but only under strictly anoxic conditions; the longer-term stability of all these phases, minerals and glasses, synthetic and natural, is extremely poor under surface conditions relative to those at depth within water-saturated rock formations (due to low-ionic-strength pore waters and not the presence or absence of oxygen): surface storage is inherently unsafe.

The basic observations underlying these conclusions are to be seen in any road cutting or excavation that cuts down through soils into bedrock and are further exemplified in the predominance of the deep geological disposal concept as the approach chosen by governments to manage higher-activity wastes into the long term.

3.11. Saline solutions and mineral stability

Curtis (1990) also calculated reaction free energies for reactions in waters of natural solute compositions rather than unit activities. These are reproduced in Table 3 and show that the same trend of relative mineral stability persists. The very aggressive nature of low-ionic-strength rainwater is seen in much more negative Gibbs Free Energy values and, as expected, the lowest values emerge from solutions that are more saline, notably seawater. These simple calculations reinforce the view that surface waters are much more aggressive than groundwaters, especially saline groundwaters. Again, the conclusion is that radioactive waste forms would be expected to be much more stable buried within rock formations well below the groundwater table than if they were at the surface. Moreover a distinctive additional mineralogical stability is imposed by highly saline groundwaters.

3.12. Direct measurement of dissolution rates

Although dissolution rates for silicate and other natural minerals in water are very slow at room temperature, some direct experimental work has been undertaken. Brady & Walther (1989), for example, discuss measured dissolution rates for a number of rock-forming minerals at 25°C and neutral pH. Rates for olivine, pyroxenes and feldspars are sequentially greater than those for quartz and kaolinite, the minerals assumed to be stable in the calculations of Tables 2 and 3. In a very limited way, therefore, direct experimental determinations of mineral breakdown rates correlate broadly with predictions made on the basis of free-energy data and field observations of long-term persistence.

3.13. Natural analogues

Savage (1995) described the study of ancient mineral deposits and their behaviour over geological time as so-called natural analogues. The very fact that ore-grade natural concentrations of uranium minerals have persisted for millions of years in the geological sub-surface surely demonstrates that conditions can prevail when no dispersive transport takes place. The practical utilization of such information is discussed by Savage (1995), including work on the long-term stability of uraninite as an analogue for spent fuel. Some uranium ore bodies offer ample evidence of mobilization by aerial oxidation, migration in aqueous solution and then re-concentration at (sometimes biologically driven – McKinley *et al.*, 1997) redox fronts–transition zones between reducing and oxidizing environments. These observations tally with deductions based on weathering-reaction energies as discussed above. Natural analogues also allow for a whole range of transport and retardation processes to be studied and the radioactive isotopes present in some can give opportunities for rates to be determined using radiometric dating.

In one sense, the entire discussion of soil-profile development can be looked upon as an analysis of a natural process where analogue materials are directly degraded by the very same processes which will eventually attack radioactive wastes.

4. Minerals within containment barrier systems

4.1. Low level wastes (LLW) and intermediate level wastes (ILW)

So far only high-level wastes and spent fuels have been considered. Although they contain much the greater part of all activity, they are generally less voluminous than the ILW and LLW waste forms. As LLW is essentially a self shielding wasteform (it is typically of large volume and consists of slightly contaminated paper, cellulosic, metallic and building rubble), it is typically disposed of in specialist engineered landfills and the discussions here will focus on ILW which yields significant external dose. Wastes arise from the operation of nuclear power and reprocessing plants – these are known as operational wastes. Ion-exchange filters (organic) from primary water systems (especially in water reactors) and the cladding removed from spent fuel during reprocessing are examples of such ILW materials. Similar materials are also generated at the power or fabrication/processing plant or during decommissioning. ILWs are quite active and may require remote handling.

The additional problem (apart from larger volumes and heterogeneity) is the fundamental instability of organic matter in the presence of any oxidant. As in landfill, microbial degradation occurs under both oxic and anoxic conditions. Two very important products are methane and carbon dioxide and repositories for these wastes have to be designed with gas escape pathways. More broadly, the impact of geomicrobiology on geological disposal of wastes has significant implications, with consumption and production of gases being possible in the case of many geological disposal facility designs.

In order to stabilize these heterogeneous wastes, they are compacted and then placed within either steel drums or concrete/steel boxes (generally a few m^3 in volume). The design is matched to particular waste-stream requirements and takes account of issues such as transport, store design and eventual disposal systems. The waste is then usually grouted into the waste container. Cementitious materials can be designed to be porous for gas release or impermeable, depending on the waste enclosed. Intermediate-level active liquids can also be solidified in this way, rendering them very much less mobile. A number of different additives are used including pulverized fly ash, bentonite and calcite. These relatively reactive materials are designed to enhance retention properties. The cement minerals themselves (hydrated potassium, sodium calcium and aluminium oxides and silicates) are all highly alkaline (pH 10–13.1). This property is an important design feature as the solubility of many oxides (therefore their potential mobility) is greatly reduced. It should be noted, however, that some important radionuclides are not immobilized in this way. This is particularly true of negatively charged radionuclides: iodine (I^-), chlorine (Cl^-) and technetium (TcO_4^-) are troublesome examples. Furthermore, the effects of complexation and colloid transport on the actinides are poorly constrained at the extremely elevated pH environments expected in ILW disposal. Indeed, it is only very recently that we have started to gain a molecular-level understanding of actinide and transuranic behaviour in cementitious matrices and more broadly in environmental matrices (*e.g.* Duff *et al.*, 1999; Heberling *et al.*, 2008; Livens *et al.*, 2004; Law *et al.*, 2010; Tits *et al.*, 2011). Other grouts have been used, notably bitumen and organic polymers. Although useful in waste-specific applications, these materials are organic and will eventually decompose to generate gases and other organic compounds. Again, the influence of microbial processes on these systems is of current interest.

4.2. Engineered barriers

Long-term waste-management designs have generally favoured deep geological disposal for HLW and ILW. Heat-generating HLW or spent fuel assemblies are encased, in most designs, within very heavy, corrosion-resistant metal canisters, either stainless steel or copper (as metallic copper is known to have survived over geological time spans in certain ore deposits). It is planned to emplace these within small tunnels or vaults and surround them with tightly packed anhydrous bentonite blocks, each engineered to leave virtually zero air space. The tunnels/vaults are then sealed with a non-alkaline backfill. Once the vaults are filled in this way, groundwater will be allowed to reclaim pore space. In the case of the bentonite, hydration causes a 50% volume expansion thereby creating an extremely impermeable, plastic barrier around the metallic canister. Further ingress and, in particular, through-flow of groundwater, will be minimized over long time scales. Finally, there is the barrier created by the rock-formation itself which contributes to physical protection of the engineered barriers as well as being itself a barrier to radionuclide escape. For ILW considered for geological disposal, the grouted wasteform is emplaced again in underground 'caverns' and backfilled. Backfill models include various materials; a cementitious environment is chemically compatible with grouted material and is the current model for backfill in the UK.

Vaults and tunnels are planned to be back-filled pending final geological disposal facility sealing. A variety of materials have been designed for this purpose including alkaline cement grouts and highly absorbing fillers. The objective is chemical as well as physical containment and chemical compatibility across the barrier system is important.

At this time, no geological disposal facility for civilian high-level wastes designed on a multiple-barrier basis has been completed. There remain uncertainties as to the performance of the engineered barriers under conditions of enhanced temperature caused by heat-generating waste forms. At the Grimsel Test Site of NAGRA (the Swiss National Co-operative for the Disposal of Radioactive Waste), a full-scale simulation of HLW geological disposal facility engineered barriers has been undertaken by ENRESA (Empresa Nacional de Residuos Radioactivos SA, Madrid), NAGRA's Spanish sister organisation. This has emplaced a canister with heating coils to simulate waste heat generation within bentonite blocks inside a purpose-drilled shaft (as described above). The aim was to test engineering procedures, monitoring, the performance of materials and models of performance (McKinley *et al.*, 1996).

5. Minerals contributing to containment in the natural environment

The geosphere (the rock formations within which a geological disposal facility is placed) provides three elements of containment: shielding; protection for the engineered barriers (see above); and long pathways for radionuclide transport to the surface.

It is certain that the escape pathway for almost all radionuclides from a geological disposal facility will be as solutes in solution, or as colloids transported in groundwater. Containment is provided when the return pathways are long and groundwater flow rates slow. For the small proportion of the total activity which is present as long-lived anthropogenic radionuclides, the time taken to reach the surface (in significant concentration) must be very long 'geological' timescales. Additionally, natural series radionuclides (^{235,238}U and their decay products) are also expected to be significant long-term risk drivers for any facility. Overall, this means a long pathway and/or relatively low mobility are needed to allow radioactive decay and dilution to reduce radiotoxicity. This applies, for example, to ^{99}Tc (half-life $= 2.13 \times 10^5$ y), ^{239}Pu (half-life $= 2.41 \times 10^5$ y), ^{237}Np (half-life $= 2.13 \times 10^6$ y) and ^{235,238}U (half-life $= 7.04 \times 10^8$ y and 4.47×10^9 y, respectively) and daughters, and regulatory scrutiny will be particularly close for these radionuclides.

Here, the balance between politics and science becomes stark; over recent decades it has become apparent that hosting a geological disposal facility is not a straightforward business of the implementers choosing a site and the facility being successfully hosted at that spot. This approach has commonly failed at the planning application level. The most advanced geological disposal facility plans are typically ones where local communities volunteer to host the geological disposal facility usually in return for some form of compensation. As volunteer communities are generally few, the choice of geological settings available to the developer is narrow. Nontheless, the technical acceptability criteria for any geological disposal facility hosted in a volunteer community will not be affected by the location choice; the facility design must fully satisfy regulatory safety requirements (over very long post-closure time frames) if it is to be implemented. Obviously some sites will be geologically and technically better than others with the appropriate safety and design criteria being more easily met and at lower cost to the nation. For example, one site may require construction at greater depth with increased engineering costs in order to achieve the necessary isolation. It may also be that the implementation process fails because no suitable site (*i.e.* one that meets regulatory safety requirements) can be found within a volunteered locality.

The last element of containment is provided by having reactive minerals along water-migration pathways as a natural component of the sub-surface environmental conditions. Clay minerals and related large-surface area phases are important here (*e.g.* Chardon *et al.*, 2008). Ion exchange, surface adsorption and mineral−water reaction all retard radionuclide migration relative to groundwater flow. In addition, the iron oxides are likely to provide potentially redox active sorption and incorporation sites for a range of relevant radionuclides.

6. Geomicrobiology and mineralogy of long-term nuclear waste management

Over recent decades, it has become clear that microbiological processes are significant in driving the geochemistry of the shallow sub-surface, including at radioactively contaminated land sites (Lloyd & Renshaw, 2005). Understanding the microbiology of extreme, deep sub-surface environments is a topic of great current interest (*e.g.* Vreeland *et al.*, 2000). In deep geological disposal for higher-activity wastes, microbial processes are likely to be significant in influencing and even controlling radionuclide behaviour *via* a range of mechanisms including biomineralization although these processes are currently relatively poorly constrained (Andersen *et al.*, 2011; Buckau & Duro, 2007). Microbial activity in higher-level waste disposal is likely as the ILW wasteform itself contains a range of the electron donors and electron acceptors that microbial processes require. In addition, the fact is that microbial activity in the deep subsurface and the extreme pH and radiation environments associated with geodisposal facilitates biogeochemical change is now acknowledged, even if these processes are relatively unconstrained. In terms of mineralogical considerations, biomineralization processes in the deep subsurface are poorly explored although it is becoming clear that they are extensive in the shallow sub-surface. For example, iron mineralization is essentially subject to biological control in many systems. Furthermore, biotransformations of key redox active radionuclides have a profound effect on their environmental behaviour in the shallow sub-surface (Lloyd & Renshaw, 2005), and are likely to influence radionuclide behaviour and mineralogical fate in a range of environmental scenarios (*e.g.* Keith-Roach & Livens, 2002 and references therein; Lloyd & Gadd, 2011 and references therein).

7. Conclusions

Mineralogy can be said to play a crucial role in radioactive waste management in that it addresses the long-term issues in a safety case that are simply not covered by engineering experience. Equally, short-term management is relatively straightforward; nuclear nations and nuclear organizations are clearly capable of managing waste in engineered structures at the present time and doing so with acceptable levels of dose to both industry workers and the public.

But what happens in 10^2, 10^3, 10^4 years and beyond? Present practice suggests that repackaging and expensive storage facility refurbishment will probably have to be redone relatively frequently (every 100+ years). Active and expensive management is required for these stores. They are susceptible to terrorism or warfare. Management regimes of this kind are susceptible to economic collapse. Will future generations be able to pay for such management? We also have a significant inventory of radioactive waste left over from the 'cold war' and from all nuclear power generation since. Is it responsible to store and bequeath the problem to future generations? Are present proposals for deep disposal (with no active management beyond the period of emplacement and closure required) acceptably safe? These are difficult questions, which certainly extend beyond mineralogy. But what does mineralogy suggest? And how confidently? The analysis here strongly suggests that mobile liquid wastes and spent metallic fuels should be conditioned to silicate glasses or synthetic ceramics. These and spent oxide fuels are calculated here to be of the same order of resistance to weathering reactions and radionuclide release as many common rock-forming minerals.

Studies of weathering reactions underline the highly aggressive nature of meteoric as opposed to saline waters and strongly suggest that weathering (and, hence, radionuclide release rates) will be very much slower in the geological subsurface. Certain key elements, notably U and Pu, would also be greatly and additionally stabilized for a different reason; namely that deep geological environments are normally completely anoxic. The role of biogeochemistry is increasingly recognized as crucial to a proper understanding of disposal systems. Overall, the mineralogical arguments here all seem to point in one direction: deep geological disposal of HLW and spent fuel. Whether or not such a solution is implemented will depend on both socio-political and technical factors. Where the site is hosted clearly must involve local communities; the technical aspects of the safety case must be robust enough to stand up to extreme external scrutiny, and must be flexible enough to include assessments of emerging topics, *e.g.* radionuclide biogeochemistry or colloid chemistry.

Acknowledgements

CDC acknowledges the wise counsel of former members of RWMAC (Radioactive Waste Management Advisory Committee) prior to publication of the original version of this paper. KM acknowledges the financial support of NERC for past studies on radionuclide biogeochemistry and NERC, EPSRC and a BNFL Endowment to The University of Manchester for ongoing geological-disposal-related research, particularly the NERC BIogeochemical Gradients and RADionuclide (BIGRAD) consortium (NE/H007768/1). The authors also acknowledge the helpful review of Prof. Francis Livens.

References

Anderson, C., Johnsson, A., Moll, H. & Pedersen, K. (2011) Radionuclide geomicrobiology of the deep biosphere. *Geomicrobiology Journal*, **28**, 540–561.

Barshad, I. (1966) The effect of variation in precipitation on the nature of clay mineral formation in soils from acid and basic igneous rocks. In: *Proceedings of the International Clay Conf*erence, *CIPEA*, (L. Heller & A.Weiss, editors). **1**, Jerusalem, Israel, pp. 167–173.

Brady, P.V. & Walther, J.V. (1989) Controls on silicate dissolution rates in neutral and basic solutions at 25°C. *Geochimica et Cosmochimica Acta*, **53**, 2823–2830.

Brookins, D.G. (1978) Retention of transuranic and actinide elements and bismuth at the Oklo natural reactor, Gabon: application of Eh–pH diagrams. *Chemical Geology*, **23**, 309–323.

Brookins, D.G. (1988) *Eh–pH Diagrams for Geochemistry*. Springer-Verlag, New York.

Buckau, G. & Duro, L.D. (2007) Fundamental processes of radionuclide migration. In: *Proceedings of the Integrated Project 6th EC FP IP FUNMIG.* Technical report TR-07-05, Swedish Nuclear Fuel and Waste Management Company (SKB), Stockholm, pp. 1–87.

Chardon, E.S., Bosbach, D., Bryan, N.D., Lyon, I.C., Marquardt, C., Romer, J., Schild, D., Vaughan, D.J., Wincott, P.L., Wogelius, R.A. & Livens, F.R. (2008) Reactions of the feldspar surface with metal ions: Sorption of Pb(II), U(VI) and Np(V), and surface analytical studies of reaction with Pb(II) and U(VI). *Geochimica et Cosmochimica Acta*, **72**, 288–297.

Craig, D.C. & Loughnan, F.C. (1964) Chemical and mineralogical transformations accompanying the weathering of basic volcanic rocks from New South Wales. *Australian Journal of Soil Research*, **2**, 218–234.

Curti, E., Crovisier, J.D., Morvan, G. & Karpoff, A.M. (2006) Long-term corrosion of two nuclear waste reference glasses (MW and SON68). A kinetic and mineral alteration study. *Applied Geochemistry*, **21**, 1152–1168.

Curtis, C.D. (1976) Stability of minerals in surface weathering reactions; a general thermochemical approach. *Earth Surface Processes*, **1**, 63–70.

Curtis C.D. (1990) Aspects of climatic influence on the clay mineralogy and geochemistry of soils, palaeosols and clastic sedimentary rocks. *Journal of the Geological Society* (London), **147**, 351–357.

DECC (2011) The management of the UK's plutonium stocks. Department of Energy and Climate Change. Report Number URN 11D/819.

Duff, M.C., Hunter, D.B., Triay, I.R., Bertsch, P.M., Reed, D.T., Sutton, S.R., Shea-McCarthy, G., Kitten, G., Eng, P., Chipera, S.J. & Vaniman, D.T. (1999) Mineral associations and average oxidation states of sorbed Pu on tuff. *Environmental Science & Technology*, **33**, 2163–2169.

Edwards, R. (2000) Hard to contain. *New Scientist.* 22nd January, p. 18.

Faure, G. (1991) *Principles and Applications of Geochemistry.* Prentice-Hall, Upper Saddle River, New Jersey, USA.

Goldich, S. (1938) A study of rock weathering. *Journal of Geology,* **46**, 17–58.

Harder, E.C. (1952) Examples of bauxite deposits illustrating variations in origin. In: *Problems of Clay and Laterite Genesis.* American Institution of Mining and Metallurgical Engineering, New York, pp. 55–64.

Haschke, J.M., Allen, T.H. & Morales, L.A. (2000) Reaction of plutonium dioxide with water: formation and properties of PuO_{2+x}. *Science*, **287**, 285–287.

Heberling, F., Denecke, M.A. & Bosbach, D. (2008) Neptunium (V) coprecipitation with calcite. *Environmental Science & Technology*, **42**, 471–476.

Huheey, J.E. (1983) *Inorganic Chemistry: Principles of Structure and Reactivity,* 3rd edition. Harper and Row, New York.

Keith-Roach, M.J. & Livens, F.R. (editors) (2002) Interactions of microorganisms in the environment. *Radioactivity in the Environment Vol 2.*, Elsevier, Oxford, UK.

Knauus, K.G., Bourcier, W.L., McKeegan, K.D., Merzbacher, C.I., Nguyen, S.N., Ryerson, F.J., Smith, D.K., Weed, H.C. & Newton, L. (2007) Dissolution kinetics of a simple analogue nuclear waste glass as a function of time, pH and temperature. *Materials Research Society Symposium Proceedings*, **176**, 371–381.

Law, G.T.W., Geissler, A., Lloyd, J.R., Livens, F.R., Boothman, C., Begg, J.D.C., Burke, I.T., Denecke, M.A., Dardenne, K., Charnock, J.M., Rothe, J. and Morris, K. (2010) Geomicrobiological redox cycling of the transuranic element neptunium. *Environmental Science & Technology*, **44**, 8924–8929.

Livens, F.R., Jones, M., Hynes, A.J., Charnock, J.M., Mosselmans, J.F.W., Hennig, C., Steele, H., Collison, D., Vaughan, D.J., Pattrick, R.A.D., Reed, W.A. & Moyes, L.N. (2004) X-ray absorption spectroscopy studies of reactions of technetium, uranium and neptunium with mackinawite. *Journal of Environmental Radioactivity,* **74**, 211–219.

Lloyd, J.R. & Gadd, G.M. (2011) The geomicrobiology of radionuclides. *Geomicrobiology Journal*, **28**, 383–386.

Lloyd, J.R. & Renshaw, J.C. (2005) Microbial transformations of radionuclides: fundamental mechanisms and biogeochemical implications. *Met Ions Biol Syst*, **43**, 205–240.

McKinley, I.G., Kickmaer, W., del Olmo, C. & Huertas, F. (1996) The FEBEX project: full scale simulation of engineered barriers for a HLW repository. *Nagra Bulletin*, **27**, 56–67.

McKinley, I.G., Hagenlocher, I., Alexander, W.R. & Schwyn, B. (1997) Microbiology in nuclear waste disposal: interfaces and reaction fronts. *FEMS Microbiology Reviews*, **20**, 545–556.

Morris, K., Law, G.T.W. & Bryan N.D. (2011) Geodisposal of higher activity wastes. In: *Issues in Environmental Science and Technology, Issue No. 32 Nuclear Power and the Environment*. Royal Society of Chemistry, Cambridge, UK, pp. 56–78.

NDA-RWMD (2010) Nuclear decommissioning authority radioactive waste management directorate (NDA-RWMD). Near field evolution status report. *NDA Report Number NDA/RWMD/033, December 2010.*

Nelson, D.L. & Lovett, M.B. (1978) Oxidation state of plutonium in the Irish Sea. *Nature*, **276**, 599–601.

Ringwood, A.E. (1981) *Safe Disposal of High Level Nuclear Reactor Wastes: A New Strategy*. Australian National University Press, Canberra, pp. 42–44.

Savage, D. (1995) *The Scientific and Regulatory Basis for the Geological Disposal of Radioactive Waste*. Wiley, Chichester, UK.

Sherman, G.D. (1952) The genesis and morphology of the alumina-rich laterite clays. In: *Problems of Clay and Laterite Genesis*. American Institute of Mining and Metallurgical Engineering, New York, pp. 154–161.

Strakhov, N.M. (1967) *Principles of Lithogenesis*. Oliver & Boyd, London.

Tits, J., Geipel, G., Mace, N., Eilzer, M. & Wieland, E. (2011) Determination of uranium(VI) sorbed species in calcium silicate hydrate phases: A laser-induced luminescence spectroscopy and batch sorption study. *Journal of Colloid and Interface Science*, **359**, 248–256.

Vreeland, R.H., Rosensweig, W.D. & Powers, D.W. (2000) Isolation of a 250 million-year-old halotolerant bacterium from a primary salt crystal. *Nature*, **407**, 897–900.

West, J.A. & McKinley, I.G. (1984) The geomicrobiology of nuclear waste disposal. *Materials Research Society Symposium Proceedings*, **25**, 487–494.

Mineralogy and cultural heritage conservation

GIACOMO CHIARI

*Getty Conservation Institute, Los Angeles, California, USA,
e-mail: gchiari@getty.edu*

A large number of instruments, as well as new analytical procedures, have been developed recently and applied to archaeometry and conservation science. Here, an effort is made to summarize the most important such developments, with particular reference to the Earth and mineral sciences. Non-invasive or micro-invasive techniques are emphasized, *i.e.* those developed with the goal of reducing, as much as possible, the impact of sampling on art objects. The examples described are taken mostly from the work of the author and his colleagues and address problems in both archaeometry and conservation. Examples of objects in collections, of building materials and of archaeological materials are all considered.

1. Introduction

Rewriting a chapter dealing with science after a ten year interval is quite an interesting challenge. Methodologies employed not that long ago have been supplanted by others which are quite different. In the first edition of the present volume, digital imaging was not mentioned because digital devices, although already on the market, were not yet competitive in resolution with the well established film-based techniques. Ten years ago, non-invasive portable instrumentation was practically non-existent. Chemical photography is now officially dead, as all the major manufacturers have stopped producing photographic paper and slide film. This is actually quite a revolution, as what I refer to as 'analytical imaging' is now a very important methodology used by conservation scientists when analysing art objects. The ability to map the chemical composition of an object using non-invasive portable techniques allows us to minimize the number of samples to be taken, and to select them in a more intelligent manner. Cultural heritage objects, therefore, can be studied and conserved in a much safer way. The interaction between mineralogy and, more generally, between science and cultural heritage, has taken giant steps over the last 50 years, from a time when very little analysis was undertaken on a serious scientific basis. In the early years, most interventions involved 'friendly scientists', who often caused problems rather than solve them (Torraca, 1982). Fortunately, in recent years things have changed, as one can see by consulting the numerous online bibliographic databases dedicated to cultural heritage conservation, such as the Getty website (http://www.getty.edu), the ICCROM (International Centre for Conservation and Restoration of Monuments, www.iccrom.org) online bibliographic database, AATA-online (Art and Archaeological Technical Abstracts), BCIN (Bibliographic Database on Conservation and Information Network),

© Copyright 2013 The J. Paul Getty Trust
DOI: 10.1180/EMU-notes.13.10

CAMEO (Conservation Art Materials Encyclopedia on-line) and many others. Many scientific laboratories are now dedicated to cultural heritage research and top-level scientists, each specializing in a specific domain, can address a whole array of problems. Each conservation laboratory maintains a website where major discoveries are shared with the conservation community, almost in real time. Very powerful search engines facilitate this task. The analytical techniques applied by conservation scientists are all those used in Earth sciences, chemistry, physics, biology and engineering. It seems that almost every day, a new method is proposed to investigate previously unsolved problems. Instrumentation is improving constantly, allowing one to carry out analyses that only a few years ago would have been impossible. The science of cultural heritage is among the fields that have profited most from the 'scientific explosion' of recent years. See, for example, the book by Artioli (2010) for a comprehensive description of scientific methods applied to this field or Rapp (2009) more specifically on archaeomineralogy.

The role of the scientist involved in any marginal field with respect to the mainstream of his or her discipline is complex. In order to take advantage of this position fully, one might make a list of the problems which still are unsolved, and then make a list of the 'dream' instruments which could, in theory, solve those problems. By cross-checking the lists, it may be possible to find someone who has produced the required instrumentation. It also sometimes happens that, in the literature, someone has described a new insight which makes a problem easier to tackle. Being on the borderline is often a good position; one where, for example, a very simple analysis can lead to important results provided that one recognizes its implications. On the one hand, the technology available has become much better; on the other hand, the ability of conservation scientists to deal with cultural heritage problems has also been refined, thanks to day-to-day practice and more frequent interaction with conservators, archaeologists and museum curators. This interaction is essential in order to have the background needed to understand the problems. In fact, to analyse a sample without knowing its historical context can easily produce false or misleading results.

As part of a team, each professional needs to contribute his expertise. However, it is often up to the scientist to clarify the issues that emerge by listening to the stories that each material has to tell, and then asking the proper questions needed to interpret these stories correctly. In fact, the materials that we are dealing with are rarely pristine. Often, many centuries have passed since they were fashioned, and this leaves its mark on the chemical and physical characteristics of the objects. This makes it more difficult to compare their characteristics with modern standard materials and with mock-ups. A profound knowledge of the field of interest is essential, and this necessitates constant interaction with conservators and art historians. Unfortunately, although in recent years the situation has greatly improved, there are still some difficulties in trying to maintain a dialogue with all the professionals involved. Scientific equations and graphs are not always welcomed by the professionals who do not have a scientific background. Therefore, it is the duty of the scientist to explain the results in as simple a manner as possible, without diminishing scientific rigour. This is not an easy task but it is essential to improving knowledge in the field.

1.1. The role of the researcher

It is not easy to define the role of the scientist in cultural heritage. In fact there are several roles which are strongly interconnected. First of all, there are two different approaches which here are called 'lab research' and 'field work'. In the 'lab research' approach, a scientist normally takes the techniques which are available, and looks for a problem that can be solved using those techniques, prepares representative samples as simplified simulations of the real objects, and almost always gets good results. If the results are not satisfactory, they are simply not published. Obviously, when well planned and executed, this type of work leads to basic knowledge that can subsequently be applied in the real world.

For 'field work', the situation is reversed. A scientist visits a restoration project and speaks with restorers, archaeologists, art historians, architects and the director overseeing the work. Here, the scientist may be besieged by problems which often, in order to be solved, need techniques that are not available in the scientist's own laboratory. If the problems are not solved, the rest of the team who were counting on 'science' to solve problems are disappointed. The techniques best suited to do the job are often out of immediate reach. It is, therefore, necessary to establish a network of alliances with several different laboratories in order to be able to deal with all or most requests. Typically, one person can deal efficiently with either Earth Science- or Life Science-related problems (and with inorganic or organic systems) but rarely with both. Even in our own area, a mineralogist cannot always provide the expertise of a petrographer and *vice versa*. However, if there is only one scientist available in the field, and if the most interesting problem concerns an area removed from his or her own expertise, he or she needs at least to be able to take samples in a proper way and pass them to a colleague who would then be able to analyse them with the appropriate instrumentation. It is also essential that the colleague be willing to do the analyses in a relatively short time. Taking samples, however, has become more and more restricted. The good news, however, is that a large number of non-invasive portable instruments have been developed and are now available for use. Most importantly, hand-held X-ray fluorescence (XRF, now used routinely in almost all laboratories) and also infrared (IR), Raman, nuclear magnetic resonance (NMR) and even X-ray diffraction and X-ray fluorescence combined (XRD/XRF) combined) instruments are available as we shall see below (see Wogelius and Vaughan, 2013, this volume, for basic information about these techniques).

The ease with which non-invasive portable instruments can produce results without touching the object has convinced most people responsible for the safeguard of such objects that taking samples is not really necessary. Furthermore, most state laws prohibit the exporting of artistic or archaeological items without making a clear distinction between an important piece of art and a few milligrams of material taken with the goal of conserving much more important and valuable objects. If the laboratory where the analyses are going to be carried out is in a different country, as often happens, a very long delay may occur before getting essential results. It is hoped that this problem may be soon solved by properly classifying scientific samples for study in a way that distinguishes them from art and archeological items.

1.2. Goal of the study

There are many possible goals when studying cultural heritage materials. Archaeometry and technical art history encompass all analytical research that sheds light on the material characterization of an object. The questions normally asked are: What is the object made of? What were the techniques used to produce it? What is its provenance? How old is the object?

Other goals might be to study how the component materials have changed over time, the state of conservation of the object, and any processes acting upon it that might cause deterioration. The goals may often include preserving it from further damage and/or restoring it to a better condition. Processes causing deterioration that took place in the past may not be active now and, therefore, would not be a cause for concern. By contrast, a deterioration process that has just started may destroy the object in a short time if not arrested immediately. To determine the rate at which any deterioration processes are active (if at all) is obviously of paramount importance.

In the literature, the great majority of papers deal with the characterization of materials while very few are dedicated to conservation issues. There is a gap between the information obtained by analysis and its practical application. In spite of this, it is quite obvious that these two general activities are almost always strongly related. For example, while a scientist is on scaffolding erected in order to study the deterioration of a mural painting, no one would refuse a request to analyse some pigments just because that is 'archaeometry' rather than 'conservation'. Similarly, while collecting pigments to ascertain the painting technique of a given civilization, no one would refuse permission to perform the analysis of an efflorescence of soluble salts because that could be regarded as 'conservation'. Good conservators, whilst doing their job of cleaning, preserving and conserving, always take advantage of being in close contact with a normally inaccessible work of art by studying it in depth. This may simply involve close examination, but the use of modern analytical techniques and the expertise of a scientist are obviously of great help.

Sometimes goals coincide. For example, mortars and plasters can be compared to establish if two construction phases of a building are coeval. This is certainly archaeometry, because the aim is to date the period of construction and to give information on construction techniques and the materials used. It is also relevant for any restoration of the building, as knowing the historical sequence of building activities is essential for the director of the work who has to keep to one or other historic period, or keep them all. The proper assessment of the chemical and physical properties of the mortar used in a structure (its composition, the amount of aggregate, the porosity, the tensile and compressive strength *etc.*) is essential when choosing the proper grouting or repointing mortar which must have properties similar to the original. Too many mistakes have been made in the past by thinking that an intervention with a stronger material would be 'better'. We now know that the best strategy is to add a weaker sacrificial layer that will deteriorate instead of the original. We also need to guarantee, if not reversibility of the treatment, at least re-treatability of the object. One example of this occurred when analysing the pigments on mortars at the decorated sculptures of Templo Mayor in

Mexico City. A substantial amount or gypsum was found mixed together with calcite. In Europe, it is not unusual to find gypsum in a mortar. Using XRD (see Wogelius and Vaughan, Chapter 2, this volume), determining the presence of gypsum is straightforward. Either the gypsum is the result of sulfuric acid attack (in which case there should be concern about the conservation of the object), or the gypsum has been added to the lime (intentionally or not) at some point. The plaster being from Mexico, this finding could be 'revolutionary', as the use of gypsum by ancient civilizations in Mesoamerica has never been documented. This is a good example of how important it is to establish contact with local experts in order to know the context of the research. The analysis of more mortar samples was needed; if the quantity of gypsum is significant and more or less constant with respect to the calcite, one can exclude a sulfuric acid attack, but there is still a question of whether the gypsum was added on purpose or not. A literature search found reports of natural outcrops of gypsum in the surroundings of the archaeological ruins (Reyna Jenaro, 1956). These gypsum deposits are always associated with calcite in beds of various thickness. On the basis of our knowledge of lime and gypsum technologies, one can establish whether or not the addition of gypsum was intentional. If one wants to use gypsum as a binder, it must not be heated above 480°C, otherwise $CaSO_4$ (α-anhydrite) is transformed into the β form which is insoluble, does not react with water and, therefore, cannot be used to prepare plaster (Turco, 1996). On the other hand, limestone has to be heated to at least 800°C in order to become CaO, and then be hydrated into slaked lime to be used as a plaster binder. Therefore, it is impossible that the gypsum found with calcite in Mexican plaster is the result of accidentally taking a rock containing both calcium carbonate and calcium sulfate in order to make lime, because the two preparation technologies are incompatible. The only possible explanation is that the gypsum, and possibly some of the calcite, was added to the lime as an aggregate. This simple example demonstrates that interdisciplinary investigation is essential in order to obtain good results.

Another example involving discrimination between the same mineral used as a binder and as an aggregate involves the large number of Roman mortars that contain pozzolanic material (red and black 'pozzolana') indistinguishable by XRD (see Chiari *et al.*, 1992, 1993, 1996a, 1996b). In many instances, the mortar contains calcite (or crushed limestone) as an aggregate, making it almost impossible to obtain a quantitative analysis and provide the matrix to aggregate ratio, an important datum for the characterization of the mortar. These problems can be solved by using thin sections (see *e.g.* Reedy, 1994) with a digital imaging program (Ruffini *et al.*, 1999) which can be used to carry out modal analysis on thin sections of mortars, and obtain the total area of parts of the sections having the same colour by counting the pixels comprised within an RGB (red, green and blue) colour range. By doing this, one can measure areas of different chemical composition on the basis of their colour. Modal analysis is certainly not a new technique. Petrographers have sometimes discarded it as insufficiently accurate, especially when dealing with fine-grained minerals. This may be because 'thin sections' 30 μm thick, observed in transmitted light, have always been used. By so doing, an opaque cubic grain with a side 3 μm long shows a cross section of 9 μm^2 that is

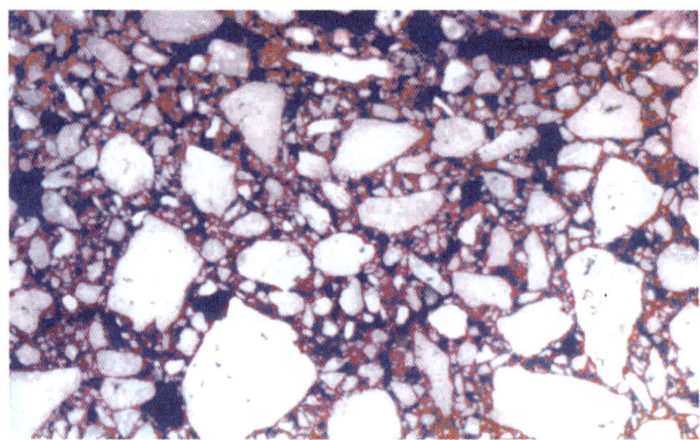

Fig. 1. Lime mortar with travertine aggregate, embedded in blue resin and stained with Alizarin Red S. The total blue area gives the porosity; the white gives the quantities of large crystals of calcite added as aggregate, and the red represents the micritic calcite as the binder.

multiplied by the entire thickness of 30 μm, thus giving a volume of 180 μm^3 instead of the real 27 μm^3 (a result over six times greater). The smaller the particle, the larger is the error. In the digital imaging approach, the cross sections are examined in reflected light, thus eliminating this systematic error. In order to become routine in mortar analysis, this procedure has to be quick, economical and reliable. Therefore, the computer revolution and the availability of cheap, high-resolution images have been essential. Porosity can also be measured in cases where simple alternatives such as measuring total water impregnation cannot be employed (*e.g.* for mud brick specimens, as the mud would dissolve completely in water within a few minutes). Here, the use of a strongly coloured resin for the impregnation of the specimen before cutting the sections can facilitate the measurement. More than one section needs to be analysed to improve the statistics, and calibration undertaken by comparing the results with those obtained by mercury porosimetry.

In mortar analysis, one of the most important variables is the binder to aggregate ratio. For lime mortars containing marble, limestone or travertine as an aggregate, it is very difficult to establish which calcite is part of the binder and which is part of the aggregate. Digital image analysis can solve this problem (Casadio *et al.*, 2005). A blue-stained resin is allowed to embed into material; it penetrates into the open porosity and is easily measured. The cross sections can then be stained with Alizarin Red S; this imparts a light red colour to the micritic calcite derived from the lime and leaves the larger crystals of the aggregate the original white colour (Figure 1).

2. Approaches, methods and applications

2.1. The sampling problem

One significant difference between cultural-heritage investigations and other fields of mineralogical research is that taking samples is often not simple. The amount of sample that one is allowed to take is often of the order of a few milligrams. When performing XRD, it is possible to obtain reasonably good data using 2–3 mg of substance, but special equipment is needed. Using other techniques such as IR or Raman spectroscopy, the amount required is even smaller. One can recover precious samples for other types of analyses after collecting the data. The zero-background plates (slices of very well formed quartz crystals cut in a way that no crystallographic planes satisfy the Bragg diffraction condition for the characteristic radiation of the tube employed) together with a monochromator on the secondary beam, are the tools that can really improve the quality of the XRD patterns when the amount of substance available is minimal. Although the monochromator reduces the intensity of the diffraction, the signal to noise ratio is greatly improved.

Modern technology helps to reduce the need for a large sample by producing more sophisticated, ever more sensitive, instrumentation. Using a Raman microscope, one can identify the composition of extremely small grains of pigment without sample preparation. There is, however, a serious limitation to the reduction of the amount of sample used, which does not depend upon the instrument. A sample needs to be representative of the object that one wants to investigate. There is a minimum volume (or mass) which is needed, and which varies not only from one object to another but may vary for the same object, depending on the particular goal of the analysis. Geologists long ago developed procedures to determine the minimum volume required, for homogenization of the sample, and use of the right amount required by the instrument. A simple way of estimating the minimum volume required of a heterogeneous material is roughly to determine the particle-size distribution, and take a volume about ten times larger than the largest particle present. This should limit the errors to $\sim 10\%$ or less. What is stated above is valid if what we wish to investigate is the overall composition of the object. Typical examples could be the pigments, the ground, or the binders in a painting, the alteration products on the surface of a stone or metal or glass object, the soluble salts crystallized on such surfaces, or the white marble of a statue. But there are other properties for which the sample required is the whole object. For example, it may be very important to know if a block of stone, perhaps a sculpture, has internal cracks. Only by studying the object as a whole can one answer this question.

One should assess what it is important to know about the object, and be aware of what its normal condition is, so as to avoid taking samples of irrelevant anomalies. There is always a temptation to sample parts which capture one's attention because they are different from the rest. If we sample those parts without clearly recording the fact that they are anomalous details, their role in the overall evaluation may be overestimated. The criteria for sampling also depend upon the goals of the investigation. If we are looking for clues concerning a deterioration process which has begun recently, a very small and apparently insignificant detail may be important as the first sign of incipient deterioration. In this case the important thing is to document all the different variations in the deteriorated area, because only when seen together may they make sense and allow us to understand the degradation mechanism.

Another problem which arises from the need to collect a small sample is the difficulty of manipulating such a small amount of substance. In the field it is sometimes almost impossible to collect, in a piece of paper, a small amount of powder scraped from a wall, *e.g.* because of exposure to wind or the need to climb a ladder to collect the sample. In order to overcome such difficulties, an instrument consisting of an aspirating device (Figure 2) has been designed which collects the powder removed from the art object directly onto a paper filter at the base of a vial. The vial is interchangeable; it can be sealed with parafilm and reopened in the laboratory (Chiari, 1999). Obviously, when the amount of substance is very small, the possibilities of contamination are greater, and particular care must be taken in cleaning all the tools used for sample collection.

Besides the small amount of sample, it often happens that the substance of interest is present only as a small percentage in the sample. A typical example would be the painted layer of a mural painting. The thickness of a painted layer is of the order of 40–90 μm. Sampling it with a scalpel, even using amplifying lenses, usually leads to the removal of the supporting layers together with the pigment we wish to analyse. A blue paint from the facade of the oldest known colonial house of all the Americas (1519–1530 A.D.), the Casa Velasquez in Santiago de Cuba, was analysed to determine whether or not the pigment was Maya Blue. The 'as collected' XRD (Figure 3) showed abundant calcite, some gypsum and traces of quartz. Unlike the sample, none of these minerals is blue; therefore, something was missing. Assuming that the blue colour is indeed Maya Blue, which is a complex material involving the clay mineral palygorskite and the organic dye indigo, a very faint peak at $\sim$8.5°2θ is seen, corresponding to the main peak of this clay. A weak acid attack (using dilute hydrochloric acid) can be used to eliminate calcite and gypsum. The pattern obtained after removal of these minerals is shown in Figure 4. Palygorskite is indeed present, together with a small quantity of albite, which had not been detected in the previous pattern. The overall quality of the pattern is much worse, however, as the quantity of sample is much less. The

Fig. 2. A small sample is removed from a Roman mural painting at Herculaneum to study pigment composition. A modified aspirator facilitates the collection of the powder, even in the presence of wind.

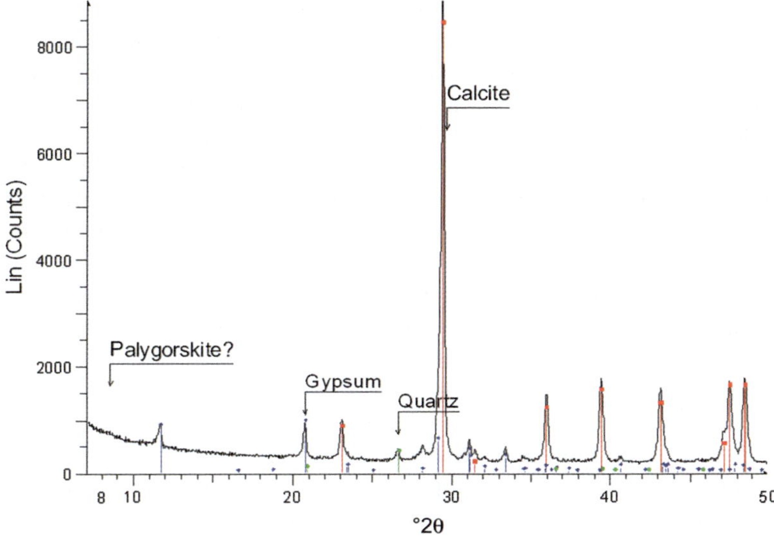

Fig. 3. XRD pattern of a blue pigment from a Cuban mural painting. Calcite, gypsum and traces of quartz are easily detectable. As none of these minerals is blue, it is logical to suspect that a small amount of blue pigment is present. If palygorskite was detected, the pigment would be Maya Blue.

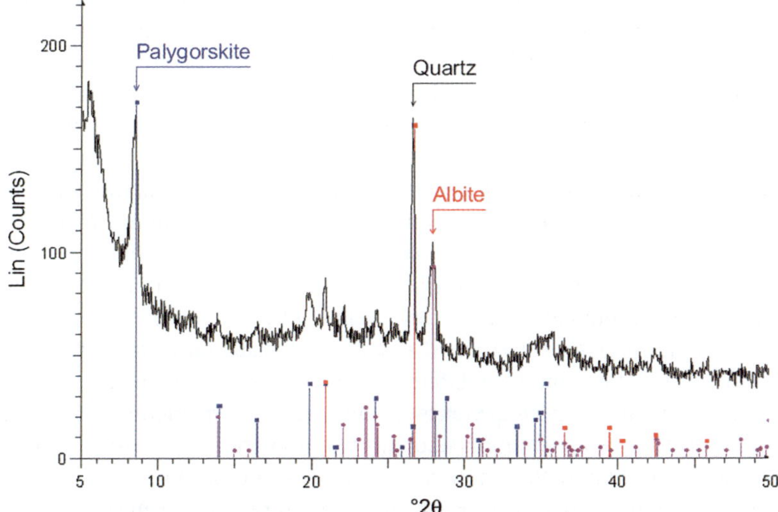

Fig. 4. XRD pattern of the same sample as in Fig. 3 but after acidic attack to remove the predominant calcite and gypsum. Palygorskite, the clay fraction of Maya Blue, is now clearly visible, together with albite (previously not detected) and quartz.

identification of the pigment is straightforward. It should be emphasized that the time of exposure was not extended to the point of obtaining a smooth background, as the presence of the three above-mentioned substances was clearly proven. In fact, the goal was to ascertain if palygorskite was present or not, and it was achieved even with a short exposure time.

As often happens, the result of the analysis solves one problem (to establish if Maya Blue was used in Santiago de Cuba as, indeed, it was) but poses a series of other problems, as the paint cannot be original (contemporary with the construction of the house). Although Maya Blue was known in Mexico at the time of the construction of the house (it was probably produced around 800 AD), at that time Mexico had not been conquered by the Spaniards (the Cortez expeditions date from 1529). So we do not know when the blue paint was applied to the wall.

2.2. Qualitative *vs.* quantitative analysis

In most cases, qualitative analysis is sufficient to solve a problem. The identification of the chemical elements or mineralogical phases in the object gives enough information to draw sound conclusions, especially in a conservation context. Often a rough estimate of the quantities involved is useful, but this can be done semi-quantitatively, *e.g.* by estimating the heights of the most intense diffraction peaks in XRD, or the absorption lines in IR, UV, Raman and other spectroscopies. Hand-held XRF instruments are very common nowadays and produce data which can be easily mistaken for quantitative analyses. This may present a problem, as a non-scientist may not be aware of the fact that, in order to obtain good quantitative data by XRF, one needs to pre-treat the sample to obtain a smooth surface, and then compare the values obtained to a series of standards. The process is long and delicate, and producing accurate and reproducible results is not trivial. A series of standard samples was sent recently to several laboratories to compare the results of their handheld XRF instruments and data treatment. This was done for canvas paints (Namowicz *et al.*, 2008) and for copper alloys (Heginbotham *et al.*, 2011). Investigations of this kind are important in avoiding incorrect interpretations of the data. In spite of the fact that newly available instruments are very user-friendly, interpretation of the results needs a person who is well trained in the relevant field.

The reason that quantitative analysis is seldom needed is, in part, due to the nature of the sample. When an object to be analysed is homogeneous, as in the case of pure pigments, glass, obsidian or most metals, it is sufficient just to identify the nature of the substance. When, on the other hand, the sample is not homogeneous, there is little purpose in carefully determining the ratios of the various components. In fact, this quantitative ratio is not the result of a conscious decision made by the manufacturer of the object. So, when studying a mortar, it is useful to know approximately the ratio between lime and aggregate, but it would be a waste of time to measure it precisely, as it can vary from one point to another in the same wall, or from one day of work to another for the same mason. Approximate methods to determine these values are sufficient, like the modal analysis mentioned above.

On the other hand, the facility to carry out quantitative analyses plays a role in the decision to carry out such analysis. X-ray diffraction full-profile fitting techniques are now routine, especially if the samples analysed are all of the same type (Altomare *et al.*, 1999). One can carry out a Rietveld refinement of many phases simultaneously in a relatively short time, as the crystallographic and structural data can be retrieved from the Inorganic Crystal Structure Database (ICSD, 2011), or from a simple internet search. The use of complementary techniques is, of course, always advisable. For example, thermogravimetry gives the precise weight of material that is released on progressively heating a sample. If the emitted gases are directed into a mass spectrometer, one can also obtain an indication of the chemical compositions of those gases. With the availability of new methods, it becomes important to determine, case by case, if a full quantitative analysis is really needed.

The situation is very different when the goal is to study the provenance of a given material, as in archaeometry studies of metals, white marble, jade, obsidian, glasses, bones or pottery (see Freestone, 1995; Quinn, 2008; Rapp, 2009; Vaughan, 1995). In this case, very precise quantitative analysis of all the chemical elements and mineralogical phases present, especially in trace amounts, and of the isotopic ratios of many of them, is essential for locating the source of the material. The task may be a difficult one. In the case of white marble, for example, XRD might indicate only the presence of pure calcite in almost all cases, with the exception of a few dolomitic marbles. Petrography may reveal the different microstructures, including grain size, shape of mineral grains, preferred orientation and so on. A chemical analysis of the elements, especially those present in trace amounts, may add more information. But only by treating all of the data in a statistical way is it possible to distinguish one marble from another (Gorgoni *et al.*, 1992; Lazzarini *et al.*, 1997; Gorgoni *et al.*, 1998; Zöldföldi & Satir, 2003). The same is true for other natural or man-made materials. In recent years the use of laser ablation inductively coupled plasma mass spectrometry (LA ICP-MS) techniques has become more common. One obtains a quantitative analysis of all elements in the same run, including both major and trace elements. Combined with statistical treatment of the data, these analyses can produce very interesting results. The same is true for the determination of isotopic ratios, especially for carbon, oxygen and lead. An interesting application of lead isotope ratios was the determination of the provenance of red lead covering a Fayum mummy (Herakleides, J.P.Getty Museum 91.AP.6). The lead was shown to have come from the copper- and silver-producing Río Tinto mine in Spain (Walton and Trentelman, 2009). Here, the litharge was a by-product of the smelting and cupellation of silver and was further treated to produce the red minium pigment, which was in strong demand in Egypt for painting the red shroud mummies.

^{14}C dating, well known to the lay public, is also a very powerful tool for organic materials. Although there is a danger of errors due to contamination, the overall impact of ^{14}C dating on archaeology has been outstanding, especially after results are corrected using data from dendrochronology. Dating is obviously of paramount importance to archaeologists, and so is the provenance of the material, as it helps to establish if a given object was produced locally or imported (Maggetti, 1981; Barra Bagnasco *et al.*, 2001; Walton and Trentelman, 2009). For an art historian, the provenance of the material can often help in the attribution or validation of a work of art. For this reason, it is essential that objects are found *in situ*. A looted specimen loses most of its scientific value.

2.3. Non-invasive and invasive methods

In recent years a large number of non-invasive, portable instruments have been produced, many of them in a cooperative effort between a conservation scientist and a commercial company. Often these instruments are adaptations of other devices developed for fields of greater economic importance and, therefore, more attractive to industrialists. This technology transfer is an important role for conservation scientists. A good example is the adaption of an XRD/XRF instrument developed by the Jet Propulsion Laboratory for a NASA mission to Mars. Not having problems of sample collection, this instrument, and its counterpart called TERRA, made for geologists (Blake, 2010), works in transmission mode. For cultural-heritage applications the geometry was changed to reflection mode, as one cannot put a detector behind a mural painting. The advent of miniaturized, Peltier cooled X-ray tubes, and of affordable miniature energy dispersive detectors was necessary for the development of such instruments.

An advantage of such instruments is that they can be taken to the object rather than the reverse. In many cases (mural paintings, rock art, *etc.*) the object cannot be moved at all. But, in general, even for easel paintings, illuminated manuscripts, polychrome statues *etc.*, it is preferable to keep the object in its original location, avoiding environmental changes and possible accidents. Furthermore, if an analysis can be carried out inside a museum without even taking the object out of its case, it is much easier to obtain permission to perform it. The results obtained by such analyses are often conclusive. Sometimes, however, the quality of the data obtained in this way is not sufficient to reach a conclusion and it is necessary to take samples or move the object to larger, non-portable instruments.

It is important to establish what we want to preserve about an object, and we need to answer this question before we can establish the criteria for justifying the intrusiveness of an investigation. For many years the external appearance of the object was the only value of importance. With new methods of analysis, other features became important. For example, the chemical composition of an object (together with structure, texture, tool marks, *etc.*) can reveal the manufacturer's technique for producing a work of art. Detailed analysis of a material, such as in determining the trace-element composition, or isotopic ratios, may help to establish the provenance of an artifact. The identification of a pigment may allow one to define or to exclude historical periods in determining the time of its production or to indicate a subsequent intervention. The presence of a given mineral may be indicative of a firing temperature in the production of pottery. The carbon isotope ratio or the thickness of tree rings may permit absolute dating of the object. So, any type of analysis or intervention which would modify one or more of these parameters, even if it leaves the object apparently unchanged, should be considered invasive because it would remove important information.

Sometimes, the information we gather from an object is not so obvious. For example, most people nowadays believe that a fresco should not be removed from its original setting unless it is in real danger (although, not long ago, it was common practice to do so, because the only value attached to the painting was aesthetic). However, a recently developed technique of dating mural paintings, based on remanent magnetization of hematite, requires the spatial orientation of the painting to be preserved and unchanged from the moment of its execution, so an important piece of information would be lost if it were to be moved (Chiari and Lanza, 1997, 1999; Zanella *et al.*, 2000).

Nowadays everyone agrees that, when dealing with art objects, it is essential to obtain information without damaging the object. However, we cannot gather information without using some sort of 'probe' which interacts with the object, is modified in some way by the object's characteristics, and comes back to us after the interaction so that we can measure these modifications. In the same way that our probe is modified by the interaction, so is the object. Therefore, strictly speaking, there is no such thing as 'non-invasive analysis'. Even the simple act of looking at an object, observing it without touching it, can be considered, in principle, an 'invasive analysis'. The light needed to see an object is absorbed by its surface (selectively absorbed, so that we see colours) and, in so doing, affects the energy levels of some electrons in the atoms present, or energy states of some molecules, changing their vibration modes or even modifying some chemical bonds. In general, these changes are reversible and when the light source is removed, the object goes back to its initial state. Although one may think that regarding as 'invasive' the mere act of looking at an object is excessive, there is now much more attention given to the fading of colours due to light. Lighting systems employed in museums are carefully modified to reduce the impact of illumination on artifacts. In recent years, the science of measuring the impact of light on light-sensitive material has come of age. The micro-fadeometers now found in many laboratories measure the damage produced on a very small spot by a light up to 50 times more intense than the sun. At the same time, these instruments collect the reflected light which is measured in a spectrophotometer and produces a spectrum, in real time, on the computer screen. A single numeric value that characterizes the magnitude of the total colour is derived from repeated spectral measurements and a plot of ΔE (colour change in CIE L*a*b* colour space) *vs.* time is produced. As soon as ΔE reaches a value of 3 (the threshold for detecting any colour difference by human eye), the measurement is stopped. No observable damage is produced by these measurements but all data needed to specify a safe time of exposure to a given illumination are collected. Old master drawings can now be seen at optimal illumination with a precisely calculated light exposure (Whitmore, 1999; Druzik, 2010). This is just one of the applications of 'preventive conservation'.

As we have seen, a non-invasive analysis (sometimes called non-destructive) is normally defined as one for which it is not necessary to remove a sample from the object. When a sample is taken, we can have various degrees of invasiveness. Some analyses are such that, at the end of the procedure, the sample no longer exists. Typical examples of this type of analysis are: Gas Chromatography-Mass Spectrometry (CG-MS; Barry and Grob, 2004); Atomic Absorption Spectrometry (AAS; Ebdon *et al.*, 1998); carbon dating (Libby, 1952); thermal analysis (Gabott, 2007) and ICP-MS (Tykot & Young, 1996; Boyd *et al.*, 2008). In other cases, some characteristics of the sample are lost, but not all of them. For example, powder XRD requires grinding of the sample in order to reduce it to powder. Any information on shape, morphology, stratigraphy or texture is lost, but the powder can be used for other bulk analyses. The sequence in which the analyses are carried out is very important, and should be planned carefully in order to gain the maximum amount of information with the minimum number of samples. In some cases, the procedure affects only the property on which the method is based, leaving everything else almost unchanged. This is the case for thermolumines-cence dating which, during the measurement, resets the age of the object to zero (Martini and Sibilia, 2001; Martini *et al.*, 2004). It should be emphasized that a technique which does not require the removal of a sample and, therefore, is normally considered non-invasive, may nevertheless be destructive because of the damage induced by, for example, electromagnetic waves. Lasers, X-rays, UV light or even visible light can be destructive to sensitive materials. Another important factor when performing non-invasive analyses is that the risk of possible accidents is always present.

2.4. Microscopic invasive analyses

We define an analysis as 'microscopic' when the amount of sample to be taken is very small. The problem is to define what we mean by 'very'. A few milligrams may seem a very small amount of material, but this is not the case if we remove it from a famous diamond or from an illuminated manuscript. It is of paramount importance that the sample be removed by an expert, a conservator, possibly aided by the scientist directly involved in the analysis, as the sampling procedure is the first step in the analysis itself. Observation of the general conditions of the object and the choice of representative material are decisions which often require a deep understanding of both the object (con-servator) and the principles and procedures involved in the analysis (scientist). One should hesitate to study samples taken by other people unless the documentation accom-panying them shows that every sampling step was properly performed by a competent person.

When dealing with micro-samples, one should also be very careful not to contaminate the material, as a modest impurity may result in a large percent error. Maximum cleanli-ness is required when handling the sample. In the case of instrumental analyses, the instruments must be calibrated to the optimum in order to reduce errors and noise. That micro-samples are representative of the whole is also of paramount importance.

2.5. Invasive analyses

An analysis can be called 'invasive' when the mark left on the object by sampling is clearly visible or detectable in other ways. Nowadays these types of analyses tend to be less frequent, but sometimes, some 'destruction' cannot be avoided. In an archaeolo-gical excavation, it is common practice to remove the top layer in order to investigate what is underneath. To assess the static properties of a pillar, it may be necessary to obtain large cores in an obviously destructive way. On the other hand, the recognition of the invasiveness of an analysis does not necessarily imply that it should be condemned. There are situations in which even invasive analyses find legitimate use.

2.6. The risk factor

Natural catastrophic events such as volcanic eruptions (Pompeii), earthquakes (San Francisco), hurricanes and floods (Katrina), or man-made destruction on a large scale through wars or vandalism (Bamiyan Buddhas) are beyond our control. Other risks are manageable. A lot of damage to objects is due to improper handling, mounting and transporting for exhibitions and carelessness that can be avoided. Fortunately, much progress has been made in packaging and mounting art for transport, but moving it remains a big hazard. The risk factor associated with a specific event depends not just on the probability of that event but also upon the amount of damage produced by the event. Insurance companies evaluate the risk factor using mathematical theories. The simplest such equation is:

$$\text{Risk Factor} = (\text{Probability of the event}) \times (\text{Damage produced})$$

Even in the case of very low probability, if the consequences of the event are catastrophic, the risk factor may be high. We need to consider this because often the damage to a work of art is the result of an accident. Consider the recording of Michelangelo's sculptures by laser scanning which generated a three-dimensional computerized record of every detail. The impact of the radiation used on the marble could have been problematic. (It was known that X-ray exposure can change the colour of white marble). Trials on small portions of the marble had to be done first. If no change occurred, the technique was considered to be non-invasive. The procedure also involved using of a lot of scaffolding for the positioning of cameras. Intense preparation and training of the operators on one-to-one scale gypsum replicas of the statues was needed. Even then, the possibility of accidental damage could not be excluded. It is up to the people who carry out a procedure to minimize the risk of accidents, but an evaluation of the risks of various types of analyses should be considered by the decision makers before any investigation is undertaken.

The amount of sample that might reasonably be removed from various types of objects has to be judged for every case on the basis of artistic and historic importance, as well as the integrity of the object. Suppose that we are studying a large wall, part of an archaeological excavation, which is already severely damaged, and of which a large percentage of material has already been lost. We want to analyse the mortar and the stone in order to determine the best conservation treatment. In this case, the removal of large pieces of material may be acceptable. Suppose again that we are studying a bronze statue, and we want to drill a large hole in order to characterize the alloy and to obtain information on its provenance. This is obviously not acceptable. We also have to consider the advances that are now being made in applying new techniques to make analytical interventions less invasive. So, even if today we do not have at our disposal a non-invasive method, such a method may be invented in the near future. We are compelled to wait to satisfy our curiosity and, for now, leave the object intact. This certainly applies to archaeometry. It may be more difficult to make a decision when the information we are looking for is vital to establishing the correct way to conserve an object which is in serious danger of destruction.

We often face situations where there are severe time constraints on the decisions to be made. During the restoration of Michelangelo's *Last Judgment*, the problem of keeping in place or removing tempera repaintings made over the centuries, in order to cover some of the nudity, was important and could not be postponed. Non-invasive analyses were attempted using a very early portable X-ray fluorescence instrument (Sciuti & Gabrielli, 1995) to see if the type of pigment employed could be used to identify the various repaintings (having decided to keep the oldest and to remove the most recent ones). The answers, in some cases, were ambiguous as, given the penetration of the beam, the technique not only analysed the superficial tempera repainting but also Michelangelo's original layer and the mortar underneath. A micro-invasive XRD analysis carried out on a few milligrams of tempera painting, which were very carefully removed, provided the necessary information (Chiari, 1999). Daniele da Volterra in 1564 used as preparation 'bianco di Sangiovanni' (*i.e.* a partially carbonated lime) and ochre as yellow pigment. Domenico Carnevali in 1568 used orpiment for the yellow. The painters of the XVIII century used lead white for preparation and 'Naples yellow' (*i.e.* lead antimonate) as yellow pigment. Therefore, it was straightforward to distinguish between the three groups of interventions. It is obvious that this information had great value insofar as it was obtained before the decision had to be made to remove some of the tempera paintings. In this case, it was not possible to wait for a non-invasive method to be invented. It is interesting to note that such an instrument now exists in the form of a portable XRD/XRF device called DUETTO, which could have solved the problem without needing to take samples at all (Chiari, 2008; Sarrazin *et al.*, 2008, see Figure 5). Other similar instruments have being produced (Uda *et al.*, 2000, 2007; Gianoncelli *et al.*, 2008; Pappalardo *et al.*, 2008; Blake, 2010; Berti *et al.*, 2011). Nakay and Abe (2012) have produced a thorough description of all these instruments, comparing their advantages and disadvantages.

Another example where the application of a non-invasive, portable XRD/XRF

Fig. 5. Non-invasive, portable XRD/XRF instrument (DUETTO), analyzing pigments within the Tutankhamen Tomb.

device has been of great utility is in the archaeometric study of Neolithic jade axes. Fifteen years ago Göbel Mirrors were used to study Neolithic objects (mainly axes) made from greenstone (Chiari *et al.*, 1996c). The goal was to characterize the rocks used to make the axes with the hope of finding their primary outcrops within the Western Alps area, and defining their provenance. A preliminary screening separating jade artifacts from those made of other less dense lithotypes was carried out non-invasively (Ricq de Bouard & Compagnoni, 1991) by using density measurements. Unfortunately, this only gives limited information. By using XRD, however, the unit-cell parameters of the pyroxenes within the jadeite–diopside series can be determined. The values of the cell parameters for jadeite and omphacite are sufficiently different to discriminate between them. Statistical analysis was used to determine the chemical composition of the mineral phases, using X-ray data (the interlayer spacings of three selected reflections). This necessitated very precise and accurate d spacing values, which could be obtained by using Göbel Mirrors and the 'detector scan' technique for data acquisition on whole axes without the need for sampling. The sample holder was removed from the diffractometer to make room for a bulky object. This approach applies only to polycrystalline samples (see Figure 6).

Microprobe data from the literature were plotted on the diopside–jadeite–aegirine compositional triangle, together with the values of the d spacings of three selected reflections: (002), ($\bar{3}$10) and (221). Lines with the same d spacing could then be added (see Figure 7). Data obtained using this procedure were compared with SEM-EDS results and showed satisfactory agreement. Today, the same data can be acquired using the non-invasive portable XRD/XRF instrument (DUETTO) mentioned above. This instrument uses a CCD as detector and records a 2D portion of the diffracted beams. The XRD pattern is obtained as the integration of an arc of the diffraction cones, not just along the equatorial line as for a conventional diffractometer. Therefore

Fig. 6. Measurement of an XRD pattern directly from a large Neolithic axe of 'greenstone'. This ceremonial axe, 34 cm long, is particularly well preserved and cannot be sampled for archaeometric purposes. The use of Göbel Mirrors permitted collection of good XRD data in a non-invasive way.

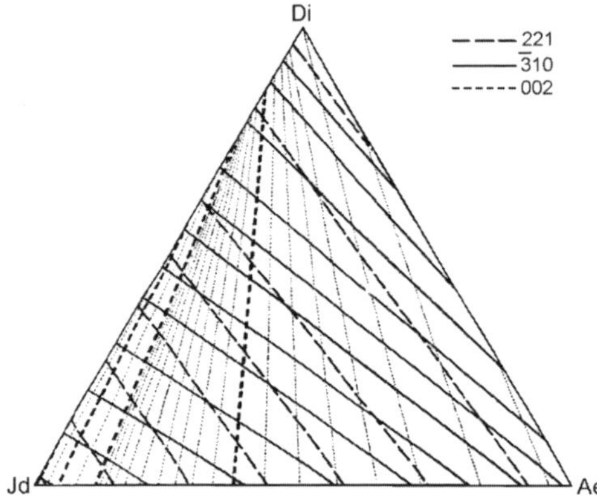

Fig. 7. Jadeite(Jd)–Diopside(Di)–Aegerine(Ac) compositional triangle with the lines of equal *d* spacing for three reflections. By measuring the *d* spacing of the three reflections in a non-invasive way, and using this chart, one can determine the chemical composition (*i.e.* the amount of Ca and Na in a given site of the pyroxene structure) to better than ±8%.

the intensity of the various diffraction peaks is, in principle, more accurate, particularly when the grain size of the material is large. In this case, the formation of dotted lines cannot be avoided as the sample is not ground, but it is useful to know that a given mineral in the sample is made of large grains. This may help in the identification and shed some light on the manufacturing technology. This is just one of many possible examples of the application of XRD with Göbel Mirrors (and now with portable XRD/XRF) in the study of cultural heritage (see Chiari *et al.*, 1996d, 1998).

2.7. 'Figure of merit' for evaluating the expediency of carrying out an analysis

The factors which influence the decision to carry out an analysis are often not all controlled by the decision makers. The optimum analytical method may not be available, or several different methods could be applied and one has to decide which one is more appropriate, given the circumstances. There are several factors which play a role in deciding if an analysis should be carried out. These factors may be brought together using a combined 'Figure of Merit' (FOM). Some factors are positive and would increase the value of the FOM, while others are negative and would tend to decrease it. The following parameters are suggested:

S = security or inverse of the risk factor (*i.e.* possibility of accident, increased by the necessity for the object to travel)

Q = quality and/or quantity and/or relevance of the information gathered

A = amount of sample to be removed and impact on the object (which includes an evaluation of the importance of the object itself)

C = cost and/or length of time needed to obtain a result

One would give a score of 1 to 4 to each of these parameters before applying a simple formula. For example, for parameter S, a value of 1 would mean very high risk of accidental damage; 2 = medium risk; 3 = low risk; 4 = no risk at all. For parameter A, one may assign a value of 1 = no sampling (non-invasive); 2 = low impact (micro-invasive); 3 = medium impact; 4 = significant damage to the object (destructive analysis). It should be noted that a value of 4 for parameter A does not immediately eliminate the possibility of carrying out the analysis. In fact, there could be reasons described by the factor Q which would redeem a low score for factor A. For example, in judging the stability of a building, it may be necessary to remove some deep cores, obviously in a very intrusive way, but justified by the fact that without that information one may not be able to save the entire building. The formula proposed is the following:

$$FOM = (S \times Q)/(A \times C)$$

Values of FOM of <1 suggest that the analysis should not be carried out. The lower the value, the more undesirable is the analysis.

Values of FOM >1 suggest that the method may be considered; the closer the FOM is to 16, the more acceptable is the proposed analysis.

Obviously, there still is a strong dependence of this FOM upon the subjective judgment of the person who uses it, but at least it represents a simple way of combining several separate evaluations. The subjectivity is embedded into the process of ranking each single parameter, and there is no obvious solution to this problem. On the other hand, the FOM may be very useful in deciding which method of analysis is best when, for some reason, not all of them can be applied. In this case, each single evaluation would be done by the same person with the same 'biases', so that the results are compared readily.

2.8. When a sample needs to be taken: stratigraphy by the use of thin sections

The study of samples in thin section, using both an optical polarizing microscope and an Environmental Scanning Electron Microscope (with Energy Dispersive X-ray analysis) (ESEM/EDS) may be of paramount importance. The spatial distribution of component materials in an object is fundamental information. Typical examples are the 'stratigraphy' of different layers in paintings (both mural and easel paintings), the shapes and dimensions of the aggregate dispersed into the fine lime matrix in a mortar, the shape and size of the filling material in a ceramic, the structure of a rock in a stone element of a building, or the thickness and structure of a superficial patina (*e.g.* a gypsum black crust or a calcium oxalate layer on stone or the oxidation patina on a metal). The ability to map elements with ESEM/EDS allows the superposition of several elements in a region to be used to suggest the minerals present. For example, a sample taken from the mural painting of the Tutankhamen tomb demonstrated the complete stratigraphy (Figure 8). Polarized light microscopy (PLM) showed the mortar, followed by a thin layer of yellow ochre and a thick greenish-blue paint (that may be interpreted as Egyptian Blue, a synthetic pigment used widely in Egyptian as well as Greek-Roman murals). Taking the section to the Environmental Scanning Electron Microscope (ESEM), one can obtain a much sharper image in backscattered mode, in which the heavier elements show lighter. If the ESEM has an energy dispersive device, one can go a step further and produce maps of the individual elements, a very useful analytical tool, and even superpose various elements at will, thus obtaining information on the way they are combined. Figure 8a–g shows these images. Onc can see that there is a good superposition of Cu, Ca and Si, the three elements constituting Egyptian Blue ($CaCuSi_4O_{10}$). The extensive presence of sodium was unexpected but gives an important clue about how the pigment was made. It is very likely that sodium nitrate

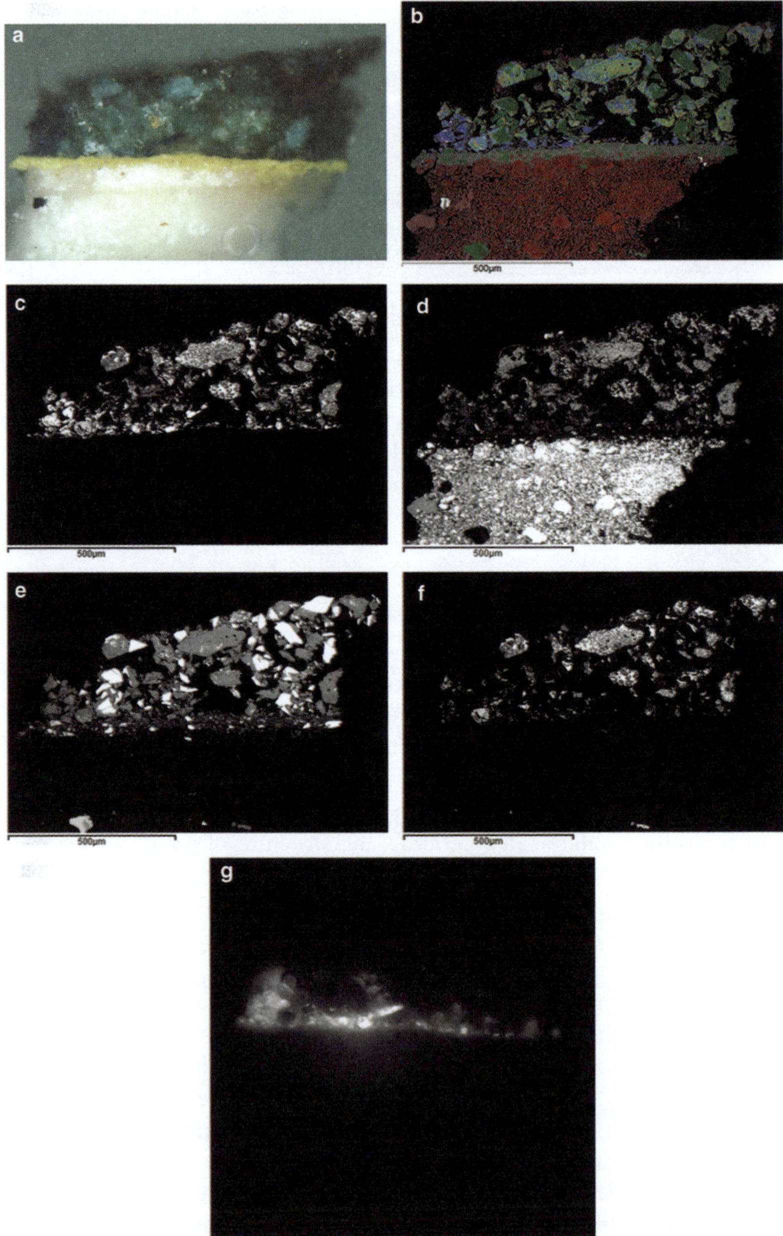

Fig. 8. ESEM/EDS maps of a cross section of a mural painting from Tutankhamen tomb. By matching the location of the various elements one can obtain information on the composition and spatial distribution of minerals. In this case the presence of Egyptian Blue ($CaCuSi_4O_{10}$) can be inferred by the superposition of the maps for Ca, Cu and Si: (**a**) PLM; (**b**) superposition of the maps for Cu, Ca and Si; (**c**) ESEM map of Cu; (**d**) map of Ca; (**e**) map of Si; (**f**) map of Na; (**g**) VIL microscopy map.

(natron) was added to the mix as a flux. The visible induced luminescence (VIL) technique is described below (in section 2.9.4). An extension of this technique has been optimized for microscopic observation of thin sections. As the luminescence depends on the crystal structure of Egyptian blue, one can even distinguish between the pigment and Egyptian Green (a glass of identical chemical composition) which is often present also (Figure 8g). Most of the pigment is clearly Egyptian Green and the combined observation of all the images gives the true distribution of the pigment, which is relevant to determining the conservative process. Once more, this case emphasizes the importance of not relying on a single technique alone, but to make use of complementary techniques. One example of the use of extremely thin sections has been the elucidation of the mysterious 'coral red' present in some Attic pottery (Walton *et al.*, 2009). Chemical analyses performed by LA ICP-MS, used to construct bi-variant plots of individual oxide components, explained some of the differences between the black gloss, the red and the coral red pigments. But transmission electron microscope (TEM) analyses of extremely thin sections (74 nm) produced and oriented using a JEOL Dual-Beam instrument was needed to understand its process of formation. The porosity of the black gloss is almost non-existent, while in the coral red it is >20%. This allows for the re-oxidation of the Fe component to produce red hematite in the coral red, whilst access by oxygen is inhibited in the black gloss, leaving black magnetite as pigment.

Although the study of thin sections using a conventional optical microscope may provide a large amount of useful information, the use of ancillary techniques can be even more informative. We have already discussed the possibility of obtaining quantitative results using Digital Imaging Analysis (DIA) based on colours reflected by a section. One can go further and use various kinds of illumination to cause emission of fluorescent light. The most common example is UV excitation, which leads to visible fluorescence. Although this may not provide specific conclusions about the composition, the presence of organic binders, especially in paintings, can be detected (Bacci, 2004).

2.9. Analyses that involve the whole object

2.9.1. *Use of synchrotron radiation for high-resolution mapping of paintings using XRF*

X-ray fluorescence analysis of paintings can show the presence of elements which may be attributed with reasonable confidence to specific pigments. This is a very powerful tool, and portable XRF instruments are now used widely. Coupled with micro-focus beams or polycapillary lenses and a sturdy x-y-z stage, this technique can produce very detailed elemental maps. The drawback is that the time needed for a 10 mm × 10 mm map can be of the order of several hours. The application of this technique is normally confined to illuminated manuscripts or to very small details of other types of objects (see Trentelman *et al.*, 2010; Szafran *et al.*, 2011). When the object is larger, the normal procedure is to measure several points that can be connected to each other visually on the basis of their colour. However, when an under-painting is present, and one is interested in mapping the colour of the figure underneath, the time involved usually becomes prohibitive. For this kind of investigation, a much more powerful and controllable source of excitation is needed (such as a synchrotron). Dik *et al.* (2008) have pioneered this technique, revealing the hidden face of a peasant in a Van Gogh painting.

2.9.2. Computerized tomography (CT scanning) of large bronze statues

In medicine we are accustomed to the imaging of parts of our bodies using X-rays and CT scans. Using a computer, 3D images obtained by adding a large number of radiographs taken at slightly different angles have proved to be very accurate, providing a great amount of information. Medical doctors have two advantages over the field of cultural heritage: human bodies are all approximately of the same size and shape (allowing the object to remain stationary and moving the rotating coupled source-detector along the body axis) and our bodies do not absorb X-rays too severely. A medical CT scan is an excellent tool to study mummies without having to unwrap them (Corcoran and Svoboda, 2010). If one wants to perform a CT scan on a bronze statue, the set-up has to be completely different. A very powerful X-ray tube (up to 450 keV) must be used to penetrate the thick sheet of metal. A flexible configuration might consist of a base that allows the object to be rotated carefully by as little as a tenth of a degree. To increase the resolution, it is good practice to take up to 1,200 radiographs. The final product is a cube in which x,y,z are the spatial coordinates and the value attached to every point is the X-ray absorption of the specific pixel. This cube can be sliced in any direction needed, and 3D reconstructions of the object can be made. The obvious advantage is that details of the otherwise invisible inner parts are detected and at high resolution. These include wax imprints, repairs, soldering, cracks and many other features of great interest to conservators and curators. Figure 9a,b shows the visible image of the Statue of Infant Cupid (Roman 1st century, Getty Museum 96AB53) and a cut through the inner part of it as obtained by CT scan. The statue is 67 cm tall, large for this application to a bronze (Casali *et al.*, 2005; Bettuzzi *et al.*, 2011).

2.9.3. Laser speckle interferometry

Cracks and plaster detachments are physical properties characterizing a whole object. They are often of paramount importance when considering conservation treatments. The use of ultrasound technology to establish the integrity of a stone statue is well known (Prassianakis, 2000). In archaeological and historic buildings, it is often a problem to determine when a mortar (perhaps holding a mural painting) is loose or not. Conservators carry out this investigation by gently tapping on the surface and listening to the sound emitted. Low frequencies (low pitch) are an indication of voids and, therefore, present potential dangers. One can intervene with an appropriate grouting which needs to be compatible with the surroundings (including material, location and weight of the loose part). Some problems arise when the detachment is located on a ceiling, perhaps several metres high. In this case, costly scaffolding needs to be erected, with all the inherent risks. Recently, a laser speckle interferometer technique has been developed (Keene and Chiang, 2009). Here, a divergent laser beam is reflected

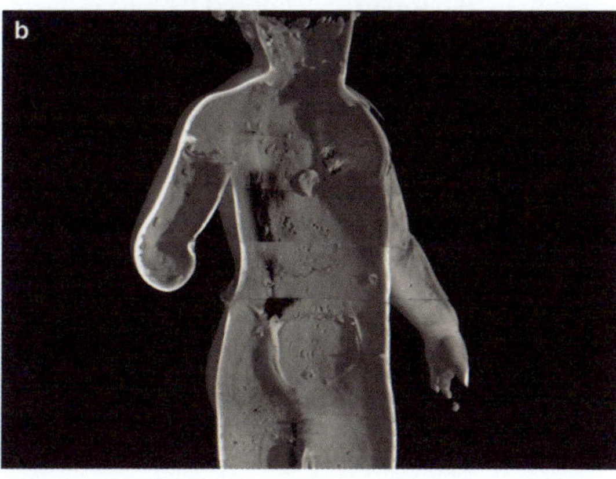

Fig. 9. (a) Visible-light image of the Statue of Infant Cupid (Roman 1st century), 67 cm in height; (b) section obtained by a CT-scan showing details of the interior.

off the wall, which is never perfectly smooth. The wall topography reflects the laser beam in many different directions, and a speckle pattern is formed by interference. A video camera can collect the speckle image in a defocused plane. This image is stored as reference points. By producing an extremely small perturbation of the object's surface and comparing the new speckle pattern with the reference pattern, one can detect loose parts as they move more freely. These changes can be produced by softly pressing, tapping, or even by sound using varying pitch produced by a loudspeaker. In this circumstance, when the resonance frequency of the detachment is reached, a clear image of the loose piece appears on the computer screen. In the thermal baths at Herculaneum, several small pieces of plaster were on the verge of falling from the ceiling, posing great risk for visitors. The voids were found using laser speckle interferometry, avoiding the need to assemble scaffolding.

2.9.4. Visible induced luminescence

Photoluminescence spectroscopy (Ajò *et al.*, 1996; Chiari *et al.*, 1999), based on laser-activated luminescence using fibre optics, has been used for the identification of compounds by point analysis but it is not portable nor does it permit mapping of large objects. It is used to study gems, but it does not have

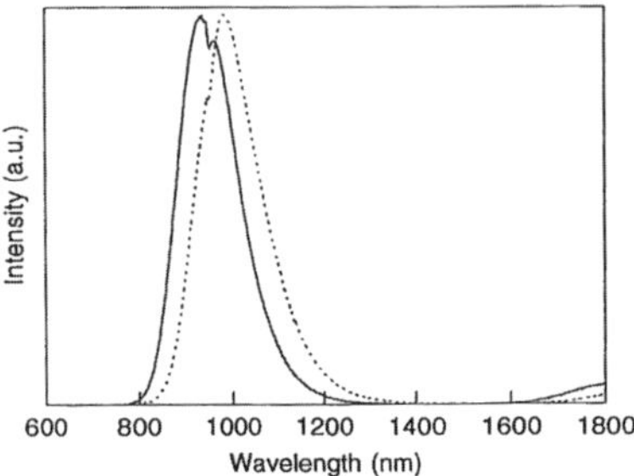

Fig. 10. Photoluminescence spectrum of the blue pigments: Egyptian Blue (solid line) and Han blue (dotted line).

widespread application for pigments. The photoluminescence spectra of several blue pigments (Lapis Lazuli, synthetic lazurite, Egyptian Blue and Han blue; see Figure 10) are remarkably different in spite of their similarity in colour. This makes it possible to determine whether a Lapis Lazuli is of natural origin or is man-made without having to take a sample; as the instrument makes use of a spectrophotometer, these various types of blues can be distinguished from one another.

A more recently developed imaging technique called Visible Induced Luminescence (VIL), which is non-invasive (Verri, 2008, 2009a, 2009b), is based on the same principle: excitation of the object using visible light and detection of the emitted luminescence in the near-infrared. Visible induced luminescence is particularly sensitive to important blue pigments such as Egyptian blue ($CaCuSi_4O_{10}$), Han blue ($BaCuSi_4O_{10}$) (see Fig. 10) and Han purple ($SrCuSi_2O_6$). In all these cases, the Cu^{2+} ion is the only luminescent centre (Pozza *et al.*, 2000). The possibility of mapping Egyptian blue is very interesting, as it was used widely in the old world, from the Assyrians (Verri *et al.*, 2009) to the Egyptians, Greeks and Romans. It was used as a blue colour, or mixed with white to make it brighter, or red to make it purple. Using this technique, Verri (2009a,b) was able to detect the presence of Egyptian blue traces on the marble of ancient statues, including some from the Parthenon, proving that they were originally polychrome. A good example of such mapping of a large mural painting is the work done in King Tutankhamen's tomb (a Getty Conservation Institute project in collaboration with the Supreme Council of Antiquities of Egypt). The camera used had the infrared (IR) filter removed and had a long pass IR filter (typically 850 nm) placed in front of the detector. The IR component of the ambient light needs to be very low as one measures the difference between the IR light fluoresced by the Egyptian blue and the one reflected by the other parts of the object. In King Tutankhamen's tomb this was not difficult to achieve, but in other situations, such as the open-air archaeological excavations at Herculaneum (Getty Conservation Institute in collaboration with Herculaneum Conservation Project) and Pompeii, it may be problematic (Figure 11). Some modifications have been made to this technique (Chiari, 2011), chiefly by using a more intense excitation source without an IR component. This can be achieved by using a normal flashlight screened with a low-passing filter (750 nm) which eliminates the IR component. In this way, the technique can be made portable; it does not require the use of a tripod, and it can be performed by one person only without the need for electrical cables (Figure 12). An extension of the technique has been optimized for microscopic observation of thin sections. In this case, an 850 nm long pass filter is used together with a normal green laser pointer for excitation. As the luminescence depends on the crystal structure of Egyptian blue, one can even distinguish between this and Egyptian green (a glass of identical chemical composition), which is often also present.

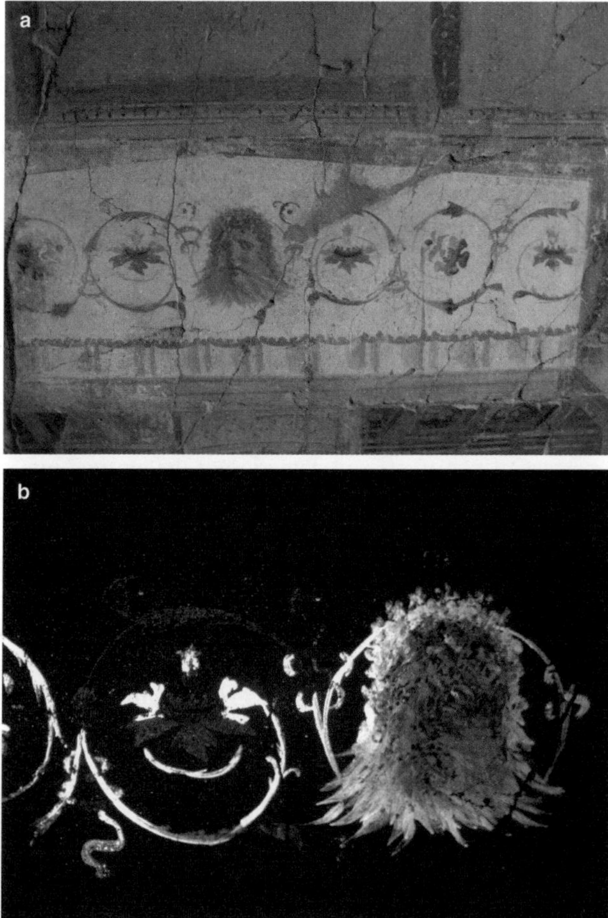

Fig. 11. VIL images from Herculaneum (**a**) visible light picture; and (**b**) VIL image in which only the Egyptian blue is visible.

2.9.5. *Polynomial texture mapping (PTM)*

Many imaging techniques useful in cultural heritage studies have been proposed recently. One worth mentioning is Polynomial Texture Mapping (PTM) introduced by Malzebender *et al.* (2001). This involves taking many (30–60) photographic images of an object using a stationary camera and a light source that moves to various angles. Knowing the direction of the light for each image, a computer is able to create an image on the screen in which, with a touch of the mouse, one can change the illumination direction, passing from raking light coming from the right, to light coming from the left or any other direction. The effects are stunning, and this technique is particularly useful for reading inscriptions on tablets or monitoring small details on

Fig. 12. Equipment used to obtain VIL images in a totally portable way including the presence of ambient light.

flaking mural paintings. The direction of the light can be controlled by geometric devices to accurately position the lamps, or can be determined by the computer by locating the reflection spot on a shiny ball inserted in the field of the images. This technique can even be extended to a microscope by using a very small ball, typically the tip of a ballpoint pen. All the programs needed for this application are available without charge at the Hewlett Packard laboratory site (http://www.hpl.hp.com/research/ptm).

2.10. Use of non-conventional techniques (at large facilities such as synchrotron and neutron sources)

In some cases, the use of large facilities such as synchrotron and neutron sources is necessary. Typically they offer very powerful tunable radiation which may be employed for many different uses, from diffraction to elemental mapping (see Wogelius & Vaughan, 2013, this volume). The study of Maya blue, a synthetic pigment developed by the Mayas, probably at around the eighth Century AD, was possible thanks to a combination of both types of radiation. The literature on Maya blue is extensive. Reyes-Valerio (1993) described many of the applications of Maya blue from the mural paintings of Bonampak to Mayan and Aztec pottery and stone sculptures. He also described in detail how to make Maya blue starting from "Sacalum" (Mexican clay, rich in palygorskite) and añil (indigo) leaves. Its properties are quite unusual, as it is extremely resistant to any sort of attack (organic solvents, or even concentrated nitric acid at 250°C).

The Maya blue crystal structure has been described in detail by Chiari *et al.* (2003) using synchrotron powder diffraction, and starting with data from Artioli *et al.* (1994) for the clay structure. It was further confirmed by neutron diffraction using Rietveld refinement of a synthetic Maya blue containing deuterated indigo (Giustetto and

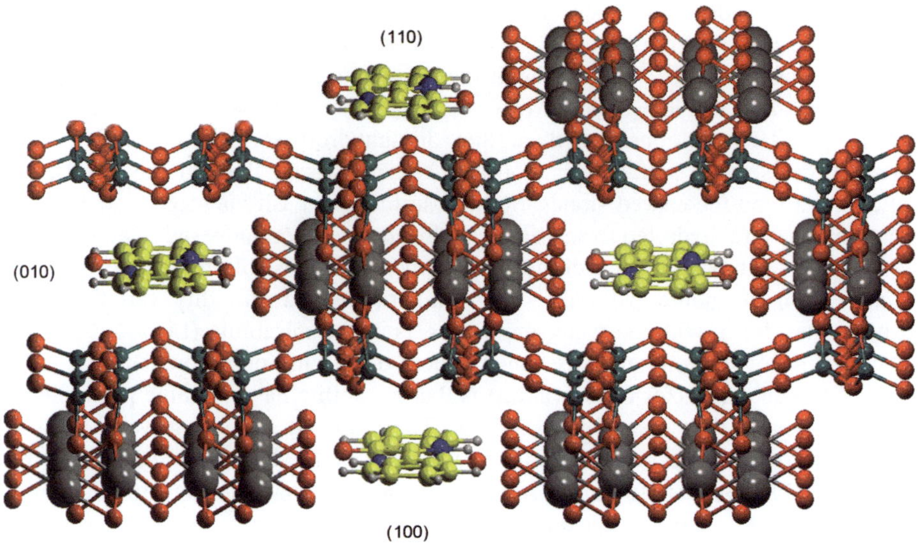

Fig. 13. A proposed model for the crystal structure of Maya Blue. The palygorskite 'grooves' along the horizontal axis are clearly visible. Their size is compatible with the dimensions of the indigo molecules, which are simultaneously and randomly inserted into the grooves on the surface of the crystals only. The penetration into the channels is energetically unfavourable.

Chiari, 2004). Good synchrotron and neutron diffraction data, which permitted the location of the deuterium atoms, gave a structural model that seemed to be correct, with indigo molecules filling the clay channels. In fact, the indigo molecule fits quite well into the channels, as can be shown by computer simulation. Nevertheless, as often happens, the use of one technique alone is not sufficient to find a complete solution. Theoretical energy calculations showed that the affinity of water for the clay is larger than that for indigo and that, therefore, the indigo molecule cannot fill the channels of the clay when these are occupied by water. Thermogravimetric measurements show that the adsorbed water is removed at $T < 120°C$ while the removal of zeolitic water in the channels takes place at higher temperatures. The 'grooves' or half channels deriving from the termination of the crystals can accommodate indigo at the low temperature of synthesis used experimentally, but the channels cannot. In this way, indigo molecules can simultaneously enter the palygorskite crystals (Figure 13) (Chiari *et al.*, 2008). This new model of the pigment structure, which explains the odd properties of this compound, could only be obtained by combining the results of several different techniques.

3. Conservation of monuments

Most of the examples discussed above address some aspect of archaeometry. Perhaps that is because the problems connected with conservation, although very important, are less exciting; indeed, there is little excitement in identifying the types of soluble salts present on a stone carving, although that piece of information may be of paramount importance for conservation. When analysing pigments on a sculpted and painted volcanic stone at the Templo Mayor in Mexico City, a white efflorescence was found (Rasmussen *et al.* 1996) to consist of thenardite, an anhydrous sodium sulfate. The likely source of the soluble sulfate was Portland cement stucco that, in the 1960s, was applied all around the carved stones. Sodium sulfate can easily absorb water from the air and form several hydrated forms, stable at different temperatures and relative humidity, all possible in a temperate climate. It is, therefore, possible to change from one hydrated form to another with the daily temperature and humidity variations. The most hydrated form, with ten water molecules, is called mirabilite (Levy & Lisensky, 1978) because of the long crystals which appear during its formation. The dangerous aspect of this material (for conservation) is that mirabilite has a volume per molecule four times greater than the anhydrous form. Therefore, by absorbing water from the air, it can exert considerable pressure inside the pores of the stone, with the possible formation of cracks. Sodium sulfate is, in fact, the substance used during simulated aging cycles in order to submit a consolidated stone to a very harsh test. Here, a modest analytical effort produced interesting results, and stimulated a lot of preservation work. Not alone was the symptom of the deterioration (the salt) identified, but the culprit (the cement) was also eliminated. Portland cement should never be used in proximity to a porous stone because of its different properties (hardness, porosity, water absorption capacity, permeability, elasticity) and, also, because it contains sulfates which can cause the above-mentioned problems of soluble salts. However, despite all the warnings, cement continues to be used for restoration work in all parts of the world.

3.1. Earthen architecture

One third of the houses in the world are made of mud. The percentage of archaeological excavations dealing with this material is even greater. Adobe (from 'al dub', which in Egyptian means 'the brick') can adsorb water into its clay component, causing dilation. If the water is abundant, it can turn into slurry. This causes enormous conservation problems. It would be difficult to list all the measures that have been adopted in order to counteract the dissolution of adobe. Two examples can be considered, one chemical and one physical. For selected parts of a structure of particular value (decorated surfaces, possibly painted or carved), it is possible to use ethyl silicate as a consolidant (Chiari, 1990). The product at first hydrolyses, generating silicic acid which can react with other silicic acid molecules, producing strong $Si-O-Si$ bonds. This polymerization creates a network of silica tetrahedra, which eventually react with the OH groups present in the clay. The very fine clay and silt particles, having a large surface:-volume ratio, act as binder of the adobe through surface bonds of electrostatic nature which are affected by water. The chemical bonds formed following ethyl silicate treatment are less susceptible to the action of water so that the clay particles can no longer be brought into suspension by liquid water, but retain their action as binding agents in the brick (Bruno *et al.*, 1968; Chiari, 1987). When an adobe monument is very large, like a

Fig. 14. Square house (1930s) built on a much older round tower in Yemen. The entire construction is made of adobe with the support of a wooden frame. The white top is a lime plaster serving both as decoration and as protection against rain.

pyramid or a ziggurat, or where entire cities are involved such as in Yemen (Jerome *et al.*, 1999; see Figure 14), the use of a surface consolidant is impossible for practical and economic reasons. Agnew (1990) has suggested using geotextiles, of the kind used to prevent landslides, in archaeological excavations. This is a good example of the exchange of ideas between applied geology and applied mineralogy for the benefit of conservation.

3.2. Inorganic consolidants

3.2.1. Barium

Another very important contribution of mineralogy to conservation, in this case the conservation of mural paintings, is the so called "barium method" originally proposed by Ferroni (1969) and refined by Matteini (1991). The principle is to eliminate gypsum contamination, often present, by first treating the plaster with ammonium carbonate which produces calcite and ammonium sulfate. An application of barium hydroxide then fixes the sulfate by forming insoluble barium sulfate. If there is an excess of barium hydroxide, the consolidation continues, as insoluble barium carbonate is formed spontaneously. The final result is a sound consolidated surface. As the treatment is irreversible, it must be carried out only by well trained specialists to avoid the formation of surface white crusts.

3.2.2. Oxalate patinas: decay or protection?

Another important topic in the conservation field concerns calcium oxalate patinas. Calcium oxalate is extremely insoluble and forms yellowish patinas of either whewellite or weddellite (the mono and di-hydrated forms) over practically every type of surface. Until recently, these patinas were ignored, but after their discovery on the Arch of Constantine in Rome (Cipriani & Franchi, 1979), researchers started to pay closer attention, first to the surfaces of marble and then to other surfaces, finding the ubiquitous presence of oxalates (on frescoes, glasses, and even on bronze). The origin of these patinas is still a subject of debate and, very likely, there is more than one mechanism at play. The important thing is that these patinas can act as protection for a fragile surface, screening it from acidic attack. On the Trajan Column, exposed to weathering for $\sim$2000 years, even the finest tool marks are preserved where oxalates are still present. This suggests that it is worth producing oxalate patinas in order to protect surfaces; an idea (Chiari *et al.*, 1989) that has been applied by Matteini & Giovannoni (1996) using ammonium oxalate with very good results. For more information on oxalate patinas and their role in cultural heritage conservation, see the publications from two international meetings on the subject (Realini and Toniolo, 1989, 1996).

3.2.3. *Nanolimes*

The basic idea common to all inorganic treatments is that by filling the voids of a deteriorated mortar or stone with an insoluble compatible compound, one can regenerate a solid, compact texture, more resistant to further deterioration. In the case of lime mortars, or marbles or limestone, the obvious choice is to fill the pores with calcite. The problem is that the solubility of portlandite ($Ca(OH)_2$) is very low (Rodriguez-Navarro *et al.*, 2005) and, therefore, the traditionally employed lime-wash is of limited use. Modern technology can produce portlandite particles at a nanometric scale (Baglioni *et al.* 2003) which can be held in suspension at high concentration. Given its nanometric dimensions, this product can easily penetrate a capillary system and consolidate loose material by crystallizing as calcite and filling the voids. We should keep in mind that any inorganic treatment will fail if the material has previously been treated with synthetic resins or polymers of the kind that were popular in the 1970s. Also, hydrophobic treatments such as the wax or paraffin, commonly used on mural paintings since the eighteenth century, inhibit the use of much better inorganic products. The lesson here is that in order to avoid future problems, we should always use a treatment compatible with the original material and that will not drastically change its physical-chemical properties.

4. Conclusions

The contribution of mineralogy and related sciences to solving problems in cultural heritage is fundamental. In order for this role to be fully recognized, it is necessary that the people involved in the research make a special effort to meet the needs of an interdisciplinary group, and to use language which can be comprehended by non-specialists. Mineralogists need to be a part of the process of studying cultural objects as a whole, not just of analysing a small detail which may not be fundamentally important. They also need to be involved in the taking of samples and, therefore, in deciding which parts of an object should be studied and with what methods. As the field is too small to generate an industry producing specialized instrumentation, it is necessary to influence the production of non-invasive portable instruments, sharing our ideas with the engineers capable of realizing them. In the debate on non-invasive and micro-invasive methods, we should contribute to the development of a series of guidelines to simplify the evaluation of the various factors which are involved. Above all, we need to develop expertise in our own field as part of cultural heritage research.

Acknowledgments

This chapter is dedicated to Professor Giorgio Torraca, who was the author's mentor, colleague and friend. The author is grateful to all the colleagues (especially those at the Getty Conservation Institute) who generously shared the results of their work. Without their help, and that of Archaeologists, Art Historians, Architects and Restorers with whom he has worked, he would not have had the opportunity to broaden his expertise. He thanks David Vaughan for reviewing the present manuscript with great care and Gary Mattison for polishing the English.

References

Agnew, N. (1990) The Getty Adobe Research project at Fort Selden. I. Experimental design for the test wall project. *ADOBE 90, 6TH International Conference on the Conservation of Earthen Architecture.* Las Cruces, New Mexico, pp. 243–249.

Ajò, D., Chiari, G., De Zuane, F., Favaro, M.L. & Bertolin, M. (1996) Photoluminescence of some blue natural pigments and related synthetic materials. *5th International Conference on Non-invasive Testing, Microanalytical Methods and Environmental Evaluation for Study and Conservation of Works of Art,* Budapest (Hungary), 24–28 September 1996.

Altomare, A., Burla, M.C., Camalli, M., Carrozzini, B., Cascarano, G.L., Giacovazzo, C., Guagliardi, A., Moliterni, A.G.G., Polidori, G. & Rizzi, R. (1999) EXPO: a program for full powder pattern decomposition and crystal structure solution. *Journal of Applied Crystallography,* **32,** 339–340.

Artioli, G. (2010) *Scientific Methods and Cultural Heritage.* Oxford University Press, Oxford, UK, 536 pp.

Artioli, G., Galli, E., Burattini, E., Cappuccio, G. & Simeoni, S. (1994) Palygorskite from Bolca, Italy: a characterization by high-resolution synchrotron radiation powder diffraction and computer modeling. *Neues Jahrbuch für Mineralogishe Monatshefte,* **H5,** 217–229.

Bacci, M. (2004) Optical microscopy and colorimetry. In: *Physics Methods in Archaeometry* (M. Martini, M. Milazzo and M. Piacentini, editors). SIF, Bologna, Italy and IOS Press, Amsterdam, The Netherlands, pp. 1–16.

Baglioni, P., Caretti, E., Dei, L. & Giorgi, R. (2003) Nanotechnology in wall painting conservation. In: *Self Assembly* (B.H. Robinson, editor), IOS Press, Amsterdam, pp. 32–41.

Barra Bagnasco, M., Casoli, A., Chiari, G., Compagnoni, R., Davit, P. & Mirti, P. (2001) Mineralogical and chemical composition of transport amphorae excavated at Locri Epizephiri (Southern Italy). *Journal of Cultural Heritage,* **2,** 229–239.

Barry, E.F. & Grob, R.L. (2004) *Modern Practice of Gas Chromatography.* Wiley-Interscience, New York.

Berti, G., De Marco, F. & Aldrighetti, S. (2011) Advances on NDT methods and technologies for early-stage diagnosis of materials. *Material Science Forum,* **681,** 461–467.

Bettuzzi, M., Casali, F., Morigi, M.P., Brancaccio, R., Chiari, G., Carson, D. & Maish, J. (2011) A new small scale computed tomographic (CT) system for museum use: case study of a Roman bronze statue of Eros. *ART'11, Florence.*

Blake, D. (2010) A historical perspective of the development of the CheMin Mineralogical instrument for the Mars Science Laboratory (MSL'11) Mission. *Geochemical News,* **144.**

Boyd, B., Bethem, R. & Basic, C. (2008) *Trace Quantitative Analysis by Mass Spectrometry.* John Wiley & Sons, New York.

Bruno, A., Chiari, G., Trossarelli, C. & Bultinck, G. (1968) Contribution to the study of the preservation of mud-brick structures. *Mesopotamia*, **III–IV**, 443–479.

Casadio, F., Chiari, G. & Simon, S. (2005) Evaluation of binder/aggregate ratios in lime mortars with carbonate aggregate: a comparative assessment of chemical, mechanical and macroscopic approaches. *Archaeometry*, **47**, 671–689.

Casali, F., Pasini, A., Bettuzzi, M., Chiari, G. & Carson, D. (2005) A new digital radiography and computed tomography system: preliminary tests on objects of artistic interest. *Art 2005, Lecce*.

Chiari, G. (1987) Consolidation of adobe with ethyl silicate: control of long term effects using SEM. *5th International Meeting of Experts on the Conservation of Earthen Architecture*. Rome, ICCROM-CRATerre, pp. 25–32.

Chiari, G. (1990) Chemical surface treatment and capping techniques of earthen structures: a long-term evaluation. In: *Adobe 90, 6TH International Conference on the Conservation of Earthen Architecture*. Las Cruces, New Mexico, pp. 267–273.

Chiari, G. (1999) Analisi dei panneggi censori. In: *Michelangelo – La Cappella Sistina – Rapporto sul Restauro del Giudizio Universale*. Istituto Geografico De Agostini, Novara, Italy, pp. 341–351.

Chiari, G. (2008) Saving art *in situ*. *Nature*, **453**, 159.

Chiari, G. (2011) Visible induced luminescence of Egyptian Blue: innovative procedures. *Art'11, 10th International Conference on Non-destructive Investigations and Microanalysis for the Diagnostics and Conservation of Cultural Heritage*. Florence, Italy.

Chiari, G. & Lanza, R. (1997) Pictorial remanent magnetization as an indicator of secular variation of the Earth's magnetic field. *Physics of the Earth and Planetary Interiors*, **101**, 79–83.

Chiari, G. & Lanza, R. (1999) Remanent magnetization of mural paintings from the *Bibliotheca Apostolica* (Vatican, Rome). *Journal of Applied Geophysics*, **41**, 137–143.

Chiari, G., Sampo', S. & Torraca, G. (1989) Formazione di ossalati di calcio su superfici marmoree da parte di funghi. *I International Symposium. Le pellicole ad Ossalati: Origine e Significato Nella Conservazione delle Opere d'arte*. Milano, pp. 85–90.

Chiari, G., Santarelli, M.L. & Torraca, G. (1992) Caratterizzazione delle malte antiche mediante l'analisi di campioni non frazionati. *Materiali e Strutture, Problemi di Conservazione*, **Anno II, N 3**, 111–137.

Chiari, G., Santarelli, M.L. & Torraca, G. (1993) Problemi nell'analisi strumentale di calcestruzzi pozzolanici antichi. *Scienza e beni Culturali*. Bressanone, **IX**, 275–286.

Chiari, G., Torraca, G. & Santarelli, M.L. (1996a) Recommendations for systematic instrumental analysis of ancient Mortars: the italian experience. In: *Standards for Preservation and Rehabilitation* (S.J. Kelley, editor). American Society for Testing Materials, pp. 275–284.

Chiari, G., Cimitan, L., Della Ventura, G., Felitici, M.G., Santarelli, M.L. & Torraca, G. (1996b) Le malte pozzolaniche del mausoleo di Sant'Elena e le pozzolane di Torpignattara. *Materiali e Strutture*, **VI**, 1–36.

Chiari, G., Giordano, A., & Menges, G. (1996c) Non invasive X-ray diffraction analyses of unprepared samples. *Science and Technology for Cultural Heritage*, **5**, 21–36.

Chiari, G., Compagnoni, R., Giustetto, R. & Ricq-de-Bouard, M. (1996d) Metodi archeometrici per lo studio dei manufatti in pietra levigata. In: *Le Vie della Pietra Verde*. Omega Edizioni, Italy, pp. 35–53.

Chiari, G., D'amicone, E. & Vigna, L. (1998) Non-invasive X-ray diffraction analysis using Göbel mirrors: an application to Egyptian pigments and glasses. In: *La couleur dans la peinture et l'Emaillage de l'Egypte anciennne* (S. Colinart, editor). Edipuglia, Bari, Italy, pp. 87–94.

Chiari, G., Ajò, D., Reyes-Valerio, C., Virdis, C., Pozza, G. & De Zuane, F. (1999) Application of photoluminescence spectroscopy to the investigation of materials used in works of art. *VI International Conference on Non-Invasive Testing and Microanalysis for the Diagnostics and Conservation of Cultural and Environmetal Heritage*. Roma, 17–19 May 1999.

Chiari, G., Giustetto, R. & Ricchiardi, G. (2003) Crystal structure refinement of palygorskite and Maya Blue from molecular modelling and powder synchrotron diffraction. *European Journal of Mineralogy*, **15**, 21–33.

Chiari, G., Giustetto, R., Druzik, J., Doehne, E. & Ricchiardi, G. (2008) Pre-colombian nanotechnology: reconciling the mysteries of the Maya blue pigment. *Applied Physics A. Materials Science & Processing*, **A90**, 3–7.

Cipriani, C. & Franchi, L. (1979) Sulla presenza di whewellite fra le croste di alterazione di monumenti Romani. *Bollettino del Servizio Geologico d'Italia,* **88**, 555–564.

Corcoran, L.H. & Svoboda, M. (2010) *Herakleides: a portrait mummy from Roman Egypt.* Paul Getty Trust, UK.

Dik, J., Janssens, K., Van Der Snickt, G., van der Loeff, L., Rickers, K. & Cotte, M. (2008) Visualization of a lost painting by Vincent van Gogh using synchrotron radiation based X-ray fluorescence elemental mapping. *Analytical Chemistry,* **80**, 6436–6442.

Druzik, J.R. (2010) Evaluating the light sensitivity of paints in selected wall paintings at the Mogao Grottoes: Caves 217, 98 and 85. In: *Conservation of Ancient Sites on the Silk Road.* Proceedings of the Second International Conference on the Conservation of Grotto Sites, Mogao Grottoes, Dunhuang, People's Republic of China (N. Agnew, editor). June 28–July 3, 2004. Getty Conservation Institute, Los Angeles, California, pp. 457–463.

Ebdon, I., Evans, E.H., Fisher, A. & Hill, S.G. (1998) *An Introduction to Analytical Atomic Spectrometry.* John Wiley & Sons, Chichester, UK.

Freestone, I. (1995) Ceramic petrography. *American Journal of Archaeology,* **99**, 111–115.

Ferroni, E., Malaguzzi-Valeri, V. & Rovida, G. (1969) Utilisation de techniques diffractométriques dans l'étude de la Conservation des fresques. 8*eme Colloque sur l'analyse de la matiére,* Florence 15–19 Septembre 1969, 4 pp.

Gabott, P. (2007) *Principles and Applications of Thermal Analysis.* Blackwell Publishing, Oxford, UK.

Gianoncelli, A., Castaing, J., Ortega, L., Dooryhee, E., Salomon, J., Walter, P., Hodeau, J.L. & Bordet, P. (2008) A portable instrument for in situ determination of chemical and phase compositions of cultural heritage objects. *X-ray Spectrometry,* **37**, 418–423.

Giustetto, R. & Chiari, G. (2004) Crystal structure refinement of palygorskite from neutron diffraction. *European Journal of Mineralogy,* **16**, 521–532.

Gorgoni, C., Lazzarini, L. & Pallante, P. (1992) Identification of ancient white marbles in Rome. I: the Arch of Titus. *Science and Technology for Cultural Heritage,* **I**, 79–86.

Gorgoni, C., Lazzarini, L., Pallante, P. & Turi, B. (1998) An updated and detailed reference database for the main Mediterranean marbles used in antiquity. *ASMOSIA 98, Fifth International Conference, Boston, Massachusetts, USA.*

Heginbotham, A., Bezur, A., Bouchard, M., Davis, J., Eremin, K., Frantz, J., Glinsman, L., Hayek, L., Hook, D., Kantarelou, V., Karydas, A., Lee, L., Mass, J., Matsen, C., McCarty, B., McGath, M., Shugar, A., Sirois, J., Smith, D. & Speakman, R. (2011) An evaluation of inter-laboratory reproducibility for quantitative XRF of historic copper alloys. In: *METAL 2010* – Proceedings of the Interim Meeting of the ICOMCC Metal Working Group, Charleston, South Carolina, 11–15 October 2010 (P. Mardikian, C. Chemello, C. Waters & P. Hull editors). Clemson University, USA, pp. 244–245.

ICSD – Inorganic Crystal Structure Database (2011) *Gmelin Institut fur Anorganische Chemie und Fachinformationszentrum,* Karsruhe, Germany, STN Service Centre, Europe.

Jerome, P., Chiari, G. & Borelli, C. (1999) The architecture of mud: construction and repair technology in the Hadhramaut region of Yemen. *APT Bulletin, Journal of Preservation Technology,* **XXX**, 39–48.

Keene, L.T. & Chiang, F.P. (2009) Real-time anti-node visualization of vibrating distributed systems in noisy environments using defocused laser speckle contrast analysis. *Journal of Sound and Vibration,* **320**, 472–481.

Lazzarini, L., Gorgoni, C., Pallante, P. & Turi, B. (1997) Identification of ancient white marbles in Rome. I: The Portico d'Octavia. *Science and Technology for Cultural Heritage,* **6**, 185–198.

Levy, A.H. & Lisensky, J.C. (1978) Crystal structure of sodium sulfate decahydrate (Glauber's salt) and sodium tetraborate decahydrate (borax) redetermination by neutron diffraction. *Acta Crystallographica,* **B**, 3502–3510.

Libby, W.F. (1952) *Radiocarbon Dating.* University of Chicago Press, Chicago, Illinois, USA.

Maggetti, M. (1981) Composition of Roman pottery from Lausanne (Switzerland). *British Museum Occasional paper,* **19**, 33–49.

Malzebender, T., Gelb, D. & Wolters, H. (2001) Polynomial Texture Maps. In: *SIGGRAPH '01* – Proceedings of the 28th Annual Conference on Computer Graphics and Interactive Techniques, New York. Association for Computing Machinery Press, USA, pp. 519–528.

Martini, M. & Sibilia, E. (2001) Radiation in archaeometry: archaeological dating. *Radiation Physics and Chemistry,* **61,** 241–246.

Martini, M., Milazzo, M. & Piacentini, M., editors (2004) *Physics Methods in Archaeometry.* SIF, Bologna, Italy.

Matteini, M. (1991) An assessment of Florentine methods of wall painting conservation based on the use of mineral treatments. *The Conservation of wall paintings – Courtauld Institute of Art and The Getty Conservation Institute, London, 13–16 July, 1987 (S. Cather editor),* pp. 137–148.

Matteini, M. & Giovannoni, S. (1996) The protective effect of ammonium oxalate treatment on the surface of wall paintings. In: *Painted Facades* (E. Mayer, editor). Osterreichische Section des IIC, Vienna, pp. 95–101.

Nakai, I. & Abe, Y. (2012) Portable X-ray powder diffractometer for the analysis of art and archaeological materials. *Applied Physics A,* **106,** 279–293.

Namowicz, C., Trentelman, K. & McGlinchey, G. (2008) XRF of Cultural Heritage materials: Round-robin IV – Paint on canvas. *Powder Diffraction Journal,* **24,** 124–129.

Pappalardo, L., Pappalardo, G., Amorini, F., Branciforti, M.G., Romano, F.P., de Saniot, J., Rizzo, F., Scafiri, E., Taormina, A. & Gatto Rotondo, G. (2008) The complementary use of PIXE-α and XRD non-destructive portable systems for the quantitative analysis of painted surfaces. *X-ray Spectrum,* **37,** 370–375.

Pozza, G., Ajò, D., Chiari, G., De Zuane, F. & Favaro, M. (2000) Photoluminescence of the inorganic pigments Egyptian Blue, Han Blue and Han Purple. *Journal of Cultural Heritage,* **1,** 393–398.

Prassianakis, I.N., Kourkoulis, S.K. & Vardoulakis, I. (2000) Marble monument examination using NDT method of ultrasounds. *15th World Conference on Non-destructive Testing,* Rome.

Quinn, P.S. (2008) The occurrence and research potential of microfossils in inorganic archaeological materials. *Geoarchaeology,* **23,** 275–291.

Rapp, G.R. (2009) *Archaeomineralogy (Natural Science in Archaeology).* Springer-Verlag, Berlin, 348 pp.

Rasmussen, S.E., Jorgensen, J.E. & Lundoft, B. (1996) Structure and phase transitions of Na_2SO_4. *Journal of Applied Crystallography,* **29,** 42–47.

Realini, M. & Toniolo, L. (1989) *International Symposium Proceedings on the Oxalate films. Origin and significance in the conservation of works of art.* Milan, Italy, 5–26 October 1989.

Realini, M. & Toniolo, L. (1996) Proceedings of the *International symposium (II) the oxalate films in the conservation of works of art, Milan, March 1996* (M. Realini & L. Toniolo, editors). Centro C.N.R. "Gino Bozza". Milano, Italy.

Reedy, C.L. (1994) Thin-section petrography in studies of cultural materials. *Journal of the American Institute for Conservation,* **33,** 115–129.

Reyes-Valerio, C. (1993) De Bonampak al Templo Mayor – El Azul Maya en Mesoamérica. *Siglo Veintiuno Editores,* 1–157.

Reyna Jenaro, G. (1956) *Riqueza minera y yacimientos minerales de México.* Departamento de Investigaciones Industriales. Banco de México, México.

Ricq-de Bouard, M. & Compagoni, R. (1991) La circulation des outils polis en éclogite Alpine au IV° millénaire. Premières observations relatives au sud-est de la France et à quelques sites plus septentrionaux. In: *Identité du Chasséen. Actes du Colloque de Nemours 1989, Memoires du Musée de Préhistoire d'Ile de France,* **4,** 273–280.

Rodriguez-Navarro, C., Hansen, E. & Ginell, W.S. (2005) Calcium hydroxide crystal evolution upon aging of lime putty. *Journal American Ceramic Society,* **81,** 3032–3034.

Ruffini, R., Borghi, A., Cadoppi, P., Cossio, R. & Dela Pierre, F. (1999) Modal determination and image analysis performed by EDS-SEM system applied to sedimentary rocks. *Proceedings Geovision 99,* Liegi, pp. 209–212.

Sarrazin, P., Chiari, G. & Gailhanou, M. (2008) A portable non-invasive XRD/XRF instrument for the study of art objects. *Advances in X-ray analysis,* **52,** 175–186.

Sciuti, S. & Gabrielli, N. (1995) Tecniche di analisi e di osservazione finalizzate al restauro di dipinti antichi. *Accademia Nazionale dei Lincei* (Roma), **91**,7–53.

Szafran, Y., Namowicz, C., Smith Patterson, C., Sciacca, C., Trentelman, K. & Turner, N. (2011) Painting on parchment and panel: an exploration of Pacino di Bonaguida's technique. *Studying Old Master Paintings Technology and practice. The national Gallery Technical Bulletin. 30th Anniversary Conference Postprints*, pp. 8–14.

Torraca, G. (1982) The scientist's role in historic preservation with particular reference to stone conservation. In: *Conservation of Historic Stone Buildings and Monuments*. National Academies Press, Washington D.C., pp. 13–21.

Trentelman, K., Bouchard, M., Ganio, M., Namowicz, C., Smith Patterson, C. & Walton, M. (2010) The examination of works of art using in situ XRF line and areas scans. *X-Ray Spectrometry*, **39**, 159–166.

Turco, T. (1996) *Il Gesso, Lavorazione Trasformazione Impieghi*. Manuale Hoepli, Milan, Italy.

Tykot, R.H. & Young, S.M.M. (1996) Archaeological applications of inductively coupled-mass spectrometry. In: *Archaeological Chemistry: Organic, Inorganic, and Biochemical Analyses* (M.V. Orna, editor). ACS Symposium Series **625**. American Chemical Society, Washington D.C., pp. 116–130.

Uda, M., Sassa, S., Taniguchi, K., Nomura, S., Yoshimura, S., Kondo, J., Iskander, N. & Zaghloul, B. (2000) Touch-free in situ investigation of ancient Egyptian pigments. *Naturwissenschaften*, **87**, 260–263.

Uda, M., Yoshimura, S., Ishizaki, A., Yamashita, D.Y. & Sakuraba, Y. (2007) Tutankhamun's golden mask investigated with XRDF. *International Journal of PIXE*, **17**, 65–76.

Vaughan, S.J. (1995) Ceramic petrology and petrology in the Aegean. *American Journal of Archaeology*, **99**, 115–117.

Verri, G. (2008) The use and distribution of Egyptian Blue: a study by visible-induced luminescence imaging. In: *Nabaum wall paintings: conservation, scientific examination and display at the British Museum* (A. Middleton & K. Uprichard, editors). Archeotype Publications, London, pp. 41–50.

Verri, G. (2009a) The spatial characterization of Egyptian Blue, Han Blue and Han Purple by photoinduced luminescence digital imaging. *Analytical and Biochemical Chemistry*, **394**, 1011–1021.

Verri, G. (2009b) Traces of paint confirmed on Parthenon sculptures. *Nature News,* **2009**, 574.

Verri, G., Collings, P., Ambers, J., Sweek, T. & Simpson, S.J. (2009) Assyrian colours: pigments on a neo-Assyrian relief of a parade horse. *Technical Research Bulletin*, **3**, 57–62.

Walton, M. & Trentelman, K. (2009) Romano-Egyptian red lead pigment: a subsidiary commodity of Spanish silver mining and refinement. *Archaeometry,* **51**, 845–860.

Walton, M., Doehne, E., Trentelman, K., Chiari, G., Maish, J. & Buxbaum, A. (2009) Characterization of coral red slips on Greek Attic pottery. *Archaeometry*, **51**, 383–396.

Whitmore, P.M., Pan, X. Pan, Xun & Bailie, C. (1999) Predicting the fading of objects: identification of fugitive colorants through direct noninvasive lightfastness measurements. *Journal of the American Institute for Conservation*, **38**, 395–409.

Wogelius, R.A. & Vaughan, D.J. (2013) Analytical, experimental and computational methods in environmental mineralogy. In: *Environmental Mineralogy* II (D.J. Vaughan and R.A. Wogelius, editors). EMU Notes in Mineralogy, **13**, European Mineralogical Union and the Mineralogical Society of Great Britain & Ireland, London, pp. 5–102.

Zanella, E., Gurioli, L., Chiari, G., Ciarallo, A., Cioni, R., De Carolis, E. & Lanza, R. (2000) Archaeomagnetic results from mural paintings and pyroclastic rocks in Pompeii and Herculaneum. *Physics of the Earth and Planetary Interiors*, **118**, 227–240.

Zöldföldi, J. & Satir, M. (2003) Provenance of white marble building stones in the monuments of the ancient Troia. In: *Troia and the Troad* (G.A. Wagner, E. Pernicka & H.P. Uerpmann, editors). Springer-Verlag, Berlin, pp. 203–223.

Minerals and human health

H. CATHERINE W. SKINNER

*Department of Geology and Geophysics, Yale University, PO Box 208109,
New Haven, CT, USA
e-mail: Catherine.skinner@Yale.edu*

Environmental Mineralogy is an area of investigation and research that applies to all the systems operating at the surface of the Earth: the atmosphere, the hydrosphere, the lithosphere and the biosphere. Minerals are basic to all the dynamic processes in each of these sectors. This chapter will focus on biominerals, a subset of mineral materials that are essential to all Earth's inhabitants as they have been since the first living species appeared over 3 billion years ago. Biominerals are those produced via biological mechanisms, a large suite of compounds, many familiar to mineralogists, but this chapter will concentrate on the biominerals found within humans. Not only are these natural, some are essential such as the mineral matter found in bones and teeth. Other biominerals found in humans are pathological and reflect disease states, like the 'stones' that occur in the kidneys of some individuals. The study of human biominerals offers a novel view of the 'geo-environment'.

1. Introduction

Humans depend on mineral-derived materials not only because they define the spaces we occupy in the global environment such as the caves occupied by the first hominids, or the houses and skyscrapers of today, but because we depend on the elements contained in minerals for many of the nutrients we require. The heterogeneous distribution of minerals in soils, described by Manning (2013, this volume) make it possible for humans to survive by eating a wide range of plants and animals that provide the food and nutrients critical to our wellbeing. In this chapter, I examine the minerals (or 'biominerals') that our bodies produce; some of these are essential for health whereas others form pathologically and indicate potential disease states. The most important biomineral is apatite, especially to all vertebrates, because it forms the skeleton. All the essential elements we extract from our food and drink, plus those minerals mined to satisfy our many other needs, have become central to the stability of worldwide populations. Acknowledging that the global environment is being changed by human activities, and that humans are now the dominant force changing the planet, we now need to improve our understanding of the systems that are critical to sustaining the benefits we enjoy.

The human 'sphere', is analogous to the other spheres that characterize our planet, and just one of the many parts of the complex 'open' systems that interact with and influence the other spheres. All life forms, from the earliest, have required a range of elements that are sequestered in minerals. Some of these elements become part of the external parts of

© Copyright 2013 the European Mineralogical Union and the Mineralogical Society of Great Britain & Ireland
DOI: 10.1180/EMU-notes.13.11

the living forms, such as the shells or carapaces of invertebrates, whereas some form internal skeletons. These are the structures that when fossilized we now employ for their identification. We study both types of biomineralized tissues. The shells of invertebrates, such as molluscs, with a range of sizes and colours, have changed over geological time. Some early life forms were, and still are, responsible for the accumulation of certain biominerals as well known geological features, *e.g.* the corals that created reefs from their 'hard parts' of $CaCO_3$. Buried over millennia, the reefs become limestones. Today limestones are often mined for construction of buildings, or for the beautiful blocks or slabs carved by artists as decorative additions to houses or monuments. This same mineral substance, $CaCO_3$, when treated and combined with other mineral materials and water is made into the concrete that contributes to the range of building styles that provide many of today's living environments.

Calcite is one of just three known minerals composed of $CaCO_3$, each with a distinct crystal structure. This common mineral is one of many identified in biomineralizing life forms (Skinner, 2005b; Table 1).

Many species of biomineralized creatures create biominerals with a distinct chemical composition, adapted to that particular living form, and often found in a specific site within the creature. Studies of biomineralized creatures, many still alive today and generating their distinctive forms in specific environments, are documented in thousands of papers that describe their unique morphologies and crystal chemistries. Over time it has been possible to define the roles that distinct biomineral materials play within these species (Lowenstam & Weiner, 1989; Skinner 2005b). Wherever these biological mineral materials are preserved they are known collectively as biominerals, and the study of their formation and growth is termed 'biomineralization'. Biominerals are mineral materials generated through biological and, to be more precise, cellular activity. Biominerals may be deposited within a species, or within its immediate surroundings or environment, as a result of the metabolism of the living creature (Skinner and Jahren, 2004).

It may come as a surprise to some readers that the human body contains its own suite of biominerals and some have functions analogous to those of other living forms. Humans, for example, do form calcium carbonate; however, it occurs in only one very specialized site, in the maculae portion of the inner human ear, as hexagonal crystallites <20 μm long referred to as statoconia or otoconia (Carlstrom *et al.*, 1953a). Human biominerals can be divided into two types: (1) those that are essential, a normal part of the expected anatomy and physiology of human systems, such as the mineral matter found in bones and teeth; and (2) those unexpected, undesired or pathologic mineral deposits. For example, we may ingest or inhale mineral particles from the environment. Asbestos, a well known example, is a natural 'hazardous' mineral. These minerals are not useful to the body and although asbestos may deposit, for example in lung tissues, the body reacts, often sealing off these interlopers *via* encapsulation within an organic layer. This organic layer may subsequently mineralize. There are several normal physiological processes that protect body sites and body systems from hazards that may inadvertently enter the human or animal. Any unwanted materials may be excluded and exit the body via sneezing or, if swallowed, pass through the digestive system and be

Table 1. List of biominerals (from Skinner, 2005b, reproduced with the permission of the Mineralogical Society of Great Britain & Ireland).

Class	Chemical designation	Example
1	Native elements/alloys	S
2,3	Sulfides/arsenides	FeS_2, pyrite, marcasite
		FeS_2S_4, greitige
		$(Fe,Ni)S_{0.9}$, mackinawite
		$Fe_{(1-x)}S$, pyrrhotite, where $x = 0{-}0.2$
		$FeS \cdot nH_2O$, hydrotroilite
		ZnS, sphalerite, wurtzite
		PbS, galena
		Ag_2S, acanthite
		Ag_2S_3, orpiment
4–8	Oxides/hydroxides	$FeFe_2O_4$, magnetite
		Fe_2O_3, amorphous
		$Fe^{2+}TiO_2$, ilmenite
		α-$FeOOH$, goethite
		γ-$FeOOH$, lepidocrocite
		$5Fe_2O_3 \cdot 9H_2O$, ferrihydrite
		Mn_3O_4, amorphous
		$(Mn^{2+}Ca,Mg)Mn_2^{4+} \cdot H_2O$, todorokite
		$Na_2Mn_2O_2 \cdot 9H_2O$, birnessite
9–12	Halogenides	CaF_2, fluorite
		K_2SiF_6, hieratite
		$Cu_2Cl(OH)_3$, atacamite
13–17	Carbonates	$CaCO_3$, calcite, aragonite, vaterite, amorphous
		$CaCO_3H_2O$, monohydrocalcite
		$(Ca_{(1-x)}Mg_x)CO_3$, magnesian calcite
		$(Ca,Mg)CO_3$, 'protodolomite'
		$Pb_3(CO_3)_2(OH)_2$, hydrocerrusite
28–32	Sulfates	$CaSO_4 \cdot 2H_2O$, gypsum
		$KFe^{3+}(SO_4)_2(OH)_6$, jarosite
		$SrSO_4$, celestite
		$CaSO_4$, baryte
37–43	Phosphate/arsenates	$Ca_5(PO_4)_3(OH,F,Cl)$, calcium apatite group
		$(Ca,X)_{10}(PO_4,CO_3)_6(OH,CO_3)_2$, bioapatite
		$\quad X =$ cations
		$CaHPO_4 \cdot 2H_2O$, brushite
		$Ca_8H_2(PO_4)_6 \cdot 6H_2O$, octacalcium phosphate
		$Ca_{18}H_2(Mg^{2+},Fe^{2+})(PO_4)_{14}$, whitlockite
		$Fe^{2+}(PO_4)_2 \cdot 8H_2O$, vivianite
Silicates		$Mg(NH_4)(PO_4) \cdot 6H_2O$, struvite
		$SiO_2 \cdot nH_2O$, opal, non-crystalline or disordered, cristobalite, tridymite

See Table 2.12 in Lowenstam and Weiner (1989), for a list of the phyla where these mineral species have been found.

naturally eliminated through normal pathways *via* the intestines to the anus or *via* the urinary system. The sequestration of any mineral within the body and the reactions we designate as pathological may lead to mineralized areas in tissues, in organs, or in some tumors (Skinner *et al.*, 1988). Inhaled or ingested particles of any natural minerals sequestered by the body defence mechanisms may become mineralized as part of normal protection reactions; any reaction products or deposits if they are part, or become part, of the human body can be thought of as biominerals.

More than twenty biominerals (Table 2) have been identified within the human body and more may well be added in future. The most common biomineral in humans is the one that forms in the skeleton. It is a calcium phosphate mineral known generally as apatite. There are, in addition, a few other phosphate species without natural mineral equivalents which are found in the body; they are also regarded as biominerals. Other biomineral species that form within us are, as one might expect, based on common chemical species found on Earth. They are carbonates, sulfates, oxides, oxalates and chlorides. The anions that give their names to these mineral groups are also characteristic of the body, and are constituents of body fluids. We humans are $>60\%$ fluids (Memmler *et al.*, 1996). Some human biominerals will be discussed in this chapter but only incomplete information is available on some of them. This chapter will

Table 2. Biominerals and pathological crystalline compounds in the human (from Skinner, 2000b).

Name	Formula
Hydroxylapatite	$Ca_5(PO_4)_3(OH)$
Fluorapatite	$Ca_5(PO_4)_3F$
Bioapatite*	$Ca_{10-y}\square Na_y(PO_4)_{6-x}(CO_3)_x(OH)_{2-x}\square$
Monetite	$CaHPO_4$
Brushite	$CaHPO_4 \cdot 2H_2O$
Octacalcium phosphate*	$Ca_2H2(PO_4)_6 \cdot 5H_2O$
Tricalcium phosphate*	$Ca_3(PO_4)_2$
Whitlockite	$Ca_9(PO_4)_3Mg(PO_3OH)$
Magnesium phosphate octahydrate (bobierrite)	$Mg_3(PO_4)_2 \cdot 8H_2O$
Magnesium hydrogen phosphate*	$MgHPO_4$ (?)
Magnesium ammonium phosphate hexahydrate*	$Mg_2(NH_4)(PO_4)_2 \cdot 6H_2O$ (?)
Diammonium calcium phosphate*	$(NH_4)_2Ca_2(PO_4)_2$ (?)
Calcium pyrophosphate dihydrate*	$Ca_2P_2O_7 \cdot 2H_2O$
Calcite	$CaCO_3$
Monohydrocalcite	$CaCO_3 \cdot H_2O$
Monosodium urate monohydrate*	$CaC_5H_3N_4O_3 \cdot H_2O$
Whewellite	$CaC_2O_4 \cdot H_2O$
Weddellite	$CaC_2O_4 \cdot 2H_2O$
Uric acid (uricite)	$C_5H_4N_4O_3$
Cystine*	$C_4H_6(NH_2)_2S_2(COOH)_2$
Struvite	$MgNH_4(PO_4) \cdot 6H_2O$

*Not considered a *bona fide* mineral.

begin with a brief overview of the methods and techniques that allow us to define bio-minerals, will outline normal apatitic biomineralization in bones and teeth, and will then discuss the crystal-chemical features of bioapatites because, not only are they essential in the human body, they are a major contributor to hazardous biomineral sequestration occurring pathologically in tissues that we do not wish to mineralize. Bioapatite deposits in the cardiovascular system, the aorta and the heart are obviously pathological, but so is the mineralization of tendons. I include brief presentations on other structures in the human body that may contain pathological but non-apatitic depos-its, such as kidney and pancreatic stones, before making a few concluding remarks about 'biominerals'.

2. Identification of biominerals

The precise chemical composition of any biomineral deposit in the human body, normal or pathologic, can be identified using generally available mineralogical techniques. Bio-minerals are characteristically exceedingly fine grained, often not well crystallized, and occur in rather minute quantities (other than the bioapatite in bone which has its own chemical and physical properties as discussed below). X-ray transmission, the method still employed to examine bones as part of the normal skeleton, such as when they have been broken, provide useful 'pictures'. This technique can also show odd materials such as swallowed pins (!) or other accumulated stomach or digestive system contents. The latest X-ray techniques include X-ray micro computerized transmission X-ray ana-lyses that allow any inorganic materials within body tissues, or specific organs, to be located rapidly using scanning observation. These techniques are successful because of the grossly different densities of the inorganic materials compared with the surrounding organic tissue materials such as proteins or lipids. The use of other energy sources for scanning, such as sound waves or any of the other relatively innocuous waves in addition to X-rays, are also used as long as they do no harm to the human body. Other, more precise, analytical techniques are discussed under the appropriate sub-sections below, especially where accurate crystal chemistry of the biomineral is desired.

Once a biomineralized site or deposit is located, the next step is accurate identification of the biomineral. This usually requires special preparation, such as excision or biopsy of the tissue to obtain samples. Initial analysis, post-detection may record the general composition and morphology of the mineral, but often more details on the species are needed. For these analyses, it is often a requirement to separate the inorganic from any potentially interfering organic molecules, those of the particular organ or those that provided encapsulation. An accurate composition and an unequivocal identification of the biomineral can be achieved using the wide range of high-resolution testing methods and techniques available on any extracted sample. Accurate biomineral identi-fication is often critically important information in defining a specific human disorder or disease, not only for diagnosis, but before undertaking any treatment. These actions should be coordinated with details on the individuals' other body systems as well as information such as the age, and genetic details of the patient if known, in order to make appropriate decisions to benefit their health. These undertakings are no less complicated for the human body than when geologists and geochemists consider the interactions in the multiple-component systems investigated in the Earth's environment.

In order to set the stage for the distinctive processes involved in biomineralization, so that the roles and contributions of our major human biomineral, bioapatite, can be appreciated, an outline of normal biomineralization processes precedes a summary review of apatite/bioapatite crystal chemistry. From this base, some of the details of the normal and major structures, bones and teeth, are presented before any abnormal or pathological situations are addressed. Pathological deposits usually result from either trauma, or localized disease processes distinctive of particular sites, and usually show specific signs, compositional markers detected generally by fluid (blood or urine) analyses, thereby providing indicators that are medically well known. A biomineral deposit, plaque formation around teeth in the oral cavity is pathological. However, since these deposits are relatively easily removed during dental cleaning, dental plaque is considered 'relatively normal' and benign. However, it is clear that mineral deposits in the pulmonary system may lead to major health issues such as when the biomineral interferes with normal bodily functions, sometimes precipitously (Skinner *et al.*, 1988). For details on pathological situations that are not presented in this chapter the reader should refer to specific treatises in the medical or dental literature. Finally, I wish to underscore that these sources rely upon well known mineralogical information and, in fact, the identical methods and techniques for mineral identification that mineralogists know and use. This statement emphasizes the crossover of mineralogy as essential to health sciences in general, and to the vast number of professionals in medicine and dentistry.

3. Normal biomineralization: bones and teeth

Humans are distinguished from all other forms of life, and other vertebrates, by the special shapes of our bones and teeth. To the casual observer, the skeleton is merely a scaffold on which to hang muscles that enable us to move. Teeth likewise seem lifeless and rock-like. Nothing could be further from the truth. Though the gross appearance of the adult skeleton appears permanent, we know that it has altered over time (Gibbons, 1997; Shubin, 2008). And today we are very conscious of variations in many parts of our bodies with age, disease and trauma, as well as genetics. The skeleton within our bodies is a living, dynamic, ever-changing structure.

There are 208 bones in the skeleton and 32 teeth in the oral cavity of a normal adult. Each one of these mineralized structures is a separate organ composed of tissues created by specialized cells. The cells first produce an extracellular matrix of connective tissues composed of intimately associated organic biomolecules that, shortly thereafter, biomineralize with calcium phosphate, bioapatite. The specific mechanisms of mineral deposition, growth, and continual repair or replacement, and the many other dynamic characteristics of the bone or tooth tissues, remain under investigation in spite of having been studied for >400 years (Glimcher and Lian, 1999; Skinner 2000a,

2005a). Today a very broad range of researchers, including biochemists, geneticists, internal medicine and orthopaedic physicians, dentists and engineers, especially those interested in prosthetics, investigate the dynamic processes that form and maintain the complex mix of biological and inorganic phases in bones and teeth (Cazalbou *et al.*, 2004a). The focus on the bio-composite nature of these structures is important, especially to an aging population predisposed to osteoporosis, or osteopenia, the diseases that affect the formation of bone tissue and cause unwanted changes and loss of function of the musculoskeletal system (Fig. 1) (Marcus *et al.*, 1996; Favus *et al.*, 2006).

All human bones begin as mesenchymal condensations in the embryo and develop in the foetus. Details of pre-natal and post-natal growth are well beyond the scope of this chapter. However, we now know that many unacceptable and unwanted effects may result from personal habits, food or drugs consumed by us, or by a mother pre-birth (Poggi and Skinner, 2012). Overviews in publications such as the series edited by Favus *et al.* (1996, 2006) and proceedings published by International Congresses in orthopaedics and dental areas are most useful as they document recent research efforts. There are also many popular books available which make more pleasant reading such as Plant and Tidey (2003) on the human body and nutrition, and Mestman and Herman (2009) on dental health. Detailed information on anatomical, physiological and clinical information can also be found on Medline, a site which has

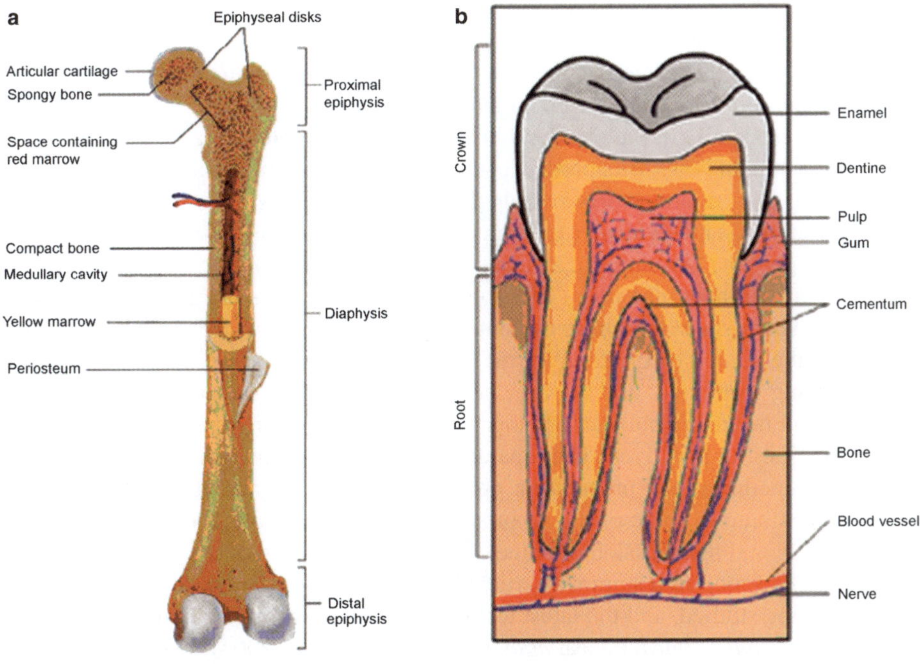

Fig. 1. (**a**) General structure of a mammalian bone. (**b**) Schematic diagram of a tooth (both images from Dorozhkin, 2007b, reproduced with the permission of Springer).

options not only for reading but for direct interactions on all manner of questions regarding normal bone growth and remodelling and the many disease states related to mineralizing systems. This internet site is probably the most up-to-date, considering the vast number of studies on patients and research undertaken all over the world.

It is worth emphasizing at this juncture that bones and teeth are composed of a complicated range of biomineralized tissues that consist of inorganic and organic components as well as multiple and distinct cell types. The next section focuses on bioapatite, the biomineral found in normal bones and teeth and that may precipitate pathologically in arteries, heart valves, joints and tendons.

4. Apatites and bioapatites: definitions and crystal chemistry

Apatite is a common mineral, found all over the world in igneous, metamorphic and sedimentary rocks; many varieties have also been synthesized (Pan & Fleet, 2002; Pasero *et al.*, 2010). The ideal formula for the calcium apatite mineral group and several subgroups can be expressed as $A_5(TO_4)_3Z$ or $A_{10}(TO_4)_6Z_2$ where A designates a site with cations such as Ca, Sr, Pb, Ba and that may contain single charged species such as Na, *etc.*; T indicates the anionic oxide tetrahedral site species, usually P, but others such as As, S, Si and V are possible; and Z is an anion species, which can be F, OH or Cl (see Table 1, p. 14, Pan & Fleet, 2002; Gaines *et al.*, 1997: *Dana's New Mineralogy* pp. 854–858, for general information on the group and subgroups based on chemical variations as well as the mineral names given to them.

Bioapatite, a separate and discrete mineral phase, is also very common. It deposits in living forms, especially in the teeth and bones of all vertebrates. It is an essential component of the human body. Bioapatite closely resembles hydroxylapatite in composition and structure; it is dominated by tetrahedral phosphorous-oxygen groups (PO_4) that bond to calcium (Ca), and always contains other cation and anion species. Bioapatites are exceedingly fine grained (size range 10–100 nm) and always occur intimately associated with organic materials (Tarasevic *et al.*, 2003). This complicates assays for the mineral which show the phase as compositionally variable. The variations depend not only on the source of the sample (the selection of specific sites in different bones and teeth being examined, and the age, gestation level, and preservation of the tissue which may not be fresh), but also whether it has been frozen or stored in formaldehyde or other solutions, or fossilized. The separation of the mineral phase by any of a number of extraction methods, and preparation processes for specific analytical investigations may also contribute to some of the reported chemical variations. The reported compositional ranges, and a discussion of some of the analytical methodology applied to bioapatites, can be found in a summary article by Elliott (2002). Some of the species and body sites being studied, and the latest techniques used in these research efforts, can be found in Fratzl *et al.* (1992), Puceat *et al.* (2004), Tang *et al.* (2004), Jager *et al.* (2006), Bertazzo and Bertran (2008), Chen *et al.* (2011) and McNally *et al.* (2012).

Because of their importance and potential use in humans, as bone graft substitutes or for dental tissue implants or replacements, investigations on bioapatite go beyond natural mineralized tissue materials and fossils into synthetics that are also usually hydroxylapatite or synthetic hydroxylapatite (Rosen *et al.*, 2002; Robinson *et al.*, 2004). Recent studies on bioapatite composition and physical properties have utilized the ultrahigh-resolution techniques now available: Atomic Force and Chemical Force Microscopy, Nuclear Magnetic Resonance Spectroscopy, small and wide angle X-ray scattering, as well as the more routine methods including Scanning Electron Microscopy (SEM), Raman Spectroscopy and Transmission Electron Microscopy (TEM) including three-dimensional TEM with options for selected area compositional analysis, and a back scattered electron imaging method described by Fratzl *et al.* (2004) to visualize the shape and size of the crystallites. For further details of many of these techniques, see Wogelius & Vaughan (2013, this volume) and other sub-sections in the present chapter.

Bioapatites were formerly considered amorphous because of their fine particle size and diffuse X-ray diffraction patterns compared with mineral hydroxylapatite (HAP), or fluorapatite (FAP), or alternatively an aggregate of several phosphate species (Neuman & Neuman, 1958; Elliot *et al.*, 1973). The calcium phosphate minerals close to hydroxylapatite are depicted in Figure 2a accompanied by diffraction patterns comparing hydroxylapatites with bioapatites.

The hexagonal morphology of hydroxylapatite and fluorapatite mineral crystals suggests that the phosphate apatites are members of space group $P6_{3/m}$ although recent investigations show that hydroxylapatite is actually monoclinic (Elliott *et al.*, 1973). Today, most investigations on bioapatites focus on the nucleation of the mineral rather than crystal-structure details (Cazalbou *et al.*, 2004b; Dorozhkin, 2007a).

Another view of bioapatite is as a mineral aggregate in which each tiny

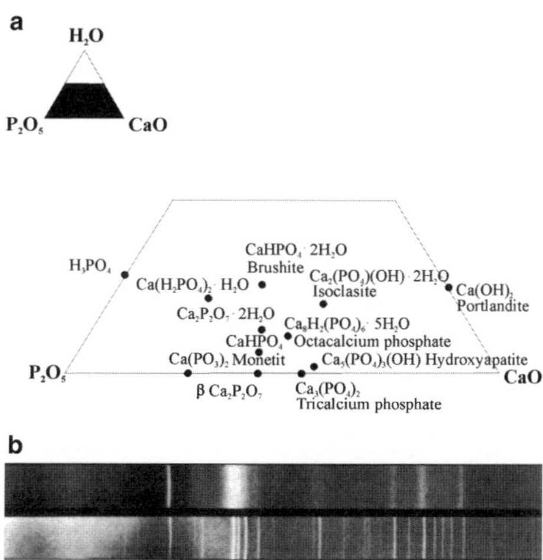

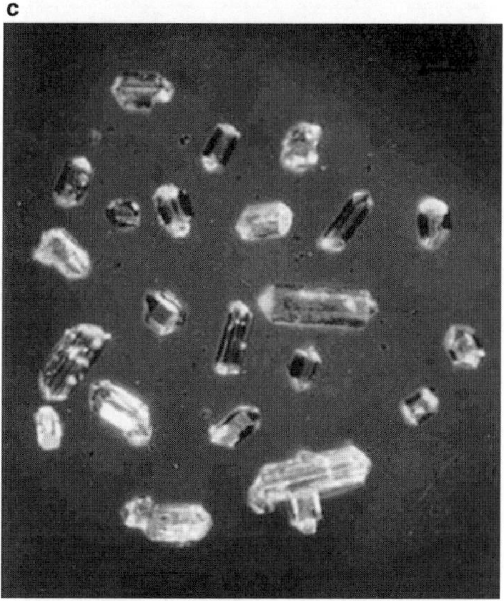

Fig. 2. (**a**) Part of the calcium-phosphate-water, CaO-P₂O₅-H₂O phase diagram showing minerals close to HA in composition (from Skinner, 2000b). (**b**) diffraction patterns of HA and bioapatite (from Skinner, 200b). (**c**) Photograph of synthetic apatite in which the crystallites are clearly hexagonal.

crystallite possesses its own unique composition, nucleated and assembled within the extracellular matrix at an advantageous site predicated by the matrix composition and structure. In addition, because of the very large surface area ($>100\ \mathrm{m^2/g}$ for bone mineral), crystallization may differ in different tissues as the protein matrix assembles and changes in the precipitation environment. Furthermore, the extracellular fluid contains not only a plethora of ions but could, and probably will, also vary compositionally over time. Therefore, compositional distinctions between cortical and trabecular bone or dentine *vs.* enamel can be seen as the response to the distinct tissue, cell type and crystallization environments (Chen *et al.*, 2011).

The following tables focus first on the compositional distinctions between geo- and bio-apatites (Table 3), the fluids in mammals from which bioapatites draw the elements for their precipitation (Table 4), and (in Table 5) the wide range of elements one might anticipate within the fluids and the bioapatite precipitate.

With the variety of sources, loci and timing of deposition, only ranges of composition for bones and teeth are usually presented, and often only a Ca/P ratio. Table 3 shows the differences between geo- and bio-apatites, Table 4 illustrates the range of elements

Table 3. Major element chemistry of geo- and bio-apatites* (from Skinner, 2005b, reproduced with the permission of the Mineralogical Society of Great Britain & Ireland).

	1	2	3	4	5
SiO_2	0.55	0.34	11.90	b.d.	b.d.
Al_2O_3	0.04	0.07	1.70	0.02–0.14	0.00–1.63
Fe_2O_3	0.14	0.06	1.10	reported as FeO	reported as FeO
FeO	0.24	0.00	b.d.	0.22–1.90	0.00–0.01
MnO	0.08	0.01	b.d.	b.d.	b.d.
MgO	b.d.	0.01	0.03	0.12–1.07	0.00–0.39
CaO	53.91	54.02	44.00	49.80–56.70	60.00–62.10
SrO	b.d.	0.07	b.d.	2.20–2.50	0.00–0.17
Na_2O	0.10	0.23	0.06	0.86–7.56	0.00–0.46
K_2O	0.01	0.01	0.05	0.06–1.04	0.00–0.01
REE_2O_3	1.69	1.43	3.75	0.00–0.23	0.00–0.37
P_2O_5	41.13	40.78	30.5	29.20–33.30	36.30–37.60
SO_3	b.d.	0.37	1.80	b.d.	b.d.
CO_2	b.d.	0.05	2.20	5.84–5.98	b.d.
F	2.93	3.53	3.10	b.d.	b.d.
Cl	0.71	0.41	0.04	b.d.	b.d.
Net	101.53	101.39	100.23	N/A	N/A
$-O=F,Cl$	1.23	1.58		N/A	N/A
Total	100.3	99.81		N/A	N/A

b.d.: below detection limit; N/A: not applicable
(1) Apatite from granitoid rock, Cougar Canyon, Elko County, Nevada (Chang *et al.*, 1996, Table 3).
(2) Fe-apatite ore, Cerro de Mercado, Durango, Mexico (Chang *et al.*, 1996, Table 4).
(3) Phosphorite (average), Phosphoria Fm., Colorado (Gulbrandsen, 1966).
(4) Ranges of typical bone bioapatite compositions (Skinner *et al.*, 1972; Rink and Schwarcz, 1995).
(5) Ranges of typical enamel bioapatite composition (Carvalho *et al.*, 2001; Brown *et al.*, 2002).
*From Skinner *et al.* (2004), reproduced with permission from Editura Fundatiel de Studii Europene, Cluj-Napoca, Romania.

Table 4. Compositions of seawater, blood, plasma, cell (from Skinner, 2000b).

	Seawater (mg/kg)		Mammalian		
		Blood (mg/l)	Plasma (mg/l)	Cell (mg/l)
Cl	19350	2900	3950	1890
Si	0.5–10	4.0	2.5	4.1
Na	10760	1990	3280	260
Ca	411	62	99	99
Sr	8	0.039	0.038	0.04
K	399	1690	170	3690
Mg	1290	41	22	61
Fe	0.002	475	1.14	1110
Cu	0.0005	1.07	1.12	0.98
Zn	0.002	6.5	1.6	12.3
B	4.5	0.13	0.17	(0.077)
F	1.3	0.36	0.28	0.43
Br	67	4.6	3.9	(5.6)
O_2	0.1–6			
O		775000	848000	689000
SO_4	2710			
S		2040	1220	3600
HCO_3	142			
C_{org} (diss.)	0.3–2	94200	40500	166000
NO_3	0.005–2			
N		33000	12000	
PO_4		0.001–0.05		
P			370	132

References: Seawater: Drever (1982), Table 10-1, p. 234; Mammalian blood/plasma/cell: Bowen (1966), Table 5.9, p. 81.

possible in precipitating environments, the bulk compositions of sea water compared with mammalian blood, plasma and cell (mg/l), whereas Table 5 lists those elements which could relatively easily, because of their ionic sizes (radii), become incorporated into the bioapatite group.

Table 6, taken from Dorozhkin (2007a), compares data for the chemical and physical variations in the mineral precipitates from different tissues. Note that the biomineral analyses include carbonate, an ion which will be discussed later in this section on bioapatites.

These analyses mostly show Ca/P ratios below that of stoichiometric hydroxylapatite (2.15 wt. ratio or 1.676 atomic), although not for bone. However, low Ca/P is the usual composition for early-forming bone, and other phosphate mineral species have also been detected [as shown in Figure 2a, and discussed by Skinner (2005a) and Dorozhkin (2007a)]. For example, octacalcium phosphate, $Ca_8H_2(PO_4)_6 \cdot 5H_2O$, a platy and metastable phosphate, has been suggested as the nucleating mineral phase of bioapatites (Brown, 1966; Elliott, 2002). Furthermore, electron microscopy studies on cross sections of human enamel crystallites, the largest of all bioapatite crystallites, show a dark central line attributed to octacalcium phosphate, and lattice defects such as screw dislocations that are parallel to the long axis (c axis) of the crystallites, as well as crystal bending. Any one, or all, of these structures could affect the chemistry,

Table 5. Ions that can substitute in calcium phosphate apatites, their charge and ionic radii depending on coordination number, i.e. VII, IX.

Ion	Ionic radius
Ca^{2+}	$1.06^{a,b}$ (VII)
	1.18^{a} (IX)
Cd^{2+}	1.14^{a} (VII)
Mg^{2+}	0.79^{a} (VII)
	0.89^{b} (VIII)
Sr^{2+}	1.21^{a} (VII)
	1.31^{b} (IX)
Ba^{2+}	1.38^{b} (VII)
	1.47^{b} (IX)
Mn^{2+}	0.90^{b} (VII)
	0.96^{b} (VIII)
Na^{+}	1.12^{b} (VII)
	1.24^{b} (IX)
K^{+}	1.46^{b} (VII)
	1.55^{b} (IX)
Pb^{2+}	1.23^{b} (VII)
	1.35^{b} (IX)
P^{5+}	$0.17^{a,b}$ (IV)
As^{5+}	0.335^{a} (IV)
Si^{4+}	$0.26^{a,b}$ (IV)
V^{5+}	0.54^{a} (VI)
S^{6+}	0.12^{b} (IV)
	0.29^{b} (VI)
Sb^{5+}	0.61^{b} (VI)
Al^{3+}	0.39^{a} (IV)
U^{3+}	0.98^{a} (VII)
Ce^{3+}	1.07^{a} (VII)
	1.146^{b} (IX)

([a]from Shannon and Prewitt, 1969; [b]from Shannon, 1976.)

crystallinity and reactivity of the mineral in biological systems (Daculsi & Kerebel, 1978; Iijima *et al.*, 1996; Elliott, 2002).

With such morphological and compositional variables, it has been suggested that bioapatite crystallites may also contain site vacancies, and ions such as HPO_4^{3-} or CO_3^{3-}. Some researchers, noting that bone bioapatite may contain a few percent of CO_3 on analysis, refer to this biomineral as 'carbonate apatite' (Skinner 2005; Wopenka & Pasteris, 2005).

The following formula for bioapatite shows it relative to hydroxylapatite:

$$(Ca,Na,H_2O,\square)_{10}(PO_4,HPO_4,CO_3)_6(OH,F,Cl,H_2O,CO_3,O,\square)_2$$

where $\square$ indicates vacancies in the lattice sites of the precipitated solid.

Table 6. Comparative composition and structural parameters of inorganic phases of adult human calcified tissues (from Dorozhkin, 2007a, reproduced with the permission of Springer).

Composition (wt.%)	Enamel	Dentin	Cementum	Bone	HA
Calcium[a]	36.5	35.1	C	34.8	39.6
Phosphorus (as P)[a]	17.7	16.9	C	15.2	18.5
Ca/P (molar ratio)[a]	1.63	1.61	C	1.71	1.67
Sodium[a]	0.5	0.6	C	0.9	–
Magnesium[a]	0.44	1.23	C	0.72	–
Potassium[a]	0.08	0.05	C	0.03	–
Carbonate (as CO_3^{2-})[b]	3.5	5.6	C	7.4	–
Flouride[a]	0.01	0.06	C	0.03	–
Chloride[b]	0.30	0.01	C	0.13	–
Pyrophosphate (as $P_2O_7^{4-}$)[b]	0.022	0.10	C	0.07	–
Total inorganic[b]	97	70	50	65	100
Total organic[b]	1.5	20	35	25	–
Water[b]	1.5	10	15	10	–
Crystallographic properties: lattice parameters (± 0.003 Å)					
a axis, Å	9.441	9.421	C	9.41	9.430
c axis, Å	6.880	6.887	C	6.89	6.891
Crystallinity index (HA = 100)	70–75	33–37	C	33–37	100
Typical crystal sizes (nm) [205, 231, 233]	100 μm × 50 nm × 50 nm	35 × 25 × 4	C	50 × 25 × 4	200–600
Ignition products (800°C)	β-TCP + HA	β-TCP + HA	β-TCP + HA	HA + CaO	HA
Elastic modulus (GPa) [607]	80	15	C	0.34–13.8	10
Tensile strength (MPa)	10	100	C	150	100

Due to the considerable variation found in biological samples, typical values are those given in these cases.
[a]Ashed samples.
[b]Unashed samples.
[c]Numerical values were not found in the literature but they should be similar to those for dentin.

Although it is not possible to show a crystal-structure figure for bioapatite, the hydroxylapatite structure should enable the reader to appreciate some of the structural issues still under investigation. Phosphate tetrahedra (Fig. 3) form the backbone of the structure and Ca occupies two distinct sites, one with 7-fold symmetry and the other 9-fold symmetry (see Fig. 3b,c) (Tang *et al.*, 2009). These two cation sites facilitate the wide range of possible cation substitutions shown in Tables 3a and 4. Note the differences between the compositions of apatite minerals, the average for phosphorites, and the ranges for bioapatites from bone, dentin, cementum and enamel (Hughes & Rakovan, 2002; Kohn & Cerling, 2002; Jager, *et al.*, 2006). It is important to emphasize that the ranges presented for bioapatites will have been taken from individual analyses representative of a specific location and that the analytical methods employed were probably not exactly the same. Clearly, presenting only a calcium:phosphorus ratio, Ca/P, for particular biomineral matter is only the beginning of any discussion of bioapatite chemistry.

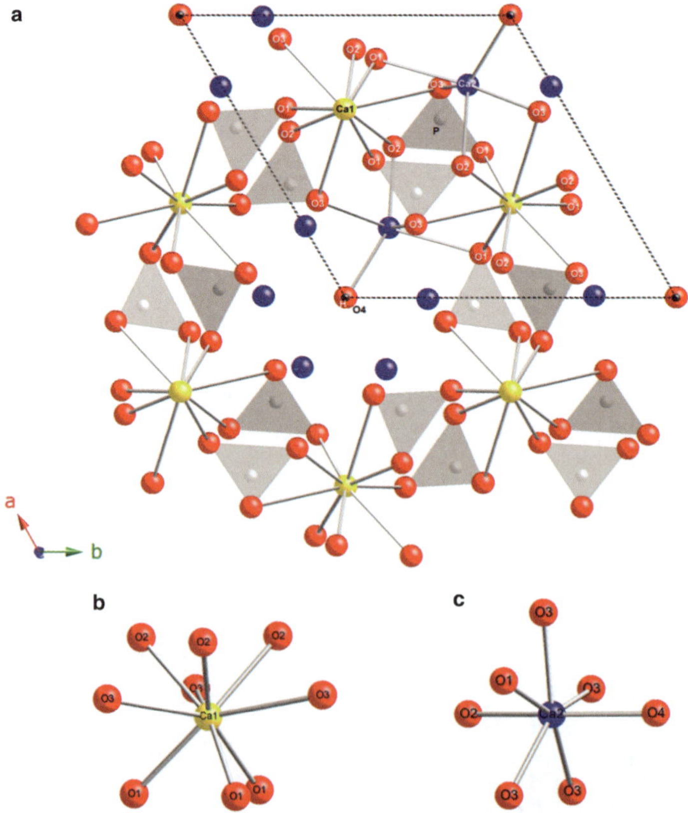

Fig. 3. Crystal structure of hydroxylapatite (from Tang *et al.*, 2009, reproduced with the permission of Elsevier).

5. The mineralizing system: CaO-P₂O₅-H₂O

Many investigations of the simple system $CaO\text{-}P_2O_5\text{-}H_2O$ have been undertaken. Experiments at 2 kbar (200 MPa) total pressure and in the temperature range 300–600°C show that hydroxylapatite is a stable solid phase over a large part of the total composition range (Skinner, 1973). However, hydroxylapatite, is usually accompanied by a second solid phase and fluid (Skinner, 1973).

The second solid phase present at equilibrium depends on the temperature of the experiments. At 300°C, monetite

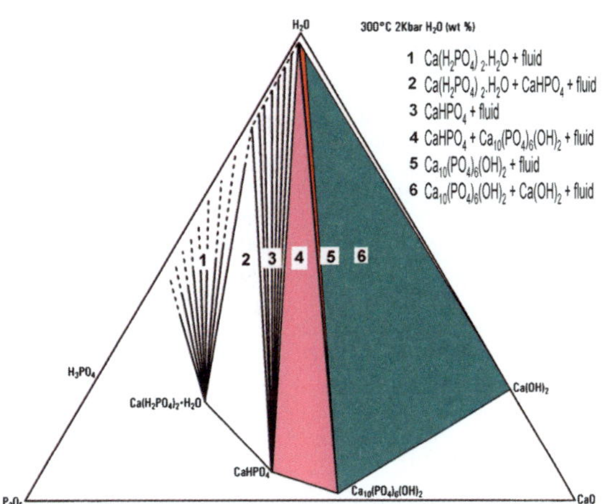

Fig. 4. Phase diagram of the system $CaO\text{-}P_2O_5$ -H_2O at 300°C (H_2O pressure) to illustrate the several calcium phosphate phases that may occur with hydroxylapatite under equilibrium conditions (from Skinner, 1973, reprinted by permission of the *American Journal of Science*).

(CaHPO, $Ca/P = 1$) is the second phase that occurs with hydroxylapatite, as seen in the lower-calcium, high-phosphorus regions of the diagram (Fig. 4) and the pH of the associated fluid in this phase field (#4) is 4. Note that $Ca_2P_2O_7 \cdot 2H_2O$, with $Ca/P = 1$ (Fig. 2a) is not a stable phase at equilibrium at these temperature and pressure conditions. At higher temperatures, 500°C, $Ca_2\,P_{27}$, an anhydrous phase with a $Ca/P = 1$ is the second phase, the stable calcium phosphate taking the place of monetite. Only in the tiny field #5, (coloured red in the figure) is hydroxylapatite the only solid phase with fluid, and that fluid has a pH of 7. In the high calcium part of the diagram (field #7) the second phase is Ca(OH) and the pH is 12. Hydroxylapatite occurs over a wide range of Ca and P concentrations and a very wide pH range of 4–12. It seems clear that hydroxylapatite may form under the variable pH conditions that are found in natural bioenvironments and within the human body.

Lattice parameters of synthetic hydroxylapatites produced in this and other experimental studies show slight variations, with the *c*-axis parameter showing the least variation for all the samples, as shown in Table 7 (Skinner, 1968, 2000b).

As the experimental study was carried out at 2 kbar H_2O pressure with identical calcium phosphate starting materials, this parameter variability is somewhat remarkable. Results from other synthetic hydroxylapatite studies show similar variations, especially when fluorine or carbonate are incorporated. The cell sizes for enamel, dentine and a wide range for bone bioapatite also show variability. These variations reflect uptake of some of the many ions that can enter the mineral and bioapatite structures, and the differences in their ionic radii (Table 5) (Veiderman *et al.*, 2005). For example, it is well known from dental studies that fluorine (F^-) can substitute for the $(OH)^-$ at the surface

Table 7. Apatite lattice parameters.

		a (Å)	c (Å)	a (Å)	c (Å)
	T at 2 kbar	Ca/P = 2.20 (wt.%/wt.%)		Ca/P = 1.95 (wt.%/wt.%)*	
Hydroxylapatites (synth.)					
Skinner (1968)	315°C	9.421(1)	6.882(1)	9.416(1)	6.883(1)
Skinner (1968)	450°C	9.418(1)	6.884(1)	9.416(1)	6.883(1)
Skinner (1968)	600°C	9.4195(8)	6.880(1)	9.4224(5)	6.8819(7)
Trautz (1955)		9.421	6.884		
Mengeot (1975)		9.4176(5)	6.8814(5)		
Fluorapatite (synth.)		9.367(1)	6.884(1)		
Carbonate apatites					
Mengeot (1975)		9.557(3)	6.872(2)		
Vignoles *et al.* (1975)		9.373(2)	6.897(2)		
Eysel & Roy (1973)		9.476	6.867		
Enamel					
Jongebloed *et al.* (1975)		9.441(6)	6.884(6)		
Driessens & Verbeek (1990)		9.440	6.881		
Young & Mackie (1980)		9.441	6.878		
Shark enamel					
Glas (1962)		9.385	6.885		
Dentine					
Glas (1962)		9.42(3)	6.84(1)		
Bone		9.29–9.50	6.687–6.696		

*For synthetic hydroxylapatites of Skinner (1968).

of enamel crystallites, forming at least a partially fluorinated hydroxylapatite. Furthermore, fluorine, along with other trace elements can be taken up in fossil bone from the burial environment during diagenesis (Koenig *et al.*, 2008; Trueman *et al.*, 2006). However, the most stable calcium apatite phase is fluorapatite (Jokl & Skinner, 1973; Moreno *et al.*, 1974; White *et al.*, 1988).

Of all the possible substituents in bioapatites, the anion groups CO_3 and HPO_4 are worthy of further discussion, as they may be critical to the roles bioapatites play in the human body.

5.1. Carbonate (CO_3) in bioapatites

The occurrence of carbonate in the apatites was suggested 60 years ago by McConnell & Gruner (1940) and is still under investigation (Simpson, 1964, 1967; Vignoles *et al.*, 1988). With high-temperature and high-pressure experiments and IR spectroscopy, Bonel & Montel (1965) produced and characterized carbonate at two different crystallographic sites in the structure. They labelled one site, in which CO_3 substituted along the c axis at the OH site in the hydroxylapatite structure, CO_3-apatite A. The other, CO_3-apatite B, has CO_3 taking the place of some PO_4 groups (Harries *et al.*, 1987; Skinner, 1989). Other early studies on synthetic carbonate apatites, as well as the apatite-group minerals francolite (carbonate fluorapatite) and dahllite (carbonate hydroxylapatite), showed variations in the unit-cell parameters of powdered samples (Table 6). As CO_2 is a usual cell metabolite, it was generally assumed that the carbonate ion would be available for incorporation during biomineralization, and located in these sites.

The carbonate ion in bioapatite continues to be investigated using synthetic preparations. For example, combining Na-free $CaCO_3$ and $CaHPO_4$ at 100°C produced a sample with 4 wt.% CO_3 (Barralet *et al.*, 1998; Morgan *et al.*, 2000). In the presence of Na, an uptake of 22% CO_3 was obtained. Many investigations have utilized Raman, Infrared and Nuclear Magnetic Resonance Spectroscopy to measure and locate the CO_3 within the apatite crystal structure (Koch *et al.*, 1997; Krajewski *et al.*, 2005; Fleet, 2009). Reitveld structure analyses on X-ray powder diffraction patterns of the synthetic samples show a reduction in apparent PO_4 volume and P occupancy for high CO_3 content samples, but not as much as expected, suggesting that the CO_3 was not occupying many of the PO_4 sites. Such substitution led to a suggestion that a second calcium phase may have been produced (Elliott, 2002). The most recent investigations have utilized X-ray diffraction and FTIR spectroscopy to confirm the two separate CO_3 locations, and were able to distinguish the orientations of CO_3 ions along and within the z-axis channel (Fleet & Liu, 2007; Peroos *et al.*, 2006; Fleet *et al.*, 2011).

Another aspect that has focused on CO_3 in biominerals is its content in fossil materials and bone (Rey *et al.*, 1989). The purpose of these efforts has been to investigate the trophic levels and diets of early omnivores. Investigations on bone samples have focused on the composite make-up of the material with C in the collagen of the matrix and C in CO_3 in the biomineral (Trueman *et al.*, 2006; van der Merwe *et al.*, 2008; van der Merwe, 2012). Problems with reproducibility of the data led the investigators to suggest that fossil samples had suffered diagenesis or were affected by the method of biomineral extraction. Further investigations, and those by other researchers, concluded that data for bone and dentine were not reliable because of the extremely small biomineral crystallite size (Trueman *et al.*, 2008; Koch *et al.*, 1997) and efforts were then concentrated on enamel for determinations of the C from CO_3 in that biomineral. Enamel, with its larger crystallites, is a more stable biomineral, and ease of extraction with a dental drill provides clean samples for analyses. The most recent C analyses on enamel samples from Pleistocene and Holocene specimens from Olduvai, Tanzania and Kenya, confirm earlier investigations on omnivore diets, identifying the plant source as C_4 plants (Lee-Thorpe and van der Merwe, 1991; Skulan & De Paolo, 1999; Kohn and Cerling, 2002; Hoppe *et al.*, 2003; van der Merwe *et al.*, 2008; Cerling, 2011). Isotopic analyses of such materials now include $^{18/16}O$ determinations on enamel bioapatite. Oxygen isotopes can also provide information on ingestion in relation to water sources. Present results suggest that the modern water source differs from that of the past. Heavy oxygen isotope $^{18/16}O$ determinations show that the rain source is now from the East, the Indian Ocean, a shorter distance for clouds to arrive at the hominid sites than from the S. Atlantic, the previous source. Presumably, weather and climate patterns have altered. The greater distance travelled from the Atlantic for the clouds means that the heavier isotope ^{18}O would fall out leaving a distinctly lighter $^{18/16}O$ signal in the environment (van der Merwe, 2012). These distinctions were not only found in the enamel, but correlate with soil determinations.

5.2. Phosphate (HPO_4^{3-}) ions in bioapatites

Another ion found in analyses of the mineral deposited in bone tissues is acidic ortho-phosphate, or HPO_4^{3-} (Wu *et al.*, 1994; George & Veis, 2008). This ion has been detected adjacent to mineral nucleation sites in the fluid phase associated with the precipitating biomineral. High P concentrations in early-formed bioapatites have usually been non-uniform with non-stoichiometric composition. Phosphorous is not only a stable elemental species with buffering capacity in the pH range of $\sim$7, but is an element essential to all living forms. Examples include the PO_4 on DNA molecules and the phosphate part of the energy system ATP-ADP-AMP. Cells accumulate phosphorus intracellularly for the energy system which operates by transferring phosphate groups from and between ATP (adenosine triphosphate), ADP (adenosine diphosphate) and AMP (adenosine monophosphate). Phosphate is therefore a cofactor in all biomolecular syntheses and metabolic pathways.

The idea that bioapatite is induced to form by phosphate groups associated with the initial collagen matrix molecules was suggested by Glimcher (1976). Other suggestions have been that brushite or octacalcium phosphate, with their greater PO_4 contents were the first phases to precipitate, or were intermediates in the mineralization process before a mature bio-mineral phase, more closely resembling hydroxylapatite, formed (Neuman & Neuman, 1958; Brown, 1966: Brown *et al.*, 2002). Even in the simplified synthetic calcium phosphate-water system described above, the other solid phases in the diagram show incongruent relationships with hydroxylapatite which are not yet well understood (Van Wazer, 1966; Omelon & Grynpus, 2008). It is known that initial precipitates, produced by adding phosphate to calcium-rich solutions, greater phosphorous (P) contents than that of stoichiometric hydroxylapatite (and therefore a lower Ca/P) so it is not surprising that the solid precipitated in mineralized tissues remains under study.

A variety of chemical techniques for estimating the Ca/P ratio in synthetic chemical systems have been devised but previous suggestions that the earliest deposited biomineral, was 'amorphous calcium phosphate' or brushite, phases which have higher phosphate (PO_4) compositions, have not been borne out in recent research (Roberts *et al.*, 1992). Attention has focused on octacalcium phosphate as an important initial or intermediate phase in the biomineralization process in bone and the enamel in teeth (Brown, 1966; Elliott, 2002).

Interestingly, a hypothesis proposed in 1970 that focused on the importance of ortho-phosphate has recently been revived (Lehninger, 1970; Omelon and Grynpas, 2008). The proposal offered a biomineralizing system that could resorb and deliver biomineral while maintaining the viability of bone and/or cartilage. It invoked a role for mitochondria, a subcellular organelle found in most body cells, as the foremost player in the process. Mitochondria are able to store polyphosphates (P-O-P-O...) and calcium, but do not form apatite except for ultrafine mineral granules occasionally detected in bacteria or within this sub-cellular component. The proposal was that in bone and in cartilage, mitochondria at sites of biomineralization, together with the actions of appropriate special enzymes, alkaline phosphateses, carbonic anhydrases and polyphosphatases, make PO_4 ions available through hydrolytic actions and assist apatitic mineral formation.

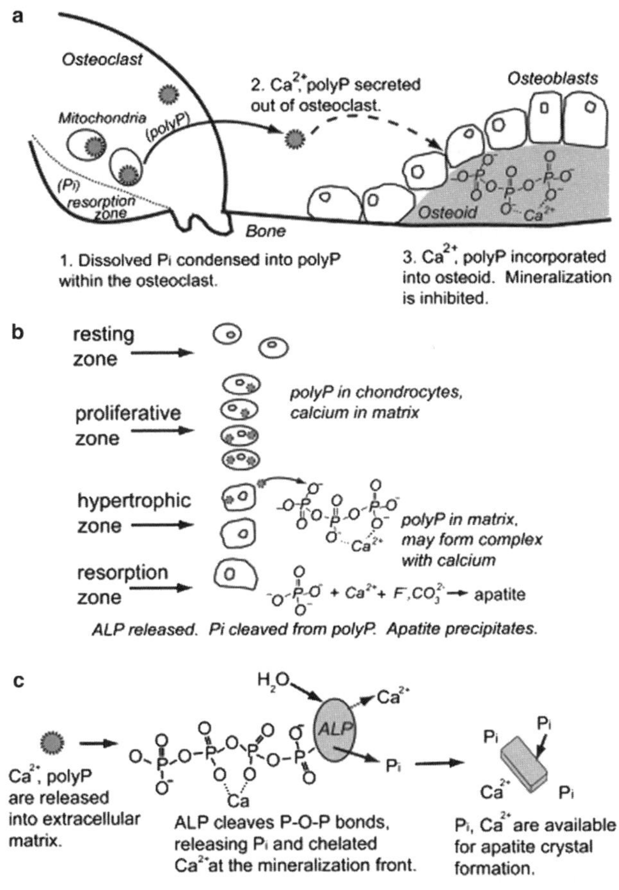

Fig. 5. Hypothetical roles of cells and polyphosphates in bone remodelling (from Omelon *et al.*, 2009). ALP: alkaline phosphatase; poly P: polyphosphate.

The system in bone would start with osteoclasis that dissolves the mineral matter by acidification, with local mitochondria through enzyme associations and activity breaking appropriate bonds, storing the phosphate PO_4 ions internally which can then be provided precisely where and when needed. Bone mineralization, therefore, could take place at some later time, perhaps days after the bone-forming cells, osteoblasts, produce the collagen matrix. Biomineralization would be assisted by localization of the PO_4 ions by the mitochondria and provide the very ion needed for an appropriately dynamic demineralizing and mineralizing system. Figure 5 shows the central importance of mitochondria, suggested by Leninger as essential in biomineralization (Omelon *et al.*, 2009).

Omelon and Grynpas (2007) gave some credence to this hypothesis by describing a high-resolution fluorescent microscopy and polyacrylic gel electrophoresis method to detect released PO_4 *via* alkaline phosphatase hydrolysis. The phosphatases break down polyphosphates, releasing PO_4 ions. The technique uses DAPI (4,6-diamidino-2-phenylindole) as a 'flagging marker' to locate phosphate ion appearance and increase in the hypertrophic zone of calcifying cartilage. By employing specific wavelength emissions, spectral shifts of enzyme hydrolysis between PO_4 arising from DNA in tissues from that of the polyphosphates can be discriminated. This technique, used to investigate a possible mitochondral mechanism, will undoubtedly be applied in studies of a wide range of creatures and sites, as phosphate and phosphatase activity are important in many biomineralizing systems.

Any phosphate made available, perhaps HPO^{3-} in the fluid phase or from the phosphate groups on biomolecules such as cell membrane phospholipids, might assist nucleation of biominerals (Chefetz & Sprecher, 2009). Other enzymes and ions, *e.g.* calcitonin or HCO^{3-} supplied to the body fluids from metabolizing cells would extend the anions available to initiate the biomineral phosphate phase. Biominerals in tissues are convenient: they are a nearby source not only of the phosphate for the energy requirements of cells, but of other elements also because the sources of phosphate, carbonate and enzymes are all generated under local cell control as is the appearance of bioapatite.

5.3. Summary

Phosphates: Geochemical, Geobiological and Materials Importance, published in 2002 (by the mineralogical Society of America) as volume **48** of the Reviews in Mineralogy and Geochemistry series provides a useful summary of data on phosphates in general, both mineral apatites and bioapatites, and, in a series of chapters, offers information on crystal structures, compositional variations, crystal growth and other issues of trace-element and ion incorporation in phosphates (see Kohn *et al.*, 2002). Some of this information is included in the present chapter. Other interesting information on apatites, such as their use in dating, in nuclear-waste disposal, as well as a review of the phosphorous cycle are recommended for general background; for the latter see Filippelli (2002).

6. Bone as a tissue

Each of the >200 bones in the human body has a separate gestation, growth and maintenance system that is initiated, modulated and repaired by specialized cells (Neuman & Neuman, 1958; Albright & Skinner 1987; Weiner & Dove, 2003; Sahai, 2005; Skinner, 2005a; Boskey, 2007; Veis, 2011). *In utero,* the initiation and growth of bone organs takes place *via* either one of two pathways: (1) that creating the appendicular skeleton, leg and arm (long) bones that start as preformed cartilaginous models of the bone structures called intra-membranous bone; and (2) a second mechanism of gestation, intra-membranous, which is typical of the skeletal flat bones as found in the skull. These two precursor pathways have phylogenetic significance but no vestige of these pathways remain by the time a bone is at the adult stage (Patterson, 1977). However, we know of these pathways because they are produced by separate cell systems, chondroblasts for the cartilaginous sequence, and osteoblasts for the other. The final results are indistinguishable (Wang & Nancollas, 2009).

The first action of the cells, either osteoblasts or chondroclasts, is to produce an extracellular organic matrix which consists predominantly of fibrous asymmetric proteins, the collagens, other globular protein moieties such as the glycoproteins, and small biomolecules that are part of the cellular metabolism (Sasaki *et al.*, 2002; Antipova & Orgel, 2010). Bone cells then make a singular and most important contribution: they aid the deposition of the biomineral that appears on, in and adjacent to the collagen fibres in the organic matrix. Whatever the mechanism, the collagen is necessary both in the accumulation of phosphate and calcium, the main elements in bioapatite to be concentrated from the surrounding extracellular fluid and the processes that localize biomineral and make bone (Canalis, 1996; Xiao *et al.*, 2007; Fratzl *et al.*, 2004; Shim *et al.*, 2009). The production of bone tissues is found in all mammals, from mice to humans, and that makes possible laboratory investigations into the gestation and growth of bones and the mineralization by bioapatite in a wide range of vertebrate species (Eagle *et al.*, 2011).

To provide some background information on the organic matrix: collagen 1 is the fibrous form in the intramembranous bones, and collagen 2 is the fibrous protein in the cartilaginous model. These two asymmetric fibrous protein biomolecules have slightly different compositions, although they both are composed of amino-acids accumulated and organized within the cells (Bornstein, 1974). Collagens are the body's major connective tissue protein found in all body systems and organs including skin. Over 23 chemical varieties of collagens have been identified and only some mineralize (Parry and Squire, 2005; Wess, 2005). As mineralization is essential for bone but pathological elsewhere, such as in the aorta, part of the cardio-vascular system, the chemical specifics of the local collagen in tissues is critical. It is also known that among different collagens there may be mutations. If there are mutations, they may limit bioapatite deposition and alter bone-tissue formation at specific sites and compromise one or more of the bone organs.

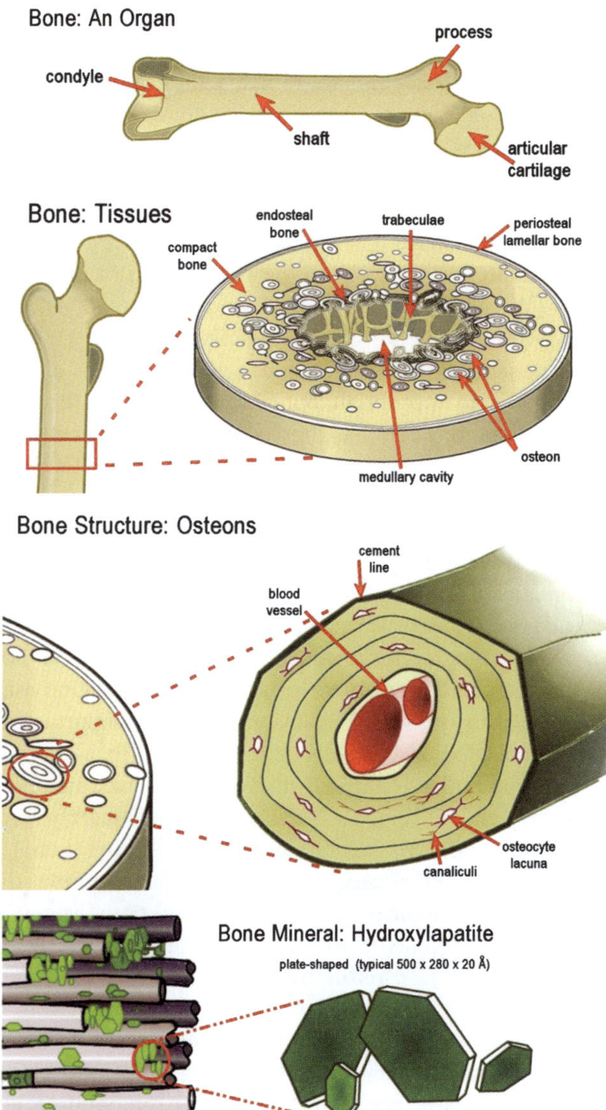

Fig. 6. Colour sketches of bone at the organ, tissue, cell and sub-micron levels to illustrate the site of the tiny apatitic crystals within the extracellular fibrous collagen (protein) matrix (images courtesy of William Straight).

Collagen 1 and 2 are both triple stranded protein molecules with a fibril size of 15 Å × 2800 Å (depicted in Fig. 6). After secretion into the local extracellular milieu, the fibrils aggregate in a special organized fashion, cross link, intertwine with other biomolecules to form fibres, in an extracellular gel-like mesh (Bornstein, 1974; Orgel *et al.* 2011). Mineral deposits on and within this mesh may be aided by specific amino acids and other characteristics of the aggregate (Skinner, 1987; Skinner, 2005a; Reddi & Reddi, 2009). The normal biomineralization progressions that have been studied demonstrate that a range of enzymes are involved and facilitate the succession from the protein complex to the localization of bioapatite, all of which remains under study (Weiner & Traub, 1991; Simon *et al.*, 2004; Xiao *et al.*, 2007; Toroian *et al.*, 2007; Roschger *et al.*, 2009; McNally *et al.*, 2012).

As bones grow, they increase in all dimensions and osteoblasts, continually created, are supplied to appropriate bone organ sites from the marrow and maintained by the circulating blood system. As they continue to produce the organic matrix that mineralizes with bioapatite, osteoblasts become surrounded, effectively embedded, but remain viable, keeping their connections to the blood circulation system (Jee, 1983; Skinner, 1987; Nancollas & Wu, 2000). The buried cells, known as osteocytes, are the bone-tissue maintenance cells. Osteocytes are connected through pseudopodia with each other and canaliculated to the arterial and venous systems which not only supply essential elements but dispose of any wastes, maintaining a dynamic and viable bone system. Oxygen arrives *via* the haemoglobin molecules in erythrocytes (red blood cells) while other elements or species, as required, move to assist the metabolizing osteocyte.

The bone system has a third cell equally important to its dynamics. Osteoclasts are multi-nucleated giant cells (50–250 nm) that literally 'chew-up' portions of the already mineralized parts of the formed tissues, including the mineral. Tissue destruction, or remodelling, is necessary to adequately sculpt each organ. Remodelling ensures that bone tissues retain a proper and functional shape as bones increase in size towards skeletal maturity. Osteoclasis, or extracellular tissue resorbtion, is also essential in the normal repair of fractured and broken bones. When osteoclasts solubilize bone tissues, the elements from the mineral and the matrix released appear to be reused for regeneration of new tissue by osteoblasts, thus retaining the viability of the organ and local function of the tissue. The rates of such ongoing reactions are remarkably rapid, taking place within seconds to minutes to weeks within all bone organs. Cells appear to be pre-programmed and linked, reacting to local and to systemic signals over the course of the multiphase bone production and remodelling activities, whether in the foetus or the adult. Distinctive histological textures in bone organs, woven, lamellar or haversian tissues express stages of bone development and reactions; see Figure 6 (Shore & Pozanski, 1996; Skinner, 2000b). Variations in the amino-acid composition and sequence in the collagen molecules may not only affect nucleation of bioapatite, but the structural integrity of the tissue and ultimately the function of the organ. The precise and coordinated mechanisms of the intricate interactive chemical shifts in this triple dynamic cell system involved in the production, maintenance and remodelling of bone draw on hormones and enzymes that may be produced elsewhere in the body, as well as locally and important and integral in other body systems (Millan, 2006).

All the data on important bone metabolites and the metabolism of the bone system have given rise to many hypotheses regarding transport of appropriate elements, and molecules, and their contributions to the formation of the complicated normal biochemical bone 'system' (Johnston *et al.*, 1996; Kacena *et al.*, 2004). Reviews of research efforts show that processes involved in bone formation and mineralization may be generally accepted, but details of how cells are programmed to consistently repair and remineralize continue to be studied (Fratzl *et al.*, 2004; Skinner, 2005a; Dorozhkin, 2007a,b; Pasteris *et al.*, 2008; Omelon *et al.*, 2009; Ono *et al.*, 2012).

Mineralization is but one facet in the production and function of normal bone tissue and organs. Studies of the importance of a particular element, such as calcium, in the diet, or in chemical supplements, or the effect of drugs or exercise regimens, all impinge on the generation, amount and distribution in the tissues and the continued bone organ vitality.

7. Teeth

Each tooth should be considered a separate organ with its own gestation, growth and mineralization schedule. There are three distinct tissues found in each tooth: enamel, dentin and cementum (Fig. 1b). Each has its own cell system and pathways of cell actions that are discrete.

Humans, of course, are not unique in having teeth. Studies of histological sections from dentine of several dinosaur genera have demonstrated similar histological structures and processes such as incremental growth lines and tooth replacement rates as found in humans (Robinson *et al.*, 2004). Dentine and cementum are physiologically similar to bone with the same amount of mineral relative to the collagen as seen in Table 6. Cementum is the more reactive tissue while dentine, once produced and fully formed, is sustained through exchange with local fluids but no new tissue is created. However, some changes may occur if the enamel is breached as a result of destruction of an individual tooth organ from external forces or a carious lesion. Enamel, on the other hand, the most highly mineralized tissue in the human body with >95% mineral, is generated by a separate group of cells as a cap, 0.5–3 mm thick, on the exterior of each tooth within the human oral cavity. The enamel cells, ameloblasts, first produce a non-collagenous organic matrix, amelogen, which directs precipitation of highly organized packets of bioapatite crystallites of much larger size than found in bone or in dentine. The cells that produce the enamel disappear once the tooth has formed and erupted. Therefore, post-eruption destruction of the enamel which includes bioapatite means there is no possibility of repair. Thus teeth, mineralized organs, have no natural way of regeneration, repair, or recovery of an individual tooth or teeth.

In humans there are usually two sets of teeth created in the jaw. 'Baby' teeth and the permanent dentition which in most adults is completed by ~18–25 years of age when the final four 'wisdom' teeth are fully erupted into the rear of the tooth rows, one each at top and bottom and the left and right in the oral cavity. Lattice dimensions for human enamel are somewhat larger than those of bone or dentine and synthetic hydroxylapatite

(Tables 4, 5). Shark enamel, which contains 3 wt.% F, compares well with synthetic fluorapatite (Young & Mackie, 1980). Dentine resembles bone; both exhibit wide variations in analytical results between samples and investigators, an expression of the difficulties of accurate measurement from a variety of sources of poorly crystalline materials. Dental plaque or calculus often shows some Mg (Leach, 1973). Brushite, octacalciumphosphate, weddelite, whewellite and calcite have also been identified in dental plaque but the dominant phase in these pathological tissues is usually apatitic (Driessens and Verbeeck, 1990, Table 15.1, p. 296). For more detail on the biomolecular components, formation and maintenance (plus any resorption or destruction of dental tissues) please consult the many volumes that have detailed these structures *e.g.* Miles (1967), for enamel, Frank *et al.* (1960) and Fernhead & Stack (1971) and a more recent paper by Brown *et al.* (2002). For substantive discussions on prevention of caries and the use of fluoride (F), the website published and updated by National Institutes of Health, National Institute of Dental and Craniofacial Research (http://www.nidcr.nih.gov/OralHealth/Topics/Fluoride) is recommended.

8. Summary

To conclude this brief background on normal biomineralizing human body tissues, bones and teeth, I would like to emphasize that all except enamel are dynamic, constantly remodelled throughout life, and will respond to the environment in which they grow, whether in the mouth or in bones. This fact must be borne in mind as we try to evaluate the remarkable amount of research that has been done and published on these tissues. Aside from fact that mineralized tissue samples are usually extracted for morphological or chemical analyses, even the immediate or modern samples will be 'out of the environment' in which they formed. Fossil materials are even more difficult to evaluate as they may have suffered changes related to the specific, possibly novel, environment in which they were found. The possibility that biomineralized tissues have been altered is great. Enamel biomineral is most likely to be selected for study because of the larger crystallites, the virtual absence of organics and greater stability. Enamel also has a shorter gestation period and cannot change, as there are no interior metabolizing cells. Isotopic and chemical measurements, however, should first be evaluated for any diagenetic effects, as they may not only influence the organic content but the composition of the mineral phase.

The study of skeletal tissues has become a way to reconstruct vertebrate history, from the earliest humans (Van der Merwe *et al.*, 2008: Cerling, 2011). This information has been used to interpret diet and rates of growth of ancient humans and can also be obtained for any modern humans or animals. Considered a 'permanent recording system', today we may use it to investigate the use and abuse of medication, or of disease in living individuals and communities all over the world. Some samples may be easily available, *e.g.* teeth fall out or can be extracted readily. Because the biominerals in teeth and bones are relatively stable, we have gained much knowledge about individuals' 'history', and fascinating insights into 'how we live' from our ancestors to the range of humans existing today, with different races, sizes, gender, geographic environments and diseases.

The latest understandings on normal biomineralization were reviewed by Gibbons (2012). Discussion focused on the growth and changes in jaw bones, tooth wear and cavity formation in addition to the evolution of human bones and teeth. Comparison was made between skulls from the ancients, Egyptians for example, and modern humans. In the former, one observes precise alignment of the jaws, upper and lower teeth, *i.e.* no over-bites, and virtually no cavities. The only tooth problems were wear with age, probably the result of chewing rough gritty probably fibrous plant materials as food. Present-day dental health problems are attributed to a modern diet that shows marked increases in ingestion of soft foods with highly refined carbohydrates and sugar that is producing prominent jaw alterations and cavities, even in young children. These foods promote the growth of *staphlococcus* bacteria in the mouth, which leads to mineral destruction and deposition of plaque. The modern diet of high carbohydrates, along with decreased physical activity, is thought to contribute not only to oral troubles and jaw changes but also the accelerated changes now encountered in the bony skeleton in general, as hominids have evolved to adopt an erect posture (Gibbons, 2012).

9. Pathological bioapatite mineralized sites in humans' arteries

Bioapatite may deposit in soft tissues or organs throughout the body, a pathologic occurrence. One area with bioapatite deposition undergoing active research today is in the cardiovascular system. Atherosclerotic plaque 'calcifications' can be visualized using echocardiography focused in the area of the heart (transthorax). These tests use high-pitched sound waves (ultrasound) to pick up echos from different-density structures to produce photographic images (Thompson, 2010; Cheitlin *et al.*, 2003). Mineralized tissues in or around the heart can be visualized because of density differences from the unmineralized tissues (Allison *et al.*, 2006; Freeman, 2012; Freeman & Otto, 2005).

An ultrasound image of calcification in the mitral valve (MAC = mitral valve calcification) and possibly in the aortic valve (AV) is shown in Figure 7. The c-shaped area segment at the base of the left ventrical (LV) of the heart and other structural areas showing similar light colour are probably regions that contain greater amounts of fibrous tissue which may appear light

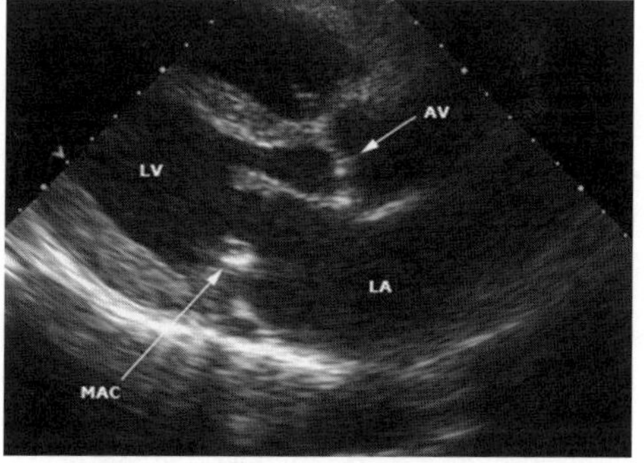

Fig. 7. Mitral annular calcification (parapsternal long-axis view) (AV: aortic valve; LV: left ventricle; MAC: mitral annular calcification; LA: left aorta). Reproduced with permission from: Freeman, R.V. Mitral Annular Calcification. In: UpToDate, Basow DS (Ed), UpToDate, Waltham, MA, 2012. Copyright © 2012 UpToDate, Inc. For more information visit www.uptodate.com.

on the image but not be calcified. Doppler echocardiograms also taken in this body site provide moving pictures that show not only valve activity and direction, but the speed of blood flow from which the volume can be calculated (Cheitlin *et al.*, 2003).

Cardiovascular testing has increased with the numbers of elderly in our society, and the appended references document details of the research and the increased survival chances of patients following identification and surgery (Harpaz *et al.*, 2001; Feindel *et al.*, 2002; Barasch *et al.*, 2006; Proudfoot and Shanahan, 2007; Arora *et al.*, 2008). Surgical intervention, which includes not only emplacement of a new valve but reconstruction of the mitral annulus and local decalcification, has been shown to be effective (Allison *et al.*, 2006; Papadopoulos *et al.*, 2009),

The differential densities of mineralized sites or, as it is referred to in most medical articles, 'calcification' or simply 'calcium' (!), is also found in the walls of different channelways in many body sites. The locations are thought to be the result of localized trauma and tissue inflammation, followed by fibroblast activity, the production of collagen fibers that may subsequently mineralize. As might be expected, considering the increase in genetic investigations worldwide, tissue 'calcinosis' in highly inbred populations has led to identification of specific gene mutations that can now be searched for and detected (Topaz *et al.*, 2006).

Mineralized tissue samples obtained during surgery or autopsy of humans show mineral aggregates which can vary from submicron size up to several millimeters in diameter (Allison *et al.*, 2006). After extraction of lipids and enzyme treatments with proteases such as pronase, papain, collagenase, or elastase that remove all the soft tissues, examination of the biomineral is possible. The following published elemental analysis of what may be labelled 'plaque' showed, as a percent of dry weight, $\sim71\%$ mineral with: Ca 29.3%, P 13.6%, weight ratio Ca/P $= 2.15$; CO_3^{2-} 8.69%, Mg 0.37%, Na 0.28%, Cl 1.08%, $<0.35\%$ F; $\sim15\%$ protein was calculated from the C, N, H determinations, and 4.5% moisture, evolved on heating to $110°C$ (Schmid *et al.*, 1980). The Ca/P ratio of the mineral fraction is exactly what one would predict for bioapatites/hydroxylapatite but the CO_3^{2-} content, $\sim9\%$, is remarkably high, not commensurate with a low-temperature, single-phase mineral. On the basis of weak, diffuse X-ray diffraction patterns, these authors concluded that the microcrystalline biomineral was apatitic, may contain an amorphous component and/or a carbonate phase. Virtually all apatites, hydroxyl- and carbonate hydroxyl-apatite minerals specifically, exhibit negative, low birefringence (0.002 and 0.009, respectively) as does calcite, $CaCO_3$, a common carbonate mineral that has large negative birefringence. These results, especially the positive optical sign, are anomalous if we are dealing with phosphatic biomineral materials (Gaines *et al.*, 1997).

Selected fragments from these samples, coated with gold-palladium, were subjected to SEM analysis and several different morphologies were noted: spherical particles with a diameter of 2.5 μm showing concentric layers, others up to 6 μm with spindle-like (openwork) centres, fibres with diameters of ~0.1 μm connected with some of the spherical particles, irregular particles to 2.5 μm, and separate flat, smooth plates 0.5 μm thick and 20 μm $\times$ 50 wide and long. These results suggest that these samples of aortic plaque were a mixture of different biomineral species. Although the investigators took pains to treat the tissue gently during extraction, it is possible some of these results are artifacts. The authors concluded that the plaque was microcrystalline, probably carbonate-apatite and, as no other mineral species was detected on diffraction analysis, was similar to the biomineral matter in bone.

Other studies of mineral deposited in cardiovascular tissues have found significantly smaller Ca/P ratios and significantly larger Mg contents. Several investigators suggested that these fine-grained phases are the mineral whitlockite, the ideal end-member composition of which is $Ca_9(PO_4)_6Mg(PO_3OH)$ and usually contains Mg and HPO_4^{2-} groups (Gopal *et al.*, 1974; Bigi *et al.*, 1980; see Tables 3a, 4).

The relatively high levels of Mg in aortic plaque, together with a reported inverse relationship between coronary heart disease with hardness of drinking water where high Mg is more likely, appears consistent. Early experiments on animals, and epidemiological studies has led to a suggestion that an increased ingestion of Mg salts is efficacious in reducing aortic mineral accumulation (Carlström *et al.*, 1953a,b; Crawford *et al.*, 1977). However, if another phosphate phase such as whitlockite forms rather than bioapatite, that is not a remedy, the biomineral is still pathological 'calcification'.

In the most recent investigations, a connection is made between valve and aortic calcification, systemic atherosclerosis, calcified deposits and age (Feindel *et al.*, 2002; Allison *et al.*, 2006; Barasch *et al.*, 2006; Arora *et al.*, 2008; Papadopolus *et al.*, 2009). The calcification appears identical to the formation of bone mineral. Indeed, lamellar bone tissue has been identified histologically in aorta samples from >1000 patients (Mohler *et al.*, 2001). These excized, histologically investigated samples showed osteoblasts and fibroblasts, bone morphogenic proteins (BMP 2, 4), calcitonin, osteopontin and other bone-related molecular species, leaving no doubt that the precipitation of biomineral is one of the stages of 'heterotrophic ossification'. This reaction appears to be the result of trauma, leading to inflammation, causing migration of pericytes from the marrow cavity to disturbed areas in the cardiovascular system. The migrated cells generate local mineralization, with many of the local components expressed in arteries and heart valves paralleling the bone system (Mohler *et al.*, 2001; Rajamannan *et al.*, 2003; Proudfoot & Shananhan, 2007).

It is interesting that the tissues of the venous system do not mineralize. Note that blood returning to the heart would be expected to have higher concentrations of CO_2 on the red blood cells and possibly more HCO_3^- in the fluid, but what effect such differences might have on venous pH, or Mg content, or whether any of these conditions modify the opportunity for future biomineral deposition in veins is unknown at present. The pathological mineral material in the aorta contains the same elements as the biomineral found in bones and teeth, carbonate-containing bioapatite.

10. Kidneys

Urine is often used to test for body metabolites in order to monitor health status. Of the several testing options there is one for calcium concentration (Skulan *et al.*, 2007). However, the kidney is another location in the body that shows pathological bioapatite

(Prie *et al.*, 2009). The disorder, acute phosphate nephropathy, was recently observed in older patients following the use of oral sodium phosphate as a purgative in preparation for colonoscopy examination. Both acute and chronic injury of the renal tubules can occur, and is often observed in older females who may have pre-existing kidney disease or other systematic disorders such as hypertension or diabetes, where the use of drugs may alter renal perfusion or colitis. Kidney injury from ingestion of the purgative in some cases has led to extensive deposition of bioapatite in necrotic tubular epithelium and/or as rounded concretions on the surface of desquamated epithelial cells in the tubular lumen. The deposition following ingestion of a large dose of sodium phosphate appears larger than accumulates in other acute kidney-injury situations, although some tubules display injury but no deposition. The purgative causes a loss of fluid and electrolytes, and an excess of phosphate in the kidney filtrates which leads to the calcium phosphate precipitation. The pathology is ascribed to 'hypovolemic induced avid proximal salt re-absorption'. Other risk factors for these patients, such as when they are undergoing treatment with certain drugs including diuretics, may also lead to kidney tubule injury and possible increased phosphate localization in the tubules. Epidemiological studies have forced the curtailment of the use of sodium phosphate (either as a solution or in tablet form) as purgatives (Markowitz and Perazella, 2009; Fogo and Kashgarian, 2012).

11. Tendons

Tendons form the attachment between muscles and bone and are composed of collagen that connects to the muscle fibrous proteins, actin and myosin. The connection and its viability are essential to body functions.

A generalized overview of the physical associations of tendons with proteinaceous soft tissue that merges into osseous tissue at the insertion sites is shown in Figure 8. Tendons complete the necessary connection of soft tissues with the hard tissues of the mineralized skeleton. Consistent stress on tendons may cause pain and the traumatic experience of breaking a tendon or pulling it out from its bony site locations causes excruciating pain, usually only tackled by sports-medicine professionals. Studies in this area of concern, utilizing animals for biomechanical testing, is localized to specific body sites. A vast literature exists about tendons, broken down to specific joints, individual physiology, disease and issues such as particular athletic activities. A summary of finger, hand and wrist tendon injuries illustrates problems in one

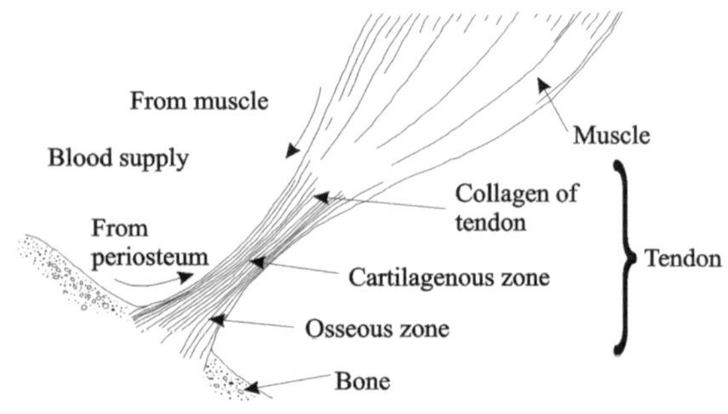

Fig. 8. Sketch of a tendon (from Skinner, 2000b).

important anatomical area (Ono *et al.*, 2012). The predominant tendon problems are pain, but this is not life threatening, a view underscored by exposing 'the myth of tendonitis' and calling it tendonopathy which can be treated with ultrasound, with rapid healing (Ebenbichler *et al.*, 1999; Khan *et al.*, 2002).

Tendons are elongated structures the lower parts of which, close to the insertions, show a combination of mineral and the collagenous fibres that feather out into proteins only (Fig. 8) Apatitic deposition can be found in peri-articular tendons as well as in the articular cartilage in joints. The deposition of bioapatitic mineral elsewhere in the tendons, a pathological, painful experience, has many analogies to apatitic calcifications in other tissues and is usually due to trauma or other disease-inducing processes. Some apatitic deposition has been observed in avascular areas, with the suggestion that changes in blood supply may affect local cell function; in general, this lowers calcification at those sites (Dieppe and Calvert, 1983, p. 201).

Pathological tendon calcification may take place and is more prevalent in men than women at ages over 60. Acute periarthritis is common in the shoulder but may occur in the hip, knees, fingers or wrist a result of minor or continued trauma, e.g. in carpel tunnel disease. The pain, stiffness, swelling and redness of the area may disappear over a few days, usually because the patient is treated with anti-inflammatory drugs (ibuprofin). However, if the area is aspirated, a milky fluid full of crystallites, usually 0.1 μm in size, but occasionally spherulitic particles (apatitic) up to 3–10 μm in diameter, or mineralized flakes, may be observed. The mineralized sites and cases are sporadic, but where multiple joints are involved, or when there is re-occurrence, there may be familial, or systemic instigators such as hormonal imbalance with genetic implications. Some treatments suggest reduction in calcium intake.

12. Other pathological biomineral deposits

12.1. Joints

Joint diseases are associated with pathological deposition of crystalline materials. Probably the best known joint disease is gout, the symptoms of which were described well by the ancient Greeks and Romans. There are other pathological mineral deposits found in joints such as in the diseases known as pseudo-gout and arthritis.

Of the three types of joints occurring in the human, the most familiar are those associated with limbs (Fig. 9) (Dieppe & Calvert, 1983). The limb bones are covered by special tissue known as articular cartilage, 2–4 mm thick. This cartilage is distinct from the cartilage described in the section on bone. It functions as the cushion for the sliding

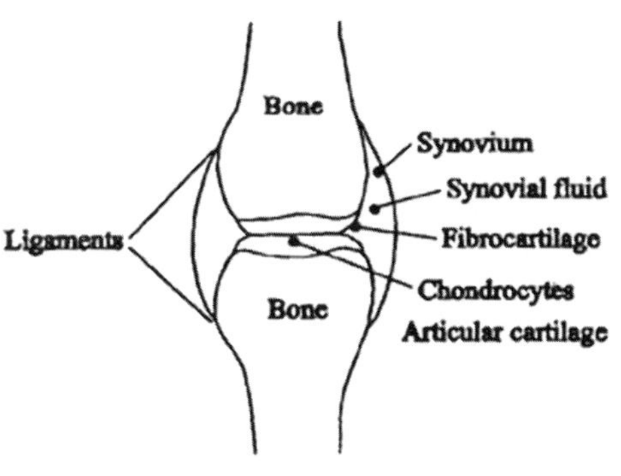

Fig. 9. Knee joint (from Skinner, 2000b).

movement of two bones meeting in a joint. Only that part of the articular cartilage closest to the bone shows blood vessels, cells (chondrocytes) and their products. The bulk of articular cartilage is intercellular ground substance, a swollen gel composed of 75% water, 15% collagen and the remainder proteoglycans (chondroitin-6-sulfate and keratan sulfate attached to a protein core in bottle-brush fashion). Where the articular cartilage on adjacent bones meets, the collagen fibres are closely packed and lie parallel to the surface. Below the surface of the cartilage, the fibres are random with variable distribution, and concentrated in the areas with most stress. Extra stability and control in some joints comes from additional fibrocartilage pads or menisci that surround the contact area within the joint space. The whole area of apposition of the two bones with their articular cartilage is encased by a cuff that defines the joint capsule. The cuff is composed of tough connective tissue thickened locally to form the ligaments that tie the bones together and control flexion and the degree of extension of the limb bones. The fluid in the joint space is confined by a thin (1–2 cell layers thick) vascular and flexible membrane that grades into a thicker intercellular matrix of ground substance or synovium. The synovial fluid has a gelatinous, egg white consistency, and is predominantly composed of protein/ polysaccharides, the main function of which is the nutrition of the avascular cartilage. Crystals appear in the synovial fluid, in the fibrocartilage, and in the capsular tissue that contains parallel bundles of collagen and a few cells (fibroblasts) that produce this protein. The general medical opinion is that pathological processes lead to the formation of crystals in joints, which then cause trauma and further damage the tissues. It is characteristic of crystal-induced inflamation that it starts suddenly, reaches a maximum, and is self-limiting within days. Crystalline materials with different compositions appear in the various joint sites mentioned above.

12.2. Gout

Gout is a systemic disease, a generalized metabolic abnormality in which crystals may be deposited anywhere in the body, but can be concentrated and directed by the local cell metabolism and physical environment (Laine *et al.*, 2010). The crystals found in gouty joints are monosodium urate monohydrate, $NaC_5H_3N_4O_3 \cdot H_2O$. They form in the joint spaces of the extremities as a result of the buildup and restricted excretion of uric acid, one product in the normal metabolism of the nucleic acid purine residues guanine and adenine. The expected and normal amount of uric acid circulating in serum is 100–400 μM (micro-molar) or ~ 1 g total in the normal human. It comes from two sources, the diet and the metabolism of normal cell systems.

Over the past 30 years specific enzyme defects in the pathways of uric acid metabolism, both synthesis and excretion, have been shown to lead to the buildup of uric acid. In extreme cases, uric acid concentration may be fifty times normal and produce accumulations of the urate monohydrate in tophi (swollen tissues that contain clusters of the crystals) at the elbow and at the joints in the toes and fingers (Richette & Bardin, 2010).

The crystals are triclinic, *P*1, platy to needle-like, and traditionally identified with polarizing optical microscopy (Mandel and Mandel, 1976). They may be up to 20 μm

long and 2 μm wide with strong negative birefringence (McCarty and Hollander, 1961). In tophi, the accumulated crystals are usually larger with smooth surfaces, whereas those from the synovial fluid often show surficial granules (probably protein) and may be found in the act of being phagocitized by leucocytes, the white cells that attack and remove foreign materials.

Urate crystals have been found in articular cartilage, peri-articular soft tissues, at the ends of the bone, around the achilles tendon, in the hands, eyes and eye lids, tongue, intestines, penis and kidney, but are characteristically absent from muscle, liver, and nervous tissues that contain little connective tissue. In cases of chronic gout, there may be extensive destruction of bone at the sites where crystals are deposited, but the mechanisms causing the destruction are unknown; certainly they are not mechanical as the crystals are soft and fracture easily.

A number of hereditary genetic disorders in the nucleic acid pathways occur and are associated with accumulations of crystals of other compositions such as the organic compounds xanthine, hypoxanthine, and oxypurinol. These crystals occur in the muscles rather than in the connective tissues typical of the urate deposition in joints.

12.3. Pseudo-gout

In a few patients with gout-like symptoms but no urate crystals in the synovial fluid, rod-shaped pyrophosphate crystallites, sometimes showing prisms and up to 10 μm long and 0.2–5 μm wide, were identified (McCarty *et al.*, 1962). X-ray diffraction analysis of crystalline materials in joint tissues over the past 40 years have identified $Ca_2P_2O_7 \cdot 2H_2O$, both monoclinic and triclinic forms, and hydroxylapatite (McCarty, 1976). (Note that the pyrophosphate materials have not been included as *bone fide* minerals.) Because of features noted on electron micrographs, the crystals were suggested to be twinned. Many fractured crystals are found in associated cartilage suggesting smaller fragments may provide additional nuclei after they reach the fluid (Pritzker, 1980).

The daily turnover of pyrophosphate as part of the cellular energy system ATP-AMP in a human is probably of the order of kilograms (total); in plasma at the sites where the action takes place, turnover of the very small amounts takes place in <2 min. The concentration of pyrophosphate in cell-free plasma for normal individuals is ∼2 μM, with excursions by a factor of two reported in analyses with cells (platelets). Normal synovial-level concentration is also ∼2 μM, and the urinary concentration is 10–100 μM. No excursions in systemic pyrophosphate concentrations have been found in patients with joint disease related to pyrophosphate crystals, referred to as pseudo-gout, although elevated amounts (about ten times normal) are present in their synovial fluids.

The formation of $Ca_2P_2O_7 \cdot 2H_2O$ crystals occurs as a local phenomenon. Disease is induced by the presence of these crystals that probably formed as the result of the metabolism of the chondrocytes in the articular cartilage. These cells, if they increase their metabolism, may release large quantities of pyrophosphate (Dieppe & Calvert, 1983, p.158). Because normal physiological calcium levels are ∼9 mg/100 ml (2.4 mM or millimolar) and 1–2.5 μM for total pyrophosphate measured at saturation level in the synovial fluid (estimated to be 40% ionized compared to 60% in plasma), the elevated pyrophosphate level measured in some individuals (5–60 μM) is very much supersaturated. The normal levels of pyrophosphate in articular cartilage, and their excursions, are not known but probably vary with the individual, with depth in the cartilage, and over time. Experimental studies have suggested that a change in pH, in the concentration of iron in the several tissues in the joint or in the cartilage, possibly aid the nucleation of the crystals.

Patients with chondrocalciosis (calcium-containing mineral deposits in cartilagenous tissues) may suffer from coincident diseases such as hyperparathyroidism which causes hypercalcemia, hypophosphatamia (low levels of the enzyme alkaline phosphatase), hypomagnesia (low Mg, a condition that could reduce the amount of Mg-pyrophosphate complexation, and increase calcium-pyrophosphate formation), haemocromotosis (a disease with excessive amounts of iron stored in connective tissues), hypothyroidism, and steroid therapy; these may induce sufficient systemic changes that local deposits form. Most of the debility associated with chondrocalcinosis increases with age, and trauma, but the onset is slower than for gout, and takes longer to resolve. In general, it is a less severe arthritis; although seen predominantly in the knee, it is also seen in ankles and wrists. The sites affected differ from those of gout-crystal deposition.

The latest view of cartilage calcium precipitation disorder, is that deposits of calcium pyrophosphate dihydrate crystals, usually in the elderly and at an average age of 72 years, increases with age and has no sexual predisposition, although men have more attacks (Becker and Ryan, 2012).

12.4. Summary

In summary, mixtures of different calcium phosphate minerals may occur in the same joint (Halverson & McCarty, 1979). In addition, bone fragments, cartilage fragments, talc or other materials from surgical operations, plant thorns, or other sharp and foreign objects such as grit, dirt and other materials from accidents that enter the capsule may be responsible for inflammation, crystal formation and the continuation of pain and difficulties. All three crystal deposits described here in joints (apatitic materials, uric acids and pyrophosphates) are associated with osteoarthritis, destruction of the articular cartilage and remodelling of the subchondral bone, and alter the normal structure of the joint. The most damaged joints usually show the greatest amount of apatitic or other crystallite mineralization. Cell-culture experiments had earlier shown that collagenase and other proteases (enzymes that attack the protein portions of tissue) arriving in the joints may also be responsible for chronic debilitating diseases (Schumacher *et al.*, 1981).

13. Stones: other pathological biomineral deposits in kidney stones

Several biomineral materials occur as calculi that form in the renal system, and are the result of supersaturation of the urine with at least one of the elements of the biomineral components. Hypercalciuria, or high levels of calcium absorbed by the intestines in some individuals, was considered the major cause of the deposition of the calcium oxalates (mineral names: whewellite and weddellite) in the kidney. These deposits are usually known as 'stones' because of their size and inorganic composition. In some stone formers, the calcium level may be brought back to normal, but the patients remain stone formers. Another group of patients with stones have normal calcium levels, while some individuals with high calcium levels do not form stones. Oxalosis, or hyperoxaluria, due to an elevation of oxalate excretion at the kidney, is also known and considered a possible reason for stone formation. Diet, fluid intake, absorption of both calcium and oxalate, and other circulating components such as sodium and organic species, and the excretion volumes and rates, are all necessary considerations in the formation of calcium oxalate crystals in the kidney.

Elevated urinary excretion of calcium and oxalate is typical of stone formers and particularly of repeat stone formers. Normal individuals excrete $\sim$5 mM (millimoles) of calcium per day, and 0.3 mM oxalate per day compared to 8–9 mM Ca and 0.4–0.5 mM oxalate for repeat stone patients (Robertson *et al.*, 1981). The increase in oxalate has a much more direct effect on the supersaturation of calcium oxalate in the urine than an increase in calcium. With the increase of oxalate/Ca ratio, larger numbers and larger crystals are formed and excreted. Larger crystals are known to increase the risk of forming stones.

In a remarkable study of 10,000 samples of urinary calculi, over twenty different components were identified (Herring, 1962) and the average sample weight was 27.5 mg. The stones were spheroidal, brown to black in colour and either smooth surfaced or with mulberry texture. Sixty one percent contained at least some well developed octahedral crystals and fine-grained calcium oxylate dihydrate, $CaC_2O_4 \cdot 2H_2O$ (weddellite), but only a few of the stones were pure; most also contained some hydroxylapatite and calcium oxylate monohydrate (whewellite). Forty-three percent were predominantly calcium oxalate monohydrate, $CaC_2O_4 \cdot H_2O$, with associated phases. The dihydrate was often identified by its octahedral appearance whereas hydroxylapatite and uric acid were noted as possible seeds. Eighty percent of the 10,000 calculi contained phosphates but only 17% showed >50% phosphates. The species listed were hydroxylapatite, carbonate apatite, calcium hydrogen phosphate dihydrate (brushite), tricalcium phosphate, calcium magnesium hydroxyl phosphate, magnesium phosphate octahydrate, magnesium ammonium phosphate hexahydrate, magnesium hydrogen phosphate, and diammonium calcium phosphate. Cystine and other organic materials were identified in <1% of the stones. The variations in crystalline species in different stones and the question of which hydrate of calcium oxalate forms has been the subject of some classic studies (Prien & Frondel, 1947; Jensen, 1940).

In recent observations, it has been reiterated that 90% of stones are calcium-containing and, therefore, treatment of nephrolithiasis should attempt to reduce circulating (plasma) calcium levels in the patient. The ultimate source of calcium, or the mechanisms that lead to hypercalcemia, are excessive resorption of bone, usually related to parathyroid levels, increased intestinal absorption that can be instigated by excessive quantities of vitamin D, or impaired renal tubule reabsorption that takes place for a variety of physical and endocrine reasons (Coe & Favus, 1986).

There are non-calcareous renal stones; $\sim 10\%$ of those causing pain and discomfort today. Struvite, $MgNH_4(PO_4) \cdot 6H_2O$, is perhaps the best understood clinically. Struvite stones are related to urinary infections with organisms such as *Escherichia coli, Pseudomonas, Klebsiella* and *Staphlococcus*. The urine becomes ammonia-rich and alkaline through urease production by the bacteria. The higher-pH solution makes more phosphate available for nucleation/precipitation and the concentration in a less than optimal performing kidney provides a local situation for crystal formation.

The preponderance of phosphatic and, particularly, apatitic components in stones is probably a direct reflection of the passage of alkaline urine. The conditions of formation and potential dissolution of the phases that occur in stones have been calculated based on the pH-dependent dissociation of other various ions, the effect of ionic strength on activity coefficients, and the composition of the urine (Pak *et al.*, 1996). The urine of stone-formers is usually supersaturated with respect to brushite and the calcium oxalates although spontaneous nucleation does not necessarily take place. After nucleation, however, growth is rapid and correlates with supersaturation. The fact that some individuals are stone formers whereas others are not has been ascribed to the presence of inhibitors to crystallization in the urine in the latter. The most recent research to minimize kidney disease and stone formation has been fostered by studying the specific composition of stones and their crystal chemistry. This understanding has led to the use of crystal inhibitors that prevent nucleation and growth and, thereby, halt pathogensis. A report on l-cysteine kidney stones illustrates this new pathway to reducing biomineral-created disease (see Rimer *et al.*, 2010; Wesson & Ward, 2006).

13.1. Pancreatic stones

Pancreatic 'stones' or calcifications, are mainly calcite with minor amounts of phosphate, magnesium and sodium along with an organic matrix (Dong *et al.*, 1972). As they occur in ducts, they are often the result of localized trauma reactions that alter the normal tissue and organ structures. The occurrence of calcite parallels the fact that the serum and blood is saturated with respect to calcite (Edmondson *et al.*, 1950). A recent review of pancreatic calcifications, detected by their dense appearance near or adjacent to the pancreas and the ducts leading to and away from this organ, leads to some insights (Lesniak *et al.*, 2002). Usually in elderly individuals, >65 years old, the ductal adenocarcinoma can be up to 8 cm wide with calcifications on or at the periphery of the pancreas. The protein, mucin, plus or minus calcific material, effectively blocks the duct. A reaction to chronic alcoholic pancreatitis, one of a variety of disease states, the obstructed ducts cause trauma, peri-ductal fibrosis and calculi formation. These biomineralized obstructions vary with duct size from submicroscopic to cm in size and may exist for years before causing pain or resulting in death (Buelow *et al.*, 1995).

Brushite ($CaHPO_4 \cdot 2H_2O$) and bioapatite have been identified in some samples, but are probably not the first formed of these mineralized materials (Rodgers & Spector, 1986). There are a number of other composition stones that appear in the bladders of animals for example (Skinner *et al.*, 1977), and many other sites and organs can mineralize. The largest organ in our bodies, the skin, has not even been mentioned in this overview but there are studies devoted to its pathological calcification (Selye, 1962).

All of these pathological mineralization phenomena can be related to the serum or fluid concentrations that are cell controlled. Trauma to cells results in reactions that are becoming more predictable from cell- and molecular-level research. From the viewpoint of biomineral deposition, it must be remembered that on death, each cell releases its accumulated contents into the surroundings and in many, many cases this is sufficient to result in nucleation, and tumors are a prime example. The human body's cycles have similarities to mineral formation on Earth; both contain a fluid phase with complicated chemistry that changes from moment to moment.

14. Concluding remarks

Deposition of bioapatite in the human has some parallels to the deposition of phosphorites, a biologically induced, constrained, localized, and time-dependent phenomenon. This is analogous to bioapatite in humans. Both are expressions of their unique environments and biogeochemical activities, and both involve the proliferation and death of living forms that use phosphorous for their energy systems and molecular pathways. It is phosphorous that allows the living to continue to thrive and reproduce, whether in the earliest species or in humans (Skinner, 2000a).

Turning on or off the supply of phosphorous or of nutrients to specific depositional sites in the human body is not yet possible, but medical and dental researchers are pursuing these possibilities. Today they are comparing and sharing information on bones, teeth and pathological conditions worldwide. At the moment there is no known way to measure out an appropriate amount of phosphate for normal bone turnover, or to restrict pathological phosphate deposition. Local cell reactions govern the sequestration and precipitation of phosphate in the human body as they do in phosphate mineral deposition on Earth. Non-bioapatite biomineral formation also draws upon the widely available elements and nutrients *via* comparable mechanisms.

This chapter on human health and disease deals with but one aspect of our environment but with many applications appropriate for mineralogists. I extoll the cross-disciplinary activities which now form an important part of our scientific endeavours. They generate novel insights aiding understanding of Earth processes and should benefit all humans, their health, welfare and ultimately, it is to be hoped, survival.

Acknowledgements

Much of the detail presented in this chapter comes from publications of investigations carried out by medical and dental researchers. I highlight seminal and original contributions where possible, although many important works in metabolic bone disease, caries research (dental), and the other disease states, have been omitted. This overview is merely an introduction to human biominerals for those mineralogists that apply a range of techniques in other fields.

I am deeply indebted to my many associates and collaborators over the years who cheerfully helped my scientific education in a diversity of fields. For this particular chapter, I thank the following for their thoughtful contributions and advice: Hugh Bradburn MD, Jonathan Samburg DDS, Patricia St. Georges, Marc Taylor MD and Shikma Zaarur.

References

Albright, J.A. & Skinner, H.C.W. (1987) Bone: structural organization and remodeling dynamics. In: *The Scientific Basis of Orthopaedics* (J.A. Albright & R. Brand, editors). Appleton and Lange, Norwalk, Connecticut, USA, pp. 165–198.

Allison, M.A., Cheung, B.S., Criqui, M.H., Langer, R.D. & Wright, C.M. (2006) Mitral and Aortic annular calcification are highly associated with systemic calcified atherosclerosis. *Circulation*, **113**, 861–866.

Antipova, O. & Orgel, J.P.R.O. (2010) *In Situ* D-periodic molecular stucture of type II collagen. *Journal of Biological Chemistry*, **258**, 7087–7096.

Arora, H., Madan, P., Simpson, L. & Stainback, R.F. (2008) Caseous calcification of the mitral annulus. *Texas Heart Institute Journal*, **35**, 211–213.

Barasch, E., Gottdiener, J.S., Marino Larson, E.K., Chaves, P.H.M. & Newman, A.B. (2006) Cardiovascular morbidity and mortality in community dwelling elderly individuals with calcification of the fibrous skeleton of the base of the heart and aortosclerosis (The Cardiovasular Health Study). *American Journal of Cardiology*, **97**, 1281–1286.

Barralet, J., Best, S. & Bonfield, W. (1998) Carbonate substitution in precipitated hydroxyapatite: An investigation into the effects of reaction temperature and bicarbonate ion concentration. *Journal of Biomedical Materials Research*, **41**, 79–86.

Becker, M.A. & Ryan, L.M. (2012) Pathogenesis and etiology of calcium pyrophosphate crystal deposition disease. *www.uptodate.com*

Bertazzo, S. & Bertran, C.A. (2008) Effect of hydrazine deproteination on bone mineral phase: a critical view. *Journal of Inorganic Biochemistry*, **102**, 137–145.

Bigi, A., Foresti, E., Incerti, A. & Ripamonti, S. (1980) Structural and chemical characterization of the inorganic deposits in calcified human aorta. *Inorganica Chimica Acta*, **55**, 81–85.

Bonel, G. & Montel, G. (1965) Studies comparing carbonate apatite obtained through different methods of synthesis. In: *Reactivity of Solids* (G.-M. Schwab, editor). Elsevier, Amsterdam, (in French), pp. 667–675.

Bornstein, P. (1974) The biosynthesis of collagen. *Annual Reviews of Biochemistry*, **54**, 567–604.

Boskey, A.L. (2007) Mineralization of bones and teeth. *Elements*, **3**, 385–391.

Bowen, H.J.M. (1966) *Trace Elements in Biochemistry*. Academic Press, New York.

Brown, C.J., Cherney, C.R.N., Smith, B., Tomkins, A., Roberts, G.J., Serimjpgi, L. & Thompson, M. (2002) A sampling and analytical methodology for dental trace elements analysis. *Analyst*, **127**, 319–323.

Brown, W.E. (1966) Crystal growth of bone mineral. *Clinical Orthopaedics and Related Research*, **44**, 205–220.

Brown, W.E., Eidelman, N. & Tomazic, B. (1987) Octacalcium phosphate as a precursor in biomineral formation. *Advances in Dental Research*, **1**, 306–313.

Buelow, P.C., Purrino, T.V., Buck, J.C. *et al.* (1995) Islet cell tumors of the pancreas: pathological image correlation size, necrosis and cysts, calcification, malignant behavior and fumginal status. *American Journal of Rheumatology*, **165**, 1175–1129.

Canalis, E. (1996) Regulation of bone remodeling. In: *Primer on the Metabolic Bone Disease and Disorders of Mineral Metabolism* (M.J. Favus, editor). Lippincott-Raven, 3rd edition, Philadephia, Pennsylvania, USA, pp. 29–34.

Carlstrom, D., Engstrom, H. & Hjorth, S. (1953a) Electron microscopy and X-ray diffraction studies of statoconia. *Larynoscope*, **63**, 1052–1057.

Carlström, D., Engfeldt, B., Engström, A. & Ringertz, N. (1953b) Studies on the chemical composition of normal and abnormal blood vessel walls. I. Chemical nature of vascular calcified deposits. *Laboratory Investigations*, **2**, 325–335.

Carvalho, M.L., Casaca, C., Marques, J.P., Pinheiro, T. & Cunha, A.S. (2001) Human teeth elemental profiles measured by synchrotron X-ray fluorescence: dietary habits and environmental influence. *X-ray Spectrometry*, **30**, 190–193.

Cazalbou, S., Combes, C., Eichert, D. & Rey, C. (2004a) Adaptive physico-chemistry of bio-related calcium phosphates. *Journal of Materials Chemistry*, **14**, 2148–2153.

Cazalbou, S., Combes, C., Eichert, D., Rey, C. & Glimcher, M.J. (2004b) Poorly crystalline apatite: evolution and maturation in vitro and in vivo. *Journal of Bone Mineral Metabolism*, **22**, 310–317.

Cerling, T. (2011) Diet of *Paranthropus boisei* in the early Pleistocene of East Africa. *Proceedings of the National Academy of Sciences, USA,* **108**, 9337–9341.

Chang, L.L.Y., Howie, R.A. & Zussman, J. (1996) *Rock-Forming Minerals, Non-Silicates: Sulphides, Carbonates, Phosphates, Halides.* Vol. **5B**, 2nd edition, Longman Scientific, Harlow, Essex, UK, 383 pp.

Chefetz, I. & Sprecher, E. (2009) Familial tumoral calcinosis and the role of O-glycosylation in the maintenance of phosphate homeostsis. *Biochimica et Biophysica Acta*, **1792**, 847–852.

Cheitlin, M.D. (editor) *et al.* (2003) ACC/AHA/ASE 2003 Guideline Update for the Clinical Application of Echocardiography: Summary Article: A Report of the American College of Cardiology/American Heart Association Task Force on Practice Guidelines (ACC/AHA/ASE Committee to Update the 1997 Guidelines for the Clinical Application of Echocardiography). *Circulation*, **108**, 1146–1162.

Chen, Po-Yu, Toroian, D. & Price, P.A. (2011) Minerals form a continuum phase in mature cancellous bone. *Calcified Tissue International*, **88**, 351–361.

Coe, F.L. & Favus, H.J. (1986) *Disorders of Stone Formation.* 2nd edition. In: *The Kidney* (R.M. Brenner & F.C. Rector, Jr., editors). Saunders, Philadelphia, Pennsylvania, USA, pp. 1403–1442.

Crawford, M.D., Clayton, D.G., Stanley, F. & Shaper, A.G. (1977) An epidemiological study of sudden death in hard and soft water areas. *Journal of Chronic Diseases*, **30**, 69–80.

Daculsi, G. & Kerebel, B. (1978) High resolution electron microscopy study of human enamel crystallites: size, shape and growth. *Journal of Ultrastructural Research*, **65**, 163–172.

Dieppe, P. & Calvert, P. (1983) *Crystals and Joint Disease.* Chapman & Hall, London.

Dong, P.K., Ballenger, G. & Leger, L. (1972) Mechanism of formation of pancreatic calculi. *Clinica Chimica Acta* (Paris), **103**, 119–126 (in French).

Dorozhkin, S.V. (2007a) Calcium orthophoshates. *Journal of Materials Science*, **42**, 1061–1095.

Dorozhkin, S.V. (2007b) A hierarchical structure for apatite crystals. *Journal of Materials Science: Materials in Medicine*, **18**, 363–366.

Drever, J.I. (1982) *The Geochemistry of Natural Waters.* Prentice-Hall, Engelwood Cliffs, New Jersey, USA.

Driessens, F.C.M. & Verbeek, R.M.H. (1990) *Biominerals.* CRC Press, Boca Raton, Florida, USA.

Eagle, R. Tutkin, T., Martin, T., Tripati, A. Fricke, H. Conneley, M., Cifelli, R. & Eiler, J. (2011) Body temperatures of dinosaurs determined from isotopic ^{13}C-^{18}O ordering in fossil biominerals. *Science*, **333**, 443–445.

Ebenbichler, G.R., Erdogmus, C.B., Resch, K.L. *et al.* (1999) Ultrasound therapy for calcific tendonitis of the shoulder. *New England Journal of Medcine*, **340**, 1533–1538.

Edmondson, H.J., Bullock, W.K. & Mehl, J.W. (1950) Chronic pancreatitis and lithiasis. *American Journal of Pathology*, **26**, 37–49.

Elliott, J.C. (2002) Calcium phosphate biominerals. In: *Phosphates: Geochemical, Geobiological and Materials Importance* (M.J. Kohn, J. Rakovan & J.M. Hughes, editors). Reviews in Mineralogy and Geochemistry, **48**. Mineralogical Society of Anmerica, Washington, D.C., pp. 427–454.

Elliott, J.C., Mackie, P.E. & Young, R.A. (1973) Monoclinic hydroxyapatite. *Science*, **180**, 1055–1057.

Favus, M.J. *et al.* (editors) (1996) *Primer on the Metabolic Bone Diseases and Disorders of Mineral Metabolism.* Lippincott-Raven, Philadelphia, Pennsylvania, USA.

Favus, M.J. *et al.* (editors) (2006) *Primer on the Metabolic Bone Diseases and Disorder of Mineral Metabolism*, 6th edition. American Society for Bone and Mineral Research, Washington, D.C.

Feindel, C.M., Tufail, Z., Ivanov, J. & Armstrong, S. (2002) Mitral valve surgery in patients with extensive calcification of the mitral annulus. *Journal of Thoracic and Cardiovascular Surgery*, **126**, 777–781.

Fernhead, R.W. & Stack, M. (editors) (1971) Tooth enamel: its composition, properties, and fundamental structure. *Proceedings, Second Internaional Symposium on the compositon properties and fundamental structues of Tooth Enamel.* London Hospital Medical College, June 1969. John Wright, Bristol.

Filippelli. G.M. (2002) The global phosporous cycle. In: *Phosphates: Geochemical, Geological and Materials Importance* (M.J. Kohn, J. Rakovan & J.M. Hughes, editors). Reviews in Mineralogy and Geochemistry, **48**, Mineralogical Society of America, Washington, D.C., pp. 391–426.

Fleet, M.E. & Liu, X. (2007) Coupled substitution of Type A and B carbonate in sodium-bearing apatite. *Biomaterials*, **28**, 916–926.

Fleet, M.E., Liu, X. & Liu, Xi (2011) Orientation of channel carbonate ions in apatite: effect of pressure and composition. *American Mineralogist*, **96**, 1148–1157.

Fogo, A.B. & Kashkarian, M. (2012) *Diagnostic Atlas of Renal Pathology: A Companion to Brenner & Rector's the Kidney.* 2nd edition. Elsevier/Saunders, Philadelphia, Pennsylvania, USA.

Frank, R.M., Sognnaes, R.F. & Kern, R. (1960) Calcification of dental tissues with special reference to enamel ultrastructure. In: *Calcification in Biological Systems* (R.F. Sognnaes, editor). American Association for the Advancement of Science, Washington, D.C., pp. 163–202.

Fratzl, P., Groschner, M., Vogt, G., Plenk, H., Jr. Eschbergert, J., Fratzl-Zelman, N., Koller, K. & Klaushofer, K. (1992) Mineral crystals in calcified tissues: a comparative study by SAXS. *Journal of Bone and Mineral Research*, **7**, 329–334.

Fratzl, P., Gupta, H.S., Paschalis, E.P. & Roscher, P. (2004) Structure and mechanical quality of the collagen-mineral nano-composite in bone. *Journal of Materials Chemistry*, **14**, 2115–2123.

Freeman, R.V. (2012) Mitral annular calcification. *UpToDate.com* (http://www.uptodate.com/contents/mitral-annular-calcification)

Freeman, R.V. & Otto, C.M. (2005) Spectrum of calcific aorta valve disease: pathogenesis, disease progression and treatment strategies. *Circulation*, **111**, 3316–3326.

Gaines, R.V., Skinner, H.C.W., Foord, E.E., Mason, B. & Rosenzweig, A. (1997) *Dana's New Mineralogy.* Wiley & Sons, New York.

George, A. & Veis, A. (2008) Phosphated proteins and control over apatite nucleation, crystal growth and inhibition. *Chemical Reviews*, **108**, 4670–4693.

Gibbons, A. (1997) Bone sizes trace the decline of man (and woman). *Science*, **276**, 896–897.

Gibbons, A. (2012) An evolutionary theory of dentistry. *Science*, **336**, 973–975.

Glas, J.E. (1962) Studies on the ultrastructure of dental enamel VI. Crystal chemistry of shark's teeth. *Odontologiskf Revy*, **13**, 315–326.

Glimcher, M.J. (1976) Composition, structure and organization of bone, other mineralized tissues, and mechanism of calcification. In: *Handbook of Physiology:Endocrinology,* Vol 7. (P.O. Creep & P. Atwood, editors). American Physiological Society, Washington, D.C., pp. 25–116.

Glimcher, M.J. & Lian, J. (editors) (1989) *The Chemistry and Biology of Mineralized Tissues.* Gordon and Breach Science Publishers, New York.

Gopal, R., Calvo, C., Ito, J. & Sabine, W.K. (1974) Crystal structure of whitlockite. *Canadian Journal of Chemistry,* **52**, 1155–1164.

Gulbrandsen, R.A. (1966) Chemical composition of phosphorites in the Phosphoria Formation. *Geochimica et Cosmochimica Acta*, **30**, 769–778.

Halverson, P.B. & McCarty, D.J. (1979) Identification of hydroxyapatite crystals in synovial fluid. *Arthritis Rheumatism*, **22**, 389–395.

Harpaz, D., Auerbach, I., Vered, Z., Motro, M., Tobar, A. & Rosenblatt, S. (2001) Caseous calcification of the mitral annulus, a neglected, unrecognized diagnosis. *Journal of the American Society of Echocardiology*, **14**, 825–831.

Harries, J.E., Hasnain, S.S. & Shah, J.S. (1987) EXAFS study of structrural disorder in carbonate-containing hydroxyapatite. *Calcified Tissue Intenational*, **41**, 346–350.

Herring, L.C. (1962) Observations on the analysis of ten thousand urinary calculi. *Journal of Urology*, **88**, 545–562.

Hoppe, K.A., Koch, P.L. & Furutani, T.T. (2003) Assessing the preservation of biogenic strontium in fossil bones and tooth enamel. *International Journal of Osteoarchaeology*, **13**, 20–28.

Hughes, J. & Rakovan, J. (2002) The crystal structure of apatite, $Ca_5(PO_4)_3(F,OH,Cl)$. In: *Phosphates: Geochemical, Geobiological and Materials Inportance* (M. Kohn, J. Rakovan & J. Hughes, editors). Reviews in Mineralogy and Geochemistry, **48**. Mineralogical Society of America, Washington, D.C., pp. 1–12.

Iijima, M., Tohda, H. & Moriwaki, Y. (1992) Growth and structure of lamella mixed crystals of octacalcium phosphate and apatite in a model system. *Journal of Crystal Growth*, **116**, 319–326.

Jager, C., Wetzel, T., Meyer-Zaoka, W. & Epple, M. (2006) A solid-state NMR investigation of the structure of nanocrystalline hydroxyapatite. *Magnetic Resonance in Chemistry*, **44**, 573–580.

Jee, W.S.S. (1983) The skeletal tissues. In: *Histology, Cell and Tissue Biology*. (L. Weiss, editor). Elsevier Biomedical, New York, pp. 200–255.

Jensen, A.T. (1940) On concrements from the urinary tract, II. *Acta Chiropracta Scandinavia*, **84**, 217–228.

Johnston, A., Jr., Conrad, C., Slemenda, C.W. & Melton, L.J., III (1996) Bone density measurement and the management of osteoporosis. In: *Primer on the Metabolic Bone Disease and Disorders of Mineral Metabolism* (M.J. Favus, editor), 3rd edition. Lippincott-Raven, Pennsylvania, Philadephia, USA, pp. 142–151.

Jokl, P. & Skinner, H.C.W. (1973) Fluoride induced changes in ashed mineral of growing sheep. *Journal of Bone and Joint Surgery*, **55A**, 761–770.

Jongeblood, W.L., Molenaar, I. & Arends, J. (1975) Morphology and size-distribution of sound and acid-treated enamel crystallites. *Calcified Tissue Research*, **19**, 109–123.

Kacena, M.A., Shivdasani, R.A., Wilson, K., Xi, Y., Nazarian, A., Gundberg, C.M., Bouxsein, M.L., Lorenzo, J.A. & Horowitz, M.C. (2004) Megakaryocyte-osteoblast interaction revealed in mice deficient in transcription factors GATA-1 and NF-E2. *Journal of Bone and Mineral Research*, **19**, 652–660.

Kahn, K.M., Cook, J.L., Kannus, N. & Bonar, S.F. (2002) Editorial: Time to abandon the 'tendinitis' myth. *British Medical Journal*, **324**, 626.

Koch, P.L., Tuross, N. & Fogel, M.L. (1997) The effects of sample treatment and diagenesis on the isotopic integrity of carbonate in biogenic hydroxylapatite. *Journal of Archaeological Science*, **24**, 417–429.

Koenig, A.E., Rodgers, R.R. & Trueman, C.N. (2008) Visualizing fossilization using laser ablation inductively coupled plasma-mass spectroscopy maps of trace elements in Late Cretaceous bones. *Geology*, **37**, 511–514.

Kohn, M. & Cerling. T.E. (2002) Stable isotope composition of biological apatite. In: *Phosphates: Geochemical, Geobiological and Materials Inportance* (M. Kohn, J. Rakovan & J. Hughes, editors). Reviews in Mineralogy and Geochemistry, **48**. Mineralogical Society of America, Washington, D.C., pp. 455–488.

Kohn, M., Rakovan, J. & Hughes, J. (editors) (2002) *Phosphates: Geochemical, Geobiological and Materials Importance*. Reviews in Mineralogy and Geochemistry, **48**. Mineralogical Society of America, Washington, D.C.

Krajewski, A., Mazzocchi, M., Buildini, P.L., Ravaglioli, A., Tinu, A. & Fagnano, C. (2005) Synthesis of carbonated hydroyapatite: efficiency of the substitution amd critical evaluation of analaytical methods. *Journal of Molecular Structure*, **744–747**, 221–228.

Laine, C., Turner, B.J., Williams, S. & Wilson, J.F. (2010) Gout. *Archives of Internal Medicine*, **152**, *In the Clinic*, Feb 2, Section **2**, 21–22.

Leach, S.A. (1973) Dental calculus. In: *Biological Mineralization* (I. Zipkin, editor). Wiley, New York, pp. 587–606.

Lee-Thorp, J.A. & Van der Merwe, N.J. (1991) Aspects of the chemistry of fossil and modern bioloical apatites. *Journal of Archaeological Science*, **18**, 343–354.

Lehninger, A.L. (1970) Mitochondria and calcium ion transport. *Biochemical Journal*, **119**, 129–138.

Lesniak, R.J., Hohenwalter, M.D. & Taylor, A.J. (2002) Spectrum of causes of pancreatic cancer. *American Journal of Roetenology*, **178**, 79–86.

Lowenstam, H.A. & Weiner, S. (1989) *On Biomineralization*. Oxford University Press, New York.

Mandel, N.S. & Mandel, G.S. (1976) Monosodium urate monohydrate, the gout culprit. *Journal of the American Chemical Society*, **98**, 2319–2323.

Manning, D.A.C. (2013) Minerals and soil development. In: *Environmental Mineralogy* II (D.J. Vaughan & R. Wogelius, editors). EMU Notes in Mineralogy, **13**. European Mineralogical Union and the Mineralogical Society of Great Britain & Ireland, pp. 103–122.

Marcus, R., Feldman, U. & Kelsey, J. (1996) *Osteoporosis*. Academic Press, New York.

Markowitz, G.S & Perazella, M.A (2009) Acute phosphate nephropathy. *Kidney International*, **76**, 1027–1034.

McCarty, D.J. (1976) Calcium pyrophosphate dihydrate crystal deposition disease. *Arthritis Rheumatology*, **19**, 275–285.

McCarty, D.J. & Hollander, J.L. (1961) Identification of urate crystals in gouty synovial fluid. *Annals of Internal Medicine*, **54**, 452–460.

McCarty, D.J., Kohn, N.N. & Faires, J.S. (1962) The significance of calcium phosphate crystals in the synovial fluid of arthritis patients: the "pseudogout syndrome": I. Clinical Aspects. *Annals of Internal Medicine*, **56**, 711–745.

McConnell, D. & Gruner, J.W. (1940) The problem of the carbonate-apatites III: the carbonate-apatite from Magnet Cove, Arkansas. *American Mineralogist*, **25**, 157–167.

McNally, E.A., Schwarcz, H., Botton, G.A. & Arsenault, A.L. (2012) A model for the ultrastructure of bone based on electron microscopy of ion-milled sections. *PloS One* 7 e29258, 12 pp.

Memmler, R.L., Cohen, B.J. & Wood, D.L. (1996) *The Human Body in Health and Disease,* 8th edition. Lippincott, Philadelphia, Pennsylvania, USA.

Mestman, S. & Herman, A. (2009) *What to do for Healthy Teeth*. Institute for Healthcare Advancement, USA.

Miles, A.E.W. (1967) *Structure and Chemical Organization of Teeth*. Academic Press, New York.

Millan, J.L. (2006) Alkaline phosphatases: structure, substrate specificity and functional relatedness to other members of a large superfamily of enzymes. *Purinergic Signaling*, **2**, 335–341.

Mohler, E.R., Gannon, F., Reynolds, C., Zimmerman, R., Keanem, M.G. & Kallan, F.S. (2001) Bone mineralization and inflamation in cardiac valves. *Circulation*, **103**, 1522–1528.

Moreno, E.C., Kresak, M. & Zahradnik, R.J. (1974) Fluorated hydroxyapatite solubility and caries formation. *Nature*, **247**, 64–65.

Morgan, H., Wilson, M., Elliott, J.C., Dowker, S.E.P. & Anderson, P. (2000) Preparation and characterization of monoclinic hydroxyapatite and its precipitated carbonate apatite intermediate. *Biomaterals*, **21**, 617–627.

Nancolas, G.H. & Wu, W. (2000) Biomineralization mechanisms: a kinetic and interfacial energy approach. *Journal of Crystal Growth*, **211**, 137–142.

Neuman, W.F. & Neuman, M.W. (1958) *The Chemical Dynamics of Bone Mineral*. University of Chicago Press, Chicago Illinois, USA.

Omelon, S.J. & Grynpas, M.D. (2007) A nonradioactive method for detecting phosphates and polyphosphates by PAGE. *Electrophoresis*, **28**, 2808–2811.

Omelon, S.J. & Grynpas, M.D. (2008) Relationshops between polyphosphate chemistry, biochemistry and apatite biomineralization. *Chemical Reviews*, **108**, 4694–4715.

Omelon, S.D., Georgiou, J., Henneman, Z.J., Wise, L.M., Sukhu, B., Hunt, T., Wynnyckyi, C., Holmyards, D., Bielecki, R. & Grynpas, M.D. (2009) Control of vertebrate skeletal mineralization by polyphosphates. *PloS ONE*, **4** issue 5, e5634.

Ono, S., Sebastin, S. & Chung, K.C. (2012) Overview of finger, hand and wrist fractures. *UpToDate.com* (http://www.uptodate.com/contents/overview-of-finger-hand-and-wrist-fractures)

Orgel, J.P.R.O., San Antonio, J.D. & Antipova, O. (2011) Molecular and structural mapping of collagen fibril interactions. *Connective Tissue Research*, **52**, 2–17.

Pak, C.Y.C., Resnick, M.I. & Preminger, G.M. (editors) (1996) *Urolithiasis. Proceedings of the 8th International Symposium on Urolithiasis*. Millet the Printer, Dallas, Texas, USA.

Pan, Y. & Fleet, M.E. (2002) Compositions of the apatite-group minerals substitution mechanisms and controlling factors. In: *Phosphates: Geochemical, Geobiological and Materials Importance* (M. Kohn,

J. Rakovan & J. Hughes, editors). Reviews in Mineralogy and Geochemistry, **48**. Mineralogical Society of America, Washington, D.C., pp. 13–50.

Papadopoulos, N., Dietrich, M., Christodoulou, T., Moritz, A. & Doss, M. (2009) Midterm survival after decalcification of the mitral annulus. *American Thoreacic Sugery*, **87**, 1143–1147.

Parry, D.A. & Squire, J.M. (2005) Fibrous proteins: new structural and functional aspects revealed. *Advances in Protein Chemistry*, **70**, 1–10.

Pasero, M., Kampf, A.R., Ferraris, C., Pekov, I.V., Rakovan, J. & White, T.J. (2010) Nomenclature of the apatite supergroup minerals. *European Journal of Mineralogy*, **22**, 163–179.

Pasteris, J.D., Wopenka, B. & Valsami-Jones, E. (2008) Bone and tooth mineralization: why apatite? *Elements*, **4**, 97–104.

Patterson, C. (1977) Cartilage bones, dermal bones and membrane bones or the exoskeleton vs. endoskeleton. In: *Problems in Vertebrate Evolution* (S.M. Andrews, R.S. Miles & A.D. Walker, editors). Linnaeus Society Symposium Series **4**. Academic Press, London, pp. 77–121.

Peroos, S., Du, Z. & De Leeuw, N.H. (2006) A computer modeling study of the uptake, structure and distribution of carbonate defects in hydroxyapatite. *Biomaterials*, **27**, 2150–2161.

Plant, J. & Tidey, G. (2003) *Understanding, Preventing and Overcoming Osteoporosis*. Virgin Books Ltd, London.

Poggi, S.H. & Skinner, H.C.W. (2012) *Environmental hazards and the human fetus*. Abstract, 34th International Geologic Congress, Brisbane, Australia.

Prie, D., Torres, P.U. & Friedlaender, G. (2009) Latest findings on phosphate homeostasis. *Kidney International*, **75**, 882–889.

Prien, E.L. & Frondel, C. (1947) Studies in urolithiasis. I: The composition of urinary calculi. *Journal of Urology*, **57**, 949–991.

Pritzker, K.P.H. (1980) Crystal-associated arthopathies; what's new in joints? *Journal of the American Geriatric Society*, **28**, 439–445.

Proudfoot, D. & Shanahan, C.M. (2007) Atherosclerosis: cellular aspects. In: *Handbook of Biomineralization* (M. Eppel & E. Bauerlein, editors). J. Wiley-VCH Verlag, Weinheim, Germany.

Puceat, E., Reynard, B. & Lecuyer, C. (2004) Can crystallinity be used to determine the degree of chemical alteration of biogneic apatites? *Chemical Geology*, **205**, 83–97.

Rajamannan, N.M., Rickard, D., Stock, S.R., Donovan, J., Springett, M., Orszulak, T., Fullerton, D.A., Tajik, A.J., Bonow, R.O. & Spelsberg, T. (2003) Human aortic valve calcification is associated with an osteoblast phenotype. *Circulation*, **107**, 2181–2184.

Reddi, A.H. & Reddi, A. (2009) Bone morphogenic proteins (BMPs): from morphogens to metabologens. *Cytokine & Growth Factor Reviews*, **20**, 341–342.

Rey, C., Collins, B., Goehl, T., Dickson, I.G. & Glimcher, M.J. (1989) The carbonate environment of bone mineral: a resolution-enhanced Fourier infrared spectroscopy study. *Calcified Tissue International*, **45**, 157–164.

Richette, P. & Bardin, T. (2010) Gout. *Lancet*, **375**, 318–328.

Rimer, J., An, Z., Zhu, Z., Lee, M.H., Goldfarb, D.S., Wesson, J.A. & Ward, M.D. (2010) Crystal gowth inhibitors for the prevention of L-cystine kidney stones throgh molecular design. *Science*, **330**, 337–341.

Rink, W.J. and Schwarcz, H.P. (1995) Tests for diagenesis in tooth enamel – ESR dating signals and carbonate contents. *Journal of Archaeological Science*, **22**, 251–255.

Robertson, W.G., Peacock, M., Ouimet, D., Heyburn, P.J. & Rutherford, A. (1981) The main risk factor for calcium oxalate stone disease in man: hypercalciuria or mild hyperoxaluria? In: *Urolithiasis: Clinical and Basic Research* (L.H. Smith, W.G. Robertson & B. Finlayson, editors). Plenum Press, New York, pp. 3–12.

Robinson, C., Connell, S., Kirkham, J., Shore, R. & Smith, A. (2004) Dental enamel – a biological ceramic: regular substructure in enamal hydroxyapatite crystals revealed by atomic force microscopy. *Journal Materials Chemistry*, **14**, 2242–2248.

Rodgers, A.L. & Spector, M. (1986) Pancreatic calculi containing brushite: Ultrastucture and pathogenesis. *Calcified Tissue International*, **39**, 342–347.

Roschger, P., Manjubala, I., Zoeger, N., Meier, F., Simon, R., Li, C., Fratzl-Zelman, N., Mishof, B.M., Paschalis, E.P., Streli, C., Fratzl, P. & Klaushofer, K. (2009) Bone material quality in transiliac bone biopsies of postmenopausal osteoporotic women after 3 years of strontium ranalate treatment. *Journal Bone and Mineral Research*, **25**, 891–900.

Rosen, V.B., Hobbs, L.W. & Spector, M. (2002) The ultrastructure of anorganic bovine bone and selected synthetic hydroxyapatites used as bone graft substitute materials. *Biominerals*, **23**, 921–928.

Sahai, N. (2005) Modeling apatite nucleation in the human body and in the geochemical environment. *American Journal of Science*, **305**, 661–672.

Sasaki, N., Tagami, A., Goto, T., Taniguchi, M., Nakara, M. & Hikichi, K. (2002) Atomic force microscopic studies on the structure of bovine cortical bone at the collagen fibril-mineral level. *Journal of Materials Science and Medicine*, **15**, 333–337.

Schmid, K., McSharry, W.O., Pameijer, C.H. & Binette, L. (1980) Chemical and physicochemical studies of the mineral deposits of the human artherosclerotic aorta. *Artherosclerosis,* **37**, 199–210.

Schumacher, H.R., Miller, J.L., Ludvico, C. & Jessar, R.A. (1981) Erosive arthritis associated with apatite crystal deposition. *Arthritis and Rheumatism*, **24**, 31–37.

Selye, H. (1962) *Calciphylaxis.* University of Chicago Press, Chicago, Illinois, USA.

Shannon, R.D. (1976) Revised effective ionic radii and systematic studies of interatomic distances in halides and chalcogenides. *Acta Crystallographica A*, **32**, 751–767.

Shannon, R.D. & Prewitt, C.T. (1969) Effective ionic radii in oxides and fluorides. *Acta Crystallogrphica B*, **25**, 925–946.

Shim, J.-H., Greenblatt, M.B., Xie, M., Schneider, M.D., Zou, W., Zhai, B., Gygi, S. & Glimcher, L.H. (2009) TAK 1 is an essential regulator of BMP signaling in cartilage. *The EMBO Journal*, **28**, 2028–2041.

Shore, R.M. & Poznanski, A.K. (1996) Radiologic evaluation of bone mineral in children. In: *Primer on the Metabolic Bone Disease and Disorders of Mineral Metabolism* (M.J. Favus, editor), 3rd edition. Lippincott-Raven, Philadephia, Pennsylvania, USA, pp. 119–133.

Shubin, N. (2008) *Your Inner Fish: a Journey into the 3.5 Billion Year History of the Human Body*. Pantheon Books, New York.

Simon, P., Carillo-Cabrera, W., Formanek, P., Gobel, C., Geiger, D., Ramlau, R., Tlatlik, H., Buder, J. & Kniep, R. (2004) On the real-structure of biomimetically grown hexagonal prismatic seeds of fluorapatite-gelatine-composites: TEM investigations along [001]. *Journal of Materials Chemistry*, **14**, 2218–2224.

Simpson, D.R. (1964) The nature of alkali carbonate apatites. *American Mineralogist*, **49**, 363–376.

Simpson, D.R. (1967) Effect of pH and solution concentration on the composition of carbonate apatite. *American Mineralogist*, **52**, 896–902.

Skinner, H.C.W. (1968) X-ray diffraction analysis techniques to monitor composition fluctuations within the mineral group apatite. *Applied Spectroscopy*, **22**, 412–414.

Skinner, H.C.W. (1973) Phase relations in the CaO-P_2O_5-H_2O system from 300–600°C at 2 kb H_2O pressure. *American Journal of Science*, **273**, 545–560.

Skinner, H.C.W. (1974) Studies in the basic mineralizing system, CaO-P_2O_5-H_2O. *Calcified Tissue Research*, **14**, 3–14.

Skinner, H.C.W. (1987) Bone: mineral and mineralization. In: *The Scientific Basis of Orthopaedics,* 2nd edition (J.A. Albright & R. Brand, editors). Appleton & Lange, Norwalk, Connecticut, USA, pp. 199–211.

Skinner, H.C.W. (1989) Low temperature carbonate phosphate materials or the carbonate apatite problem. A review. In: *Origin, Evolution and Modern Aspects of Biomineralization in Plants and Animals* (R. Crick, editor). Plenum Press, New York, pp. 251–264.

Skinner, H.C.W. (2000a) In praise of phosphates or why vertebrates chose apatite to mineralize their skeletons. *International Geology Review*, **42**, 232–240.

Skinner, H.C.W. (2000b) Minerals and Human Health. In: *Environmental Mineralogy* (D.J. Vaughan & R.A. Wogelius, editors). European Union Notes in Mineralogy, **2**. Eötvös University Press, Budapest, Hungary, pp. 383–412.

Skinner, H.C.W. (2005a) Mineralogy of Bone. In: *Essentials of Medical Geology* (O. Sellinus, editor). Elsevier, New York, pp. 667–693.

Skinner, H.C.W. (2005b) Biominerals. *Mineralogical Magazine*, **69**, 621–641.

Skinner, H.C.W. & Jahren, A.H. (2004) *Biomineralization.* In: *Treatise on Geochemistry*, Volume **8** (H.D. Holland & K.K. Turekian, editors). Elsevier, Pergamon Press, Amsterdam, pp. 117–184.

Skinner, H.C.W., Kempner, E.S., & Pak, C.Y.C. (1972) Preparation of the mineral phase of bone using ethylenediamine extraction. *Calcified Tissue Research,* **10**, 257–268.

Skinner, H.C.W., Osbaldistan, G. & Wilner, A. (1977) Monohydrocalcite in a guinea pig bladder stone: a novel occurrence. *American Mineralogist*, **62**, 273–277.

Skinner, H.C.W., Ross, M. & Frondel, C. (1988) *Asbestos and other Fibrous Materials: Mineralogy, Crystal Chemistry and Health Effects.* Oxford University Press, New York.

Skinner, H.C.W., Nicolescu, S. & Raub, T.D. (2004) A tale of two apatites. In: *Environment and Progress* (I. Petrescu, editor). Editura Fundatiel de Studii Europene, Cluj-Napoca, Romania, pp. 283–288.

Skulan, J., Bullen, T., Anbar, A.D., Puzas, J.E., Shackelford, L., LeBlaine, A. & Smith, S.M. (2007) Natural calcium isotope composition of urine as a marker of bone mineral balance. *Clinical Chemistry*, **53**, 1155–1158.

Skulan, J. & DePaolo, D.J. (1999) Calcium isotope fractionation between soft and mineralized tissues as a monitor of calcium use in vertebrates. *Proceedings of the National Academy of Sciences,* **96**, 13709–13713.

Tang, R., Wang, L. & Nancolas, G.H. (2004) Size effects in the dissolution of hydroxyapatite: an understanding of biological mineralization. *Journal of Materials Chemistry*, **14**, 2341–2346.

Tang, Y., Chappell, H.F., Dive, M.T., Reeder, R.J. & Lee, Y.J. (2009) Zinc incorporation into hydroxyapatite. *Biomaterials*, **30**, 2864–2872.

Tarasevic, R.J., Chusuei, C.C. & Allara, D.I. (2003) Nucleation and growth of calcium phosphate from physiological solutions onto self-assembled templates by a solution-formed nucleus mechanism. *Journal of Physical Chemistry*, **107**, 10367–10377.

Toroian, D., Lim, J.E. & Price, P.A. (2007) The size exclusion characteristics of type 1 collagen. *Journal of Biological Chemistry*, **282**, 22437–22447.

Topaz, O., Indelman, M., Chefetz, I., Geiger, D., Metzker, A., Alischuler, Y., Choder, M., Bercovich, D., Utto, J., Bergman, R., Richard, G. & Sprecher, E. (2006) A deleterious mutation in SAMD9 causes normophosphatemic familial tumoral calcinosis. *American Journal of Human Genetics*, **79**, 750–764.

Trautz, O.R. (1955) X-ray diffraction of biologic and synthetic apatite. *Annals of the New York Academy of Science*, **60**, 696–712.

Trueman, C.N., Behrensmeyer, A.K., Potts, R. & Tuross, N. (2006) High-resolution records of location and stratigraphic provenance from the rare earth element composition of fossil bones. *Geochimica et Cosmochimica Acta*, **70**, 4343–4355.

Trueman, C.N., Palmer, M.R., Field, J., Privat, K., Ludgate, N., Chavagnae, V., Edert, D.A., Sifelli, R. & Rogers, R.R. (2008) Comparing rates of recrystallization and the potential for preservation of biomolecules from the distribution of trace elements in fossil bones. *Comptes Rendus Palevol*, **7**, 745–756.

Van der Merwe, N.J. (2012) Oxygen isotopes and the rain on Olduvai: climatic change in Tanzania since the emergence of homo sp. *Archaeometry* (in press).

Van der Merwe, N.J., Masao, F.T. & Bamford, M.K. (2008) Isotopic evidence for contrasting diets of early hominins *Homo habilis* and *Australopithecus boisei* of Tanzania. *South African Journal of Science*, **102**, 153–155.

Van Waser, J.R. (1966) *Phosphorous and its Compounds.* Interscience Publishers, John Wiley & Sons, New York.

Veiderman, M., Tonsuaadu, K., Knubovets, R. & Peld, M. (2005) Impact of anionic substitutions on apatite structure and properties. *Journal of Organometallic Chemistry*, **690**, 2638–2643.

Veis, A. (2011) A letter from the Editor. *Connective Tissue Research*, **52**, 1.

Vignoles, M., Bonel, G., Holcomb, D.W. & Young, R.A. (1988) Influence of preparation conditions on the composition of the type B carbonate hydroxyapatite and on the localization of the carbonate ions. *Calcifed Tissue International*, **43**, 33–40.

Wang, L. & Nancollas, G.M. (2009) Calcium orthophosphates: crystallizaion and dissolution. *Chemical Reviews*, **108**, 4628–4669.

Weiner, S. & Dove, P.M. (2003) An overview of mineralization processes and the problem of the vital effect. In: *Biomineralization* (P.M. Dove, J.J. DeYoreo & S. Weiner, editors). Reviews in Mineralogy and Geochemistry, **48**, Washington, D.C., pp. 1–30.

Weiner, S. & Traub, W. (1991) Bone structure from Angstrom to microns. *FASB*, **6**, 879–885.

Wess, T.J. (2005) Collagen form and function. *Advances in Protein Chemistry*, **70**, 342–374.

Wesson, J.A. & Ward, M.D. (2006) Pathological biominerlization of kidney stones. *Elements*, **3**, 413–421.

White, D.J., Bowman, W.D., Faller, R.V., Mobley, M.J., Wolfgang, R.A. & Yesinowski, J.P. (1988) ^{19}F MAS-NMR and solution chemical characterization of the reactions of fluoride with hydroxyapatite and powdered enamel. *Acta Odontologica Scandinavia*, **46**, 375–389.

Wogelius, R.A. & Vaughan, D.J. (2013) Analytical, experimental and computational methods. In: *Environmental Mineralogy* II (D.J. Vaughan & R. Wogelius, editors). EMU Notes in Mineralogy, **13**. European Mineralogical Union and the Mineralogical Society of Great Britain & Ireland, pp. 5–102.

Wopenka, B. & Pasteris, J.D. (2005) A mineralogical perspective on the apatite in bone. *Materials Science and Engineering,* **C 25**, 131–143.

Wu, Y., Glimcher, M.J., Rey, C. & Ackerman, J.L. (1994) A unique protonated phosphate group in bone mineral not present in synthetic calcium phosphate. Identification by phosphorus-31 solid state NMR spectroscopy. *Journal of Molecular Biology*, **244**, 423–435.

Xiao, Y-T., Xiang, L-X. & Shao, J-Z. (2007) Bone morphological protein. *Biochemical and Biophysical Research Communications*, **362**, 550–553.

Young, R.A. & Mackie, P.E. (1980) Crystallography of human tooth enamel: initial structure refinement. *Materials Research Bulletin*, **15**, 17–29.

Subject and Author Index

Numbers refer to the page where a definition or explanation of, and/or a *figure* or a **table** for, a given subject is found. Author names are given with the number of the first page of the paper in which they are involved.

© Copyright 2013 the European Mineralogical Union and the Mineralogical Society of Great Britain & Ireland
DOI: 10.1180/EMU-notes.13.index